The Chemistry of Natural Products

The Chemistry of Natural Products

Edited by

R.H. THOMSON, D.Sc.
Professor of Organic Chemistry
University of Aberdeen

Blackie

Glasgow and London
Distributed in the USA by
Chapman and Hall
New York

Blackie & Son Limited
Bishopbriggs
Glasgow G64 2NZ

Furnival House
14–18 High Holborn
London WCIV 6BX

Distributed in the USA by
Chapman and Hall
in association with Methuen, Inc.
733 Third Avenue, New York 10017.

British Library Cataloguing in Publication Data

Recent advances in the chemistry of natural
  products
  1. Natural products
  I. Thomson, R.H.
  547'.7   QD415
  ISBN 0-216-91595-3

Library of Congress Cataloging in Publication Data
Main entry under title:

The Chemistry of natural products.

  Includes bibliographical references and index.
  1. Natural products.   I. Thomson, R.H.
QD415.C483 1984   547.7   84-9412
ISBN 0-412-00551-4

Printed in Great Britain by
Thomson Litho Ltd, East Kilbride, Scotland

Preface

The rapid growth of the study of natural products in recent years has been accompanied by the publication of numerous specialist monographs on alkaloids, carbohydrates, coumarins, acetylenes, terpenes, etc., and there are several on biosynthesis. In contrast general texts covering the whole field no longer exist, and a comprehensive work would be enormous. This volume aims to partly fill the gap in a modest way by describing what has been happening in the main areas of natural products research during approximately the last ten years. The emphasis is entirely on the structure, chemistry, and synthesis of natural products with only passing reference to biosynthesis.

R.H. Thomson

Contributors

1 **Carbohydrates**
Professor J.S. Brimacombe, Department of Chemistry, University of Dundee.

2 **Aliphatic compounds**
Dr E.J. Thomas, Dyson Perrins Laboratory, University of Oxford.

3 **Aromatic compounds**
Dr T.J. Simpson, Department of Chemistry, University of Edinburgh.

4 **Terpenoids**
Dr J.R. Hanson, School of Chemistry and Molecular Sciences, University of Sussex.

5 **Steroids**
Dr B.A. Marples, Department of Chemistry, University of Technology, Loughborough.

6 **Amino acids, peptides and proteins**
Professor B.W. Bycroft and Dr A. Higton, Department of Pharmacy, University of Nottingham.

7 **Alkaloids**
Dr I.R.C. Bick, Department of Chemistry, University of Tasmania (Hobart)

8 **Nucleosides, nucleotides and nucleic acids**
Dr J.B. Hobbs, Department of Chemistry, City University, London.

9 **Porphyrins and related compounds**
Professor A.H. Jackson, Department of Chemistry, University College, Cardiff.

Contents

Abbreviations

ABIBN	azobisisobutyronitrile
Ac	acetyl
BDMS	*n*-butyldimethylsilyl
BOP-Cl	*N*,*N*-bis[2-oxo-3-oxazolidinyl]phosphorodiamidic chloride
Bz	benzoyl
Bzl	benzyl
Cb$_3$	benzyloxycarbonyl
DBN	1,5-diazabicyclo[4,3,0]non-5-ene
DBU	1,8-diazabicyclo[5,4,0]undec-7-ene
DCC	dicyclohexylcarbodiimide
DDQ	2,3-dichloro-5,6-dicyano-1,4-benzoquinone
DHP	dihydropyran
DIBAL	diisobutylaluminium hydride
DME	1,2-dimethoxyethane
DMF	*N*,*N*-dimethylformamide
DMS	dimethyl sulphide
DMSO	dimethyl sulphoxide
Fuc	fucose
Gal	galactose
Glc	glucose
GlcNAc	2-acetamido-2-deoxyglucose
HMDS	hexamethyldisilazane
HMPA = HMPT	
HMPT	hexamethylphosphoric triamide
LAH	lithium aluminium hydride
LDA	lithium diisopropylamide
Man	mannose
MCPBA	*m*-chloroperbenzoic acid
MEM	methoxyethoxymethyl
MOM	methoxymethyl

Ms	methanesulphonyl
MSA	mesitylenesulphonic acid
MTHP	4-methoxytetrahydropyranyl
NBS	*N*-bromosuccinimide
PCC	pyridinium chlorochromate
Piv	pivaloyl
py	pyridine
Rha	rhamnose
TBDMS	*t*-butyldimethylsilyl
TFA	trifluoroacetic acid
TFAA	trifluoroacetic anhydride
THF	tetrahydrofuran
THP	tetrahydropyranyl
Tm	mesitylenesulphonyl
TMS	trimethylsilyl
TPS-Cl	2,4,6-tri-isopropylbenzenesulphonyl chloride
Ts	toluene-*p*-sulphonyl
Tr	trityl

1 Carbohydrates

J.S. BRIMACOMBE

During the past decade there has been unprecedented growth in syntheses that involve carbohydrates. This upsurge of activity can be attributed to a number of factors. First, the discovery of sugars of unusual structures, for example, L-evernitrose[1] (1) and D-aldgarose[2] (2), as components of antibiotic substances has presented the carbohydrate chemist with unusually difficult synthetic targets. Second, there has been exceptional activity in the total synthesis of other classes of natural product (for example pheromones and macrolide antibiotics) using 'chiral templates' derived from carbohydrates.[3] Third, the oligosaccharide chains of glycoconjugates, which include glycolipids and glycoproteins, are now known[4] to have important roles in cellular biology, including, among others, intercellular recognition, the transportation of proteins between cells, the specificity of the immune reaction, and as receptors for enzymes, hormones, proteins, and viruses. Since biogenic material is often difficult to obtain, considerable efforts[5,6] are now being directed towards the synthesis of part or whole of the oligosaccharide chains of glycoconjugates in order that their biological functions can be studied in depth. One of the aims of this chapter is to give a broad impression of what has been achieved in these areas.

Carbohydrates possess a higher density of functional groups than any other class of compound, so that protection of one or more of these groups (usually hydroxyl groups) is of fundamental importance to any synthetic strategy. Temporary protecting groups developed for use with other hydroxylic compounds are used increasingly in carbohydrate chemistry,[7] but, as will be seen later, others (for example, allyl and related ethers[8]) have been developed specifically to endow an added flexibility to syntheses involving carbohydrates.

The protection of carbohydrates as cyclic acetals is of long-standing and

(1)   (2)   (3)

enduring importance in carbohydrate chemistry.[9] It is easy to see why, for example, 1,2:5,6-di-O-isopropylidene-α-D-glucofuranose (3), which is readily prepared by acid-catalysed acetonation of D-glucose,[9] has been such a popular starting material for the synthesis of many other sugars.[3,10] The isolated hydroxyl group on C-3 of 3 can be protected or modified prior to exposure of the hydroxyl groups on C-5 and C-6 by selective hydrolysis of the 5,6-O-isopropylidene group with acid. Differences between the reactivities of the primary hydroxyl group on C-6 and the secondary hydroxyl group on C-5 can then be exploited in effecting further modifications at these positions. More vigorous acidic hydrolysis removes the 1,2-O-isopropylidene group, thereby exposing the hydroxyl group on C-2 and, if the molecule reverts to a pyranose ring, the hydroxyl group on C-4. Such procedures, in which cyclic acetals have fulfilled the fundamental role of a protecting group, are commonplace in syntheses involving carbohydrates. During the past few years, procedures for the regioselective (even regiospecific) cleavage of cyclic acetals in a synthetically useful way have been introduced into carbohydrate chemistry, so that these groups assume a much more active role in the route to the target molecule—in this context, cyclic acetals may be regarded as functional groups.[11] Some aspects of the chemistry of cyclic acetals, which have had a decisive influence on the structural modification of carbohydrates, and other useful reactions and protecting groups are discussed in the following sections.

1.1 Cyclic acetals as functional groups

1.1.1 *Halogenation*

An important method for the structural modification of carbohydrates is founded on the cleavage of O-benzylidene acetals by N-bromo-succinimide.[12,13] Thus, treatment of methyl 4,6-O-benzylidene-α-D-glucopyranoside (4, R = H) and its derivatives 4 (R = Bz or Ms) with N-bromosuccinimide in refluxing carbon tetrachloride, in the presence of barium carbonate, gave the corresponding methyl 4-O-benzoyl-6-bromo-6-deoxy-α-D-glucopyranoside (5) regiospecifically and in good yield. This reaction can be conducted in the presence of a wide range of other groups (O-mesyl, -tosyl,

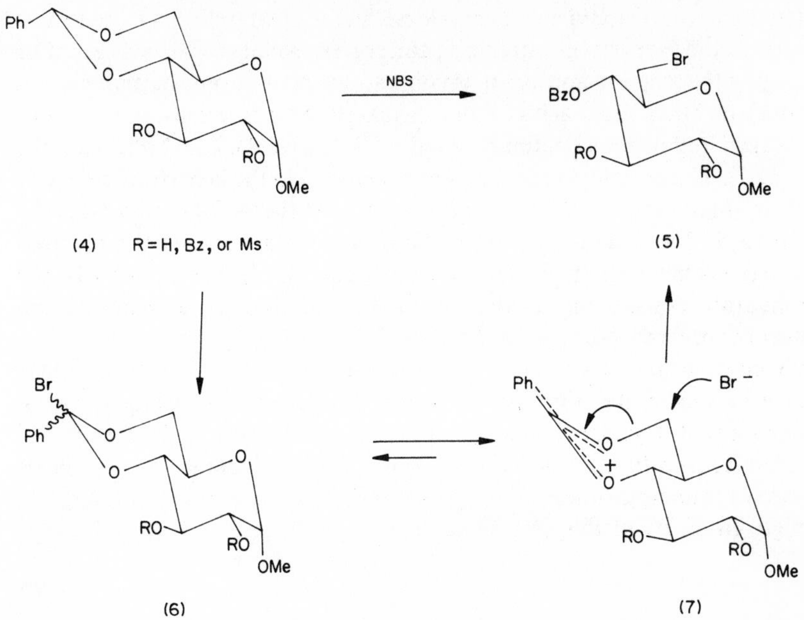

(4) R = H, Bz, or Ms (5)

(6) (7)

Scheme 1

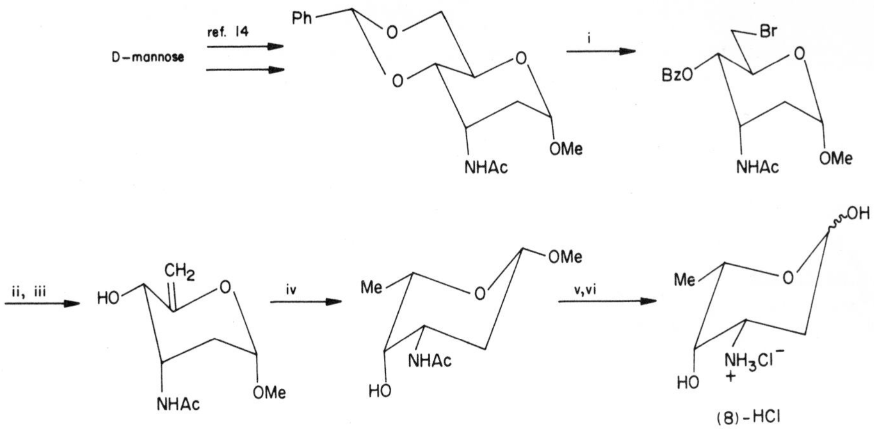

D−mannose

Reagents: i, NBS; ii, AgF; iii, NaOMe; iv, H$_2$-Pd; v, Ba(OH)$_2$; vi, H$_3$O$^+$

(8)−HCl

Scheme 2

-acetyl, -benzoyl, *N*-acetyl, and α-epoxides), and is often followed by reductive cleavage of the 6-bromo group to give the corresponding 6-deoxy sugar. The first step of the reaction appears to involve attack by *N*-bromosuccinimide (or bromine) on the acetal carbon atom, probably by a free-radical process (a free-radical initiator is sometimes included[13]), to give the *gem*-bromoacetal 6 (Scheme 1). This is followed by rearrangement of 6 to the benzoxonium ion 7, which is then attacked by bromide ion to give the 4-*O*-benzoyl-6-bromo derivative 5. The reaction provided the means for generating the required L-*lyxo* stereochemistry in an efficient synthesis of L-daunosamine (8), the carbohydrate constituent of the antitumour antibiotics adriamycin and daunorubicin, from D-mannose (Scheme 2).[14]

N-Bromosuccinimide reacted with the L-rhamnoside 2,3-*O*-benzylidene acetal 9 to yield[15] the 3-bromo derivative 11, despite the strong *syn*-axial interaction that develops during the attack of bromide ion at C-3 of the 2,3-benzoxonium ion 10. Conformational factors or the reluctance of pyranoside derivatives to undergo nucleophilic attack at C-2 have been invoked to explain the regiospecificity of this and related reactions.[16]

(9)   (10)   (11)

On the basis of these results, the diacetal 12 (R = Me) might be expected to react with *N*-bromosuccinimide to give the 3,6-dibromo derivative 15, via the 2,3-benzoxonium ion 13. In fact, the isomeric dibromo derivatives 16 and 17 were also formed,[17] in all likelihood by ring-opening of the 3,4-benzoxonium ion 14 resulting from rearrangement of 13 (Scheme 3). The dibromo derivatives 15–17 can be reduced collectively to a separable mixture of the dideoxy sugars 18 (methyl α-tyveloside) and 19,[15] so that an effective procedure for the deoxygenation of methyl α-D-mannopyranoside at positions 3 and 6 is available.

Regiospecificity was observed in the reaction of the diacetal 12 (R = Me) with triphenylmethyl fluoroborate (a strong hydride-acceptor), the only ion formed being 20.[18] On the addition of tetraethylammonium bromide, 20 underwent regiospecific ring-opening to give the 3-bromo compound 21 in 50% yield. Other examples of the formation of halogenated carbohydrates from 1,3-dioxolanylium ions have been comprehensively reviewed.[11]

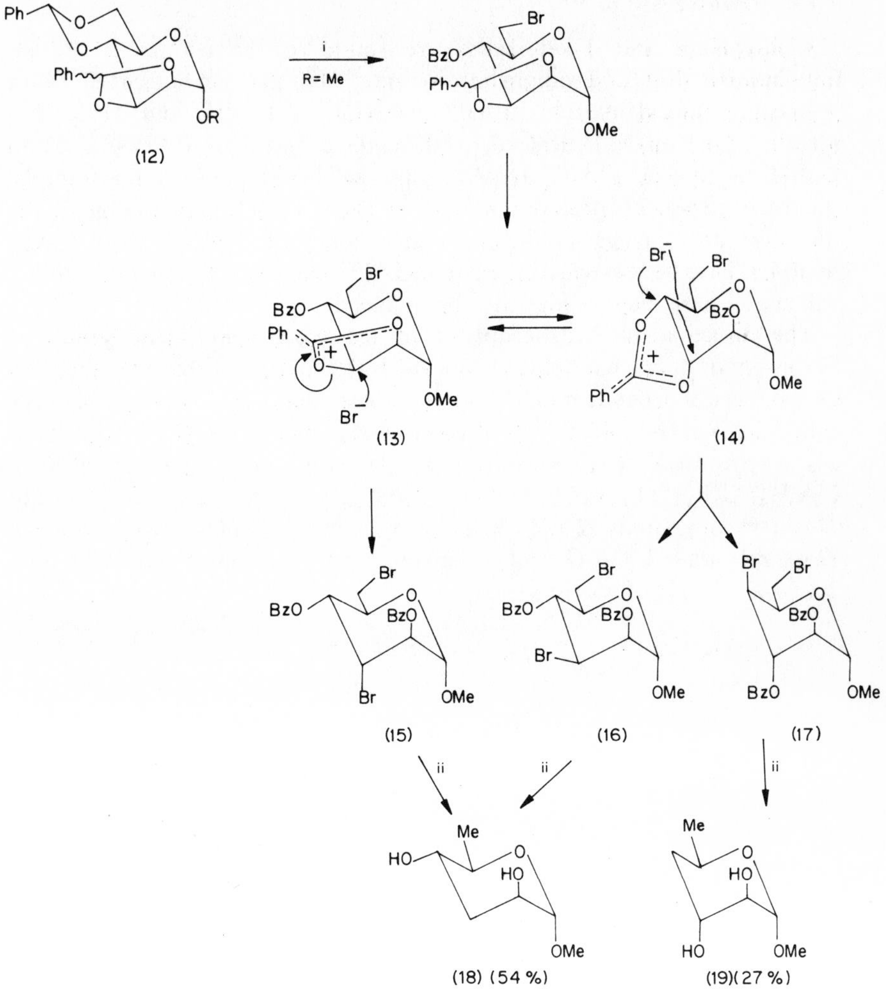

(12)

i
R = Me

(13) (14)

(15) (16) (17)

ii ii ii

(18) (54 %) (19)(27 %)

Scheme 3
Reagents: i, NBS; ii, LiAlH₄

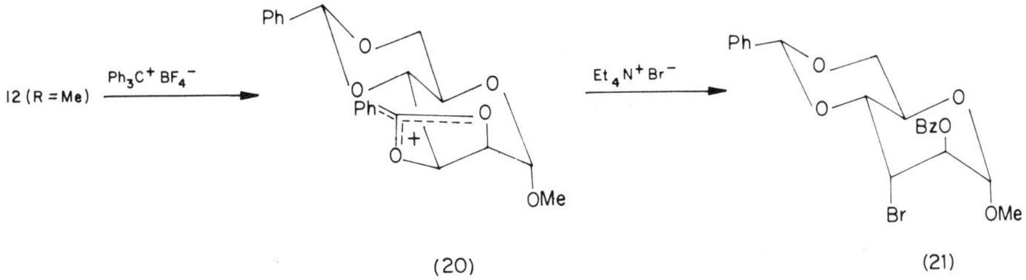

12 (R = Me) →[Ph₃C⁺ BF₄⁻]→ (20) →[Et₄N⁺Br⁻]→ (21)

1.1.2 *Hydrogenolysis*

1,3-Dioxolanes and 1,3-dioxanes are stable to the action of lithium aluminium hydride and sodium borohydride,[19] but they can be cleaved[20] with a so-called 'mixed hydride', usually a mixture of $LiAlH_4$ and $AlCl_3$. The identity of the 'mixed hydride' depends on the proportions of Lewis acid and hydride used; when $LiAlH_4$ and $AlCl_3$ are used in a ratio of 1:1, for example, the reactive species is probably AlH_2Cl.[21] The polar effects that influence the direction of cleavage of 1,3-dioxolanes and 1,3-dioxanes with 'mixed hydrides' have been extensively examined,[20,21] but, as the following examples will show, steric factors may also be involved.

The direction of hydrogenolysis of the 4,6-*O*-benzylidene group of hexopyranosides is determined primarily by the nature of the substituent on O-3, but is not dependent on the anomeric configuration or the nature of the substituents at O-1 and -2.[22] Hydrogenolysis of methyl 4,6-*O*-benzylidene-α-D-glucopyranoside **4** (R = H), for example, with a one-molar equivalent of $LiAlH_4$-$AlCl_3$ (1:1 ratio) in an inert solvent gave[23] a mixture of the 4- and 6-*O*-benzyl compounds **22** and **23**, respectively, in a ratio of 3:2, whereas the 3-*O*-benzyl- and 2,3-di-*O*-benzyl-D-gluco- and -D-manno-pyranoside deri-

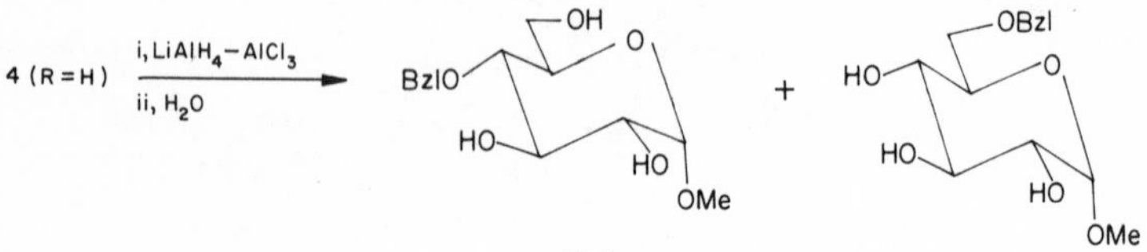

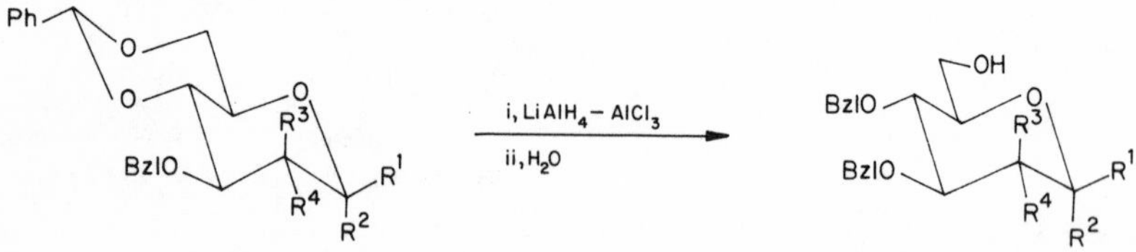

	R¹	R²	R³	R⁴
β–D–*gluco*	OBzl	H	H	OBzl
α–D–*gluco*	H	OPh	H	OBzl
	H	OMe	H	OBzl
	H	OMe	H	OH
α–D–*manno*	H	OBzl	OBzl	H

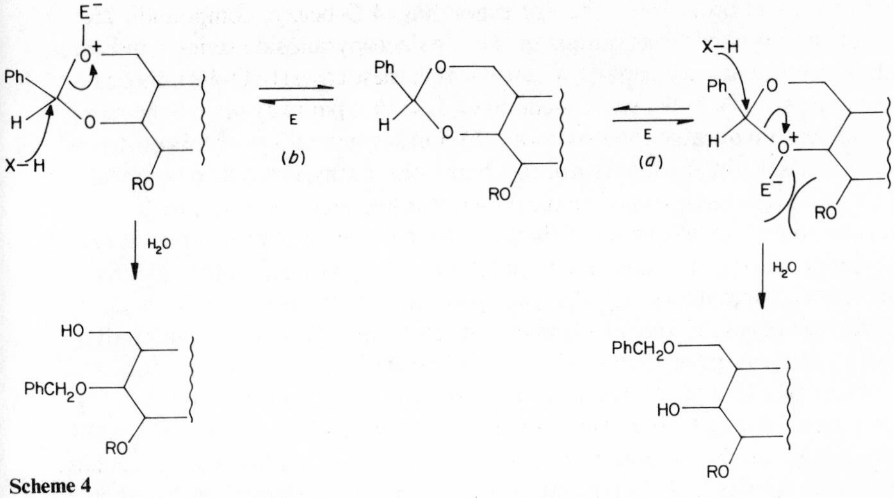

Scheme 4

E is AlH$_2$Cl or a related species

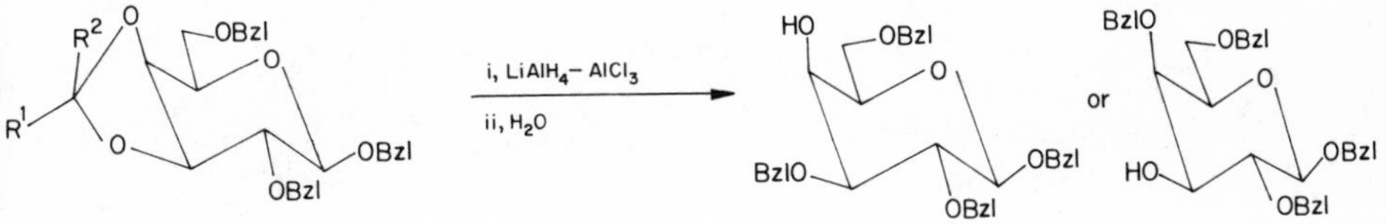

(26)

β-D-*galacto*

exo R^1 = Ph, R^2= H

endo R^1 = H, R^2= Ph

(27)   (28)

100%	0%
0%	100%

(29) R$_{endo}$ R$_{exo}$

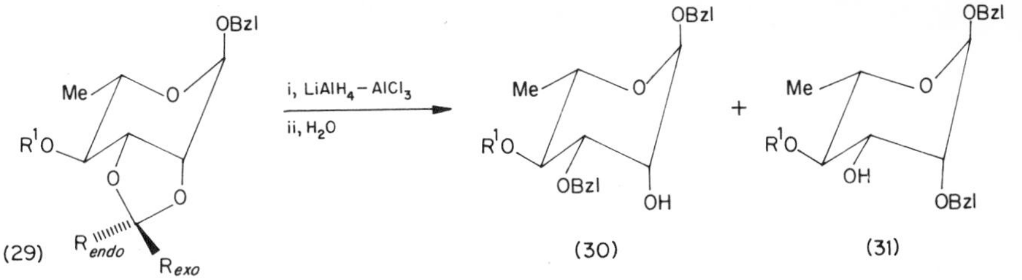

α-L-*rhamno*

(30)   +   (31)

R$_{exo}$	R$_{endo}$	R^1	%	%
Ph	H	H	98	2
Ph	H	Bzl	94	6
Ph	H	CH$_2$CH=CH$_2$	85	15
H	Ph	H	2	98
H	Ph	Bzl	18.5	81.5

vatives **24** afforded only the corresponding 4-*O*-benzyl compounds **25**.[22] These and related observations in the D-galactopyranoside series[22] indicate that access of the electrophile (a Lewis acid in these cases) to O-4 is impeded by the presence of a bulky substituent on O-3, so that pathway (*a*) in Scheme 4 is much less favourable than pathway (*b*). Only when O-3 is unsubstituted or carries a substituent that is not too bulky can pathway (*a*) also compete.

Extensive investigations[11] have shown that five-membered acetals are more susceptible to hydrogenolysis than are six-membered acetals, and that the direction of ring-cleavage of 3,4- and 2,3-*O*-benzylidene acetals of aldopyranosides is determined by the configuration of the acetal carbon atom. Hydrogenolysis of the *exo*-isomers of **26**[24] and **29**[25] with the LiAlH₄-AlCl₃ reagent gave exclusively or preferentially the equatorial *O*-benzyl compounds **27** and **30**, respectively, whereas the alternative mode of ring-cleavage, leading to the axial *O*-benzyl compounds **28** and **31**, was mainly observed with the *endo*-isomers. The reactions of related benzylidene acetals with the LiAlH₄-AlCl₃ reagent follow a similar pattern,[26] although the

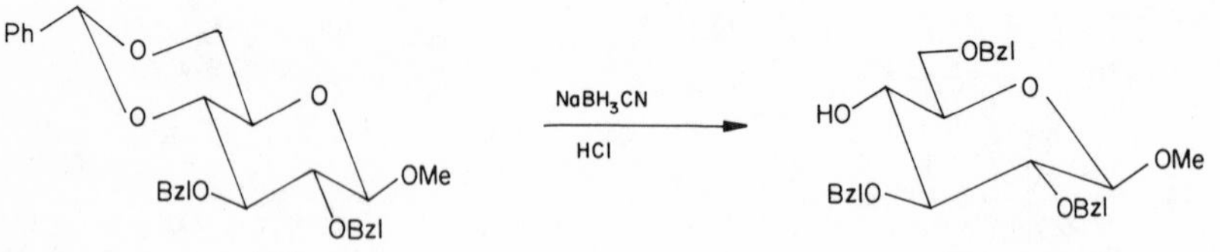

(32)   → NaBH₃CN / HCl →   (33) (87%)

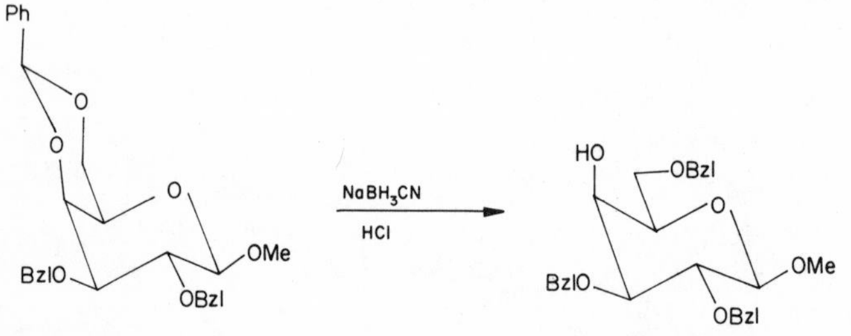

(34)   → NaBH₃CN / HCl →   (35) (81%)

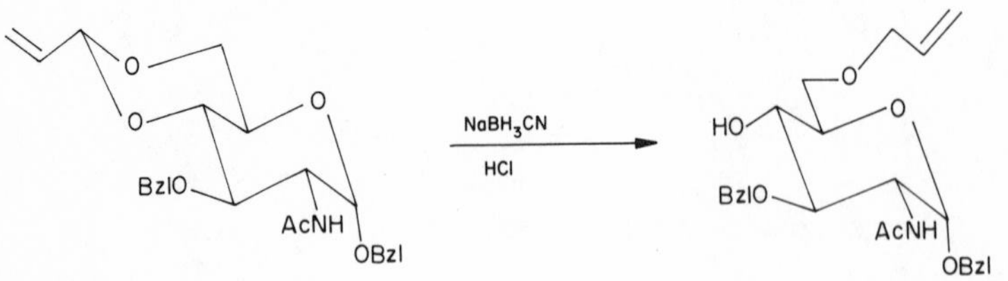

(36)   → NaBH₃CN / HCl →   (37) (74%)

factors governing the direction of ring-cleavage are not fully understood at present. An example of the usefulness of this type of reaction in oligosaccharide synthesis is described in a later section.

A facile, highly regioselective, reductive opening of 4,6-*O*-benzylidene acetals of hexopyranosides, to give *O*-benzyl ethers, is brought about by sodium cyanoborohydride-hydrogen chloride in an inert solvent.[27] The regioselectivity is opposite to that generally observed with the LiAlH$_4$-AlCl$_3$ reagent.[22] Thus, reduction of **32** and **34** in diethyl ether using sodium cyanoborohydride-hydrogen chloride furnished the compounds **33** and **35**, respectively, in which the newly-produced benzyl group is located on O-6 and the 4-OH group is unsubstituted.[27] The reaction is compatible with the presence of ester and amide groups in the starting material. In these reductions, the steric requirement of the proton is much smaller than that of a Lewis acid, so that the direction of the equilibrium (Scheme 4) appears to be governed chiefly by the relative acidities of O-4 and O-6, consequently favouring pathway (*a*). The regioselectivity observed[27] in the reduction of diastereoisomeric 2,3-*O*-benzylidene acetals of glycosides related to **29** is the same as that observed with the LiAlH$_4$-AlCl$_3$ reagent.

A useful variation in the range of substituents on partially protected hexopyranosides can be acquired by using 4,6-*O*-(prop-2-enylidene) acetals, such as **36**, which are reduced with sodium cyanoborohydride—hydrogen chloride in tetrahydrofuran to 6-*O*-allyl ethers (for example, **37**).[28]

1.1.3 *Photolysis*

Light-sensitive protecting groups are also potentially useful in the preparation of partially protected carbohydrates required for oligosaccharide and other syntheses. U.v.-irradiation of the *endo*- and *exo*-diastereoisomers of **38**, either

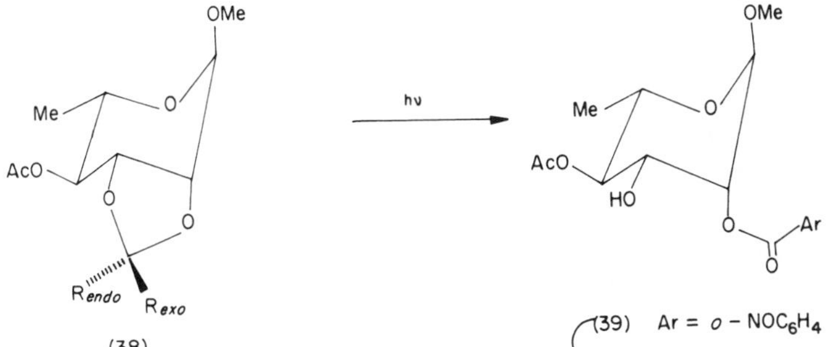

R$_{exo}$	R$_{endo}$
o-NO$_2$C$_6$H$_4$	H
H	o-NO$_2$C$_6$H$_4$

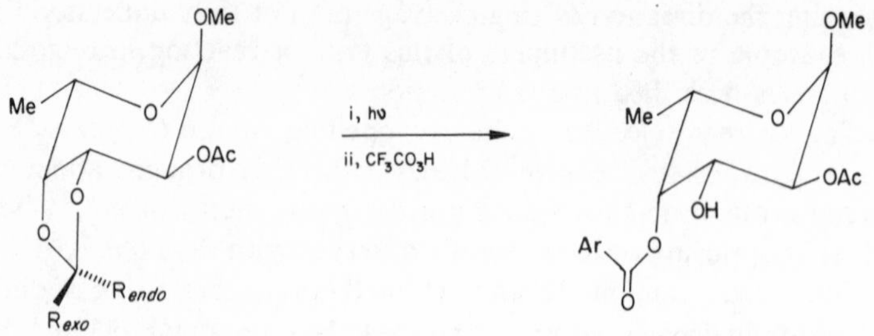

(41)

R_{exo}	R_{endo}
$o-NO_2C_6H_4$	H
H	$o-NO_2C_6H_4$

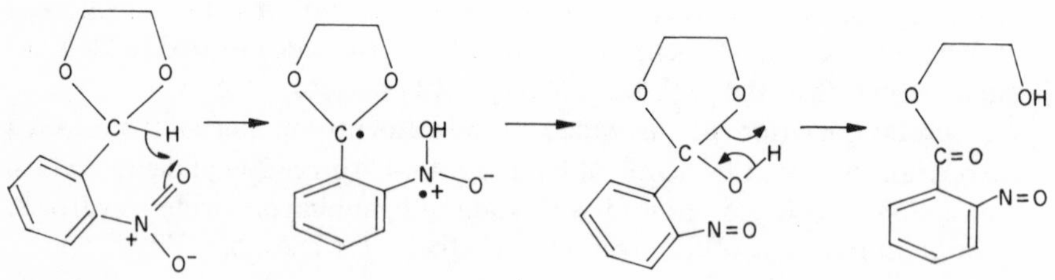

(42) $Ar = o-NO_2C_6H_4$

Scheme 5

separately or in a mixture, as approximately 1% solutions in methanol gave a compound that was assumed to be the hydroxynitrosobenzoate **39**, since on oxidation with trifluoroperoxyacetic acid it afforded the axial 2-*O*-(*o*-nitrobenzoate) **40** in 95% overall yield.[29] The oxidation step is usually performed immediately after irradiation in order to avoid further reaction or degradation of the photoproducts. U.v.-irradiation and subsequent oxidation of such 3,4-*O*-(*o*-nitrobenzylidene)hexopyranosides as **41** also furnished the axial regio-isomer **42** almost exclusively.

The fact that photochemical rearrangement does not occur with either *meta*- or *para*-nitrobenzylidene derivatives[30] is consistent with the mechanism set out in Scheme 5. The high regioselectivity ($\geq 95\%$) of ring-opening can then be ascribed[29] to the influences of conformational and stereoelectronic effects in the breakdown of the orthoacid intermediate, analogous to those operating in the hydrolysis of cyclic orthoesters,[31] which favour cleavage of the bond between the equatorial oxygen atom and the acetal carbon atom. A similar mechanistic rationale may be used to explain the regioselectivity observed in photochemical rearrangements of methyl 4,6-*O*-(*o*-

nitrobenzylidene)hexopyranosides, which invariably produce a mixture of the corresponding 4- and 6-O-(o-nitrosobenzoates).[30]

1.1.4 Acetal elimination

The fragmentation of 2-phenyl-1,3-dioxolanes brought about by butyllithium yields either an alkene and the benzoate anion (Scheme 6, pathway (a)) or benzaldehyde and an enolate anion (pathway (b)).[32] Pathway (b) was markedly favoured by the di-O-benzylidene derivative **12** (R = Me; reacted as

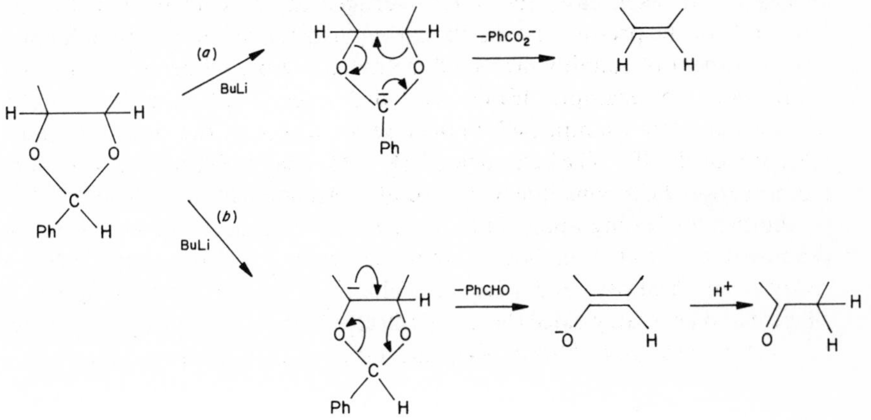

Scheme 6

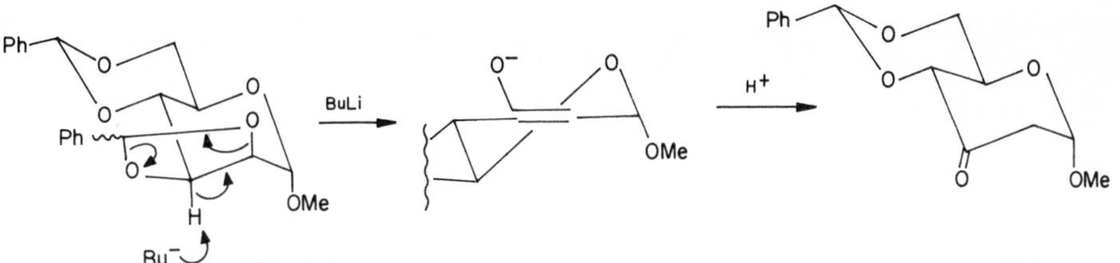

12 (R=Me)

(43)

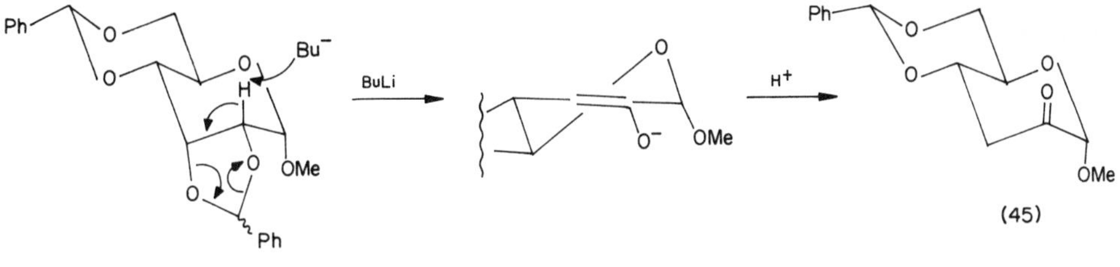

(44)

(45)

a mixture of diastereoisomers), which gave a high yield of the 3-keto sugar **43**, but only a trace of the 2,3-unsaturated sugar, on treatment with two molar equivalents of butyl-lithium in tetrahydrofuran at −30°C and subsequent quenching of the enolate anion.[14,33] The stability of the carbanion formed presumably determines which of the two pathways is followed. Further insight into the dependence of the regioselectivity on the configuration of the starting material was provided by the formation of the 2-keto sugar **45** (86% yield) when the di-O-benzylidene derivative **44** reacted with butyl-lithium.[34] The direction of cyclo-elimination is apparently determined by the configuration of the carbon atom bearing a hydrogen atom that could be abstracted. In each case, the *axial* hydrogen atom is abstracted, leading ultimately to the product having the carbonyl group at this position. Similar base treatment of suitably protected methyl 2,3-O-benzylidene-α-L-rhamno-pyranosides (for example, **46**) has provided access[35] to synthetically useful intermediates (for example, **47**) in the L-series, although the yield of product seldom exceeds 40%. The keto sugar **47** (R = Me) has been transformed[36] into L-evernitrose (**1**), a constituent of several everninomicin antibiotics.[1]

Another interesting application of acetal elimination from sugars is that promoted by lithium in liquid ammonia from 2,3-O-isopropylidene derivatives of furanosyl and pyranosyl chlorides.[37] The reaction, which is efficient and general, yielded the glycals **49** and **51** from 2,3-O-isopropylidene-β-D-erythrofuranosyl chloride (**48**) and 6-deoxy-2,3-O-isopropylidene-4-O-

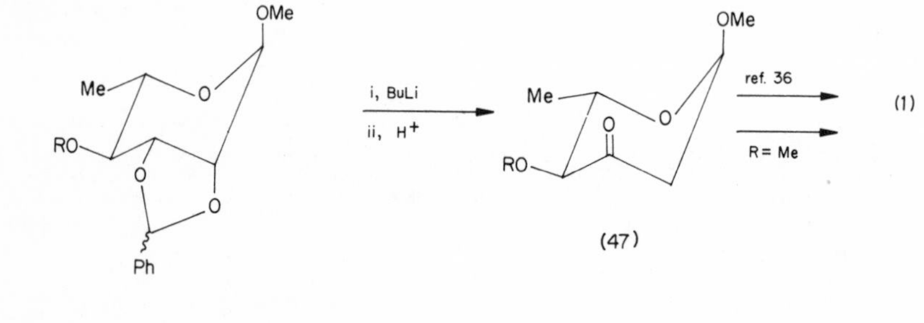

(46)   R = Me, MOM, MEM, or THP

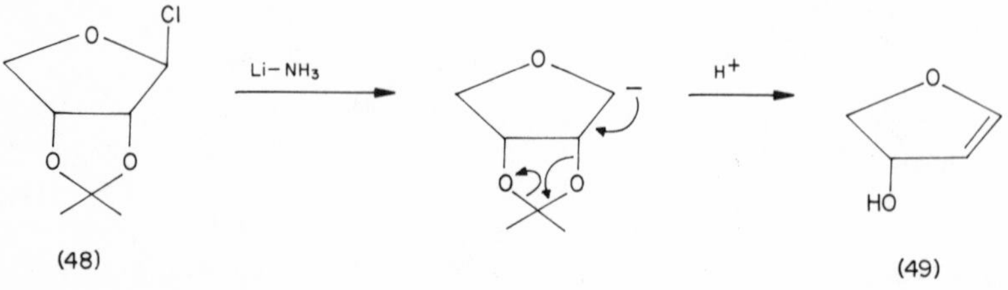

(50) (51)

methyl-β-L-gulopyranosyl chloride (**50**), respectively. The reduction of related furanosyl and pyranosyl halides with sodium naphthalide in tetrahydrofuran also gave furanoid and pyranoid glycals,[38] which are useful precursors in synthetic carbohydrate chemistry.[39] Application of the enolate Claisen rearrangement to furanoid glycals has also been used to control the stereochemistry of the side-chain in chiral syntheses of (−)- and (+)- nonactic acids.[40]

1.2 Other useful protecting groups and reactions

1.2.1 *Allyl and substituted-allyl groups*

Allyl and substituted-allyl groups have been used increasingly as temporary protecting groups in carbohydrate chemistry over recent years.[8,41] Allyl ethers **52** are readily prepared and are stable to aqueous alkali and moderately acidic conditions, but they undergo isomerization to *cis*-prop-1-enyl ethers **53** under certain basic conditions.[42] As the latter compounds are vinyl ethers, they can be cleaved under mildly acidic conditions. Procedures for removing prop-1-enyl groups under essentially non-acidic conditions include treatment with a mercury(II) chloride-mercury(II) oxide reagent[43] or with palladium-on-charcoal in a hydroxylic solvent containing a *trace* of acid.[44] The isomerization **52** → **53** is usually carried out with potassium *tert*-butoxide in dimethyl sulphoxide,[45] but it is also accomplished in other ways, including catalysis by tris(triphenylphosphine)rhodium(I) chloride under neutral conditions.[46] Isomerization of an *O*-allyl group by the rhodium catalyst produces a mixture of *cis*- and *trans*-prop-1-enyl ethers,[47] but this does not cause undue problems since the prop-1-enyl group is usually removed straight away.

$$\text{ROCH}_2\text{CH}{=}\text{CH}_2 \xrightarrow{\text{base}} \quad \overset{\overset{\displaystyle H}{|}}{\underset{\underset{\displaystyle RO}{|}}{C}}{=}\overset{\overset{\displaystyle H}{|}}{\underset{\underset{\displaystyle Me}{|}}{C} \quad \xrightarrow{\text{H}_3\text{O}^+} \text{ROH} + \text{CH}_3\text{CH}_2\text{CHO}}$$

(**52**) (**53**)

The reactivities of allyl and substituted-allyl ethers towards the actions of potassium *tert*-butoxide in dimethyl sulphoxide and the rhodium catalyst are

Table 1.1 Properties of allyl and related ethers

Name and formula	Stability to acid and base[a]	$KOBu^t/DMSO$	$(Ph_3P)_3RhCl$[b]	References
allyl ether $ROCH_2CH=CH_2$	acid and base stable	rapid isomerization to $ROCH=CHMe$[c]	rapid isomerization (1h) to $ROCH=CHMe$	43, 45, 46
prop-1-enyl ether $ROCH=CHMe$	acid labile base stable	stable	stable	43, 45, 46
2-methylallyl ether $ROCH_2\overset{\mid}{C}=CH_2$ Me	acid and base stable	slow isomerization to $ROCH=CMe_2$	rapid isomerization to $ROCH=CMe_2$[f]	46, 48
but-2-enyl ether $ROCH_2CH=CHMe$[d]	acid and base stable	rapid cleavage to ROH[e]	slow isomerization (24h) to $ROCH=CHCH_2CH_3$	46, 49
3-methylbut-2-enyl ether $ROCH_2CH=CMe_2$	acid and base stable	rapid cleavage to ROH[e]	stable (after 24h)	47

[a] Refers to moderately acidic conditions and, usually, to the basic conditions used for O-benzylation;
[b] in refluxing water-ethanol-benzene;
[c] at elevated temperatures;
[d] a mixture of trans- and cis-isomers in a ratio of c. 95:5;
[e] at or above room temperature;
[f] at a slightly lower rate than an allyl group.

shown in Table 1.1. In contrast to the rapid isomerization of an O-allyl group with the basic reagent at elevated temperatures, that of an O-(2-methylallyl) group is very much slower, whereas O-(but-2-enyl) and O-(3-methylbut-2-enyl) groups are rapidly cleaved at room temperature, at comparable rates, to give the alcohol directly. Since an O-(but-2-enyl) group is cleaved more rapidly than an O-allyl group is isomerized, the 3-O-(but-2-enyl) group was completely removed from the allyl α-D-galactoside **54** by the action of potassium *tert*-butoxide in dimethyl sulphoxide at room temperature to give **55**, with little isomerization to the prop-1-enyl glycoside **56**.[49] The latter compound was obtained from **55** by using the basic reagent at a higher temperature.

The versatility of allyl and substituted-allyl groups is enhanced by significant differences in their rates of isomerization by tris(triphenyl-phosphine)rhodium(I) chloride. An O-(but-2-enyl) group is isomerized much more slowly with the rhodium catalyst than are O-allyl and O-(2-methylallyl) groups,[46] whereas an O-(3-methylbut-2-enyl) group is relatively stable.[47] Thus, brief treatment of **57** with the rhodium catalyst gave the prop-1-enyl

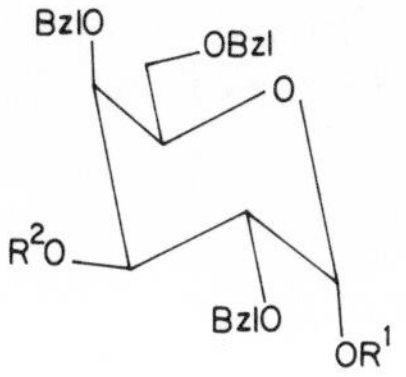

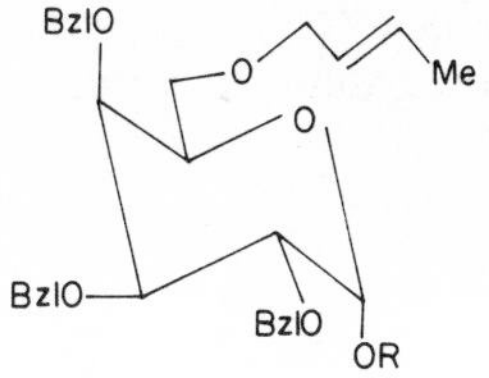

(54) $R^1 = CH_2CH=CH_2$,
      $R^2 = CH_2CH=CHMe$

(55) $R^1 = CH_2CH=CH_2$, $R^2 = H$

(56) $R^1 = CH=CHMe$, $R^2 = H$

(57) $R = CH_2CH=CH_2$

(58) $R = CH=CHMe$

(59) $R = H$

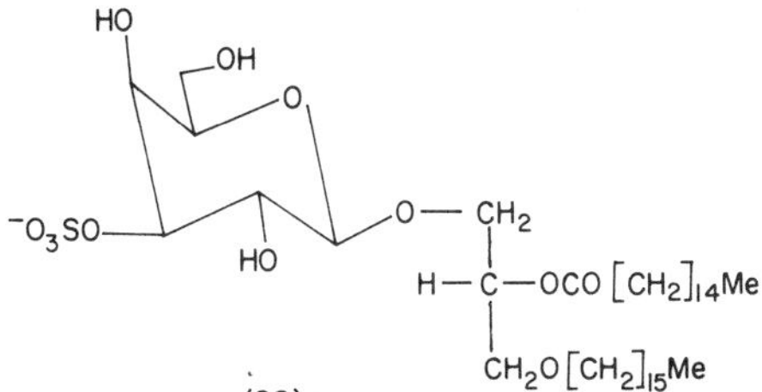

(60)

glycoside **58**, which on hydrolytic removal of the prop-l-enyl group with mercury (II) chloride in aqueous acetone afforded **59**.[50] The ready hydrolysis of prop-l-enyl glycosides like **58** has been used to advantage in the preparation of *O*-benzylated monosaccharides[45] in order to overcome the problem of benzyl ether cleavage that may accompany[51] the removal of other glycosidic substituents under acidic conditions.

An impressive illustration of the usefulness of allyl and allied ethers as temporary protecting groups can be found in a recent synthesis[52] of 'seminolipid' **60**, the major species of glycolipid present in mature mammalian testes and spermatozoa.

1.2.2   *The tetraisopropyldisiloxane-1,3-diyl group*

A novel approach to the synthesis of partially protected ribonucleosides is based[53] on the use of a bifunctional silylating agent, namely 1,3-dichloro-1,1,3,3-tetraisopropyldisiloxane (**61**). Guanosine (**62**), for example, reacted with **61** to give the 3′,5′-*O*-(tetraisopropyldisiloxane-1,3-diyl) derivative **63**, in which the 2′-OH group can be protected prior to removal of the silyl group with base. Equally interesting is the observation that, on treatment with mesitylenesulphonic acid or pyridine hydrochloride in anhydrous *N,N*-

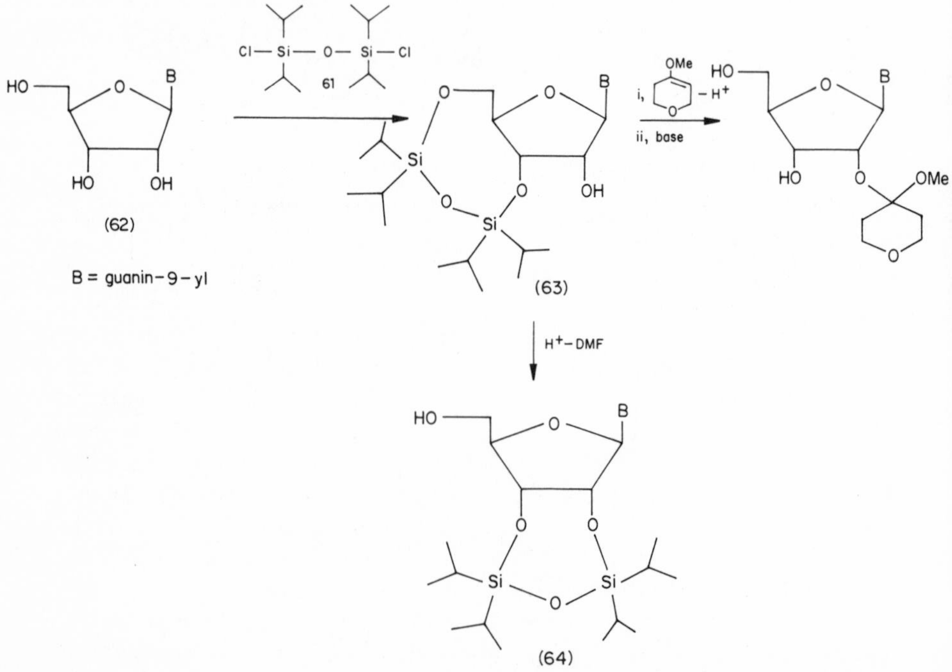

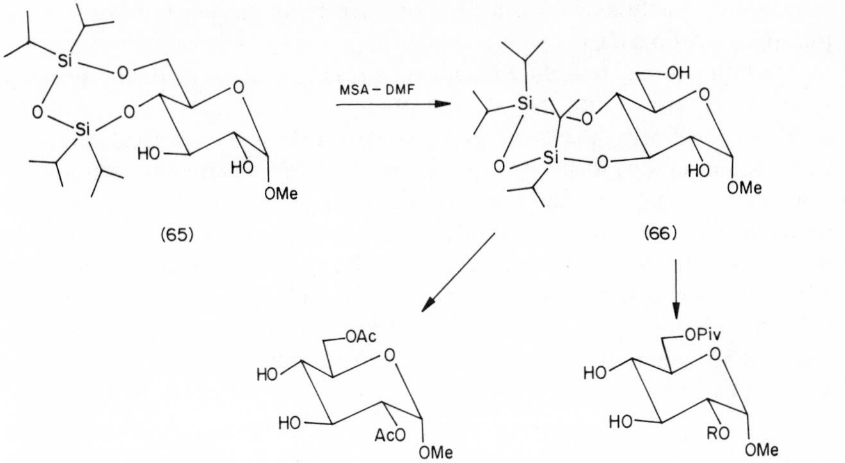

(65)

MSA – DMF

(66)

(67)

(68) R = H or MTHP

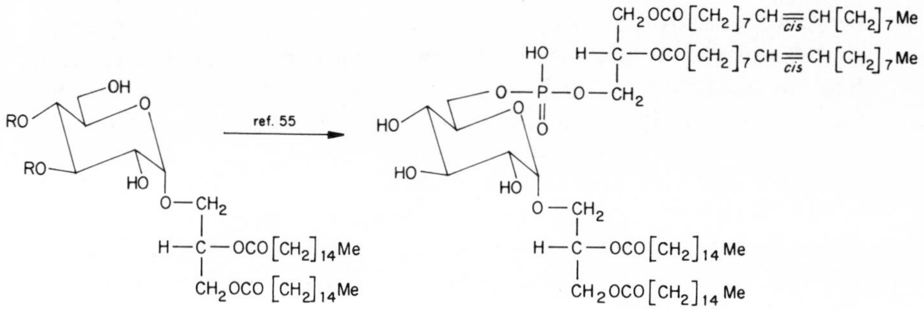

(69) R = H

(70) R,R = $Pr_2^i SiOSiPr_2^i$

(71)

ref. 55

dimethylformamide, **63** isomerized to the 2′,3′-*O*-(tetraisopropyldisiloxane-1,3-diyl) derivative **64**.[54] This acid-catalysed isomerization appears to occur only when *N,N*-dimethylformamide is used as the solvent.

Methyl α-D-glucopyranoside reacted with **61** in anhydrous pyridine to give the crystalline derivative **65** in 60% yield.[54] Isomerization of **65** by the action of mesitylenesulphonic acid in *N,N*-dimethylformamide yielded the 3,4-*O*-silylated derivative **66**, which can be converted into a variety of potentially useful 2,6-di- and 6-substituted derivatives of the parent glycoside (for example, **67** and **68**) by esterification and subsequent removal of the silyl group with tetrabutylammonium fluoride. The conversion of the glyceride **69** into the 3,4-*O*-(tetraisopropyldisiloxane-1,3-diyl) derivative **70**, by a similar sequence of reactions, constituted a key step in an elegant synthesis of the phosphoglycolipid **71**.[55]

The bifunctional reagent **61** appears to react initially with the primary hydroxyl group, so that the position of ring-closure will be decided by the ability of the newly introduced group to span two positions. Acid-catalysed ring-opening of the cyclic derivative (for example **65**) presumably occurs at the primary position, so that the position of any new ring-closure will be determined by the ability of the disiloxane-1,3-diyl group to span the distance between the new site and the original secondary position. Hence, the location of the primary hydroxyl group within the molecule may ultimately influence the structure of the rearranged product. Despite their obvious promise, there has been no systematic study of the reactions of bifunctional reagents, such as **61**, with carbohydrates.

1.2.3   Organotin derivatives

Although a large body of information[56] exists on the relative reactivities of carbohydrate hydroxyl groups towards electrophilic reagents, differences between the reactivities of the secondary hydroxyl groups are generally small, so that attempts at selective substitution often give more than one product. Recently increasing use has been made of organotin derivatives to activate selectively one of the hydroxyl groups of pyranosides towards electrophilic reagents, thereby providing ready access to a number of valuable derivatives that in the past have been difficult to obtain.

The reaction of 1,2-diols with dibutylstannyl bisalkoxides[57] or oxide[58] suspended in benzene, with azeotropic removal of either the alcohol or water, gives the so-called 'dibutylstannylenes'.[59] From their general properties, dibutylstannylenes appear to exist in non-polar solvents as dimers **72** of the acetal-like compounds **73**.[60] The ^{119}Sn chemical shifts in the n.m.r. spectra of deuteriochloroform solutions of carbohydrate stannylenes are found[61] in the range characteristic of penta-coordinated tin, and the dibutylstannylene derivative of methyl 4,6-O-benzylidene-α-D-glucopyranoside is known[62] to exist as a dimer **74** in the solid state. Simply as a matter of convenience, carbohydrate stannylenes will usually be represented as uncoordinated, monomeric units, although the presence of uncoordinated, five-membered ring stannylene derivatives is unlikely in view of the probable strain associated with the O-Sn-O angle[62] (c. 80°), which is better accommodated in a trigonal bipyramid. Moreover, dibutylstannylene derivatives of carbohydrates have sometimes been reacted in good coordinating solvents, often in the presence of an excess of an amine, so that complexes of monomeric units with solvent, amine, or even reagents may be present.

The dibutylstannylene derivative **74** reacted with acyl and sulphonyl

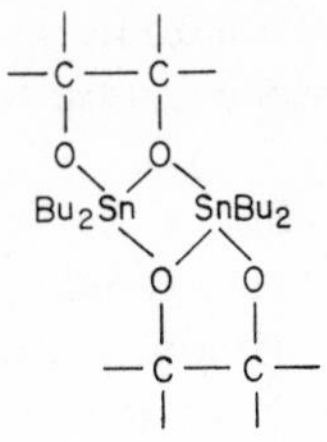

(72)

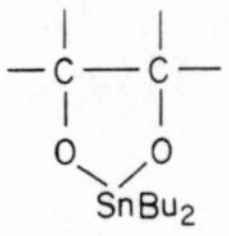

(73)

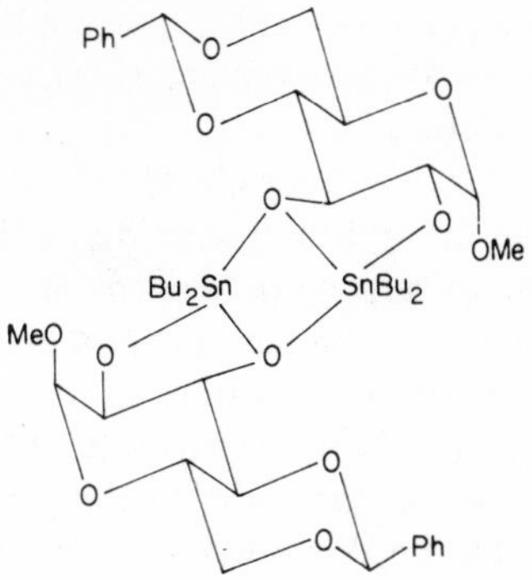

(74)

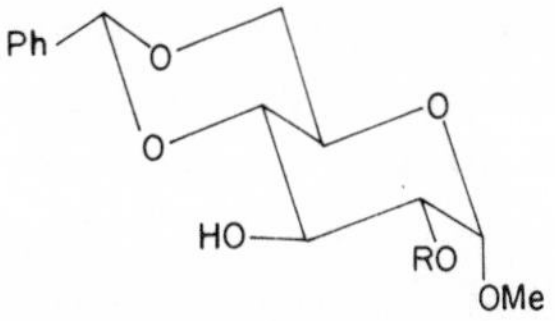

(75) R = Bz, CO[CH₂]₁₀Me, or Ts

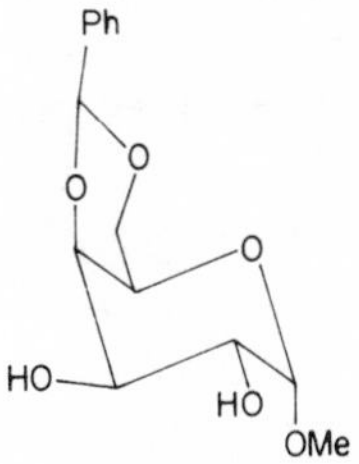

(76)

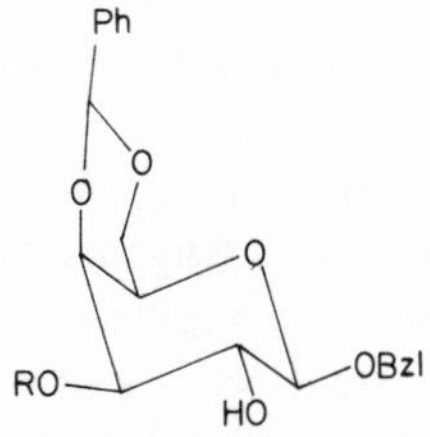

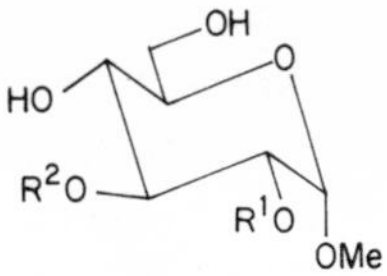

(77) R¹ = R² = H
(78) R¹ = Bz or Ts, R² = H
(79) R¹,R² = SnBu₂

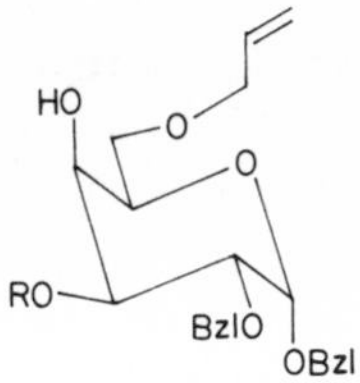

(80) R = H
(81) R = Bz

(82) R = H
(83) R = Bzl

chlorides in 1,4-dioxane, in the presence of triethylamine, at room temperature to give the 2-esters **75** in good yield.[63] Selective esterification of position-2 of the 2,3-O-(dibutylstannylene) derivatives of the α-D-galactopyranoside **76** and methyl 4,6-O-benzylidene-α-D-allopyranoside also occurred under these conditons, whereas the β-glycoside corresponding to **74** and the 2,3-O-(dibutylstannylene) derivative of methyl 4,6-O-benzylidene-α-D-manno-pyranoside afforded mixtures containing both 2- and 3-esters. Thus, it is clear that the configuration at the anomeric centre of such derivatives as **74** has an important bearing on the regioselectivity of the reaction, possibly involving coordination of the glycosidic oxygen atom to the tin atom of the active species. An even more impressive regioselectivity is observed with the dibutylstannylene derivative of methyl α-D-glucopyranoside(**77**), which again yielded the 2-esters **78** on treatment with acyl and sulphonyl halides.[63] The fact that 2-esters are formed in the presence of the more reactive primary hydroxyl group indicates that the dibutylstannylene formed from **77** has the structure **79**.

A subsequent study[64] has shown that benzoylation of the 2-oxygen atom of the dibutylstannylene derivative **74** takes place very rapidly (< 5 min) and quantitatively in benzene at room temperature without added bases, which might change the properties of the active species by coordination. Under analogous conditions, the dibutylstannylene derivative of the diol **80** afforded exclusively the 3-benzoate **81** (cf. the regioselectivity observed with related β-D-glycopyranosides in the presence of triethylamine[63]).

In the dimeric structure **74**,[62] which probably persists in weakly polar solvents, the 3-oxygen atom of one monomeric unit is coordinated to the tin atom of the other, and it has been suggested[64] that this may result in a general deactivation towards electrophilic attack at this oxygen atom. Similar arguments[64] have been advanced to account for the regioselectivity observed with other carbohydrate dibutylstannylenes in their reactions with electrophilic reagents. Since it is not known whether the dimers are involved in the rate-determining step, it is safer to make the more general assumption that the regioselectivity observed in weakly polar solvents is governed by the abilities of the stannylene oxygen atoms to coordinate with diorganotin compounds in the medium.

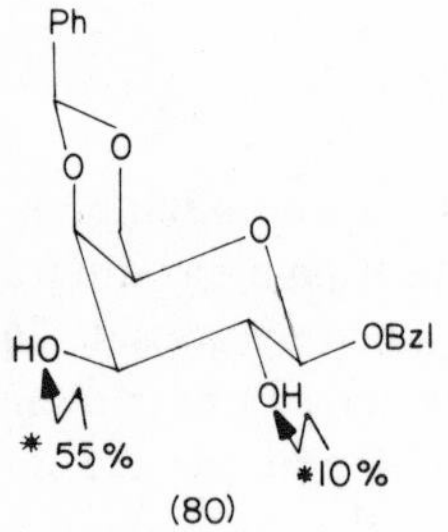

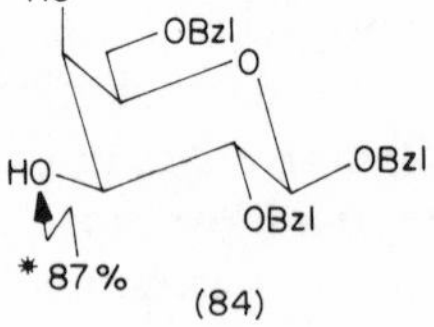

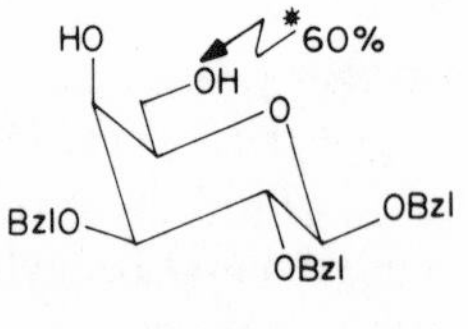

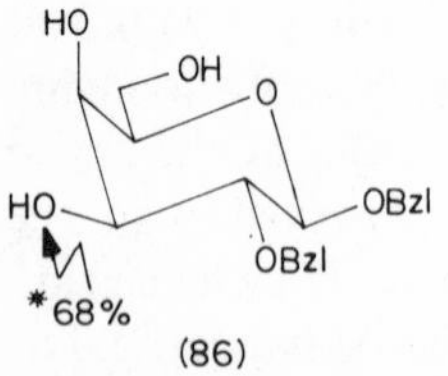

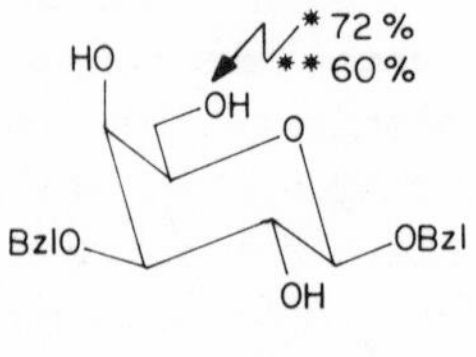

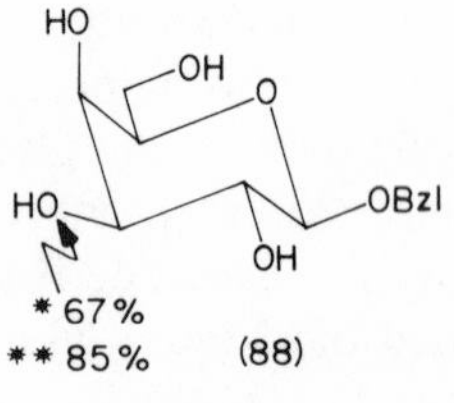

By contrast, the reactions of carbohydrate stannylenes with benzyl and allyl bromides normally proceed at insignificant speed in refluxing benzene solution, and it has been necessary to conduct them in N,N-dimethylformamide at 100°C.[65] Under these conditions, the crude dibutylstannylene derived from **82** reacted with benzyl bromide to give only the 3-substituted product **83**. It is noteworthy that partial benzylation of α-D-galactopyranosides related to **82** in *N,N*-dimethylformamide normally occurs at position 4.[66] Recent investigations[61] have shown that the reactions of active bromides with carbohydrate stannylenes in refluxing benzene are markedly accelerated in the presence of quaternary ammonium halides. Thus, in the presence of a one-molar equivalent of tetrabutylammonium iodide in refluxing benzene, the dibutylstannylene derivatives of the partially protected benzyl β-D-galactopyranosides **80** and **84–88** reacted with an excess of benzyl(∗) or allyl(∗∗) bromide at the positions, and in the yields, indicated. Only the diol **80** afforded a mixture of products, otherwise regiospecific substitution was observed. The synthesis of benzyl 3-*O*-benzyl-β-D-galactopyranoside (**87**) from the tetraol **88** would normally require several steps, whereas it is achieved by a one-pot synthesis via the dibutylstannylene derivative, although it is necessary to remove organotin by-products at the end of the reaction by chromatography. The effect of quaternary ammonium iodides cannot simply be due to the replacement of bromine by iodine in the active halide, since quaternary ammonium bromides are also effective catalysts. A more likely explanation[61] is that coordination of the halide ion to the tin atom enhances the nucleophilicity of one of the bound oxygen atoms.

Some general comments[61] can be made on the substitution patterns of benzyl β-D-galactopyranoside and its derivatives on the basis of the above results. Substitution of a stannylene involving the primary position always occurs at that position. However, five-membered stannylene rings are formed in preference to six-membered stannylene rings. When a five-membered stannylene ring spans the 3- and 4-positions substitution always occurs on O-3, whereas the outcome of the reaction may be more complex when it spans the 2- and 3-positions.

As well as enhancing the nucleophilicity of oxygen atoms towards electrophilic reagents, carbohydrate stannylenes are rapidly oxidized with bromine at room temperature to hydroxy-ketones[64] (Scheme 7). The oxidation, which is carried out on solutions of the crude stannylene derivatives in benzene in the presence of 4 Å molecular sieves or tributyl-stannyl methoxide, is regiospecific, giving only one of the two possible

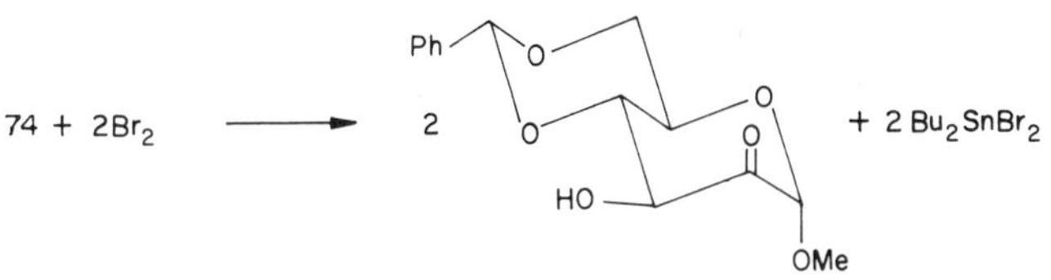

$$74 + 2Br_2 \longrightarrow 2 \quad (89) \quad + 2 Bu_2SnBr_2$$

(89)

Scheme 7

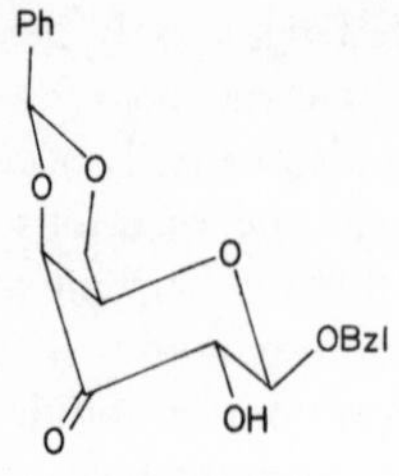

(90)

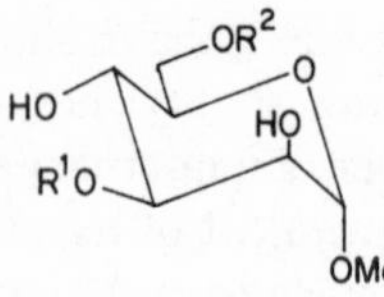

(91) $R^1 = R^2 = H$

(92) $R^1 = R^2 = CH_2CH=CH_2$

(93) $R^1 = CH_2CH=CH_2$, $R^2 = H$

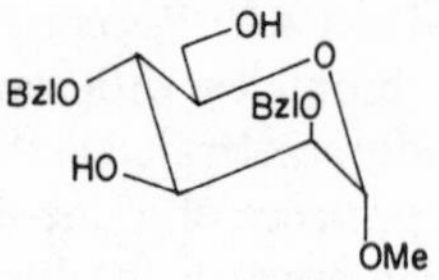

(94)

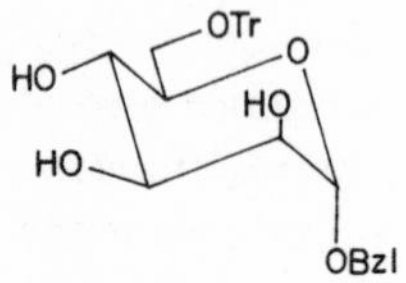

(95)

hydroxy-ketones. Thus, brominolysis converted **74** and the 2,3-*O*-(dibutyl-stannylene) derivative of **80** into **89** and **90**, respectively. The preferred site of oxidation is the same as that of benzoylation of these derivatives in benzene solution in the absence of a base.

Regioselective enhancement of the nucleophilicity of hydroxyl groups can also be achieved by means of trialkylstannylation. As an illustration, treatment of methyl α-D-mannopyranoside (**91**) with 1.5 molar proportions of bis(tributylstannyl) oxide in toluene at 140°C, with continuous removal of water, gave a partially stannylated product, which on heating with allyl bromide afforded the 3,6-di- and 3-*O*-allyl derivatives, **92** and **93**, in 71 and 13% yield, respectively.[67] Benzylation of **92** and subsequent removal of the allyl groups yielded methyl 2,4-di-*O*-benzyl-α-D-mannopyranoside (**94**), a potentially useful intermediate for the synthesis of branched-chain oligosaccharides of biological interest. Tetrabutylammonium bromide efficiently catalysed the monoallylation of a partially tributylstannylated derivative of **95**, the reaction with a moderate excess of allyl bromide yielding the corresponding 3- and 2-*O*-allyl ethers in 62 and 15% yield, respectively.[68]

1.2.4 Deoxygenation

The need to replace a secondary hydroxyl group by hydrogen is frequently encountered in the synthesis of the rare sugars that are found as components of antibiotics and other classes of natural product. Furthermore, the selective removal of secondary hydroxyl groups may confer enhanced biological activity on many polyhydroxylated antibiotics. In general, the primary hydroxyl groups of sugars are easily converted into toluene-p-sulphonates and methanesulphonates, etc., which are equally easily reduced either directly or via appropriate halide or sulphur displacements followed by reduction, to the corresponding methyl compounds.[69] The reduction of sulphonic esters at the secondary positions of sugars generally results in S–O bond-cleavage, with regeneration of the alcohol,[69] although displacement procedures can be extended to secondary alcohols where the leaving group is attached to a carbon atom at which S_N2 processes take place readily.[70] Use of the readily displaceable trifluoromethanesulphonate (triflate) leaving group substantially improves the prospects of achieving an S_N2 process at secondary positions, as in the case of the triflate ester **96**, which reacted with tetrabutylammonium iodide in refluxing benzene to give the iodo derivative **97** in 87% yield.[71] Subsequent reduction of **97** afforded **98**, a deoxygenated analogue of **3**. 2-Alkoxy-N,N'-diphenyl-1,3,2-diazaphospholanes[72] (**99**) are also readily transformed into chlorides (by the action of sulphuryl chloride at 0°C), bromides (by the action of bromine at 0°C), and iodides (by the action of methyl iodide in toluene at 85°C), with inversion of configuration, and appear to offer an attractive alternative to triflate esters for the preparation of halogenated and deoxy sugars.

Access to deoxygenated analogues of carbohydrates, via halogenated derivatives, is also provided by other phosphorus-based reagents that do not require prior derivatization of the hydroxyl group undergoing replacement. Isolated primary and secondary hydroxyl groups of carbohydrate derivatives

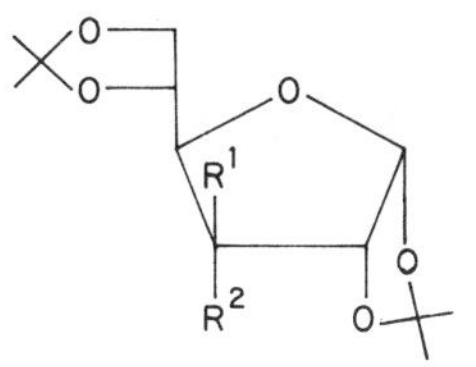

(96) $R^1 = OSO_2CF_3$, $R^2 = H$
(97) $R^1 = H$, $R^2 = I$
(98) $R^1 = R^2 = H$

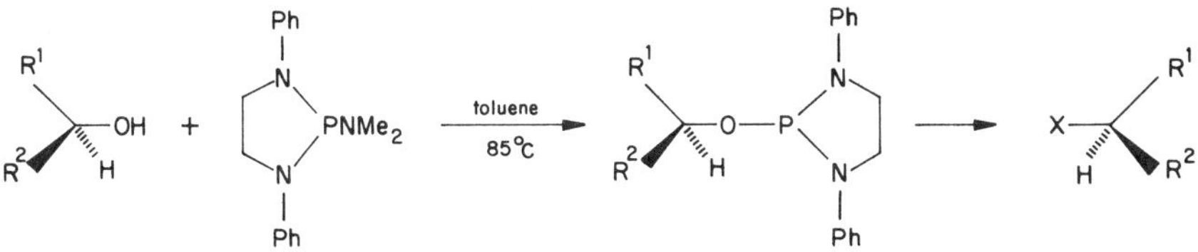

(99)

X= Cl, Br, or I

are transformed into iodo groups, with inversion of configuration, on treatment in toluene at elevated temperatures with either triphenylphosphine, iodine, and imidazole or triphenylphosphine and tri-iodoimidazole.[73] These systems are heterogeneous and the carbohydrate derivative does not have to be soluble in toluene. Thus, **3** reacted with triphenylphosphine, iodine, and imidazole in toluene at 120°C to give the 3-iodide **97** in 60% yield; under identical conditions, a 78% yield of **97** was obtained using triphenylphosphine and tri-iodoimidazole. This novel reaction, which is related to the Rydon reaction,[74] is likely to proceed via the intermediates shown in Scheme 8. Since S_N2 displacements are normally difficult to effect at the 2-position of α-glycosides,[75] the facile conversion of **100** and **101** into the 2-iodoglycosides **102** and **103**, respectively, in yields of 82–87% by this procedure is particularly noteworthy.

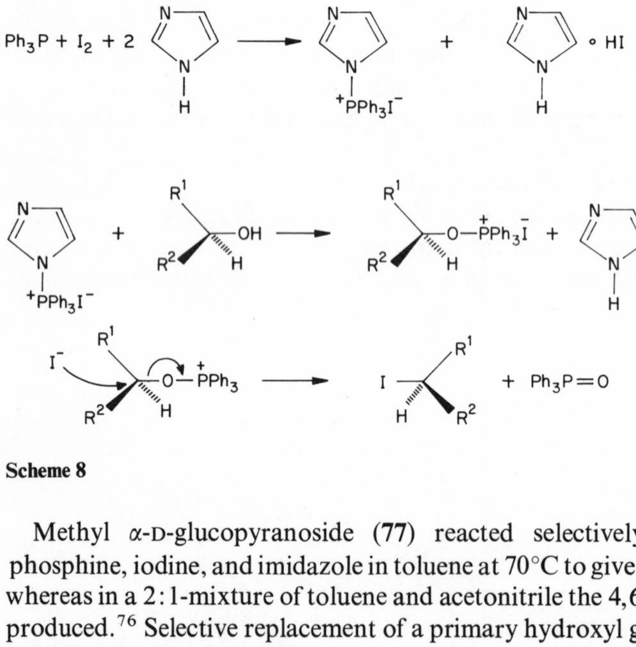

Scheme 8

Methyl α-D-glucopyranoside (**77**) reacted selectively with triphenylphosphine, iodine, and imidazole in toluene at 70°C to give the 6-iodide **104**,[73] whereas in a 2:1-mixture of toluene and acetonitrile the 4,6-di-iodide **105** was produced.[76] Selective replacement of a primary hydroxyl group by a halogen atom in the presence of unprotected secondary hydroxyl groups is also possible using triphenylphosphine in combination with either an N-halosuccinimide[77] or carbon tetrahalides.[78]

Somewhat unexpectedly, methyl α-D-glucopyranoside (**77**) gave the 3,6-dibromide **106** on treatment with triphenylphosphine, bromine, and imidazole in toluene at elevated temperature.[79] The dibromide **106** was readily reduced to methyl α-paratoside (**107**), so that an extremely short route to this 3,6-dideoxy sugar from methyl α-D-glucopyranoside (**77**) is now available.

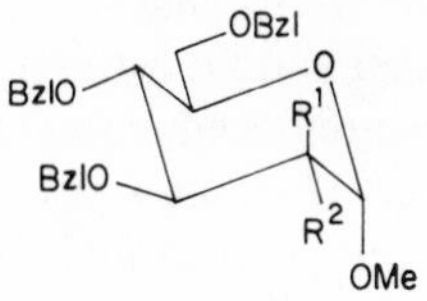

(100) R¹ = H, R² = OH

(100) R^1 = H, R^2 = OH
(101) R^1 = OH, R^2 = H
(102) R^1 = I, R^2 = H
(103) R^1 = H, R^2 = I

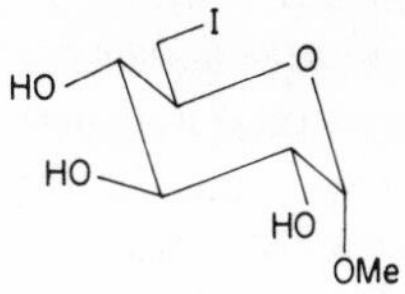

(104)

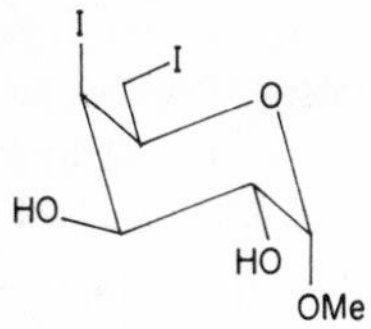

(105)

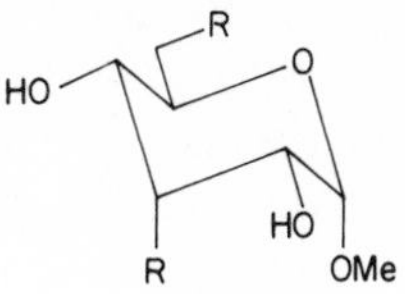

(106) R = Br
(107) R = H

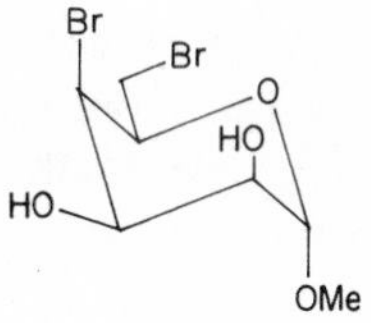

(108)

Similar dibromination of methyl α-D-mannopyranoside (**91**) furnished the 4,6-dibromide **108** in good yield, but neither methyl α- nor β-D-galacto-pyranoside afforded dibrominated products.[79] This novel bromination system also effected the replacement of an hydroxyl group by bromine, with inversion of configuration, at the 2-position of otherwise protected methyl α-D-gluco- and -manno-pyranosides (for example, **100** and **101**).

Despite the scope of the phosphorus-based reagents described above, there are still situations[75] where the secondary hydroxyl groups of sugars are attached to carbon atoms at which S_N2 processes are hindered. In such cases, deoxygenation may be achieved by homolytic cleavage of the carbon-oxygen bond to give carbon radicals which are quenched by hydrogen-atom transfer from a suitable donor. In practice this is accomplished by the reduction of thioesters or S-methyl dithiocarbonates with tributylstannane, according to the general mechanism shown in Scheme 9, whereby the conversion of a thiocarbonyl group to a carbonyl group provides the driving force for the

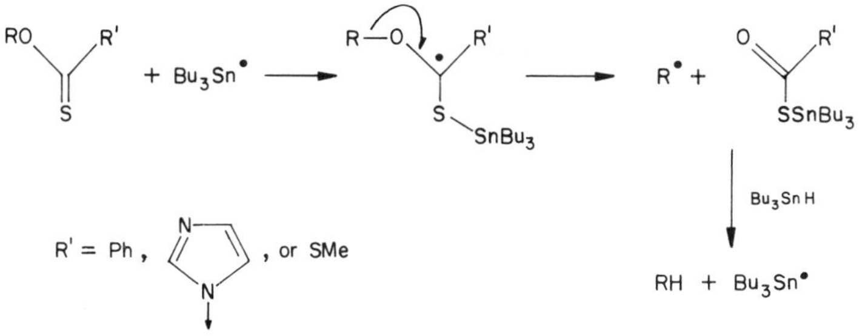

Scheme 9

reaction.[80] The reaction takes place under neutral conditions, usually in refluxing toluene or other aromatic solvents, and is compatible with the presence of a wide range of functional and protecting groups. Reduction of the *S*-methyl dithiocarbonate **109**[80] and the thiobenzoates **111**[80] and **113**,[81] for example, gave, after removal of organotin by-products by chromatography, the deoxygenated analogues **110**, **112**, and **114**, respectively.

Occasional use has been made of the photochemical cleavage of carboxylic esters[82] to prepare deoxy sugars. A simple synthesis of methyl α-D-amicetoside (**116**) was based on u.v.-irradiation of a solution of methyl 2,3,6-tri-*O*-pivaloyl-α-D-glucopyranoside (**115**) (or the corresponding triacetate) in aqueous hexamethylphosphoric triamide.[83] In general, the yields of deoxygenated products are satisfactory when photolysis is confined to a single site.

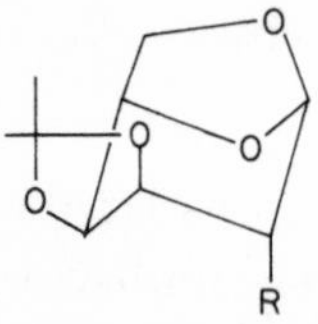

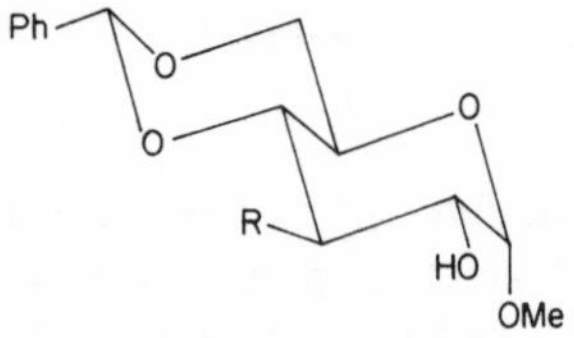

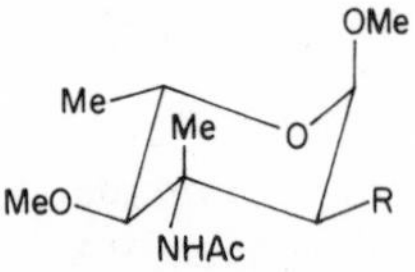

(109)   R = O·CS·SMe   (111)   R = O·CS·Ph   (113)   R = O·CS·Ph
(110)   R = H   (112)   R = H   (114)   R = H

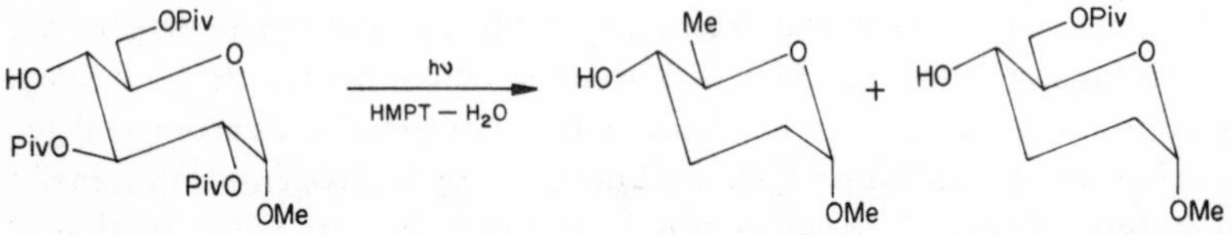

(115)   (116) (20%)   (10%)

1.3   The synthesis of antibiotic sugars

Most sugar components of antibiotics possess unusual structural features that are seldom, if ever, found in the sugar components of animals and plants. They include L-evernitrose[1] (**1**), D-aldgarose[2] (**2**), L-daunosamine[84] (**8**), L-nogalose[85] (**117**), L-sibirosamine[86] (**118**), D-hikosamine[87] (**119**), and D-rubranitrose[88] (**120**)—but there are many more with equally unusual and fascinating structures. The synthesis of antibiotic sugars may be undertaken with a view to establishing their structures or as part of the total synthesis of the antibiotics of which they form a part. The desire to prepare optically pure forms of antibiotic sugars has meant that most syntheses are effected from

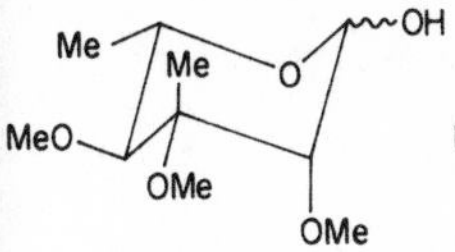

(117) 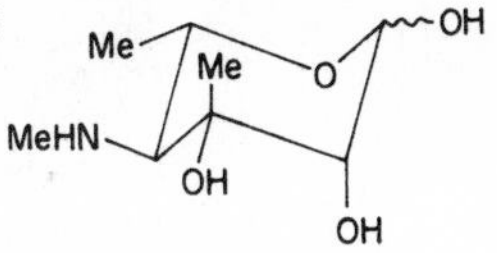　(118) 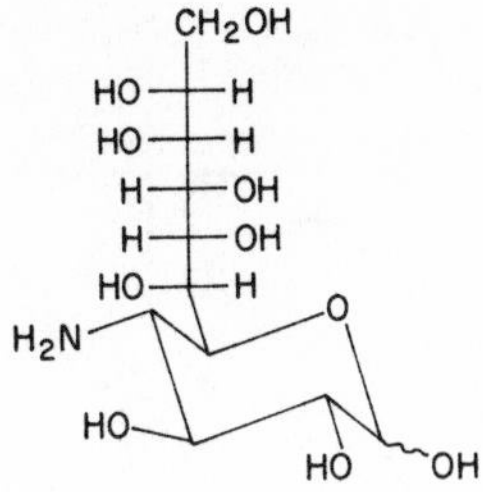　(119) 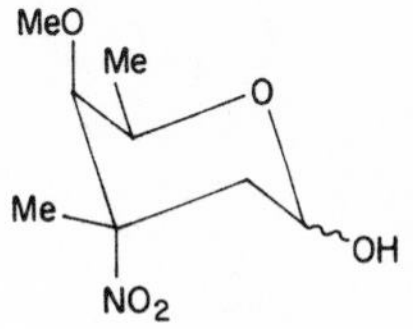　(120)

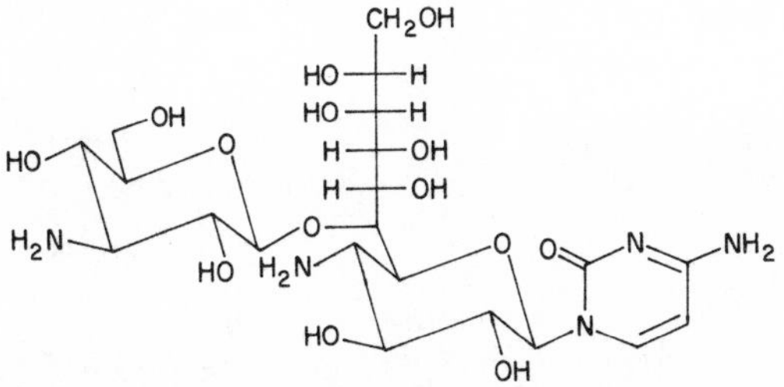

(121)

readily available sugars like D-glucose, D-galactose, D-mannose, and L-rhamnose, etc. Consequently, many synthetic operations are concerned with replacing or removing hydroxyl groups and in achieving the desired stereochemistry. Two examples should suffice to indicate the lines along which such syntheses may proceed.

The nucleoside antibiotic hikizimycin[87] (or anthelmycin) (121), isolated from *Streptomyces longissimus* and *Streptomyces A-5*,[89] is a member of a rare class of naturally-occurring compounds that contain a long-chain sugar component. The long-chain sugar component of hikizimycin is an undecose 119, referred to as hikosamine. The strategy used in assembling the eleven-carbon chain of hikosamine was based[90] on a Wittig reaction between the ylid derived from the five-carbon phosphonium iodide 124 and the six-carbon aldehyde 127, thereby placing in position eight of the ten consecutive chiral centres (see Scheme 10). The other two chiral centres were then elaborated from the newly formed olefinic linkage.

The phosphonium salt 124 was prepared[90] in a straightforward manner from 2,3:4,5-di-O-cyclohexylidene-D-arabinitol[91] (122), via the iodide 123. Its partner for the Wittig reaction was obtained from methyl 2,3-di-O-benzyl-α-D-galactopyranoside[92] (125), which was first converted into the 4-mesylate 126 by a sequence of conventional reactions. An azide displacement on the

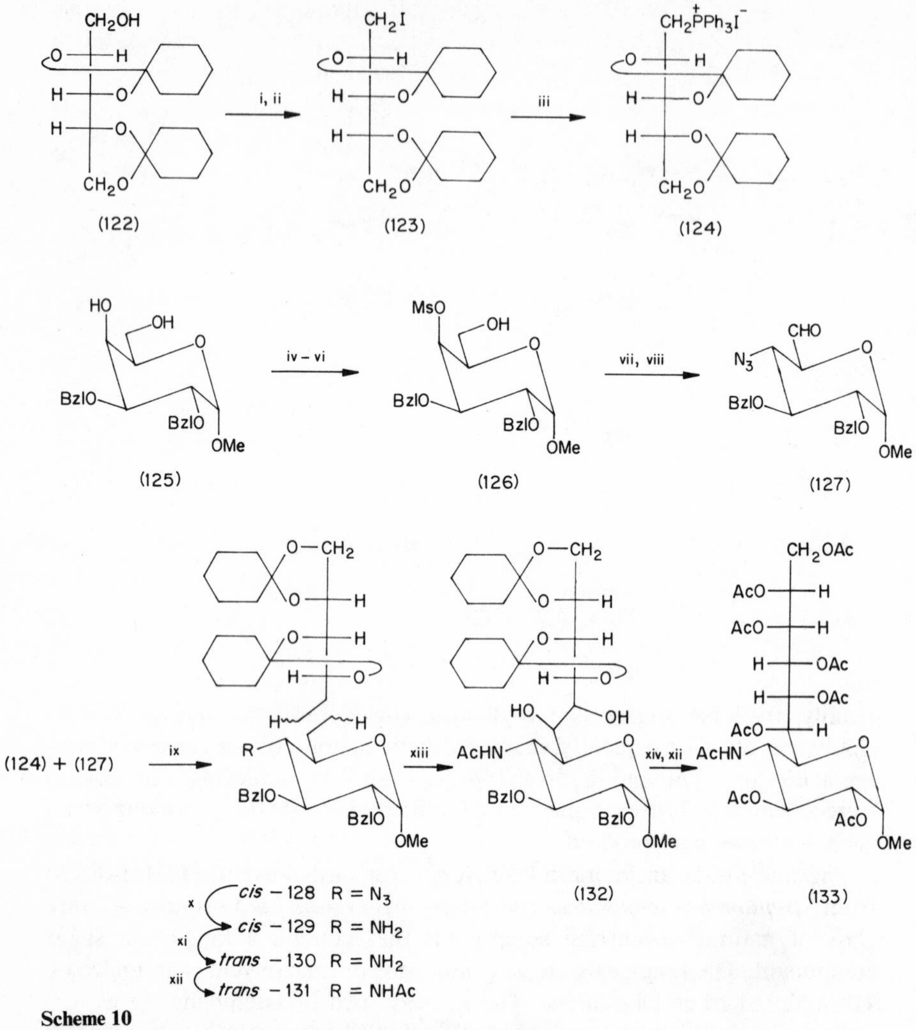

Scheme 10

Reagents: i, TsCl-py; ii, NaI-DMF; iii, Ph$_3$P; iv, Ph$_3$CCl-py; v, MsCl-py; vi, H$_3$O$^+$; vii, NaN$_3$-DMF; viii, DMSO - C$_6$H$_{11}$N=C=NC$_6$H$_{11}$; ix, BuLi-THF-HMPT; x, LiAlH$_4$; xi, hν; xii, Ac$_2$O-py; xiii, OsO$_4$; xiv, H$_2$-Pd/C-HCl.

latter compound and oxidation of the primary hydroxyl group furnished the desired aldehyde **127**. A Wittig reaction between **124** and **127** gave[90] only the *cis*-olefin **128**, which was required to undergo bishydroxylation of the carbon-carbon double bond in the *trans* fashion for proper development of the 6- and

7-OH groups of hikosamine. However, neither of the two possible epoxides obtained on treatment of the *cis*-olefin **128** with *m*-chloroperoxybenzoic acid could be opened with oxygen nucleophiles, presumably for steric reasons. This problem was overcome by *cis*-bishydroxylation of the *trans*-olefin **131** resulting from photochemical isomerization of *cis*-**129** and *N*-acetylation of the *trans*-photoproduct **130**. Only the required *cis*-diol **132** was formed on treatment of the *trans*-olefin **131** with a catalytic amount of osmium tetroxide and *N*-methylmorpholine *N*-oxide in aqueous tetrahydrofuran at room temperature. Simultaneous hydrogenation and hydrolysis of **132** gave, after acetylation, the peracetylated derivative **133**, which was indistinguishable from that derived from methyl α-hikosaminide.[93]

The principal concern in any synthesis of rubranitrose (**120**), a component of the antibiotic rubradirin[88] isolated from *Streptomyces achromogenes*,[94] is in placing an axial nitro group at the tertiary centre. This may be achieved in a number of ways, but most satisfactorily by making use of the reactions outlined in Scheme 11. This approach envisages[95] the transformation of a spiro-epoxide into a spiro-aziridine, which then undergoes hydrogenolysis *in situ* to produce the readily oxidizable tertiary-alkylamine. Since the configuration at the tertiary centre is inverted in going from the spiro-epoxide to the spiro-aziridine, the stereochemistry of the latter compound is determined by that of the spiro-epoxide. Thus, a key compound in the synthesis[95] (Scheme 12) of rubranitrose (**120**) was the spiro-epoxide[96] **135**, which was prepared by peroxyacid oxidation of the olefin **134** resulting from a Wittig reaction on the keto sugar[14] **43**. The methyl-branched amino sugar **136** was obtained in acceptable overall yield when the sequence of reactions outlined in Scheme 11 was applied to **135**. For practical reasons, oxidation of the amino group to a nitro group was postponed until later in the synthesis. Instead, **136** was converted into its *N*-acetyl derivative **137** and then, by the action of *N*-bromosuccinimide in refluxing carbon tetrachloride, into the 6-bromide **138**.

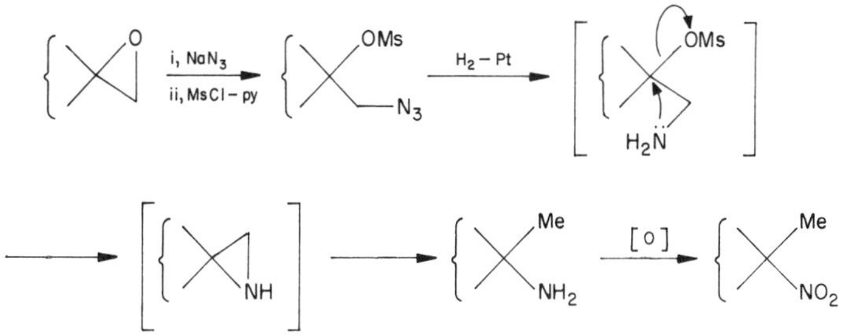

Scheme 11

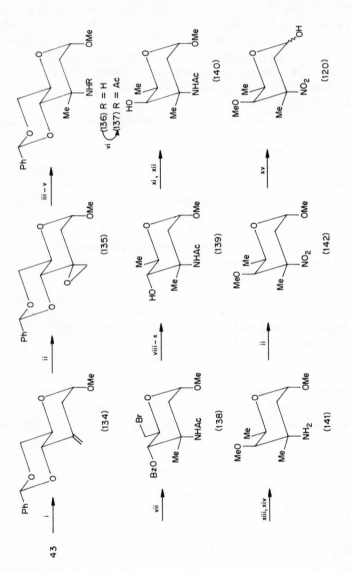

Scheme 12

Reagents: i, $Ph_3P^+MeBr^- BuLi$; ii, m-$ClC_6H_4CO_3H$; iii, NaN_3; iv, MsCl-py; v, H_2-Pt; vi, Ac_2O; vii, NBS; viii, NaI; ix, H_2-Pd/C; x, NaOMe; xi, PCC; xii, L-Selectride; xiii, MeI; xiv, Ca-liq. NH_3; xv, H_3O^+.

The latter compound was readily transformed into the 6-deoxy sugar **139**, which, in turn, was converted into its 4-epimer **140** on oxidation followed by reduction of the resulting keto sugar with L-Selectride.[97] Methylation of the axial 4-OH group and *N*-deacetylation (with calcium in liquid ammonia[98]) then furnished the amino sugar **141**, which was oxidized to the corresponding nitro sugar **142** with *m*-chloroperoxybenzoic acid in refluxing dichloromethane. Rubranitrose (**120**) was obtained[95] on acidic hydrolysis of the latter compound, establishing that it belongs to the D-series and not, as originally assigned,[88] to the L-series.

The keto sugar **43** originates[14] from methyl 2,3:4,6-di-*O*-benzylidene-α-D-mannopyranoside (**12**, R = Me) (see p. 11), which, at first sight, seems an unlikely starting material for a synthesis of D-rubranitrose. However, as we have seen already, the acetal groups can be treated as functional groups and their removal used to functionalize the molecule in a synthetically useful way.

1.4 'Chiral templates' derived from carbohydrates

Over the past decade or so, monosaccharides have been used increasingly as starting materials for the synthesis of optically pure forms of other classes of natural product, because, above all else, the functional groups of monosaccharides can be manipulated with relative ease and a high degree of regio- and stereochemical control. The first indication that monosaccharides could be used in this way was given in 1957 by Hardegger and Lohse,[99] who transformed L-arabinose into the alkaloid (+)-muscarine (**143**). Many classes of natural product discovered in the intervening years, especially those containing chiral tetrahydrofuranyl and tetrahydropyranyl rings, are capable of elaboration from such readily available sugars as D-glucose and D-mannose. The advantages that accrue from using sugars as 'chiral templates' have been underlined in recent reviews.[3]

In considering a recent synthesis[100] of thromboxane B$_2$ (**150**) from methyl α-D-glucopyranoside (**77**) (Scheme 13), the relation between the starting material and the target molecule is readily apparent from examination of the two structures. The heavy lines in formula **150** indicate the carbon atoms of the sugar precursor that remained after modification. Such modifications that are needed to append the appropriate functionalities and to achieve the

(143)

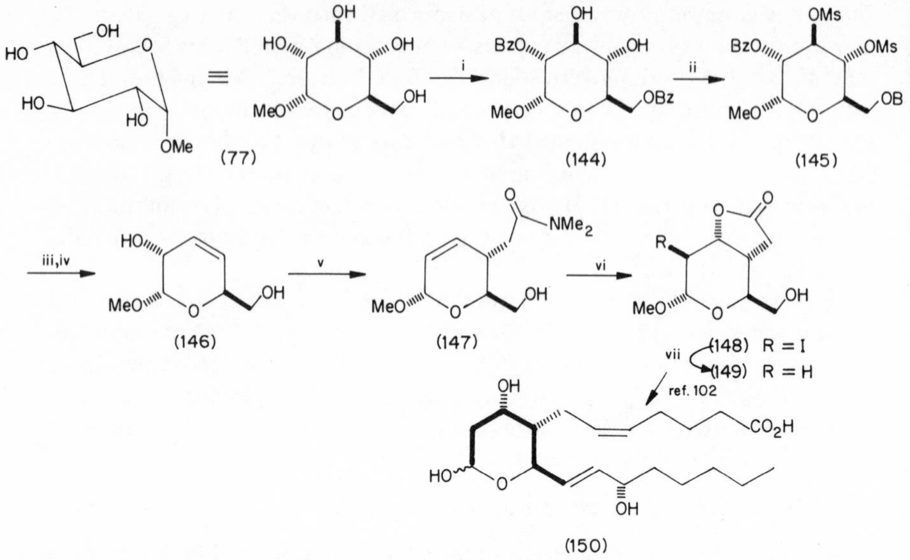

(77) ≡ (144) →i (145) →ii

(146) →iii,iv (147) →v (148) R = I / (149) R = H →vi

vii (148) R = I
    (149) R = H
ref. 102

(150)

Scheme 13

Reagents: i, BzCl (2 equiv.) at − 30°C; ii, MsCl-py; iii, NaI-Zn-DMF; iv, Et$_3$N-aq. MeOH; v, CH$_2$=C(OMe)NMe$_2$-diglyme, Δ; vi, I$_2$-aq. THF; vii, Bu$_3$SnH.

correct stereochemistry are taken into account in the overall synthetic planning, which, in this case, required the conversion of methyl α-D-glucopyranoside (77) into the allylic alcohol[101] 146, via the known compounds 144 and 145. The allylic alcohol 146 was then transformed[100] stereospecifically into the dimethylamide 147 by Claisen rearrangement. Treatment of 147 with iodine in aqueous tetrahydrofuran afforded the iodo-lactone 148, which was smoothly deiodinated, with tributylstannane, to give the key hydroxy-lactone 149. Standard methodology developed[102] with the racemic form of 149 could then be applied[100] in completing a total synthesis of optically pure thromboxane B$_2$.

A total synthesis of the aglycone moiety of the sixteen-membered macrolide antibiotics leucomycin A$_3$ and carbomycins A and B was based[103] on the three strategic bond-disconnections indicated on the key cyclic intermediate 151. These disconnections led progressively to simpler intermediates, one of which was the Michael acceptor 152. After unravelling the stereochemistry, it turned out that a synthesis of 152 might be realized by way of a Wittig reaction on a six-carbon aldehyde derived from D-glucose. In proceeding along the synthetic route (Scheme 14), D-glucose was first converted into the terminal olefin[104] 153, which furnished[105] the primary alcohol 154 on hydroboration. Benzylation of 154 and subsequent deprotection under acidic conditions

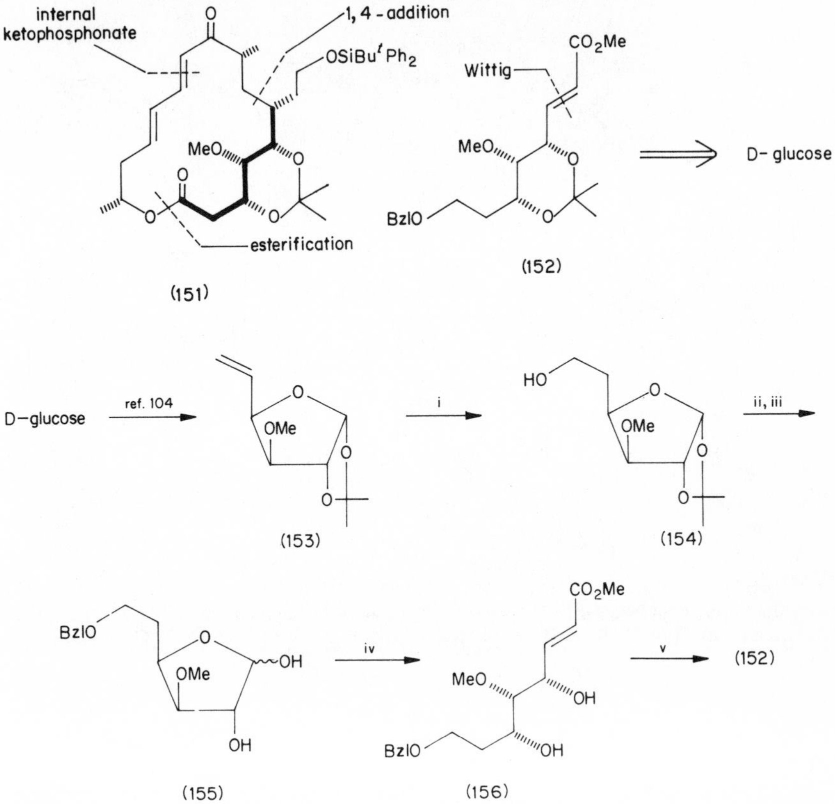

Scheme 14

Reagents: i, $(iso\text{-}C_5H_{11})_2BH\text{-}THF$; ii, $PhCH_2Br\text{-}NaH\text{-}DME$; iii, H_3O^+; iv, $Ph_3P =$
$CHCO_2Me$; v, $Me_2C(OMe)_2\text{-}H^+$

produced the lactol **155**, which reacted in its *aldehydo*-form with (carbo-
methoxymethylene)triphenylphosphorane to give the *trans*-olefin **156**.[103] The
desired intermediate **152** was obtained on acetonation of **156**. Obviously there
is still a long way to go before the synthesis of **151** is completed,[103] but
subsequent operations do not affect the integrity of the three consecutive
chiral centres introduced via D-glucose.

It may not always be quite so obvious how the chiral segment of other
natural products might be obtained from a carbohydrate precursor. The
leukotriene A_4 (LTA_4) methyl ester **164** provides a good example. The heavy
lines again indicate the carbon atoms that remained after asymmetric
modification of the carbohydrate precursor, in this case '2-deoxy-D-ribose'
(157). In practice, the synthesis[106] was stereospecific and required only seven

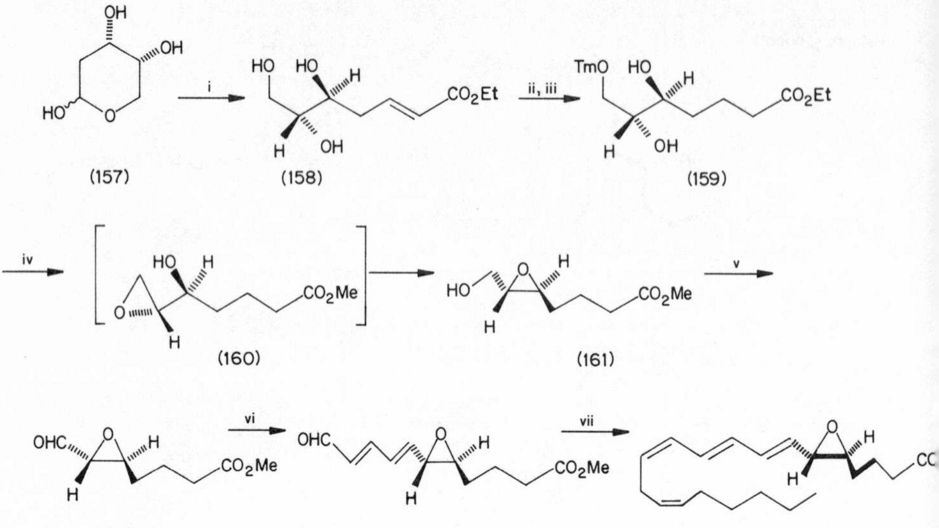

(157) (158) (159)

(160) (161)

(162) (163) (164)

Scheme 15

Reagents: i, $Ph_3P{=}CHCO_2Et$; ii, H_2-Pd/C; iii, mesitylenesulphonyl chloride-py; iv, NaOMe (1.1 equiv.)-MeOH; v, CrO_3; vi, $Ph_3P{=}CHCHO$ (2equiv.); vii, $Ph_3P{=}CHCH_2CH{=}CHC_5H_{11}$

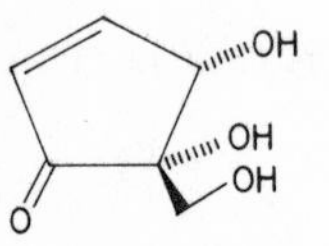

(165) 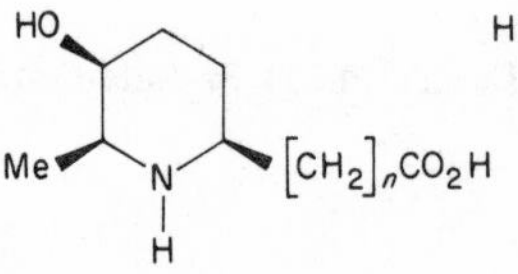 (166) 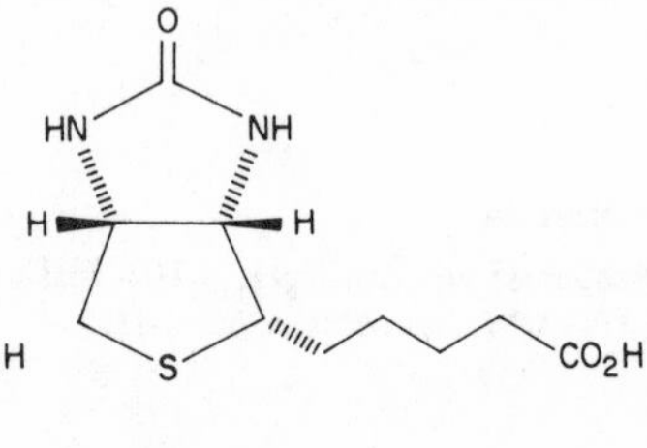 (167)

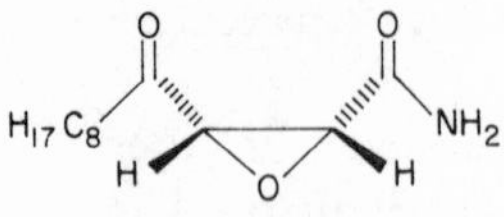

(168) 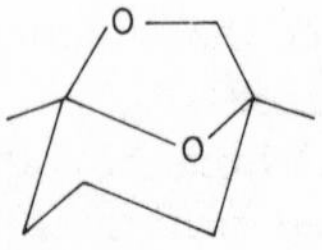 (169) 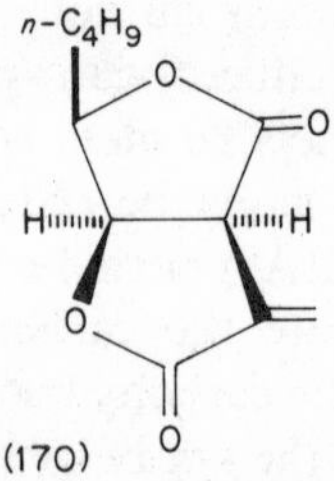 (170) 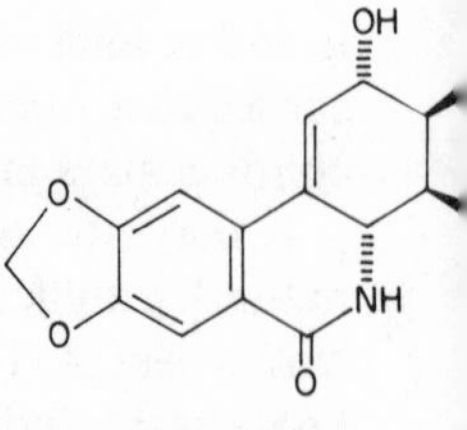 (171)

steps (Scheme 15). The reaction of **157** with (carboethoxymethylene)triphenylphosphorane gave the triol **158**, which, on hydrogenation and selective mesitylenesulphonylation, afforded **159**. Treatment of **159** with 1.1 equivalents of sodium methoxide in methanol gave, via rearrangement of the terminal epoxide **160**, the epoxy-alcohol **161**. Collins oxidation of **161** then afforded the (5S,6R)-epoxy-aldehyde **162**, which was converted into LTA$_4$ methyl ester **164** by a Wittig reaction with two equivalents of (formylmethylene)triphenylphosphorane, to give the diene aldehyde **163**, followed by a second Wittig reaction with the ylid derived from triphenyl[(Z)-non-3-en-1-yl]phosphonium chloride.

Some measure of the success of chiral syntheses developed from monosaccharides can be gauged from the fact that D-glucose alone has been converted into an impressive array of natural products that includes (−)-pentenomycin[107] (**165**), (+)-azimic acid[108] (**166**, $n = 5$), (+)-carpamic acid[108] (**166**, $n = 7$), (+)-biotin[109] (**167**), (+)-tetrahydrocerulenin[110] (**168**), (−)-frontalin[111] (**169**), (−)-canadensolide[112] (**170**), and (+)-lycoricidin[113] (**171**).

1.5 The synthesis of complex oligosaccharides

An oligosaccharide such as **172** possesses a unique arrangement of monosaccharide residues and intersaccharide linkages which must be reproduced in its synthesis. In the synthesis of an oligosaccharide, two polyfunctional reaction partners must be coupled. The most successful couplings in the synthesis of complex oligosaccharides involve the reaction of a glycosyl halide with a suitably protected saccharide (represented simply as R^{2}OH in Scheme 16). Depending on the relation between the groups at C-1 and C-2, the coupling step can furnish linkages of the 1,2-*cis* or 1,2-*trans* type in the glycoside

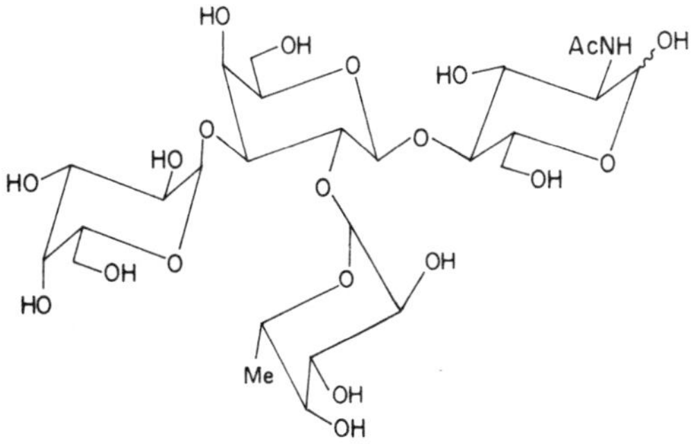

(172) ≡   α – D – Gal*p* – (1 → 3)
          α – L – Fuc*p* – (1 → 2)   β – D – Gal*p* – (1 → 4) – D – GlcNAc*p*

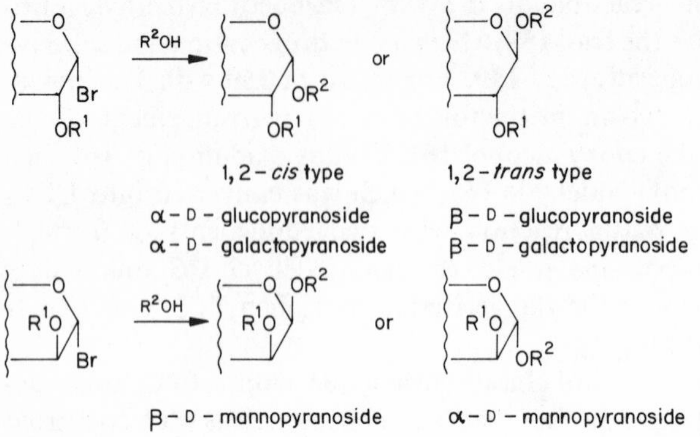

1,2 – *cis* type

α – D – glucopyranoside
α – D – galactopyranoside

1,2 – *trans* type

β – D – glucopyranoside
β – D – galactopyranoside

β – D – mannopyranoside

α – D – mannopyranoside

Scheme 16

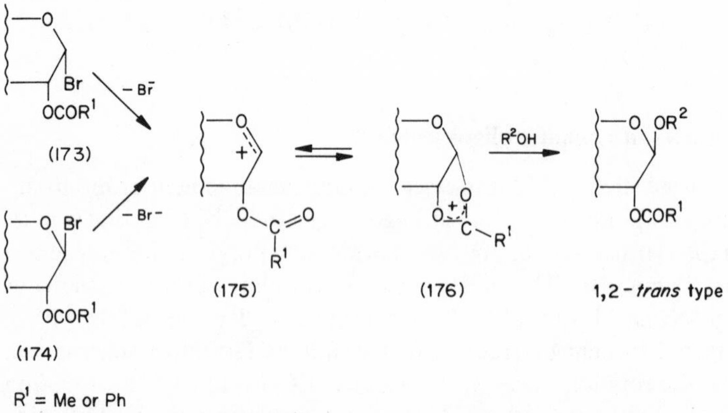

(173)

(174)

(175)

(176)

1,2 – *trans* type

R^1 = Me or Ph

produced. The overall efficiency of an oligosaccharide synthesis will depend crucially on the stereoselectivity that can be achieved in the coupling reactions.

An ester substituent on C-2 of the glycosyl halide that is capable of neighbouring-group participation can be used to control the stereochemical outcome of the coupling reaction. Irrespective of whether the α-form **173** or the β-form **174** of the halide is used, it reacts, via an oxocarbenium ion **175**, to give the more stable cyclic ion **176**.[114] Nucleophilic ring-opening of **176** with the hydroxylic component then leads, for steric reasons, to glycosides of the 1,2-*trans* type. Only with moderately reactive hydroxylic components do mixtures of both types of product result as a consequence of attack on the oxocarbenium ion **175**. By utilizing the neighbouring-group activity of a substituent on C-2, it is generally straightforward to prepare β-glycopyranosides in the *gluco* and *galacto* series and α-glycopyranosides in the *manno*

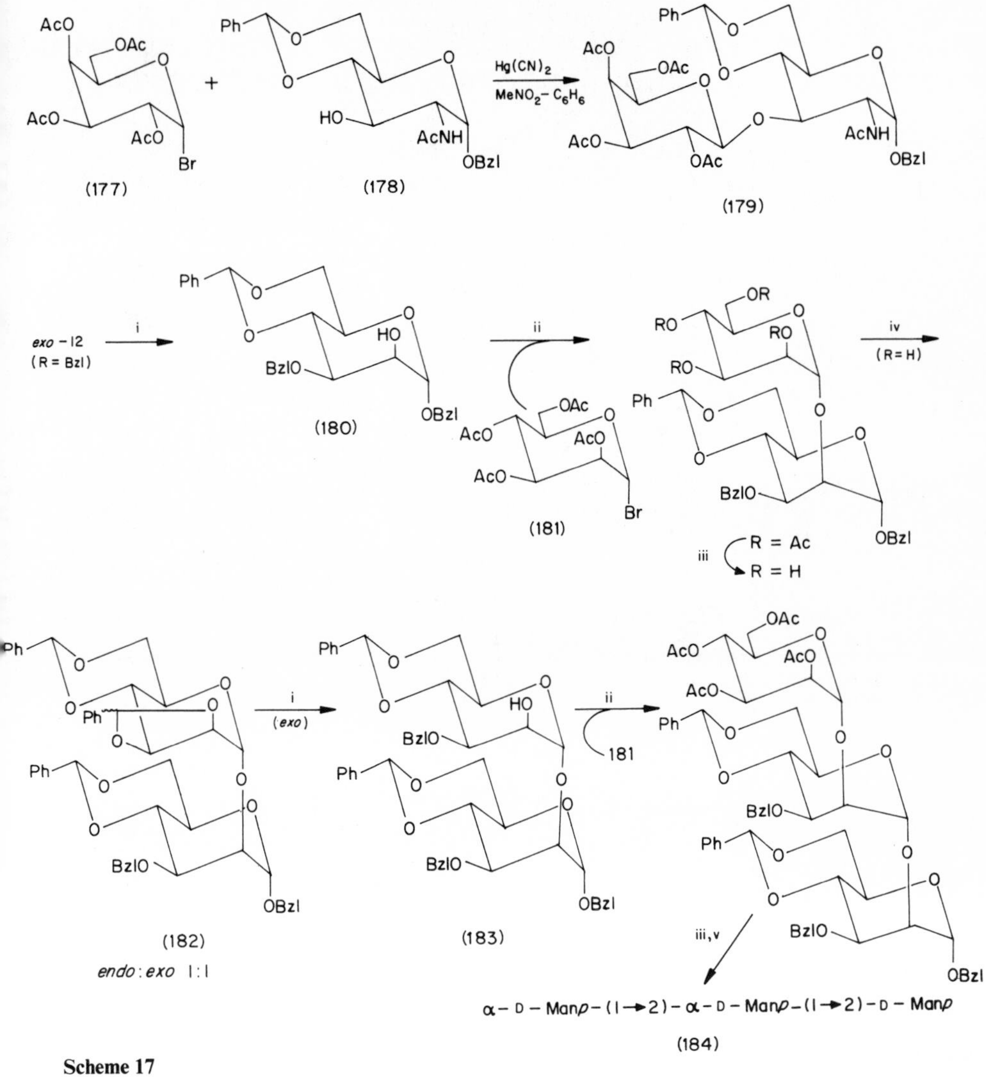

Scheme 17

Reagents: i, LiAlH$_4$-AlCl$_3$; ii, Hg(CN)$_2$; iii, NaOMe; iv, PhCHO-ZnCl$_2$; v, H$_2$-Pd/C

series. The coupling reaction between acetobromogalactose **177** and **178** (protected so that only the 3-OH group reacted) proceeded by way of an acetoxonium ion **176** (R^1 = Me) to give the β-linked disaccharide **179** in 81% yield.[115] The synthesis of the α-(1→2)-linked D-*manno*-trisaccharide **184** (Scheme 17), which represents the repeating-unit of the 08-antigen of *Escherichia coli*,[116] provides another example. In this case, the key hydroxylic

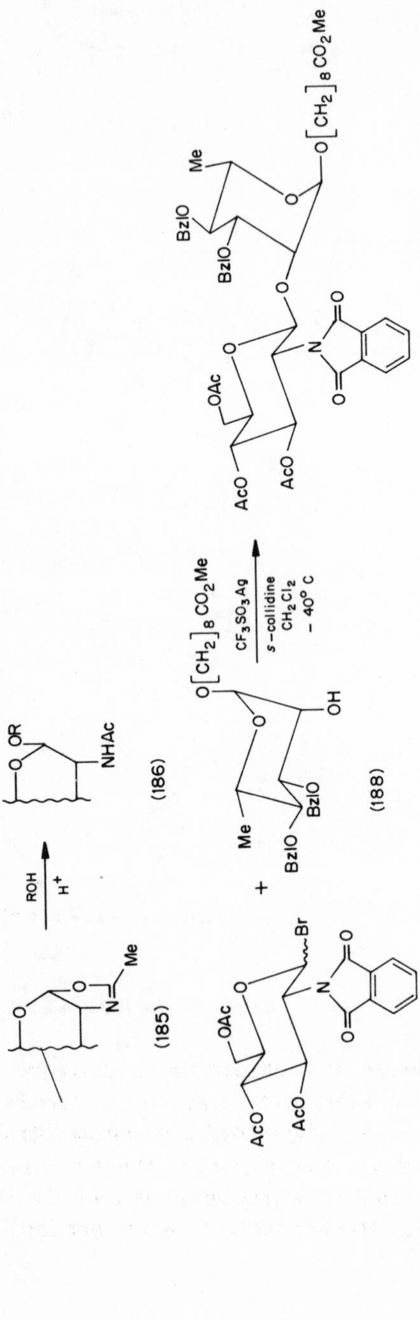

(189)

(186)

(185)

(188)

(187)

AcO—, OAc, O
AcO—, Cl
RO

+

HO—, OAc, O
AcO—, OAc
OAc

$\xrightarrow[\substack{s\,-\,\text{collidine} \\ \text{Et}_2\text{O (or CH}_2\text{Cl}_2)}]{\text{AgClO}_4}$

AcO—, OAc, O
AcO—
RO
O
AcO
OAc
AcO—, O
OAc

(190) R = Bzl or Cl$_3$CCO   (191)   (192)

components **180** and **183** were obtained[117] by preferential and regiospecific hydrogenolysis of the *exo*-substituted dioxolane rings of **12** (R = Bzl) and **182** with the LiAlH$_4$-AlCl$_3$ reagent (see section 1.1.2).

Effective syntheses of 1,2-*trans*-glycosides (**186**) from 2-amino sugars can be achieved by the ring-opening of oxazolines (**185**)[118] or by making use of the neighbouring-group activity of a phthalimido group.[119] Thus, condensation of 3,4,6-tri-*O*-acetyl-2-deoxy-2-phthalimido-D-glucopyranosyl bromide (**187**) with the partially benzylated L-rhamnopyranoside **188** gave[120] the β-linked disaccharide **189** in 70% yield. Cleavage of the phthalimido group may be accomplished[120] with hydrazine at elevated temperature, but this treatment may cause problems if the oligosaccharide is base-sensitive.

The formation of linkages of the 1,2-*cis* type (an α-glycosidic linkage in the *gluco* and *galacto* series and a β-glycosidic linkage in the *manno* series) is a much more exacting task. A prerequisite for the synthesis of this type of linkage is that the glycosyl halide must carry a non-participating group at C-2. This requirement is met by the β-D-glucopyranosyl chlorides **190** coupled to **191** in the synthesis of the α-(1 → 4)-linked disaccharide derivatives **192**.[121] These reactions must be conducted in such a way that they proceed, as completely as possible, with inversion of the configuration at the reaction centre. Unfortunately, this approach is possible only with relatively stable, isolable β-halides. In most cases, the β-halide is labile and can be used, if at all, only with difficulty. Because of this, the synthesis of 1,2-*cis* α-glycopyranosides is usually approached from the more accessible and stable α-glycosyl halides.[5] The higher stability of α-glycosyl halides has its origin in the anomeric effect,[122] which is particularly strong for halogeno substituents.

The approach used is based on halide-catalysed, *in-situ* anomerization of the α-halide **193** to the less stable β-halide **196**, via the various intermediates shown in Scheme 18.[5,123] It is assumed that the intimate ion pairs **194** and **195** which precede ion-triplet formation are structurally well-defined and different, and undergo attack on the opposite side of

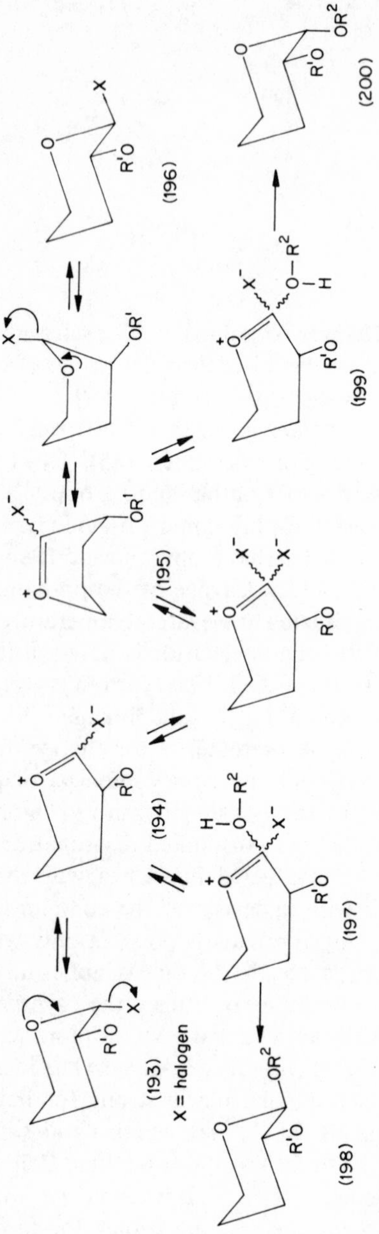

Scheme 18

the anomeric centre much more rapidly than the intially formed ions can separate. Although there is a higher proportion of α-halide **193** than β-halide **196** at equilibrium, the latter is more reactive[123] towards the hydroxylic component, so that formation of the α-glycoside (via **196 → 195 → 199 → 200**) is faster than that of the β-glycoside (via **193 → 194 → 197 → 198**). The higher reactivity of β-halides must be related to the stereoelectronic requirements for their reaction, as well as to the fact that β-halides are thermodynamically less stable and, therefore, closer to their transition states than are α-halides. Provided the equilibrium **193 ⇌ 196** is established quickly enough, the difference between the rates of reaction of **193** and **196** can be exploited in the selective synthesis of α-glycosides from α-halides. In order to do this, it is essential to have as much information as possible about the reactivities of the reactants and the catalyst. The reaction, termed halide-catalysed glycosidation,[123] has to be carried out in a solvent of low polarity such as dichloromethane (sometimes mixed with toluene), since its α-stereoselectivity is considerably reduced in polar solvents. Tetra-alkylammonium halides were originally used[123] as catalysts, but the conversion of α-halides **193** into α-glycosides **200** can also be carried out in the presence of other soluble catalysts such as mercury(II) cyanide, mercury(II) bromide, silver perchlorate, and silver triflate.[124,125] These catalysts are much more reactive than tetra-alkylammonium halides, so that less reactive glycosyl halides react under very mild conditions in their presence. The precise structures of the ion pairs (see Scheme 18) formed with these catalysts are not known, but the preferential formation of **199**, with $X^- = BrHg(CN)_2^-$, $HgBr_3^-$, ClO_4^-, or $CF_3SO_3^-$, can be invoked to explain the stereoselectivities observed.

As already indicated, the stereoselectivity of oligosaccharide syntheses utilizing *in situ* anomerization can be influenced by the reactivities of the glycosyl halide, the hydroxylic component, and the catalyst. Valuable information on all of these factors has been provided by Paulsen and co-workers,[5,125] and some of the results from their extensive studies are summarized in Table 1.2. In this particular example, the glycosyl halide used

Table 1.2 Reactivities of glycosyl halides, catalysts, and alcohols in oligosaccharide synthesis.

Glycosyl halide	Catalyst	Alcohol
 R ~ Bzl > Bz > Ac X ~ I > Br > Cl	Et_4NBr-4Å molecular sieves $Hg(CN)_2$ $Hg(CN)_2$-$HgBr_2$ $HgBr_2$-4Å molecular sieves $AgClO_4$-Ag_2CO_3 Ag triflate-Ag_2CO_3	MeOH ≫ HOCH$_2$R > 6-OH 6-OH ≫ 3-OH > 2-OH > 4-OH

in the coupling reaction is substituted on C-2 by the non-participating azido group, which can be reduced to an amino group after the glycosidic linkage has been formed—this so-called 'azide method' offers distinct advantages in the synthesis of α-glycosides from 2-amino sugars.[5] Reference to Table 1.2 shows that, in the coupling reaction, benzyl-substituted glycosyl halides are more reactive than their acyl analogues; glycosyl bromides are more reactive than glycosyl chlorides, although they are less stable; there is a range of catalytic reactivities that can be called upon; and the primary hydroxyl group of a pyranos(id)e is considerably more reactive than are the secondary hydroxyl groups, which also exhibit a range of reactivities. The generally moderate reactivities of the secondary hydroxyl groups of pyranos(id)es are well suited to reactions with reactive glycosyl halides under the conditions of *in situ* anomerization, and it is possible to obtain high α-stereoselectivities.

While the synthesis of 1,2-*cis*-α-glycosidic linkages may offer some scope for modifying the reactivity of the glycosyl halide (irrespective of their positions, benzyl groups increase, and acetyl groups decrease, its reactivity, whereas an additional glycosyl residue has a slightly weaker activating effect than a benzyl group), there is usually less scope for modifying the reactivity of the hydroxylic component, which, depending on the requirements of the synthesis, may have to be protected in a predetermined way so that other hydroxyl groups can be selectively exposed for subsequent coupling. However, the reactivity of the catalyst can be varied at will and, as the following examples show, the choice of catalyst has a significant bearing on the stereoselectivity of the coupling reaction and the yield of coupled product, although sometimes a compromise must be reached between them.

Whereas the glycosyl bromide **201** reacted with the L-rhamnopyranoside derivative **202** in the presence of $HgBr_2$ to give only the α-linked disaccharide **203**, the β-linked disaccharide **205** was the principal product (**205**:**206**, 81:19) when the more reactive glycosyl bromide **204** was used.[126] If the milder catalyst tetraethylammonium bromide was used instead of $HgBr_2$, the reaction of **202** with **204** yielded almost exclusively the α-linked disaccharide **206**, although the yield was reduced.[5] Interestingly, the reactivity of the 4-OH group of the 2,2,2-trichloroethyl glycoside **207** was suppressed to such an extent that the $HgBr_2$-catalysed reaction with the more reactive glycosyl bromide **204** now favoured the formation of the corresponding α-linked disaccharide.[126]

In the presence of tetraethylammonium bromide, the L-fucosyl bromide **208** (a reactive halide) coupled with the moderately reactive disaccharide derivative **209** (prepared from **179**) to give the α-glycoside **210** in 80% yield.[127] Deprotection of **210** by standard procedures yielded **211**, the trisaccharide determinant of blood-group substance H (type 1). The 2-OH group of the non-reducing residue of a similar disaccharide **212** is much less reactive, so

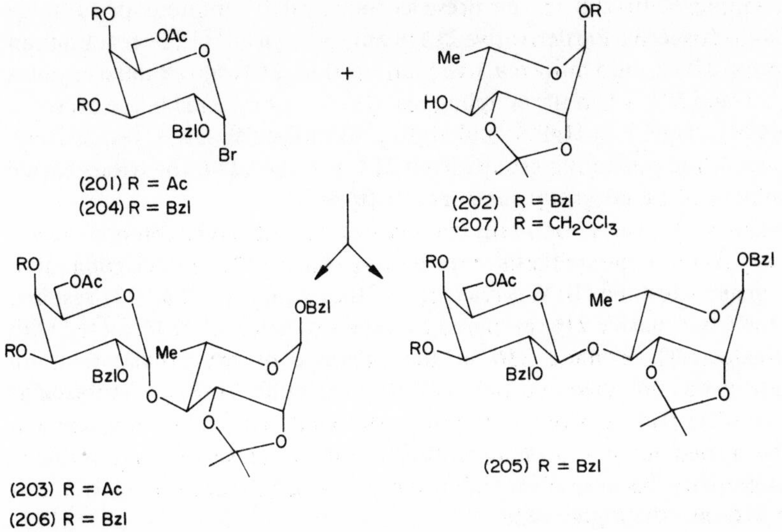
(201) R = Ac
(204) R = Bzl
(202) R = Bzl
(207) R = CH_2CCl_3
(203) R = Ac
(206) R = Bzl
(205) R = Bzl

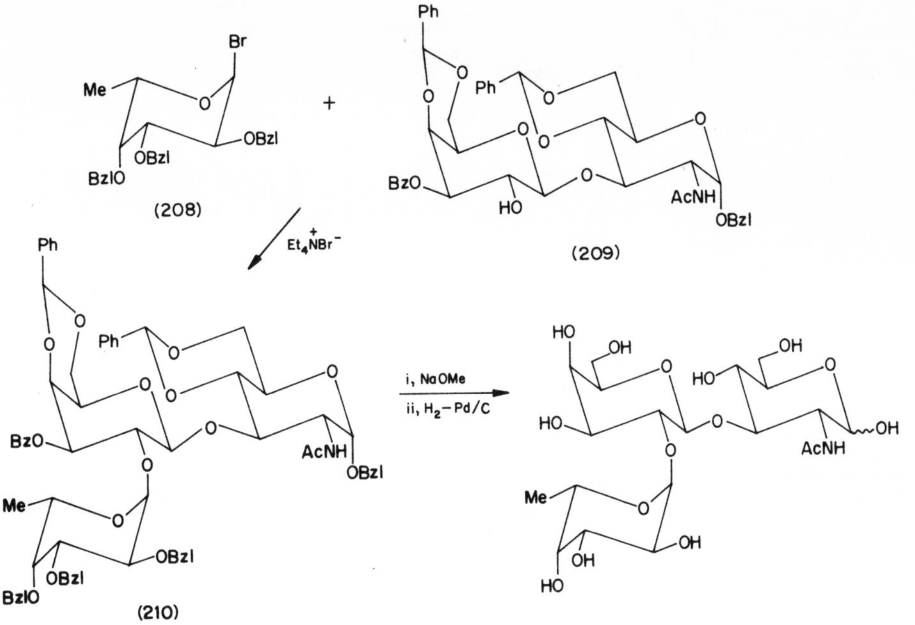
(208)
(209)
$Et_4\overset{+}{N}Br^-$
i, NaOMe
ii, H_2–Pd/C
(210)

α-L-Fuc*p*-(1→2)-β-D-Gal*p*-(1→3)-D-GlcNAc*p*

(211)

that coupling with **208** in the presence of tetraethylammonium bromide yielded the trisaccharide derivative **213** in only 30% yield.[125] The reaction can be improved by using a more reactive catalyst: thus, **213** was obtained in yields of 40, 50, and 80% when the coupling reaction was conducted in the presence of $Hg(CN)_2$, $Hg(CN)_2$-$HgBr_2$, and $HgBr_2$-4Å molecular sieves, respectively. Removal of the protecting groups from **213** furnished **214**, the trisaccharide determinant of blood-group substance H (type 2).

The choice of catalyst to suit the reactivity of the hydroxylic component was also a key factor in the synthesis of the tetrasaccharide **172**, the determinant of blood-group substance B. Whereas little of the coupled product **217** resulted when the trisaccharide **215** (prepared by debenzoylation of **213**) reacted with the D-galactosyl bromide **216** in the presence of tetraethylammonium bromide, it was obtained in a yield of 80% using either $HgBr_2$-4Å molecular sieves or silver triflate-silver carbonate as the catalyst.[125] The latter reaction must be carried out at a low temperature in order to preserve a favourable α-stereoselectivity, because silver triflate also catalysed the conversion of the α-halide **216** into the β-glycoside.

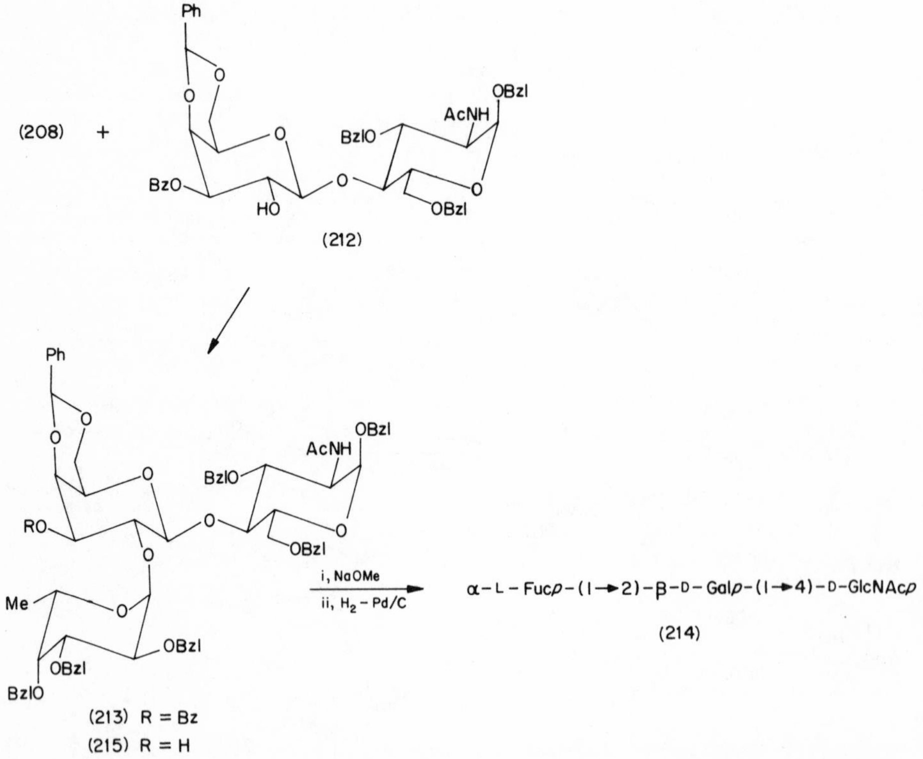

α-L-Fuc$p$-(1→2)-β-D-Gal$p$-(1→4)-D-GlcNAcp

(214)

Ph

BzIO
OBzI
(215) + 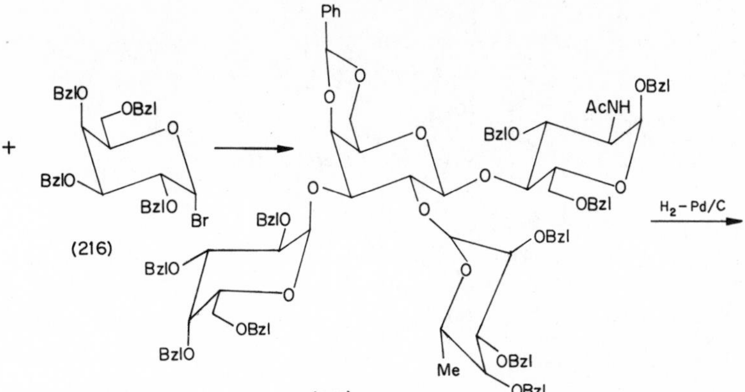
BzIO
BzIO
Br
(216)

OBzI
AcNH
BzIO
OBzI
OBzI
H₂-Pd/C
(172)

BzIO
BzIO
BzIO
OBzI
BzIO
O
Me OBzI
OBzI
(217)

The α-stereoselectivity decreases when compounds with very reactive hydroxyl groups (for example, the 6-OH group of hexopyranosides) are coupled with glycosyl halides by *in situ* anomerization techniques. This effect can sometimes be countered by using a mild catalyst,[50,128] although at other times it may be necessary to isolate and react the β-halide.

1.5.1 Block synthesis

It is generally advantageous in the synthesis of long oligosaccharide chains to link together two prefabricated oligosaccharide units (or blocks). An α-glycosidic linkage is often chosen for the block-coupling step, as in the synthesis[129] of the pentasaccharide chain **221** of the Forssman antigen (Scheme 19). Special methods are needed for the preparation of sensitive trisaccharide halides like **218**, which must be coupled as soon as possible using highly reactive catalysts. Bearing in mind the low reactivity exhibited by the 4'-OH group of the lactose derivative **219** in reactions with simple glycosyl halides, a 28% yield of the pentasaccharide derivative **220** from the coupling reaction between **218** and **219** is an encouraging result.

Other biologically important oligosaccharides prepared by block synthesis include **222**[130] (the repeating-unit of the O-specific chain of the lipopolysaccharide from *Escherichia coli* O 75), **223**[131] (the branched pentasaccharide unit of the O-specific chain of the lipopolysaccharide from *Shigella dysenteriae*, serotype 2), and the branched heptasaccharide **224**[5] (a key structure in the carbohydrate moiety of glycoproteins).

The final words in this section must be accorded to Paulsen,[5] who comments, 'Although we have now learned to synthesize oligosaccharides, it should be emphasized that each oligosaccharide synthesis remains an

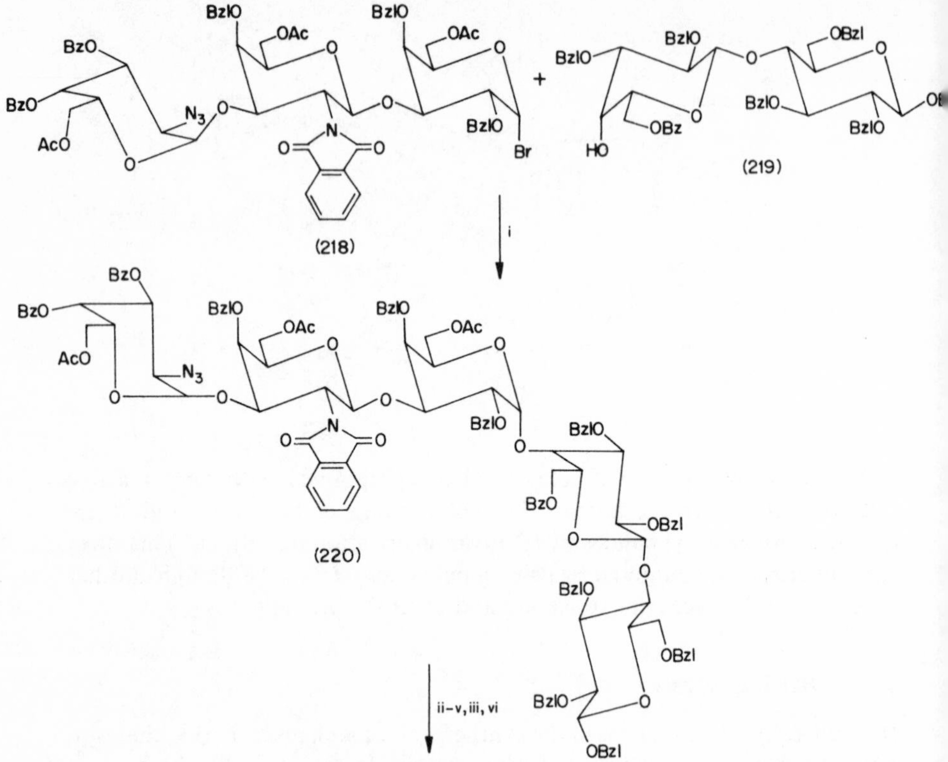

(218)

(219)

i

(220)

ii–v, iii, vi

α-D-GalNAcp-(1→3)-β-D-GalNAcp-(1→3)-α-D-Gal$p$-(1→4)-β-D-Gal$p$-(1→4)-D-Glc$p$

(221)

Scheme 19

Reagents: i, AgClO$_4$-Ag$_2$CO$_3$; ii, NiCl$_2$-NaBH$_4$; iii, Ac$_2$O; iv, NaOMe; v, NH$_2$NH$_2$-EtOH; vi, H$_2$-Pd/C

independent problem, whose resolution requires considerable systematic research and a good deal of know-how.'

1.6 Polysaccharides

Investigations of the primary structures of polysaccharides are generally concerned with establishing the nature and sequence of the monosaccharide residues and the positions and anomeric configurations of the linkages in the repeating-unit. A complete knowledge of the primary structure and conformation of a polysaccharide is indispensable to an understanding of the relationship between its structure and specific biological role.[132]

β- D - Manp - (I → 4)
　　　　　　　　　＼
　　　　　　　　　　α- D- Galp - (I → 4) - L - Rhap
　　　　　　　　　／
β- D - GlcNAcp - (I → 3)

(222)

α- D - GalNAcp - (I → 3)
　　　　　　　　　＼
　　　　　　　　　　α- D- GalNAcp - (I → 4) - α - D - Glcp - (I → 4) - D - Galp
　　　　　　　　　／
α- D - GlcNAcp - (I → 4)

(223)

β- D - Galp - (I → 4) - β - D - GlcNAcp - (I → 2) - α - D - Manp - (I → 6)
　　　　　　　　　　　　　　　　　　　　　　　　　　　　　　　　　　＼
　　　　　　　　　　　　　　　　　　　　　　　　　　　　　　　　　　　D - Manp
　　　　　　　　　　　　　　　　　　　　　　　　　　　　　　　　　　／
β- D - Galp - (I → 4) - β - D - GlcNAcp - (I → 2) - α - D - Manp - (I → 3)

(224)

Methylation analysis has stamped itself as an important method in structural polysaccharide chemistry.[133] It involves complete methylation of the polysaccharide and hydrolysis of the methylated polysaccharide to a mixture of methylated sugars, which are then separated, identified, and quantitatively estimated. The positions of glycosidic linkages in the polysaccharide correspond to the positions of unsubstituted hydroxyl groups in the methylated monosaccharides. Gas-liquid chromatography is extensively employed nowadays for the analysis of mixtures of methylated sugars obtained on acidic hydrolysis of methylated polysaccharides. By reduction of methylated monosaccharides (usually obtained as a mixture of both anomers), a single alditol derivative is obtained from each sugar, which is generally converted into the corresponding acetate for identification by g.l.c., often combined with mass spectrometry.[133] Mass spectrometry has proved particularly effective in determining the substitution patterns of partially methylated alditol acetates, either by comparing the spectra with previously determined spectra or by a careful analysis of the primary fissions that occur between adjacent carbon atoms in the chain.

Methylation analysis gives no information on the relative order of monosaccharide residues in the repeating-unit of a polysaccharide or on their anomeric configurations. Thus, determination of the complete primary structure of a polysaccharide requires complementary analyses, among the most important of which are graded hydrolysis by acids[134] or enzymes,[135] followed by isolation and identification of the oligosaccharides released, and various modifications of periodate oxidation.[134,136] Often, the assignment of the anomeric configurations of the linkages presents difficulties. Enzymic[135] and chemical[134] methods may sometimes yield information on the anomeric

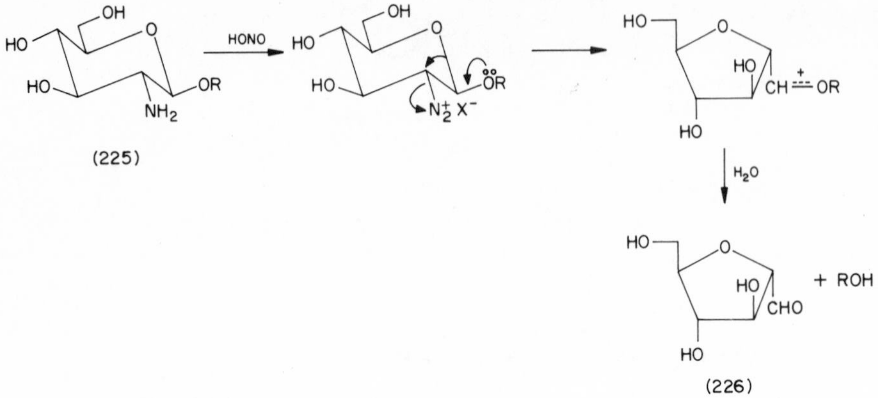

(225)   (226)

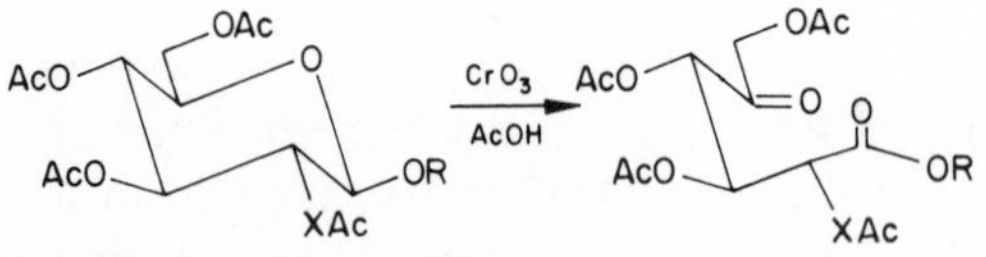

(227) X = O or NH   (228)

configurations, but, nowadays, high-resolution $^1H^{137}$ and ^{13}C n.m.r. spectroscopy[138] are used increasingly for this purpose.

Some of the methodologies that can be brought to bear in determining the structures of polysaccharides can best be demonstrated by considering a specific example. But before embarking on a detailed structural investigation, mention should be made of two reactions, originally developed at the monomer level, that feature prominently in it. First, 2-amino-2-deoxy-D-glucopyranosides 225 undergo a ring-contraction on deamination to give 2,5-anhydro-D-mannose (226), with scission of the glycosidic linkage.[139] Second, fully acetylated β-D-hexopyranosides such as 227 are oxidized by chromium trioxide in acetic acid at room temperature to acetylated 5-hexulosonates 228, whereas the corresponding α-D-hexopyranosides are not oxidized in the same way.[140]

Structural investigations on the polysaccharide O-antigen from *Shigella flexneri* variant Y have indicated[141] that it has the basic repeating-unit 229, namely → 2)-α-L-Rhap-(1 → 2)-α-L-Rhap-(1 → 3)-α-L-Rhap-(1 → 3)-β-D-GlcNAcp-(1 →, with, additionally, two of the hydroxyl groups acetylated. Perhaps the wisest course would be to assume this structure and to see how the various pieces of information can be reconciled with it. Acidic hydrolysis of the polysaccharide liberated principally L-rhamnose and 2-acetamido-2-

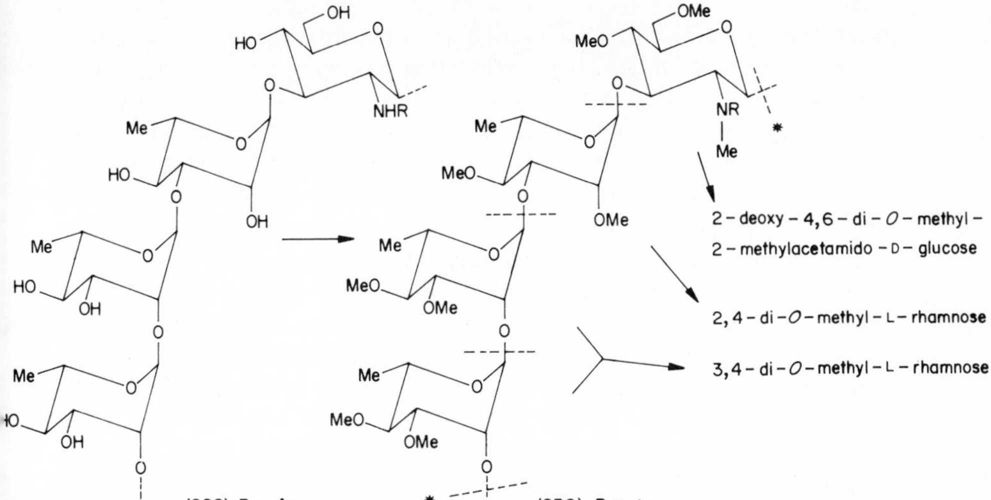

2 – deoxy – 4,6 – di – O – methyl –
2 – methylacetamido – D – glucose

2,4 – di – O – methyl – L – rhamnose

3,4 – di – O – methyl – L – rhamnose

(229) R = Ac   (230) R = Ac
(231) R = H   (232) R = H

deoxy-D-glucose, which were analysed by g.l.c.-m.s. of their alditol acetates. The ^{1}H n.m.r. spectrum of the polysaccharide (dissolved in deuterium oxide) revealed the presence of N-acetyl (δ 2.06) and O-acetyl (δ 2.15) groups in the ratio of ~ 1:2. The O-deacetylated polysaccharide gave a better-resolved spectrum which contained, inter alia, signals for the methyl protons of the L-rhamnose residues (δ 1.25–1.32) and N-acetyl group (δ 2.06), and for the anomeric protons (δ 4.75–5.15) in the proportions 9:3:4, respectively. These data are consistent with a repeating-unit containing one 2-acetamido-2-deoxy-D-glucose and three L-rhamnose residues. As the anomeric protons of α- and β-L-rhamnopyranosyl residues should display small couplings ($J_{1,2} \leq 4\,\mathrm{Hz}$), the presence of a signal for one anomeric proton at δ 4.75, with $J_{1,2}$ 7Hz, signified that the 2-acetamido-2-deoxy-D-glucopyranosyl residue is β-linked. Furthermore, the chemical shifts of the other anomeric protons, at δ 4.80 and 5.15 (2 protons), suggested that at least two of the three L-rhamnopyranose residues are α-linked.

The polysaccharide was next subjected to exhaustive methylation, with the aim of methylating all the free hydroxyl groups. Since strongly basic conditions were employed for methylation of the polysaccharide, O-acetyl groups would also be replaced by O-methyl groups. Acidic hydrolysis of the methylated polysaccharide (230) yielded 3,4-di-O-methyl-L-rhamnose and 2,4-di-O-methyl-L-rhamnose (determined as their alditol acetates[133]) in the proportions 2:1, and also the 4,6-di-O-methyl ether of 2-deoxy-2-N-methyl-acetamido-D-glucose, thereby establishing that two of the L-rhamnose residues are linked through O-2 while the other L-rhamnose residue and the 2-acetamido-2-deoxy-D-glucose residue are linked through O-3. Additional information was obtained by methylation analysis of the N-deacetylated polysaccharide (231), which was prepared from 229 by the action of sodium hydroxide-sodium thiophenolate in aqueous dimethyl sulphoxide. Acidic hydrolysis of the N-deacetylated, methylated polysaccharide (232) liberated equimolar proportions of 3,4- and 2,4-di-O-methyl-L-rhamnose, because the glycosidic linkage (marked*) of the modified amino sugar was not significantly hydrolysed due to the proximity of the $-\overset{+}{N}H_2Me$ group formed under the conditions of hydrolysis. Consequently, the 2-acetamido-2-deoxy-D-glucose residue is linked to O-2 of one of the L-rhamnose residues.

Deamination of the N-deacetylated polysaccharide (231), by treatment with sodium nitrite in aqueous acetic acid, yielded the tetrasaccharide 233 (cf. 225 → 226). Part of 233 was reduced with sodium borodeuteride to give the 2,5-anhydro-D-[1 − ^{2}H]mannitol derivative 234 {[α]$^{25}_{578}$ − 60°}, which, as expected, furnished 2,5-anhydro-1,4,6-tri-O-methyl-D-[1 − ^{2}H]mannitol (identified by mass spectrometry), 2,4- and 3,4-di-O-methyl-L-rhamnose, and 2,3,4-tri-O-methyl-L-rhamnose following methylation and acidic hydrolysis.

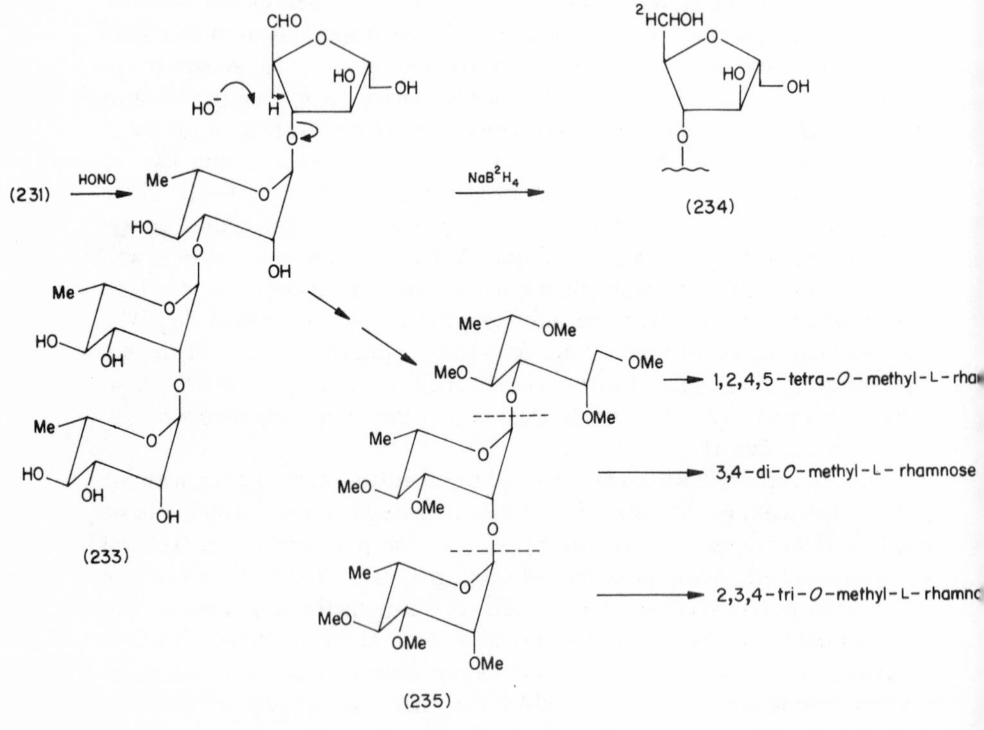

Another part of the tetrasaccharide **233** was treated with dilute sodium hydroxide solution, thereby causing β-elimination of the trisaccharide side-chain linked to O-3 of the anhydro-sugar residue. The trisaccharide was reduced to its alditol $\{[\alpha]_{578}^{25} - 44^\circ\}$, which then gave **235** on methylation. The formation of 3,4-di-O-methyl-L-rhamnose and 1,2,4,5-tetra-O-methyl-L-rhamnitol on acidic hydrolysis of **235** demonstrated that the L-rhamnose residue linked to the 2-acetamido-2-deoxy-D-glucose residue is also substituted on O-3. The values for the optical rotations of **234** and the trisaccharide alditol further indicated that all the L-rhamnose residues are α-linked. This was confirmed by oxidation of the fully acetylated polysaccharide with chromium trioxide in acetic acid.[140] Since only the 2-acetamido-2-deoxy-D-glucopyranose residues of the acetylated polysaccharide were affected by this treatment (cf. **227** → **228**), they are the only residues involved in β-linkage. All the evidence[141] therefore points to the structure **229** for the basic repeating-unit of the *Sh. flexneri* O-antigen, although the locations of the O-acetyl groups, which may be immunologically significant, have still to be determined.

In principle, any reaction that can be performed at the monomer level should be applicable to polysaccharides. Encouraging progress has already been made in the range of methods available for the specific degradation of polysaccharides,[134] some of which are discussed above, and further developments are awaited with interest.

Acknowledgements

I am indebted to Professor Hans Paulsen (University of Hamburg) for permission to reproduce Table 1.2 and Scheme 18, and to Dr. Roy Gigg (M.R.C., Mill Hill) for comments on Table 1.1.

References

1. A.K. Ganguly, O.Z. Sarre, A.T. McPhail, and K.D. Onan, 1977, *J. Chem. Soc. Chem. Comm.*, 313; D.E. Wright, 1979, *Tetrahedron*, 35, 1207.
2. G.A. Ellestad, M.P. Kunstmann, J.E. Lancaster, L.A. Mitscher, and G. Morton, 1967, *Tetrahedron*, 23, 3893.
3. S. Hanessian, 1979, *Acc. Chem. Res.*, 12, 159; B. Fraser-Reid, 1975, *ibid.*, 8, 192; T.D. Inch, 1972, *Adv. Carbohydrate Chem. Biochem.*, 27, 191.
4. S. Hakomori and A. Kobata, 1974, in 'The Antigens' (M. Sela, ed.), Vol. 2, Academic Press, New York, p. 79; R.C. Hughes, 1976, 'Membrane Glycoproteins', Butterworths, London; J. Montreuil, 1980, *Adv. Carbohydrate Chem. Biochem.*, 37, 157; N. Sharon and H. Lis, 1981, *Chem. Eng. News*, 58, 21.
5. H. Paulsen, 1982, *Angew. Chem. (Int. Edn.)*, 21, 155.
6. R.U. Lemieux, 1978, *Chem. Soc. Rev.*, 7, 423.
7. A.H. Haines, 1981, *Adv. Carbohydrate Chem. Biochem.*, 39, 13.
8. R. Gigg, 1977, *Amer. Chem. Soc. Symp. Ser.*, 39, 253.
9. A.N. de Belder, 1965, *Adv. Carbohydrate Chem.*, 20, 219; 1977, *Adv. Carbohydrate Chem. Biochem.*, 34, 179.

10. J.S. Brimacombe, A.J. Rollins, and S.W. Thompson, 1973, *Carbohydrate Res.*, 31, 108; J.S. Brimacombe, J.G.H. Bryan, A. Husain, M. Stacey, and M.S. Tolley, 1967, *ibid.*, 3, 318; K.C. Nicolaou, M.R. Pavia, and S.P. Seitz, 1982, *J. Amer. Chem. Soc.*, 104, 2027 provide some examples.
11. J. Gelas, 1981, *Adv. Carbohydrate Chem. Biochem.*, 39, 71.
12. S. Hanessian, 1966, *Carbohydrate Res.*, 2, 86; 1968, *Adv. Chem. Ser.*, 74, 159.
13. D.L. Failla, T.L. Hullar, and S.B. Siskin, 1966, *J. Chem. Soc. Chem. Comm.*, 716.
14. D. Horton and W. Weckerle, 1975, *Carbohydrate Res.*, 44, 227.
15. C. Monneret, J.-C. Florent, N. Gladieux, and Q. Khuong-Huu, 1976, *Carbohydrate Res.*, 50, 35.
16. J.-C. Florent and C. Monneret, 1980, *Carbohydrate Res.*, 85, 243.
17. J. Thiem and J. Elvers, 1978, *Carbohydrate Res.*, 60, 63.
18. S. Jacobsen and C. Pedersen, 1974, *Acta Chem. Scand. Ser. B*, 28, 1024.
19. N.G. Gaylord, 1956, 'Reduction with Complex Metal Hydrides', Interscience, New York.
20. E.L. Eliel, V.G. Badding, and M.N. Rerick, 1962, *J. Amer. Chem. Soc.*, 84, 2371; B.E. Leggetter and R.K. Brown, 1963, *Can. J. Chem.*, 41, 2671; 1964, *ibid.*, 42, 990, 1005.
21. M.N. Rerick, 1968, in 'Reduction' (R.L. Augustine, ed.), Dekker, New York, pp. 2–94; E.C. Ashby and J. Prather, 1966, *J. Amer. Chem. Soc.*, 88, 729.
22. A. Lipták, I. Jodál, and P. Nánási, 1975, *Carbohydrate Res.*, 44, 1.
23. S.S. Bhattacharjee and P.A.J. Gorin, 1969, *Can. J. Chem.*, 47, 1195.
24. A. Lipták, 1976, *Tetrahedron Lett.*, 3551.
25. A. Lipták, P. Fügedi, and P. Nánási, 1978, *Carbohydrate Res.*, 65, 209.
26. P. Rollin and P. Sinaÿ, 1977, *C.R. Acd. Sci. Ser. C.*, 284, 65; P. Fügedi, A. Lipták, P. Nánási, and A. Neszmélyi, 1980, *Carbohydrate Res.*, 80, 233.
27. P.J. Garegg, H. Hultberg, and S. Wallin, 1982, *Carbohydrate Res.*, 108, 97.
28. P.J. Garegg, H. Hultberg, and S. Oscarson, 1982, *J. Chem. Soc. Perkin Trans. 1*, 2395.
29. P.M. Collins and N.N. Oparaeche, 1975, *J. Chem. Soc. Perkin Trans. 1*, 1695.
30. P.M. Collins, N.N. Oparaeche, and V.R.N. Munasinghe, 1975, *J. Chem. Soc. Perkin Trans. 1*, 1700.
31. J.F. King and A.D. Allbutt, 1970, *Can. J. Chem.*, 48, 1754; P. Deslongchamps, R. Chênevert, R.J. Taillefer, C. Moreau, and J.K. Saunders, 1975, *ibid.*, 53, 1601.
32. J.N. Hines, M.J. Peagram, G.H. Whitham, and M. Wright, 1968, *J. Chem. Soc. Chem. Commun.*, 1593.
33. A. Klemer and G. Rodemyer, 1974, *Chem. Ber.*, 107, 2612.
34. D. Horton and W. Weckerle, 1977, *Amer. Chem. Soc. Symp. Ser.*, 39, 22.
35. D.M. Clode, D. Horton, and W. Weckerle, 1976, *Carbohydrate Res.*, 49, 305; A. Klemer and D. Balkau, 1978, *J. Chem. Res. (S)*, 303; (*M*), 3823; J.S. Brimacombe, R. Hanna, M.S. Saeed, and L.C.N. Tucker, 1982, *J. Chem. Soc. Perkin Trans. 1*, 2583.
36. J. Yoshimura, M. Matsuzawa, K. Sato, and Y. Nagasawa, 1979, *Carbohydrate Res.*, 76, 67.
37. R.E. Ireland, C.S. Wilcox, and S. Thaisrivongs, 1978, *J. Org. Chem.*, 43, 786; R.E. Ireland, S. Thaisrivongs, N. Vanier, and C.S. Wilcox, 1980, *ibid.*, 45, 48.
38. S.J. Eitelman, R.H. Hall, and A. Jordaan, 1978, *J. Chem. Soc. Perkin Trans. 1*, 595.
39. R.J. Ferrier, 1965, *Adv. Carbohydrate Chem.*, 20, 67; 1969, *Adv. Carbohydrate Chem. Biochem.*, 24, 199.
40. R.E. Ireland and J.-P. Vevert, 1980, *J. Org. Chem.*, 45, 4259.
41. R. Gigg, A.A.E. Penglis, and R. Conant, 1977, *J. Chem. Soc. Perkin Trans. 1*, 2014.
42. T.J. Prosser, 1961, *J. Amer. Chem. Soc.*, 83, 1701; C.C. Price and W.H. Snyder, 1961, *ibid.*, 83, 1773.
43. R. Gigg and C.D. Warren, 1968, *J. Chem. Soc. C*, 1903.
44. R. Boss and R. Scheffold, 1976, *Angew. Chem. (Int. Edn.)*, 15, 558.
45. J. Gigg and R. Gigg, 1966, *J. Chem. Soc. C*, 82.
46. P.A. Gent and R. Gigg, 1974, *J. Chem. Soc. Comm.*, 277.
47. R. Gigg, 1980, *J. Chem. Soc. Perkin Trans. 1*, 738.
48. P.A. Gent, R. Gigg, and R. Conant, 1973, *J. Chem. Soc. Perkin Trans. 1*, 1858.
49. P.A. Gent, R. Gigg, and R. Conant, 1972, *J. Chem. Soc. Perkin Trans. 1*, 1535.

50. P.A. Gent and R. Gigg, 1974, *J. Chem. Soc. Perkin Trans. 1*, 1835.
51. C.M. McCloskey, 1957, *Adv. Carbohydrate Chem.*, 12, 137.
52. R. Gigg, 1979, *J. Chem. Soc. Perkin Trans. 1*, 712.
53. W.T. Markiewicz, 1979, *J. Chem. Res. (S)*, 24; (*M*), 181.
54. C.H.M. Verdegaal, P. L. Jansse, J.F.M. de Rooij, and J.H. van Boom, 1980, *Tetrahedron Lett.*, 1571.
55. C.A.A. van Boeckel and J.H. van Boom, 1980, *Tetrahedron Lett.*, 3705.
56. A.H. Haines, 1976, *Adv. Carbohydrate Chem. Biochem.*, 33, 11; J.M. Sugihara, 1953, *Adv. Carbohydrate Chem.*, 8, 1.
57. J.C. Pommier and J. Valade, 1965, *Bull. Soc. chim. France*, 1257; R.C. Mehrotra, and V.D. Gupta, 1965, *J. Organometallic Chem.*, 4, 145; D. Wagner, J.P.H. Verheyden, and J.G. Moffatt, 1974, *J. Org. Chem.*, 39, 24.
58. W.J. Considine, 1966, *J. Organometallic Chem.*, 5, 263.
59. J.C. Pommier and M. Pereyre, 1976, *Adv. Chem. Series*, 157, 82.
60. P.J. Smith, R.F.M. White, and L. Smith, 1972, *J. Organometallic Chem.*, 40, 341; J.C. Pommier and J. Valade, 1968, *ibid.*, 12, 433; J. Bornstein, B. LaLiberté, T.M. Andrews, and J. Montermoso, 1959, *J. Org. Chem.*, 24, 886.
61. S. David, A. Thieffry, and A. Veyrières, 1981, *J. Chem. Soc. Perkin Trans. 1*, 1796.
62. S. David, C. Pascard, and M. Cesario, 1979, *Nouveau J. Chim.*, 3, 63.
63. R.M. Munavu and H.H. Szmant, 1976, *J. Org. Chem.*, 41, 1832.
64. S. David and A. Thieffry, 1979, *J. Chem. Soc. Perkin Trans. 1*, 1568.
65. C. Augé, S. David, and A. Veyrières, 1976, *J. Chem. Soc. Chem. Comm.*, 375.
66. H.M. Flowers, 1975, *Carbohydrate Res.*, 39, 245.
67. T. Ogawa and M. Matsui, 1978, *Carbohydrate Res.*, 62, C1.
68. J. Alais and A. Veyrières, 1981, *J. Chem. Soc. Perkin Trans. 1*, 377.
69. R.S. Tipson, 1953, *Adv. Carbohydrate Chem.*, 8, 107; D.H. Ball and F.W. Parrish, 1968, *ibid.*, 23, 233.
70. D.H. Ball and F.W. Parrish, 1969, *Adv. Carbohydrate Chem. Biochem.*, 24, 139.
71. R.W. Binkley and D.G. Hehemann, 1978, *J. Org. Chem.*, 43, 3244.
72. S. Hanessian, Y. Leblanc, and P. Lavallée, 1982, *Tetrahedron Lett.*, 23, 4411.
73. P.J. Garegg and B. Samuelsson, 1979, *J. Chem. Soc. Chem. Comm.*, 978; 1980, *J.C.S. Perkin Trans. 1*, 2866.
74. S.R. Landauer and H.N. Rydon, 1953, *J. Chem. Soc.*, 2224.
75. A.C. Richardson, 1969, *Carbohydrate Res.*, 10, 395.
76. P.J. Garegg, R. Johansson, C. Ortega, and B. Samuelsson, 1982, *J. Chem. Soc. Perkin Trans. 1*, 681.
77. S. Hanessian, M.M. Ponpipom, and P. Lavallée, 1972, *Carbohydrate Res.*, 24, 45.
78. A.K.M. Anisuzzaman and R.L. Whistler, 1978, *Carbohydrate Res.*, 61, 511.
79. B. Classon, P.J. Garegg, and B. Samuelsson, 1981, *Can. J. Chem.*, 59, 339.
80. D.H.R. Barton and S.W. McCombie, 1975, *J. Chem. Soc. Perkin Trans. 1*, 1574.
81. J.S. Brimacombe and A.S. Mengech, 1980, *J. Chem. Soc. Perkin Trans. 1*, 2054.
82. H. Deshayes, J.-P. Pete, C. Portella, and D. Scholler, 1975, *J. Chem. Soc. Chem. Comm.*, 439.
83. P.M. Collins and V.R.Z. Munasinghe, 1977, *J. Chem. Soc. Chem. Comm.*, 927.
84. F. Arcamone, G. Cassinelli, P. Orezzi, G. Franceschi, and R. Mondelli, 1964, *J. Amer. Chem. Soc.*, 86, 5335; F. Arcamone, F. Franceschi, S. Penco, and A. Selva, 1969, *Tetrahedron Lett.*, 1007.
85. P.F. Wiley, D.J. Duchamp, V. Hsiung, and C.G. Chidester, 1971, *J. Org. Chem.*, 36, 2670.
86. K.A. Parker and R.E. Babine, 1982, *J. Amer. Chem. Soc.*, 104, 7330.
87. M. Vuilhorgne, S. Ennifar, B.C. Das, J.W. Paschal. R. Nagarajan, E.W. Hagaman, and E. Wenkert, 1977, *J. Org. Chem.*, 42, 3289.
88. H. Hoeksema, S.A. Mizsak, L. Baczynskyj, and L.M. Pschigoda, 1982, *J. Amer. Chem. Soc.*, 104, 5173.
89. R.L. Hamill and M.M. Hoehn, 1964, *J. Antibiot., Ser. A*, 17, 100; K. Uchida, T. Ichikawa, Y. Shimauchi, T. Ishikura, and A. Ozaki, 1971, *J. Antibiot.*, 76, 259.

90. J.A. Secrist, III, and K.D. Barnes, 1980, *J. Org. Chem.*, 45, 4526.
91. H. Zinner and J. Milbradt, 1966, *Carbohydrate Res.*, 2, 470.
92. J. Kiss and T. Burkhardt, 1970, *Helv. Chim. Acta*, 53, 1000.
93. K. Uchida and B.C. Das, 1973, *Biochemie*, 55, 635.
94. B.K. Bhuyan, S.P. Owen, and A. Dietz, 1965, *Antimicrob. Agents Chemother. 1964*, 91; H. Hoeksema, C. Lewis, S.A. Mizsak, J.A. Shiley, D.R. Wait, H.A. Whaley, and G.E. Zurenko, 1978, *J. Antibiot.*, 31, 945.
95. J.S. Brimacombe and K.M.M. Rahman, 1983, *Carbohydrate Res.*, 113, C6; 1983, *ibid.*, 114, C1.
96. M. Funabashi, N. Hong, H. Kodama, and J. Yoshimura, 1978, *Carbohydrate Res.*, 67, 139.
97. S. Krishnamurthy, 1974, *Aldrichim. Acta*, 7, 55.
98. G. Stork, S.D. Darling, I.T. Harrison, and P.S. Wharton, 1962, *J. Amer. Chem. Soc.*, 84, 2018; A.J. Pearson and D.C. Rees, 1982, *J. Chem. Soc. Perkin Trans. 1*, 2467.
99. E. Hardegger and F. Lohse, 1957, *Helv. Chim. Acta*, 40, 2383.
100. E.J. Corey, M. Shibasaki, and J. Knolle, 1977, *Tetrahedron Lett.*, 1625.
101. N.L. Holder and B. Fraser-Reid, 1973, *Can. J. Chem.*, 51, 3357.
102. N.A. Nelson and R.W. Jackson, 1976, *Tetrahedron Lett.*, 3275.
103. K.C. Nicolaou, M.R. Pavia, and S.P. Seitz, 1981, *J. Amer. Chem. Soc*, 103, 1224.
104. J.S. Josan and F.W. Eastwood, 1968, *Carbohydrate Res.*, 7, 161.
105. K.C. Nicolaou, M.R. Pavia, and S.P. Seitz, 1979, *Tetrahedron Lett.*, 2327.
106. J. Rokach, R. Zamboni, C.-K. Lau, and Y. Guindon, 1981, *Tetrahedron Lett.*, 22, 2759.
107. J.P.H. Verheyden, A.C. Richardson, R.S. Bhalt, B.D. Grant, W.L. Fitch, and J.G. Moffatt, 1978, *Pure Appl. Chem.*, 50, 1363.
108. S. Hanessian and R. Frenette, 1979, *Tetrahedron Lett.*, 3391.
109. T. Ogawa, T. Kawano, and M. Matsui, 1977, *Carbohydrate Res.*, 57, C31.
110. H. Ohrui and S. Emoto, 1978, *Tetrahedron Lett.*, 2095; J.-R. Pougny and P. Sinaÿ, 1978, *ibid.*, 3301.
111. S. Jarosz, D.R. Hicks, and B. Fraser-Reid, 1982, *J. Org. Chem.*, 47, 935.
112. R.C. Anderson and B. Fraser-Reid, 1978, *Tetrahedron Lett.*, 3233.
113. H. Paulsen and M. Stubbe, 1982, *Tetrahedron Lett.*, 3171.
114. K. Igarashi, 1977, *Adv. Carbohydrate Chem. Biochem.*, 34, 243.
115. H. Paulsen, Č. Kolář, and W. Stenzel, 1978, *Chem. Ber.*, 111, 2370.
116. K. Reske and K. Jann, 1972, *Eur. J. Biochem.*, 31, 320.
117. A. Lipták. J. Imre, and P. Nánási, 1983, *Carbohydrate Res.*, 114, 35.
118. A.Y. Khorlin and S.E. Zurabyan, 1975, *Recent Dev. Chem. Nat. Carbon Prod.*, 6, 137.
119. R.U. Lemieux, T. Takeda, and B.Y. Chung, 1976, *Amer. Chem. Soc. Symp. Ser.*, 39, 90.
120. D.R. Bundle and S. Josephson, 1979, *Can. J. Chem.*, 57, 662.
121. K. Igarashi, J. Irisawa, and T. Honma, 1975, *Carbohydrate Res.*, 39, 341; B. Helferich, W.M. Müller, and S. Karbach, 1974, *Justus Liebigs Ann. Chem.*, 1514.
122. R.U. Lemieux, 1964, 'Molecular Rearrangements' (P. de Mayo, ed.), Wiley-Interscience, New York, p. 735; J.F. Stoddart, 1971, 'Stereochemistry of Carbohydrates', Wiley-Interscience, New York, p. 72.
123. R.U. Lemieux, K.B. Hendriks, R.V. Stick, and K. James, 1975, *J. Amer. Chem. Soc.*, 97, 4056.
124. H. Paulsen and O. Lockhoff, 1978, *Tetrahedron Lett.*, 4027.
125. H. Paulsen and Č. Kolář, 1981, *Chem. Ber.*, 114, 306.
126. H. Paulsen and O. Lockhoff, 1981, *Chem. Ber.*, 114, 3079.
127. H. Paulsen and Č. Kolář, 1979, *Chem. Ber.*, 112, 3190.
128. P.A. Gent and R. Gigg, 1975, *J. Chem. Soc. Perkin Trans. 1*, 1521.
129. H. Paulsen and A. Bünsch, 1982, *Carbohydrate Res.*, 100, 143.
130. H. Paulsen and O. Lockhoff, 1981, *Chem. Ber.*, 114, 3115.
131. H. Paulsen and H. Bünsch, 1981, *Chem. Ber.*, 114, 3126.
132. D.A. Rees, 1977, 'Polysaccharide Shapes', Chapman and Hall, London.
133. H. Björndal, C.G. Hellerqvist, B. Lindberg, and S. Svensson, 1970, *Angew. Chem. Int. Edn.*, 9, 610.

134. H.O. Bouveng and B. Lindberg, 1960, *Adv. Carbohydrate Chem.*, 15, 53; B. Lindberg, J. Lönngren, and S. Svensson, 1975, *Adv. Carbohydrate Chem. Biochem.*, 31, 185.
135. J.J. Marshall, 1974, *Adv. Carbohydrate Chem. Biochem.*, 30, 257.
136. J.M. Bobbitt, 1956, *Adv. Carbohydrate Chem.*, 11, 1; R.D. Guthrie, 1961, *ibid.*, 16, 105; R.L. Whistler and J.N. BeMiller, 1958, *ibid.*, 13, 289.
137. G.G.S. Dutton and K.L. Mackie, 1977, *Carbohydrate Res.*, 55, 49; G.G.S. Dutton and M.-T. Yang, 1977, *ibid.*, 59, 179.
138. P.A.J. Gorin, 1981, *Adv. Carbohydrate Chem. Biochem.*, 38, 13; A.S. Perlin, 1976, *MTP Internat. Rev. Sci. Org. Chem., Ser. Two*, 7, 1.
139. J.M. Williams, 1975, *Adv. Carbohydrate Chem. Biochem.*, 31, 9.
140. S.J. Angyal and K. James, 1970, *Aust. J. Chem.*, 23, 1209.
141. L. Kenne, B. Lindberg, K. Petersson, and E. Romanowska, 1977, *Carbohydrate Res.*, 56, 363.

2 Aliphatic compounds

E.J. THOMAS

2.1 Introduction

In this chapter a few of the recent developments in the chemistry of aliphatic natural products are discussed with the emphasis on synthetic aspects. The coverage, which is not meant to be comprehensive, parallels that in the Royal Society of Chemistry's Specialist Periodical Reports on 'Aliphatic and Related Natural Product Chemistry,' except for acyclic terpenoids which are discussed in chapter 4.

2.2 Fatty acids and derivatives

There is still considerable interest in the isolation and characterization of polyacetylenic compounds from plants, most of the new compounds being found in species within the family Compositae. However, more recently this area has become one of consolidation, with interest changing towards chemotaxonomic implications and biosynthesis.

A useful introduction to fatty acid chemistry has been published, and regular reviews discuss progress in this area (see reviews listed at the end of this chapter). Of particular interest recently was the isolation of endiandric acids A (1; $n = 0$), B(1: $n = 1$), and C(2) from the dried leaves and leaf stems of *Endiandra introrsa* (Lauraceae), a large tree occurring in the rain forests of northern New South Wales and Southern Queensland.[1] Since these acids were found to be racemic, it was suggested that they were derived from open-chain polyenes by the series of thermally allowed, pericyclic reactions shown in Scheme 1. Thus sequential conrotatory and disrotatory ring closure of the all-*cis*-or the *trans, cis, cis, trans-conjugated* tetraenes (3) or (4) would give the cyclohexadienes (7) and (8). Endiandric acids A and B (1; $n = 0$, 1) could then be formed by an intramolecular Diels Alder reaction of cyclohexadienes (7;

56

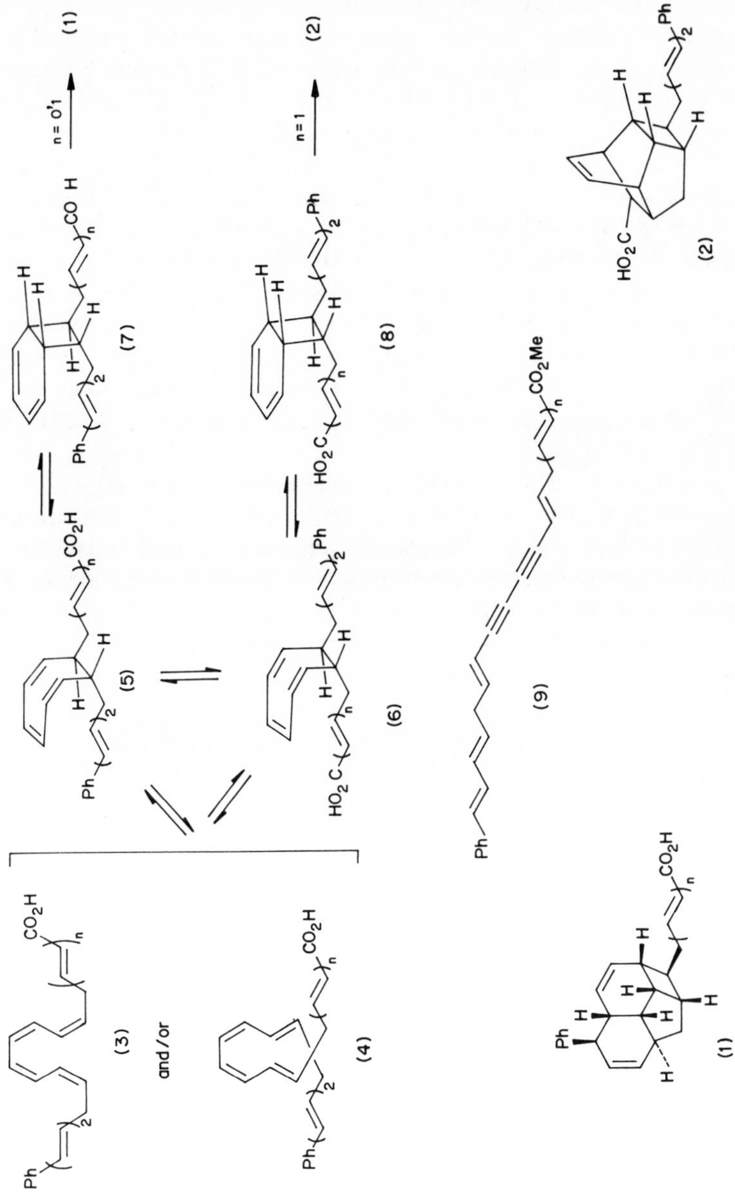

Scheme 1

$n = 0, 1$), whereas cyclohexadiene ($8; n = 1$) would yield endiandric acid C (2).

This biosynthetic proposal has been supported by an elegant series of biomimetic syntheses reported by Nicolaou.[2,3] Amongst the syntheses he developed are two based upon hydrogenation of the acetylenic esters ($9; n = 0$, 1). Selective hydrogenation at room temperature of the tetraene bis-acetylene ($9; n = 0$) in the presence of a Lindlar catalyst, gave a mixture of the methyl esters of cyclohexadienes ($7; n = 0$) and ($8; n = 0$) which presumably had been formed via the methyl ester of the *trans, cis, cis, trans*-conjugated tetraene ($4; n = 0$) as shown in Scheme 1. These cyclohexadiene methyl esters were found to interconvert at 70°C, with isomer ($7; n = 0$) undergoing an irreversible intramolecular cycloaddition to give endiandric acid A methyl ester in 30% overall yield. Cyclohexadiene ($8; n = 0$) has since been isolated as a natural product, endiandric acid D. Similarly selective hydrogenation of pentaene bis-acetylene ($9; n = 1$) at room temperature gave a mixture of the methyl esters of cyclohexadienes ($7; n = 1$) and ($8; n = 1$) which were converted by brief thermolysis in toluene into the methyl esters of endiandric acids B ($1; n = 1$) and C (2).

Of the oxygenated acids isolated during the last few years, the pseudomonic acids A, B, and C (10)–(12) have attracted considerable attention because of their antimicrobial activity.[4] The pseudomonic acids A, B, are unstable in the presence of acid and base because of epoxide opening with participation of the neighbouring hydroxyl groups; however, pseudomonic acid C is more stable. Aspects of the chemistry of these acids have been studied, for example, monic acid A (13) was obtained from pseudomonic acid A (10) and esterified with a series of alcohols to give a range of analogues.

An outline of the first published synthesis of ($\pm$)-pseudomonic acid C (12) is given in Scheme 2.[5] Oxidative elimination of the major (2.5:1) benzyloxy-selenide prepared by treatment of dihydropyran (14) with phenylselenenyl chloride and benzyl alcohol in THF[6] gave the *cis*-2, 5-disubstituted dihydro-pyran (15) which was oxidized stereoselectively to give hemiacetals ($16; \alpha/\beta$ anomers 85:15) after hydrogenolysis and protection. These hemiacetals were subjected to a Wittig reaction with acetylmethylenetriphenylphosphorane to give a mixture of tetrahydropyrans (17) and (18), ratio 2.5:1. The major 'kinetic' product (17) was treated with the sodium salt of ethyl diethyl-

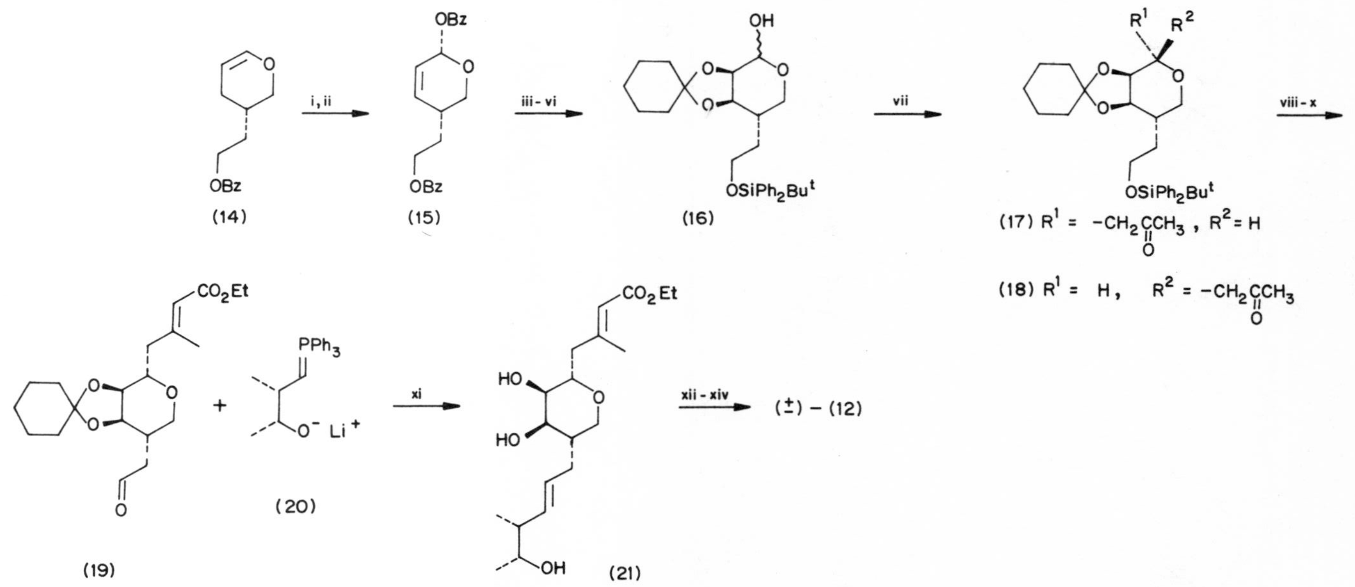

Scheme 2

Reagents: i, PhSeCl, PhCH$_2$OH; ii, NaIO$_4$, NaHCO$_3$, MeOH, then CCl$_4$, reflux; iii, OsO$_4$, N-methylmorpholine N-oxide; iv, cyclohexanone, H$^+$; v, H$_2$, Pd, C; vi, Ph$_2$Bu^{t}Cl, imidazole; vii, MeCOCH=PPh$_3$; viii, (EtO)$_2$P(O)CHCO$_2$Et; ix, F$^-$; x, PCC; xi, HOAc; xii, NaOH; xiii, methyl 9-iodononanoate; xiv, KOH, NaHCO$_3$

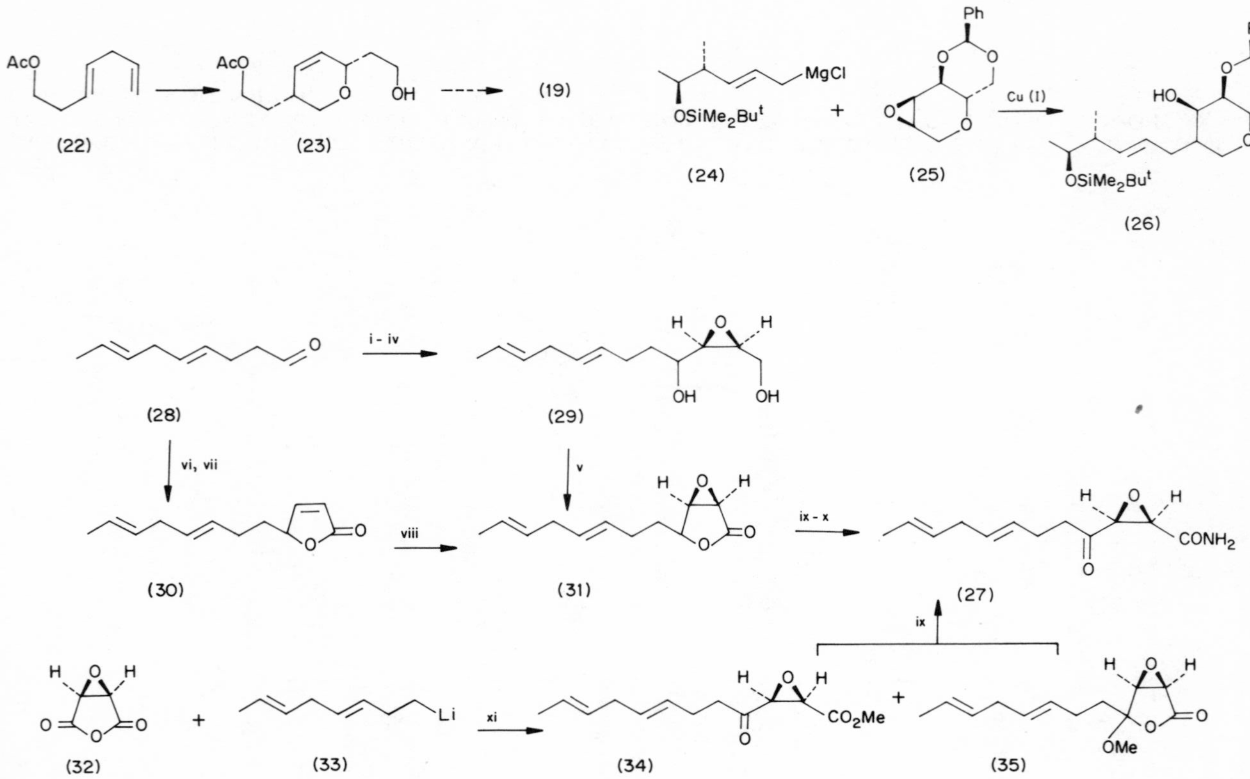

Scheme 3

Reagents: i, LiC≡CCH₂OTHP; ii, TsOH; MeOH; iii, Pd, BaSO₄, quinoline, H₂; iv, VO(acac)₂, Bu'OOH; v, Ag₂CO₃-celite; vi, BrMgC≡CCO₂MgBr; vii, H₂, Lindlar; viii, NaOCl, py; ix, NH₄OH; x, PCC or Collins; xi, CH₂N₂

phosphonoacetate, and the major product (selectivity 4:1) deprotected and oxidized to aldehyde (19) which was treated with an excess of the racemic phosphorane (20) derived from tiglic acid. The Wittig products were hydrolysed, and one of the two major products isolated was identified as the desired diastereoisomer (21). Saponification of ester (21), and alkylation using methyl 9-iodononanoate gave (±)-pseudomonic acid C (12) after hydrolysis. Stereoselective conversions of pseudomonic acid C into pseudomonic acid A are known, and so this work also constitutes a synthesis of racemic pseudomonic acid A.[7]

A shorter synthesis of the key intermediate (17) has since been developed which involves a one-pot conversion of 7-acetoxyhepta-1, *trans*-4-diene (22) into the *cis*-2,5-disubstituted dihydropyran (23) using formaldehyde and ethylaluminium dichloride.[8] This transformation involves an 'ene' reaction followed by a quasi-intramolecular Lewis acid-catalysed Diels Alder reaction.

Several attractive approaches to the synthesis of optically active pseudomonic acids using carbohydrates as starting materials are being developed,[9] and the first such completed synthesis has recently been published.[10] The key step in this synthesis was the copper (I) catalysed reaction between the allylic Grignard reagent (24) and epoxide (25) which are available from D-glucose and D-xylose respectively.

Cerulenin (27), isolated from *Cephalosporium caerulens*, exhibits both antibiotic and antifungal activity. It is a potent inhibitor of fatty acid synthesis, and in *E. coli* has been shown to inactivate the enzyme β-keto-acyl carrier protein synthetase. Scheme 3 outlines three syntheses of (±)-cerulenin reported in 1977. In Boeckman's synthesis epoxylactone (31) was obtained from nonadienal (28) via selective oxidation of the epoxydiol (29),[11] whereas Tishler epoxidized butenolide (30) using sodium hypochlorite-pyridine.[12] In both cases ammonolysis of epoxylactone (31), and further oxidation, gave racemic cerulenin (27). In Corey's synthesis, the epoxyanhydride (32) was treated with octadienyllithium (33) to give, after treatment with diazomethane, a mixture of keto-ester (34) and pseudolactone (35). Ammonolysis then gave racemic cerulenin (27).[13] More recently two syntheses of optically pure lactone (31) from glucose were reported, so providing (+)-cerulenin (27), which was found to be the natural enantiomer, and corrected the previously assigned absolute configuration.[14]

(36) X = O
(37) X = <OH, H
(38) X = <OH, H
(39) X = O
(40)

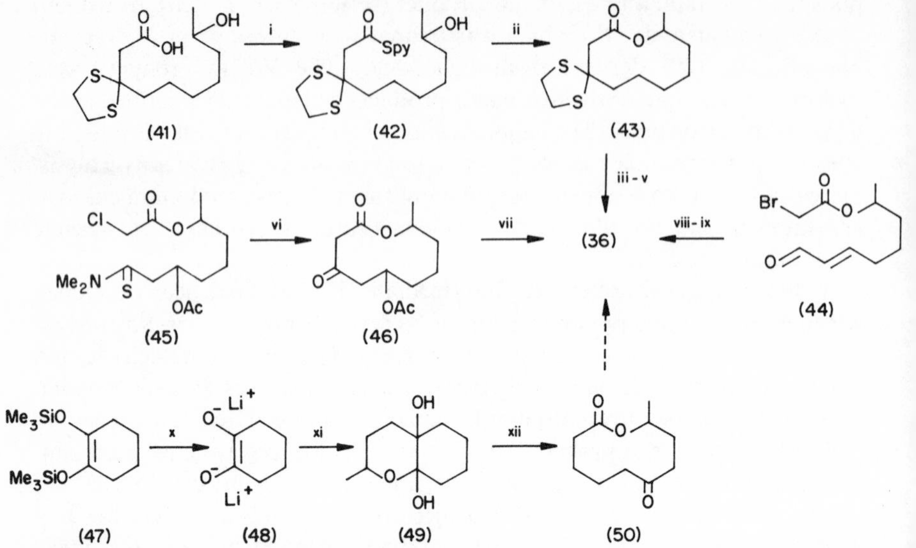

Scheme 4

Reagents: i, $(PyS)_2,Ph_3P$; ii, xylene, heat; iii, NBS; iv, LDA,PhSeBr; v, H_2O_2; vi, NaI, $P(OEt)_3$, Pr_2^iNEt; vii, Pr_2^iNEt, heat; viii, Et_2AlCl, Zn; ix, MnO_2; x, MeLi; xi, $MeCHOHCH_2CH_2I$; xii, $Pb(OAc)_4$

Medium and large-ring lactones derived from long-chain hydroxy-acids have attracted considerable interest recently. The macrolide antibiotics will be discussed later, but of present interest are the much simpler lactones, the diplodialides A–D (36)–(39)[15] and recifeiolide (40).[16]

Four syntheses of diplodialide A (36), which shows significant inhibitory activity against progesterone 11α-hydroxylase in vegetable cell cultures of *Rhizopus stolonifer*,[15] are outlined in Scheme 4. These syntheses illustrate some of the modern methods which have been developed for the synthesis of medium-ring lactones. Thus the ring-forming step in Ishida and Wada's synthesis was the lactonization of hydroxy-thioester (42) according to the procedure developed by Corey.[17,18] In contrast a carbon-carbon bond forming reaction, namely an intramolecular Reformatski reaction of bromoacetate (44) was used to close the ring in a second synthesis.[19] During studies of the efficiency of Eschenmoser's sulphur extrusion reaction for macrolide formation, Ireland cyclized chloroacetate (45) to give the diplodialide A precursor (46) in modest yield.[20] Finally a ring-cleavage approach to the diplodialides has been developed which involves treatment of enediolate (48) with 4-iodobutan-2-ol to give the bicyclic diols (49). These were oxidized by lead tetraacetate to give keto-lactone (50), a precursor of both diplodialides A and C.[21]

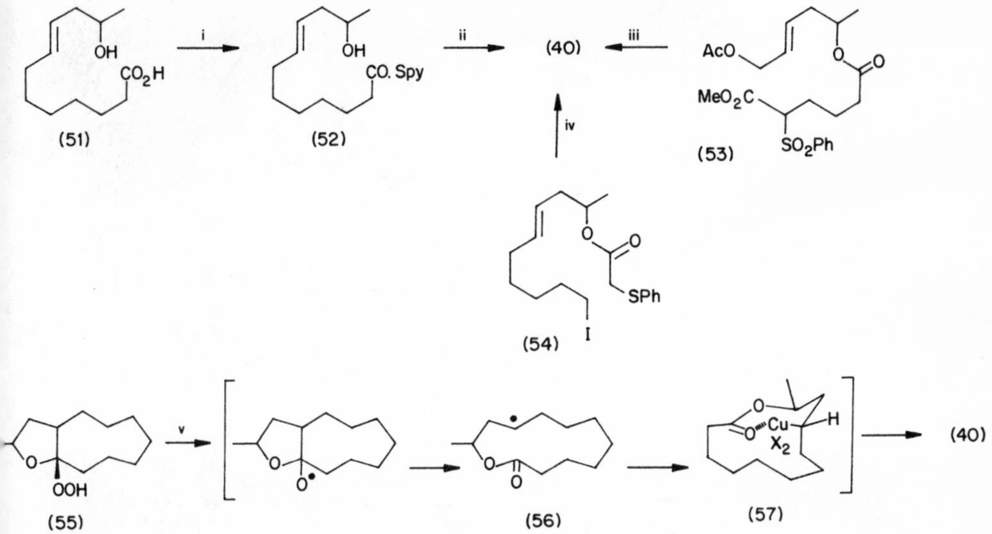

Scheme 5

Reagents: i, $(pyS)_2$, Ph_3P; ii, heat, xylene, or $AgClO_4$; iii, NaH, then $Pd(Ph_3P)_4$ followed by decarboxymethylation and desulphonylation; iv, $KN(SiMe_3)_2$ then Raney Ni; v, $FeSO_4$, $Cu(OAc)_2$, MeOH

Similarly recifeiolide (**40**), a macrolide isolated from the fungus *Cephalosporium recifei*,[16] has been synthesized several times with different procedures being used to close the 12-membered ring, see Scheme 5. Thus the seco-acid (**51**) has been lactonized via the pyridine thio-ester (**52**),[22] and carbon-carbon bond forming reactions which have been used to close the macrocycle include the organopalladium catalysed cyclization of the sodium anion of sulphone-acetate (**53**),[23] and the intramolecular alkylation of the phenylthioacetate (**54**).[24] A novel stereoselective synthesis of (±)-recifeiolide has also been reported in which alkylhydroperoxide (**55**), available in two steps from cyclononanone, was subjected to an iron (II) induced fragmentation in the presence of copper (II) acetate, which gave recifeiolide (**40**) in 96% yield.[25] The alkyl copper intermediate (**57**), formed via an oxidative coupling of radical (**56**) with copper (II) acetate, may be involved. The anomeric effect and intramolecular co-ordination of the copper species control the selectivity of the final β-elimination process.

The 16-membered dilactones pyrenophorin (**58**) and vermiculene (**67**) have also attracted considerable attention. Pyrenophorin (**58**), isolated from plant pathogenic fungi shows both antifungal and cytostatic activity. The first published synthesis of (±)-pyrenophorin involved cyclization of the hydroxy-acylimidazolide (**60**) which had been prepared stepwise from al-

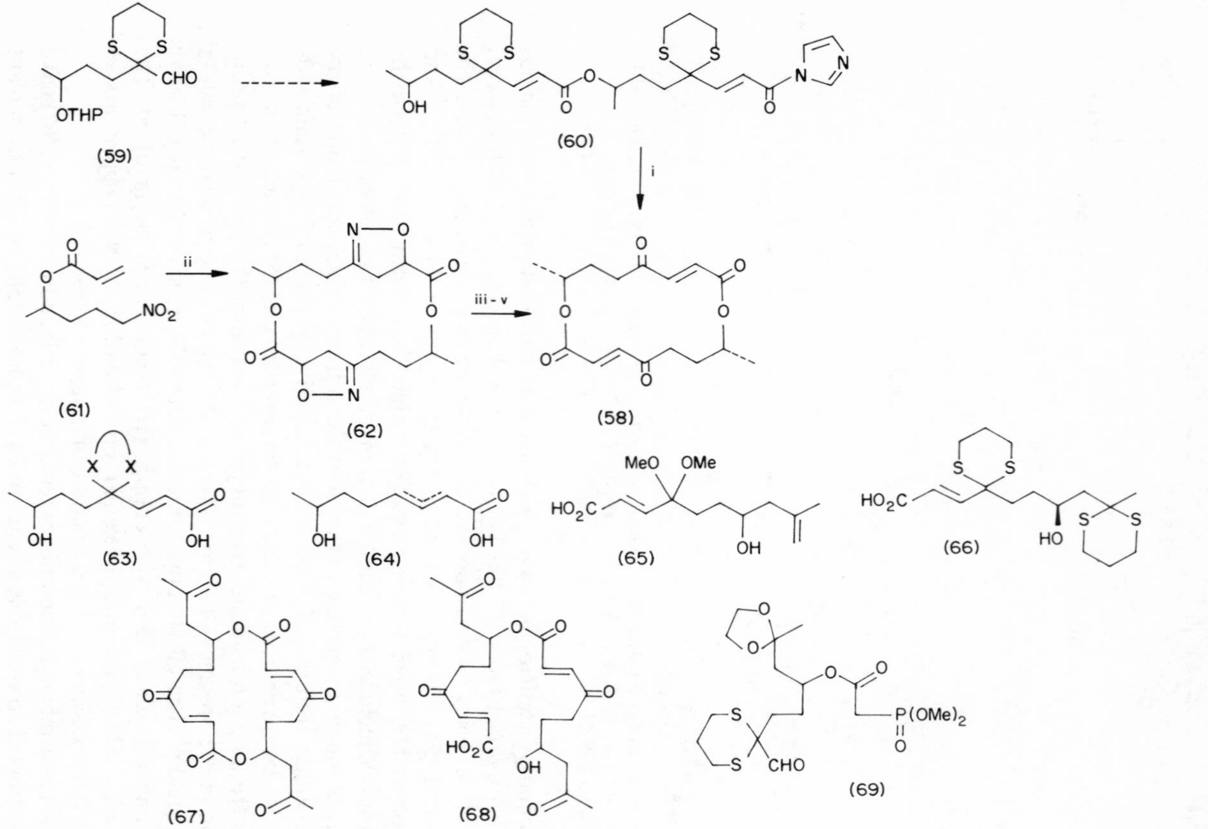

Scheme 6

Reagents: i, DBN; ii, Me$_3$SiCl, Et$_3$N, 37°C, 3 days, then d.HCl; iii, H$_2$, Pd/C, EtOH, HOAc, H$_2$O; iv, TFAA-CH$_3$CN; v, Et$_3$N

dehyde (**59**) (Scheme 6). Cyclization of imidazolide (**60**) gave a 60% yield of a separable mixture of (±)-pyrenophorin (**58**) and its meso-diastereoisomer after deprotection.[26] Other approaches to pyrenophorin have involved dimerization of hydroxyacids (**63**) and (**64**).[27,28] The Mitsunobu method (diethyl azodicarboxylate-triphenylphosphine) is the method of choice for these dimerizations. An alternative approach to the 16-membered ring of pyrenophorin which involves 1,3-dipolar cycloaddition of silyl nitronates has been described. Thus nitro-acrylate (**61**) gave dilactone (**62**) (85%) on treatment with chlorotrimethylsilane and triethylamine followed by dilute HCl. Hydrolytic hydrogenolysis and dehydration then gave (±)-pyrenophorin (**58**) together with its meso-isomer.[29]

The first published synthesis of (±)-vermiculene (**67**) involved the thio-ester induced dimerization of hydroxy-acid (**65**).[30] Later the absolute configuration of (−)-vermiculene (**67**) was established by a synthesis which involved the Mitsunobu dimerization of the optically active hydroxy-acid (**66**).[28] This procedure was also used to cyclize the vermiculene seco-acid (**68**).[31] Finally dimerization of aldehyde-phosphonate (**69**) has also been used to assemble the dilactone ring of (−)-vermiculene (**67**).[32]

Brefeldin A (**70**), a fungal metabolite isolated from a range of organisms, exhibits a broad spectrum of biological activity including antibiotic and antiviral effects. During cyclization studies, Corey showed that the 2-pyridine thiol ester (**71**), with the natural configuration at C(15), cyclized on heating in anhydrous xylene much faster than its epimer (**72**).[33] This observation was interpreted in terms of steric hindrance between the C(15) methyl group and one of the substituents at C(1) in the tetrahedral intermediate for lactone formation for the unnatural epimer, and simplified stereochemical problems involved in synthesis. This cyclization rate difference was not found however when other methods of lactonization were used. It was also found that sodium borohydride reduction of a C(4) carbonyl group was sensitive to the nature of the C(7) substituent, the MEM-ether (**75**) giving the 4α-alcohol, whereas the 4β-alcohol was obtained preferentially from the C(7) alcohol (**76**).[33]

Several syntheses of brefeldin A (**70**) have been described which involve formation of the C(9)–C(10) bond by conjugate addition of a cuprate to a cyclopentenone (Scheme 7). In Corey's modified synthesis, addition of the racemic cuprate (**78**) to cyclopentenone (**77**) gave the *trans*-adduct which was converted into a mixture of the C(15) epimers of aldehyde (**79**). The upper side chain was then added using a vinyl lithium reagent to give seco-acid (**80**) after protection-deprotection. Lactonization of the 2-pyridine thiol ester was selective for the natural C(15) diastereoisomers as discussed above, and provided the MEM-ketone (**75**) after selective hydrolysis of C(4)-THP ether and oxidation. Sodium borohydride reduction and deprotection then gave (±)-brefeldin (**70**).[34] Two later syntheses which involved conjugate addition

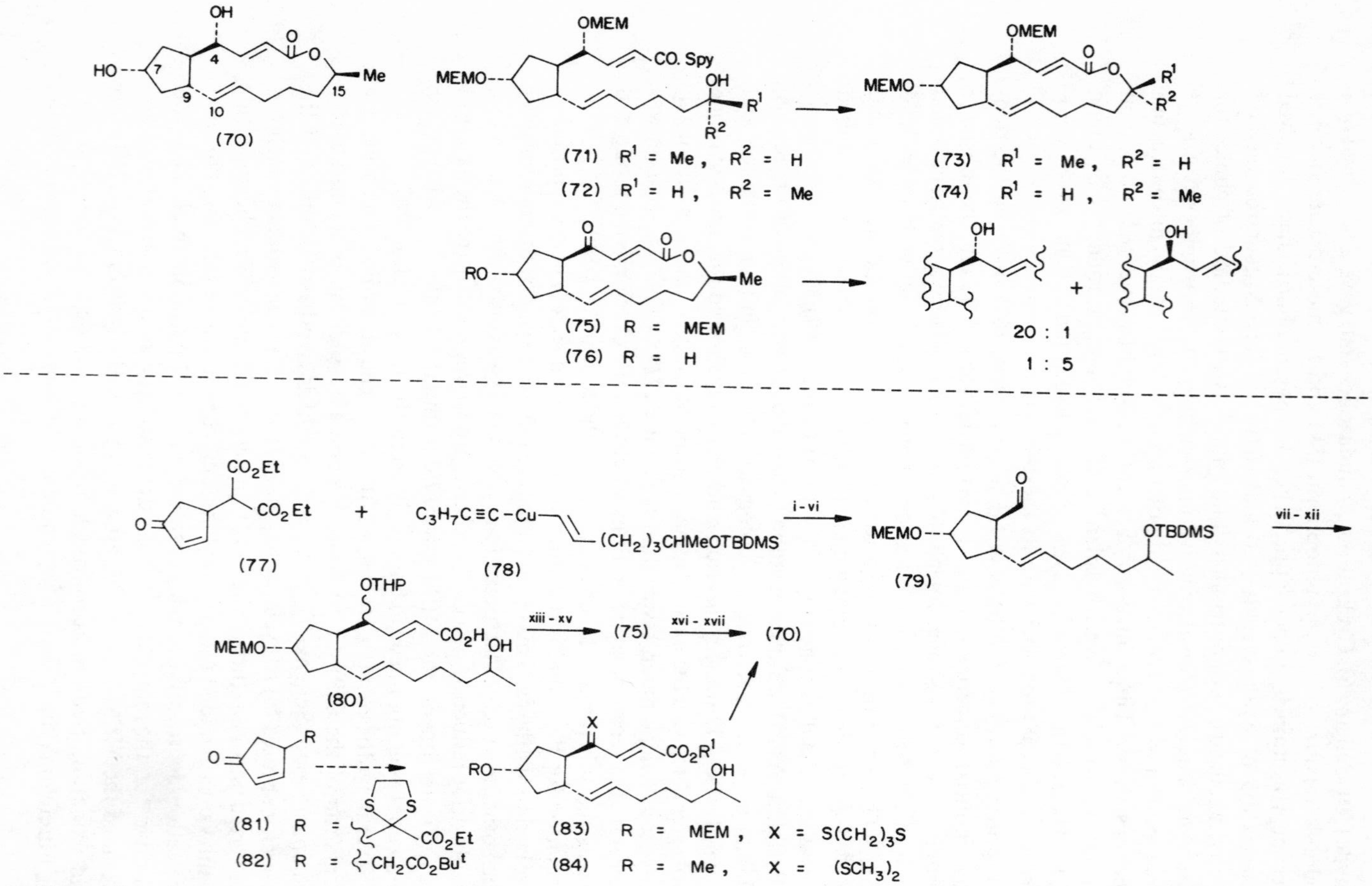

Scheme 7

Reagents: i, $LiBH_4$; ii, MEMCl; iii, NaOH; iv, py, heat; v, LDA, O_2, then $(MeO)_3P$; vi, $Pb(OAc)_4$; vii, $LiCH{=}CHCH_2OCH_2SCH_3$; viii, DHP, H^+; ix, $HgCl_2$; x, Collins; xi, Ag_2O; xii, F^-; xiii, $(pyS)_2$, Ph_3P; heat; xiv, HOAc; xv, Collins; ...

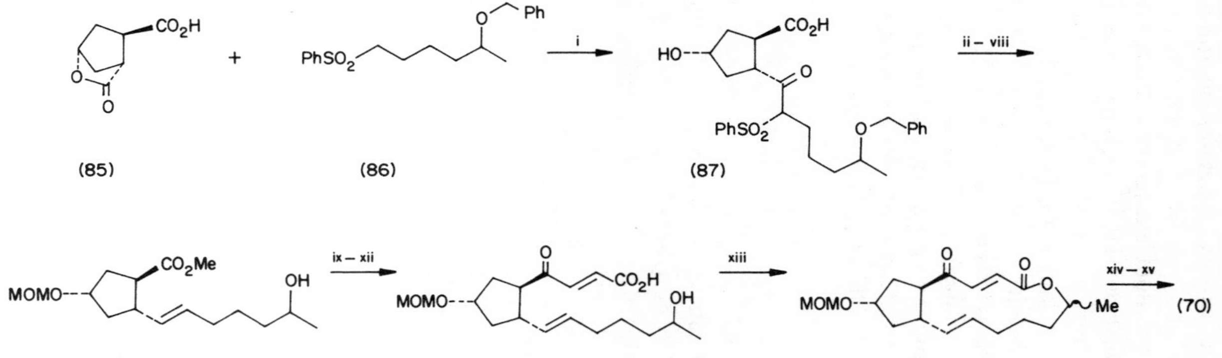

(85) (86) (87) (88) (89) (90)

Scheme 8

Reagents: i, LDA, *n*-BuLi; ii, CH_2N_2; iii, $CH_2(OMe)_2$; iv, $(PhO)_2P(O)Cl$, DMAP; v, Na-Hg; vi, OH^-; vii, Na-NH_3; viii, CH_2N_2; ix, NaH, DMSO; x, $BrCH_2CO_2Me$, $KOBu^t$; xi, heat; xii, OH^-; xiii, 2-chloro-1-methylpyridinium tetrafluoroborate, Et_3N; xiv, $NaBH_4$; xv, TFA, EtOH

of an organo-cuprate to an optically active cyclopentenone have been described. Thus cyclopentenones (81) and (82) were converted into adducts (83) and (84) in which the sensitive γ-keto-$\alpha\beta$-unsaturated ester moiety [C(1) − C(4)], was developed as a γ-thioketal group. Saponification and cyclization, using either Corey's double-activation procedure, or by using Mukaiyama's α-chloro-N-methylpyridinium iodide procedure, gave lactones which were converted into brefeldin A(70).[35]

In Bartlett's synthesis of brefeldin A (70), Scheme 8, the bicyclic lactone (85) was treated with the α-lithium derivative of sulphone (86) to give the keto-sulphone (87).[36] This was converted into *trans*-alkene (88) via an intermediate alkyne,[37] and development of the γ-keto-$\alpha\beta$-unsaturated ester moiety, was followed by saponification and lactonization using Mukaiyama's procedure[39] to give brefeldin A (70) after reduction at C(4) and deprotection.[36] Several other syntheses of brefeldin A (70) have been described.[40]

2.3 Leukotrienes

The leukotrienes are a group of long-chain carboxylic acid derivatives biosynthesized from arachidonic acid. In particular the pathway shown in Scheme 9 has been elucidated by a combination of biological and chemical methods.[41] Thus it is now believed that 5-hydroperoxyeicosatetraenoic acid (5-

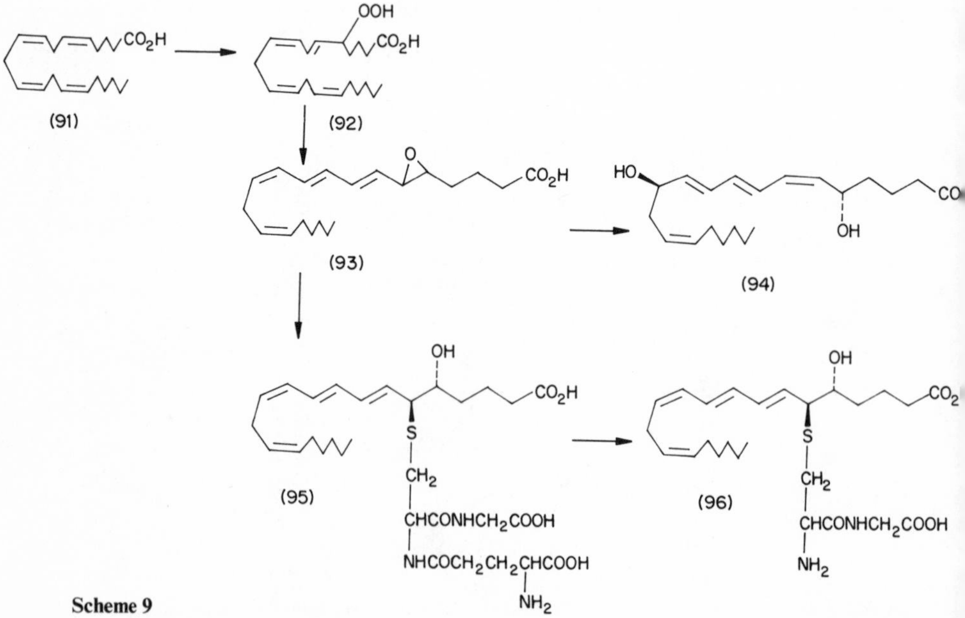

Scheme 9

HPETE) (92), formed by a lipoxygenase mediated oxidation of arachidonic acid (91), is the biosynthetic precursor of leukotriene A_4 (LTA_4) (93).[42] This unstable epoxide is either hydrolysed to leukotriene B_4 (94), or is opened by sulphur-containing peptides to give a series of peptolipids. Thus glutathione gives leukotriene C_4 (95) which is then converted into the cysteinylglycine dipeptide derivative, leukotriene D_4 (96). These natural products were isolated in trace amounts from murine mastocytoma cells treated with the calcium ionophore A-23187, and were identified from spectroscopic data and by comparison with synthetic material. In particular synthesis was necessary to establish stereochemistry and absolute configuration. The thio-ethers have been found to have biological activity of 'slow releasing substances' (SRS's), and it has been suggested that the 'slow releasing substance of anaphylaxis' (SRS-A) derived from human lung is comprised mainly of LTD_4 (96) together with LTC_4 (95), although it is likely that the exact composition of SRS-A depends upon its origin and mode of generation. Since 1979, there has been a vast amount of work on the synthesis of the leukotrienes and their analogues, in the hope of finding an SRS-A agonist which would be useful for the treatment of asthma and other allergies. Some of this work has been reviewed, and only a brief outline of the early synthetic work will be given here.

Aspects of the chemistry of arachidonic acid (91) were studied by Corey to provide a framework for synthetic work in this area, see Scheme 10. Thus arachidonic acid (91) was converted into its 5,6-epoxide (98) by iodolactonization and treatment with base, and into its 15,16-epoxide (100) by an ingenious intramolecular epoxidation involving perarachidonic acid (99).[43] Iodolactone (97) was also treated with DBU to give methyl 5-hydroxyeicosanoate (101) after methanolysis.[44] Mesylation of alcohol (101), and displacement of the mesylate with hydrogen peroxide gave 5-HPETE (92) after saponification.[44] The 15,16-epoxide (100) was isomerized into the 15-hydroxy acid (102) using isopropyl cyclohexylamine-methylmagnesium bromide (MICA), and converted into the 11,12-epoxide (104) by epoxidation-reduction of bromohydrin (103). Isomerization of the 11,12-epoxide (104) using MICA gave a mixture of the 11- and 12-hydroxy acids (105) and (106).[45]

The first synthesis of leukotriene A_4 (93) published in 1979 by Corey used a sulphonium ylid to form the 5,6-epoxide connectively, and was non-stereoselective in that an approximately 1:1 mixture of cis- and trans-epoxides was obtained.[46] Other syntheses have since been described that similarly use sulphonium ylids to form the epoxide,[47] and it has been suggested that arsenic ylids would be more stereoselective.[48] Corey's second synthesis of LTA_4 (93) used D-ribose to establish the absolute configuration of the epoxide and is outlined in Scheme 11.[49] In this synthesis, 2,3,5-tribenzoyl-D-ribose was treated with ethoxycarbonylmethylenetriphenylphosphorane to give (107) which was converted into the key hydroxy-epoxide (109). Treatment of the

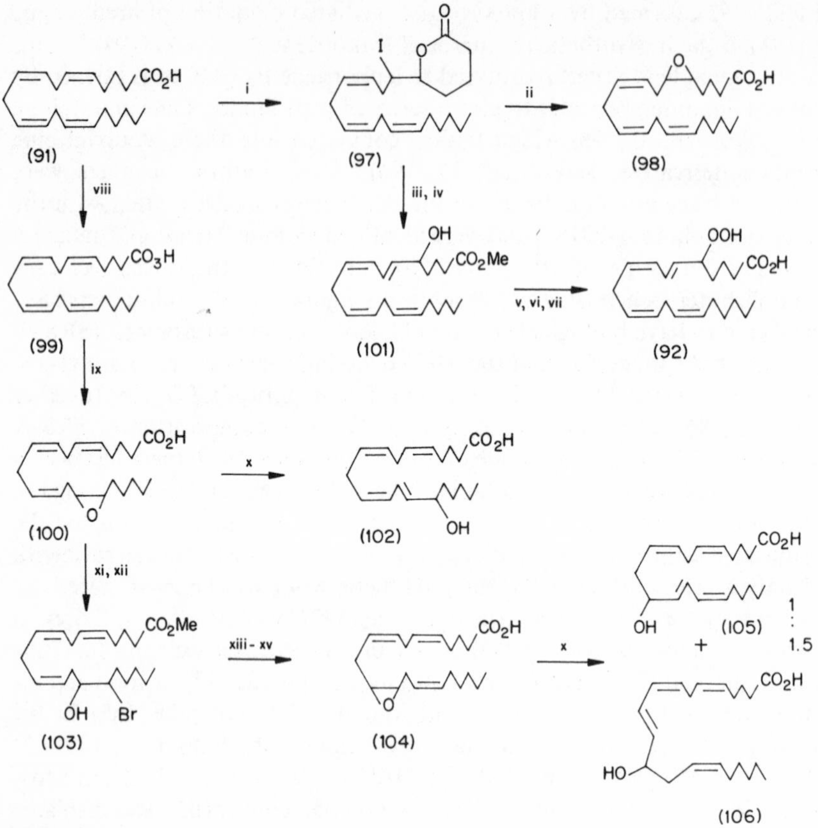

Scheme 10

Reagents: i, $KI, I_2, KHCO_3$; ii, OH^-; iii, DBU; iv, Et_3N, MeOH; v, $MeSO_2Cl, Et_3N$; vi, H_2O_2; vii, OH; viii, $(Im)_2CO, H_2O_2$, ix, 25°C; x, $Pr^iNHC_6H_{11}$, $MeMgBr$, xi, CH_2N_2; xii, KBr,HOAc; xiii, Bu^tOOH, $VO(acac)_2$; xiv, $(CF_3SO_2)_2O$,py; xv, LiOH.

aldehyde formed by Collins oxidation of the hydroxy-epoxide (109) with 1-lithio-4-ethoxybutadiene gave dienal (110) after mesylation-hydrolysis. A final Wittig reaction then gave LTA_4 methyl ester (111). Treatment of epoxide (111) with N-trifluoroacetylglutathione dimethyl ester followed by selective hydrolysis gave LTC_4 (95) identical with the natural product so confirming the structure of the latter.[49] Subsequently the stereoselectivity of the final Wittig was varied, so affording the 11,12-cis-isomers of LTA_4 and LTC_4.[50]

Several alternative routes to the key epoxy alcohol (109) and the corresponding aldehyde have been described,[51] with perhaps the most efficient being based upon the Sharpless chiral epoxidation procedure. For example, treatment of 8-methylnona-trans-2,7-dien-1-ol (112) with t-butyl hydrope-

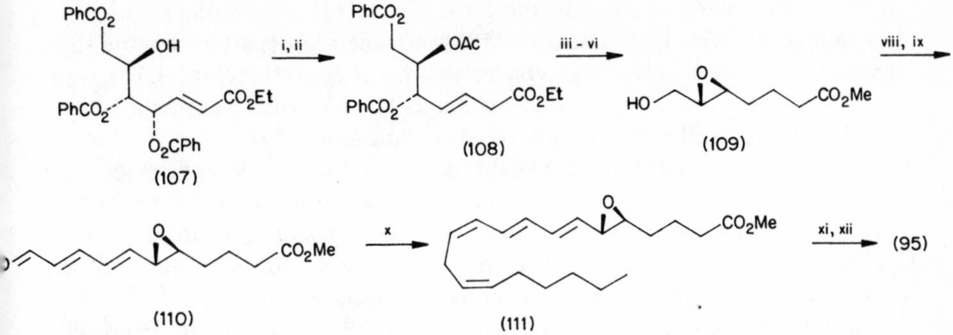

Scheme 11
Reagents: i, Ac_2O; ii, Zn,Hg; iii, H_2,Pd/C; iv, HCl,MeOH; v, TsCl, py; vi, K_2CO_3,MeOH vii, Collins; viii, LiCH=CH—CH=CHOEt; ix, $MeSO_2Cl$,Et_3N; x, $C_5H_{11}CH=CHCH_2CHPPh_3$; xi, N-trifluoroacetylglutathione dimethyl ester, Et_3N; xii, K_2CO_3, H_2O

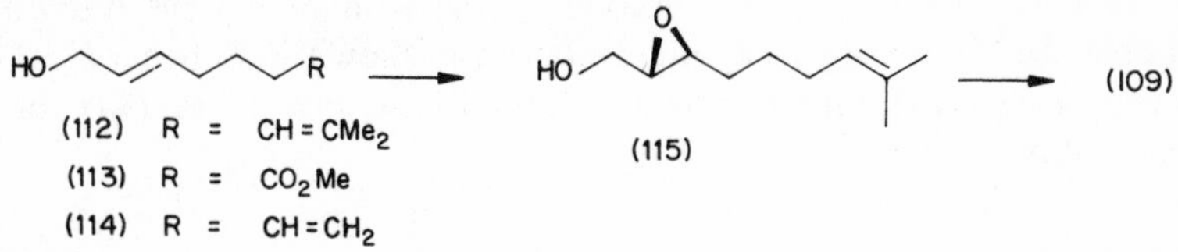

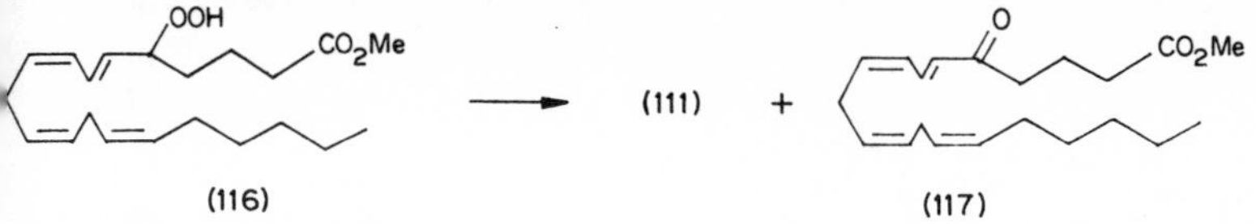

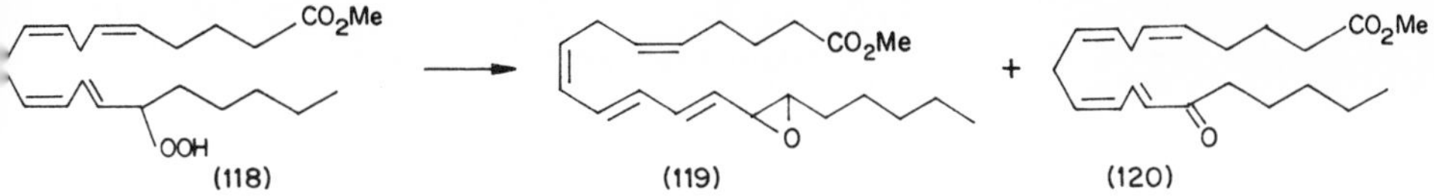

roxide in the presence of L(+)-diethyl tartrate and titanium isopropoxide gave the 2,3-epoxide (115) (75%). Acetylation, ozonolysis, and Jones oxidation then gave epoxide (109) after esterification with diazomethane and deacetylation (93%ee).[52] Direct epoxidation of ester (113) was less efficient.[52,53] An analogous route from octa-2,7-dien-l-ol (114) has also been described.[53]

The methyl ester of LTA_4 (111) has also been prepared by a biomimetic route. In this synthesis, the methyl 5-hydroperoxyeicosatetraenoate (116) was converted into a mixture of LTA_4 methyl ester (111) and the isomeric ketone (117) by treatment with triflic anhydride and 1,2,2,6,6-pentamethylpiperidine. After reduction with sodium borohydride to reduce the ketone to the more easily separated alcohol, the LTA$_4$ ester (111) was isolated in 25% overall yield.[54] The 14,15-epoxide (119) was similarly prepared from the 15-hydroperoxy ester (118), and converted into a regioisomer of LTC$_4$ by treatment with glutathione.[55] The analogous 11,12-epoxide was prepared by total synthesis, and similarly treated with glutathione.[55]

Recently some doubt has been expressed as to the stereochemistry of the epoxides formed in these hydroperoxide reactions. Sih found that epoxides (111) and (119) when formed from the corresponding hydroperoxides contained appreciable amounts of their *cis*-isomers. However the ratio of *trans*- to *cis*-epoxides is dependent upon the reaction conditions.[56]

Initially the biological role of leukotriene B$_4$ (94) was unclear. However more recently it has been shown to be chemotactic for macrophages and neutrophils, and as a result very inflammatory. It has been detected in the synovia of patients with rheumatoid arthritis, and may have a major role in the mechanisms of inflammatory and allergic states. Several synthesis of LTB$_4$ have been described,[57] Corey's second synthesis involving the interesting intramolecular base catalysed fragmentation is shown in Scheme 12. Thus epoxybenzoate (121), on treatment with mild base gave LTB$_4$ (94) via the fragmentation shown.[58]

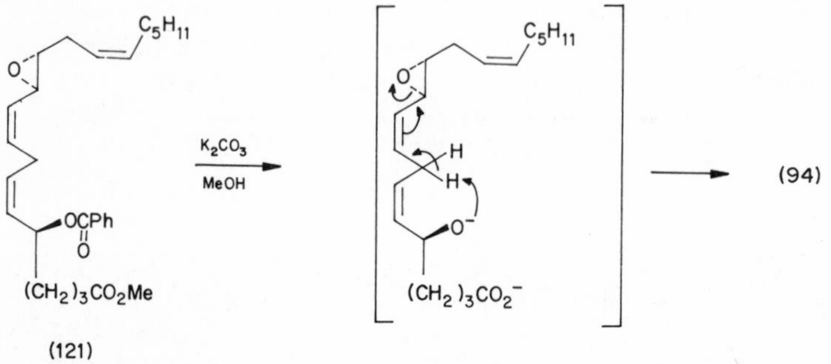

Scheme 12

Finally several reports have been published concerning the synthesis of potential inhibitors of leukotriene biosynthesis,[59] and it was found that 4,5-dehydroarachidonic acid irreversibly inhibits the 5-lipoxygenase from rat basophilic leukemic cells in the presence of oxygen.[60]

2.4 Marine natural products

Compounds representing a wide range of structural types have been isolated from marine organisms. For example several polysulphide antibiotics were isolated from *Chondria californica*, of which sulphone (122) showed the most activity,[61] and a cyclic peroxyketal, chondrillol (123), was isolated from a

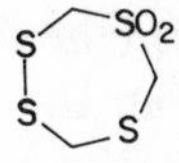

(122) 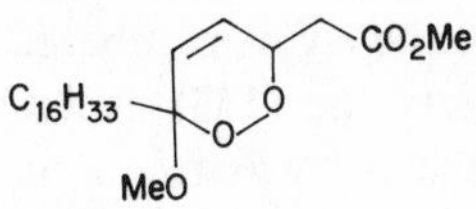　　　(123) 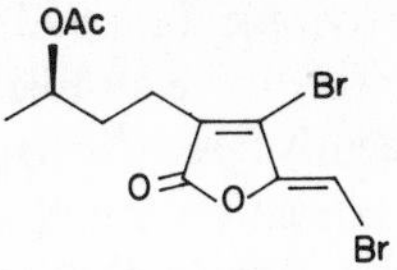　　　(124)

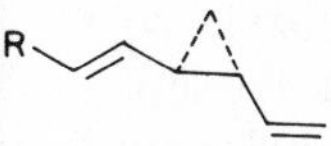

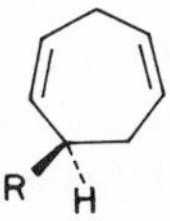

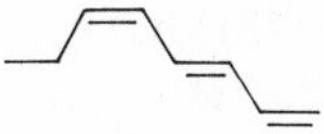

(125)　R = n-Bu—　(127)　R = n-Bu—　　　(129)　　　(130)
(126)　R = C₂H₅CH=CH—　(128)　R = C₂H₅CH=CH—

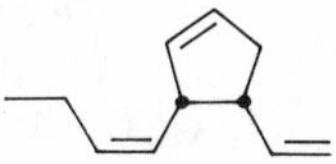

sponge of the genus *Chondrilla*.[62] Studies of the constituents of the red alga *Asparagopsis taxiformis* led to the isolation of a range of simple halogenated metabolites including bromoform, iodoform, several bromopropenes, and iodoacetone.[63] Several halogenated γ-lactones, fimbrolides, the major component being identified as butenolide (124), have been isolated from Australian and Antarctic *Delisea fimbriata*, and are responsible for the antibiotic activity of the algae.[64]

The chemistry of volatile compounds derived from the Hawaiian species of the brown seaweed *Dictyopteris plaiogramma* or *D. australis* has been studied, with the *trans*-disubstituted cyclopropanes (125) and (126) being the major constituents, together with cycloheptadienes (127) and (128), and several acyclic undecapolyenes, as minor components.[65] The synthesis and Cope rearrangements of these divinylcyclopropanes have been studied; their biosynthesis is of current concern. Fucoserratene (129) and multifidene (130) have been isolated as sperm attractants in brown algae.[66,67]

A range of halogenated enynes has been isolated from red algae of the genus *Laurencia*, e.g. laurencin (131).[68] Structural assignment in this series was particularly difficult, and several of the original structural assignments made

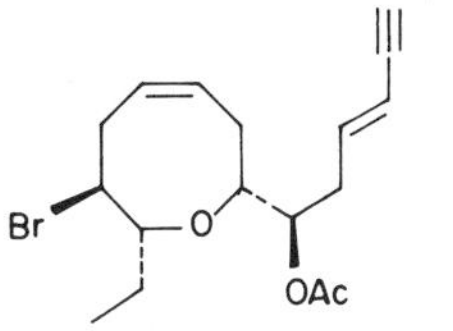

(131) 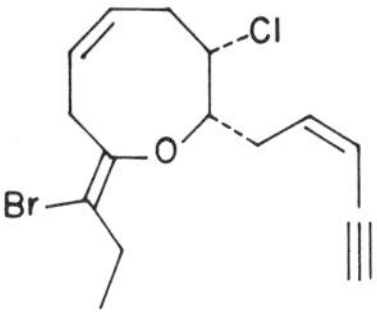　　　(132) 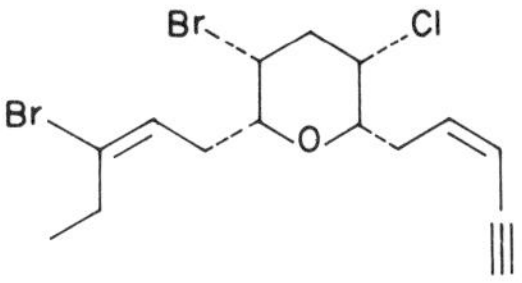　　　(133) 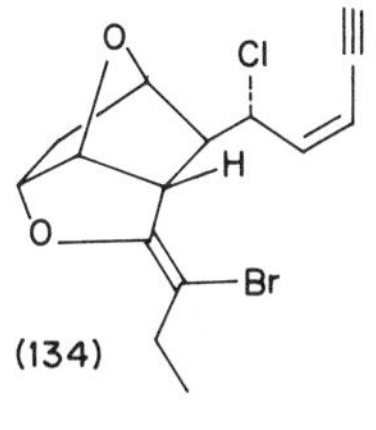　　　(134)

on the basis of spectroscopic data had to be revised when X-ray data became available. Rhodophytin, for example, was originally described as a cyclic peroxide, but has recently been shown to be the eight-membered cyclic ether (132) by an X-ray diffraction study.[69] Dactylyne (133), isolated from the opisthobranch mollusc *Aplysia dactylomela*, was the first *Laurencia* acetylene to be identified which had a six- rather than an eight-membered ring.[70] Tricyclic halogenated acetylenes isolated from *Laurencia* include the manneonenes, e.g. (134),[71] and other novel halogenated ethers include obtusin,[72] obtusenyne,[73] and a bromoallene.[74]

Laurencin (131) has been synthesized from 5-ethylfuran-2-carboxylic acid via the dihydro-2H-oxocin-3-one (135). The ene-yne side-chain was developed by the stereoselective reaction between aldehyde (135) and dimethyloxosulphonium methylide which gave epoxide (136), followed by epoxide opening using 2-lithio-1,3-dithiane to give the acetoxyaldehyde (137) after acetylation and hydrolysis. Introduction of the ene-yne unit via a Wittig reaction was then followed by conversion of the 3-keto group into the desired bromine substituent.[75] More recently synthetic interest has centred upon biomimetic formation of the halogenated dihydro-2H-oxocin-3-one ring.

Aplasmomycin (152) is a novel boron containing antibiotic isolated from a cultured marine strain of *Streptomyces griseus* which exhibits activity against gram-positive bacteria and plasmodia. It has recently been synthesized by Corey, whose synthesis is outlined in Scheme 13.[76] In this synthesis critical stereochemical control is achieved by the selective hydroxylation of unsaturated ketone (139), available in two steps from (+)-pulegone, on the *si*-face of the double bond to give hemiacetal (140). Selective protection and Baeyer-Villiger oxidation then gave lactone (141) which with propane-1,3-dithiol and trimethylaluminium was converted into ketenethioacetal (142).

Ozonolysis and reprotection, followed by conversion of the vicinal diol into the inverted epoxide, then gave the key intermediates (143) and (144). These were coupled with the vinylstannane (145), which had been prepared in optically active form from D-mannose, to give adducts, e.g. (146) from (143). Transprotection of thioether (146) gave ester (148) after carboxymethylation, and the ester was selectively deprotected and coupled to give the open-chain dimer (149). Further selective deprotection and cyclization gave the complex

(135)   (136)   (137)   (138) $\xrightarrow{vii-ix}$ (13

Reagents: i, Me$_2$SOCH$_2$,DMSO; ii, 1,3-dithiane, Bu^{n}Li; iii, Ac$_2$O,py; iv, HgO,BF$_3$.Et$_2$O; v, Ph$_3$P≡CHC≡CTMS; vi, F$^-$; vii, TsOH; viii, NaBH$_4$; ix, Ph$_3$P, CBr$_4$

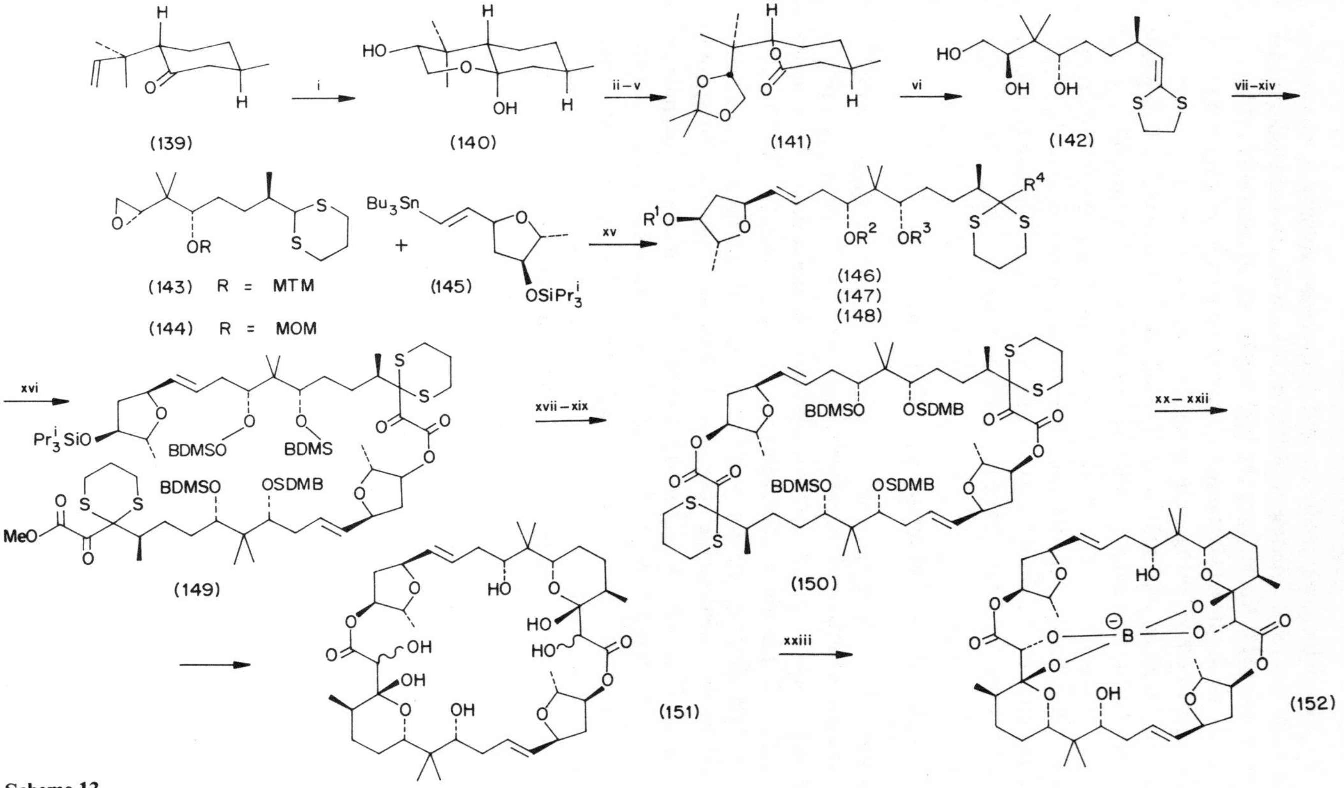

Scheme 13

Reagents: i, OsO_4, N-methylmorpholine N-oxide; ii, $LiAlH_4$; iii, acetone, H^+; iv, PCC; v, MCPBA; vi, $HS(CH_2)_3SH$, Me_3Al; vii, O_3, DMS; viii, $HS(CH_2)_3SH$, $BF_3.Et_2O$; ix, 2,2-dimethoxypropane; x, either DMSO, HOAC, $Ac_2O \rightarrow$ (143), or $ClCH_2OMe$, $Et_3N \rightarrow$ (144); ix, AcOH, H_2O; xii, PhCOCN, Et_3N; xiii, MsCl, Et_3N; xiv, $Cu_2(CN)_2$, n-BuLi; xv, Transprotection, then (147) + (148), N,N-bis[2-oxo-3-oxazolidinyl]-phosphorodiamidic chloride (BOP chloride) xvii, Li1; xviii, F^-; xix, BOP chloride, Et_3N; xx, $NaBN_4$; xxi, $HF\cdot CH_3CN$; xxii, $HgCl_2$, $CaCO_3$; xxiii, $(MeO)_3B$, MeOH (146), R' $SiPr^i_3$, R^3=MTM, R^2=R^4=H; (147), R'=$SiPr^i_3$, R^2=R^3=TBDMS, R^4=$COCO_2H$; (148), R'=H, R^2=R^3= TBDMS, R^4=$COCO_2Me$

dilactone (**150**) which was reduced and hydrolysed to give a mixture of diastereoisomers (**151**). Treatment of this mixture with trimethylborate gave aplasmomycin (**152**). Apart from the high degree of stereocontrol shown in this synthesis, the high yield obtained in the coupling and cyclization steps (98 and 71%, respectively), using BOP-chloride,[77] are notable.[76]

Palytoxin, the toxic principle isolated from marine soft corals of the genus *Palythoa*, is one of the most poisonous substances known to date. Its structure has recently been established by Kishi who used both spectroscopic and synthetic techniques to assign the stereochemistry at each of palytoxin's sixty-four chiral centres. Palytoxin turned out to be highly polyhydroxylated aliphatic compound whose synthesis is now awaited.[78]

2.5 Insect pheromones and related natural products

Insect chemistry is now an active area of research; because of the large number of pheromones isolated and characterized over the last few years, no attempt will be made to list all the new compounds here. The pheromones from moths and butterflies (Lepidoptera) have been particularly well studied, and it is now recognized that these pheromones are generally multicomponent in nature, most of their constituents being unsaturated alcohols, aldehydes, and acetates of C_{12}, C_{14}, C_{16}, and C_{18} carbon chains. The ratio of the different components can have a marked effect; for example females of *G. molesta* from the U.S.A. produce four bio-active compounds, (*Z*)-and (*E*)-8-dodecenyl acetate, (*Z*)-8-dodecenol, and dodecanol.[79] As the ratio of the two acetates is varied, other European species of the same genus are attracted, but each

(153)

(154) $R^1 = C_{14}H_{29}$, $R^2 = C_{18}H_{37}$

(155) $R^1 = R^2 = C_{16}H_{33}$

(156)

(157)

(158)

(159) (160) (161) R = Me (162) R = H

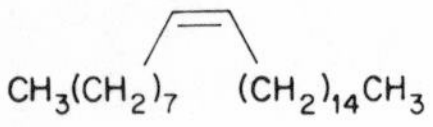

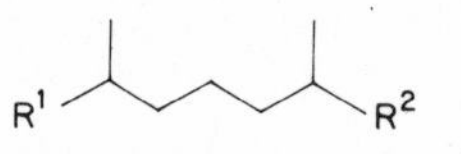

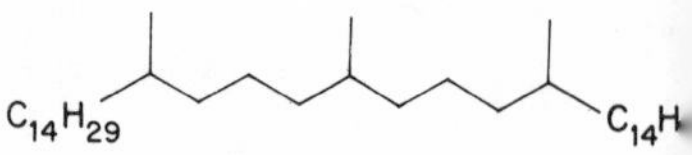

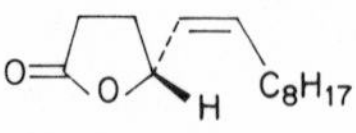

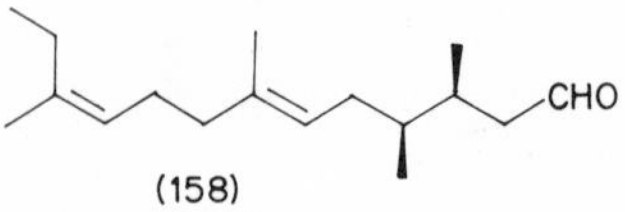

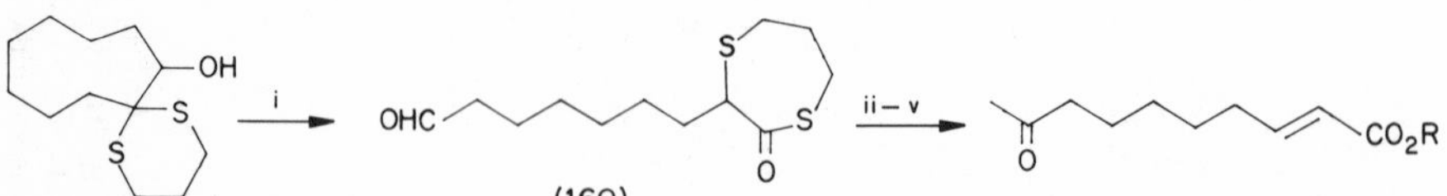

Reagents: i, $Pb(OAc)_4$; ii, MeOH, I_2; iii, $NaIO_4$, then heat; iv, MeMgBr; v, Jones

species responds optimally to a different ratio.[80] Individuals of *G. molesta* collected in France do not produce the same ratio of compounds as the species given the same name in the New World.[81]

Communication between the Diptera (flies) generally involves long-chain alkanes and alkenes as contact pheromones, e.g. (Z)-9-pentacosene (153) is a mating stimulant for the little housefly *Fannia canicularis*,[82] and the three most active compounds isolated from the cuticle of the tsetse flies *Glossina morsitans* are the alkanes (154)–(156).[83] The sex pheromone of the female Japanese beetle *Popillia japonica* has been identified as the chiral lactone (157).[84] In this case the racemic pheromone was inactive as the enantiomer of the natural pheromone was an inhibitor. The aldehyde (158) has been identified as a component of the trail pheromone of the Pharaoh's ant *Monomorium pharaoris*, a pest and public health problem in many countries.[85]

Synthesis is important for insect pheromone work both as an aid in structure determination, particularly for stereochemical assignments, when only trace quantities of the natural pheromone are available, and also to provide pheromones in sufficient amounts for field trials. One interesting synthesis of the queen substance, (E)-9-oxo-2-decenoic acid (162) involves oxidative cleavage using lead (IV) acetate of the hydroxydithiane (159), followed by oxidative-elimination and addition of methylmagnesium bromide to give ester (161) after Jones oxidation.[86] Syntheses of two of the components, (167) and (170), of the sex attractant of the California red scale, a pest of citrus groves, are outlined in Scheme 14. Thus alkylation of alcohol (164), prepared from bromide (163), using iodomethyl tri-*n*-butylstannane, gave the alkoxystannane (165) which was treated with *n*-butyllithium to generate the intermediate organolithium species (166). This rearranged stereoselectively to give the (Z)-alkene (167) after acetylation.[87] The stereoselectivity of this [2,3]-rearrangement was explained in terms of a preferred transition state in which the alkyl substituent adopts an axial position to avoid torsional interaction with the vinylic methyl group. The key carbon-carbon bond forming step in the synthesis of acetate (170) was the aluminium trichloride catalysed 'ene' reaction between methyl propiolate and citronellyl acetate (168) which gave adduct (169). This was then transformed into the pheromone component (170).[88] Other sigmatropic shifts used in insect pheromone synthesis include the [3,3]-rearrangements of allylic thionocarbonates.[89]

Several novel approaches to the stereoselective synthesis of trisubstituted alkenes have been applied to pheromone syntheses. For example the unsaturated acid (171), produced by the male monarch butterfly, has been synthesized by a route which uses the reaction between lithium dimethylcuprate and the enol phosphate of an $\alpha\beta$-ketoester, to control the stereochemistry of the trisubstituted double-bonds.[90] The decadienol (172), a component of

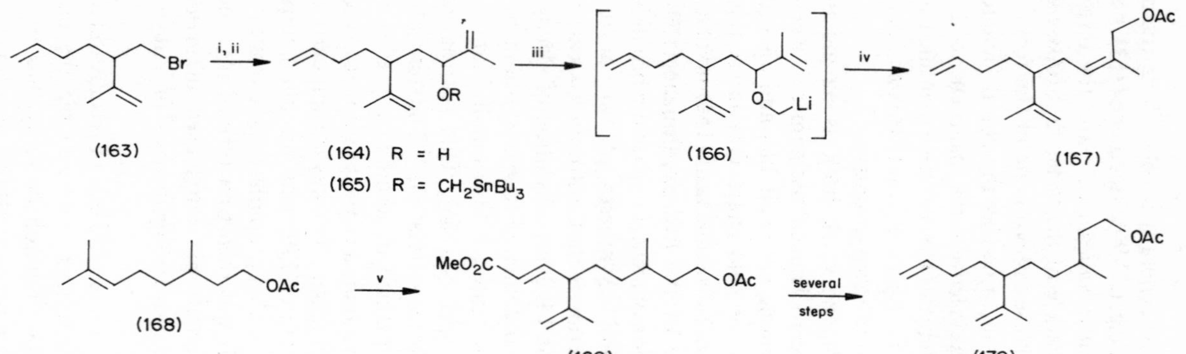

(163)

(164) R = H

(165) R = CH₂SnBu₃

(166)

(167)

(168)

(169)

(170)

(171)

(172)

(173)

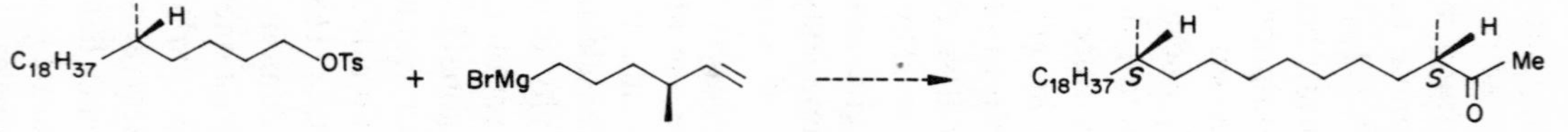

(174)

(175)

(176)

Scheme 14

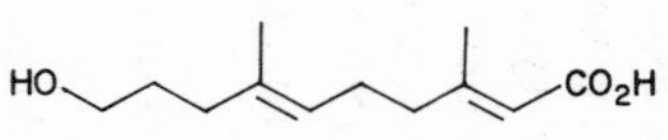

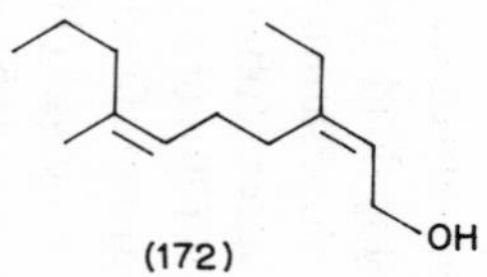

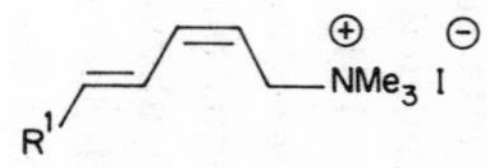

Reagents: i, Mg, CH₂=C(Me)CHO; ii, KH, Bu₃SnCH₂I; iii, BuLi; iv, Ac₂O, py; v, HC≡CCO₂Me, AlCl₃

the pheromone of the codling moth, has been synthesized by the addition of an alkylcopper reagent to a terminal acetylene followed by capture of the resulting vinyllithium species.[91]

Many new syntheses of conjugated dienes have been developed and applied to pheromone synthesis. These include the use of dienylammonium salts (173) obtained from 2-alkylpyridines,[92] the lithium tetrachlorocuprate catalysed coupling of a dienyl acetate with a Grignard reagent,[93] the reduction of conjugated ene-ynes,[94] and the coupling of alkenyl halides with either (E)-1-alkenyl-zirconium compounds or with (E)-1-alkenylalanes, derived from alkynes.[95,96] A 1,4-dehydration of an allylic alcohol was used in a synthesis of (9E)-9,11-dodecadienyl acetate.[97]

Of particular interest recently has been the development of syntheses of optically pure pheromones. For example, syntheses of all possible stereo-isomers of *erythro*-3,7-dimethylpentadec-2-yl acetate and propionate have been reported to determine the configuration of the natural material isolated from pine sawflies. These syntheses involved reactions between chiral cuprates derived from optically active monoterpenes, and chiral epoxides prepared from optically active tartaric acid.[98] Similarly the configuration of the 3,11-dimethyl 1,2-nonacosanone which is the female sex pheromone of the German cockroach, was established by synthesis of all four possible stereoisomers; the natural material was identified as (176) being obtained by coupling the optically active toluene-*p*-sulphonate (174) with the optically active Grignard reagent (175).[99]

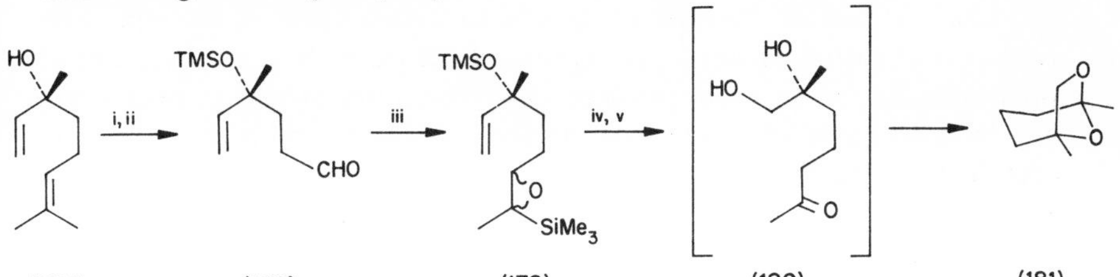

(177)   (178)   (179)   (180)   (181)

Scheme 15

Reagents: i, Me₃SiCl; ii, O₃,Me₂S; iii, Me₃SiCLiClMe; iv, O₃, NaBH₄; v, BF₃.OEt₂

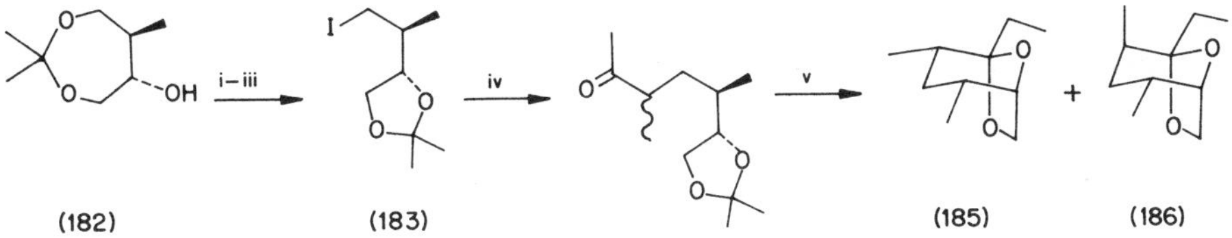

(182)   (183)   (185)   (186)

(184)

Reagents: i, HCl; ii, TsCl,py; iii, NaI; iv, EtCOEt, LDA; v, HCl, MeCN, H₂O

There has been considerable activity in the synthesis of optically active bicyclic acetal pheromones. Thus (−)-frontalin, the aggregation pheromone of the southern pine beetle, has been synthesized several times including two syntheses from D-glucose which established the absolute configuration of the natural product.[100] One short synthesis of the enantiomer of the natural material (181) from (−)-(3R)-linalool (177) is indicated in Scheme 15, and involved treatment of aldehyde (178) with lithiated α-chloroethyl trimethylsilane to give the silylepoxide (179). This was oxidized to give the diol (180) which spontaneously cyclized to the desired acetal (181).[101]

The configuration of (−)-α-multistriatin (185), a pheromone of the European elm bark beetle, was established as (1S, 2R, 4S, 5R) by synthesis of all possible stereoisomers and comparison of these with the natural material.[102] In one synthesis of (−)-α-multistriatin (185), the optically active acetal (182) was converted via iodide (183) into a mixture of diastereoisomeric ketones (184). These on treatment with acid were converted into (−)-α-multistriatin (185) and its γ-isomer (186).[103] Various multistriatin isomers have also been obtained from carbohydrate precursors.[104]

2.6 Prostaglandins, thromboxanes, and analogues

Since they were first characterized, a vast programme of research has been concerned with developing the chemistry of the prostaglandins in the hope that new compounds of therapeutic use would be discovered. The isolation and identification of the prostacyclins and thromboxanes in the mid-1970s further accelerated research in this area. Because of the large amount of prostaglandin research that has been described in the literature, no attempt will be made here to review prostaglandin chemistry with any degree of completeness. Rather just a few of the more recent developments will be mentioned to indicate the flavour of present research.

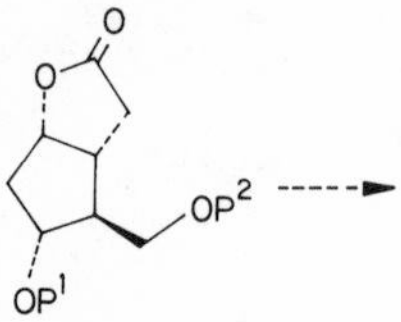

(187) 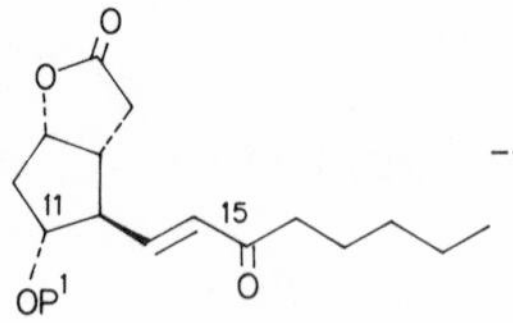   (188)

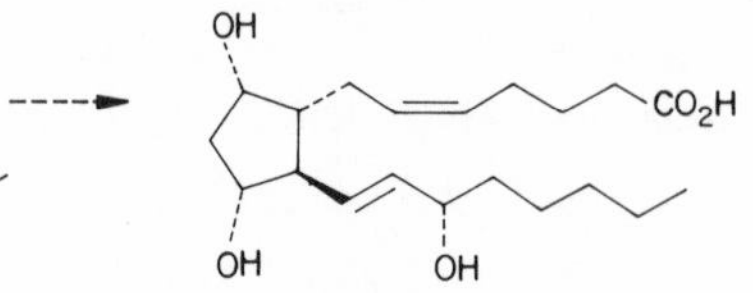

(189)

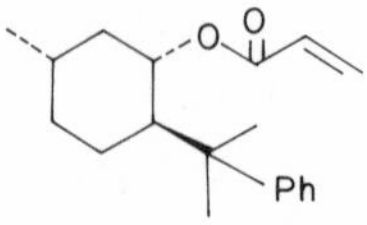

(190)

The pioneering synthetic work of Corey in the late 1960s and early 1970s established lactones (187; P^1, P^2 are protecting groups) as key intermediates in prostaglandin synthesis since the lower side-chain could be introduced by deprotection, oxidation, and a Wittig reaction, and the upper side-chain could be introduced by DIBAL reduction of the lactone followed by a second Wittig, see (187)–(189).[105] Many modifications to the original syntheses have since been reported, for example, the problem of the stereoselectivity of reduction of the C(15) ketone, see (188), was solved by using a bulky protecting group for the C(11) hydroxyl in conjunction with a hindered reducing agent,[106] and optically active lactones (187) have been prepared using the optically active acrylate (190) for the Diels Alder step.[107] Other dienophiles which have been used include nitroethylene.[108]

Amongst new syntheses of the Corey lactone are two based upon a Prins reaction of norbornadiene.[109] Sutherland's approach involves treatment of norbornadiene with formaldehyde and formic acid to give the tricyclic formate (191) which on oxidation and HBr cleavage gave lactone (192) after a Baeyer-Villiger oxidation. Hydrolysis and relactonization followed by reduction then gave the alcohol-lactone (193) which was subsequently converted into prostaglandin (PG) F_{2a} (189) using Corey's methodology.[109]

Another short synthesis of a Corey lactone used a Diels Alder reaction between 6-acetoxyfulvene and α-chloroacrylonitrile to give adduct (194), a precursor of the acetal ketone (195). Resolution of this ketone followed by a Baeyer-Villiger oxidation and iodolactonization gave lactone (196) which was converted into several prostaglandins via the lactone-aldehyde (197).[110]

A third recent synthesis of a Corey lactone is based upon the [2 + 2]

(191)   (192)   (193)

(194)   (195)   (196)   (197)

Reagents: i, HCO_2H, HCHO; ii, Jones; iii, HBr, HOAc; iv, $MeCO_3H$; v, $NaHCO_3$; vi, PhSH, py, DCC; vii, Raney Ni; viii, d.HCl, then HCl, 84°C; ix, $HC(OMe)_3$, TsOH; x, KOH, DMSO; xi, H_2O_2, OH^-; xii, KI, I_2; xiii, PhC_6H_4COCl; xiv, Bu_3^nSnH; xv, HCl, H_2O

cycloaddition of chloromethoxycarbonylketene to cyclopentadiene to give adduct (198) after Bu_3SnH reduction. Cyclobutanone cleavage, saponification, and iodolactonization, then gave iodolactone (200) which was converted into the desired lactone (202) via bromohydrin (201).[111] A related route has been developed using trimethylsilylcyclopentadiene.[112]

An extensive series of prostaglandin syntheses have been reported which use the cycloadduct (203) of cyclopentadiene and dichloroketene as starting material. More recently this work has been developed to include syntheses of optically active prostaglandins. Thus reductive removal of the chlorine substituents from (203) gave ($\pm$)-cyclobutanone (204) which was reduced enzymatically to give a separable mixture of the optically active (S)-alcohols (205) and (206). Reoxidation of the separated alcohols gave both of the optically pure enantiomers of cyclobutanone (204) which were then used in prostaglandin syntheses.[113] One example of their use is given in Scheme 16 where both enantiomers are used separately in syntheses of PGE_2 (211).[114] Thus the ($-$)-enantiomer of cyclobutanone (204) was converted via its protected bromohydrin into the tricyclic ketone (207). This underwent regioselective cleavage on treatment with a vinylic cuprate to give norbornanone (208) which on Baeyer-Villiger oxidation and *trans*-lactonization gave lactone (210). Selective reduction of this lactone, followed by treatment of the derived aldehyde with a phosphorane gave PGE_2 (211) after deprotection. The ($+$)-enantiomer of cyclobutanone (204) was protected as its acetal, and converted into the more hindered epoxide (209) via a bromohydrin. The lower side-chain was then introduced through epoxide cleavage, and the cyclobutanone moiety oxidized to give lactone (210). Many variations on

(198)   (199)   (200)   (2

(202

Reagents: i, cyclopentadiene, then Bu_3SnH, AIBN; ii, $NABH_4$, then NaOMe; iii, HCl, $HC(OMe)_3$; iv, NaOH; v, $K1,1_2$; vi, DBU; vii, $MeCONHBr.H_2O$; viii, PhC_6H_4COCl, py; ix, Bu_3^nSnH, AIBN

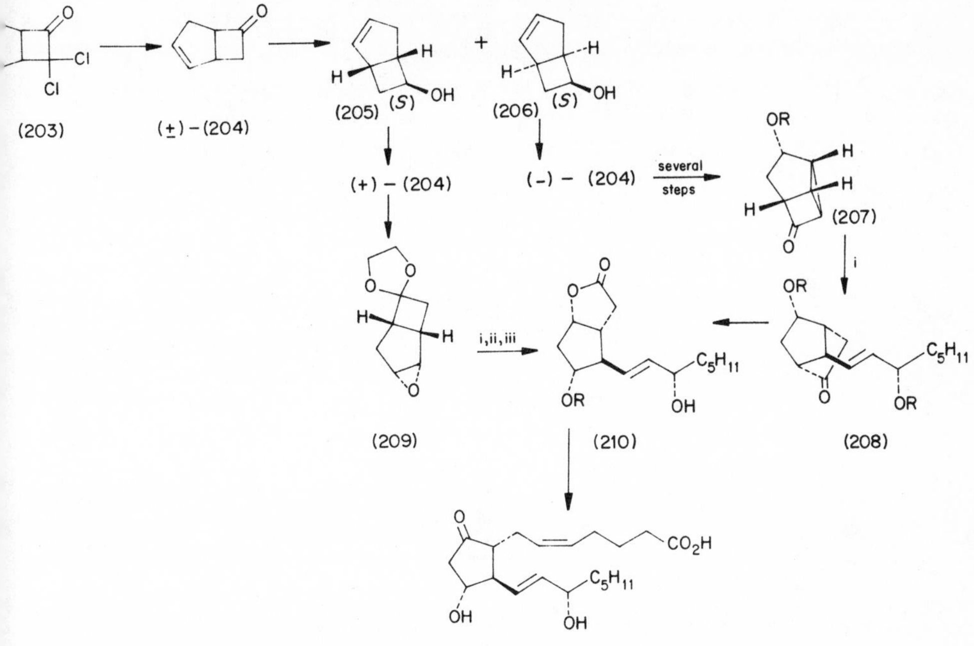

Scheme 16

Reagents: i, $LiCu(C_5H_7)CH{=}CHCH(OSiBu^tMe_2)C_5H_{11}$; ii, H^+; iii, $MeCO_3H$:

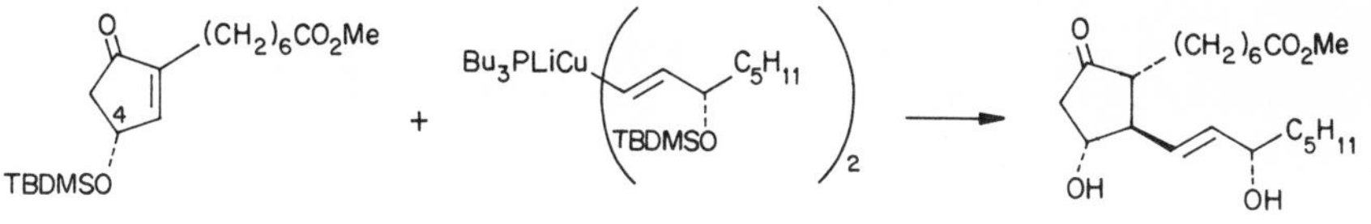

(212)   (213)   (214)

these syntheses have been reported including syntheses of PGA_2, PGD_2, and $PGF_{2\alpha}$.[115]

The conjugate addition of a cuprate to an $\alpha\beta$-unsaturated cyclopentenone has been widely used in prostaglandin synthesis. In one early example, addition of cuprate (213) to the optically active cyclopentenone (212) gave $(-)$-PGE_1 methyl ester (214) after deprotection.[116] In this synthesis the incoming cuprate is directed to the opposite face of the cyclopentenone ring by the protected 4-hydroxyl substituent, and the two vicinal alkyl chains prefer to be trans to each other. The efficiency of this conjugate addition depends upon cuprate structure,[117] and has been carried out using alananes[118] and copper(I) catalysed Grignard reagents.[119]

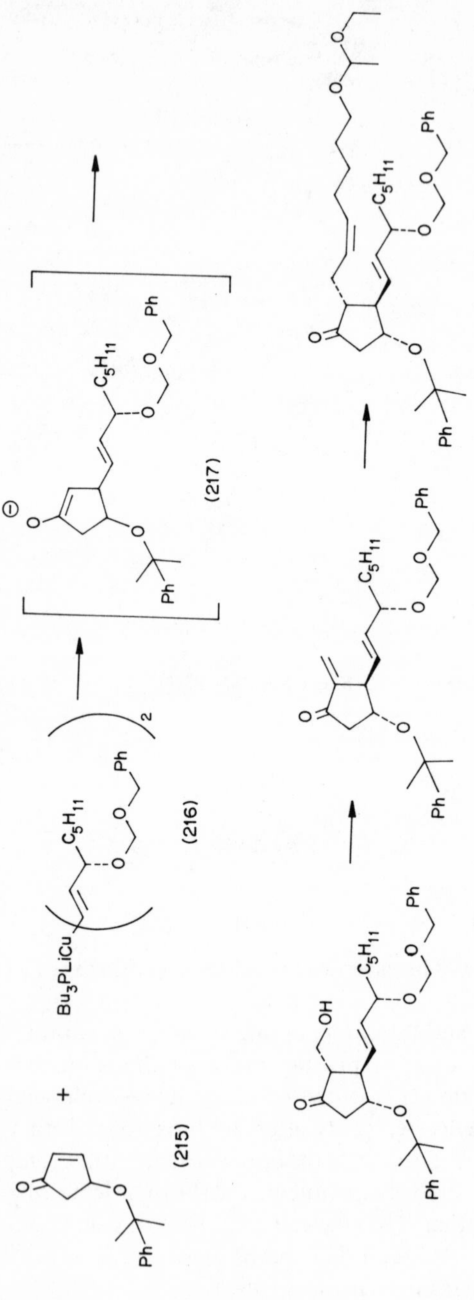
(215)
+
Bu_3PLiCu
C_5H_{11}
Ph
2
(216)
(217)
(218)
OH
(219)
(220)
O
Ph

   Although 11-deoxyprostaglandins are available by conjugate addition of vinyl cuprates to cyclopentenone followed by trapping the intermediate enolate with an alkyl halide, this direct approach could not be used for the synthesis of prostaglandins with the 11-hydroxyl group present.[120] However in these cases it was possible to trap the intermediate enolate with a more reactive electrophile, and then develop the upper side-chain subsequently. For example addition of the optically active vinyl cuprate (216) to the racemic cyclopentenone (215) gave a mixture of hydroxyketone (218) and the enantiomer of the 15-epi-isomer, after trapping the intermediate enolates (217) with formaldehyde. Dehydration and a further conjugate addition gave the PGE_2 precursor (220).[121]

   Other interesting syntheses of prostaglandins include Stork's synthesis from carbohydrate precursors, and Woodward's synthesis.[122]

   Since its isolation, there has been considerable interest in the synthesis of prostacyclin, PGI_2 (221), because of its potent vasodilatory and anti-etherification of $PGF_{2\alpha}$ derivatives followed by elimination, e.g. dehydrobromination of the major bromo-ether (222) prepared by N-bromosuccinimide cyclization of $PGF_{2\alpha}$ bis-tetrahydropyranyl ether and deprotection, gave PGI_2 (221).[123] The analogous iodo-ether (223) similarly gave PGI_2 methyl ester on elimination,[124] whereas oxidative elimination from the phenylselenenylether (224) gave the isomeric alkene (225).[125] A large range of PGI_2 analogues have been prepared.

   Recently a total synthesis of PGI_2 has been reported. The key step involved the aldol condensation of the lithium enolate of cyclopentanone with aldehyde (226) to give a mixture consisting mainly of the threo-adducts (227) and (228). These were converted separately into PGI_2 (221).[126]

   Evidence for the importance of thromboxane A_2 (229), a potent platelet

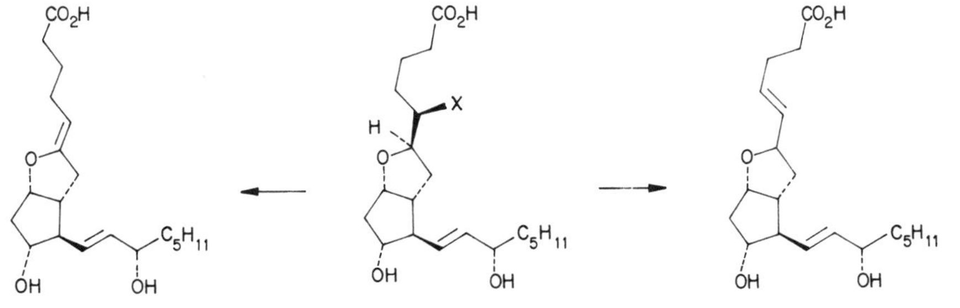

(221)

(222)   X = Br, R = H

(223)   X = I , R = Me

(224)   X = PhSe , R = Me

(225)

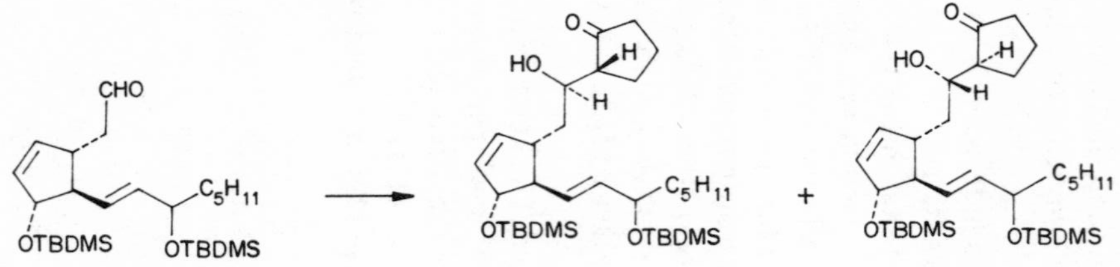

(228)

+

(227)

(226)

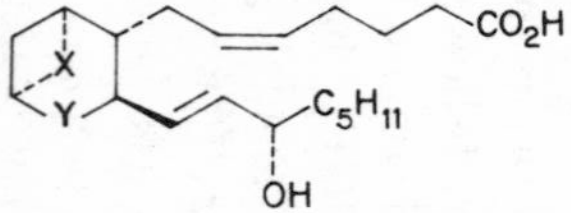

(229) X = Y = O

(230) X = Y = CH$_2$

(231) X = CMe$_2$, Y = CH$_2$

(232) X = CH$_2$, Y = O

(233) X = O Y = CH$_2$

(234) X = CH$_2$, Y = CH$_2$

(235) X = S , Y = CH$_2$

(236) X = Y = S

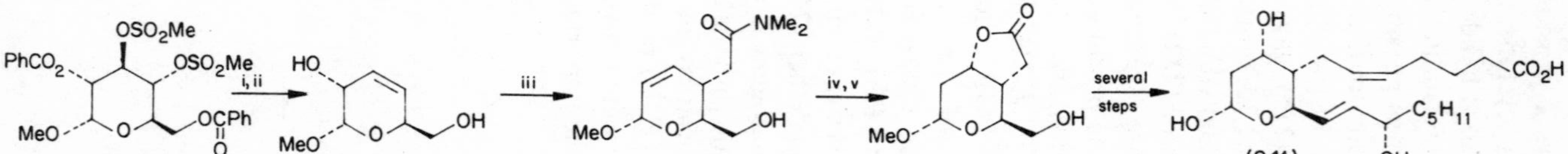

(241)

several steps

(240)

iv, v

(239)

iii

(238)

i, ii

(237)

Reagents: i, NaI, Zn(Cu), DMF; ii, MeOH, Et$_3$N; iii, MeC(OMe)$_2$NMe$_2$, 160°; iv, l$_2$, THF, H$_2$O v, Bu$_3$SnH

aggregating agent, emerged in the mid 1970s. Although thromboxane A_2 (229) itself has not been isolated or synthesized because of its instability ($t\frac{1}{2} = 36$ seconds in water at 37°C), many of its analogues have been prepared by total synthesis. These include the carbocyclic, monomethylene, monothia-monomethylene, and the dithia-analogues (230)–(236).[127] However, several syntheses of the more stable, but biologically less active, thromboxane B_2 (241) have been described,[128] and optically active thromboxane B_2 (241) has been prepared from other prostaglandins[129] and carbohydrates[130] as starting materials. In one such synthesis, the bis-mesylate (237), available from methyl β-glucoside, was reduced and hydrolysed to give diol (238). Claisen rearrangement of this diol gave amide (239) which was iodolactonized and reduced to give lactone (240), a precursor of thromboxane B_2 (241).[131]

2.7 Polyether antibiotics and related ionophores

A group of macrotetrolide antibiotics exemplified by nonactin (242) have been isolated from various actinomycetes, and found to complex selectively with many alkali and alkaline earth metals. Nonactin is composed of four

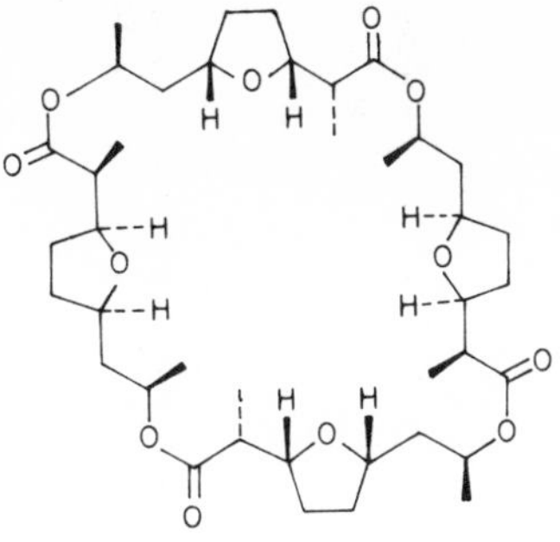

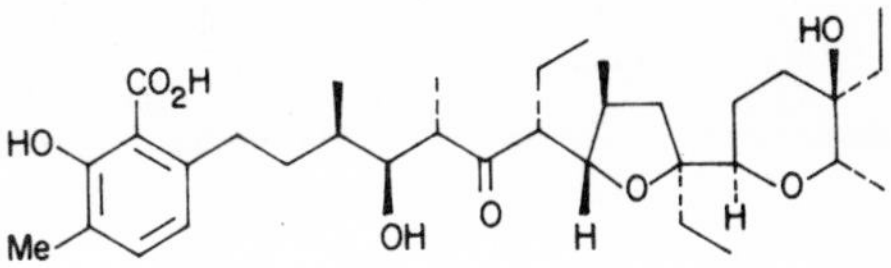

(243)

(242)

nonactic acid units and is overall a *meso* compound due to the alternation of (+)-and (−)-nonactic acid units around the periphery.[132] Several syntheses of nonactic acid have been reported, both of the racemic and optically active material, and several procedures have been described for the oligomerization of nonactic acid into nonactin.[133,134]

The polyether antibiotics are a group of structurally complex natural products which are characterized by an oxygen bridged acyclic carbon framework containing tetrahydrofuran and tetrahydropyran units, together with spiroacetals. They exhibit remarkable cation complexing ability, for example lasalocid (243) complexes and transports all alkali and alkaline earth metals, lanthanides, and organic amines. Because of their complex structures the polyethers have represented a formidable target for synthetic organic

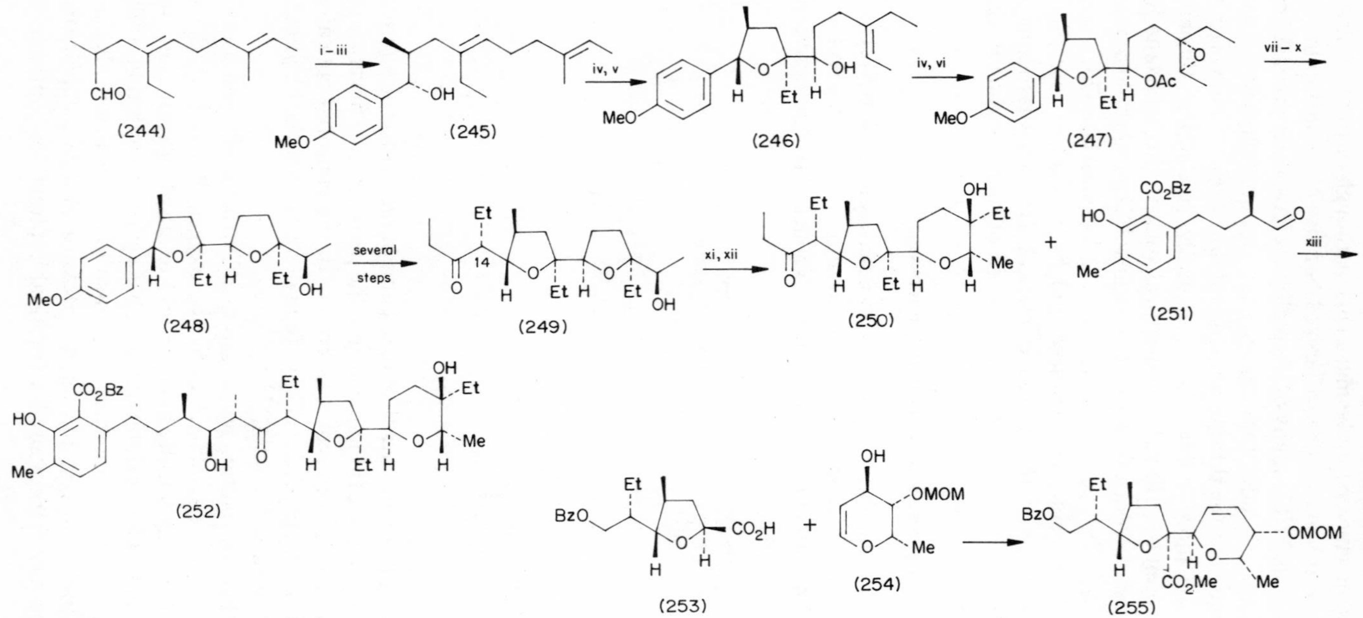

Scheme 17

Reagents: i, MeOC₆H₄MGBr; ii, Jones; iii, LiAH₄-dl-2-(o-toluidinomethyl)pyrrolidine; iv, Bu'OOH, VO(acac)₂; NaOAc; v, AcOH; vi, Ac₂O, py; vii, d, H₂SO₄; viii, TsCl, py; ix, K₂CO₃, MeOH; x, AcOH; xi, MsCl, py; xii, Ag₂CO₃, H₂O; xiii, LDA, ZnCl₂; xiv, H₂,Pd/C

chemists, and it wasn't until 1978 that the first synthesis of a polyether antibiotic, namely lasalocid (243), was reported by Kishi.[135]

An outline of Kishi's synthesis of lasalocid A (243) is given in Scheme 17. Thus the racemic aldehyde (244) was converted into the *threo*-alcohol (245) by treatment with *p*-methoxyphenyl magnesium bromide, oxidation, and stereoselective reduction. This alcohol was resolved, and the laevorotatory enantiomer epoxidized using *t*-butyl hydroperoxide-VO(acac)$_2$ to give tetrahydrofuran (246) after treatment with acid. This epoxidation procedure was then repeated, but with an acetic anhydride-pyridine work-up, to give epoxide (247).

These stereoselective epoxidations of bis-homoallylic alcohols (245) and (246) are a key feature of this synthesis, and have been studied as a general procedure.[136]

The epoxide (247) was then inverted by a three-step sequence, and the isomeric epoxide treated with acid to give the bis-tetrahydrofuran (248). The anisole ring was then reduced and converted by an eight step sequence into the ethyl ketone (249). During this transformation the configuration at C(14) had to be established by a procedure involving equilibration and recovery of the unwanted isomer. Ring expansion of the hydroxyalkyltetrahydrofuran by mesylate solvolysis then gave (250) which was condensed as its zinc enolate with the optically active aldehyde (251) to give a mixture of aldol products from which the major isomer (252) was isolated and deprotected to give lasalocid A (243).[135] Kishi subsequently published a second synthesis of the intermediate ketone (249).[137]

A convergent approach to lasalocid A (243) has been described by Ireland.[138] The key coupling step in Ireland's synthesis involved an enolate Claisen rearrangement of the ester derived from acid (253) and alcohol (254), both of which had been prepared optically pure from carbohydrate precursors. The enolate Claisen rearrangement gave a mixture of the *syn*- and *anti*-products, ratio 76:24, from which the desired *syn*-isomer (255) was isolated and converted through to ethyl ketone (250) (Scheme 17) so completing a formal synthesis of lasalocid A. Ireland subsequently prepared the optically active aldehyde (251) from (−)-citronellene and found his product to be identical to that obtained by degradation.[139]

Perhaps the most complex polyether antibiotic synthesized to date is monesin (256). Monesin was isolated in 1967 from *Streptomyces cin-*

(256)

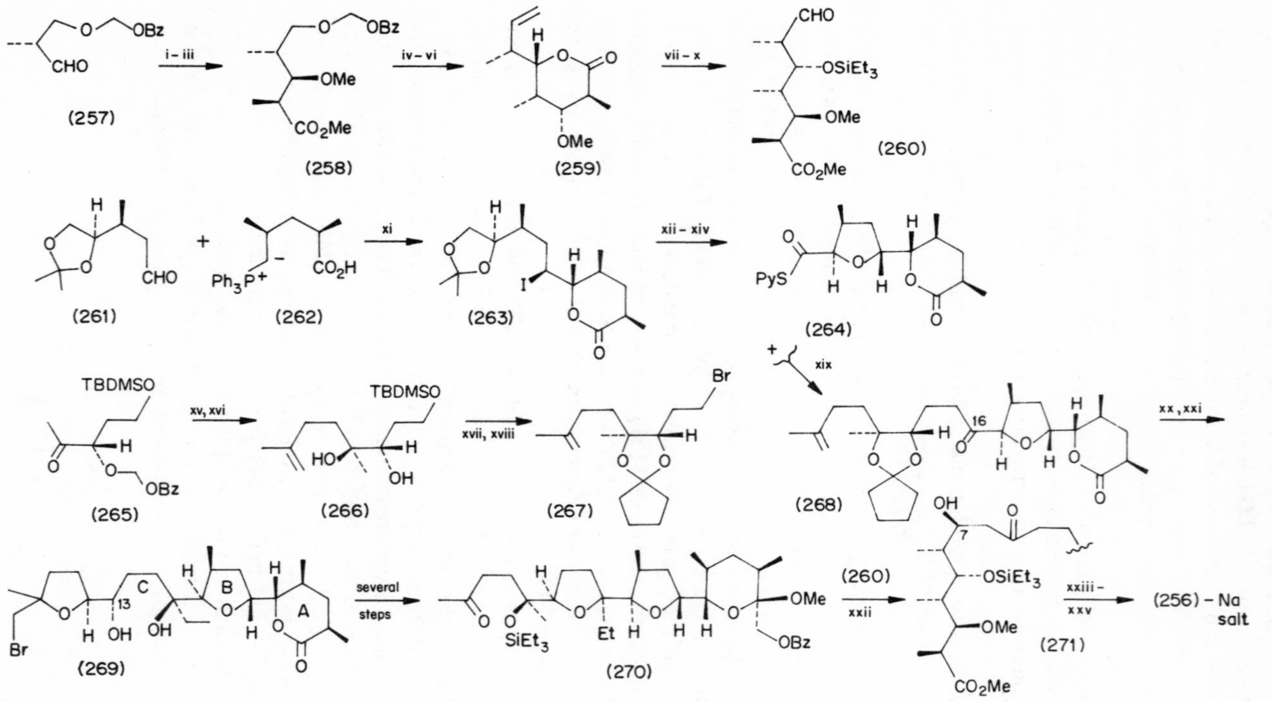

Scheme 18

Reagents: i, EtCOC(CH$_3$)$_2$OSiMe$_3$, LDA, MgBr$_2$; ii, H$_5$IO$_6$; iii, KN(TMS)$_2$, Me$_2$SO$_4$; iv, H$_2$, Pd/C; v, CrO$_3$, py; vi, Et$_2$AlCH$_2$CH=CHMe; vii, LiOH; viii, CH$_2$N$_2$; ix, Et$_3$SiOClO$_3$; x, O$_3$,Me$_2$S; xi, KI$_3$, NaHCO$_3$; xii, AgO$_2$CCF$_3$; xiii, Jones; xiv, pySH, COCl$_2$,Et$_3$N; xv, CH$_2$= C(Me)CH$_2$CH$_2$MgBr; xvi, Li,NH$_3$; xvii, cyclopentanone, H$^+$; xviii, Ph$_3$P, NBS; xix, Mg, CuI.Bu$_3$P; xx, EtMgBr; xxi, NBS, TsOH; xxii, LDA; xxiii, H$_2$, Pd/C; xxiv, TsOH; xxv, OH$^-$

namonensis and is used to control coccida infections in poultry and as a feed additive for cattle. The first synthesis of monensin (**256**) was described by Kishi in 1979 and made use of the selective epoxidation of bis-homoallylic alcohols, as discussed above, and the stereoselective hydroboration of trisubstituted double-bonds, to control the stereochemistry at several of monensin's seventeen chiral centres.[140] This synthesis was followed by a second described in 1980 by W.C. Still, and outlined in Scheme 18.[141] Still's synthesis is convergent, the three fragments (**260**), (**264**), and (**267**) being prepared from optically active components, and illustrates several modern techniques for acyclic stereochemical control, notably the use of the chelation-controlled addition of a Grignard reagent to an α-alkoxyketone.[142] The left-hand fragment (**260**) was prepared from the protected (R)-β-hydroxyisobutyraldehyde (**257**) via a chelation-controlled, erythro-selective aldol condensation to give ester (**258**) after periodate cleavage and methylation. Hydrogenolysis and oxidation, was followed by a Cram-selective addition of *cis*-but-2-enyldiethylaluminium to give lactone (**259**). Methanolysis and ozonolysis then gave fragment (**260**). Aldehyde (**261**) was prepared from (*R*)-citronellic acid and condensed with phosphorane (**262**) to give lactone (**263**) after iodolactonization. Hydrolysis and cyclization then gave the thio-ester (**264**) after Jones oxidation and treatment of the derived acid chloride with 2-mercaptopyridine. This thio-ester was treated with the Grignard reagent prepared from bromide (**267**), which had been obtained from (*S*)-(−)-malic acid via the chelation-controlled addition of 3-methyl-3-butenylmagnesium bromide to α-alkoxyketone (**265**), to give the key ketone (**268**). Chelation-controlled addition of ethylmagnesium bromide followed by treatment with *N*-bromosuccinimide in acid gave the bromomethyltetrahydrofuran (**269**). This was converted into intermediate (**270**) by cyclization, with inversion at C(13), to form the intact C ring, development of the A ring, and reductive opening-ozonolysis of the bromomethyltetrahydrofuran. Finally condensation of this methyl ketone with aldehyde (**260**) gave an alcohol with the correct configuration at C(7), which on deprotection, spiroacetalization, and saponification, gave the sodium salt of monensin.[141]

2.8 Complex macrocyclic compounds including the macrolide antibiotics

The macrolide antibiotics are a commercially important group of natural products which have been studied widely since the early 1950s. The synthesis of these compounds has been under vigorous investigation in recent years, and this aspect of their chemistry will be briefly discussed in this section. Two of the major problems associated with the synthesis of these complex macrocyclic compounds are the efficient closure of the large lactone ring, and the efficient control of configuration at the many chiral centres. Most of the ring

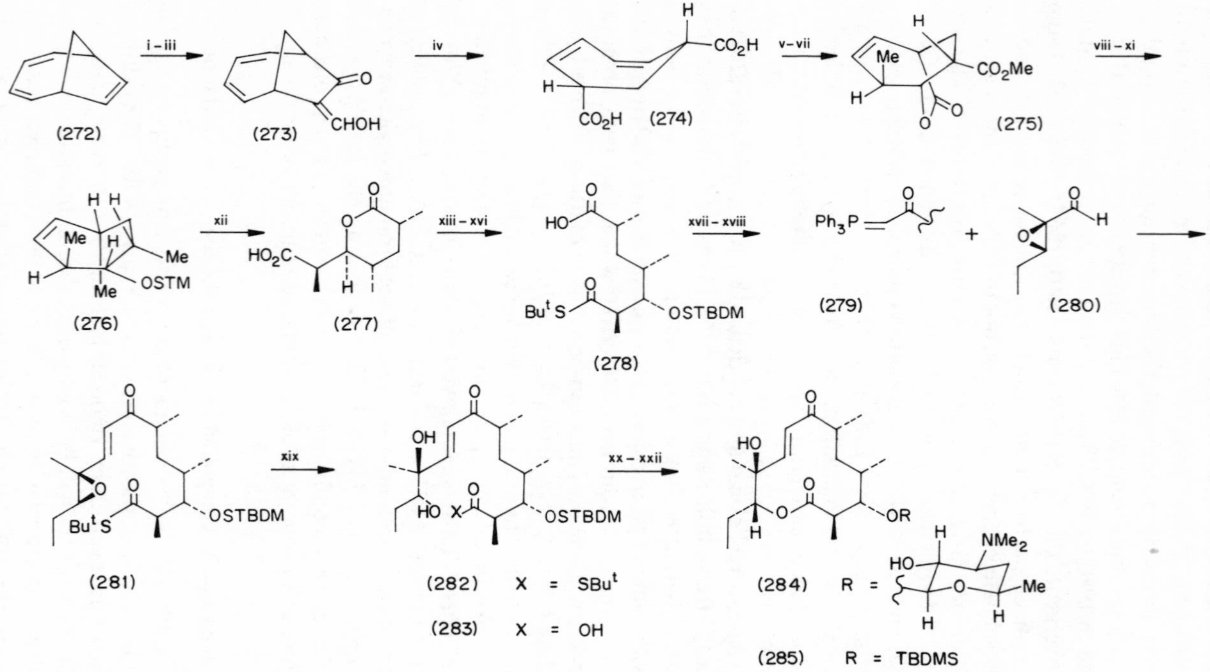

Scheme 19

Reagents: i, (Me₂CHOHCMe)₂BH, H₂O₂,OH⁻; ii, Al(OBuⁱ)₃, 1,4-benzoquinone; iii, formylation; iv, NaIO₄; v, MCPBA; vi, CH₂N₂; vii, LiCuMe₂; viii, LiAlH₄; ix, TsCl, py; x, Me₃SiCl; xi, LiCuH₂; xii, KMnO₄,NaIO₄; xiii, BuⁱSTl on acid chloride; xiv, KOH; xv, BuᵗMe₂Si Im, DMF; xvi, KOH; xvii, Im₂CO; xviii, Ph₃P=CH₂; xix, H⁺; xx, (CF₃CO₂)₂Hg; xxi, CF₃CO₂H, THF, H₂O; xxii, 1α-bromo-2-acetyldesosamine hydrobromide

closing reactions that have been used have involved lactonization of a suitably activated seco-acid, and these would appear to be facilitated by the many substituents present which limit the number of degrees of freedom of the open-chain seco-acid. Thio-esters in particular have been widely used for macrolide formation. During the course of these syntheses, many efficient techniques have been developed to control stereochemistry along an acyclic carbon chain. Of particular and general importance are the developments in the aldol condensation, which can be carried out to provide either *erythro- or threo-*adducts. Other syntheses have been based upon the use of carbohydrates as optically-active building blocks.

Methymycin (**284**), a 12-membered macrolide antibiotic, was the first member of this class to be synthesized. An outline of the synthesis reported in 1975 by Masamune is given in Scheme 19.[143] In this synthesis selective cleavage of bicyclo [4,2,1]nona-2,4,7-triene (**272**) gave the cycloheptadiene carboxylic acid (**274**). Monoepoxidation and regioselective cleavage of the epoxide, using lithium dimethyl cuprate, gave lactone (**275**), which was reduced to the protected trimethyl cycloheptenol (**276**), and then oxidized to give Prelog-Djerassi lactone (**277**). Conversion of this to its *t*-butyl thioester, saponification, and silylation, gave the open-chain acid (**278**), which was condensed via phosphorane (**279**) with the optically active epoxyaldehyde (**280**) to give the desired Wittig product (**281**). Hydrolysis of the epoxide, and cyclization of the seco-acid derivative (**282**), gave a moderate yield of lactone (**285**). This was deprotected, and treated with 1α-bromo-2-acetyldesosamine hydrobromide, to give methymycin (**284**).[143]

Since this synthesis was first reported, many other syntheses of Prelog-Djerassi lactone (**277**) have been described, and Grieco has reported an alternative synthesis of methynolide seco-acid (**283**) starting from 7-methyl-bicyclo [2,2,1] heptenone.[144]

The erythromycins, e.g. A and B (**286**) and (**287**) which are produced by the

L – cladinose =

D – desosamine =

(286) R¹ = L – cladinose , R² = D – desosamine , X = OH

(287) R¹ =   "   R²   "   X = H

(288) R¹ = R² = H X = OH

(289) R¹ = R² = X = H

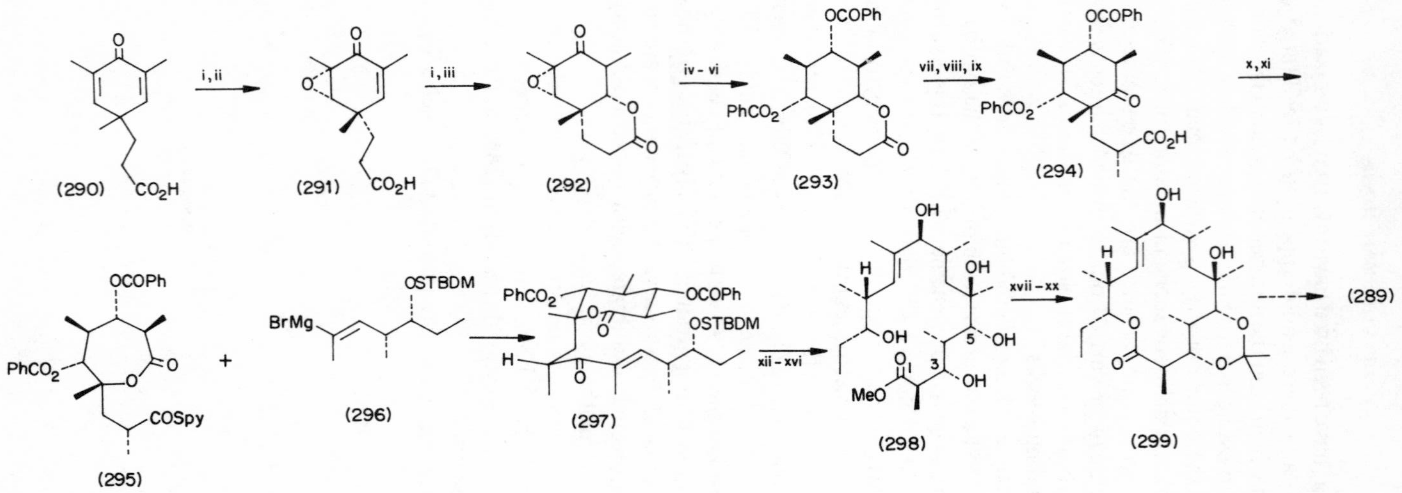

Scheme 20

Reagents: i, Br$_2$, KBr; ii, KOH; iii, Bu$_3$SnH; iv, Al-Hg; v, H$_2$, Raney Ni; vi, PhCOCl, py; vii, LDA, MeI; viii, LiOH; ix, Jones; x, CH$_3$CO$_3$H; xi, (PyS)$_2$Ph$_3$P; xii, Zn(BH$_4$)$_2$; xiii, AcOH; xiv, H$_2$O$_2$, LiOH; xv, CH$_2$N$_2$; xvi, KOH; xvii, 2-methoxypropene; xviii, H$^+$; xix, KOH; xx, 4-t-butyl-N-isopropyl-2-mercaptoimidazole

fungus *Streptomyces erythreus*, are a widely-used group of antibiotics. To date one total synthesis of erythromycin A (**286**) has been reported, and several groups have described syntheses of the erythronolides A and B (**288**) and (**289**), the aglycones of the parent antibiotics lacking the sugar moieties.

The first synthesis of the aglycone of erythromycin B (**289**), was reported by Corey in 1978.[145] In this synthesis, outlined in Scheme 20, bromolactonization and saponification of cyclohexadienone-acid (**290**) gave epoxide (**291**) which was resolved and bromolactonized a second time to give epoxylactone (**292**) after reduction using tri-*n*-butyl tin hydride. Reductive opening of the epoxide followed by hydrogenation and benzoylation gave the benzoate (**293**) which was stereoselectively methylated and saponified to give keto-acid (**294**) after an acidic work-up. This keto-acid was subjected to a Baeyer-Villiger oxidation, and treated, as thio-ester (**295**), with the vinylic Grignard reagent (**296**), to give keto-lactone (**297**). Reduction, debenzoylation and esterification using diazomethane, gave the open-chain ester (**298**). Selective protection of the 3- and 5-hydroxyls, and cyclization via a thio-ester then gave the macrolide (**299**) which was converted into erythronolide B (**289**) in five steps.

This approach was subsequently developed and led to the first total synthesis of erythronolide A (**288**). Erythronolide A is identical to erythronolide B apart from an additional hydroxyl substituent at C(12). However, despite this structural similarity, the presence of this additional hydroxyl group introduced many problems to the synthesis.[146]

The first asymmetric total synthesis of the complete antibiotic erythromycin A (**286**), including incorporation of the unusual sugars L-cladinose and D-desosamine, was published by Woodward's group in 1981.[147] The stereochemical problems were solved in this synthesis by using the optically active bicyclic thioketal **300** (see Scheme 21), to control the stereochemistry of the C(3)–C(8) and C(9)–C(13) portions of the seco-acid precursor. Thus thioketal (**300**) was converted into aldehyde (**301**) and ketone (**302**), which were condensed to give diketone (**303**) after oxidation at C(9). Further elaboration gave the seco-acid thioester (**304**). It was found that efficient macrolactonization of seco-acids related to thioester (**304**) was very dependent both upon the configuration at C(9), the *S*-configuration being preferred, and the presence of cyclic protecting groups linking the C(3) and C(9) hydroxyls with the C(5) and C(11) hydroxyls, respectively. Thus the conversion of thioester (**304**) into (**305**) (the configuration at C(9) is irrelevant since it ends up as a ketone), was followed by cyclization in toluene (110°C) to give the corresponding macrolide (**306**), the opportunity being taken during this sequence to change the C(9) substituent into an amino function. Deprotection to give triol (**307**), was followed by the selective introduction of D-desosamine at C(5) and L-cladinose at C(3), the correct anomer being obtained in each case. Deprotection and conversion of the C(9) amino group into a ketone, then gave erythromycin A (**286**).[147]

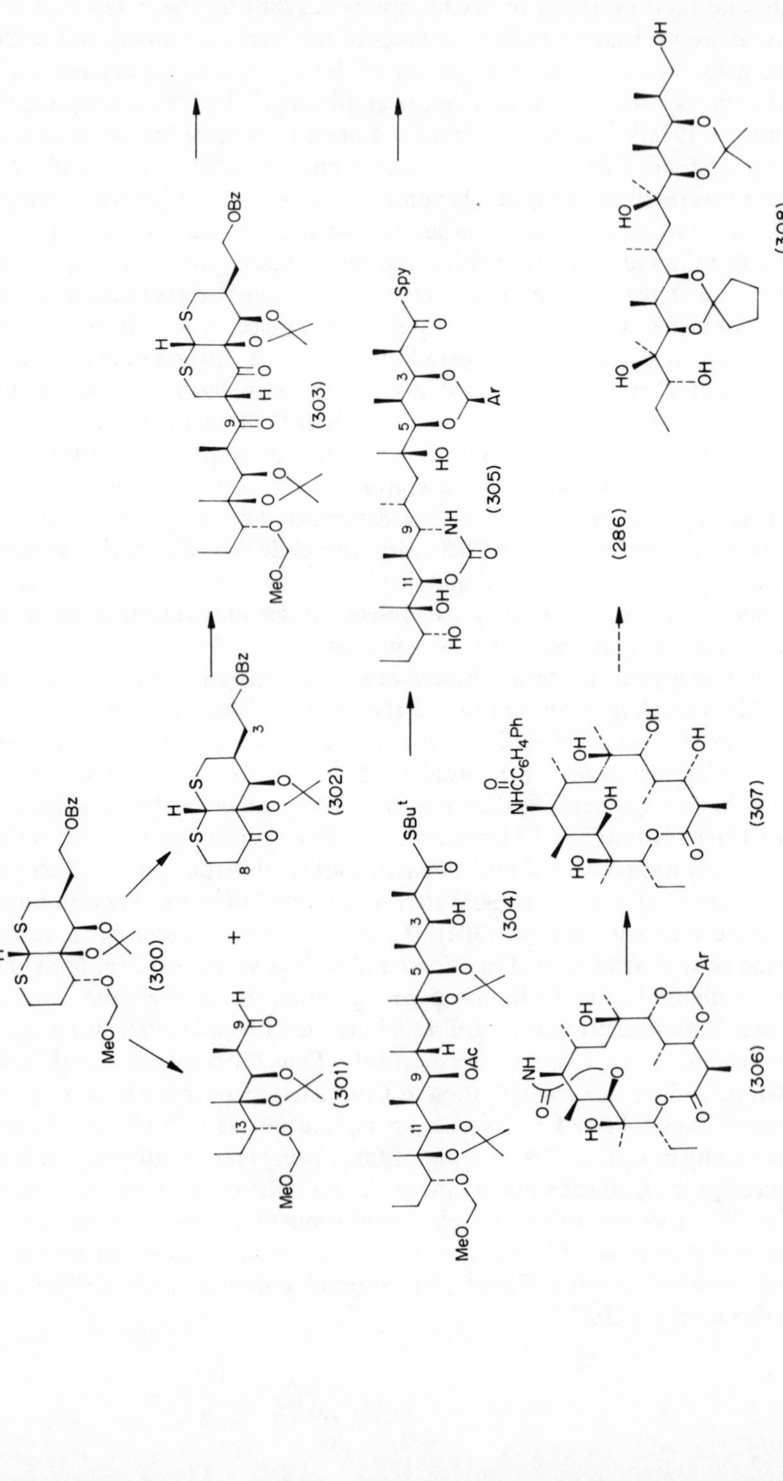

(300)
(301)
(302)
(303)
(304)
(305)
(306)
(307)
(286)
(308)
OBz
MeO
SBu^t
Spy
OAc
Ar
NHCC$_6$H$_4$Ph

Scheme 21

Other approaches to erythromycin synthesis include a synthesis of 6-deoxyerythronolide B using chiral boron enolates to effect stereoselective aldol condensations reported by Masamune,[148] a stereoselective synthesis of the open-chain precursor (308) described by Stork,[149] and a synthesis of an erythronolide A precursor based on the use of optically active carbohydrate building blocks, reported by Hanessian.[150]

(309)

Tylosin (309) is a complex 16-membered macrolide antibiotic presently used for the treatment of chronic respiratory disease in chickens. A considerable effort has been directed towards tylosin synthesis, and the aglycone tylonolide has been prepared from carbohydrate precursors,[151] and from 7-methylbicyclo [2,2,1] heptenone.[152] The synthesis outlined in Scheme 22 was described by Masamune in 1982, and illustrates some of the recent developments in aldol methodology.[153] Thus condensation of the optically active boron enolate (310) with racemic aldehyde (311) gave a 1:1 mixture of the optically active diastereoisomeric adducts (312) and (313), which were separated and the desired adduct (312) triethylsilylated and stereoselectively hydroborated using (–)-bis (isopinocamphenyl) borane, to yield alcohol (314). Selective oxidation and lactonization gave (315), the precursor of aldehyde (316), which was condensed with the optically active boron enolate (317) to give a 4:1 mixture of adducts containing the desired isomer (318) as the major component. Aldol condensations of methyl ketones generally show poor stereoselectivity, and the 4:1 selectivity reported here was only obtained after considerable experimentation. Conversion of ketone (318) into aldehyde-thioester (319) was followed by treatment with acidic methanol, to give acid-acetal (320). Treatment of the derived 2-mercaptopyridine thio-ester with lithium bis [(trimethylsilyl)methyl] cuprate gave the α-trimethylsilylmethylketone (321) which was condensed with the optically active aldehyde (322) to give seco-acid (323) after deprotection. Cyclization using a phosphoric acid mixed anhydride procedure gave the target aglycone (324) after deprotection.[153]

Synthetic studies have also been reported which are directed towards the synthesis of other 16-membered macrolide antibiotics, including carbomycin B and leucomycin A_3.[154]

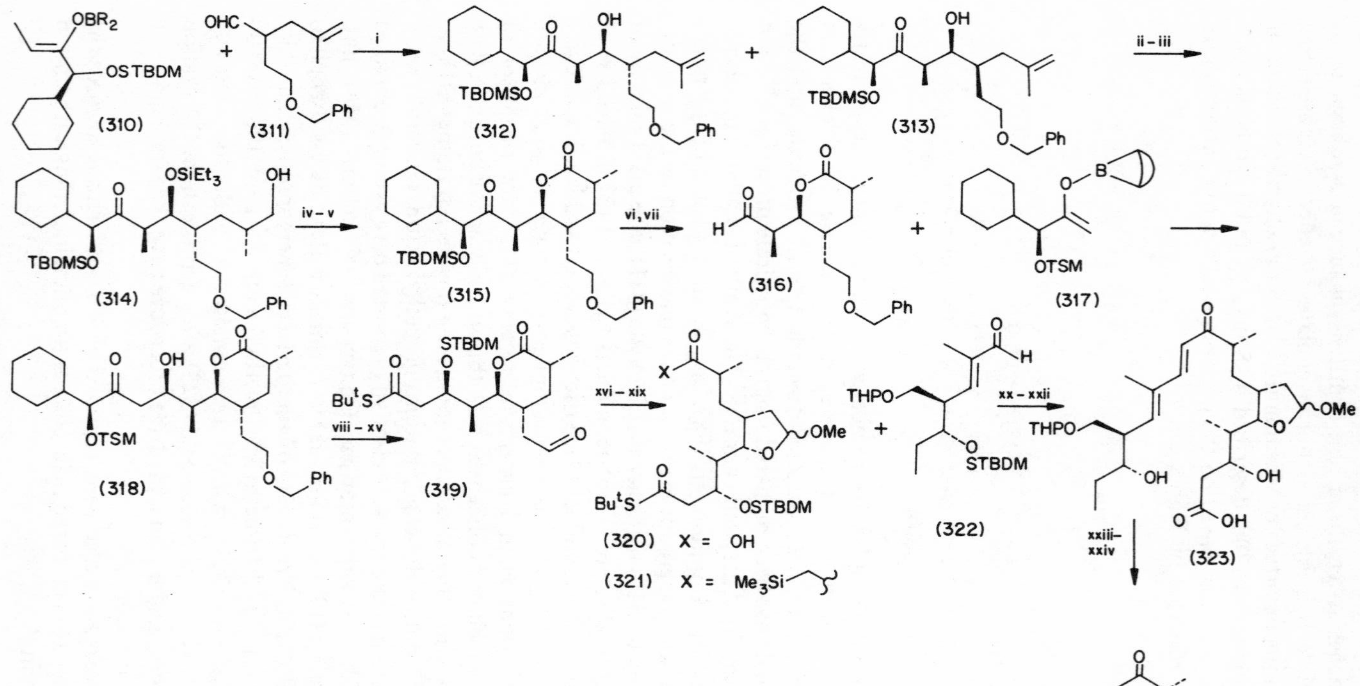

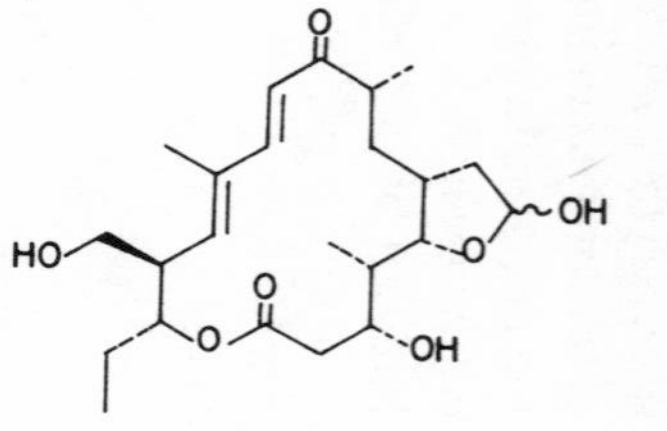

Scheme 22

Reagents: i, R=n-Bu, CH$_2$Cl$_2$, 17h; ii, Et$_3$SiOTf, 2,6-lutidine; iii, (−)-(IPC)$_2$BH, then MCPBA; iv, HOAc; v, Ag$_2$CO$_3$-Celite; vi, HF=CH$_3$CN; vii, NaIO$_4$, viii, F$^-$, ix, NaIO$_4$; x, H$_2$,Pd/C; xi, Bu^{t}Me$_2$SiCl, imid.;xii, ClCO$_2$Et; xiii, Bu^{t}STl; xiv, AcOH; xv, CrO$_3$,py; xvi, HOAc; xvii, HC(OMe)$_3$,TsOH; xviii, (PyS)$_2$,Ph$_3$P; xix, LiCu(Me$_3$SiCH$_2$)$_2$; xx, BuLi;(TMS)$_2$NH; xxi, (CF$_3$CO$_2$)$_2$Hg; xxii, HF,py; xxiii, (PhO)$_2$POCl, Et$_3$N; xxiv, AcOH.

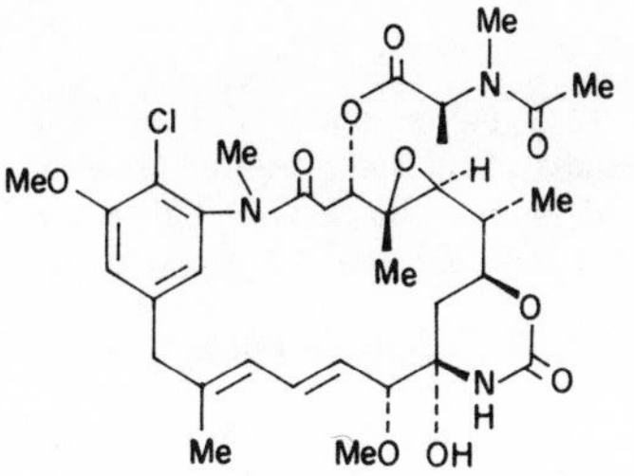

(325)

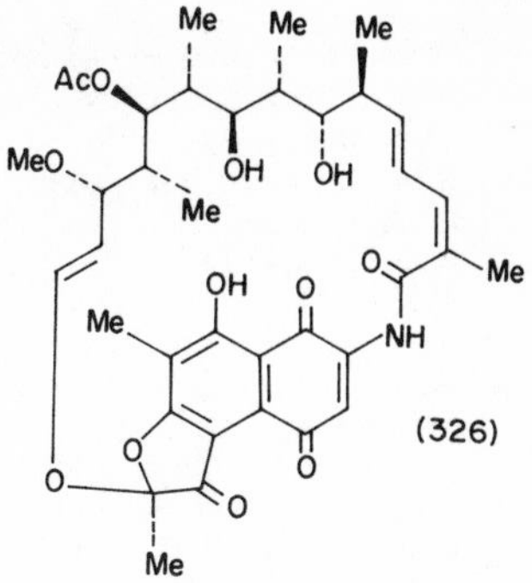

(326)

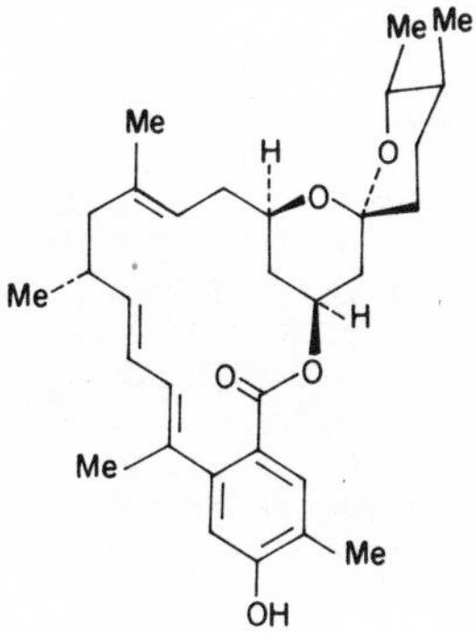

(327)

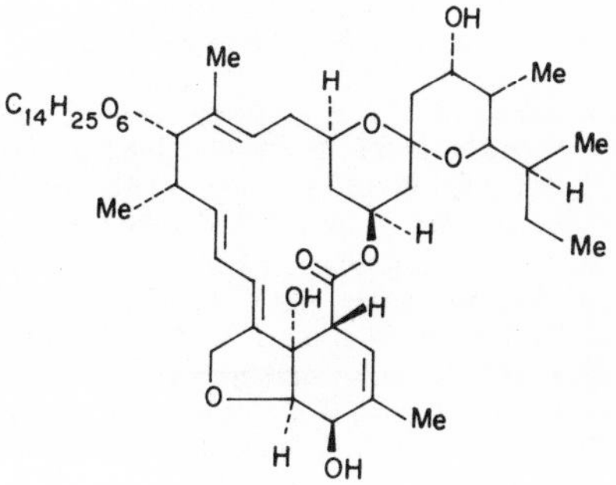

(328)

Finally, there has been considerable activity directed toward the synthesis of other macrocyclic natural products. In particular, because of its anti-tumour activity, maytansine (325) was the object of considerable synthetic work which culminated in successful syntheses of maytansine[155] and related maytansinoids.[156] Another ansa macrolide which has now been synthesized is rifamycin S (326). Kishi reported the first synthesis of this compound in 1980,[157] and has since provided details of an improved procedure.[158] A synthesis of the ansa chain of rifamycin S using stereoselective aldol condensations has been described by Masamune,[159] and a synthesis using optically active carbohydrate-derived building blocks has been achieved by Hanessian.[160]

Other compounds presently attracting the interest of synthetic chemists include the milbemycins, e.g. milbemycin β_3 (327),[161] and the avermectins, e.g. avermectin B_{2a} (328),[162] which are of interest because of their potent biological activity. Two syntheses of milbemycin β_3 (327) have been reported to date, and more syntheses in this area are to be expected.[163]

100   THE CHEMISTRY OF NATURAL PRODUCTS

References

1. W.M. Bandaranayake, J.E. Banfield, D. St. C. Black. G.D. Fallon, and B.M. Gatehouse, 1980, *J. Chem. Soc., Chem. Commun.*, 162; W.M. Bandaranayake, J.E. Banfield, and D. St. C. Black, *ibid.*, p. 902; W.M. Bandaranayake, J.E. Banfield, D. St. C. Black. G.D. Fallon, and B.M. Gatehouse, 1981, *Aust. J. Chem.*, 34, 1655; W.M. Bandaranayake, J.E. Banfield, and D. St. C. Black, 1982. *Aust. J. Chem.*, 35, 557.
2. K.C. Nicolaou, N.A. Petasis, R.E. Zipkin, and J. Uenishi, 1982, *J. Amer. Chem. Soc.*, 104, 5555, 5557.
3. K.C. Nicolaou, R.E. Zipkin, and N.A. Petàsis, 1982, *J. Amer. Chem. Soc.*, 104, 5558, 5560.
4. E.B. Chain and G. Mellows, 1977, *J. Chem. Soc., Perkin I*, 294; T.C. Feline, R.B. Jones, G. Mellows, and L. Phillips, *ibid*, p. 309; E.B. Chain and G. Mellows, *ibid.*, p. 318; R.G. Alexander, J.P. Clayton, K. Luk, N.H. Rogers, and T. King, 1978, *J. Chem. Soc., Perkin I*, 561; J.P. Clayton, K. Luk, and N.H. Rogers, 1979, *J. Chem. Soc., Perkin I*, 308; J.P. Clayton, R.S. Oliver, N.H. Rogers, T.J. King, *ibid.*, p. 838; S. Coulton, P.J. O'Hanlon, and N.H. Rogers, 1982, *J. Chem. Soc., Perkin I*, 729; J.P. Clayton, P.J. O'Hanlon, N.H. Rogers, and T.J. King, *ibid.*, p. 2827.
5. A.P. Kozikowski, R.J. Schmiesing, and K.L. Sorgi, 1980, *J. Amer. Chem. Soc.*, 102, 6577.
6. A.P. Kozikowski, K.L. Sorgi, and R.J. Schmiesing, 1980, *J. Chem. Soc., Chem. Comm.*, 477,
7. J.P. Clayton, P.J.O'Hanlon, and N.H. Rogers, 1980, *Tetrahedron Lett.*, 21, 881; A.P., Kozikowski, R.J. Schmiesing, and K.L. Sorgi, 1981, *Tetrahedron Lett.*, 22, 2059.
8. B.B. Snider and G.B. Phillips, 1982, *J. Amer. Chem. Soc.*, 104, 1113.
9. G.W.J. Fleet and M.J. Gough, 1982, *Tetrahedron Lett.*, 23, 4509, G.W.J. Fleet and C.R.C. Spensley, *ibid*, p. 109; A.P. Kozikowski and K.L. Sorgi, *ibid*, p. 2281; D.P. Curran, *ibid.*, p. 4309; R.A. Raphael, J.H.A. Stibbard, and R. Tidbury, *ibid.*, p. 2407; B. Schönenberger, W. Summermatter, and C. Ganter, 1982, *Helv. Chim. Acta*, 65, 2333.
10. J.M. Beau, S. Aburaki, J.R. Pougny, and P. Sinaÿ, 1983, *J. Amer. Chem. Soc.*, 105, 621.
11. R.K. Boeckman, Jr., and E.W. Thomas, 1977, *J. Amer. Chem. Soc.*, 99, 2805; 1979, 101, 987.
12. A.A. Jakubowski, F.S. Guziec, and M. Tishler, 1977, *Tetrahedron Lett.*, 2399.
13. E.J. Corey and D.R. Williams, 1977, *Tetrahedron Lett.*, 3847.
14. N. Sueda, H. Ohrui, and K. Kuzuhara, 1979, *Tetrahedron Lett.*, 20, 2039; N. Pietraszkiewicz and P. Sinaÿ, *ibid.*, p. 4741.
15. K. Wada and T. Ishida, 1979, *J. Chem. Soc., Perkin I*, 1154.
16. R.F. Vesonder, F.H. Stodola, L.J. Wickerham, J.J. Ellis, and W.K. Rohwedder, 1971, *Can J. Chem.*, 49, 2029; R.F. Vesonder, F.H. Stodola, and W.K. Rohwedder, 1972, *Can J. Biochem.*, 50, 363.
17. T. Ishida and K. Wada, 1979, *J. Chem. Soc., Perkin I*, 323.
18. E.J. Corey and K.C. Nicolaou, 1974, *J. Amer. Chem. Soc*, 96, 5614; E.J. Corey, D.J. Brunelle, and P.J. Stork, 1976, *Tetrahedron Lett.*, 3405.
19. J. Tsuji and T. Mandai, 1978, *Tetrahedron Lett.*, 1817.
20. R.E. Ireland and F.R. Brown, Jr., 1980, *J. Org. Chem.*, 45, 1868.
21. T. Wakamatsu, K. Akasaha, and Y. Ban, 1979, *J. Org. Chem.*, 44, 2008.
22. E.J. Corey, P. Ulrich, and J.M. Fitzpatrick, 1976, *J. Amer. Chem. Soc.*, 98, 222; K. Narasaka, M. Yamaguchi, and T. Mukaiyama, 1977, *Chem. Lett.*, 959; J. Tsuji, T. Yamakawa, and T. Mandai, 1978, *Tetrahedron Lett.*, 565; H. Gerlach, K. Oertle and A. Thalmann, 1976, *Helv. Chim Acta*, 59, 755; K. Utimoto, K. Uchida, M. Yamaya, and H. Nozaki, 1977, *Tetrahedron Lett.*, 3641.
23. B.M. Trost and T.R. Verhoeven, 1978, *Tetrahedron lett.*, 2275.
24. T. Takahashi, S. Hashiguchi, K. Kasuga, and J. Tsuji, 1978, *J. Amer. Chem. Soc.*, 100, 7424.
25. S.L. Schreiber, 1980, *J. Amer. Chem. Soc.*, 102, 6163.
26. E.W. Colvin, T.A. Purcell, and R.A. Raphael, 1976, *J. Chem. Soc., Perkin I.*, 1718.
27. H. Gerlach, K. Oertle, and A. Thalmann, 1977, *Helv. Chim. Acta*, 60, 2860; P. Bakuzis, M.L.F. Bakuzis, and T.F. Weingartner, 1978, *Tetrahedron Lett.*, 2371; M. Asaoka, N. Yanagida, N. Sugimura, and H. Takei, 1980, *Bull. Chem. Soc. Japan*, 53, 1061; T.A. Hase, A. Ourila, and C. Holmberg, 1981, *J. Org. Chem.*, 46, 3137; B.M. Trost and F.W. Gowland, 1979, *J. Org. Chem.*, 44, 3448.

28. D. Seebach, B. Seuring, H.-O. Kalinowski, W. Lubosch, and B. Renger, 1977, *Angew. Chem. (Int Edn.)*, 16, 264; B. Seuring and D. Seebach, 1978, *Annalen*, 2044; R.S. Mali, M. Pohmakotr, B. Weidmann, and D. Seebach, 1981, *Annalen*, 2272.
29. M. Asaoka, T. Mukuta, and H. Takei, 1981, *Tetrahedron Lett.*, 735.
30. E.J. Corey, K.C. Nicolaou, and T. Toru, 1975, *J. Amer. Chem. Soc.*, 97, 2287.
31. Y. Fukuyama, C.L. Kirkemo, and J.D. White, 1977, *J. Amer. Chem. Soc.*, 99, 646.
32. K.F. Burri, R.A. Cardone, W.Y. Chen, and P. Rosen, 1978, *J. Amer. Chem. Soc.*, 100, 7069.
33. E.J. Corey, K.C. Nicolaou, and L.S. Melvin, 1975, *J. Amer. Chem. Soc.*, 97, 654; E.J. Corey and R.H. Wollenberg, 1976, *Tetrahedron Lett.*, 4701.
34. E.J. Corey and R.H. Wollenberg, 1976, *Tetrahedron Lett.*, 4705; E.J. Corey, R.H. Wollenberg, and D.R. Williams, 1977, *Tetrahedron Lett.*, 2243.
35. Y. Köksal, P. Raddatz, and E. Winterfeldt, 1980, *Angew. Chem. (Int. Edn.)*, 19, 472; A.E. Greene, C. Le Drian, and P. Crabbé, 1980, *J. Amer. Chem. Soc.*, 102, 7583.
36. P.A. Bartlett and F.R. Green, III, 1978, *J. Amer. Chem. Soc.*, 100, 4858.
37. P.A. Bartlett, F.R. Green, III and E.H. Rose, 1978, *J. Amer. Chem. Soc.*, 100, 4852.
38. P.A. Bartlett, 1976, *J. Amer. Chem. Soc.*, 98, 3305.
39. T. Mukaiyama, M. Usui, and K. Saigo, 1976, *Chem. Lett.*, 49.
40. R. Baudony, P. Crabbe, A.E. Greene, C. Le Drian, and A.F. Orr, 1977 *Tetrahedron Lett.*, 2973; T. Livinghouse and R.V. Stevens, 1978, *J. Chem. Soc., Chem. Comm.*, 754; M. Honda, K. Hirata, H. Sueoka, T. Katsuki, and M. Yamaguchi, 1981, *Tetrahedron Lett.*, 2679; T. Kitahara, K. Mori, and M. Matsui, 1979, *Tetrahedron Lett.*, 3021.
41. R.C. Murphy, S. Hammarström, and B. Samuelsson, 1979, *Proc. Natl. Acad. Sci*, 76, 4275; P. Borgeat and B. Samuelsson, *ibid.*, 3213; H.R. Morris, G.W. Taylor, P.J. Piper, and J.R. Tippins, 1980, *Nature*, 285, 104; S. Hammarström, B. Samuelsson, D.A. Clark, G. Goto, A. Marfat, C. Mioskowski, and E.J. Corey, 1980, *Biochem. Biophys. Res. Comm.*, 92, 946.
42. B. Samuelsson and S. Hammarström, 1980, *Prostaglandins*, 19, 645.
43. E.J. Corey, H. Niwa, and J.R. Falck, 1979, *J. Amer. Chem. Soc.*, 101, 1586.
44. E.J. Corey, J.O. Albright, A.E. Barton, and S.-i. Hashimoto, 1980, *J. Amer. Chem. Soc.*, 102, 1435.
45. E.J. Corey, A. Marfat, J.R. Falck, and J.O. Allbright, 1980, *J. Amer. Chem. Soc.*, 102, 1433.
46. E.J. Corey, Y. Arai, and C. Mioskowski, 1979, *J. Amer. Chem. Soc.*, 101, 6748.
47. M. Rosenberger and C. Neukom, 1980, *J. Amer. Chem. Soc.*, 102, 5425; J. Rokach, Y. Girard, Y. Grindon, J.G. Atkinson, M. Larue, R.N. Young, P. Masson, and G. Holme, 1980, *Tetrahedron Lett.*, 1485.
48. W.C. Still and V.J. Novack, 1981, *J. Amer. Chem. Soc.*, 103, 1283.
49. E.J. Corey, D.A. Clark, G. Goto, A. Marfat, C. Mioskowski, B. Samuelsson, and S. Hammarström, 1980, *J. Amer. Chem. Soc.*, 102, 1436, 3663.
50. E.J. Corey, D.A. Clark, A. Marfat, and G. Goto, 1980, *Tetrahedron Lett.*, 3143.
51. D.P. Marriott and J.R. Bantick, 1981, *Tetrahedron Lett.*, 3657; N. Cohen, B.L. Banner, and R.J. Lopresti, 1980, *Tetrahedron Lett.*, 4163; J. Rokach, R.N. Young, M. Kakushima, C.-K. Lau, R. Seguin, R. Frenette, and Y. Guindon, 1981, *Tetrahedron Lett.*, 979; J.G. Gleason, D.B. Bryan, and C.M. Kinzig, 1980, *Tetrahedron Lett.*, 1129.
52. E.J. Corey, S.-i. Hashimoto, and A.E. Barton, 1981, *J. Amer. Chem. Soc.*, 103, 721.
53. B.E. Rossiter, T. Katsuki, and K.B. Sharpless, 1981, *J. Amer. Chem. Soc.*, 103, 464.
54. E.J. Corey, A.E. Barton, and D.A. Clark, 1980, *J. Amer. Chem. Soc.*, 102, 4278.
55. E.J. Corey, A. Marfat, and G. Goto, 1980, *J. Amer. Chem. Soc.*, 102, 6607.
56. V. Atrache, J.-K. Pai, D.-E. Sok, and C.J. Sih, 1981, *Tetrahedron Lett.*, 3443.
57. E.J. Corey, A. Marfat, G. Goto, and F. Brion, 1980, *J. Amer. Chem. Soc.*, 102, 7984; E.J. Corey, P.B. Hopkins, J.E. Munroe. A. Marfat, and S.-i. Hashimoto, *ibid.*, p. 7986; Y. Guindon, R. Zamboni, C.-K. Lau, and J. Rokach, 1982, *Tetrahedron Lett.*, 739.
58. E.J. Corey. A. Marfat, J. Munroe, K.S. Kim, P.B. Hopkins, and F. Brion, 1981, *Tetrahedron Lett.* 1077.
59. K.C. Nicolaou, N.A. Petasis, and S.P. Seitz, 1981, *J. Chem. Soc., Chem. Comm.*, 1195; E.J. Corey, H. Park, A.E. Barton, and Y. Nii, 1980, *Tetrahedron Lett.*, 4243.
60. E.J. Corey, S.S. Kantner, and P.T. Lansbury, Jr., 1983, *Tetrahedron Lett.*, 265.
61. S.J. Wratten and D.J. Faulkner, 1976, *J. Org. Chem.*, 41, 2465.

62. N.J. Wells, 1976, *Tetrahedron Lett.*, 2637.
63. B.J. Burreson, R.E. Moore, and P.P. Roller, 1975, *Tetrahedron Lett.*, 473; F.X. Woolard, R.E. Moore, and P.P. Roller, 1976, *Tetrahedron*, 32, 2843; O.J. McConnell and W. Fenical, 1977, *Phytochemistry*, 16, 367; W. Fenical, 1974, *Tetrahedron Lett.*, 4463.
64. J.A. Pettus, Jr., R.M. Wing, and J.J. Sims, 1977, *Tetrahedron Lett.*, 41; R. Kazlauskas, P.T. Murphy, R.J. Quinn, and R.J. Wells, 1977, *ibid.*, p. 37.
65. R.E. Moore, 1977, *Acc. Chem. Res.*, 10, 40.
66. L. Jaenicke, D.G. Muller, and R.E. Moore, 1974, *J. Amer. Chem. Soc.*, 96, 3324; L. Jaenicke and K. Seferiadis, 1975, *Chem. Ber.*, 108, 225.
67. L. Jaenicke and W. Boland, 1976, *Annalen*, 1135.
68. A.F. Cameron, K.K. Cheung, G. Ferguson, and J.M. Robertson, 1969, *J. Chem. Soc. (B)*, 559.
69. B.M. Howard, W. Fenical, K. Hirotsu, B. Solheim, and J. Clardy, 1980, *Tetrahedron*, 36, 171.
70. F.J. McDonald, D.C. Campbell, D.J. Vanderah, F.J. Schmitz, D.M. Washecheck, J.E. Burns, and D. van der Helm, 1975, *J. Org. Chem.*, 40, 665.
71. S.M. Waraszkiewicz, H.H. Sun, and K.L. Erickson, 1976, *Tetrahedron Lett.*, 3021; H.H. Sun, S.M. Waraszkiewicz, and K.L. Erickson, *ibid.*, p. 4227.
72. B.M. Howard, W. Fenical, E.V. Arnold, and J. Clardy, 1979, *Tetrahedron Lett.*, 2841.
73. T.J. King, S. Imre, A. Öztunc, and R.H. Thomson, 1979, *Tetrahedron Lett.*, 1453.
74. A. Fukuzawa and E. Kurosawa, 1979, *Tetrahedron Lett.*, 2797.
75. T. Masamune, H. Murase, H. Matsue, and A. Murai, 1979, *Bull. Chem. Soc. Japan*, 52, 135; T. Masamune, H. Matsue, and H. Murase, 1979, *ibid.*, p. 127.
76. E.J. Corey, B.-C. Pan, D.H. Hua, and D.R. Deardorff, 1982, *J. Amer. Chem. Soc.*, 104, 6816; E.J. Corey, D.H. Hua, B.-C. Pan and S.P. Seitz, *ibid.*, p. 6818.
77. J. Diago-Meseguer, A.L. Palomo-Coll, J.R. Fernandez-Lizarbe, and A. Zugaza-Biblao, 1980, *Synthesis*, 547.
78. J.K. Cha, W.J. Christ, J.M. Finan, H. Fujioka, Y. Kishi, L.L. Klein, S.S. Ko, J. Leder, W.W. McWhorter, Jr., K.-P. Pfaff, and M. Yonaga, 1982, *J. Amer. Chem. Soc.*, 104, 7369.
79. A.M. Carde, T.C. Baker, and R.T. Carde, 1979, *J. Chem. Ecol.*, 5, 423.
80. G. Biwer and C. Descoins, 1978, *C.R. Hebd. Séances Acad. Sci., Ser., D.*, 286, 875.
81. G. Biwer, C. Descoins, and M. Gallois, 1979, *C.R. Hebd. Séances Acad. Sci., Ser. D*, 288, 413.
82. E.C. Uebel, P.E. Sonnet, R.E. Menzer, R.W. Miller, and W.R. Lusby, 1977, *J. Chem. Ecol.*, 3, 269.
83. D.A. Carlson, P.A. Langley, and P. Huyton, 1978, *Science*, 201, 750.
84. J.H. Tumlinson, M.G. Klein, R.E. Doolittle, T.L. Ladd, and A.T. Proveaux, 1977, *Science*, 197, 789.
85. F.J. Ritter, I.E.M. Brüggemann-Rotgans, P.E.J. Verwiel, C.J. Persoons, and E. Talman, 1977, *Tetrahedron Lett.*, 2617.
86. B.M. Trost, T.N. Salzmann, and K. Hiroi, 1976, *J. Amer. Chem. Soc.*, 98, 4887; B.M. Trost and K. Hiroi, *ibid.*, p. 4313.
87. W.C. Still and A. Mitra, 1978, *J. Amer. Chem. Soc.*, 100, 1927.
88. B.B. Snider and D. Rodini, 1978, *Tetrahedron Lett.* 1399.
89. T. Nakai, T. Mimura, and T. Kurokawa, 1978, *Tetrahedron Lett.*, 2895.
90. F.W. Sum and L.S. Weiler, 1978, *J. Chem., Soc., Chem. Comm.*, 985.
91. A. Marfat, P.R. McGuirk, and P. Helquist, 1979, *J. Org. Chem.*, 44, 1345, 3888.
92. G. Decodts, G. Dressaire, and Y. Langlois, 1979, *Synthesis*, 510.
93. D. Samain, C. Descoins, and A. Commercon, 1978, *Synthesis*, 388; H.J. Bestmann, J. Suss, and O. Vostrowsky, 1978, *Tetrahedron Lett.* 3329.
94. G. Cassani, P. Massardo, and P. Piccardi, 1979, *Tetrahedron Lett.*, 633; D. Samain and C. Descoins, 1979, *Bull. Soc. Chim. Fr.*, Part 2, 71; E-i. Negishi and A. Abramovitch, 1977, *Tetrahedron Lett.*, 411.
95. N. Okukado, D.E. VanHorn, W.L. Klima and E-i. Negishi, 1978, *Tetrahedron Lett.*, 1027.
96. S. Baba and E-i. Negishi, 1976, *J. Amer. Chem. Soc.*, 98, 6729.

97. J.H. Babler and B.J. Invergo, 1979, *J. Org. Chem.*, 44, 3723.
98. K. Mori and S. Tamada, 1979, *Tetrahedron*, 15, 1279; K. Mori, S. Tamada, and M. Matsui, 1978, *Tetrahedron Lett.*, 901.
99. K. Mori, T. Suguro, and S. Masuda, 1979, *Tetrahedron Lett.*, 3447.
100. K. Mori, 1975, *Tetrahedron*, 31, 1381; H. Ohrui and S. Emoto, 1976, *Agric. Biol. Chem.*, 40, 2267; D.R. Hicks and B. Fraser-Reid, 1976, *J. Chem. Soc., Chem. Comm.*, 869.
101. P. Magnus and G. Roy, 1978, *J. Chem. Soc. Chem. Comm.*, 297.
102. G.T. Pearce, W.E. Gore, and R.M. Silverstein, 1976, *J. Org. Chem.*, 41, 2797; K. Mori, 1976, *Tetrahedron*, 32, 1979; C.J. Cernigliaro and P.J. Kocienski, 1977, *J. Org. Chem.*, 42, 3622.
103. W.J. Elliott, G. Hromnak, J. Fried, and G.N. Lanier, 1979, *J. Chem. Ecol.*, 5, 279.
104. P.-E. Sum and L. Weiler, 1978, *Can. J. Chem.*, 56, 2700; K. Mori and H. Iwasawa, 1980, *Tetrahedron*, 36, 87.
105. E.J. Corey, N.M. Weinshenker, T.K. Schaaf, and W. Huber, 1969, *J. Amer. Chem. Soc.*, 91, 5675; E.J. Corey, T. Ravindranathan, and S. Terashima, 1971, *J. Amer. Chem. Soc.*, 93, 4326; E.J. Corey, S.M. Albonico, U. Koelliker, T.K. Schaaf, and R.K. Varma, *ibid.*, 1491.
106. E.J. Corey, K.B. Becker, and R.K. Varma, 1972, *J. Amer. Chem. Soc.*, 94, 8616.
107. E.J. Corey and H.E. Ensley, 1975, *J. Amer. Chem. Soc.*, 97, 6908.
108. S. Ranganathan, D. Ranganathan, and A.K. Mehrotra, 1974, *J. Amer. Chem. Soc.*, 96, 5261; P.A. Bartlett, F.R. Green, III, and T.R. Webb, 1977, *Tetrahedron Lett.*, 331.
109. N.R.A. Beeley, P. Peel, J.K. Sutherland, J.J. Holohan, K.B. Mallion, and G.J. Sependa, 1981, *Tetrahedron*, (Suppl. 1) 37, 411; J.S. Bindra, A. Grodski, T.K. Schaaf, and E.J. Corey, 1973, *J. Amer. Chem. Soc.*, 95, 7522.
110. E.D. Brown, R. Clarkson, T.J. Leeney, and G.E. Robinson, 1974, *J. Chem. Soc., Chem. Comm.*, 642; E.D. Brown, R. Clarkson, T.J. Leeney, and G.E. Robinson, 1978, *J. Chem. Soc., Perkin I*, 1507; E.D. Brown and T.J. Lilley, 1975, *J. Chem. Soc., Chem Comm.*, 39.
111. S. Goldstein, P. Vannes, C. Houge, A.M. Frisque-Hesbain, C. Wiaux-Zamar, L. Ghosez, G. Germain, J.P. Declercq. M. van Meerssche, and J.M. Arrieta, 1981, *J. Amer. Chem. Soc.*, 103, 4616.
112. I. Fleming and B.-W. Au-Yeung, 1981, *Tetrahedron*, (Suppl. 1), 37, 13.
113. R.F. Newton, J. Paton, D.P. Reynolds, S. Young, and S.M. Roberts, 1979, *J. Chem. Soc., Chem. Comm.*, 908.
114. J. Davies, S.M. Roberts, D.P. Reynolds, and R.F. Newton, 1981, *J. Chem. Soc., Perkin I*, 1317.
115. C.B. Chapleo, M.A.W. Finch, T.V. Lee, S.M. Roberts, and R.F. Newton, 1980, *J. Chem. Soc., Perkin I*, 2084; M.A.W. Finch, S.M. Roberts, G.T. Woolley, and R.F. Newton, 1981, *J. Chem. Soc., Perkin I*, 1725; R.F. Newton, D.P. Reynolds, C.F. Webb, and S.M. Roberts, *ibid.*, p. 2055.
116. C.J. Sih, J.B. Heather, G. Peruzzotti, P. Price, R. Sood, L.F. Hsu Lee, 1973, *J. Amer. Chem. Soc.*, 95, 1676; C.J. Sih, J.B. Heather, R. Sood, P. Price, G. Peruzzotti, L.F. Hsu Lee, and S.S. Lee, 1975, *J. Amer. Chem. Soc.*, 97, 865; G. Stork and T. Takahashi, 1977, *J. Amer. Chem. Soc.*, 99, 1275; G.R. Kieczykowski, C.S. Pogonowski, J.E. Richman, and R.H. Schlessinger, 1977, *J. Org. Chem.*, 42, 175.
117. A.F. Kluge, K.G. Unteh, and J.H. Fried, 1972, *J. Amer. Chem. Soc.*, 94, 7827; 9256; J.G. Miller, W. Kurz, K.G. Unteh, and G. Stork, 1974, *J. Amer. Chem. Soc.*, 96, 6774.
118. M.B. Floyd and M.J. Weiss, 1973, *Prostaglandins*, 3, 921.
119. K.F. Bernady and M.J. Weiss, 1973, *Prostaglandins*, 3, 505.
120. J.W. Patterson, Jr., and J.H. Fried, 1974, *J. Org. Chem.*, 39, 2506.
121. G. Stork and M. Isobe, 1975, *J. Amer. Chem. Soc.*, 97, 4745, 6260.
122. G. Stork, T. Takahashi, I. Kawamoto, and T. Suzuki 1978, *J. Amer. Chem. Soc.*, 100, 8272; G. Stork and S. Raucher, 1976, *J. Amer. Chem. Soc.*, 98, 1583; R.B. Woodward, J. Gosteli, I. Ernest, R.J. Friary, G. Nestler, H. Raman, R. Sitrin, Ch. Suter, and J.K. Whitesell, 1973, *J. Amer. Chem. Soc.*, 95, 6853; I. Ernest, 1976, *Angew. Chem. (Int. Edn.)*, 15, 207.
123. E.J. Corey, G.E. Keck, and I. Székely, 1977, *J. Amer. Chem. Soc.*, 99, 2006; E.J. Corey, H.L. Pearce, I. Székely, and M. Ishiguro, 1978, *Tetrahedron Lett.*, 1023.

104 THE CHEMISTRY OF NATURAL PRODUCTS

124. K.C. Nicolaou, W.E. Barnette, G.P. Gasic, R.L. Magolda, and W.J. Sipio, 1977, *J. Chem. Soc., Chem. Comm.*, 630; R.A. Johnson, F.H. Lincoln, J.L. Thompson, E.G. Nidy, S.A. Mizsak, and U. Axen, 1977, *J. Amer. Chem. Soc.*, 99, 4182.
125. K.C. Nicolaou and W.E. Barnette, 1977, *J. Chem. Soc., Chem. Comm.*, 331.
126. R.F. Newton, S.M. Roberts, B.J. Wakefield, and G.T. Woolley, 1981, *J. Chem. Soc., Chem. Comm.*, 922.
127. S. Ohuchida, N. Hamanaka, and M. Hayashi, 1979, *Tetrahedron Lett.*, 3661; K.C. Nicolaou, R.L. Magolda, and D.A. Claremon, 1980, *J. Amer. Chem. Soc.*, 102, 1404; M.F. Ansell, M.P.L. Caton, M.N. Palfreyman, and K.A.J. Stuttle, 1979, *Tetrahedron Lett.*, 4497; E.J. Corey, J.W. Ponder, and P. Ulrich, 1980, *Tetrahedron Lett.*, 137; K.M. Maxey and G.L. Bundy, *ibid.*, p. 445; S. Kosuge, N. Hamanaka, and M. Hayashi, 1981, *Tetrahedron Lett.*, 22, 1345; S. Okuchida, N. Hamanaka, and M. Hayashi, *ibid.*, p. 1349; S. Okuchida, N. Hamanaka, and M. Hayashi, 1981, *J. Amer. Chem. Soc.*, 103, 4597.
128. N.A. Nelson and R.W. Jackson, 1976, *Tetrahedron Lett.*, 3275; R.C. Kelly, I. Schletter, and S.J. Stein, *ibid.*, 3279; E.J. Corey, M. Shibasaki, J. Knolle, and T. Sugahara, 1977, *Tetrahedron Lett.*, 785.
129. W.P. Schneider and R.D. Morge, 1976, *Tetrahedron Lett.*, 3283.
130. A.G. Kelly and J.S. Roberts, 1980, *J. Chem. Soc., Chem. Comm.*, 228; S. Hanessian and P. Lavalle, 1981, *Can. J. Chem.*, 59, 870.
131. E.J. Corey, M. Shibasaki and J. Knolle, 1977, *Tetrahedron Lett.*, 1625; O. Hernandez, 1978, *Tetrahedron Lett.*, 219.
132. J. Dominquez, J.D. Dunitz, H. Gerlach, and V. Prelog, 1962, *Helv. Chim. Acta*, 45, 129; H. Gerlach and V. Prelog, 1963, *Justus Liebigs Ann. Chem.*, 669, 121.
133. G. Beck and E. Henseleit, 1971, *Chem. Ber.*, 104, 21; M.J. Arco, M.H. Trammell, and J.D. White, 1976, *J. Org. Chem.*, 41, 2075; P.A. Bartlett and K.K. Jernstedt, 1980, *Tetrahedron Lett.*, 1607; R.E. Ireland and J.-P. Vevert, 1980, *J. Org. Chem.*, 45, 4259; R.E. Ireland and J.-P. Vevert, 1981, *Can. J. Chem.*, 59, 572.
134. H. Gerlach and H. Wetter, 1974, *Helv. Chim. Acta*, 57, 2306; H. Gerlach, K. Oertle, A. Thalmann, and S. Servi, 1975, *Helv. Chim. Acta*, 58, 2036; H. Zak and U. Schmidt, 1975, *Angew. Chem. (Int. Edn.)*, 14, 432; J. Gombos, E. Haslinger, H. Zak, and U. Schmidt, 1975, *Tetrahedron Lett.*, 3391; U. Schmidt, J. Gombos, E. Haslinger, and H. Zak, 1976, *Chem. Ber.*, 109, 2628.
135. T. Nakata, G. Schmid, B. Vranesic, M. Okigawa, T. Smith-Palmer, and Y. Kishi, 1978, *J. Amer. Chem. Soc.*, 100, 2933.
136. T. Fukayama, B. Vranesic, D.P. Negri and Y. Kishi, 1978, *Tetrahedron Lett.*, 2741.
137. T. Nakata and Y. Kishi, 1978, *Tetrahedron Lett.*, 2745.
138. R.E. Ireland, S. Thaisrivongs, and C.S. Wilcox, 1980, *J. Amer. Chem. Soc.*, 102, 1155.
139. R.E. Ireland, G.J. McGarvey, R.C. Anderson, R. Badoud, B. Fitzsimmons, and S. Thaisrivongs, 1980, *J. Amer. Chem. Soc.*, 102, 6178.
140. G. Schmid, T. Fukuyama, K. Akasaka, and Y. Kishi 1979, *J. Amer. Chem. Soc.*, 101, 259; T. Fukuyama, C.-L. J. Wang, and Y. Kishi, *ibid.*, p. 260; T. Fukuyama, K. Akasaka, D.S. Karanewsky, C.-L. J. Wong, G. Schmid, and Y. Kishi, *ibid.*, p. 262.
141. D.B. Collum, J.H. McDonald, III, and W.C. Still, 1980, *J. Amer. Chem. Soc.*, 102, 2117, 2118, 2120.
142. W.C. Still and J.H. McDonald, III, 1980, *Tetrahedron Lett.*, 1031.
143. S. Masamune, C.U. Kim, K.E. Wilson, G.O. Spessard, P.E. Georghiou, and G.S. Bates, 1975, *J. Amer. Chem. Soc.*, 97, 3512; S. Masamune, H. Yamamoto, S. Kamata, and A. Fukuzawa, *ibid.*, p. 3513.
144. P.A. Grieco, Y. Ohfune, K. Yokoyama, and W. Owens, 1979, *J. Amer. Chem. Soc.*, 101, 4749.
145. E.J. Corey *et al.*, 1978, *J. Amer. Chem. Soc.*, 100, 4618, 4620.
146. E.J. Corey, P.B. Hopkins, S. Kim, S.-e. Yoo, K.P. Nambiar, and J.R. Falck, 1979, *J. Amer. Chem. Soc.*, 101, 7131.
147. R.B. Woodward *et al.*, 1981, *J. Amer. Chem. Soc.*, 103, 3210, 3213, 3215.
148. S. Masamune, M. Hirama, S. Mori, Sk. A. Ali and D.S. Garvey, 1981, *J. Amer. Chem. Soc.*, 103, 1568.
149. G. Stork, I. Paterson, and F.K.C. Lee, 1982, *J. Amer. Chem. Soc.*, 104, 4686.

150. S. Hanessian, G. Rancourt, and Y. Guindon, 1978, *Can. J. Chem.*, 56, 1843.
151. K.C. Nicolaou, M.R. Pavia, and S.P. Seitz, 1982, *J. Amer. Chem. Soc.*, 104, 2027, 2030; K. Tatsuta. Y. Amemiya, Y. Kanemura, and M. Kinoshita, 1981, *Tetrahedron Lett.*, 3997.
152. P.A. Grieco, J. Inanaga, N.-H. Lin, and Y. Yanami, 1982, *J. Amer. Chem. Soc.*, 104, 5781.
153. S. Masamune, L.D.-L.Lu, W.P. Jackson, T. Kaiho, and T. Toyoda, 1982, *J. Amer. Chem. Soc.*, 104, 5523; S. Masamune, Y. Hayase, W.K. Chan, and R.L. Sobczak, 1976, *J. Amer. Chem. Soc.*, 98, 7874.
154. K.C. Nicolaou, S.P. Seitz, and M.R. Pavia, 1981, *J. Amer. Chem. Soc.*, 103, 1222; K.C. Nicolaou, M.R. Pavia and S.P. Seitz, *ibid.*, 1224; K.C. Nicolaou, M.R. Pavia and S.P. Seitz, 1979, *Tetrahedron Lett.*, 2327.
155. E.J. Corey, L.O. Weigel, A.R. Chamberlin, H. Cho, and D.H. Hua, 1980, *J. Amer. Chem. Soc.*, 102, 6613; E.J. Corey, L.O. Weigel, A.R. Chamberlin, and B. Lipshutz, *ibid.*, 1439.
156. A.I. Meyers, P.J. Reider, and A.L. Campbell, 1980, *J. Amer. Chem. Soc.*, 102, 6597; M. Isobe, M. Kitamura, and T. Goto, 1982, *J. Amer. Chem. Soc.*, 104, 4997; A.I. Meyers, D.L. Comins, D.M. Roland, R. Henning, and K. Shimizu, 1979, *J. Amer. Chem. Soc.*, 101, 7104; A.I. Meyers, D.M. Roland, D.L. Comins, R. Henning, M.P. Fleming, and K. Shimizu, *ibid.*, p. 4732.
157. H. Nagaoka, W. Rutsch, G. Schmid, H. Iio, M.R. Johnson, and Y. Kishi 1980, *J. Amer. Chem. Soc.*, 102, 7962; H. Iio, H. Nagaoka, and Y. Kishi, *ibid.*, p. 7965.
158. H. Nagaoka and Y. Kishi, 1981, *Tetrahedron*, 37, 3873.
159. S. Masamune, B. Imperiali, and D.S. Garvey, 1982, *J. Amer. Chem. Soc.*, 104, 5528.
160. S. Hanessian, J.-R. Pougny. and I.K. Boessenkool, 1982, *J. Amer. Chem. Soc.*, 104, 6164.
161. Y. Takiguchi, H. Mishima, M. Okuda, M. Terco, A. Johi, and R.T. Fukade, 1980, *J. Antibiotics*, 33, 1120.
162. G. Albers-Schönberg *et al.*, 1981, *J. Amer. Chem. Soc.*, 103, 4216.
163. A.B. Smith, III, S.R. Schow, J.D. Bloom, A.S. Thompson, and K.N. Winzenberg, 1982, *J. Amer. Chem. Soc.*, 104, 4015; D.R. Williams, B.A. Barmer, K. Nishitani, and J.G. Phillips, *ibid.*, p. 4708.

Reviews

Fatty Acids and Derivatives

1. F.D. Gunstone in 'Comprehensive Organic Chemistry', volume 5, ed. E. Haslam, Pergamon, 1979, p. 587.
2. F.D. Gunstone, Fatty Acids and Glycerides, in 'Aliphatic and Related Natural Product Chemistry,' ed. F.D. Gunstone, Royal Society of Chemistry; 1983, volume 3, p. 209; 1981, volume 2, p. 194; 1979, volume 1, p. 236.
3. C.M. Scrimgeour, Natural Acetylenic and Olefinic Compounds, in 'Aliphatic and Related Natural Product Chemistry', ed. F.D. Gunstone, Royal Society of Chemistry; 1983, volume 3, p. 1; 1981, volume 2, p. 1.
4. V. Thaller, Natural Acetylenic and Olefinic Compounds, in 'Aliphatic and Related Natural Product Chemistry,' ed. F.D. Gunstone, Royal Society of Chemistry, 1979, volume 1, p. 1.

Leukotrienes

5. J. Ackroyd and F. Scheinmann, 1982, *Chem. Soc. Reviews*, 11, 321; Leukotriene Synthesis.
6. P.J. Piper (ed.), 'SRS-A and Leukotrienes,' J. Wiley and Sons, 1981.

Marine Natural Products

7. R.E. Moore, Marine Aliphatic Natural Products, in 'Alphatic and Related Natural Product Chemistry,' ed. F.D. Gunstone, Royal Society of Chemistry, 1979, volume 1, p. 20.
8. R.E. Moore, 1977, *Acc. Chem. Research*, 10, 40; Volatile Compounds from Marine Algae.
9. D.J. Faulkner, 1977, *Tetrahedron*, 33, 1421; Marine Natural Products.
10. C. Christophersen and N. Jacobsen, 1979, 'Marine Natural Products', in Annual Reports, Section B, Royal Society of Chemistry, 76, 433.

Insect Pheromones and Related Natural Products

11. R. Rossi, *Synthesis*, 1978, 413; Synthesis of Chiral Components of Insect Pheromones.

12. C.A. Henrick, 1977, *Tetrahedron*, 33, 1845; Aspects of Pheromone Synthesis.
13. K. Mori, Insect Pheromone Synthesis, in 'The Total Synthesis of Natural Products,' ed. J. ApSimon, Wiley-Interscience, 1981, volume 4, p. 1.
14. J.M. Brand, J. Chr. Young, and R.M. Silverstein, 1979' Progress in the Chemistry of Organic Natural Products', *37*, 1; critical review on insect pheromones.
15. R. Baker and J.W.S. Bradshaw, Insect Pheromones and Related Natural Products, in 'Aliphatic and Related Natural Product Chemistry,' ed. F.D. Gunstone, Royal Society of Chemistry, 1983, volume 3, p. 66; 1981, volume 2, p. 46.
16. R. Baker and D.A. Evans, Insect Pheromones and Related Behaviour-Modifying Chemicals, in 'Aliphatic and Related Natural Product Chemistry,' ed. F.D. Gunstone, Royal Society of Chemistry, 1979, volume 1, p. 102.
17. R. Baker and J.W.S. Bradshaw, Insect Chemistry, in Annual Reports, Section B, Royal Society of Chemistry, 1979, 76, 404.

Prostaglandins
18. S.M. Roberts and R.F. Newton (ed), 'Prostaglandins and Thromboxanes,' Butterworths, 1982.
19. J.S. Bindra, The Synthesis of Prostaglandins, in 'The Total Synthesis of Natural Products, ed. J. ApSimon, Wiley-Interscience, 1981, volume 4, p. 353.
20. A. Mitra, 'The Synthesis of Prostaglandins,' Wiley, 1977.
21. J.S. Bindra and R. Bindra, 'Prostaglandin Synthesis,' Academic Press, 1977.
22. S.M. Roberts and F. Scheinmann, 'Recent Synthetic Routes to Prostaglandins and Thromboxanes,' Academic Press, 1982.
23. R.F. Newton and S.M. Roberts, 1980, *Tetrahedron*, 36, 2163; Aspects of Prostaglandin Synthesis.
24. I. Ernest, 1976, *Angew. Chem. (Int. Edn)*, 15, 207; Woodward's synthesis discussed.
25. M.P.L. Caton, 1979, *Tetrahedron*, 35, 2705; Prostaglandin synthesis.
26. K.C. Nicolaou, G.P. Gasic, and W.E. Barnette, 1978, *Angew. Chem. (Int. Edn.)*, 17, 293; synthesis of prostacyclin and analogues.
27. S.M. Roberts and F. Scheinmann (eds), 'Chemistry, Biochemistry, and Pharmacological Activity of Prostanoids,' Pergamon, 1979.
28. R.F. Newton, S.M. Roberts, R.H. Green, and P.F. Lambeth, Prostaglandins and Leukotrienes, in 'Aliphatic and Related Natural Product Chemistry,' ed. F.D. Gunstone, Royal Society of Chemistry, 1983, volume 3, p. 107.
29. P.R. Marsham, Prostaglandins, in 'Aliphatic and Related Natural Product Chemistry', ed. F.D. Gunstone, Royal Society of Chemistry 1981, volume 2, p. 125; 1979, volume 1, p. 170.
30. R.F. Newton and S.M. Roberts, Prostaglandins, in Annual Reports, Section B, Royal Society of Chemistry, 1981, 78, p. 347.

Polyethers
31. W. Wierenga, The Total Synthesis of Ionophores, in 'The Total Synthesis of Natural Products,' ed. J. ApSimon, Wiley-Interscience,1981, volume 4, p. 263.
32. W. Keller-Schierlein and H. Gerlach, 1968, *Progress in the Chemistry of Organic Natural Products*, 26, 161; macrotetrolides.

Macrolide Antibiotics
33. S. Masamune, G.S. Bates, and J.W. Corcoran, 1977, *Angew. Chem. (Int. Edn.)*, *16*, 585; macrolides, recent progress.
34. K.C. Nicolaou, 1977, *Tetrahedron*, 33, 683; macrolide synthesis.
35. T.G. Beck, 1977, *Tetrahedron*, 33, 3041; macrolide synthesis.
36. S. Masamune, 1978, *Aldrichim. Acta*, 11, 23; recent progress in macrolide synthesis.
37. W. Keller-Schierlein, 1973, *Progress in the Chemistry of Organic Natural Products*, 30, 314; macrolide antibiotics.
38. R.C.F. Jones, Olefinic Microbial Compounds, Including Macrocyclic Compounds, in 'Aliphatic and Related Natural Product Chemistry,' ed. F.D. Gunstone, Royal Society of Chemistry, 1983, volume 3, p. 288; 1981, volume 2, p. 76; 1979, volume 1, p. 128.

3 Aromatic compounds

T.J. SIMPSON

Aromatic compounds can be isolated from all possible natural sources—micro-organisms, plants, insects, mammals and marine organisms. In any discussion of them a major, and in many ways unresolvable, problem arises in trying to subdivide the huge variety of compounds and structural types which occur, and in this chapter the choice of compounds is necessarily both highly selective and highly subjective. As many reviews and books have appeared on plant phenolic substances in recent years, their coverage here bears no relation to the size and importance of the field but is limited simply to allow coverage of areas which have been less adequately treated. In general this chapter will discuss compounds in order of increasing number of aromatic rings and increasing condensation of rings.

Aromatic compounds are formed by several biosynthetic routes, in particular the polyketide and shikimate pathways, but also by the terpenoid pathway, and by combinations of some or all of these. These have been thoroughly reviewed in a number of publications[1] and so biosynthetic aspects will not be covered here in any detail, but it is important to bear in mind that many of the structural studies described have been carried out along with biosynthetic studies.

Major advances in natural product chemistry in recent years have undoubtedly been the development of ^{13}C n.m.r. spectroscopy,[2] advances in multipulse sequences in both ^{1}H and ^{13}C n.m.r. for structural studies,[3] and associated isotopic labelling methodology for studying both the structures and biosynthesis of metabolites.[4] Thus ^{13}C n.m.r. studies will be mentioned where appropriate.

3.1 Benzenoids

A large number of compounds containing a benzene ring continue to be isolated from many sources. The number, size and disposition of substituents

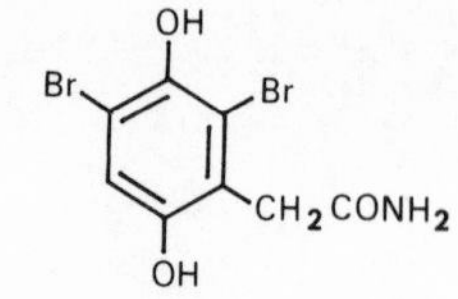

(1)

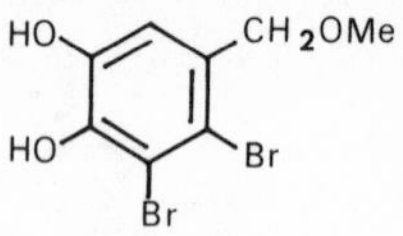

(2)

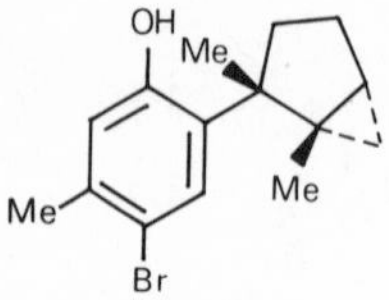

(3)

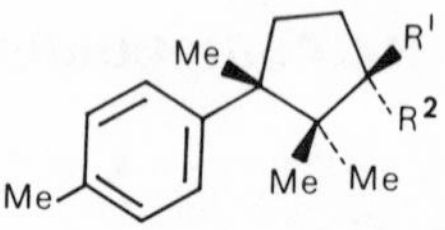

(4) R^1 = Br, R^2 = H

(5) R^1 = H, R^2 = Br

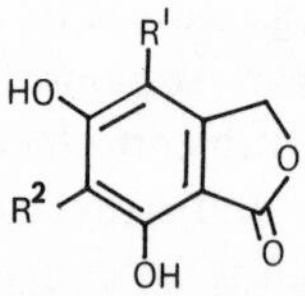

(6) R^1 = Me, R^2 = H

(7) R^1 = H, R^2 = Me

(8) R^1 = Me, R^2 = farnesyl

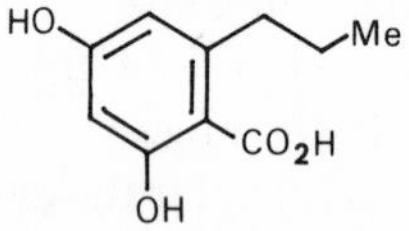

(13)

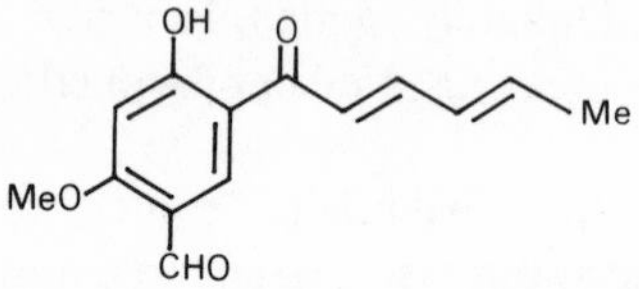

(14)

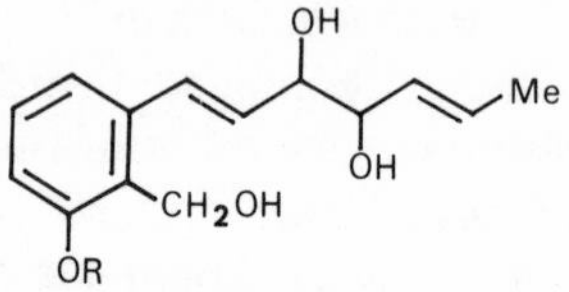

(15) R = H

(16) R = Me

on the aromatic ring varies widely and some illustrative examples are given below.

Marine organisms are a source of characteristic bromine-containing compounds.[5] The dibromo-amide (1) has been isolated from a sponge, *Verongia aurea*,[6] whereas the red alga *Rhodomela subfusca* produces 2,3-dibromo-4,5-dihydroxybenzyl methyl ether (2).[7] A number of phenolic sesquiterpenoids, e.g. laurinterol (3)[8] are responsible for the antibiotic activity[9] of *Laurencia* (red algal) species. α-Bromocuparene (4) and α-isobromocuparene (5) have been isolated from *Laurencia glandulifera* and *L. nipponica*[10] and it is suggested that these are the precursors of the other aromatic sesquiterpenes from *Laurencia* species.

Fungi are a particularly prolific source of benzenoid compounds, many of

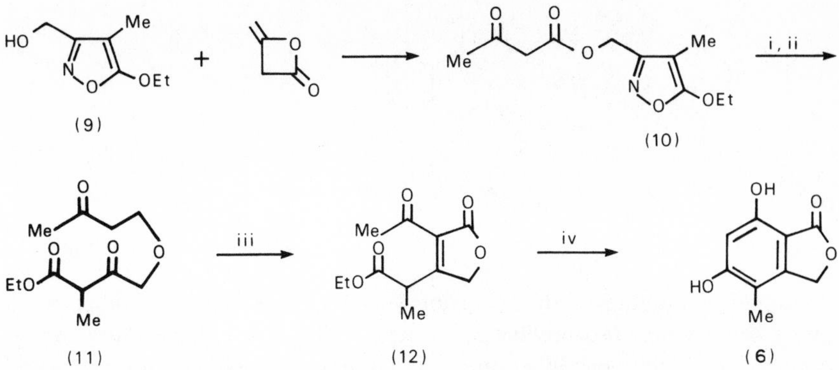

Scheme 1 Reagents: i, H_2, Pd-Lindlar; ii, H^+; iii, NaOEt; iv, BuLi, $-70°C$.

polyketide origin, and a comprehensive listing of compounds isolated up to 1982 has recently appeared.[11] The isomeric phthalides (**6**) and (**7**) have been isolated from *Alectoria nigricans*[12] and *Aspergillus flavus*,[13] respectively. 5,7-Dihydroxy-4-methylphthalide (**6**) is also a key intermediate in the biosynthesis of the antibiotic mycophenolic acid in *Penicillium brevi-compactum* where it is first converted to the 6-farnesyl compound (**8**).[14] An interesting synthesis of (**6**) proceeds (Scheme 1) via the isoxazole (**9**). Reaction of (**9**) with diketene gives the acetoacetyl ester (**10**); reduction and hydrolysis unmasks the latent polyketide-type precursor (**11**) which is converted first to the butenolide (**12**) and is then cyclized to the phthtalide (**6**).[15]

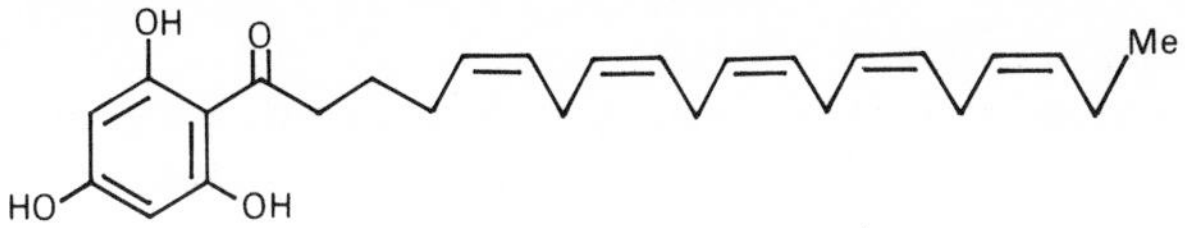

(17)

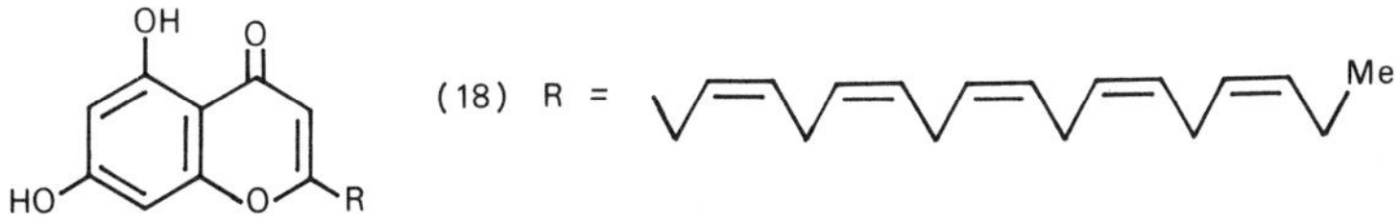

(19) R = $(CH_2)_{14}Me$

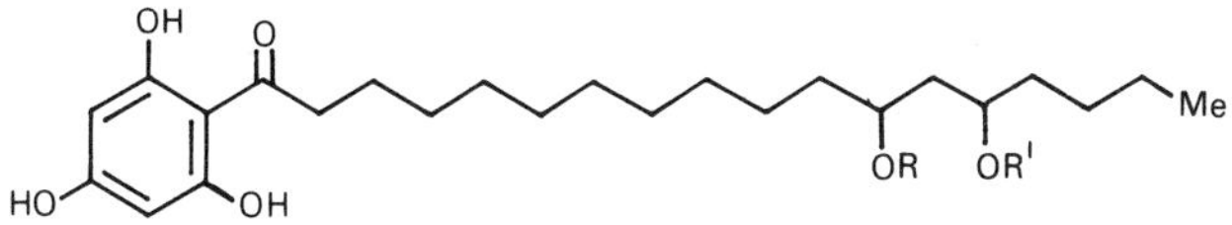

(20) R^2 = H, R^1 = mannosyl

(21) R^1 = H, R^2 = mannosyl

(23)   (24)   (25)

Fungal metabolites with increasing length of side chain are represented by 2,4-dihydroxy-6-propylbenzoic acid (13) from *Penicillium brevicompactum*;[16] the sorbophenone (14) from a *Scytalidium* species;[17] and 2-(3,4-dihydroxyhepta-1,5-dienyl)-6-hydroxybenzyl alcohol (15) from *Pyricularia oryzae*.[18] The corresponding methyl ether (16) has also been isolated from *Aspergillus variecolor*.[19]

Long-chain phenolic substances have been isolated from many sources. The polyunsaturated acylphloroglucinol (17) has been obtained from the brown alga *Zonaria tournefortii* along with the related unsaturated and saturated chromones (18) and (19).[20] Fruiting bodies of the basidiomycete *Byssomerulius corium* contain large quantities of the byssomeruliols, e.g. (20) and (21).[21] Lasolicid A (22) is a benzoic acid with a long polyoxygenated chain containing both tetrahydrofuran and tetrahydropyran rings isolated from *Streptomyces lasaliensis*.[22] It was one of the first of the polyether antibiotics, which act as ionophores,[23] to be isolated. A recent comprehensive review discusses the isolation, structures, synthesis and properties of the long-chain phenolic substances of plant origin.[24]

Many of the more interesting benzene-containing compounds have substituents of isoprenoid origin. The isoprenoid moieties can themselves be highly modified. *Aspergillus duricantis* produces several hydroxyphthalides, e.g. (23) and (24).[25] The disubstituted dihydropyran (chroman) ring of (24) is presumably formed by an intramolecular 'ene' reaction between the dimethylallyloxy and formyl substituents on (23). A similar system is found in tajixanthone (123). Pergillin (25) is a plant-growth retarding substance isolated from *Aspergillus ustus* which contains a 2-isopropyliden-benzofuran-3-one system.[26] *Stereum frutulosin* produces an antibiotic frutulosin (26). Its structure has been confirmed by synthesis.[27] An alternative synthesis[28] is outlined in Scheme 2. 2,5-Dimethoxybenzyl alcohol protected as the alkyl ether (27) was metallated with butyl lithium and iodinated to give (28). Deprotection and oxidation to the aldehyde (29) was followed by coupling with the copper acetylide derived from 3-methylbut-3-en-1-yne.

The plant pathogenic fungus *Colletotrichum nicotiniae* produces a series of prenylated benzaldehydes including colletochlorin D (30)[29] and the geranyl

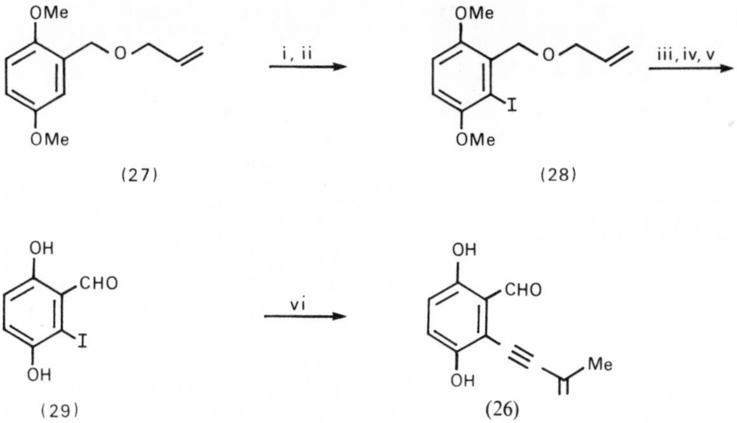

(27)   (28)

(29)   (26)

Scheme 2   Reagents: i, $BuLi/C_5H_{12}/0°C$; ii, ICH_2CH_2Cl; iii, H^+; iv, PCC; v, BBr_3; vi, $Cu-C\equiv C-C(Me)=CH_2/DMF$

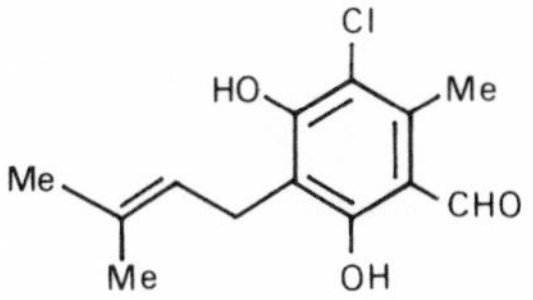

(30) 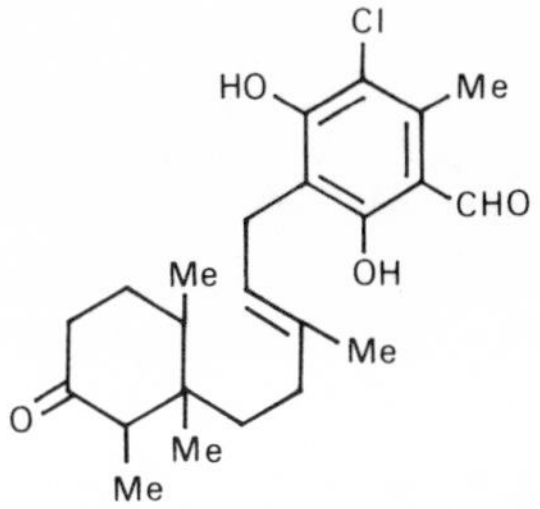   (31)

homologue, colletochlorin B.[30] These are closely related to ascochlorin (**31**), a metabolite of *Ascochyta vicae*,[31] and similar substances which have a variety of biological activities, in particular antiviral, which has stimulated much recent interest in their synthesis. A recent synthesis of colletochlorin B is summarized in Scheme 3. Birch reduction of O-dimethylorcinol gave the cyclohexadiene (**32**) which was lithiated and alkylated with geranyl bromide. Chlorination and aromatization gave the phenol (**33**) which was formylated to give collectotrichin B (**34**).[32] Other synthetic approaches to ascochlorin[33] and colletochlorins[34] have appeared.

K-76 (**35**), an inhibitor of the complement system, has been isolated from *Stachybotrys complementi*.[35] It has a bicyclofarnesyl moiety attached to a bis-aldehyde to give what is formally a dihydrobenzofuran. *Aspergillus ustus* produces a group of mycotoxins with novel complex structures, the austalides, e.g. austalide A (**36**).[36] It is likely that the farnesyl-substituted phthalide (**8**) is a key intermediate in the biogenesis of the austalides. However the farnesyl

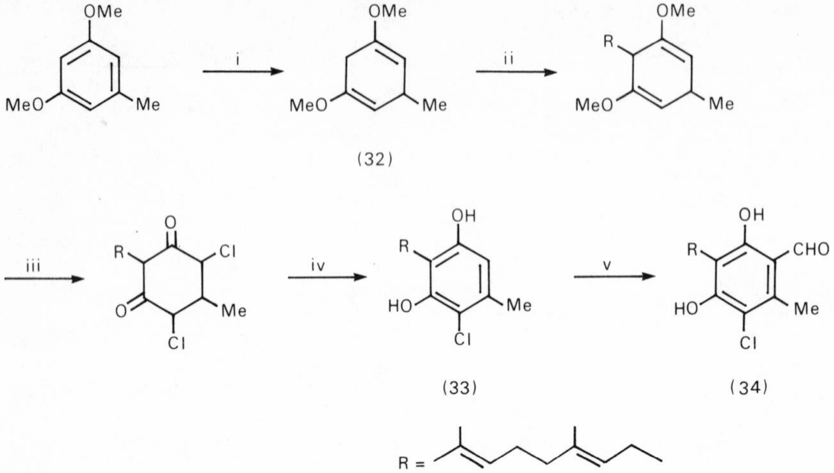

(32)

(33)   (34)

$$R = \text{\raisebox{0pt}{\includegraphics{}}}$$

Scheme 3 i, Li/NH$_3$/$^+$BuOH/THF; ii, $^+$BuLi/HMPA/geranyl bromide; iii, NCS; iv, DBU; v, hexamethylenetetramine/AcOH.

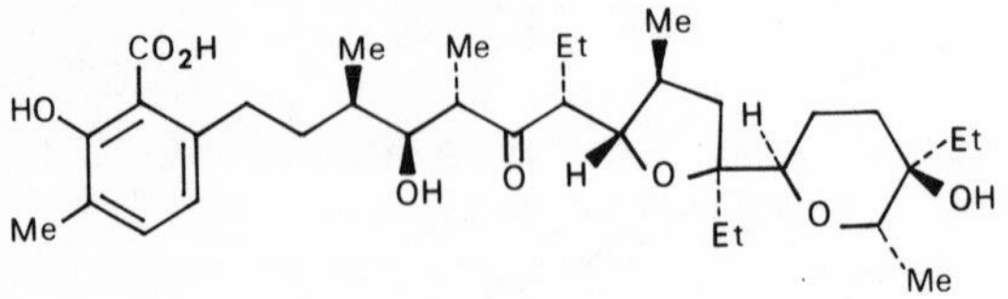

(22)

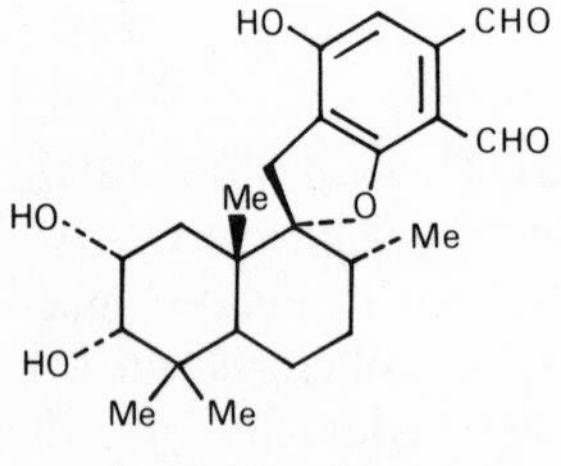

(35)

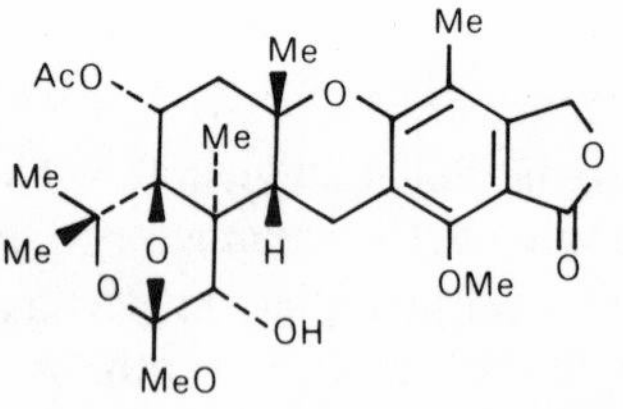

(36)

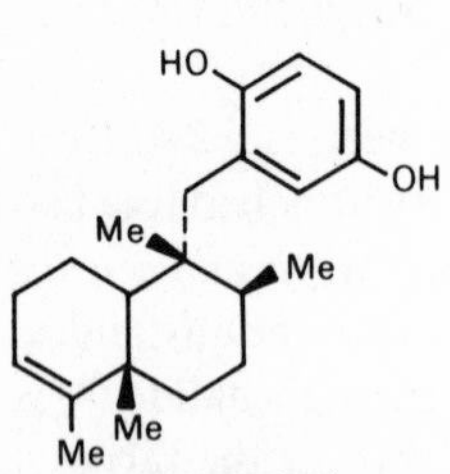

(37)

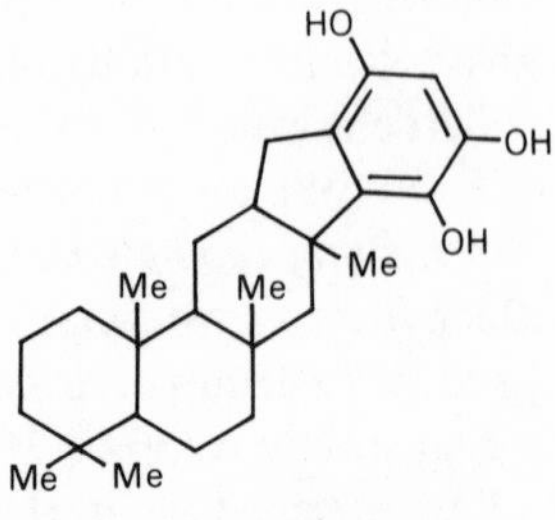

(41)

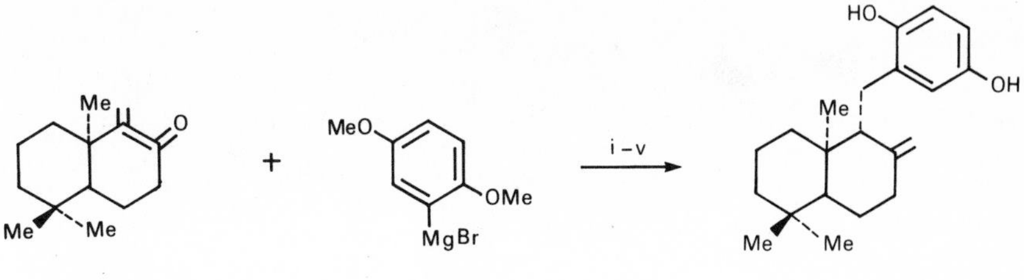

(40) (39) (38)

Scheme 4 i, cat. CuI; ii, Ac$_2$O; iii, KOH/CH$_3$OH; iv, Ph$_3$P$^+$CH$_2$Br$^-$/NaH/DMSO; v, LiSBu/HMPA.

moiety has been highly cyclized and oxidized. Note that this compound is formally a chroman.

Sesquiterpenoid-substituted benzoquinones and quinols have been isolated from both sponges and brown algae. Avarol (**37**) has been isolated from the sponge *Disidea avaria*[37] and zonarol (**38**) from the alga *Dictyopteris undulata*.[38]

The key feature of a synthesis[39] of zonarol is the conjugate addition of 2,5-dimethoxyphenylmagnesium bromide (**39**) to the enone (**40**) as shown in Scheme 4. Disidein (**41**) is unique in having a sesterterpenoid-derived moiety fused to a quinol. It is a metabolite of *Disidea pallescens*.[40]

3.2 Coumarins

A substantial monograph on coumarins is available.[41] The ^{13}C n.m.r. spectra of coumarin and substituted coumarins have been analysed.[42] Siderin (**42**) is a polyketide-derived compound which has been isolated both from plant (*Sideritis romana*[43] and *Cedrela toona*[44]) and fungal (*Aspergillus variecolor*[45]) sources. Its structure has been confirmed by synthesis.[43,45] Kotanin (**43**), a dimer of siderin, has been isolated from a toxigenic strain of *Aspergillus glaucus* and the structure confirmed by synthesis.[46] The most widely studied group of coumarins are the aflatoxins, e.g. aflatoxin B$_1$ (**44**), a group of mycotoxins produced by the common moulds *Aspergillus parasiticus* and *A. flavus*. Their biosynthesis,[47] analysis[48] and toxicology[49] have been extensively studied. Much work has been done on chemical methods of detoxification of aflatoxin-infected foodstuffs.[50] Aflatoxin M$_1$ (**45**), a metabolite of aflatoxin B$_1$, is commonly detected in milk, and its synthesis has been reported.[51]

3.3 Isocoumarins

A number of biologically active isocoumarins have been isolated from micro-organisms. The bacterium *Bacillus pumulus* produces a gastroprotective

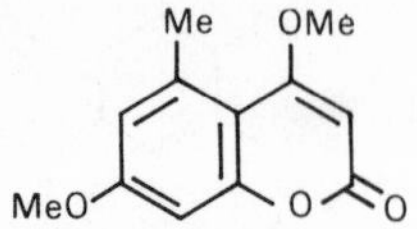

(42)

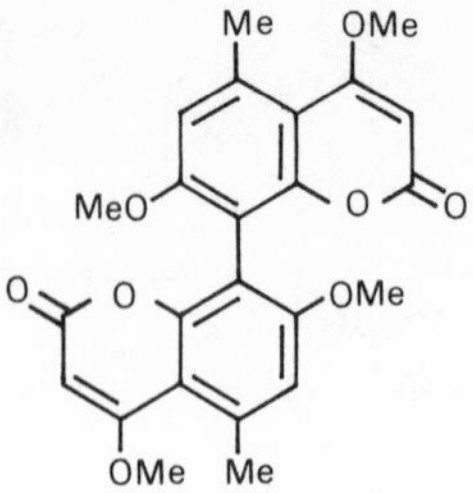

(43)

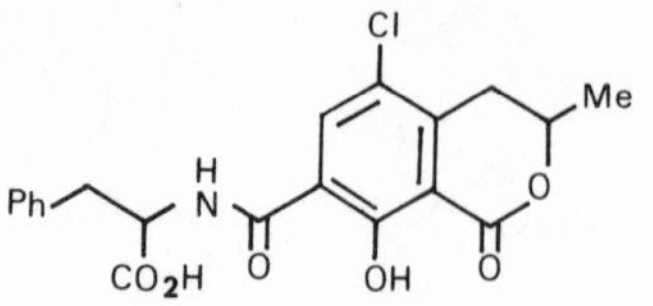

(47)

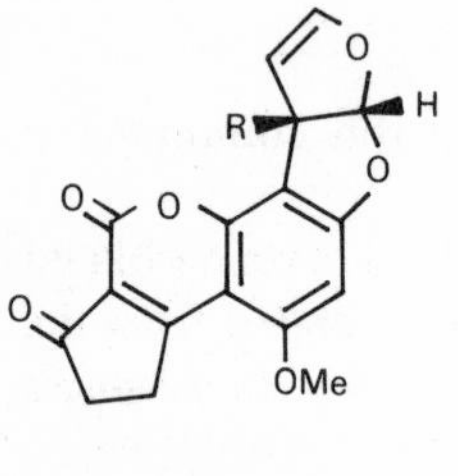

(44)   R = H
(45)   R = OH

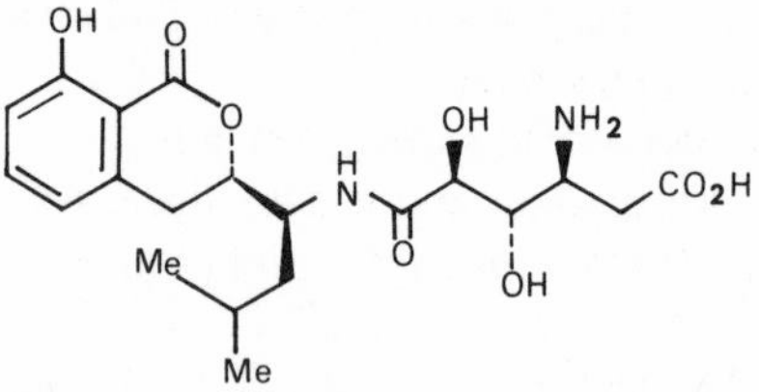

(46)

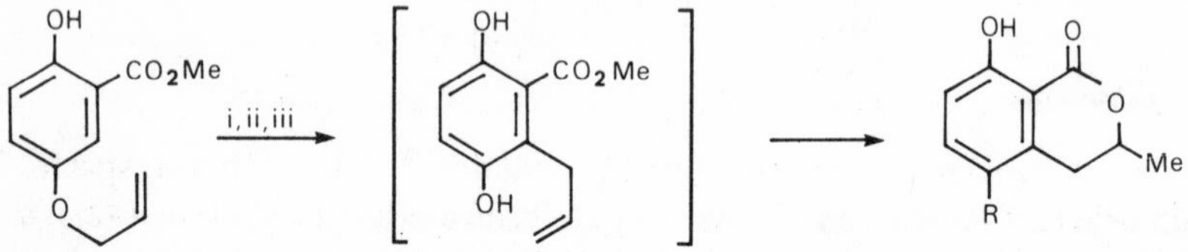

(49)

(50) R = OH
(48) R = H

Scheme 5   i, CF_3CO_2H; ii, $MeSO_2Cl-C_5H_5N$; iii, H_2/Pt.

substance A1-77-B (**46**).[52] Ochratoxin A (**47**), a metabolite of *Aspergillus ochraceus*, is an important mycotoxin responsible for kidney damage in pigs.[53] Its synthesis has recently been reported.[54] A facile synthesis of mellein (**48**), a dihydroisocoumarin which is a common fungal metabolite, is summarized in Scheme 5. 5-Allyloxy-2-hydroxybenzoates are smoothly converted, presumably via (**49**), to 3,4-dihydro-5,8-dihydroxy-3-methyli-socoumarin (**50**). Selective mesylation and reductive cleavage of the sulphonate-carbon bond gave mellein.[55] 3,4-Dihydroisocoumarins have been identified as intermediates in the biosynthesis of the fungal cyclopentenones terrein[56] and cryptosporiopsinol.[57] They have been synthesized in isotopi-

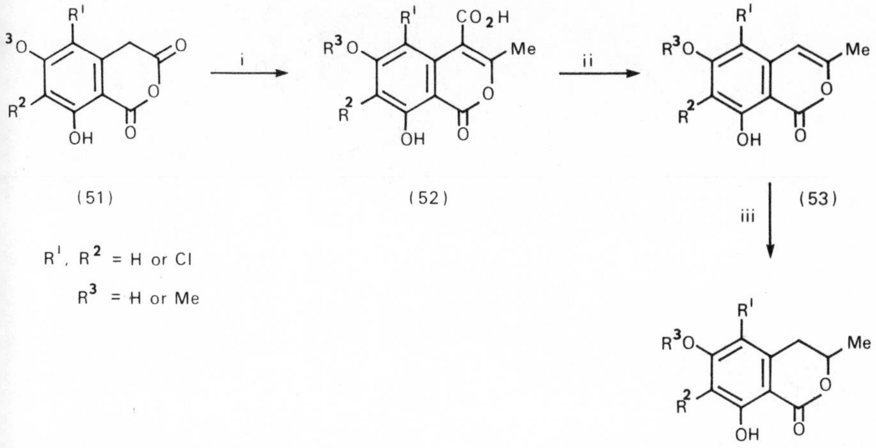

Scheme 6 i, Ac$_2$O; ii, NaOH; iii, NaBH$_4$.

cally labelled form by conversion of the homophthalates (51) with acetic anhydride to give the 3-methyl-4-carboxyisocoumarins (52), which are then decarboxylated to the 3-methylisocoumarins (53) and finally reduced to the 3-methyl-3,4-dihydroisocoumarins (Scheme 6).[57,58]

Fomajorin S(54) and fomajorin D(55) are isocoumarins of terpenoid origin produced by the common wood-rotting fungus *Fomes annosus*.[59] Stellatin (56), produced by *Aspergillus stellatus*, is unusual among fungal dihydroisocoumarins in lacking a substituent on either C-2 or C-3. Its structure was defined by a complete analysis of long-range (i.e. greater than one bond) ^{1}H–^{13}C couplings in its fully ^{1}H-coupled ^{13}C n.m.r. spectrum.[60] The 2- and 3-bond couplings detected in a series of selective low-power decoupling experiments are indicated on structure (56). This is a very powerful technique for assigning the substitution pattern of highly substituted benzenoid compounds (see also naphthoquinones below). Monocerin (57), a metabolite of *Helminthosporium monoceras* and other fungi,[61] has an unusual fused dihydrofurobenzopyrone ring system. It has been isolated along with the closely related fusarentins as an insecticidal metabolite of *Aspergillus parvulus*.[61]

Sclerin (60) is a highly substituted homophthalic anhydride metabolite of *Sclerotinia sclerotiorum*[62] and *Aspergillus carneus*[63] which stimulates plant root formation. It has been synthesized (Scheme 7) by condensation of the bis-trimethyl-silyl ether (58) of methyl 3-oxopentanoate with methyl orthoacetate in the presence of titanium tetrachloride. The resulting homophthalate (59) is then methylated and hydrolysed to produce sclerin.[64] The elaboration of (59) can be considered as a controlled condensation of diketide and triketide moieties and so constitutes a biomimetic synthesis.

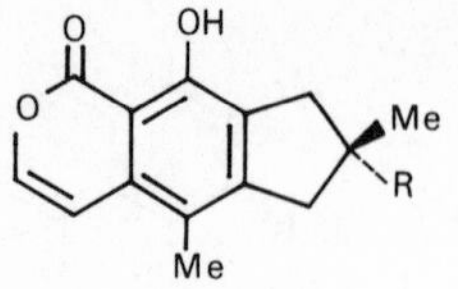

(54)   R = Me

(55)   R = CO$_2$H

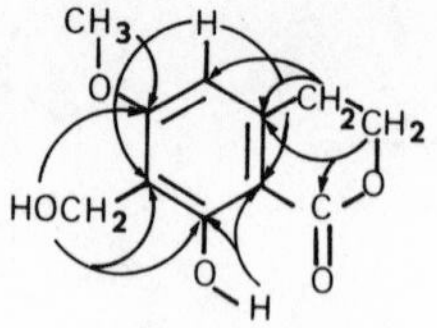

(56)

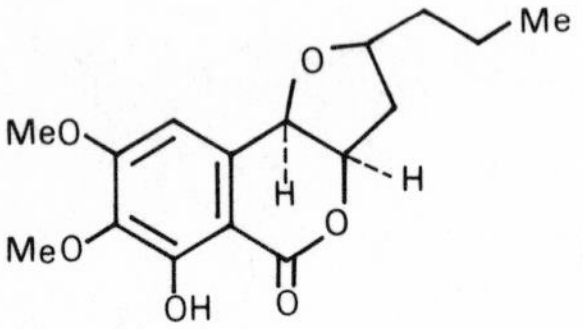

(57)

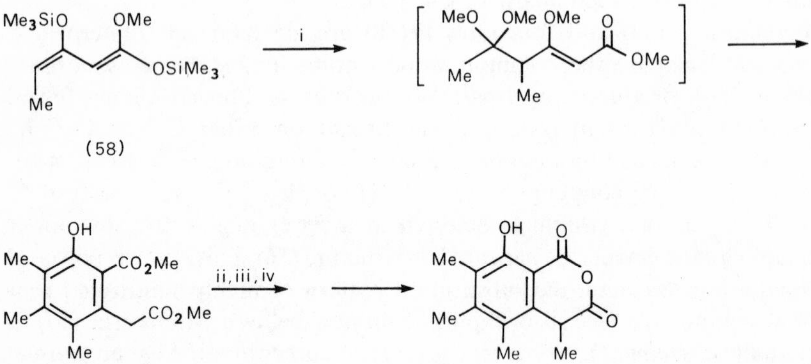

(58)

Scheme 7   i, MeC(OMe)$_3$/TiCl$_4$; ii, LDA/MeI; iii, NaOH; iv, H$^+$.

(59)   (60)

3.4   Chromanones and chromones

LL-D253α (62), an antibiotic metabolite of *Phoma pigmentivora*,[65] *P. violacea*,[66] and *Sclerotinia fructigena*,[66] was originally assigned structure (61), 5-hydroxy-6-(2′-hydroxyethyl)-7-methoxy-2-methylchromanone.[65] However its structure has been revised to (62) by analysis of the fully [1]H-coupled

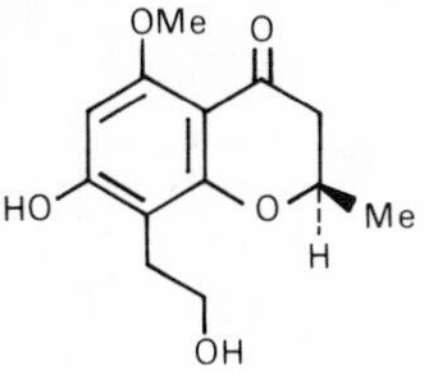

(62)

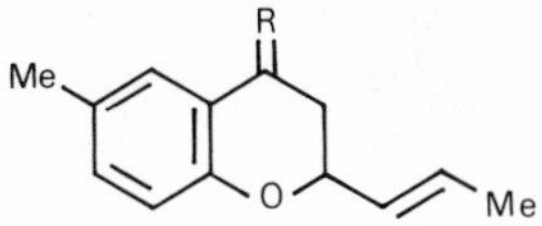

(63) R = H, OH

(64) R = O

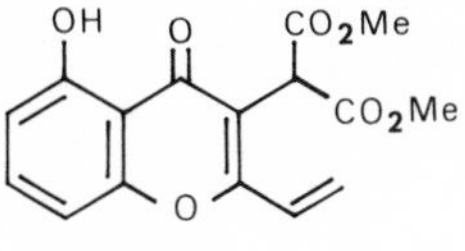

(65)

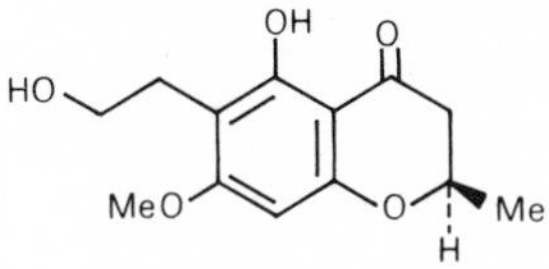

(61)

[13]C n.m.r. spectrum and by unambiguous synthesis of both structures (61) and (62).[67] Mycochromanol (63) and mycochromanone (64) are co-metabolites in *Myrothecum roridum*.[68] Mycochromone (65) has been isolated from the phytotoxic fungus *Mycosphaerella rosigena* which is responsible for leaf spot of greenhouse roses.[69] It co-occurs with mycoxanthone (126) which suggests it is formed from (126) by oxidative degradation of the methoxylated benzenoid ring. Chromenes, chromones, and chromanones have been reviewed in a recent text.[70]

3.5 Cannabinoids

Cannabinoids can be regarded as chroman derivatives. Their chemistry, analysis and synthesis have been extensively reviewed.[71] Two of the more interesting recent syntheses are described in Schemes 8 and 9. The key step in a synthesis of cannabinol (70) makes use of the remarkably smooth displacement of methoxyl groups in *O*-methoxy-aryloxazolines described by Meyers.[72] Thus reaction of oxazoline (66) with the aryl Grignard reagent (67) gives the sterically hindered biphenyl intermediate (68) which on hydrolysis and deprotection is converted to lactone (69). Methylation with methylmagnesium iodide completes the synthesis of cannabinol.[73] In a second biogenetically patterned approach, condensation of 1,3-bis-trimethyl-silyloxy-1-methoxybutadiene (71) with the acid chloride (72) gave methyl olivetolate (73) which on condensation under carefully controlled conditions with (+)-*p*-mentha-2,8-dien-1-ol gave methyl Δ'-tetrahydrocannabinolate (74). Hydrolysis and decarboxylation then gave Δ'-tetrahydrocannabinol (75).[74]

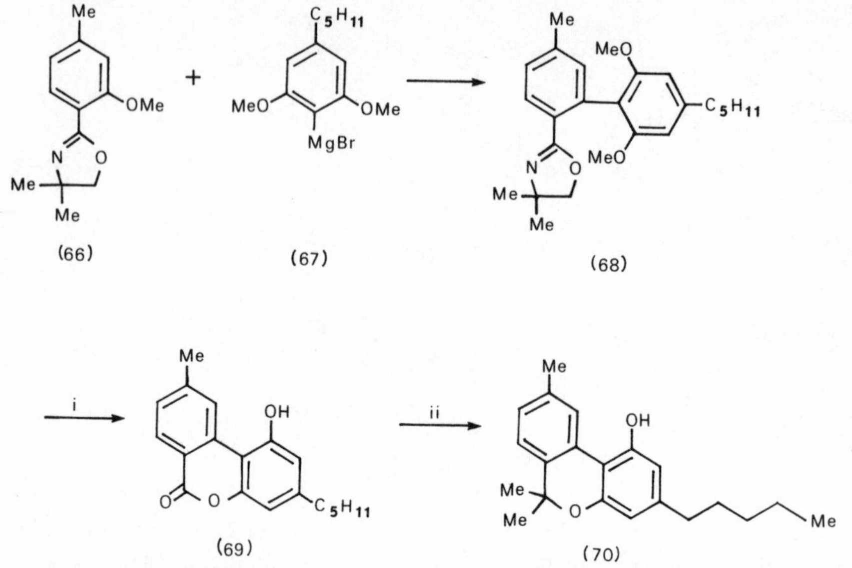

(66)　　　　　(67)　　　　　　(68)

(69)　　　　　　　(70)

Scheme 8　i, HI/Ac$_2$O; ii, MeMgI.

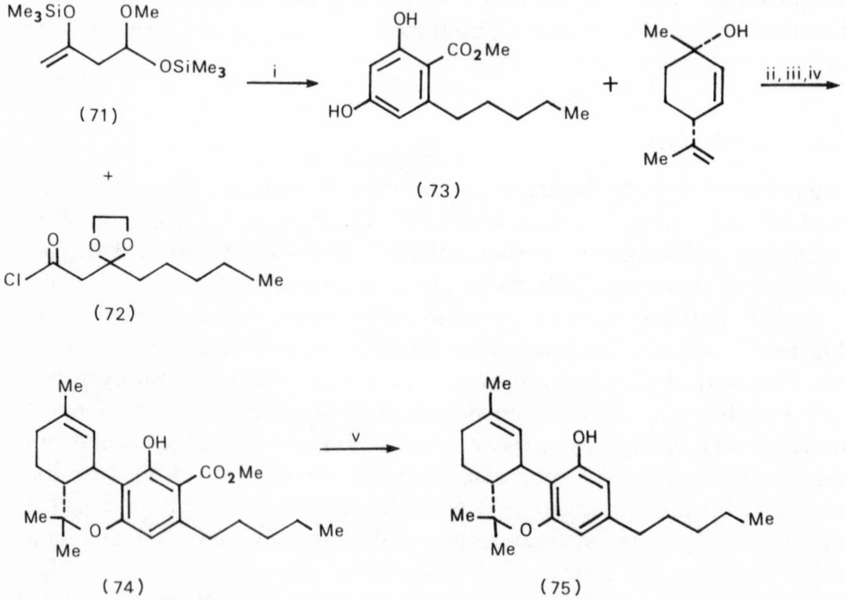

(71)

(73)

(72)

(74)　　　　　　　　(75)

Scheme 9　i, TiCl$_4$/CH$_2$Cl$_2$; ii, MgSO$_4$; iii, BF$_3$.Et$_2$O; iv, NaHCO$_3$; v, NaOH.

3.6 Macrocyclic lactones

Zearalenone (80), a mycotoxin isolated from *Fusarium* fungi, is notable because of its oestrogenic and anabolic activity in animals.[75] A large number of related compounds has been isolated from fungal sources[76] and all feature a macrocyclic lactone attached to a benzene ring. Zeanalenone has been synthesized as shown in Scheme 10 via a palladium-catalysed carbonylation of the iodothioether (76) in the presence of the iodoalcohol (77) to give the ester

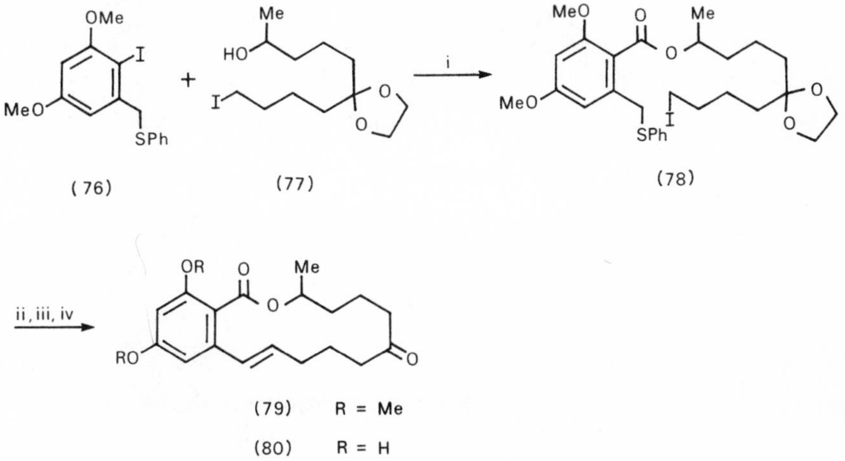

(76)   (77)   (78)

(79)   R = Me

(80)   R = H

Scheme 10   i, $PdCl_2/K_2CO_3/C_6H_6/120°/CO$; ii, $KN(SiMe_3)_2$; iii, $NaIO_4$; iv, TsOH.

(78). This was cyclized using potassium hexamethyldisilazide as base to give, after oxidation and elimination, zearalenone dimethyl ether (79).[77] A similar approach has been used to synthesize curvularin[78] and lasiodiplodin.[79] Lasiodiplodin (81), first obtained from the fungus *Lasiodiplodia theobromae*,[80] has recently also been isolated from a plant source, *Euphorbia splendens*,[81] to join the list of metabolites produced by both lower and higher organisms.

3.7 Pyrones and butenolides

A number of biologically interesting compounds contain a benzenoid ring linked to a pyrone. The nitrobenzene moiety is found in luteoreticulin (82) and aureothin (83), toxic metabolites of *Streptomyces luteoreticuli*[82] and *S. thioluteus*[83] respectively; and in spectinabilin (84), an antibiotic metabolite of *S. spectabilis*.[84] A new kawain derivative (85) has been isolated from the roots of *Piper sanctum* and its structure confirmed by synthesis.[85] Piperolide (87) is a closely related metabolite,[86] containing the 4-hydroxybutenolide (tetronic

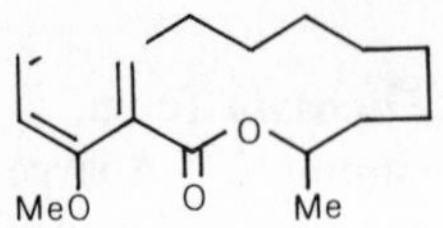

(81)

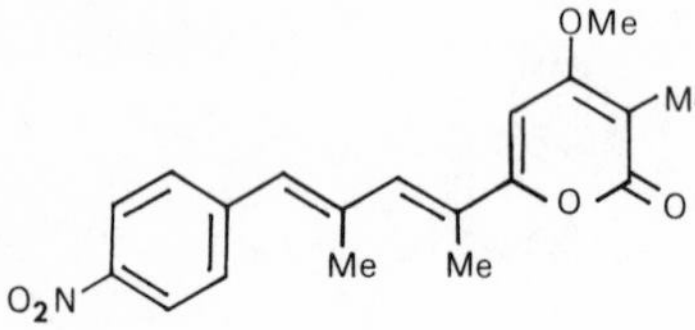

(82)

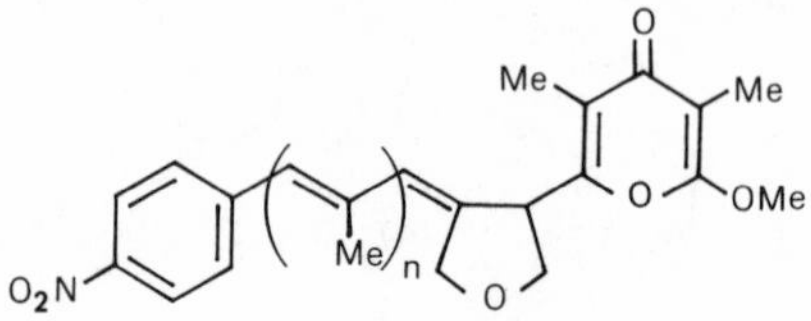

(83) n = 1

(84) n = 3

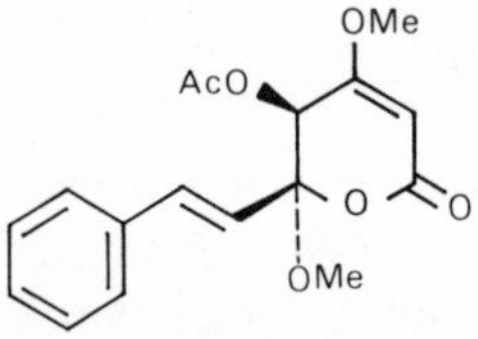

(85)

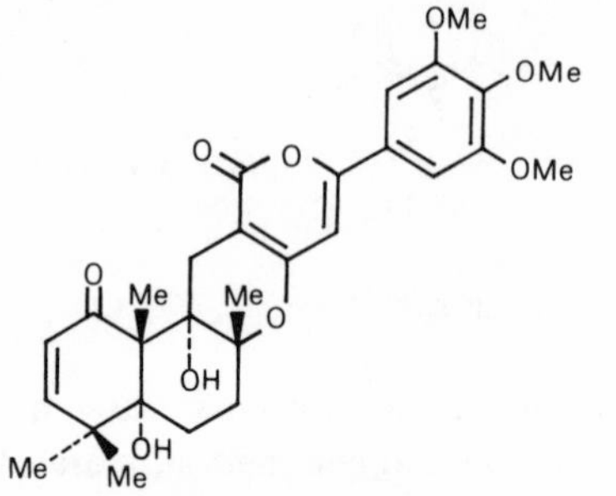

(91)

acid) moiety. It has been synthesized (Scheme 11) by reaction of the anion of 4-methoxytetronate (86) with 3-phenylpropanal. Methylation followed by allylic bromination and elimination as shown gave piperolide;[87] presumably the co-occurring pyrone and butenolide metabolites have a common biogenetic precursor, probably formed from cinnamic acid and two acetate units (Ar-C$_7$). Fadyenolide (90), isolated from *Piper fadyenii*, has two carbons (one acetate) less than piperolide.[88] It has been synthesized (Scheme 12) by conversion of 4-methoxytetronate to 2-trimethylsilyloxy-4-methoxyfuran (88) followed by condensation with methyl orthobenzoate to give (89) followed by base-catalysed elimination of methanol to give fadyenolide.[89]

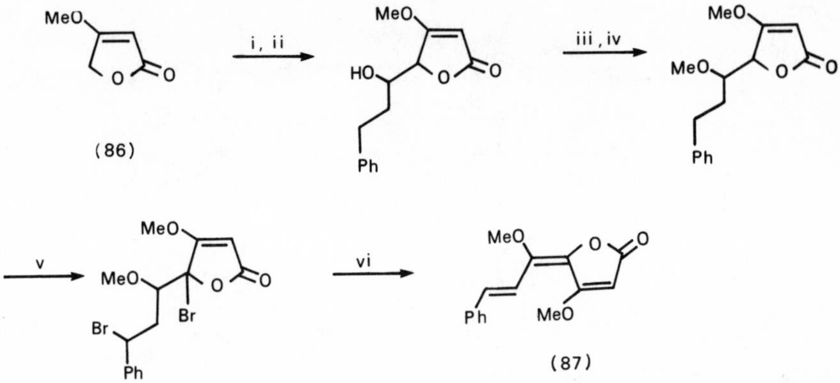

Scheme 11 i, LDA/HMPA/THF; ii, ArCH$_2$CH$_2$CHO; iii, BuLi/HMPA/ $-78°$; iv, MeI; v, NBS/CCl$_4$; vi, DBU/C$_6$H$_6$.

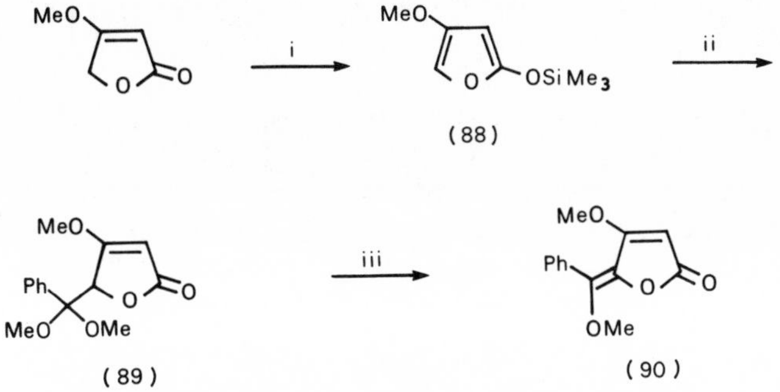

Scheme 12 i, BuLi/Me$_3$SiCl; ii, PhC(OMe)$_3$/ZnBr$_2$; iii, Bu^{t}Li.

A similar Ar-C$_5$ derived pyrone moiety is present in the territrems, e.g. territrem A (**91**), tremorgenic mycotoxins isolated from *Aspergillus terreus*.[90] These have a bicyclofarnesyl-derived moiety substituted on to the pyrone ring to form a chroman ring.

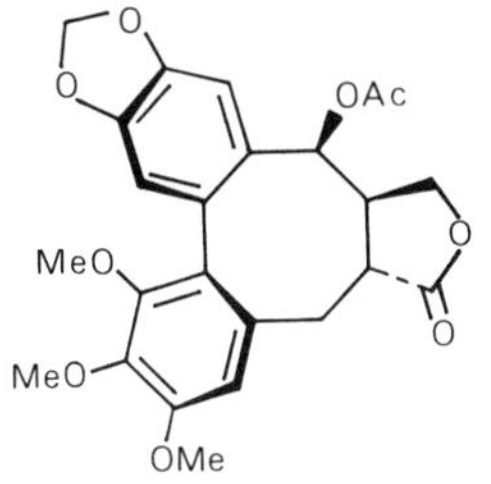

(92)

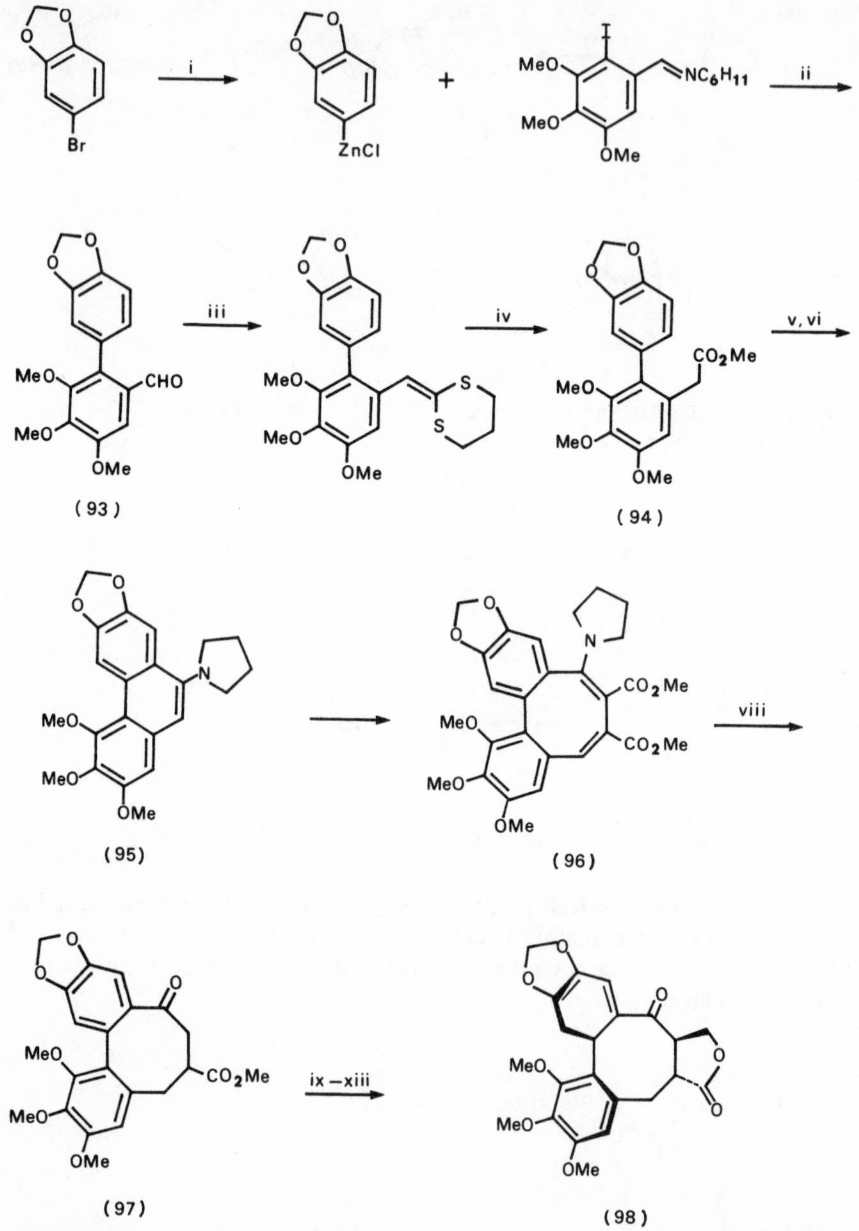

(93)

(94)

(95)

(96)

(97)

(98)

Scheme 13 i, Bu^{n}Li/THF/ $-78°$/ZnCl$_2$; ii, Ni(acac)$_2$/Bu$_2^i$AlH/THF/Ph$_3$P/ $-20°$C; iii, Bu^{n}Li/S$-$(CH$_2$)$_3$$-S-$CHSiMe$_3$/THF/ $-20°$C; iv, HgCl$_2$/HCl/MeOH; v, pyrrolidine; vi, PCl$_3$; vii, MeO$_2$CC≡CCO$_2$Me; viii, HCl/MeOH; ix, H$_2$/Ra$-$Ni; x, LiOH; xi, resolution; xii, KOH/HCHO; xiii, xylene/reflux.

3.8 Lignans

Steganacin (**92**), isolated from the Ethiopian plant *Steganotaenia araliacea*, belongs to a novel class of dibenzocyclooctadiene lignan lactones which display significant antileukaemic properties.[91] The first synthesis was reported in 1976.[92] More recently a synthesis of enantiomerically pure (−)steganone (**98**) has been described.[93] This proceeds (Scheme 13) via the biphenyl (**93**) formed by coupling of 1-bromo-3,4-methylenedioxybenzene with the cyclohexylimine of 2-iodo-3,4,5-trimethoxybenzaldehyde. Conversion of (**93**) via the dithian gave the aryl acetate (**94**), which was ring closed leading to the enamine (**95**). Cycloaddition and ring expansion gave the cyclooctatetraene (**96**) which was converted by hydrolysis and reduction to the ester (**97**) which was resolved via the (S)-2-amino-3-phenylpropan-1-ol amide derivative of the acid. Further elaboration of the keto-acid gave (+)-isosteganone which on heating isomerized to (−)-steganone (**98**).

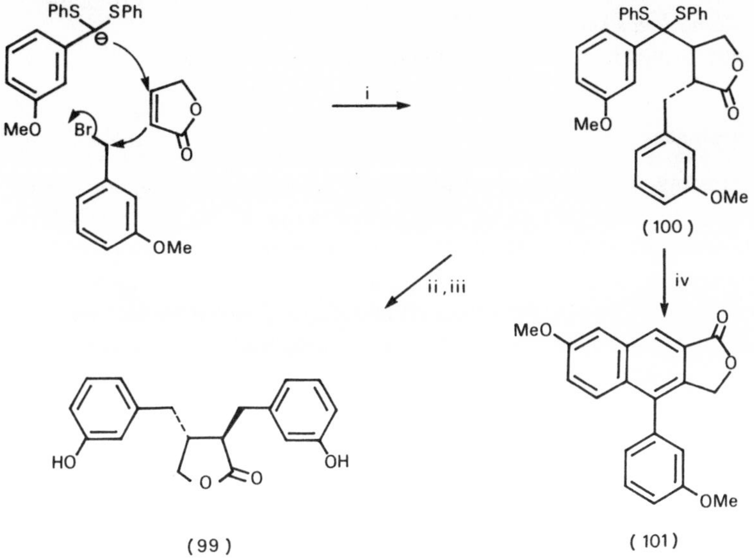

Scheme 14   i, Bu^{n}Li/THF/ − 78 °C; ii, Ra—Ni/EtOH; iii, BBr$_3$; iv, Hg(O$_2$CCF$_3$)$_2$/O$_2$.

In the first reported isolation of lignans from humans, the *trans*-dibenzylbutyrolactone (**99**) was found to be excreted by females during the luteal phase of the menstrual cycle and during early pregnancy.[94] The structure was proved by synthesis.[95] A further interesting synthesis (Scheme 14) used a tandem conjugate addition involving Michael addition of the anion of an aryldiphenylthiomethane to butenolide at − 78 °C, followed by trapping of the intermediate with a benzyl bromide. Raney nickel treatment of the in-

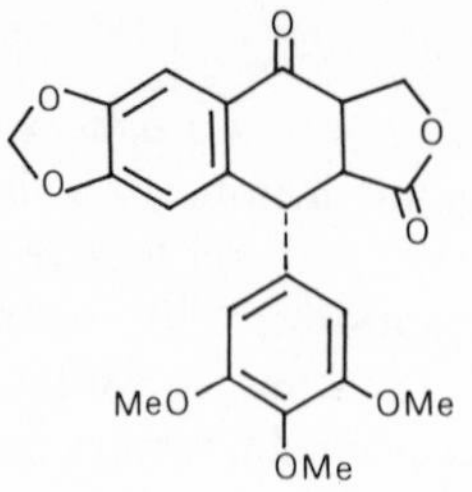

(102)

termediate (100) gave the desired butyrolactone (99) after deprotection. When treated with mercuric trifluoroacetate, compound (100) gave the arylnaphthalene lignan (101).[96] It had been hoped to stop this reaction at the dihydronaphthalene oxidation level which would have provided a good route to another class of lignans with significant antitumour activity, e.g. isopicropodophyllone (102) which has been isolated from the roots and rhizomes of *Podophyllum pleianthum*.[97]

Synthetic routes to a variety of lignans have been reviewed.[98]

3.9 Benzofurans

Two interesting approaches to the related benzofurans (104) and (105), isolated from *Sophora tomentosa*, have been described.[99] In the first (Scheme 15) the heterocycle was constructed via an intramolecular Wittig reaction of the phenolic ester (103). The second (Scheme 16) proceeds by condensation of an *ortho*-iodophenolic acetate with a copper acetylide itself formed from the corresponding methyl ketone via the hydrazone and vinyl iodide. These

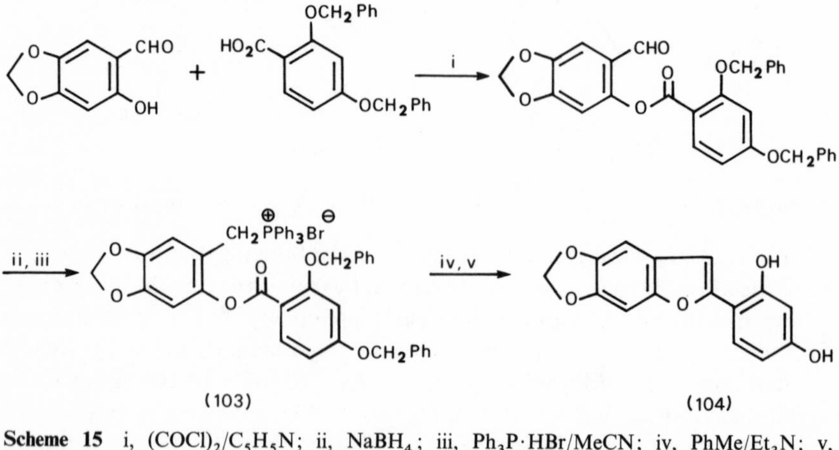

(103)   (104)

Scheme 15   i, $(COCl)_2/C_5H_5N$; ii, $NaBH_4$; iii, $Ph_3P \cdot HBr/MeCN$; iv, $PhMe/Et_3N$; v, $Pd/C-H_2$.

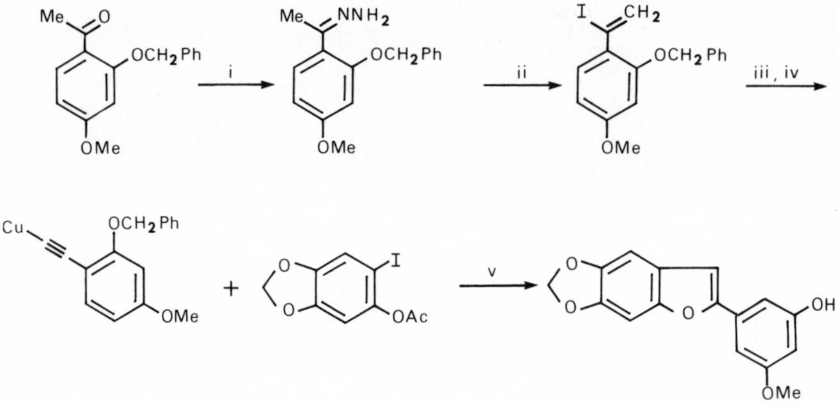

(105)

Scheme 16 i, Et_3N/H_2NNH_2 ; ii, $I_2/Et_3N/THF$; iii, NaH/THF; iv, $CuSO_4/HONH_2$. HCl aq. NH_3 ; v, $C_5H_5N/heat$

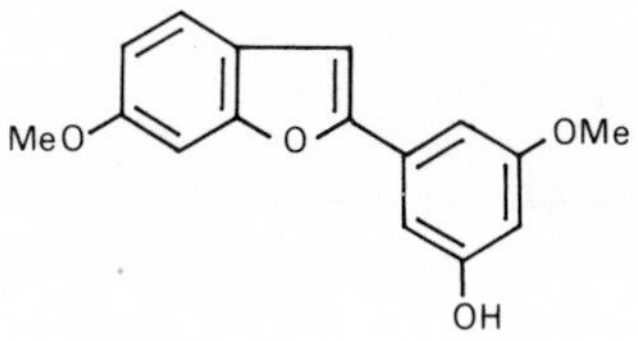

(106)

compounds are of increasing importance as several are antifungal phytoalexins, e.g. vignafuran (**106**) which is the phytoalexin of cow pea (*Vigna unguicalata*) leaves infected with *Colletotrichum lindemuthianum*.[100]

3.10   Terphenyls

Many diphenylbutenolides and related systems are formed via diphenylbenzoquinones and the related quinols. The large variety of structures and structural types have been reviewed[101] so a few representative examples only are given here. All are fungal or lichen metabolites. In corticin A (**107**), a metabolite of *Corticium caeruleum*,[102] the three phenolic rings are linked by ether bridges, whereas pigment C_2 (**108**), isolated from *Suillius grevillei*,[103] has only one ether linkage. The central ring can be subjected to a variety of oxidative changes and rearrangement processes to give metabolites such as atromentic acid (**109**), a tetronic acid isolated from several fungal sources,[104] and aspulvinone I (**110**), one of a series of prenylated butenolides produced by *Aspergillus terreus*.[105] The structures and syntheses of these butenolides have

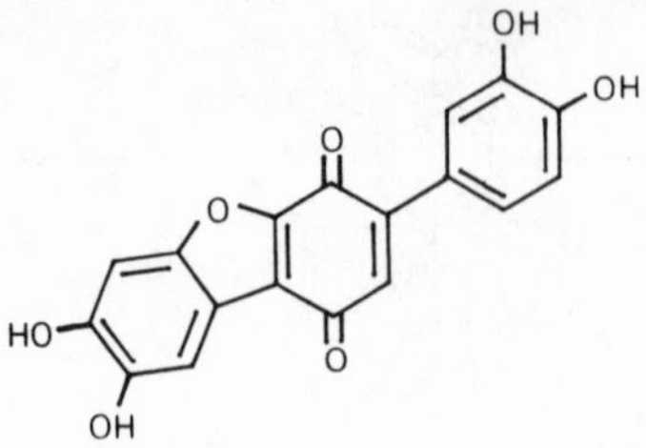

(108)

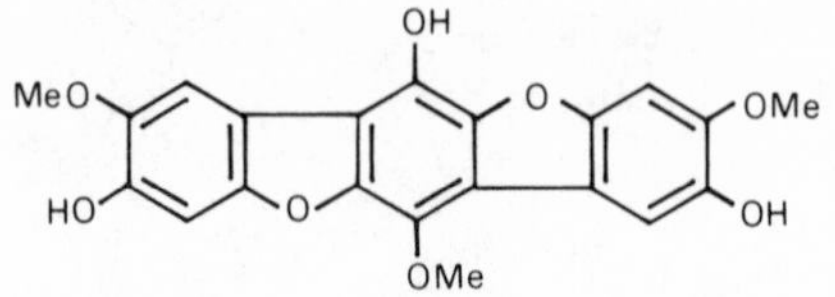

(107)

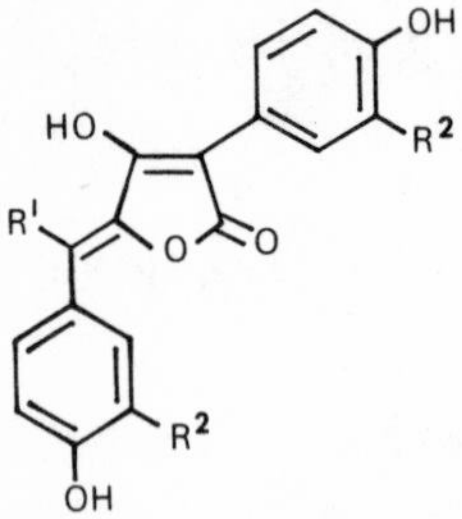

(109) $R^1 = CO_2H$, $R^2 = H$

(110) $R^1 = H$, $R^2 = $ prenyl

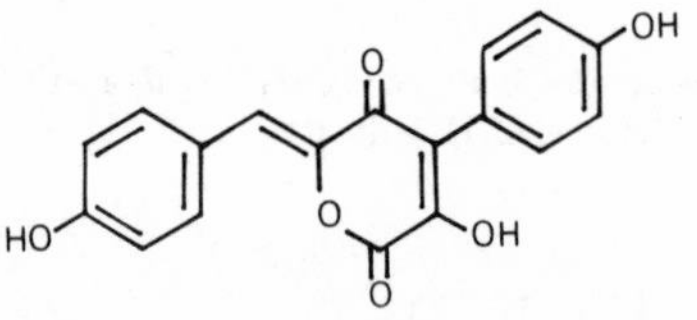

(111)

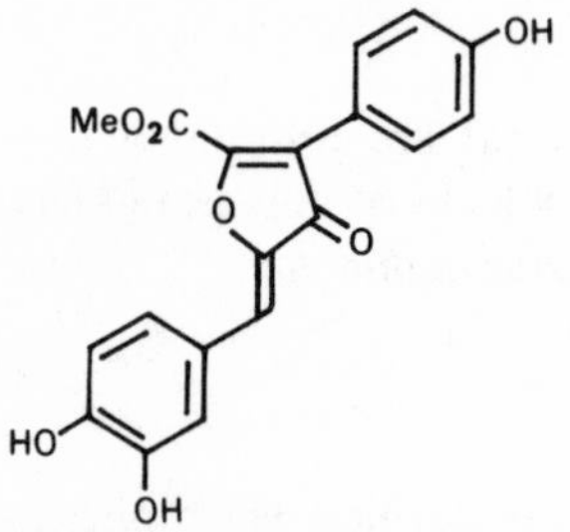

(112)

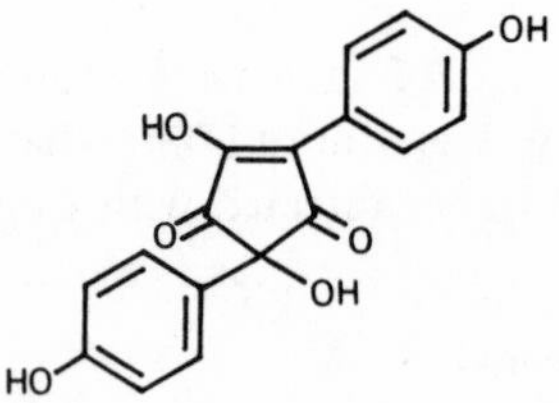

(113)

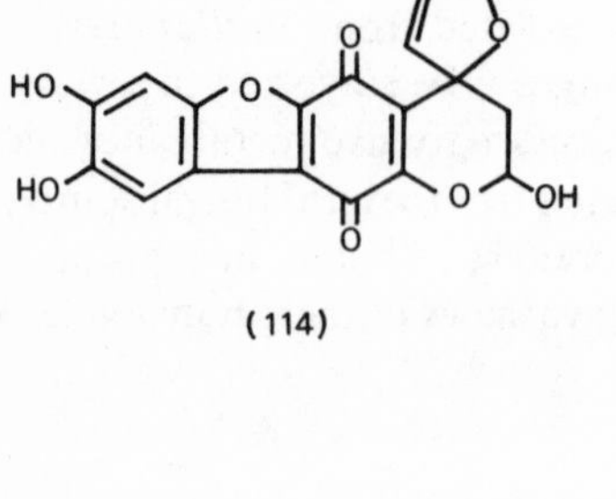

(114)

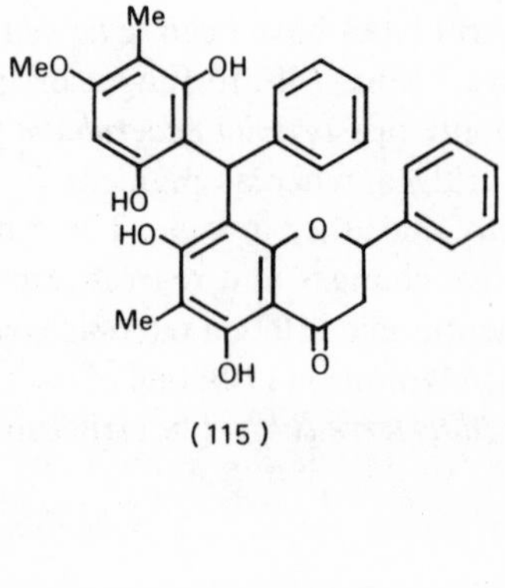

(115)

been the subject of a good comprehensive review.[106] They can be regarded as being derived formally by a cleavage of the bond between C-2 and C-3 of the central ring of a terphenyl precursor. Grevillin A has the pyrone structure (111)[107] which can be formed via a formal fission of the bond between C-1 and C-2 of a terphenyl, whereas pigment A (112), also from *S. grevillei*,[108] represents an alternative ring closure of the same intermediate to a furan-3-one. In gynosporin (113), a metabolite of (*inter alia*) *Chamonixia caespitosa*,[109] the central terphenyl ring has been contracted to form a furan-1,2,4-trione. Finally, in hydnuferrugin (114), produced by *Hydnellum ferrugineum*, one of the outer benzenoid rings has been cleaved to form a spiro-fused dihydropyranylbutenolide moiety.[110]

3.11 Flavanoids

An authoritative book covering many aspects of flavanoid isolation, structure and synthesis has appeared.[111] Two more recent books[112,113] cover plant phenolics, the shikimate pathway, phenolic acids, phenylpropanoids, lignin, flavanoids and tannins. In addition very comprehensive reviews have appeared on isoflavanoids, including rotenoids and aryl benzofurans,[114] homoisoflavones,[115] neolignans,[116] tannins,[117] and gallic acid metabolites and ellagitannins.[118]

Of the many unusual structures to appear in this area, two are worthy of special mention. The isolation and synthesis of melanervin (115) from *Maleuca quincinerva*[119] has been reported. This is the first naturally occurring compound with a triphenylmethane structure. The synthesis of scillescillin (119), the only naturally-occurring benzocyclobutene, isolated from the bulbs of *Scilla scilloides*,[120] has been reported (Scheme 17).[121] The benzocyclobutene nitrile (116) was converted to the phenylketone (117). Introduction of

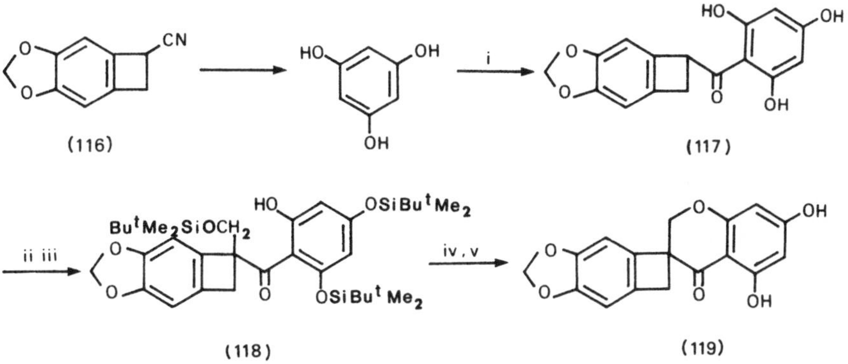

(116)   (117)

(118)   (119)

Scheme 17 i, HCl/Et$_2$O/ZnCl$_2$/0°C; ii, Bu^{t}Me$_2$SiCl/DMF/imidazole; iii, Bu^{t}OK/HCHO; iv, HF/MeCN; v, TsOH/C$_6$H$_6$.

the hydroxymethyl substituent necessitated prior protection of the phenolic hydroxyls as their silyl ethers. Surprisingly, the product of the reaction with formaldehyde was the silyl ether (118) in which migration of a silyl group had occurred. Deprotection and cyclization produced scillescillin (119).

3.12　Xanthones, benzophenones and grisans

Major reviews of xanthones from plant sources have appeared.[122] A systematic ^{13}C n.m.r. study of substituted xanthones allows the effect of substituents on chemical shifts to be deduced.[123] Although many compounds

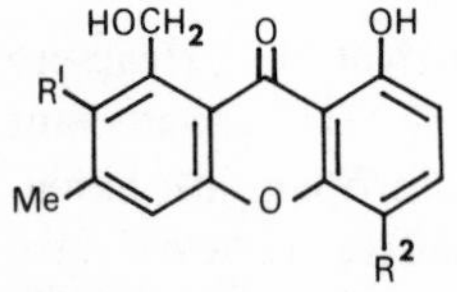

(120)　$R^1 = R^2 = H$
(121)　$R^1 = OCH_2CH = CMe_2$,　$R^2 = H$
(122)　$R^1 = OCH_2CH = CMe_2$,　$R^2 = CH_2CH = CMe_2$

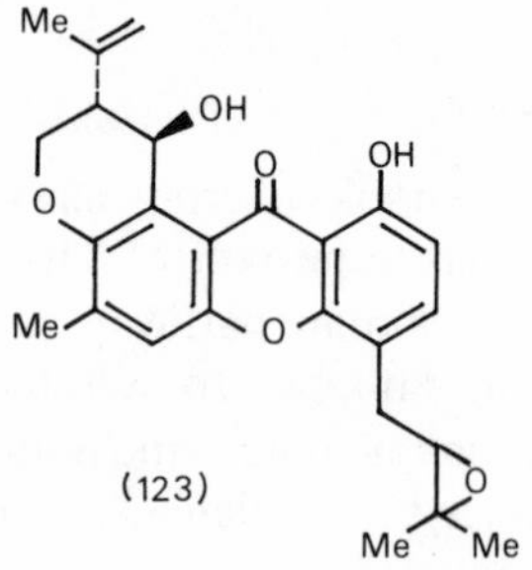

(123)

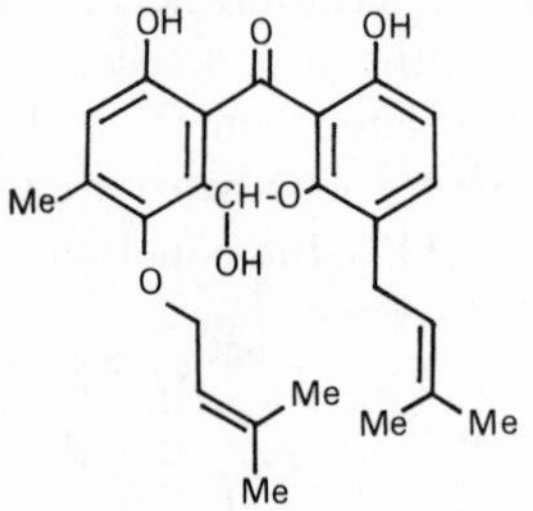

(124)

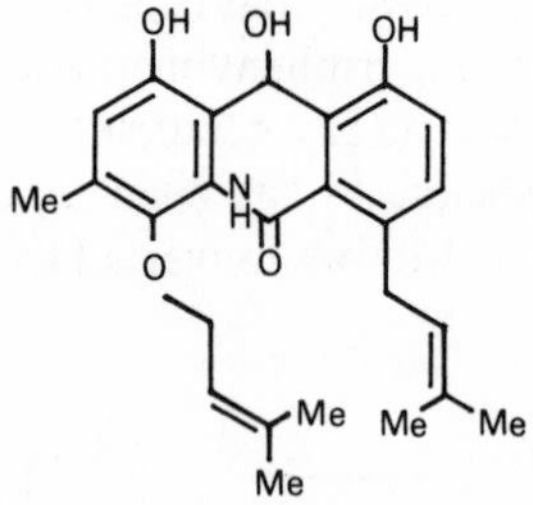

(125)

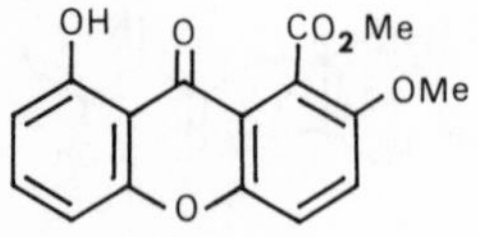

(126)

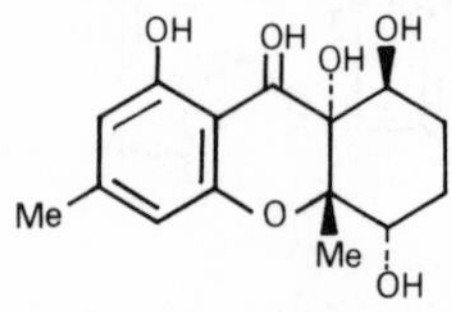

(129)

have been isolated from plant sources, fungi have also proved a source of several interesting systems in recent years. A relatively simple compound (120) has been isolated from the bird's nest fungus *Cyathus intermedius*.[124] A group of biogenetically related xanthones and benzophenones have been isolated from *Aspergillus variecolor, A. rugulosus*, and *A. nidulans*. These include the xanthones varieioxanthone A (121) and varieixanthone B (122),[125] tajixanthone (123),[126] and the benzophenone arugosin A (124) which contains a hemi-acetal linkage between the two aromatic rings.[127] A closely related benzhydrol, silvaticamide (125) has been isolated from *Aspergillus silvaticus*.[128] This has an unusual amide linkage between the rings. All these compounds are apparently formed via anthraquinonoid precursors.[129] Mycoxanthone (126) has been isolated along with mycochromone (65).[70]

The ergochromes are dimeric tetrahydroxanthones.[130] Further compounds related to the ergochromes have been isolated. These include the secalonic acids, e.g. (127), isolated from several fungi;[131] the eumetrins, e.g. (128), isolated from the lichen *Usnea baylei*;[132] and the monomeric diversonol (129), a metabolite of *Penicillium diversum*.[133] The eumetrins are of particular

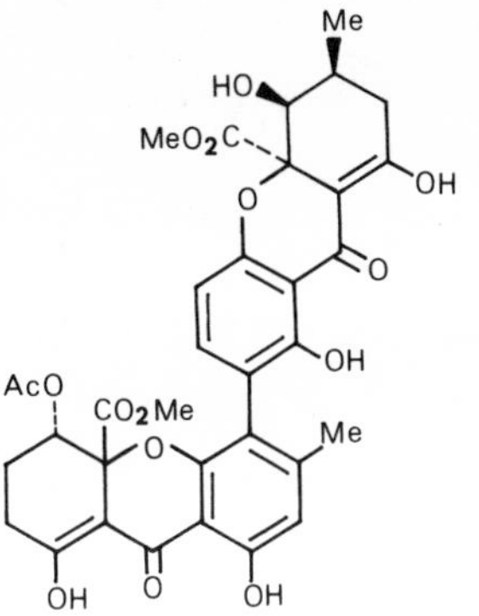

(128)

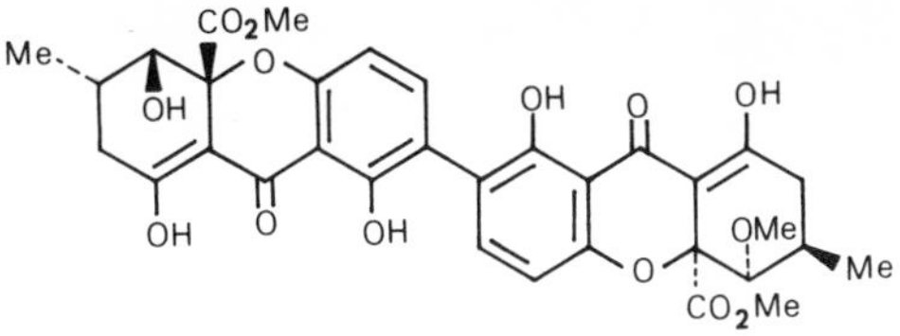

(127)

biosynthetic interest because the monomeric units present represent alternate modes of cleavage of the quinonoid ring of the presumed anthraquinonoid precursor. Some biomimetic synthetic studies on the ergochromes have been reviewed.[134]

Bikaverin (132), a red pigment with specific antiprotozoal activity, has been isolated from several fungi including *Fusarium oxysporum*.[135] It has been synthesized as shown in Scheme 18. The intermediate xanthone (130) was oxidized to the quinone (131) which was then selectively demethylated with

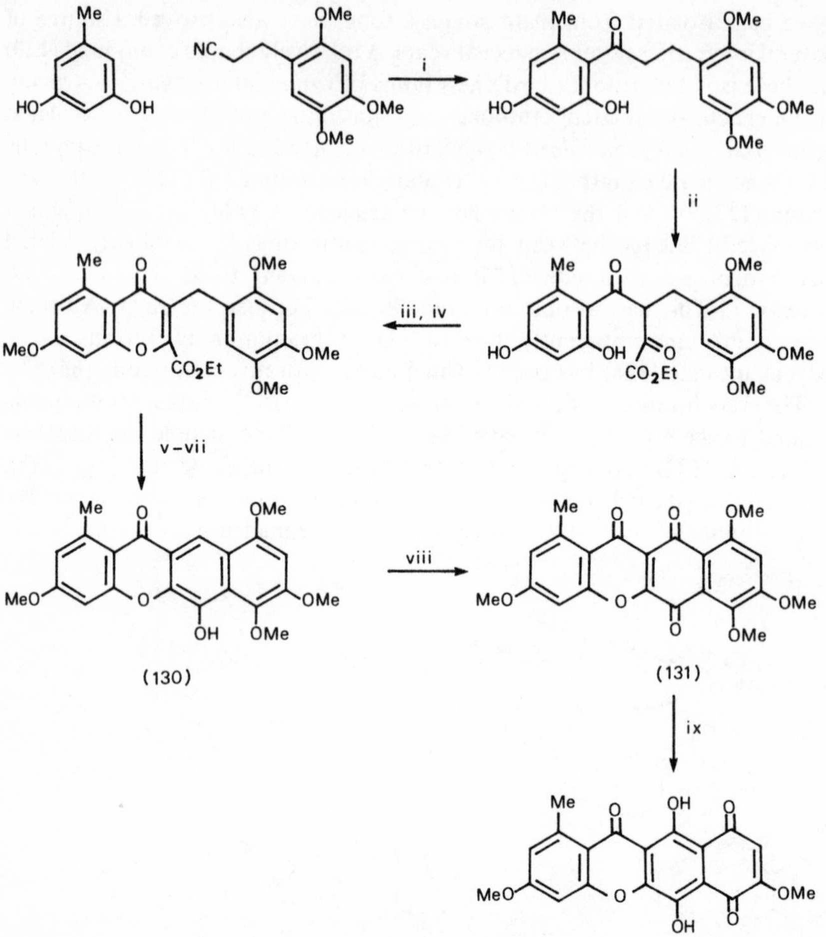

(130)

(131)

(132)

Scheme 18 i, $ZnCl_2/HCl$; ii, $(EtO_2C)_2/NaOEt$; iii, TsOH; iv, Me_2SO_4; v, KOH/EtOH; vi, $SOCl_2$; vii, BF_3; viii, $K_2Cr_2O_7/AcOH$; ix, $LiI/MeCOCMe_3$.

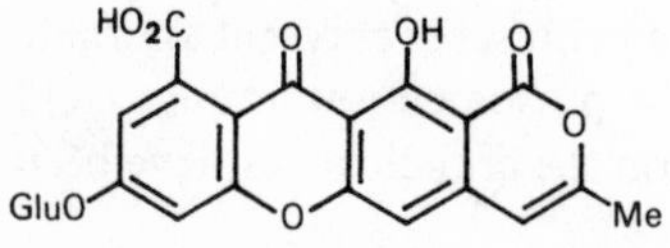

(133)

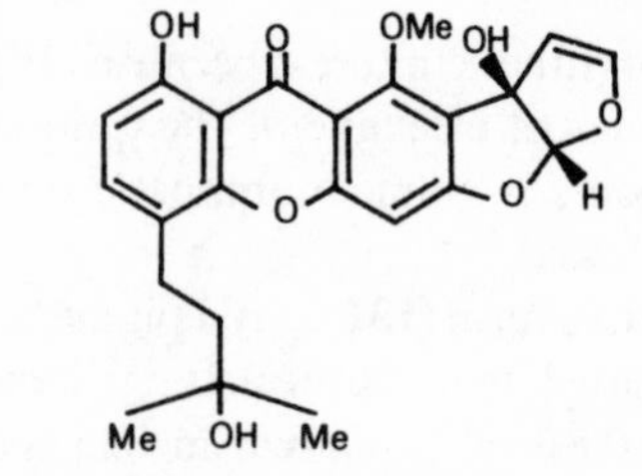

(134)

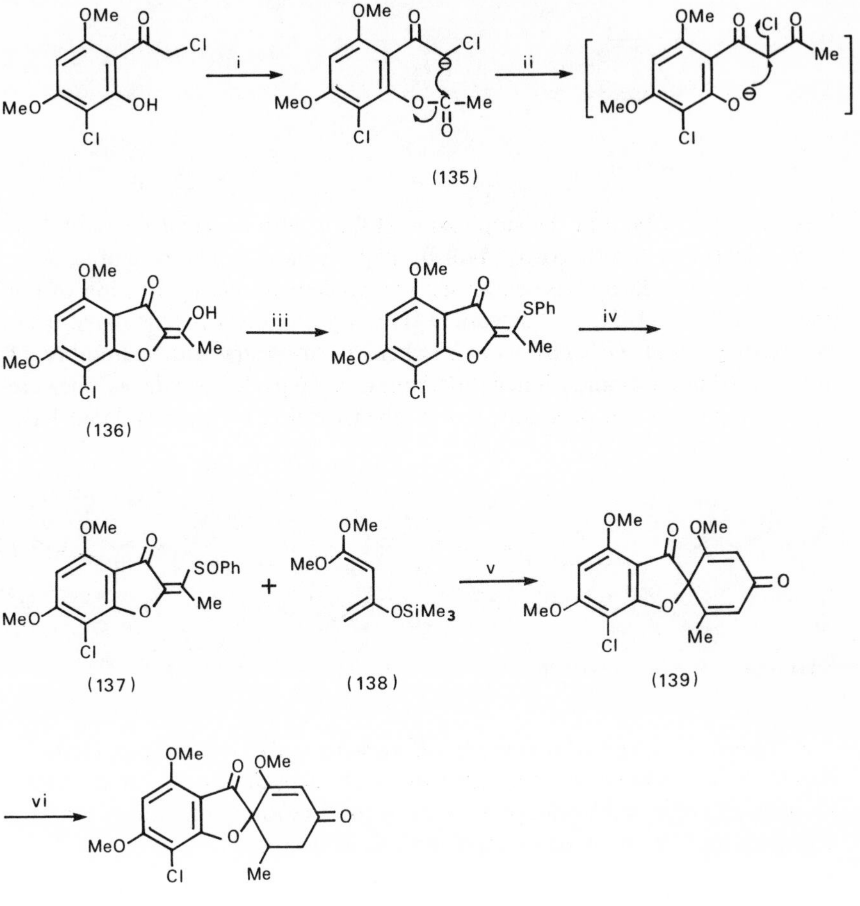

(135)

(136)

(137)   (138)   (139)

(140)

Scheme 19 i, Ac$_2$O/HClO$_4$; ii, NaH/THF/HMPA/55°C; iii, TsOH/PhSH; iv, MCPBA; v, PhMe/135°C; vi, Pd/C—H$_2$.

iodide ion.[136] Leprocybin, a glucoside responsible for the yellow fluorescence of the fruiting bodies under u.v. light, has been isolated from *Cortinarius cotoneus* and related *Leprocybes*.[137] It has the unusual pyranoxanthone structure (133). A number of xanthones containing a bis-furano side chain and modified prenyl substituents have been isolated from a toxigenic strain of *Aspergillus ustus*. These are the austocystins, e.g. austocystin E (134); their structures have been assigned by extensive ^{13}C n.m.r studies.[138]

A new total synthesis of griseofulvin (140) has been described (Scheme 19). The key step is a cycloaddition between 1,1-dimethoxy-3-trimethylsilyloxy-

(141)

1,3-butadiene (138) and the sulphoxide (137) to afford dehydrogriseofulvin (139) which was converted to (140) by hydrogenation. The benzofuranone (136) was prepared by base-catalysed rearrangement and cyclization of the phenolic acetate (135).[139] Thelepin (141) is a grisan derivative isolated from the marine worm *Thelepus setosus*, where it co-occurs with a number of brominated phenols and diphenylmethanes.[140] Depsides and depsidones are predominantly lichen products. Comprehensive lists of those isolated have been prepared.[141]

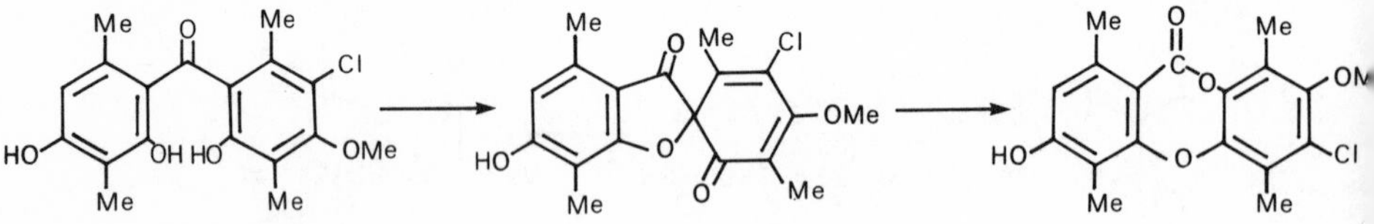

Scheme 20   i, $K_3Fe(CN)_6/H_2O/K_2CO_3$.

It is generally accepted that depsidones are derived from depsides. However the facile conversion of benzophenones to depsidones via grisandiones (Scheme 20) reported by Sargent[142] raises the interesting possibility that this sequence may be involved in depsidone biosynthesis.

3.13   Naphthalenes and naphthoquinones

The isochroman quinones are a class of antibiotic substances isolated mainly from various streptomyces.[143] Ventilagone and the eleuthrins are members of the class which had previously been found in higher plants.[143] The simplest class are the nanomycins, e.g. nanomycin D (142) isolated from *S. rosa*.[144] The griseusins, e.g. (143), a metabolite of *S. griseus*, have a more complex highly oxygenated side chain which forms a tri-substituted spiro-tetra-hydropyran ring.[145] A synthetic approach to the griseusins and their absolute configuration has been reported.[146] The granaticins, e.g. (144), metabolites of *S. olivaceus*, have a six-carbon substituent derived from glucose fused to the quinone moiety unusually through two carbon-carbon bonds.[147] Dimeric compounds also occur. The actinorhodins[148] and phenocyclinone[149] are produced by *S. coelicolor*. The most complex structures however are found in

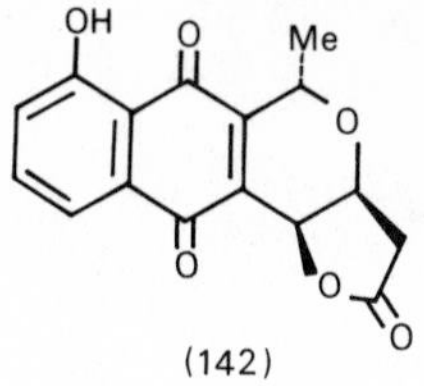

(142)

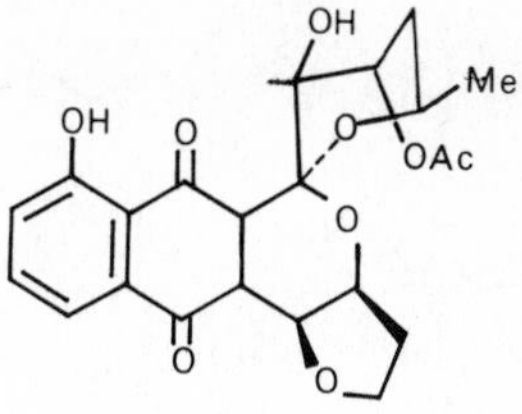

(143)

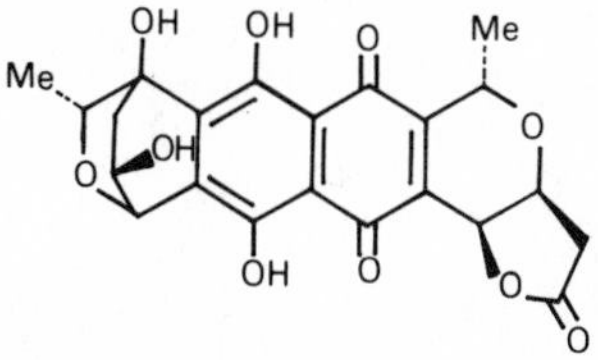

(144)

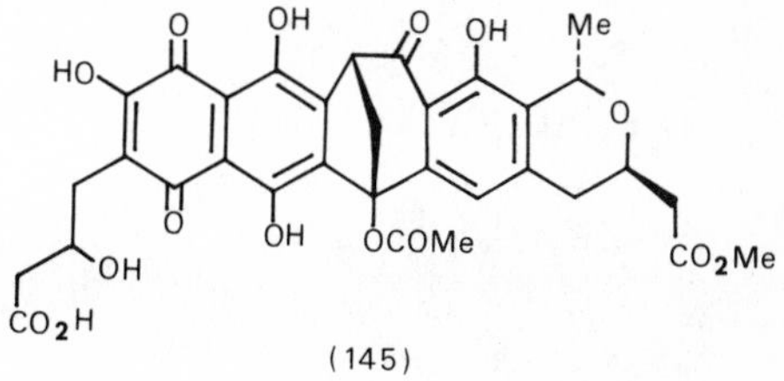

(145)

the naphthocyclinones, e.g. α-naphthocylinone (**145**), metabolites of *S. avenae*.[150] In these compounds the dimerization results in the modification of one of the isochroman quinone units to an aryl ketone and a two-carbon unit has been lost from the other in (**145**).

The unusual polychlorinated dimeric naphthoquinone (**146**) has been isolated from an unidentified soil fungus.[151] The naphthoquinone moiety in viomellein (**147**) and related metabolites was originally assigned an angular structure, but detailed [13]C n.m.r. studies and in particular the observation of a 3-bond [1]H-^{13}C coupling between the aromatic hydrogen and a quinonoid carbonyl indicated that the linear structure was correct.[152] Viomellein and related naphthalenoid dimers have been isolated from several fungi, *Aspergillus sulphureus*,[153] *A. melleus*, *Penicillium citreo-viride*,[154] *P. viridicatum*,[155] and *Microsporum cookei*.[156] The monomeric unit semi-vioxanthin (**148**) has been isolated from *P. citreo-viride*.[154] Xanthoviridicatum D (**149**), also from *P. viridicatum*,[157] has a semi-vioxanthin moiety linked to 2-methoxy-5-hydroxy-1,4-naphthoquinone.

Another synthesis of the spinochrome (**152**) from methoxynaphthazarin (**150**) has been described.[158] This uses (Scheme 21) a photo-Fries rearrangement of the diacetate (**151**) to introduce the acetyl moiety. 7-Methyljuglones bearing 4-hydroxy-5-methylcoumarin-3-yl units at C-2 and both C-2 and C-3, e.g. (**153**), have been isolated from *Diospyros ismailii*, and 2-methyljuglone with the coumarinyl unit at C-3 has been isolated from the bark of *Diospyros*

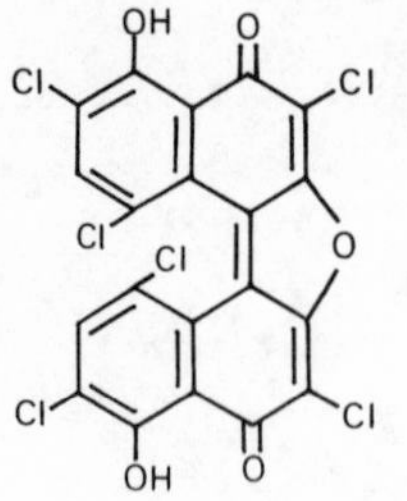

(146)

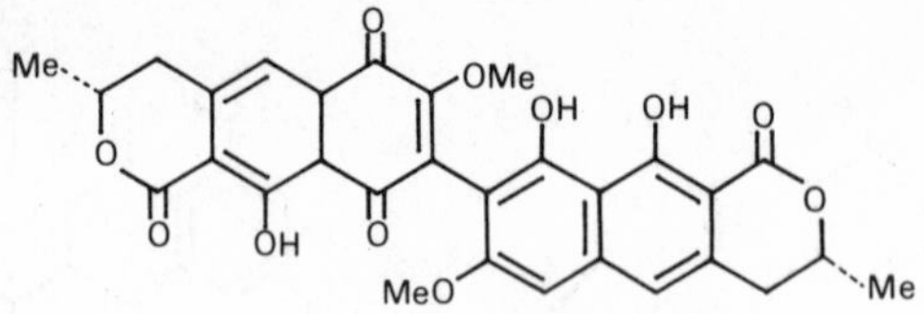

(147)

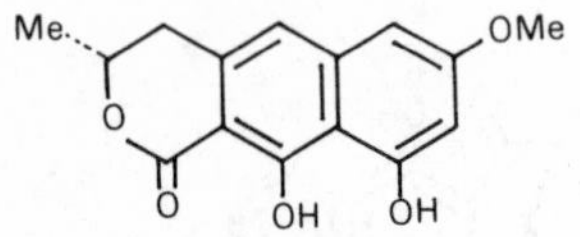

(148)

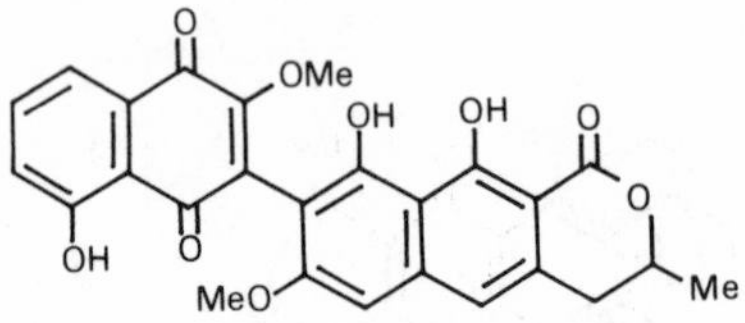

(149)

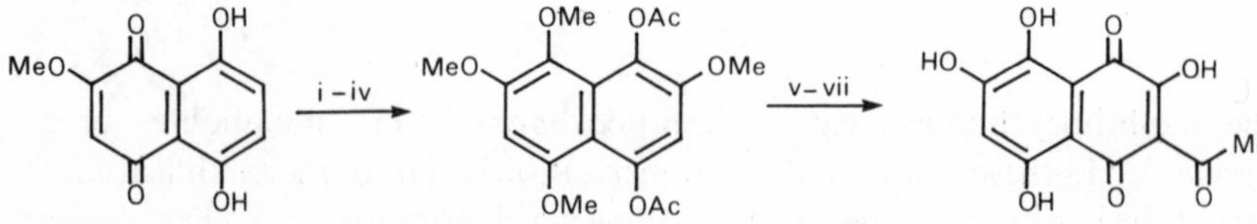

(150)　　　　　　　　　(151)　　　　　　　　(152)

Scheme 21　i, Pb(OAc)$_4$; ii, H$_2$SO$_4$; iii, MeI/Ag$_2$O; iv, Ac$_2$O/Zn; v, hν; vi, AlCl$_3$; vii, HCl/H$_2$O

---

canaliculata. The structures have been confirmed by synthesis.[159] The trypethelones, e.g. (**154**), are a group of 1,2-naphthoquinones isolated from the fungal mycosymbiont of the tropical lichen *Trypethelium eleuteria*.[160] Their structures suggest that they may be formed by degradation of the phenalenone, deoxyherqueinone.[210] Trichione (**155**, $n = 2$) is the main red pigment found in the fruiting bodies of the slime mould *Trichia floriformis,* whereas homotrichone (**155**, $n = 4$) is found in *Metatrichia vesperum.*[161]

Isohemigossypol (**157**) is a phytoalexin isolated from a cotton plant infected with *Verticillium dahliae.*[162] The asparvenones, e.g. (**158**) and (**159**), are tetralone metabolites of *Aspergillus parvulus.*[163] Systematic ^{13}C n.m.r. studies of substituted naphthalenes[164] and substituted naphthoquinones[165] have been reported. The effect of substituents on the chemical shifts of the quinonoid carbonyls and observation of long-range ^{1}H–^{13}C couplings provide valuable methods for obtaining the relative dispositions of substituents in the benzenoid and quinonoid rings.

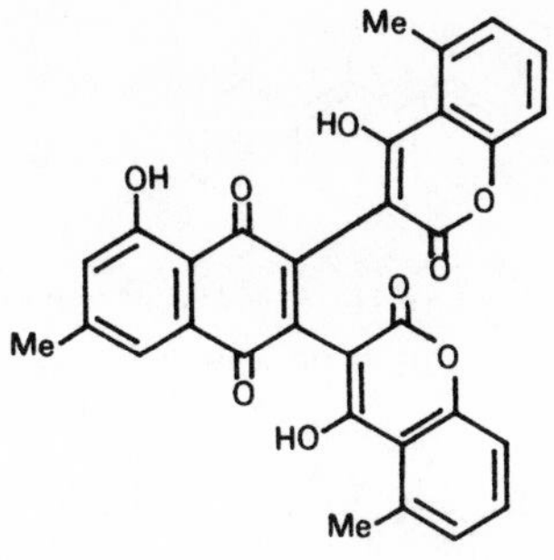

(153)

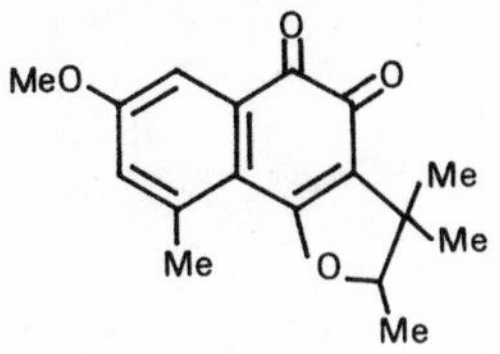

(154)

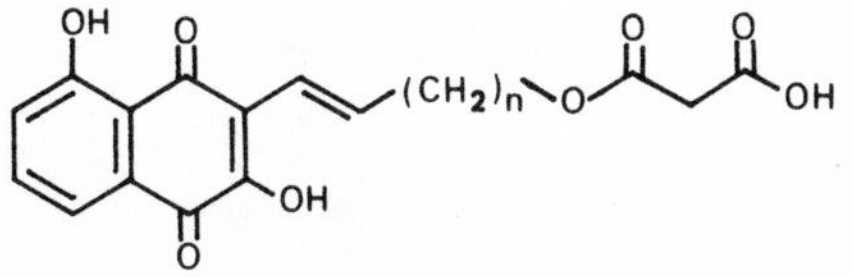

(155) n = 2
(156) n = 4

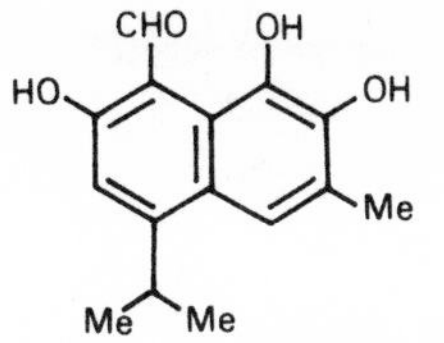

(157)

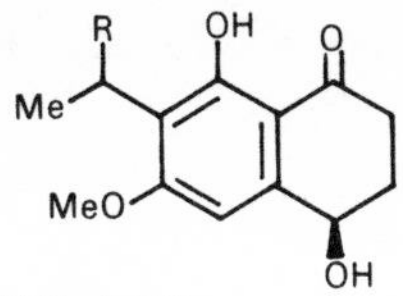

(158) R = H
(159) R = OMe

3.14 Anthraquinones and anthracenes

A number of prenylated anthracenes, e.g. ferruginin A (160), have been isolated from the berries of *Vismia* species. [166] ^{13}C n.m.r. was used to establish their structures and to study the keto-enol tautomerism of ring C. [166] The mycelium of *Aspergillus cristatus* produces the pigments viocristin (161) and isoviocristin (162). These are the first 1,4-anthraquinones to be found in nature. [167] Phomazarin is an aza-anthraquinone present in *Phoma terrestris*, the fungus responsible for 'pink root' disease of onions. Its structure has undergone many mutations over the years. Detailed ^{13}C n.m.r. studies finally established its structure as (163). [168] A feature of this study was the biosynthetic incorporation of ^{15}N and (1,2-^{13}C$_2$)acetate to enable ^{13}C–^{15}N, ^{13}C–^{13}C, in addition to ^{13}C–^{1}H couplings to be found, which greatly facilitated the structure determination. Note that the molecule exists in the pyridol rather than pyridone tautomeric system.

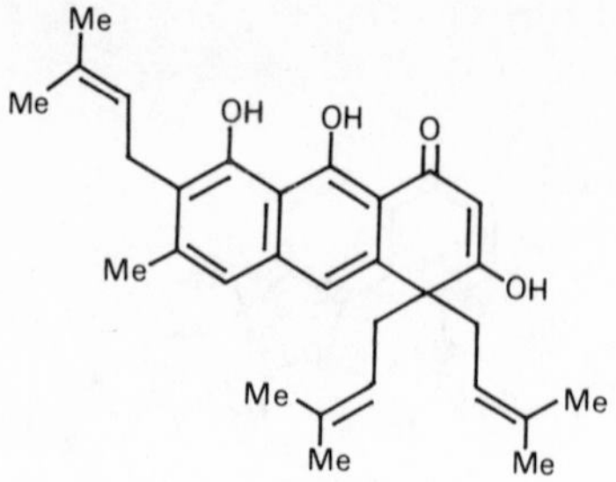

(161) R^I = Me, R^2 = H

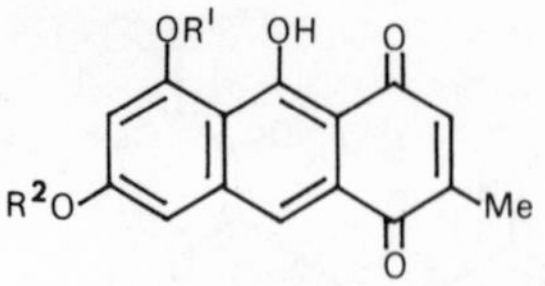

(162) R^I = H, R^2 = Me

(160)

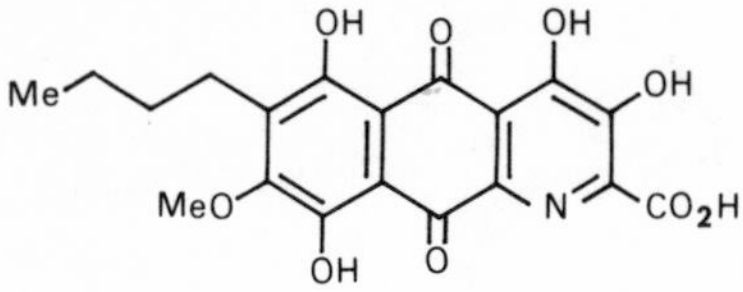

(163)

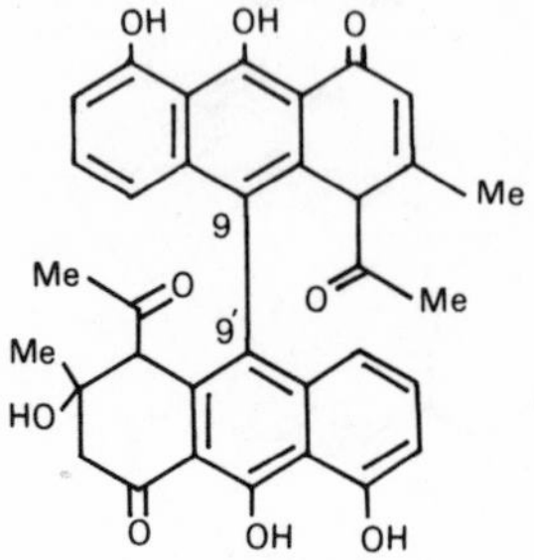

(169)

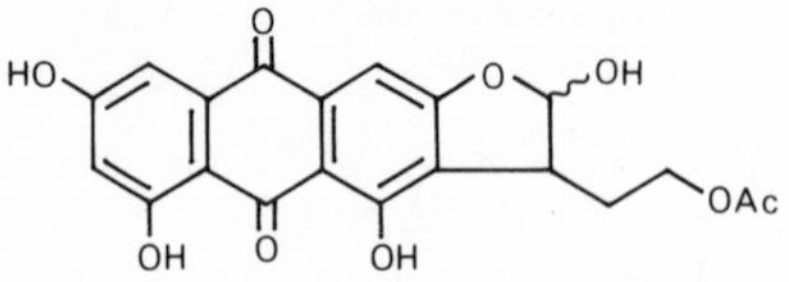

(164)

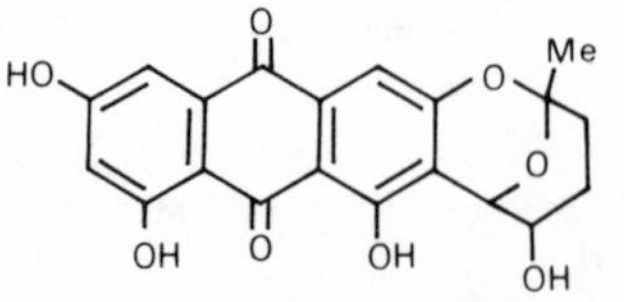

(165)

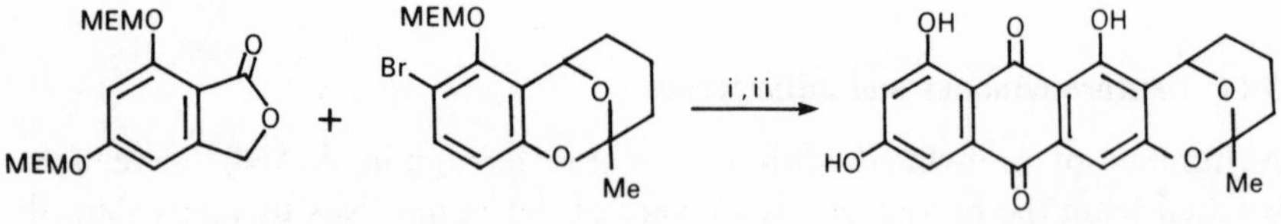

(166)   (167)   (168)

Scheme 22   i, lithium 2,2,6,6-tetramethylpiperidide/THF/ − 60°C; ii, H$_2$SO$_4$/AcOH/H$_2$O.

Treatment of aflatoxin-producing cultures of *Aspergillus parasiticus* with the insecticide dichlorvos inhibits the production of aflatoxin B$_1$ and causes accumulation of the anthraquinone versiconal acetate (**164**).[169] Nidurufin (**165**), which is also believed[170] to be on the biosynthetic pathway to aflatoxin B$_1$, has been isolated from *Aspergillus nidulans*,[171] *Dothistroma pini*,[172] and as the 6,8-di-*O*-methylether from *Aspergillus versicolor*,[173] though none of these fungi are themselves aflatoxin producers. Averufin (**168**) is another anthraquinoid precursor of the aflatoxins. Townsend has developed a synthesis of averufin (Scheme 22) in which the key steps are (i) the regiospecific coupling of the anion of phthalide (166) and the benzyne derived from the aryl bromide (167) to give the desired anthraquinone, and (ii) the use of methoxymethyl (phenol) protecting groups for regiospecific aryl metallation and the introduction of electrophiles to selectively elaborate simple oxygenated benzenoid precursors.[174] This synthesis has been used to prepare a variety of specifically (^{13}C and ^{2}H) labelled averufins which have been used for biosynthetic studies.[175]

Setomimycin (169) is a novel 9,9'-bianthryl antibiotic isolated from *Streptomyces pseudovenezuelae.* Its structure was deduced largely by ^{13}C n.m.r. and established by X-ray analysis. An interesting feature is the non-aromaticity of the cyclohexadienone ring, which presumably prefers to have C-1 sp^3 hybridized for steric reasons.[176]

Much effort has gone into developing new routes to anthraquinones, stimulated in part by their close structural relationship to the anthracyclinones (see below). Reaction of 1,1-dimethoxyethene with stypandrone (170) (Scheme 23) gave the 1:2 addition product which on deprotection gave the insect pigment acetylemodin (171).[177] Similar reaction with juglone gave a

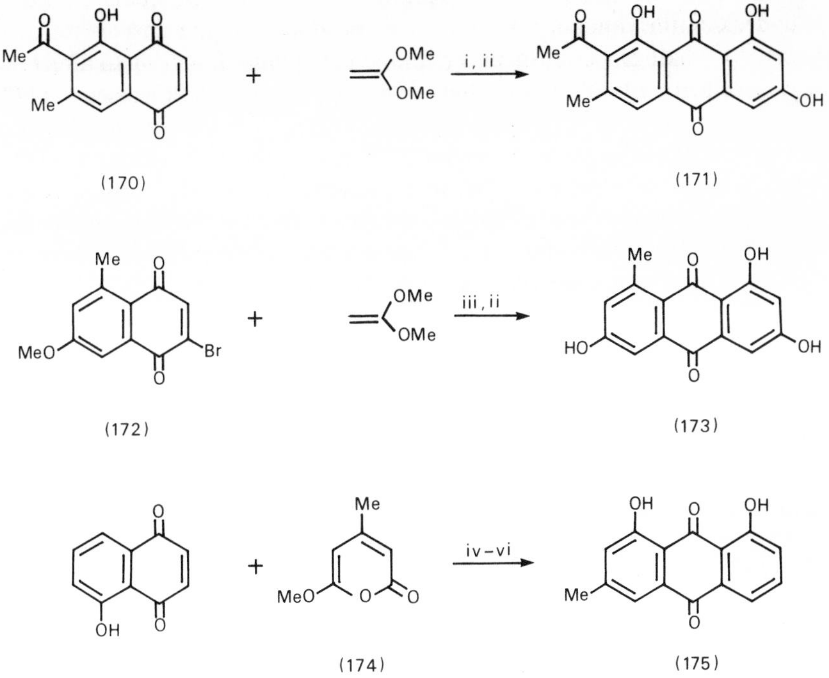

(170)   (171)

(172)   (173)

(174)   (175)

Scheme 23   i, DMSO;   ii, AlCl$_3$/NaCl;   iii, DMF;   iv, PhMe/heat;   v, Ag$_2$O/MgSO$_4$; vi, HBr/HOAc.

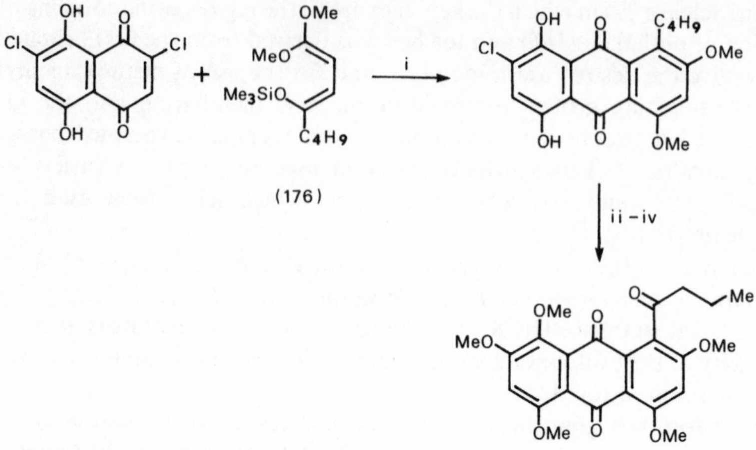

(176)

ii − iv

(177)

Scheme 24   i, C_6H_6/heat; ii, NaOMe/CuI/MeOH/DMF; iii, Me_2SO_4/K_2CO_3; iv, O_2/EtOH/hv

good yield of emodin.[178] It appears that the presence of a 5-hydroxyl in these naphthoquinones has a strong directing effect so that the required 1,3,8-trihydroxyanthraquinones are the main products. In the absence of a 5-hydroxyl, a halogen substituent is required in the quinonoid ring to direct the regiospecificity of addition. Thus synthesis of deoxyerythrolaccin (173) required the use of the bromo-quinone (172).[179] Jung has developed a convenient synthesis of chrysophanol (175) by reaction of the readily available 6-methoxy-3-methylpyrone (174) with juglone.[180] A further cycloaddition approach to the synthesis of anthraquinones uses 1,1-dimethoxybutadienes, e.g. (176), and naphthoquinones. Thus the permethylated crinoid pigment (177) has been synthesized by the route shown in Scheme 24.[181]

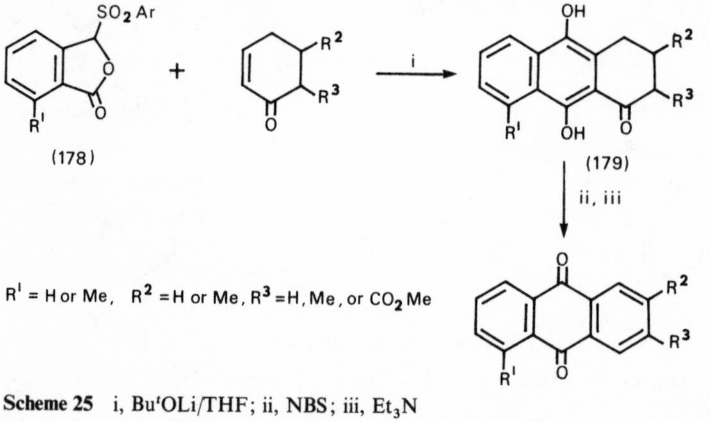

(178)   (179)

ii, iii

R^1 = H or Me, R^2 = H or Me, R^3 = H, Me, or CO_2Me

Scheme 25   i, Bu^{t}OLi/THF; ii, NBS; iii, Et_3N

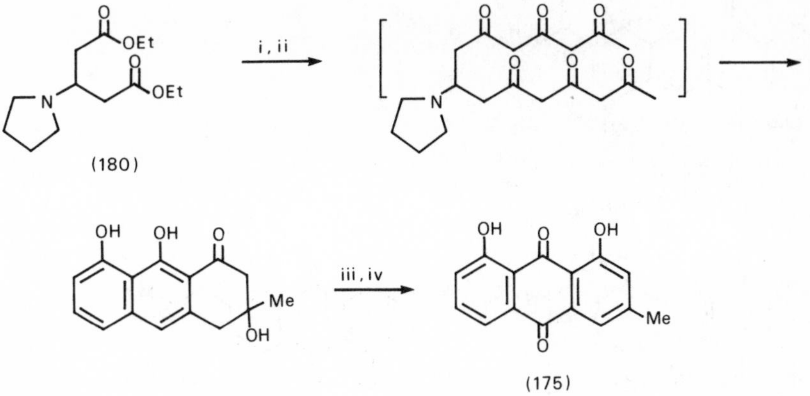

(180)

(175)

Scheme 26   i, CH_2=C(OLi)CH=C(OLi)Me; ii, NaOH; iii, HCl/HOAc; iv, CrO_3/HOAc.

The tetrahydroanthracenones (179) prepared by the addition of the anions of the sulphones (178) to substituted cyclohexenones (Scheme 25) can be aromatized and oxidized to give a simple route to 1-hydroxy-anthraquinones.[182] Harris has reviewed the use of masked polyketide synthons in the biomimetic synthesis of a wide range of aromatic compounds.[183] The approach is exemplified by the synthesis of chrysophanol (175) from the glutarate derivative (180) as shown in Scheme 26.[184]

3.15   Anthracyclinones

One of the main areas of activity in natural products chemistry in the past decade has been the study of the anthracycline group of antitumour antibiotics. They are all glycosides of tetracyclic anthraquinones and are produced by a variety of actinomycetes.[185] In addition to isolation, structure determination and biosynthetic studies,[186,187] there has been an enormous amount of work done on the synthesis of these compounds, particularly to find analogues which retain the desired antitumour activity but have lower toxicity to non-cancerous mammalian cells. Four main aglycones are known—daunomycinone (181), its 14-hydroxy derivative adriamycinone (182), aklavinone (183), and steffimycinone (184). The recently isolated 11-deoxydaunomycin[188] and aclacinomycin[189] have particularly low toxicity and this has prompted several recent syntheses of the corresponding aglycones, 11-deoxydaunomycinone[190] and aklavinone.[191] Other earlier work on the synthesis of the anthracyclinones has been reviewed,[185,192] and some more recent approaches are described below.

In Hauser's approach (Scheme 27) to 7,9-dideoxy daunomycinone (190), the anion of sulphone (185) is condensed with 2-ethoxybutenolide leading to (186)

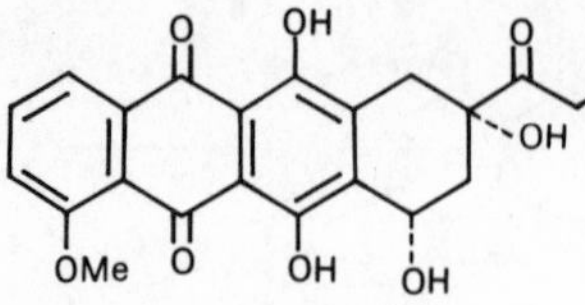

(181) R = H
(182) R = OH

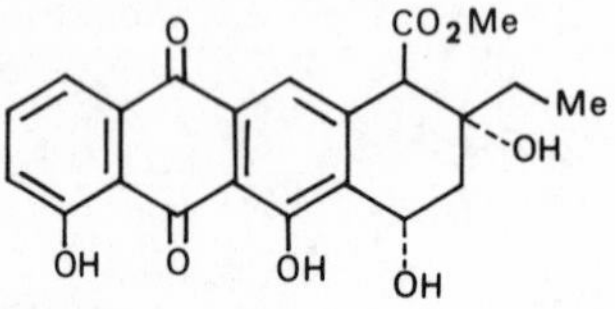

(183)

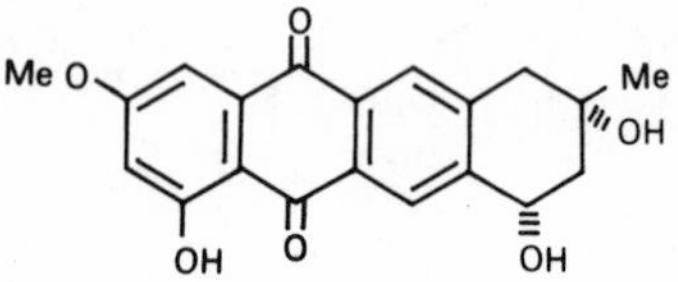

(184)

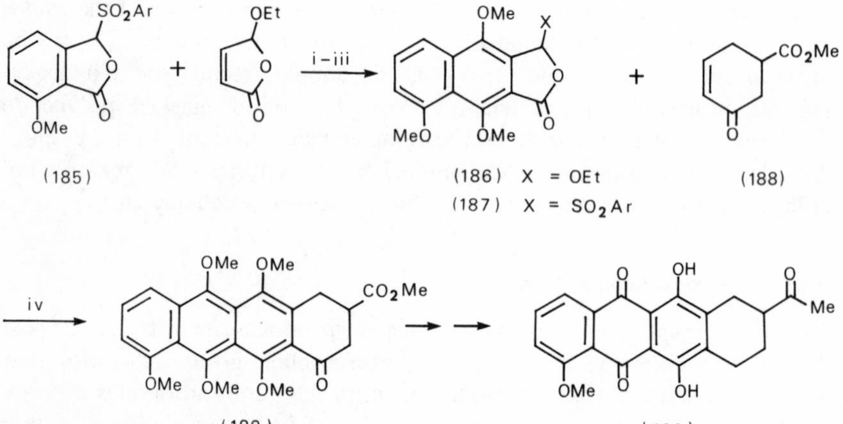

(185)

(186) X = OEt
(187) X = SO$_2$Ar

(188)

(189)

(190)

Scheme 27   i, LDA/THF/ − 78°C; ii, TsOH, PhSH; iii, MCPBA; iv, Bu^{t}OLi/THF.

which is converted to the sulphone (187). The anion of (187) is then condensed with the cyclohexanone (188) to give the heterocyclic intermediate (189) which is converted using established chemistry to (190).[193]

Most other approaches utilize cycloaddition reactions in the key steps. A flexible route which gives daunomycinone, adriamycinone, and their 6-deoxy analogues, uses the diethoxycyclobutene (191) prepared as shown in Scheme 28. On heating (191) with 5-methoxy-1,4-naphthoquinone, the tetracyclic ketone (192) is formed and this again is converted to the anthracyclinones using standard methodology.[194] In a related approach to the synthesis of 11-deoxydaunomycinone the diene (193) was prepared as shown in Scheme 29

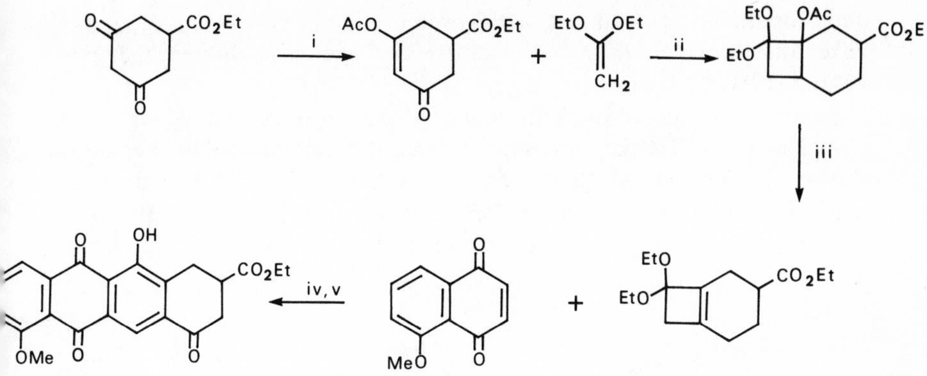

(192)

(191)

Scheme 28 i, Ac$_2$O; ii, hν; iii, LiOEt; iv, heat; v, NaOEt/O$_2$.

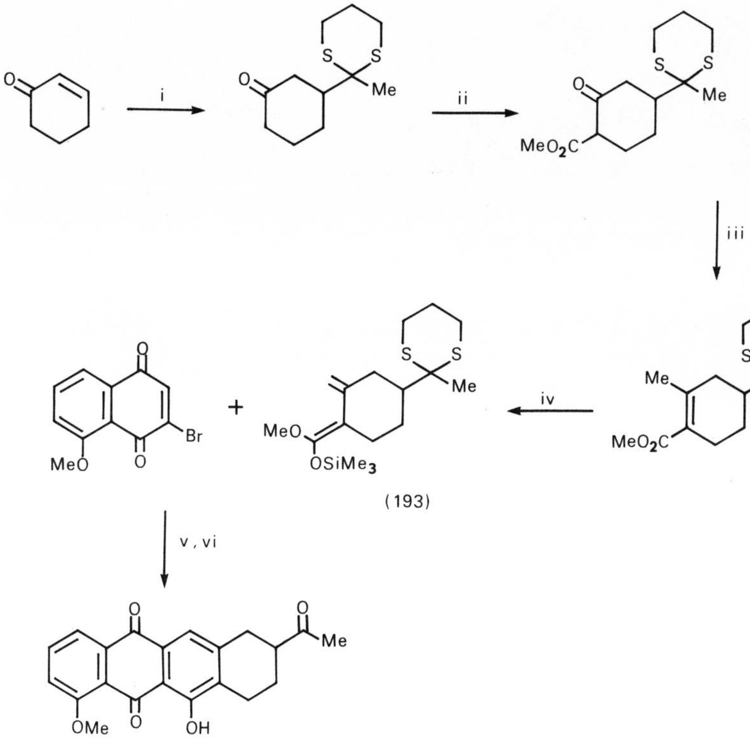

(193)

(194)

Scheme 29 i, Me(Li)CS(CH$_2$)$_3$S/THF/HMPA; ii, CO(OMe)$_2$/NaOH/C$_6$H$_6$; iii, NaH/THF/(EtO)$_2$POCl/Me$_2$CuLi; iv, LDA/Me$_3$SiCl; v, THF/20°C; vi, HgCl$_2$/CaCO$_3$/MeCN/H$_2$O.

and subsequent cycloaddition with 3-bromo-5-methoxy-1,4-naphthoquinone gave, after acidic work-up and oxidative cleavage of the dithian, the known ketone (194).[195]

In an extension of previous work, Kishi has prepared optically pure (+)aklavinone. The key intermediate was the aldehyde (199) prepared as shown in Scheme 30. Cycloaddition between 3-bromo-5-methoxy-1,4-naphthoquinone and the triene (195) led to the anthraquinone (196) which was converted to the allyl derivative (197). Palladium-catalysed cyclization

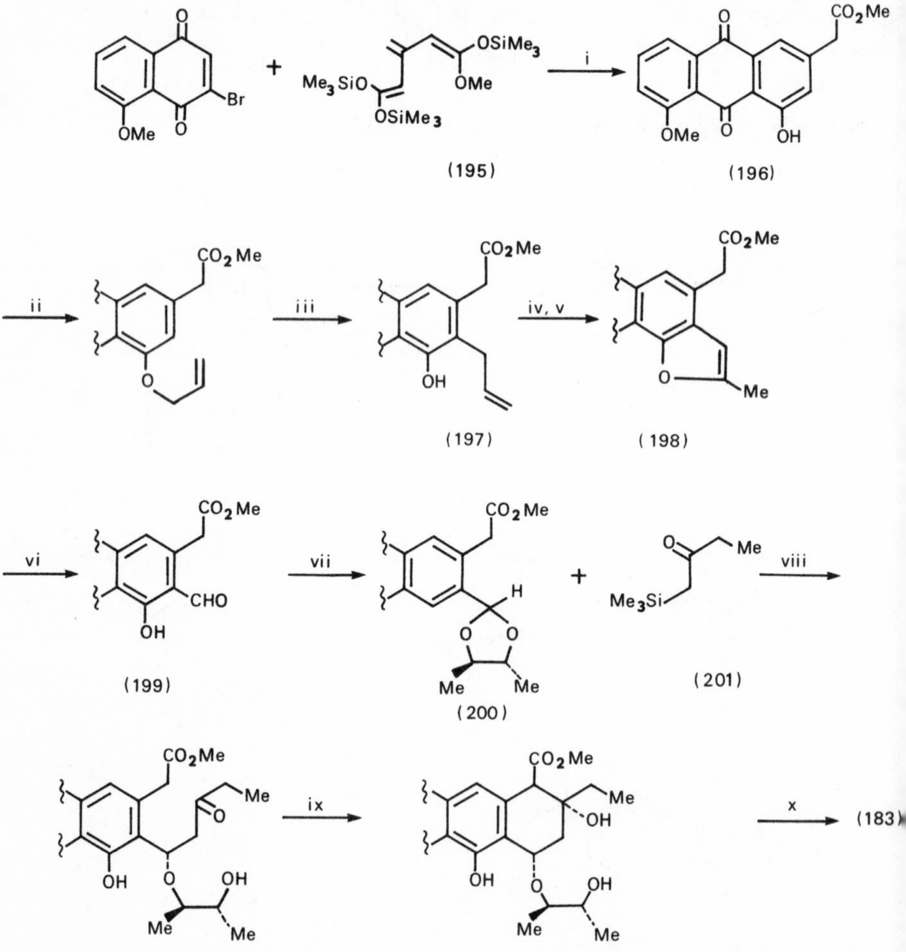

Scheme 30 i, $Et_2O/CH_2Cl_2/-40°C$; ii, $CH_2{=}CHCH_2Br/Ag_2O$; iii, $o\text{-}C_6H_4Cl_2$/reflux; iv, $PdCl_2(C_6H_5CN)_2/C_6H_6$/reflux; v, BCl_3; vi, $O_3/CH_2Cl_2/-78°C$; vii, $D(-)$-2,3-butandiol/TsOH; viii, $SnCl_4/MeCN/-20°C$; ix, $K_2CO_3/MeOH$; x, $CF_3CO_2H/-78°C$.

gave the benzofuran (198) which was converted by ozonolysis to aldehyde (199). This yielded the optically active acetal (200) with D(−) butan-2,3-diol. Condensation with the silylketone (201) followed by further cyclization and deprotection gave (+)aklavinone (183).[196]

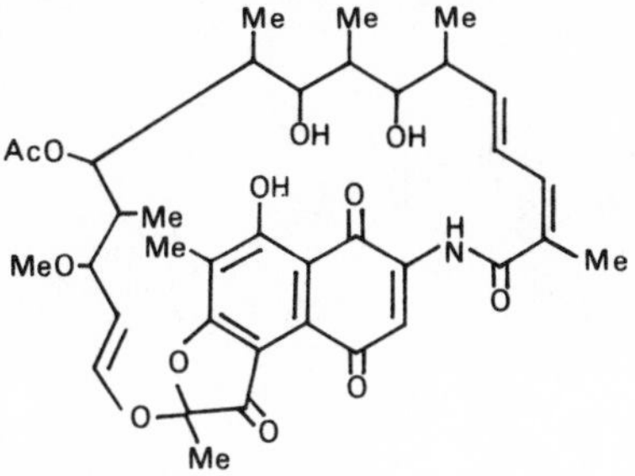

(202)

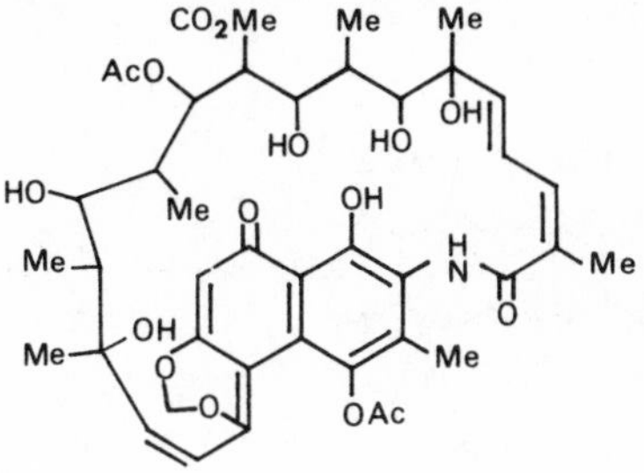

(203)

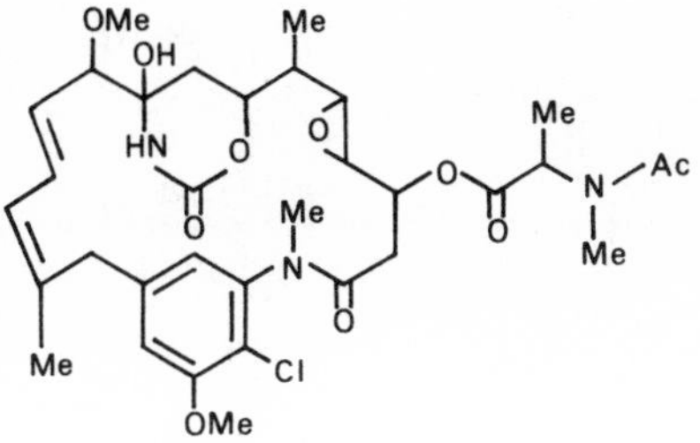

(204)

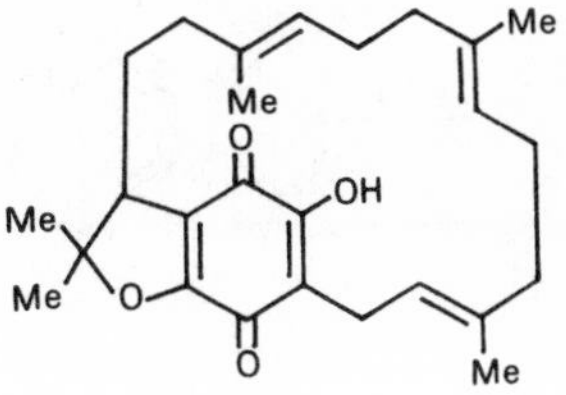

(205)

3.16 Ansamycins

Another major area involving aromatic compounds has been the study of the ansamycin antibiotics which display important antibacterial, antiviral, and antitumour activites.[198] They are characterized by having either a benzenoid or a naphthalenoid chromophore spanned by a long (polyketide-derived) aliphatic bridge which usually contains an amide linkage. The major groups are the rifamycins, e.g. (202) produced by *Nocardia mediterranei*, the streptovaricins, e.g. (203) produced by *Streptomyces spectabilis*, and naphthomycin, produced by *Streptomyces collinus*. These all contain naphthalenoid chromophores.[198] However, geldanamycin, produced by *Streptomyces hygroscopicus*,[198] and the recently isolated mycotrienins (*Streptomyces rishiniensis*),[199] macebins (a *Nocardia* species),[200] and ansatrienins (*Streptomyces collinus*)[201] all contain a benzoquinone or benzenoid chromophore. Perhaps the most interesting of all are the maytansinoids, e.g. (204). These were first isolated from the higher plants *Maytenus ovatus, M. buchanii*,

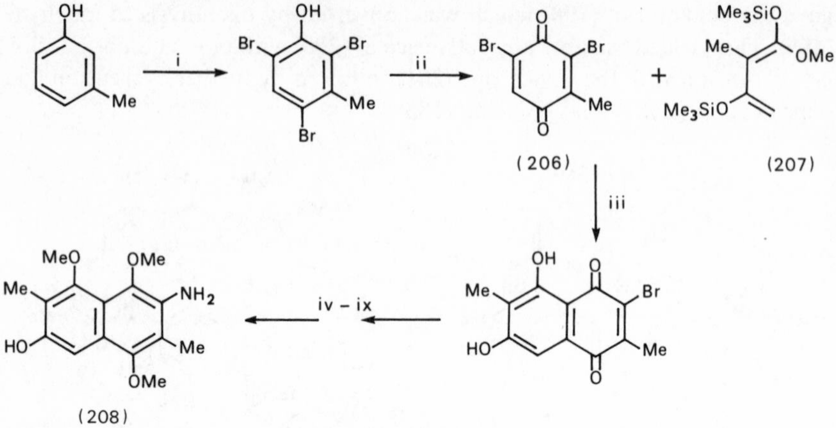

(206)   (207)

Scheme 31 i, Br$_2$/HOAc; ii, CrO$_3$/HOAc; iii, heat; iv, CH$_2$(OMe)$_2$/P$_2$O$_5$; v, NaBH$_4$/CeCl$_3$; vi, Me$_2$SO$_4$/NaOH; vii, Bu^{n}Li/THF/ MgBr$_2$; viii, PhSCH$_2$N$_3$; ix, KOH.

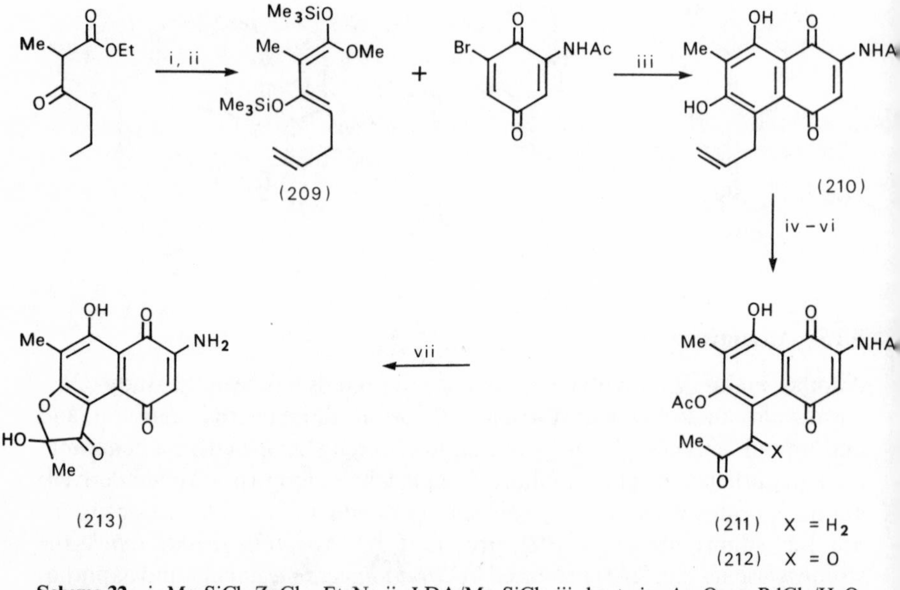

(209)   (210)

(213)

(211) X = H$_2$
(212) X = O

Scheme 32 i, Me$_3$SiCl, ZnCl$_2$, Et$_3$N; ii, LDA/Me$_3$SiCl; iii, heat; iv, Ac$_2$O; v, PdCl$_2$/H$_2$O; vi, SeO$_2$; vii, H$^+$/H$_2$O

and *Colubrina texensis*,[198] but have recently been isolated from a micro-organism, a *Nocardia* species,[202] which suggests that these compounds may not be true plant metabolites but are produced by symbiotic micro-organisms. Tridentoquinone **(205)**, produced by the fungus *Suillius tridentinus*, has an ansa-type chain of terpenoid origin.[203]

A five-step synthesis of the naphthalene nucleus (**208**) of streptovaricin D uses phenylthiomethyl azide for the introduction of the amino group (Scheme 31). The naphthalene system is built up using a cycloaddition between the benzoquinone (**206**) and 1,3-bis(trimethylsilyloxy)-methoxy-2-methyl-butadiene (**207**).[204] A similar regiospecific cycloaddition-based synthesis of the naphthofuranone chromophore of the rifamycins has been described (Scheme 32). In this the trimethylsilyl ether (**209**) is reacted with 2-bromo-6-acetamido-1,4-benzoquinone. Palladium-catalysed conversion of (**210**), derived from the adduct, to the methyl ketone (**211**) and selenium dioxide gave (**212**) which on hydrolysis produced (**213**).[205] A total synthesis of rifamycin has been developed[206] but this used a naphthalene precursor derived from the natural product, and a total synthesis of maytansine (**204**) has been described.[207]

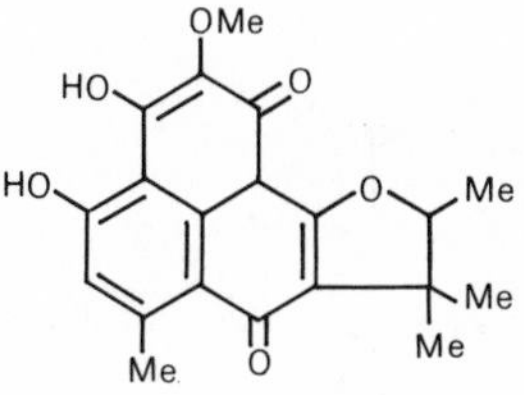

(214)

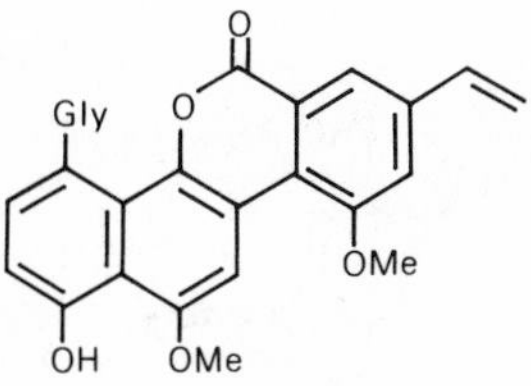

(215)

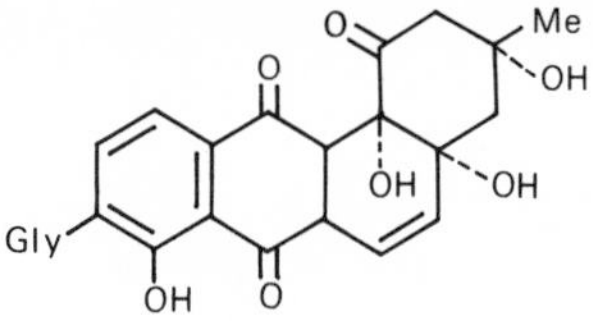

(216)

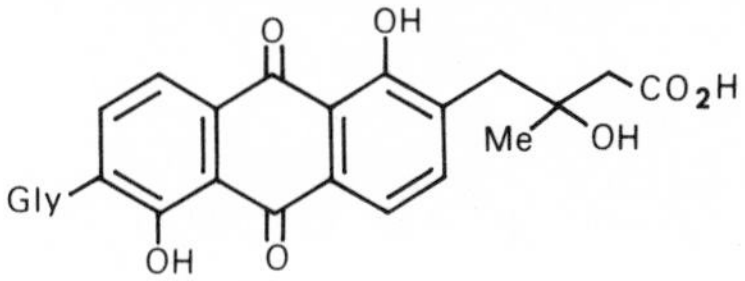

(217)

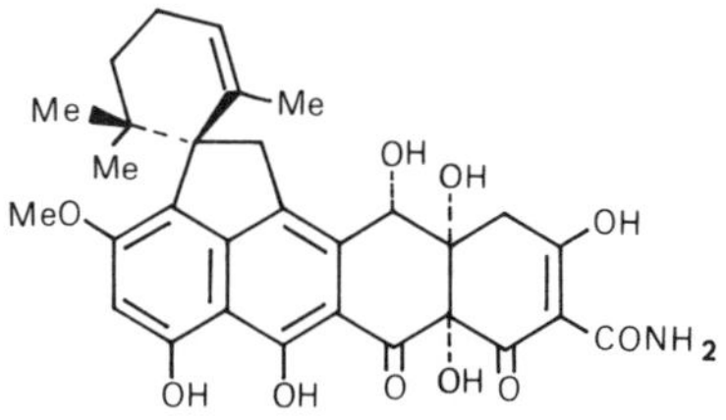

(218)

3.17 Some other polycyclic antibiotics

Phenalenones can be isolated from many fungal and plant sources and a comprehensive review of the area has appeared.[208] Herqueichrysin (**214**), a metabolite of *Penicillium herquei*, has the dihydrofuran ring in the opposite orientation to the other members of the group. Its structure has been established by synthesis[209] and detailed ^{13}C n.m.r. studies which provided information on the tautomerism occurring in these systems.

Ravidomycin (215), recently isolated from *Streptomyces ravidus*,[211] represents the newest member of the chrysomycin family of antitumour antibiotics.[212] These compounds have now been found in a number of organisms. They are apprently formed by cleavage of a benzphenanthrene precursor.[213] This skeleton is found intact in vineomycin A$_1$ (216) produced by *Streptomyces matensis*.[214] Its co-metabolite vineomycin B$_2$ (217) is presumably formed by an alternative cleavage pathway.

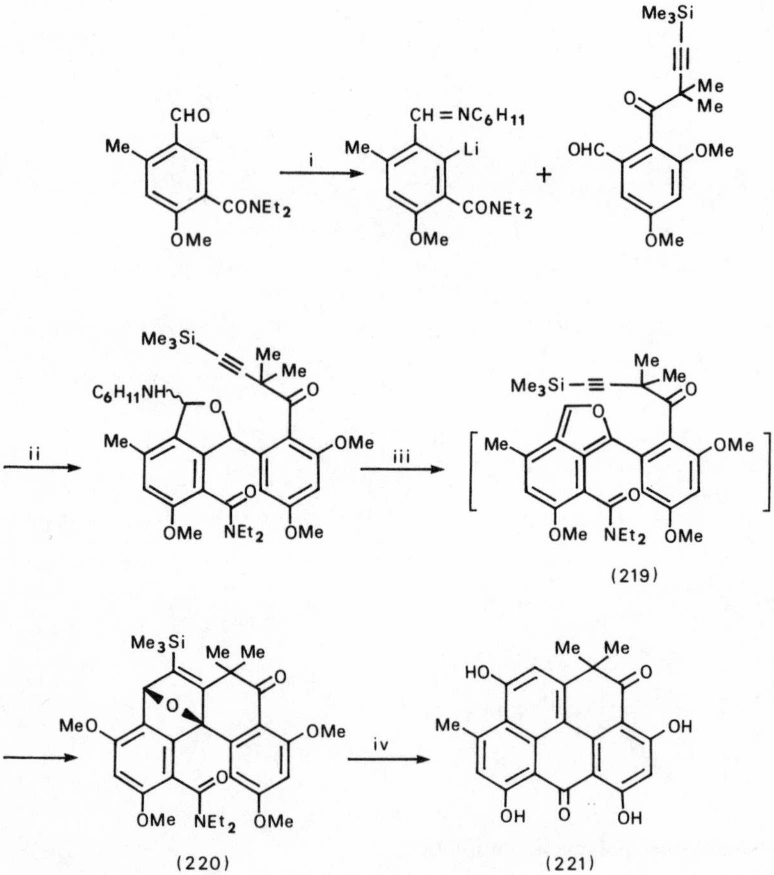

(219)

(220)   (221)

Scheme 33   i, cyclohexylamine; ii, BuLi/TMCDA/THF; iii, ICH$_2$CO$_2$H/C$_5$H$_5$N/C$_6$H$_6$/heat; iv, C$_5$H$_5$N.HCl/heat.

(222)

Viridicatum toxin (218) is a mycotoxin produced by the fungus *Penicillium viridicatum*.[215] It has a structure closely related to that of the tetracycline antibiotics. Resistomycin (221) is a polycyclic antibiotic produced by *Streptomyces griseoflavus*. It has been synthesized for the first time by an intramolecular cycloaddition of the isobenzofuran (219) which was itself synthesized in masked form as shown in Scheme 33. The Diels-Alder adduct (220) was converted to (221) in high yield in a remarkable one-pot desilylation, aromatization, demethylation and cyclization sequence using pyridinium hydrochloride.[216]

Ristocetin is a glycopeptide elaborated by *Nocardia lurida*. It is a member of the vancomycin class of antibiotics which inhibit bacterial cell wall synthesis.[217] The structure of the aglycone has been established as (222), mainly by detailed n.m.r.[218] and degradative studies.[219] Containing as it does a biphenyl, a diphenyl ether and a phenoxydiphenyl ether it is perhaps one of the most complex 'aromatic' natural products to be isolated so far!

References

1. For comprehensive reviews, see 'Biosynthesis' (Specialist Periodical Reports), The Chemical Society, London, Vols 1–7. Several relevant books have appeared: 'Biosynthesis of Natural Products', P. Mannitto, Ellis-Horwood, Chichester, 1981. 'The Biosynthesis of Secondary Metabolites', R.B. Herbert, Chapman and Hall, London, 1981; 'Secondary Metabolism', J. Mann, Clarendon Press, Oxford, 1980; 'Natural Product Chemistry', K.H.G. Torssell, Wiley, Chichester, 1983.
2. J.B. Stothers, 'Carbon-13 NMR Spectroscopy', Academic Press, New York, 1972; F.W. Wehric, and T. Nishida, 1979, *Proc. Chem. Org. Nat. Prod.*, 36, 1; H. Gunther, 'N.M.R. Spectroscopy', Wiley, Chichester, 1980.
3. R. Benn, and H. Günther, 1983, *Angew. Chem.*, 22, 350.
4. T.J. Simpson, 1975, *Chem. Soc. Revs.*, 4, 497; M.J. Garson, and J. Staunton, 1979, *Chem. Soc. Revs.*, 8, 539; J.C. Vederas, 1982, *Can. J. Chem.*, 60, 1637.
5. D.J. Faulkner, 1977, *Tetrahedron*, 33, 1421; 'Marine Natural Products', Vols. I to IV., P.J. Scheur, ed., Academic Press, New York, 1972.
6. G.E. Krejcarek, R.H. White, L.P. Hager, W.O. McClure, R.D. Johnson, Jr. K.L. Rinehart, Jr. P.D. Shaw, and R.C. Brusca, 1975, *Tetrahedron Lett.*, 508.

7. K. Kurata, and T. Amiya, 1975, *Nippon Suisan Gakkaishi*, 41, 657, (*Chem. Abs*, 1975, 83, 144 489).
8. T. Irie, M. Susuki, E. Kurosawa, and T. Masamune, 1970, *Tetrahedron*, 26, 3271.
9. J.J. Sims, M.S. Donnell, J.V. Leavy, and G.H. Lacy, 1975, *Antimicrobial Agents and Chemotherapy*, 7, 320.
10. T. Suzuki, M. Suzuki, and E. Kurosawa, 1975, *Tetrahedron Lett.*, 3057.
11. W.B. Turner, and D.C. Aldridge, 1983, 'Fungal Metabolites II', Academic Press, London.
12. Y. Solberg, 1975, *Acta Chem. Scand., B29*, 145.
13. J.F. Grove, 1972, *J. Chem. Soc. Perkin Trans. 1*, 2406.
14. L. Canonica, W. Kroszcynski, B.M. Ranzi, B. Rindone, and C. Scolastico, 1971, *J. Chem. Soc. Chem. Comm.*, 257.
15. S. Arrichio, A. Ricca, and O.V. de Pava, 1983, *J. Org. Chem.*, 48, 602.
16. M. Nukina, and S. Marumo, 1977, *Agric. Biol. Chem.*, 41, 717.
17. J. Geigert, F.R. Stermitz, and H.A. Schroeder, 1973, *Tetrahedron*, 29, 2343.
18. S. Iwasaki, H. Muro, Sasaki, S. Nozol, and S. Okuda, 1973, *Tetrahedron Lett.*, 3537.
19. A.W. Dunn, and R.A.W. Johnstone, 1979, *J. Chem. Soc. Perkin Trans 1*, 2122.
20. U. Amies, R. Currenti, G. Oriente, M. Piattelli and G. Trigali, 1981, *Phytochemistry*, 20, 1457; G. Tringali, and M. Piatelli, 1982, *Tetrahedron Lett.*, 1509'
21. M.C. Lunel, J. Favre-Bonvin, J. Bernillon, and N. Arpin, 1982, *Tetrahedron*, 38, 1235.
22. S.M. Johnson, J. Herrin, S.J. Liu, and I.C. Paul, 1970, *J. Amer. Chem. Soc.*, 92, 4428.
23. J.W. Westley, 1981, in 'Antibiotics IV', (J.W. Corcoran, ed.) Springer-Verlag, Heidelberg, p. 41.
24. J.H.P. Tyman, 1979, *Chem. Soc. Revs.*, 8, 499.
25. H. Achenbach, A. Mühlenfeld, B. Weber, and G.U. Brillinger, 1982, *Tetrahedron Lett.*, 4659.
26. H.G. Cutler, F.G. Crumley, J.P. Springer, R.H. Cox, R.J. Cole, J.W. Dorner, and J.H. Thean, 1980, *J. Agric. Food Chem.*, 28, 989.
27. A.F. Orr, 1979, *J. Chem. Soc. Chem. Comm.*, 40.
28. R.C. Ronald, J.M. Lansinger, T.S. Lillie, and C.J. Wheeler, 1982, *J. Org. Chem.*, 47, 2541.
29. Y. Kosuge, A. Suzuki, and S. Tamura, 1974, *Agric. Biol. Chem.*, 38, 1553.
30. Y. Kosuge, A. Suzuki, and S. Tamura, 1984, *Agric. Biol. Chem.*, 38, 1265.
31. Y. Nawata, K. Ando, G. Tamura, K. Arima, and Y. Haka, 1969, *J. Antibiotics*, 22, 511.
32. K. Mori, and K. Sato, 1982, *Tetrahedron*, 38, 1221.
33. K. Mori, and T. Funjioka, 1982, *Tetrahedron Lett.*, 5443.
34. A.E. Guthrie, J.E. Semple, and M.M. Joullie, 1982, *J. Org. Chem.*, 47, 2369; K.M. Chen, and M.M. Joullie, 1982, *Tetrahedron Lett.*, 4567.
35. H. Kaise, M. Shinohara, W. Miyazaki, T. Izawa, Y. Wakano, M. Sugawara, K. Saguira, and K. Sasaki, 1979, *J. Chem. Soc. Chem. Comm.*, 726.
36. R.M. Horak, P.S. Steyn, P.H. Van Rooyen, R. Vleggaar, and C.J. Rabie, 1981, *J. Chem. Soc. Chem. Comm.*, 1265.
37. L. Minale, R. Riccio, and G. Sodano, 1974, *Tetrahedron Lett.*, 3401.
38. W. Fenical, J.J. Sims, D. Squatrito, R.M. Wing, and P. Radlick, 1973, *J. Org. Chem.*, 38, 2383.
39. S.C. Welch, and A.S.C. Prakasa Rao, 1978, *J. Org. Chem.*, 43, 1957.
40. G. Cimino, P. de Luca, S. de Stefano, and L. Minale, 1975, *Tetrahedron*, 31, 271.
41. R.D.H. Murray, J. Méndez, and S.A. Brown, 1982, 'The Natural Coumarins,' Wiley, Chichester.
42. N.J. Cussans, and T.N. Huckerby, 1975, *Tetrahedron Lett.*, 2445; *Tetrahedron*, 31, 2587, 2591; H. Gunther, J. Prestian, and P. Joseph-Nathan, 1975, *Org. Magn. Resonance*, 7, 339; S.A. Sojka, 1975, *J. Org. Chem.*, 40, 1175.
43. P. Venturella, A. Bellino, and F. Piozzi, 1974, *Tetrahedron Lett.*, 979.
44. B.A. Nagasampagi, F.H. Stodola, L.J. Wickerham, J.J. Ellis, and W.K. Rohwedder, 1971, *Can. J. Chem.*, 49, 2029.
45. K.K. Chexal, C.J. Fouweather, and J.S.E. Holker, 1975, *J. Chem. Soc. Perkin Trans. 1*, 554.
46. G. Buchi, D.H. Klaubert, R.C. Shank, S.M. Weinreb, and G.N. Wogan, 1971, *J. Org. Chem.*, 36, 1143.

47. P.S. Steyn, R. Vleggaar, and P.L. Wessels, 1980, in 'The Biosynthesis of Mycotoxins', (P.S. Steyn, ed.) Academic Press, London, p. 105.
48. A.D. Campbell, 1979, *Pure and Appl. Chem.*, 52, 205.
49. A.C. Ciegler, 1975, *Lloydia*, 38, 21.
50. A.C. Beckwith, R.F. Vesonder and A. Ciegler, 1976, in 'Mycotoxins and Other Fungal Related Food Problems', (J.V. Rodricks, ed.) *Advances in Chemistry Series, Amer. Chem. Soc.*, p. 58.
51. G. Buchi, M.A. Francesco, J.M. Leisch, and P.F. Schuda, 1981, *J. Amer. Chem. Soc.*, 103, 3497.
52. Y. Shimojima, H. Hayashi, T. Ooka, M. Shibuhama, and Y. Litaka, 1982, *Tetrahedron Lett.*, 5434.
53. M.D. Northolt, H.P. van Egmond, and W.E. Paulsch, 1979, *J. Food Prod.*, 42, 485.
55. L.M. Horwood, 1982, *J. Chem. Soc. Chem. Comm.*, 1120.
56. R.A. Hill, P.H. Carter, and J. Staunton, 1981, *J. Chem. Soc. Perkin Trans. 1*, 2570.
57. G.B. Henderson, and R.A. Hill, 1981, *J. Chem. Soc. Perkin Trans. 1*, 3037.
58. G.B. Henderson, and R.A. Hill, 1982, *J. Chem. Soc. Perkin Trans. I*, 1111.
59. D.M.X. Donnelly, J. O'Reilly, J. Polonsky, and G.W. van Eijk, 1982, *Tetrahedron Lett.*,5451.
60. T.J. Simpson, 1978, *J. Chem. Soc. Chem. Comm.*, 627.
61. D.C. Aldridge, and W.B. Turner, 1970, *J. Chem. Soc. (C)*, 2598; D.T. Robeson, and G.A. Strobel, 1982, *Agric. Biol. Chem.*, 46, 2681; J.F. Grove, and M. Pople, 1979, *J. Chem. Soc. Perkin Trans. 1*, 2048.
62. S. Marukawa, S. Funakawa, and Y. Satomura, 1975, *Agric. Biol. Chem.*, 39, 645.
63. M.M. Chien, Jr. P.L. Schuff, D.J. Slatkin, and J.E. Knapp, 1977, *Lloydia*, 40, 301.
64. T.H. Chan, and P. Brownbridge, 1981, *J. Chem. Soc. Chem. Comm.*, 20.
65. W.J. McGahren, G.A. Ellestad, G.O. Morton, and M.P. Kunstmann, 1972, *J. Org. Chem.*, 37, 1636.
66. G.C. Crawley, and C.J. Strawson, unpublished results cited in ref. 11, p. 98.
67. C.R. McIntyre, and T.J. Simpson, 1984, *J. Chem. Soc. Chem. Comm.*, 704.
68. Ch. Tamm, B. Böhner, and W. Zürcher, 1972, *Helv. Chem. Acta.*, 55, 510.
69. G. Assante, L. Camarda, L. Merlini, and G. Nasini, 1979, *Phytochemistry*, 18, 311.
70. G.P. Ellis, (ed.), 1977, 'Chromenes, Chromones, and Chromanones', Wiley, New York.
71. L. Crombie, and W.M.L. Crombie, 1976, in 'Cannabis and Health', (J.D.P. Graham, ed.), Academic Press, London, p. 43.
72. A.I. Meyers, R. Gabel, and G.D. Mihelich, 1978, *J. Org. Chem.*, 43, 1372.
73. J. Novak, and C.A. Salemink, 1982, *Tetrahedron Lett.*, 253.
74. T.H. Chan, and T. Chaly, 1982, *Tetrahedron Lett.*, 2935.
75. C.J. Mirocha, C.M. Christensen, and G.H. Nelson, 1967, *Appl. Microbiol.*, 15, 497.
76. Ref. 11, p. 176.
77. T. Takahashi, T. Nagashima, and J. Tsuji, 1980, *Chem. Lett.*, 369.
78. T. Takahashi, H. Ikeda, and J. Tsuji, 1980, *Tetrahedron lett.*, 3885.
79. T. Takahaski, K. Kasuga, and J. Tsuji, 1978, *Tetrahedron Lett.*, 4917.
80. D.C. Aldridge, S. Galt, D. Giles and W.B. Turner, 1970, *J. Chem. Soc. (C)*, 539.
81. K.H. Lee, N. Hayashi, M. Okano, I.R. Hall, R.Y. Wu and A.T. McPhail, 1982, *Phytochemistry*, 21, 1119.
82. E. Suzuki, and S. Inone, 1976, *J. Chem. Soc. Perkin Trans. 1*, 404.
83. R. Cardillo, C. Fuganti, D. Ghiringhelli, D. Giangrano, P. Graselli, and A. Santopietro-Amisano, 1974, *Tetrahedron*, 30, 459.
84. K. Kakinuma, C.A. Hanson, and K.L. Rinehart, 1976, *Tetrahedron*, 32, 217.
85. R. Hansel, A. Pelter, J. Schulz, and C. Hille, 1976, *Chem. Ber.*, 109, 1617.
86. R. Reinhardt, and R. Hansel, 1977, *Z. Naturforsch*, 32c, 290.
87. A. Pelter, M.T. Ayoub, J. Schulz, R. Hansel, and D. Reinhardt, 1979, *Tetrahedron Lett.*, 1627.
88. A. Pelter, R. Al-Bayati, R. Hansel, H. Dinter, and B. Burke, 1981, *Tetrahedron lett.*, 1545.
89. A. Pelter, R. Al-Bayati, and W. Lewis, 1982, *Tetrahedron Lett.*, 353.
90. K.H. Lin, C.K. Yang, and F.T. Peng, 1979, *Appl. Environ. Microbiol.*, 37, 355.

91. S.M. Kupchan, R.W. Britton, M.F. Ziegler, C.J. Gilmore, R.J. Restino, and R.F. Bryan, 1973, *J. Amer. Chem. Soc.*, 95, 1335.
92. A.S. Kende, and L.S. Leibeskind, 1976, *J. Amer. Chem. Soc.*, 98, 267.
93. E.R. Larson, and R.A. Raphael, 1982, *J. Chem. Soc. Perkin Trans. 1*, 521.
94. S.R. Stitch, J.K. Toumba, M.B. Groen, C.W. Funke, J. Leemhuis, J. Vink, and G.F. Woods, 1980, *Nature*, 287, 738.
95. M.B. Groen, and J. Leemhuis, 1980, *Tetrahedron Lett.*, 5043.
96. A. Pelter, P. Satyanarayana, and R.S. Ward, 1981, *Tetrahedron Lett.*, 1549.
97. F.C. Chang, C.-K. Chiang, and V.N. Aiyer, 1975, *Phytochemistry*, 14, 1440.
98. K.M. Smith, 1982, *Chem. Soc. Revs.*, 11, 75.
99. B.A. McKittrick, R.T. Scannell, and R. Stevenson, 1982, *J. Chem. Soc. Perkin Trans. 1*, 2017.
100. K. Chamberlain, and R.A. Skipp, 1975, *Phytochemistry*, 14, 1843.
101. Ref. 11, p. 14.
102. L.H. Briggs, R.C. Cambie, I.C. Dean, R. Hodges, W.B. Ingram, and P.S. Rutledge, 1976, *Aust. J. Chem.*, 29, 179.
103. R.L. Edwards, and M. Gill, 1975, *J. Chem. Soc. Perkin Trans. 1*, 351.
104. P. Singh, and M. Anchel, 1971, *Phytochemistry*, 10, 3259.
105. N. Ojima, I. Takahashi, K. Ogura, and S. Seto, 1976, *Tetrahedron Lett.*, 1013.
106. G.A. Pattenden, 1978, *Progr. Chem. Org. Nat. Prod.*, 35, 133.
107. W. Steglich, H. Besl, and A. Prox, 1972, *Tetrahedron Lett.*, 4895.
108. R.L. Edwards, and M. Gill, 1973, *J. Chem. Soc. Perkin Trans. 1*, 1921.
109. W. Steglich, A. Thillmann, H. Besl, and A. Bresinsky, 1977, *Z. Naturforsch*, 32c, 46.
110. J. Gripenberg, 1974, *Tetrahedron Lett.*, 619.
111. J.B. Harborne, T.J. Mabry, and H. Mabry, 1975, 'The Flavanoids', Chapman and Hall, London, Vols, I and II.
112. T. Swain, J.B. Harborne, and C.F. van Sumere, (eds)., 1979, 'Recent Advances in Phytochemistry', Vol. 12, Plenum Press, New York.
113. P.K. Stumpf, and E.E. Lonn, (eds), 1981, 'The Biochemistry of Plants', Vol. 7, Academic Press, New York.
114. J.L. Ingham, 1983, *Prog. Chem. Org. Nat. Prod.*, 43, 1.
115. W. Heller, and Ch. Tamm, 1981, *Prog. Chem. Org. Nat. Prod.*, 40, 105.
116. O.R. Gottlieb, 1978, *Prog. Chem. Org. Nat. Prod.*, 35, 1.
117. D.G. Roux, and D. Ferreira, 1982, *Prog. Chem. Org. Nat. Prod.*, 41, 47.
118. E. Haslam, 1982, *Prog. Chem. Org. Nat. Prod.*, 41, 1.
119. S. Antus, E. Schindlbeck, S. Ahmad, O. Sehgmann, V.M. Chari, and H. Wagner, 1982, *Tetrahedron*, 38, 133.
120. I. Kuono, T. Komori, and T. Kawasaki, 1973, *Tetrahedron Lett.*, 4569.
121. W.H. Rawal, and M.P. Cava, 1982, *Tetrahedron Lett.*, 5581.
122. M.U.S. Sultanbawa, 1980, *Tetrahedron*, 36, 1465; M. Afzal, and J.M. Al-Hassan, 1980, *Heterocycles*, 14, 1173.
123. P.W. Westerman, S.P. Gunasekara, M.U.S. Sultanbawa, and R. Kazlauskas, 1977, *Org. Magnetic Res.*, 9, 631.
124. W.A. Ayer, and D.R. Taylor, 1976, *Can. J. Chem.*, 54, 1703.
125. K.K. Chexal, J.S.E. Holker, T.J. Simpson and K. Young, 1975, *J. Chem. Soc. Perkin Trans. 1*, 543.
126. K.K. Chexal, C. Fouweather, J.S.E. Holker, T.J. Simpson and K. Young, 1974, *J. Chem. Soc. Perkin Trans. 1*, 1584.
127. J.A. Ballantine, D.J. Francis, C.H. Hasall, and J.L.C. Wright, 1970, *J. Chem. Soc. (C)*, 1175.
128. M. Yamazaki, H. Fujimoto, Y. Ohta, Y. Iitaka, and A. Itai, 1981, *Heterocycles*, 15, 889.
129. J.S.E. Holker, R.D. Lapper, and T.J. Simpson, 1974, *J. Chem. Soc. Perkin Trans. 1*, 2135.
130. B. Franck, and H. Flasch, 1973, *Prog. Chem. Org. Nat. Prod.*, 30, 151.
131. C.C. Howard, and R.A.W. Johnstone, 1973, *J. Chem. Soc. Perkin Trans. 1*, 2440; I. Kurobane, L.C. Vining, and A.G. McInnes, 1978, *Tetrahedron Lett.*, 4633.

132. D.M. Yang, N. Takeda, Y. Iitaka, U. Sankawa, and S. Shibata, 1973, *Tetrahedron*, 29, 519.
133. W.B. Turner, 1978, *J. Chem. Soc. Perkin Trans. 1*, 1621.
134. B. Franck, 1980 in 'The Biosynthesis of Mycotoxins', (P.S. Steyn, ed.)., Academic Press, New York, p. 157.
135. J.W. Cornforth, G. Ryback, P.M. Robinson, and D. Park, 1971, *J. Chem. Soc. (C)*, 2786.
136. D.H.R. Barton, L. Cottier, K. Freund, F. Luini, P.D. Magnus, and I. Salazar, 1976, *J. Chem. Soc. Perkin Trans. 1*, 499.
137. L. Kopanski, M. Klaar, and W. Steglich, 1982, *Annalen*, 1280.
138. P.S. Steyn, and R. Vleggaar, 1974, *J. Chem. Soc. Perkin Trans. 1*, 2250.
139. S. Danishefsky, and F.J. Walker, 1979, *J. Amer. Chem. Soc.*, 101, 7018.
140. T. Higa, and P.J. Scheuer, 1975, *Tetrahedron*, 2379.
141. C.F. Culberson, 1969, 'Chemical and Botanical Guide to Lichen Products', University of N. Carolina Press, Chapel Hill; C.F. Culberson, 1970, 'Supplement to Chemical and Botanical Guide to Lichen Products', *Bryologist*, (1977), 73, 177; C.F. Culberson, W.L. Culberson, and A. Johnson, 'Second Supplement to Chemical and Botanical Guide to Lichen Products', The American Bryological and Lichenological Society, St. Louis.
142. T. Sala, and M.V. Sargent, 1981, *J. Chem. Soc. Perkin Trans. 1*, 855.
143. H.G. Floss, 1981, in 'Antibiotics IV', (J.W. Lorcoran, ed.), Springer-Verlag, Heidelberg, p. 215.
144. S. Omura, H. Tanaka, Y. Koyama, N. Oiwa, M. Katagiri, and T. Hata, 1974, *J. Antibiotics*, 27, 363.
145. N. Tsuji, M. Kobayashi, Y. Terui, and K. Tori, 1976, *Tetrahedron*, 32, 2207.
146. Kometani, T., Takeuchi, Y. and Yoshii, E., 1982, *J. Org. Chem.*, 47, 4725.
147. C.J. Chang, H.G. Floss, P. Soong, and C.T. Chang, 1975, *J. Antibiotics*, 28, 156.
148. A. Zeeck, and P. Christiansen, 1969, *Annalen*, 772; C.P. Gorst-Allman, B.A.M. Rudd, C.J. Chang, and H.G. Floss, 1981, *J. Org. Chem.*, 46, 455.
149. H. Brockmann, and P. Christiansen, 1970, *Chem. Ser.*, 103, 708.
150. A. Zeeck, H. Zähner, and M. Marden, 1974, *Annalen*, 1100.
151. D.W. Cameron, and M.D. Sidell, 1978, *Aust. J. Chem.*, 31, 1323.
152. G. Höfle, and K. Roser, 1978, *J. Chem. Soc. Chem. Comm.*, 611.
153. R.C. Durley, J. MacMillan, T.J. Simpson, A.T. Glen, and W.B. Turner, 1975, *J. Chem. Soc. Perkin Trans. 1*, 163.
154. A. Zeeck, P. Rub, H. Laatsch, W. Loeffler, H. Wehrle, H. Zähner, and H. Holst, 1979, *Chem. Ber.*, 112, 957.
155. M.E. Stack, R.M. Eppley, P.A. Dreifuss, and A.E. Pohland, 1977, *Appl. Environ. Microbiol.*, 33, 351.
156. T. Akita, K. Kawai, H. Shimonada, Y. Nozawa, Y. Ito, S. Nishibe, and Y. Ogihara, 1975, *Shinkin to Shikinsho*, 16, 177.
157. M.E. Stack, E.P. Mazzola, and R.M. Eppley, 1979, *Tetrahedron Lett.*, 4989.
158. F. Farina, R. Martinez-Utrilla, and M. Paredes, Carmen, 1982, *Tetrahedron*, 38, 1531.
159. J.A.D. Jeffreys, M. bin, Zakaria, P.G. Waterman, and S. Zhong, 1983, *Tetrahedron*, 1085.
160. A. Mathey, and W. Steglich, 1980, *Annalen*, 779.
161. L. Koranski, G.R. Li, H. Besl, and W. Steglich, 1982, *Annalen*, 1722.
162. A.S. Sandykov, L.V. Metlitskii, A.K. Karimdzhanov, A.I. Ismailov, R.A. Mukhamedova, M.K.H. Avazkhadzhaev, and F.G. Kamaev, 1974, *Doklady Akad. Nauk SSSR*, 218, 1472 (*Chem. Abs.*, 1975, 82, 82996).
163. P.D. Chao, P.L. Schuff, Jr. D.J. Slatkein, and J.E. Knapp, 1979, *J. Chem. Res. (M).*, 2685; (S) 236.
164. J. Serta, J. Sandstrom, and T. Drakemberg, 1978, *Org. Magnetic Res.*, 11, 329; W. Kitching, M. Bulpitt, D.M. Doddrell, and W. Adcock, 1974, *ibid.*, 6, 289; P.E. Hauser, 1979, *ibid.*, 12, 109.
165. G. Höfle, 1977, *Tetrahedron*, 33, 1963; G. Castillo, S.J. Ellames, A.G. Osborne, and P.G. Sammen, 1978, *J. Chem. Res. (S)*, 45, (M), 833; B.F. Bowden, D.W. Cameron, M.J. Crossley, S.I. Feutrill, P.G. Griffiths, D.P. Kelley, 1979, *Aust. J. Chem.*, 32, 769; I.G. MacDonald, A.F. Sierakowski, and T.J. Simpson, 1977, *Aust. J. Chem.*, 30, 1727.

152   THE CHEMISTRY OF NATURAL PRODUCTS

166. M. Nicoletti, G.B. Marini-Bettolo, F. Delle Monache, and G. Delle Monache, 1982, *Tetrahedron*, 38, 3686.
167. H. Laatsch, and H. Anke, 1982, *Annalen*, 2189.
168. A.J. Birch, D.N. Butler, R. Effenberger, R.W. Richards, and T.J. Simpson, 1979, *J. Chem. Soc. Perkin Trans. 1*, 807.
169. D.P.H. Hseih, R.C. Yao, D.L. Fitzell, and C.A. Reece, 1976, *J. Amer. Chem. Soc.*, 98, 1020.
170. C.A. Townesend, S.B. Christenson, J.C. Link, and C.P. Lewis, 1981, *J. Chem. Soc.*, 103, 6885.
171. P.J. Aucamp, and C.W. Holzapfel, 1970, *J. S. Afr. Chem. Inst.*, 23, 40.
172. A.V. Danks, and R. Hodges, 1974, *Aust. J. Chem.*, 27, 1603.
173. D.G.I. Kingston, P.N. Chen, and J.R. Vercellotti, 1976, *Phytochemistry*, 15, 1037.
174. C.A. Townesend, S.B. Christenson, and S.G. Davies, 1982, *J. Amer. Chem. Soc.*, 104, 6152; 6154.
175. C.A. Townesend and G.B. Christenson, 1983, *Tetrahedron*, 39, 3575.
176. K. Kakinuma, N. Imamura, N. Ikekawa, H. Tanaka, S. Minami, and S. Omura, 1980, *J. Amer. Chem. Soc.*, 102, 7493.
177. D.W. Cameron, M.J. Crossley and G.I. Feutrill, 1976, *J. Chem. Soc. Chem. Comm.*, 275.
178. D.W. Cameron, M.J. Crossley, J. Maxwell, G.I. Feutrill, and P.G. Griffiths, 1978, *Aust. J. Chem.*, 31, 1335.
179. D.W. Cameron, M.J. Crossley, J. Maxwell, G.I. Feutrill, and P.G. Griffiths, 1978, *Aust. J. Chem.*, 31, 1363.
180. M.E. Jung and J.A. Lowe, 1978, *J. Chem. Soc. Chem. Comm.*, 95.
181. J.L. Grandmaison, and P. Brassard, 1978, *J. Org. Chem.*, 43, 1435.
182. F.M. Hauser, and S. Prasanna, 1982, *J. Org. Chem.*, 384.
183. T.M. Harris, and C.M. Harris, 1977, *Tetrahedron*, 37, 2159.
184. T.M. Harris, A.D. Webb, C.M. Harris, P.J. Wittek, and T.P. Murray, 1976, *J. Amer. Chem. Soc.*, 98, 6065.
185. F. Arcamone, 1978, in 'Topics in Antibiotic Chemistry', (P.G. Sammes, ed.), Ellis-Horwood, Chichester.
186. C.R. Hutchinson, 1981, in 'Antibiotics IV', (J.W. Corcoran, ed.), Springer-Verlag, Heidelberg, p. 1.
187. F. Arcamone, G. Cassinelli, F. Di Malteo, S. Forenza, M.C. Ripamonti, G. Rivola, A. Vigevani, J. Clardy, and T. McCabe, 1980, *J. Amer. Chem. Soc.*, 102, 1462.
188. T. Oki, 1980, in 'Anthracyclines. Current Status and New Developments', (S.T. Crooke, and S.D. Reich, eds.), Academic Press, New York.
189. T. Oki, I. Kitamura, Y. Matsuzawa, N. Shibamoto, Y. Ogasawara, A. Yoshimoto, T. Inui, H. Naganawa, T. Takeuchi and H. Umezawa, 1979, *J. Antibiotics*, 32, 801.
190. A.S. Kende and J.P. Rizzi, 1981, *J. Amer. Chem. Soc.*, 103, 4247; B.A. Pearlman, J.M. McNamara, I. Heison, S. Matakeyama, M. Sekizaki, and Y. Kishi, 1981, *J. Amer. Chem. Soc.*, 103, 4228; P.N. Confalone, S.P. Pizzolato, 1981, *J. Amer. Chem. Soc.*, 103, 4251; T.T. Li, and Y.L. Wu, 1981, *J. Amer. Chem. Soc.*, 103, 7007.
191. S.D. Kimball, D.R. Walt, and F. Johnson, 1981, *J. Amer. Chem. Soc.*, 103, 1561; For a synthesis of 11-deoxy-carminomycinone see A.S. Kende, and S.D. Boetther, 1981, *J. Org. Chem.*, 46, 2799; for a synthesis of 4-demethoxy-11-deoxydaunomycin and adriamycin see H. Umezawa, M. Konoshita, and H. Naganawa, 1980, *J. Antibiotics*, 33, 1581; for a synthesis of 7,11-dideoxydaunomycone see J. Yadav, P. Corey, C.T. Hsu, K. Perlman, and C.J. Sih, 1981, *Tetrahedron Lett.*, 811.
192. T.R. Kelley, 1979, *Ann. Rep. Med. Chem.*, 14, 288; T. Kametani, and K. Fukumoto, 1981, *Med. Res. Rev.*, 1, 23; P. Cava, 1981, *J. Amer. Chem. Soc.*, 103, 1992.
193. F.M. Hauser, and S. Prassanna, 1981, *J. Amer. Chem. Soc.*, 103, 6378.
194. R.K. Boekmann, Jr. and S.H. Cheon, 1983, *J. Amer. Chem. Soc.*, 105, 4112.
195. J.P. Gerson, and M. Mondon, 1982, *J. Chem. Soc. Chem. Comm.*, 421.
196. J. McNamara, and Y. Kishi, 1982, *J. Amer. Chem. Soc.*, 104, 7371.
197. H. Sekizaki, M. Jung, J. McNamara, and Y. Kishi, 1982, *J. Amer. Chem. Soc.*, 104, 7372.
198. M. Brufani, 1977, in 'Topics in Antibiotic Chemistry', (P.G. Sammes, ed.), Ellis Horwood, Chichester, vol. 1, p. 91.

199. M. Sugita, K. Furihata, H. Seto, N. Otake, and T. Sasaki, 1982, *Agric. Biol. Chem.*, 46, 1111.
200. K. Hatano, M. Muroi, E. Higashide, and M. Yoneda, 1982, *Agric. Biol. Chem.*, 46, 1699.
201. M. Damberg, P. Russ, and A. Zeeck, 1982, *Tetrahedron Lett.*, 59.
202. E. Higashide, M. Asai, K. Ootsu, S. Tanida, Y. Kozai, T. Hasegawa, T. Kishi, Y. Sugino and M. Yoneda, 1977, *Nature*, 270, 721.
203. H. Besl, H.-J. Hecht, P. Luger, V. Pasupathy and W. Steglich, 1975, *Chem. Ber.*, 108, 3675.
204. B.M. Trost, and W.H. Pearson, 1983, *Tetrahedron Lett.*, 269.
205. T.R. Kelley, M. Behlorouz, A. Echevarren, and J. Vanya, 1983, *Tetrahedron Lett.*, 2331.
206. H. Iio, H. Nagaska, and Y. Kishi, 1980, *J. Amer. Chem. Soc.*, 102, 7965.
207. E.J. Corey, L.O. Weigel, A.R. Chamberlin, H. Cho, and D.H. Hua, 1980, *J. Amer. Chem. Soc.*, 102, 6613; A.I. Meyers, P.J. Reider, and A.L. Campbell, 1980, *J. Amer. Chem. Soc.*, 102, 6597.
208. R.G. Cooke, and J.M. Edwards, 1981, *Progr. Chem. Org. Nat. Prod.*, 40, 153.
209. D.A. Frost, and G.A. Morrison, 1977, *J. Chem. Soc. Perkin Trans. 1*, 2443.
210. T.J. Simpson, 1979, *J. Chem. Soc. Perkin Trans. 1*, 1233.
211. S.N. Seghal, H. Czerkawski, A. Kudelski, K. Pandev, R. Saucier, and C. Vezina, 1983, *J. Antibiotics*, 36, 355.
212. U. Weiss, K. Yoshira, R.J. Highet, R.J. White and T.T. Wei, 1982, *J. Antibiotics*, 35, 1194.
213. S.T. Carter, A.A. Fantini, J.C. James, D.B. Borders, and R.J. White, 1984, *Tetrahedron Lett.*, 255.
214. N. Imamura, K. Kakinuma, N. Ikekawa, H. Tanaka, and S. Omura, 1982, *J. Antibiotics*, 35, 602.
215. C. Kabuto, J.V. Silverton, T. Akiyama, U. Sankawa, R.D. Hutchinson, P.S. Steyn, and R. Vleggaar, 1976, *J. Chem. Soc. Chem. Comm.*, 728.
216. B.A. Keay, and R. Rodrigo, 1982, *J. Amer. Chem. Soc.*, 104, 4725.
217. D.H. Williams, V. Rajananda, M.P. Williamson, Williamson and G. Bojesen, 1980, in 'Topics in Antibiotic Chemistry', Vol. 5, (P.G. Sammes, ed.), Ellis Horwood, Chichester, p. 119.
218. N.R. Kalman, and D.H. Williams, 1980, *J. Amer. Chem. Soc.*, 102, 897.
219. C.M. Harris, and T.M. Harris, 1982, *J. Amer. Chem. Soc.*, 104, 363.

4 Terpenoids

J.R. HANSON

4.1 Introduction

This chapter describes some of the advances that have been made in the study of the terpenoids during the past ten years.[1] Whilst the selection of topics is inevitably subjective and limited by space, nevertheless I hope that it affords an indication of the direction in which progress is being made. Each group of terpenoids will be discussed in terms of novel natural products, biosynthesis, synthesis and chemistry.

Apart from the continuing search for novel substances from plants and fungi, marine organisms, and to lesser extent insects such as termites, have provided a variety of novel terpenoid skeleta. Many of these studies have been stimulated by the wide range of biological activities shown by terpenoids. In structure elucidation the impact of ^{13}C n.m.r. and high-field ^1H n.m.r. has been immense over the past ten years and these technical developments have also shown themselves in the results of biosynthetic studies concerned with the generation of the basic skeleta. Routine X-ray crystallography, particularly in the elucidation of structures lacking a heavy atom, has also played an increasing role. Nevertheless there is still an unfortunate tendency to assume an underlying carbon skeleton and absolute stereochemistry often on the grounds of co-occurrence in the same or a related species and then to demonstrate that the spectral data for a novel compound are compatible with this. Whilst in the majority of cases this is probably sufficient for structural proof, there are an increasing number of situations, particularly in the sesqui- and diterpenoids, where the skeleta are sufficiently similar for this assumption to be strictly untenable and chemical interrelationships with known substances are required.

In biosynthesis there have been several major technical advances. The use of ^{13}C-labelled, particularly doubly-labelled, substrates and the analysis of the

154

resultant enrichment and coupling patterns in terpenoid metabolites have enabled the formation of various skeleta to be defined. On paper it is often possible, particularly where rearrangements are involved, to generate a carbon skeleton by several different plausible foldings of the acyclic (e.g. farnesyl) prenyl precursor. The coupling and enrichment patterns observed in material biosynthesized from variously labelled acetates and mevalonates have provided a distinction between these. An increasing number of hydrogen rearrangements have been observed and it is apparent that a secondary methyl group often marks the terminus of such a rearrangement. Cell-free systems have been established from a number of fungal and higher plant species to enable the elucidation of detailed, particularly stereochemical, features of terpenoid biosynthetic cyclizations to be made. As a result a number of pathways such as those involving some sesquiterpenoids and the diterpenoid gibberellin plant hormones are well defined.

In synthetic studies there have been a number of major achievements such as the total synthesis of gibberellic acid. The stereochemical control over ring junction formation, developed with six-membered rings in the previous decades, has been extended and new methodology developed to deal with five- and seven-membered rings. Together with novel methods for constructing butenolides, this has had considerable ramifications in sesquiterpenoid synthesis.

The chain branching inherent in polyprenyl structures provides many opportunities for skeletal rearrangements and these have continued to provide a range of mechanistic problems during the decade with the unravelling, for example, of some deep-seated rearrangements in the chemistry of the sesquiterpene longifolene. Another aspect of prenyl chemistry to attract interest is that of biomimetic cyclizations as in those of various geranyl derivatives. Hence over the past ten years the terpenoids have revealed a plethora of fascinating structural, synthetic, biosynthetic and mechanistic problems with compounds that have the added interest of a wide range of biological activities.

4.2 Monoterpenoids

The advent of routine gas chromatography: mass spectrometry, and in particular glass capillary columns, has led over the past ten years to a thorough re-analysis of many essential oils with the result[2] that seasonal and geographical as well as infra-species variations in monoterpenoid content have been recorded for a number of plants including several *Mentha, Pinus* and *Salvia* species. Because of their commercial importance, plant genetics are now being studied in relation to monoterpenoid production. Although iridoid glycosides have been known for a long time, an increasing number of simpler monoterpenoids have been found as their glycosides and this may represent both a storage and a transport mechanism within the plant. Surprisingly, cyclic monoterpenoids are quite rare as fungal metabolites. However a *p*-menthane triol (1) has been isolated from *Fusicoccum amygdali* and its biosynthesis has been studied.

A number of monoterpenoids have been isolated as insect pheromones.[3] Considerable effort, for example, has been expended in studies on the production of monoterpenoid bark beetle pheromones. It has been suggested that some of these, e.g. verbenone (2), arise through oxidation of the host-wood monoterpene hydrocarbons, e.g. α-pinene by the gut bacteria within the beetle. Grandisol (3), which is a major pheromone released by the male boll

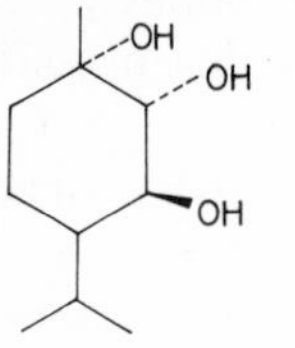

(1) 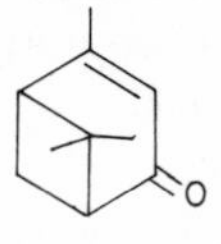 (2) 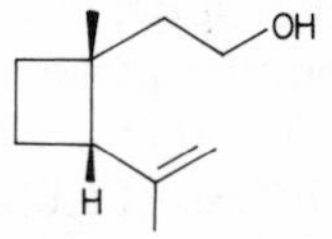 (3)

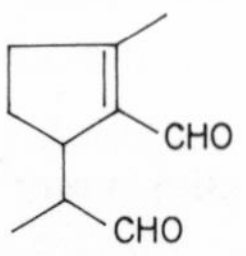

(4) 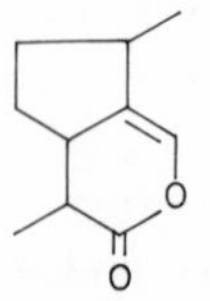 (5) 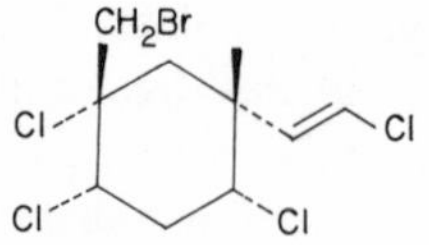 (6)

weevil *Anthonomus grandis*, has obvious monoterpenoid origins. Chrysomelidial (4) and the related lactone, plagiolactone (5), are components of the larval defence secretions of the beetle *Plagiodera versicolora*, whilst other iridoid monoterpenoids such as iridomyrmecin and dolichodial are found as components of the defensive secretions of ants.

Throughout the terpenoid area marine organisms have provided[4] sources of novel compounds. *Aplysia, Chondrococcus, Microcladia* and *Plocamium* species have all yielded halogenated monoterpenoids of unusual structure which are exemplified by violacene (6) and chondrocolactone (7). In a number of instances these structures have been confirmed by X-ray crystallographic studies. Not unexpectedly these compounds have attracted considerable interest in view of their biological, particularly anti-microbial, activity.

Another group of naturally occurring monoterpenoids whose numbers have burgeoned over the past few years are the highly oxygenated iridoids.[1]H and [13]C n.m.r. methods of structural elucidation have had an obvious impact in this area.[5] Typical examples are lamioside (8) and the secoiridoid, xylomollin (9). The complex series of bio-transformations via 10-hydroxy-

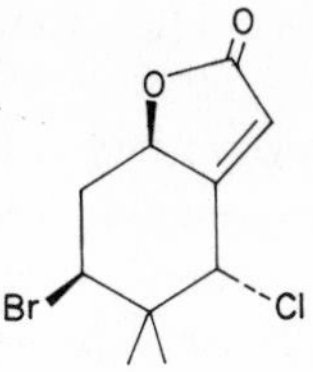

(7) 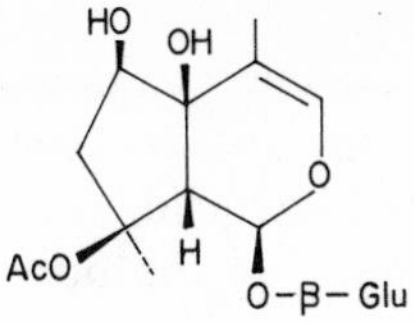   (8) 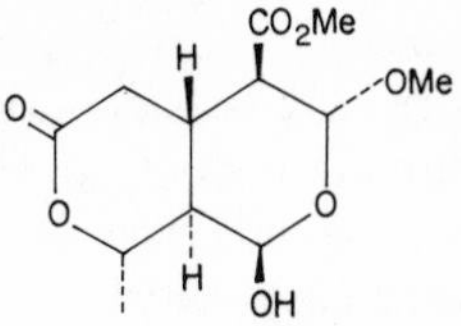   (9)

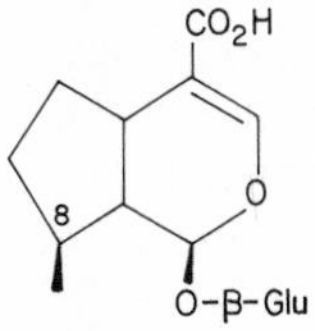

(10) 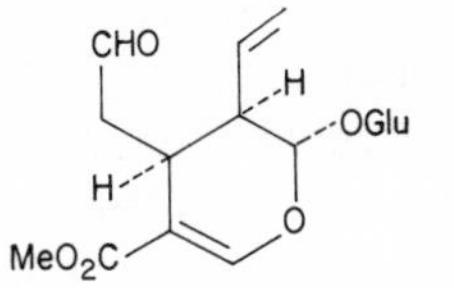   (11) 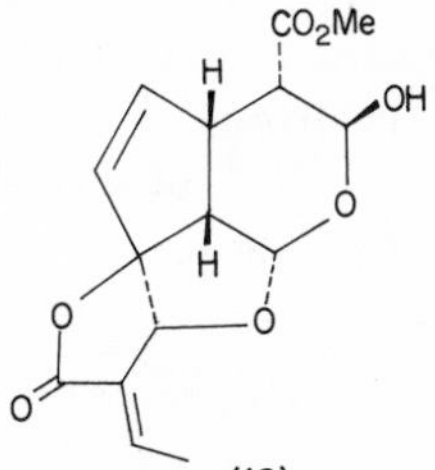   (12)

geraniol which leads to the formation of the iridoids is being gradually
unravelled. There appear to be two series of compounds, one derived from
deoxyloganic acid (10) and the other from its C-8 epimer. Whilst some of
these, such as secologanin (11), have attracted interest as precursors of the
monoterpenoid portion of the indole alkaloids, others have aroused interest
through their biological activity—for example, allamandin (12) was isolated
as a compound with anti-leukaemic activity.

A number of new natural products with irregular monoterpenoid skeleta
have been isolated from *Artemisia* species. Thus methyl santolinate (13), the
epoxide (14) and the santolinolide lactone (15) were found in the sagebrush,
Artemesia tridentata.

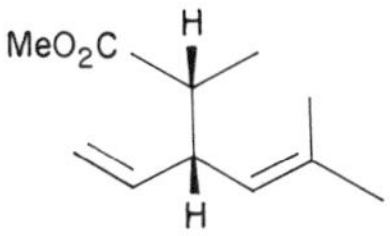

(13) 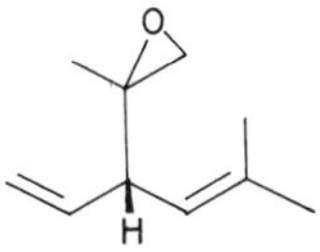   (14) 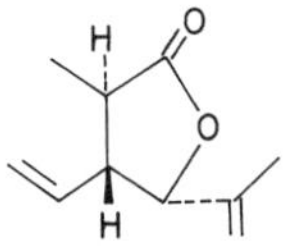   (15)

The cannabinoids are meroterpenes which are found in both fresh and dried
Cannabis sativa. A number of these substances including Δ^1-tetrahydro-
cannabinol (Δ^1THC) (16) and cannabidiol (17) have attracted interest as a
result of investigations into the psychotomimetic activity of cannabis. The
synthesis of these compounds by the acid-catalysed reaction of olivetol with

various monoterpenes such as p-mentha-2,8-dien-1-ol and verbenol has been described.

There have been considerable advances in the study of the biosynthesis of the monoterpenoids over the past decade. The asymmetric labelling of monoterpenoids from 2-^{14}C mevalonic acid has received some attention.[6,7] In many cases more label is found in the portion derived directly from isopentenyl pyrophosphate as opposed to the prenyl unit derived via dimethylallyl pyrophosphate. This has been interpreted in terms of pools of differing size, and there is some evidence that the dimethylallyl pyrophosphate pool is partially amino-acid derived. Whereas geranyl pyrophosphate appears in many cases to be converted directly into cyclic monoterpenoids without the loss of label from C-1, the free alcohol, geraniol, is converted to nerol with the loss of label. The isomerization of geraniol to nerol involves the loss of the pro-1(S)-hydrogen atom whereas the reverse isomerization involves the loss of the pro-1(R) atom. The incorporation of geraniol and geranyl pyrophosphate

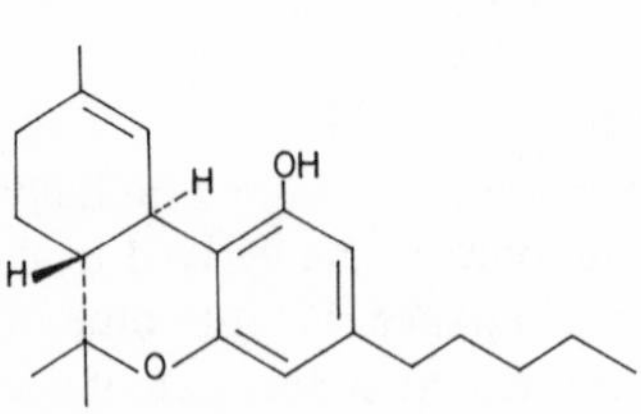

(16) 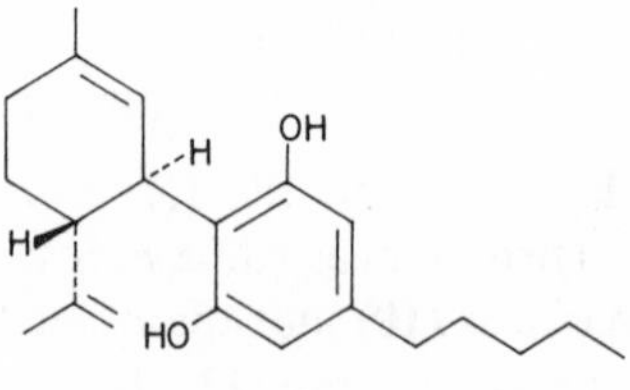   (17)

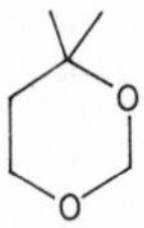

(18) 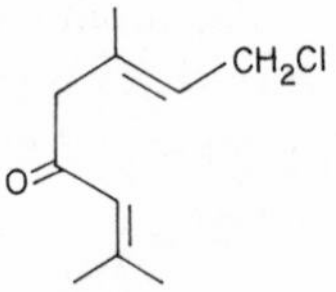   (19) 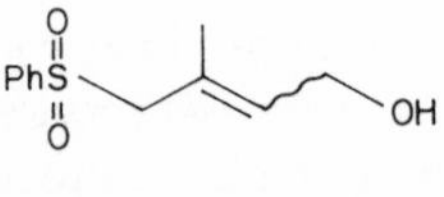   (20)

into the cyclic monoterpenoids has been studied and a number of cell-free systems have been obtained from *Salvia* species which will mediate steps in the biosynthesis of the cyclic monoterpenoids. One of the most interesting observations made with the cell-free systems has been the isolation of bicyclic alcohols such as borneol as their pyrophosphates, and the demonstration that the pyrophosphate group has its origin in the parent geranyl pyrophosphate implying the existence of a close ion-pair throughout the cyclization. The preferential participation of linaloyl pyrophosphate rather than neryl pyrophosphate has been demonstrated in the biosynthesis of some cyclic monoterpenoids such as α-terpineol and carvone.

Because of the commercial importance of monoterpenoids, the synthesis of these compounds has always received considerable attention.[8] Thus the dimerization of isoprene units to afford monoterpenoid raw materials continues to be studied. The dioxin (18), obtained along with isoprene from the reaction of isobutene with formaldehyde, affords a useful series of isopentenols on reaction with oxalic acid. The problem with many catalytic procedures is achieving a head-to-tail rather than a head-to-head dimerization. Various ways of overcoming this have been described. Acylation of isoprene with senecioyl chloride in the presence of stannic chloride affords the expected derivative (19). Other isoprenoid synthons of note which have been developed recently include the sulphone (20) and some derivatives which have been utilized in syntheses of geraniol.

A simple synthesis of artemisia ketone (21) utilizes the sequence shown in Scheme 1. Methyl santolinate has been synthesized by a Cope rearrangement of the silyl enol ether (22 – 23).

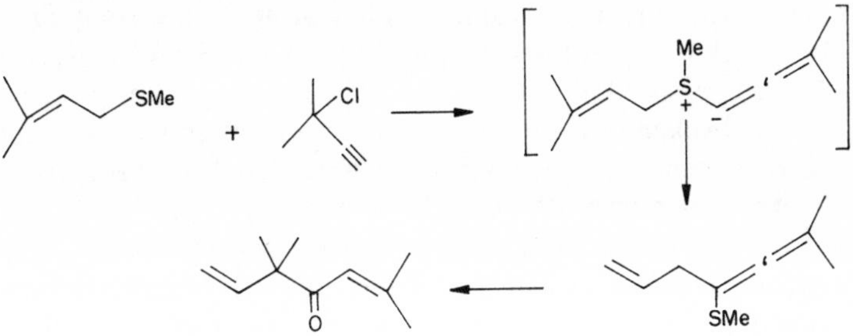

(21)

Scheme 1

A substantial literature has developed on syntheses of chrysanthemic acid and its relatives. Although strictly outside the scope of this chapter, the pyrethrins, which are chrysanthemic acid esters, have attracted considerable attention as insecticides.[9] The naturally occurring pyrethrins and cinerins are exemplified by pyrethrin 1 (24). Modification of the cyclopentane portion eventually led to bioresmethrin (25). This compound whilst having high potency lacked photochemical stability and was readily degraded. Various structural modifications particularly involving the chrysanthemic acid por-

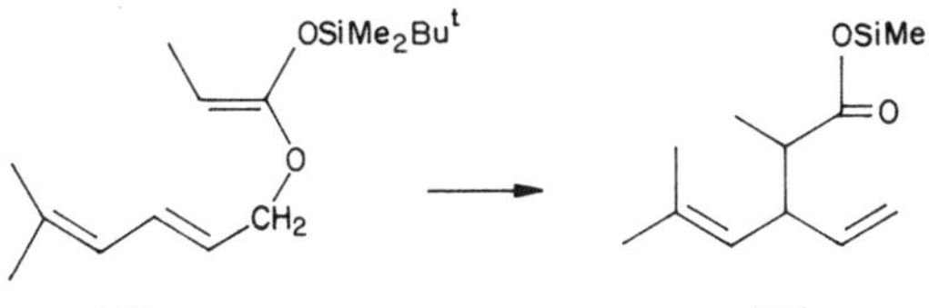

(22)   (23)

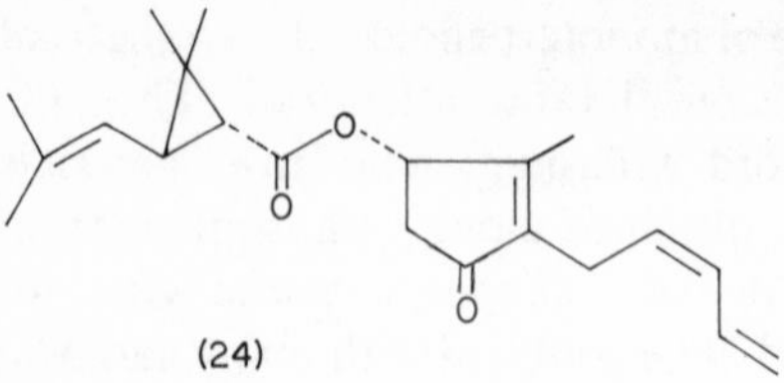

(24)

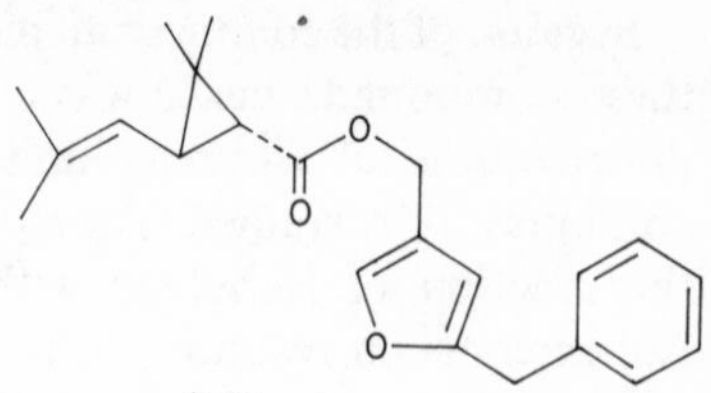

(25)

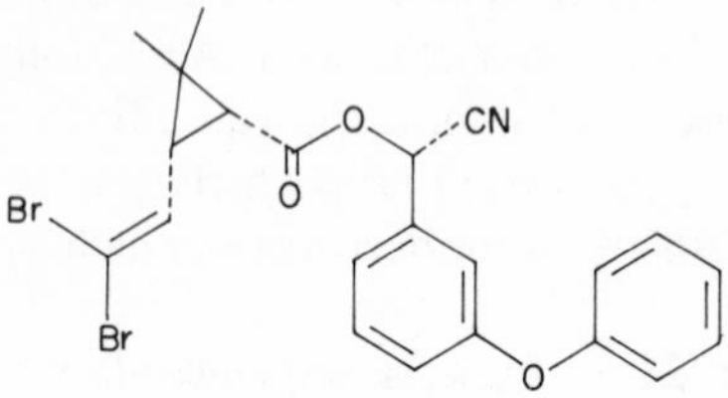

(26)

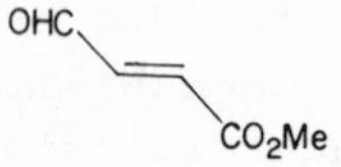

(27)

tion gave material with enhanced stability culminating in the compound NRDC 161 (26) which is one of the most powerful insecticides known for a wide range of insect pests. Although it is a photostable pyrethroid, it is nevertheless quickly broken down by soil micro-organisms and so does not persist in the environment. The early syntheses were based on reactions of diazoacetates with 2,5-dimethyl-2,4-hexadiene and these have been re-examined with the object of generating optically active chrysanthemates using asymmetric copper complexes as catalysts. Other new syntheses have been based on the aldehyde (27) obtained from sorbic acid. A non-photochemical synthesis (Scheme 2) is exemplified by the reaction of the monoepoxide (28) of 2,5-dimethyl-2,4-hexadiene with the diethyl malonate carbanion. Hydrolysis and decarboxylation then gave the lactone, pyrocine (29) which was converted via a chloro-ester to ethyl *trans*-chrysanthemate (30).

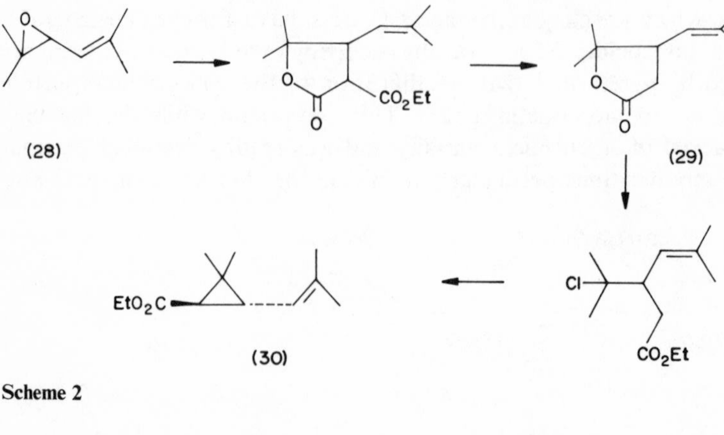

(28)

(29)

(30)

Scheme 2

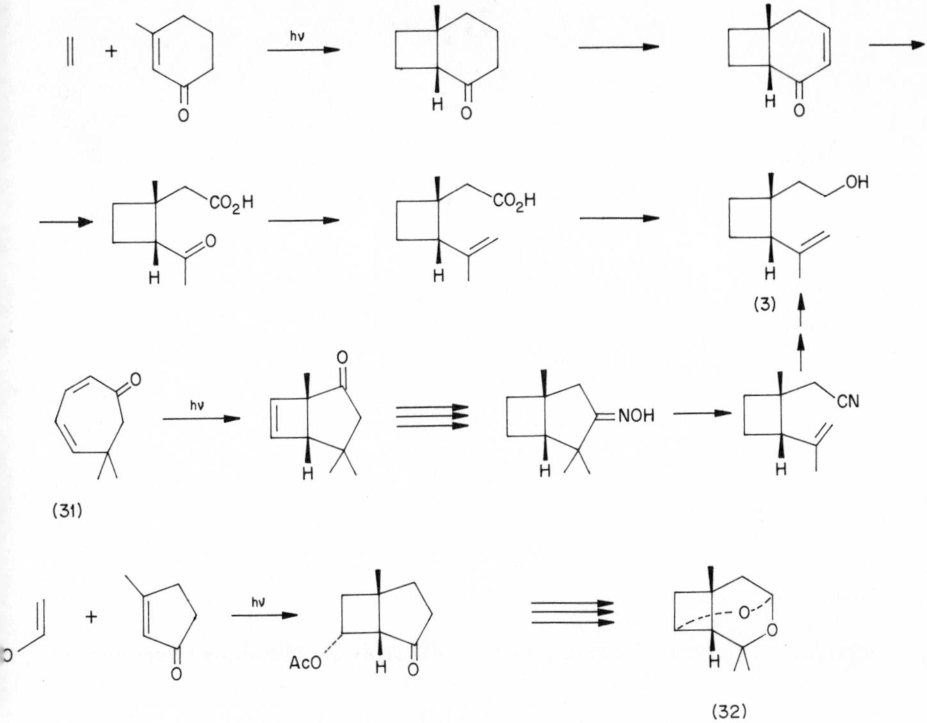

Scheme 3

The control of insect population has led to many syntheses of the monoterpenoid insect pheromones.[10] These are exemplified by syntheses of the unusual cyclobutane, grandisol (3), which is part of the pheromonal system of the male boll weevil, *Anthonomus grandis*. Several syntheses, such as that shown in Scheme 3, have been based on photochemical cycloaddition to generate the cyclobutane ring. The photocyclization of eucarvone (31) provided another approach to this system. Similar photochemical cyclo-addition reactions were used to generate the cyclobutane ring of lineatin (32) which is an attractant produced by female beetles of *Trypodendron lineatum* when feeding on Douglas fir.

As the understanding of terpenoid biosynthesis has progressed so there has been interest in biomimetic chemistry. There has been considerable interest for many years in the mechanism of solvolysis of neryl (33) and geranyl (34) derivatives. In general neryl derivatives such as the chloride, phosphate and pyrophosphate readily give cyclic products such as α-terpineol and limonene (35) with much smaller amounts of open-chain products such as linalool. On the other hand predominantly open-chain products such as linalool (36) are

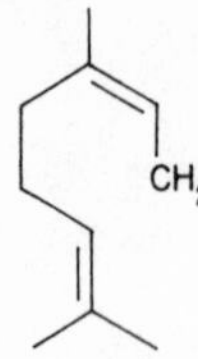

(33) 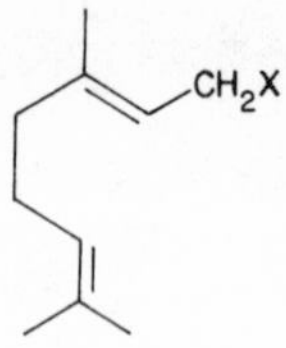   (34) 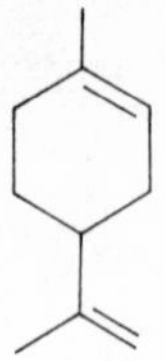   (35) 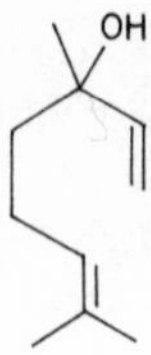   (36)

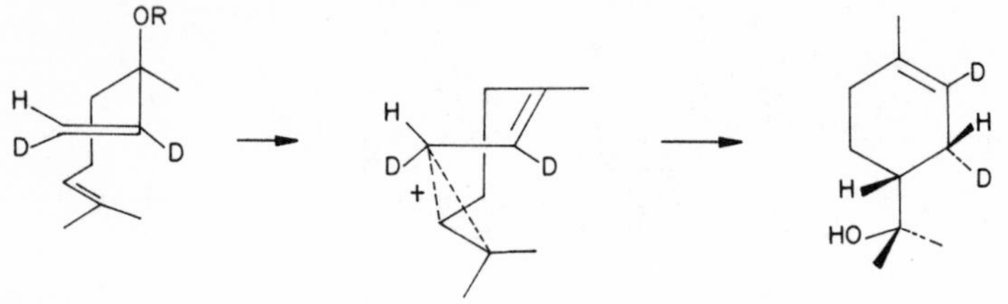

(37)   (38)

Scheme 4

obtained from geranyl derivatives.[11] The solvolysis of linalyl p-nitrobenzoate (37) to form α-terpineol (38) has been shown by labelling studies to be consistent with an anti-process (see Scheme 4).

Further cyclization of geraniol or cyclogeraniol in fluorosulphonic acid in $SO_2:CS_2$ at $-78°C$ leads to extensive rearrangement as shown in Scheme 5. The cyclization of limonene (35) with phosphoric acid has been studied and bicyclic products, exemplified by (39) and its double bond isomers, were detected.

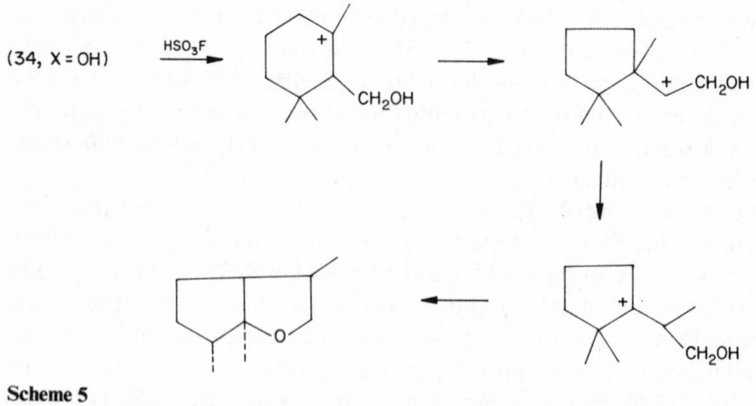

(34, X = OH) $\xrightarrow{HSO_3F}$

Scheme 5

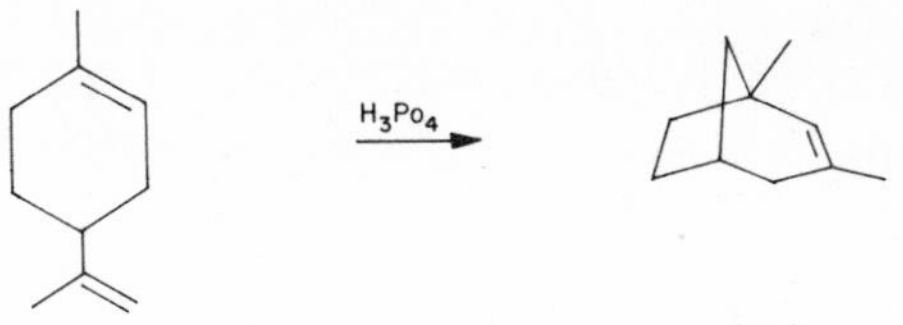

Investigations into the heterolysis of the C–O bond of the thujan-3-ols and their derivatives (**40** and **41**) show that, depending on the stereochemistry of the leaving group, either an undelocalized cation or a trishomocyclopropyl cation is formed first (Scheme 6). A subsequent 1,2-hydride shift may then convert this into the bishomocyclopropyl cation from which *p*-menthane products are obtained.

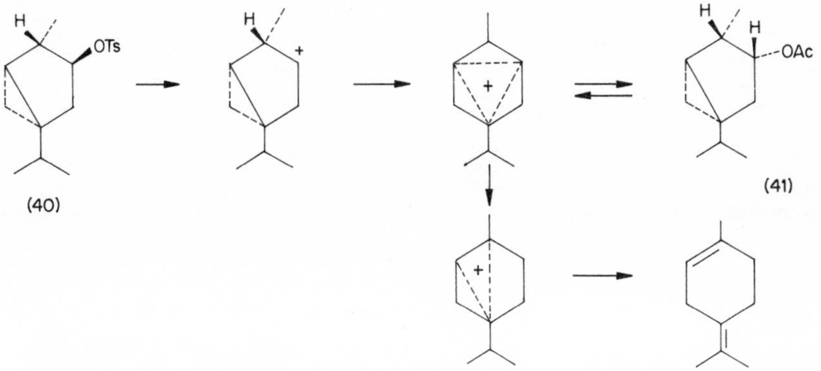

Scheme 6

The relationship between the pinane, bornane, camphane and fenchane series has been extensively studied over many years. The following scheme (Scheme 7) summarizes the fate of the pinyl carbonium ions derived from α- and β-pinene (**42** and **43**). The influence of substituents on pinyl carbonium ions has been investigated. An alkyl substituent at C-3 has quite a marked effect. If it is *cis* to the methyl groups on the bridge, bornanes tend to predominate, whilst if it is *trans*, fenchanes are formed. Low temperature n.m.r. studies have been used on a number of occasions to define carbonium ion intermediates in for example the Nametkin rearrangement of the camphene cation (**44**). The *syn* addition of deuterium chloride to the less hindered face of α- and β-pinenes (**42** and **43**) has been demonstrated using a combination of carbon-13 and deuterium n.m.r. spectroscopy. The initial tertiary chloride that is formed first rapidly isomerizes to bornyl chloride.

Studies of the solvolysis of chrysanthemyl systems have been made as

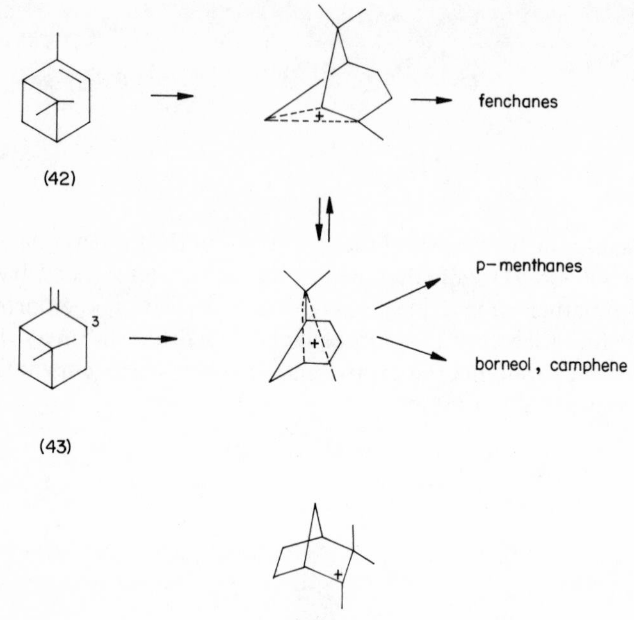

(42) → → fenchanes

⇅

(43) → → p-menthanes

borneol, camphene

(44)

Scheme 7

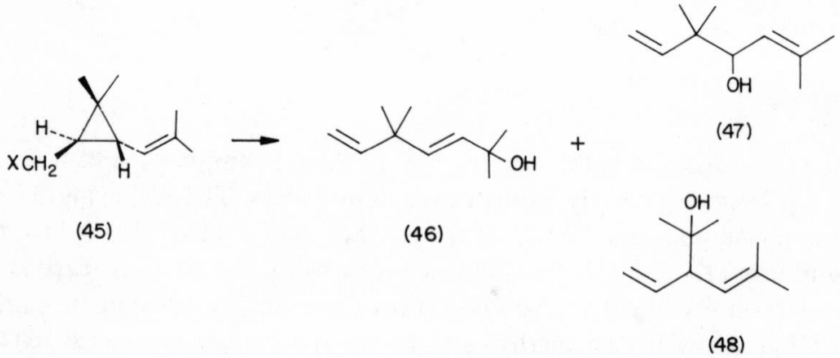

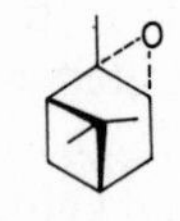

(45) → (46) + (47)

(48)

(49) 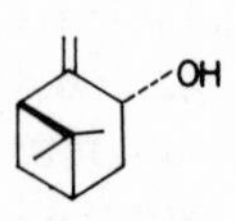   (50) 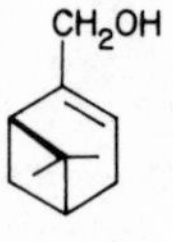   (51)

models for the formation of the irregular monoterpenoids. Thus yomogi alcohol (46), artemisia alcohol (47) and santolinyl alcohol (48) have been obtained from several chrysanthemyl esters (45,X = ODNB, OP, OMs).

A number of monoterpene epoxides react with base to form allylic alcohols (e.g. $2\alpha,3\alpha$-epoxypinane 49 → trans-pinocarveol, 50). The conditions for the allylic rearrangement of these, exemplified by the conversion of trans-pinocarveol (50) to myrtenol (51) have been examined.

Monoterpenoids have provided the chiral starting materials for a number of sesquiterpenoid syntheses. For example the synthesis of the highly oxygenated compound picrotoxinin (52) starts from carvone, whilst the

(52)　　　　　　　　　(53)

photochemical adduct (53) of methyl cyclobutene and L-piperitone is a key intermediate in the synthesis of (−)-shyobunone and its elemene isomers. A number of reagents employed for asymmetric induction in synthesis are also derived from monoterpenoids. Thus the mono and di-isopinocamphenyl-boranes, which are useful reagents for hydroboronation, have played a useful role in a number of chiral syntheses.

4.3 Sesquiterpenoids

A vast array of novel sesquiterpenoids has been described over the last ten years. Many of these have been found as a result of an extensive phyto-chemical survey of the Compositae. A recent review of sesquiterpenoid lactones lists 924 naturally-occurring compounds.[12] The major skeleta found amongst the sesquiterpenoid lactones are shown in Scheme 8. The largest number of these fall into the germacranolide and guaianolide types. The germacranolides are further subdivided into four groups based on the geometry of the cyclodecadiene ring. Many of these lactones contain hydroxyl groups which are esterified by acids such as isobutyric, angelic, epoxyangelic and tiglic acids. Some recent isolates revealing the variety of oxygenation pattern include tagitinin A (54), mikanokryptin (55) and hyporadiolide (56). The biological activity of these unsaturated lactones has stimulated synthetic endeavours which are discussed later.

Another group that has afforded a number of novel sesquiterpenoids is the liverworts (Hepaticae).[13] An interesting feature of these compounds is that

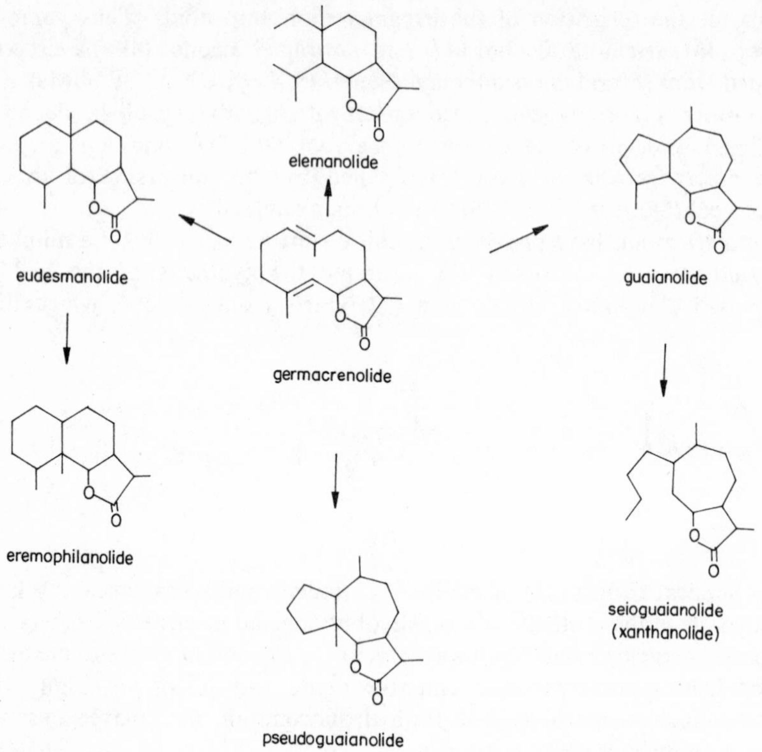

elemanolide

eudesmanolide

germacrenolide

guaianolide

eremophilanolide

pseudoguaianolide

seioguaianolide
(xanthanolide)

Scheme 8

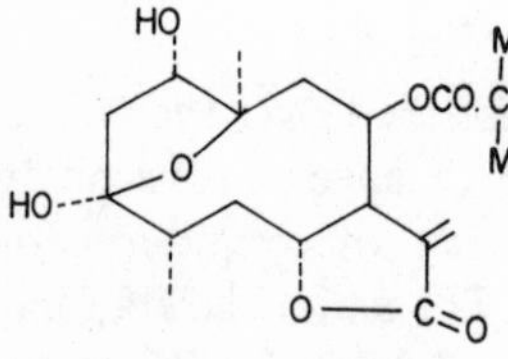

(54)

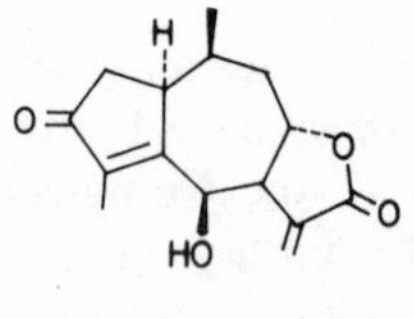

(55)

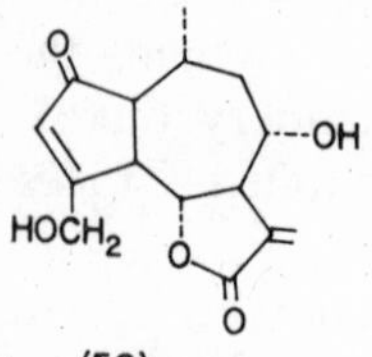

(56)

many possess skeleta that are enantiomeric with those found in higher plants. Eudesmane-type sesquiterpenoids (e.g. α-selinene, **57**) are quite widespread. Other novel hydrocarbons include β-barbatene (**58**, R = H) and β-bazzanene (**59**). The alcohol, gymnomitrol (**58**, R = OH), has attracted some synthetic

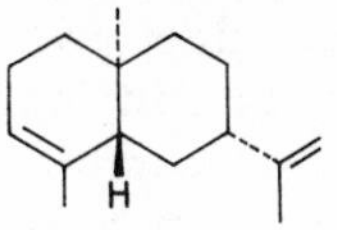

(57) 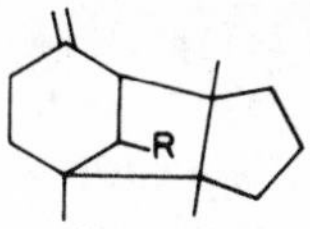   (58)

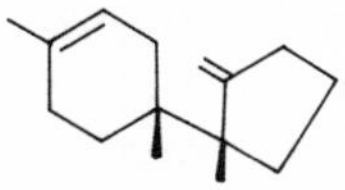

(59) 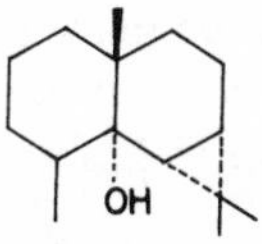   (60) 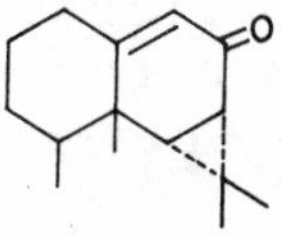   (61)

interest. Compounds of the maalane (e.g. **60**) and aristolane (e.g. **61**) skeleta have been found whilst a number of 2,3-secoaromadendrane derivatives (e.g. plagiochiline A,**62**) are amongst the more highly-oxygenated compounds that have been isolated.

Studies on the stress metabolites of the Solanaceae have yielded a number of interesting compounds such as oxylubimin (**63**) and rishitin (**64**). The features that stimulate the production of these compounds and the biosynthetic relationships between them have been the subject of a number of studies over the past ten years.

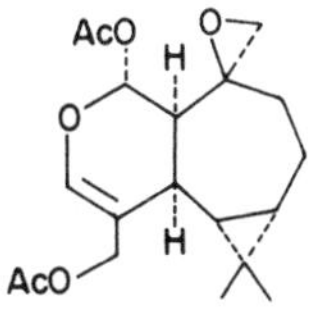

(62) 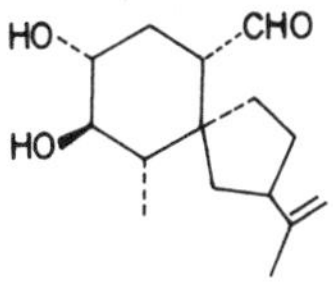   (63) 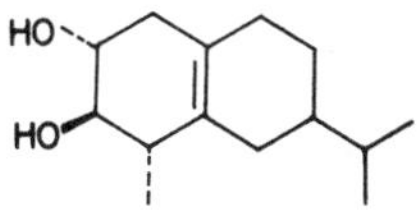   (64)

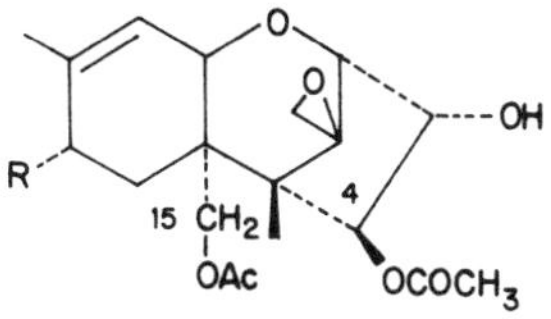

(65) 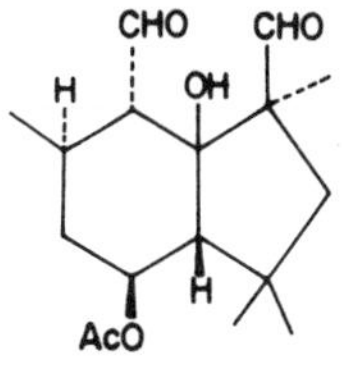   (66)

In contrast to the monoterpenoids, there are a large number of fungal sesquiterpenoids many of which have been discovered in the last ten years. Considerable interest has been shown in the trichothecenes[14] as a result of their biological activity both as mycotoxins (e.g. T-2 toxin, **65**,R = OCOCH$_2$CH(Me$_2$)) and as tumour-inhibitory agents (diacetoxy-scirpenol, anguidine, **65**,R = H). The trichothecenes fall into two classes—those which although highly hydroxylated contain simple esterifying groups such as T-2 toxin, and those with esters linking C-4 and C-15 to form a macrocycle such as the verrucarins and roridins. These compounds have achieved some notoriety as the alleged constituents of 'yellow rain' which was reputedly used in chemical warfare in South East Asia. A number of sesquiterpenoid fungal metabolites are phytotoxic—an aspect revealed by botrydial (**66**) which is a metabolite of the plant pathogen, *Botrytis cinerea*. The majority of new compounds derived from the Basidiomycetes over the last few years have been sesquiterpenoid.[15] Many of the skeleta, exemplified

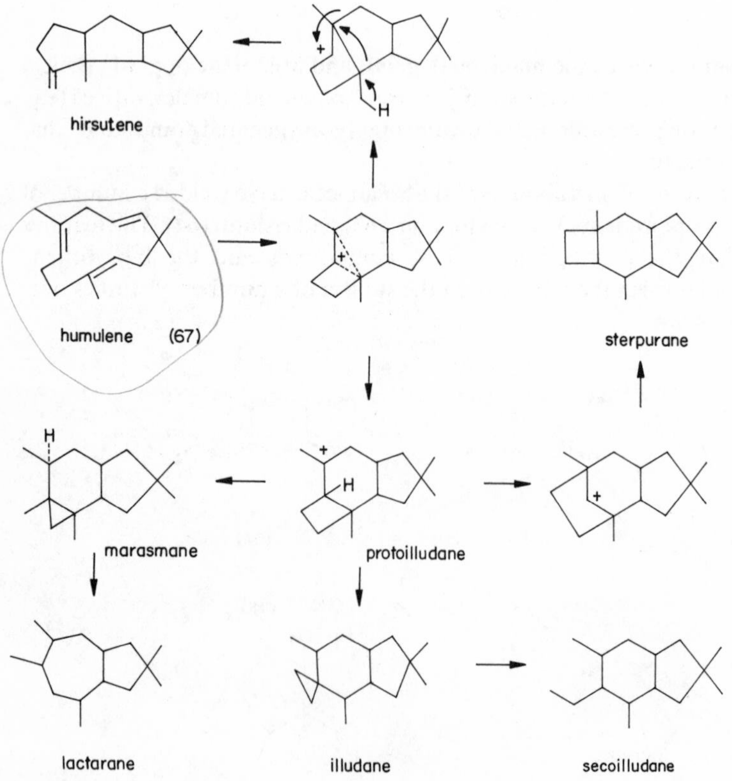

hirsutene

humulene   (67)

sterpurane

marasmane

protoilludane

lactarane

illudane

secoilludane

Scheme 9

by those in Scheme 9, can be derived formally by the cyclization of humulene (67). Typical examples of structures that have attracted interest are the coriolins (e.g. 68) (from *Coriolus consors*), fomannosin (69) and fomajorin D (70) (from *Fomes annosus*), lactarorufin A (71) (from *Lactarius rufus*) and alliacolide (72) (from *Marasmius alliaceus*). The pentalenolactones (e.g. 73) are an interesting group of biologically active compounds isolated from a streptomycete which are clearly derived by a cyclization of humulene.

There have been extensive studies on the biosynthesis of the sesquiterpenoids[16] in which particular attention has been paid to stereochemical aspects.[17,18] In order to form a number of cyclic sesquiterpenoids, the Δ^2-double bond of farnesyl pyrophosphate has to 'formally' adopt a *cis* geometry.

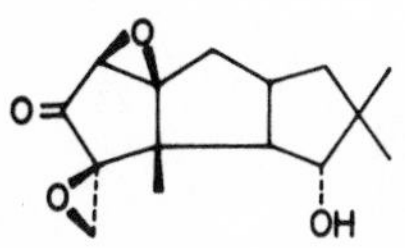

(68) 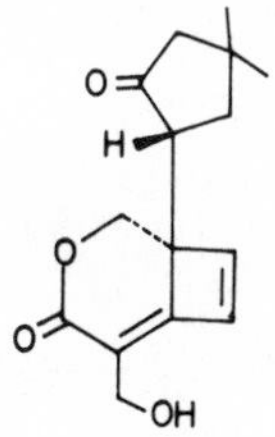   (69) 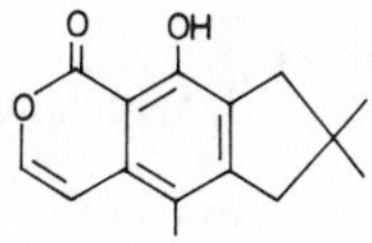   (70)

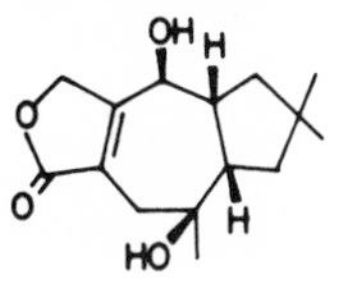

(71) 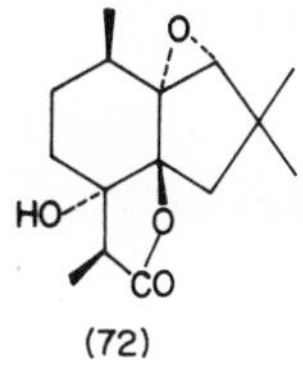   (72) 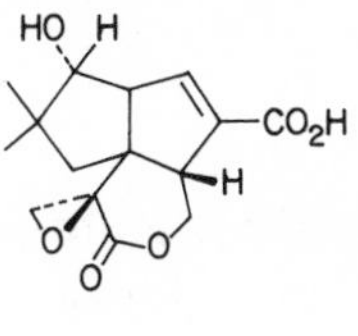   (73)

Although interconversion of the farnesols has been observed to proceed with the stereospecific loss of hydrogen from C-1 via a redox process, the prior isomerization to nerolidyl pyrophosphate and subsequent cyclization of the latter does not involve loss of hydrogen from C-1.

[13]C n.m.r. studies, using both enrichment and [13]C–[13]C coupling patterns from acetate and mevalonate feeding experiments, have been used to define the prenyl units and the folding of the farnesyl chain to generate various sesquiterpenoid skeleta. These are exemplified by trichothecolone (74) and dihydrobotrydial (75). In the case of capsidiol (76) and the *Penicillium roquefortii* metabolite, PR-toxin, these studies support the Robinson theory that the eremophilane skeleton arises by a 1,2-shift of a methyl group from a eudesmane. The enzymatic conversion of farnesyl pyrophosphate (77) and the

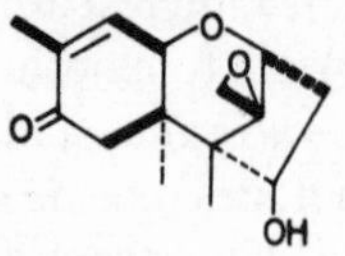

(74) 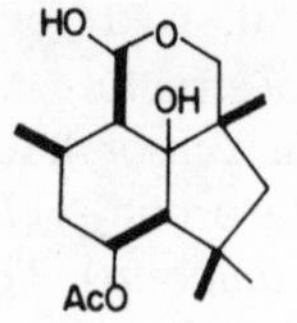   (75) 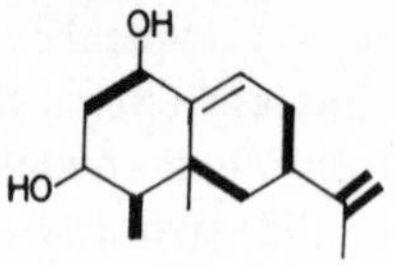   (76)

— $= {}^{13}C - {}^{13}C$ Coupling from $[1,2-{}^{13}C_2]$ - Acetate

(77)   (78)

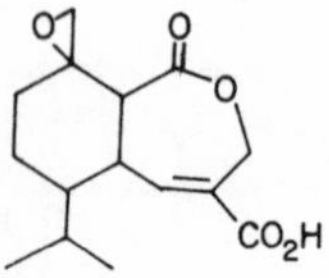

(79)

cyclization of the latter to cyclonerodiol (78) has been studied leading to the following proposal for the cyclization.

Many of the cyclizations of farnesyl pyrophosphate are accompanied by hydrogen shifts. In those biosyntheses that involve initial attack of C-1 of farnesyl pyrophosphate on the distal double bond they serve to transfer a cationic centre from C-10 or C-11 back to C-1 to initiate further isomerization and secondary cyclizations of the ten or eleven-membered ring.[18] The mevalonoid labelling pattern of avocettin (79) implied that a γ-cadinene was probably a precursor and that a 1,3-hydrogen shift of the pro-5(S) mevalonoid hydrogen from C-1 of farnesyl pyrophosphate to the isopropyl unit had occurred. A detailed study (see Scheme 10) of the formation of the tricyclic hydrocarbons (−)-sativene (80) and ent-longifolene (81) via putative ten or eleven membered ring intermediates revealed that in the case of (−)-sativene, the pro-5(R)-hydrogen (H$_A$) of mevalonate is involved in the migration whereas in the case of ent-longifolene, the pro-5(S)-hydrogen (H$_B$) is involved.

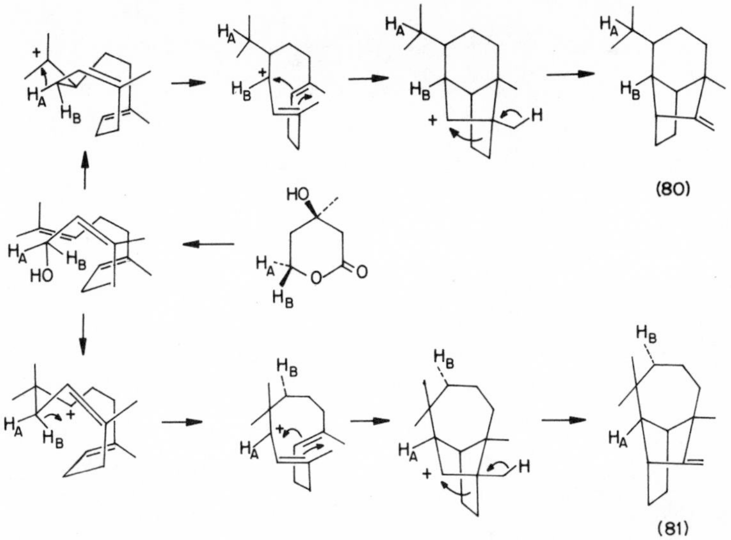

(80)

(81)

Scheme 10

The variety of sesquiterpenoid structures, and in particular the number of contiguous asymmetric centres coupled with ring systems of various sizes, has made the sesquiterpenoids prime targets for synthetic studies. As an added stimulus a number of these compounds have an interesting biological activity. The syntheses have generated novel solutions to the formation of cyclopentane rings and α-methylene lactones which have had application in other areas. Although it is not possible in a short chapter to describe all of these syntheses, the following selection reveals some of the salient steps. the antitumour activity of the elemanolides, vernolepin (**82**) and vernomenin (**83**) has stimulated several syntheses, one of which is reproduced in Scheme 11(a). Interesting features of this synthesis are the use of organoselenium reagents and the manner in which the *cis* fused ring junction is established using the angular methoxymethyl group present in the starting material.

A completely contrasting series of synthetic challenges are presented by the fused cyclopentanes of the coriolins (**84**). There have been several solutions to this problem of which one is reproduced in Scheme 11(b). Apart from the control of ring junction stereochemistry, the sequence was notable for the use of hydroxyl groups to direct the stereochemistry of epoxidation.

The interrelationships between the germacrene and elemene sesquiterpenoids have been explored on a number of occasions. In general thermal rearrangements of *trans*, *trans*-cyclodeca-1,5-diene sesquiterpenoids proceed in a highly stereospecific manner through a chair-like transition state resulting

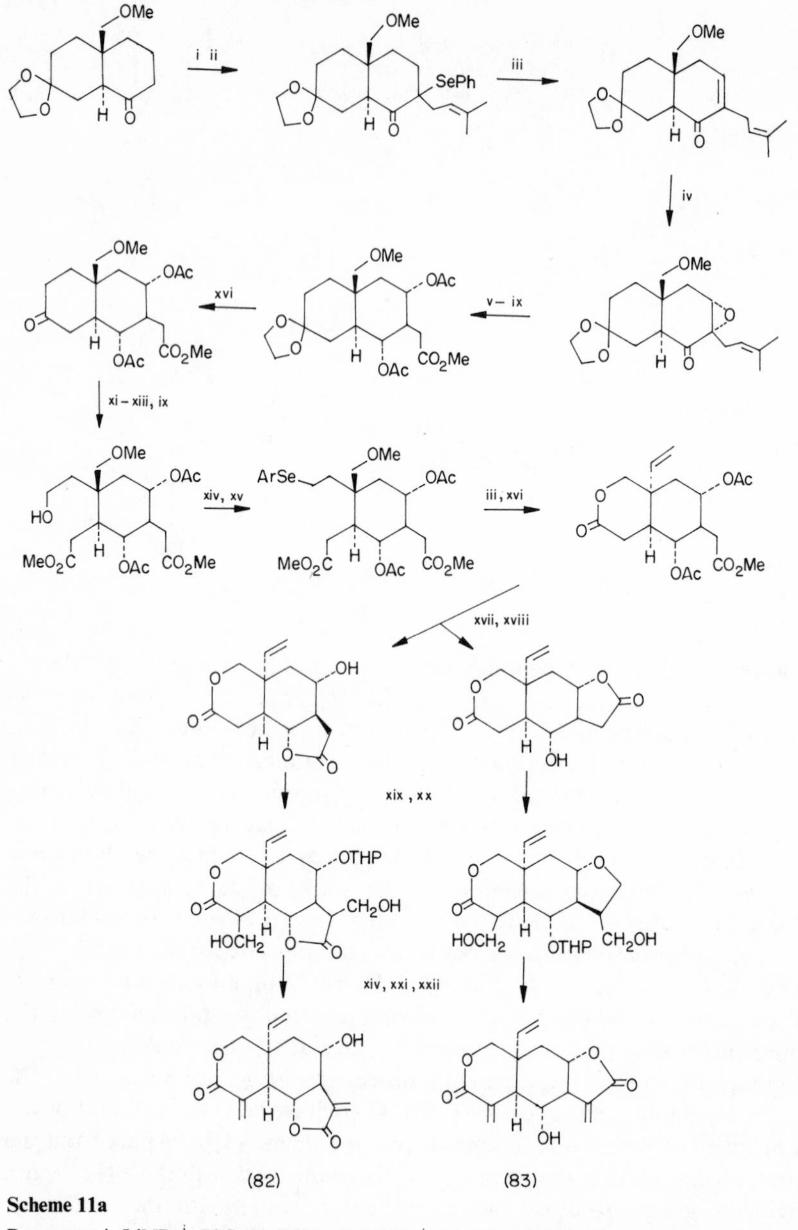

(82)   (83)

Scheme 11a

Reagents: i, LiNPr$_2^i$, PhSeCl, THF; ii, LiNPr$_2^i$; Me$_2$C=CHCH$_2$Br; iii, H$_2$O$_2$; iv, Bu^{t}OOH, triton B; v, Li,NH$_3$; vi, Ac$_2$O,pyr; vii, CrO$_3$; ix, CH$_2$N$_2$; x, HCl; xi, CH$_2$=C(Me)OAc, TsOH; xii, O$_3$,CH$_2$Cl$_2$; xiii, NaBH$_4$; xiv, MsCl, pyr; xv, o-O$_2$N.C$_6$H$_4$SeCN, NaBH$_4$, DMF; xvi, BBr$_3$, CH$_2$Cl$_2$; xvii, K$_2$CO$_3$,MeOH; xviii, TsOH; xix, DHP, H$^+$; xx, LiNPr$_2^i$, HCHO; xxi, DBU; xxii, HOAc, H$_2$O

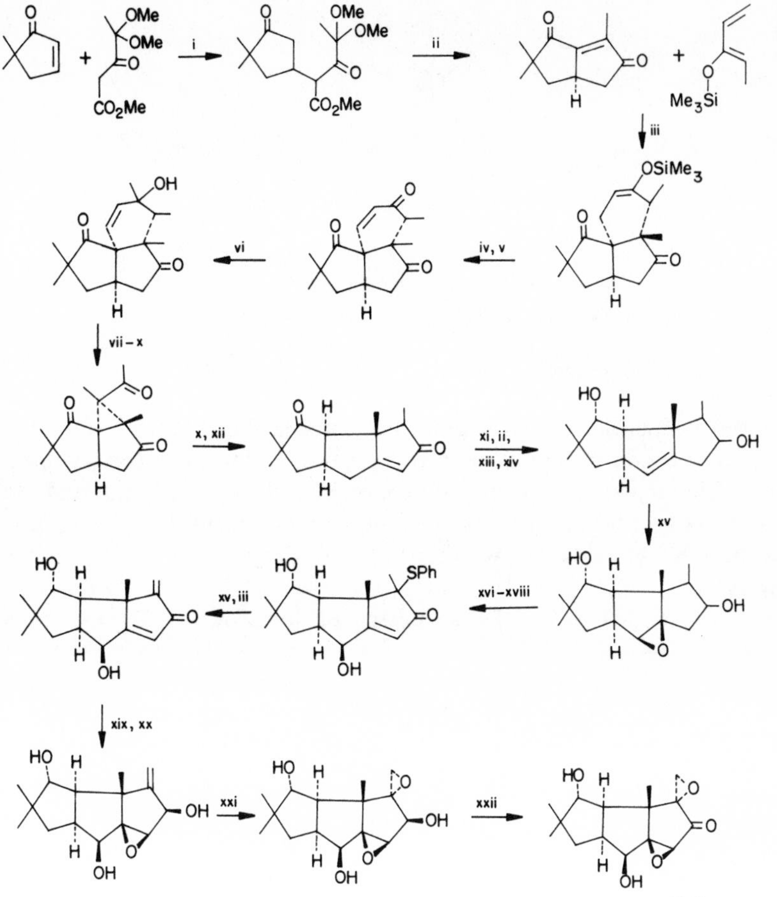

(84)

Scheme 11b

i, NaOMe; ii, H$^+$; iii, Δ; iv, PhSeCl; v, H$_2$O$_2$; vi, MeLi; vii, O$_3$; viii, CrO$_3$; ix, Ba(OH)$_2$; x, Pb(OAc)$_4$; xi, KOBu$^+$; xii, pTsOH; xiii, Bu$^i_2$AlH; xiv, Lin HN$_3$, MeOH; xv, ClC$_6$H$_4$CO$_3$H; xvi, PCC; xviii, PhSSO$_2$Ph; xix, H$_2$O$_2$, NaHCO$_3$; xx, NaBH$_4$; xxi, Bu^{t}OOH, VO(acac)$_2$; xxii, CrO$_3$—pyr.

in a divinylcyclohexane system. Thus a Cope rearrangement links the elemene skeleton of elemol (**85**) and hedycaryol (**86**). More recent examples include the thermal isomerization of dihydrotamaulipin acetate and the relationship between linderalactone and isolinderalactone.

A number of biomimetic studies have explored aspects of the chemistry of humulene (**87**) which is a possible precursor of polycyclic sesquiterpenoids. The relationship between the preferred conformations of humulene and its

(85)   (86)

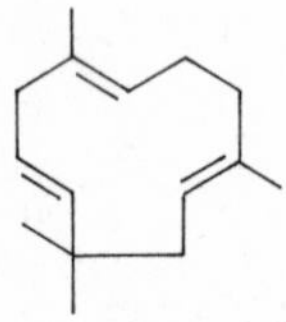

(87)

cyclization products have been examined by force-field calculations whilst the chemical conversion of humulene into the proto-illudane skeleton has been reported. The acid-catalysed cyclization of humulene 4,5-epoxide (88) to compounds (89) of the africanol type has been reported whilst rearrangement of humulene 8,9-epoxide (90) with tin(IV) chloride gives rise to the bicyclic alcohol (91).

Over the last ten years the chemistry of longifolene (92) has been

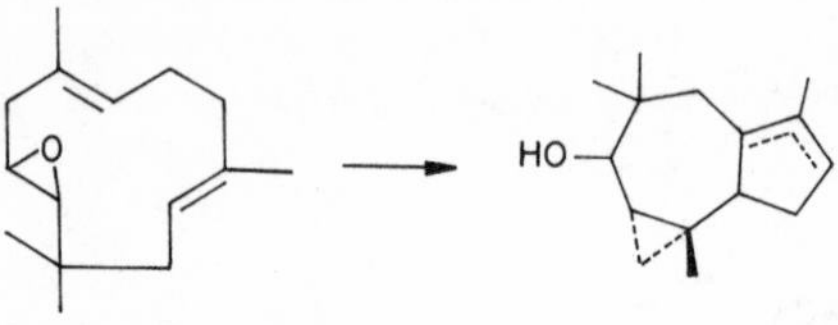

(88)   (89)

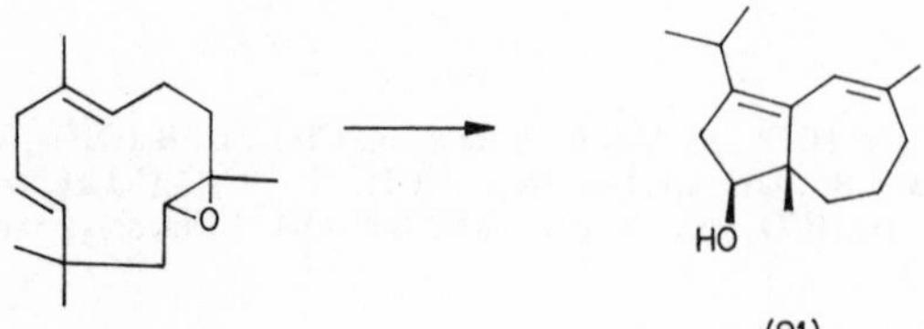

(90)   (91)

(92)   (93)

investigated further[19] particularly with regard to the deep-seated rearrangements involved in the conversion to isolongifolene (**93**). Deuterium labelling studies have been particularly informative in defining this pathway.

4.4 Diterpenoids

The number of known diterpenoids has increased substantially over the last ten years.[20] The biogenesis of the cyclic terpenoids reveals two fundamentally different types of cyclization reaction. In the first, characteristic of the mono- and sesquiterpenes, the pyrophosphate acts as a leaving group to initiate the primary cyclization by generating a carbocation which then alkylates one of the prenyl double bonds. In the other, characteristic of the higher terpenes, cyclization is initiated by protonation of a double bond or its corresponding epoxide. In both instances these primary cyclizations are often followed by a series of secondary cyclization and rearrangement reactions. The diterpenoids reveal examples of both pathways. Over the last ten years the number of diterpenoids apparently derived by the formation of a macrocyclic system followed by its secondary cyclization has increased enormously.[21]

The diversity of diterpenoid skeleta now parallels that of the sesquiterpenoids. An interesting cyclic ether, zoapatanol (**95**), has been obtained from a Mexican plant, *Montanoa tomentosa*, which is used as an abortifacient. The phytochemical examination of the Labiatiae has afforded various novel bicyclic and tetracyclic diterpenoids.[22,23] A number of the bicyclic clerodanes have attracted interest as insect anti-feedants. There has been some dispute

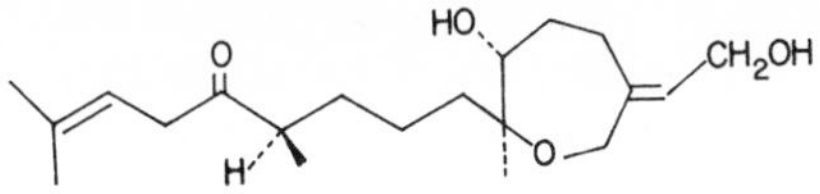

(95)

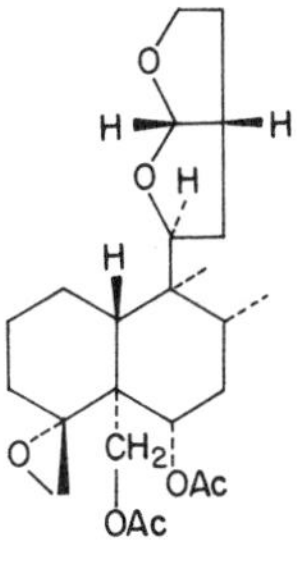

(96)

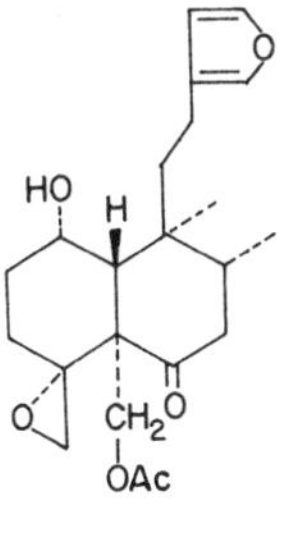

(97)

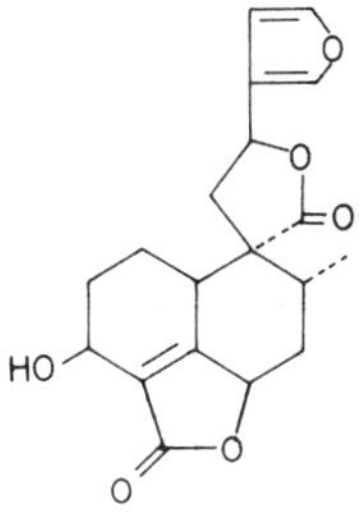

(98)

over the absolute stereochemistry of these compounds which has now been resolved. Clerodin has the absolute stereochemistry shown in (96). Examination of *Ajuga, Salvia* and *Teucrium* species has afforded compounds exemplified by (97) and (98).

Amongst the tricyclic diterpenoids a number of cytotoxic lactones have been isolated from *Podocarpus* species. These podolactones, some of which also have plant growth regulatory properties, are exemplified by inumakilactone A (99). *Coleus* and related *Plectranthus* species have also been a rich source of highly oxidized diterpenoid quinones and quinone-methides exemplified by barbatusin (100). Triptolide (101,R = H) and triptodiolide

(99)   (100)   (101)

(101,R = OH) are two novel anti-leukaemic diterpenoid triepoxides which have been isolated from *Tripterygium wilfordii* and, because of their unique structure and biological activity, have attracted some synthetic interest.

A large number of tetracyclic kaurenoid diterpenoids have been isolated from various members of the Compositae and the Labiatae such as *Sideritis* and *Isodon* species. The majority of these possess the *ent*-kaurenoid configuration with an antipodal A/B stereochemistry. Some of the more highly hydroxylated diterpenoids obtained from *Isodon* (*Rhabdosia*) species, such as (103), possess tumour-inhibitory properties. Another interesting diterpenoid which possesses the normal (steroid-like) A/B stereochemistry is the anti-tumour antibiotic, aphidicolin (102) obtained from the fungus, *Cephalosporium aphidicola*. This compound is biosynthetically interesting not only because of the unusual C/D ring junction, but also because the hydrogen atom at C-8 which originates from C-9 lies on the same face of the molecule as the methyl group at C-10.

The gibberellin plant hormones are tetracyclic diterpenoids. Sixty-six gibberellin plant hormones are now known.[24,25] These compounds fall into two series, a C_{20} family represented by gibberellin A_{13}, and a C_{19} series exemplified by gibberellic acid. Gibberellins have been found in many higher plants. Recent examples are gibberellin A_{54} (104) and A_{60} (105) which are found in wheat. A second fungal source, *Sphaceloma manihoticola*, has recently been discovered. The combination of gas chromatography: mass

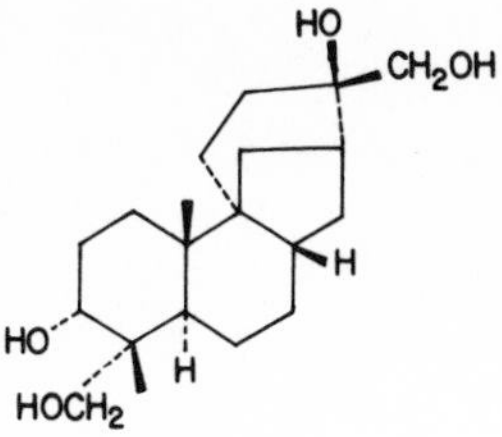

(102)

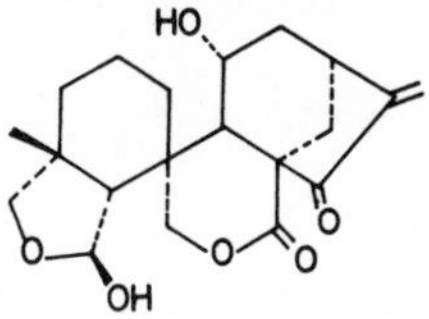

(103)

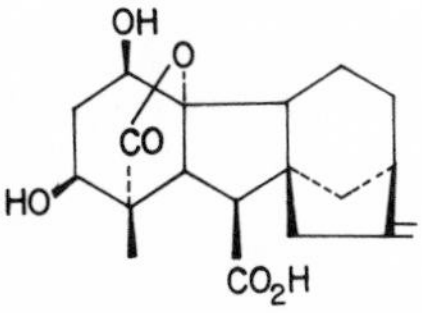

(104)

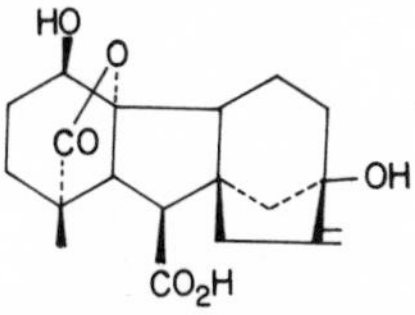

(105)

spectrometry has played a major role in the detection of these substances. Considerable effort has been expended in the preparation of the rare plant gibberellins from the more readily available fungal gibberellins such as gibberellic acid and gibberellin A_{13}. The stereochemistry of deuteration of the gibberellins has been examined with a view to the preparation of suitably-labelled substrates for metabolic studies. Several other aspects of gibberellin chemistry including the rearrangements of ring A, the photochemistry of the gibberellins and the replacement of the hydroxyl groups (including the bridgehead group) by halogen have also received attention.

An increasing number of diterpenoids have been isolated with structures based on the macrocyclic cembrane skeleton. Marine organisms have provided a number of these diterpenoids. A range of cembranolides, exemplified by sinularin (**106**), has been isolated from *Sinularia flexibilis* and related species. Some of the diterpenoids obtained from marine organisms resemble prenylated sesquiterpenoids, such as hydroxydilophol (**107**) and pachydictyol (**108**). As with the monoterpenoid examples, diterpenoids from marine sources can contain halogen, often at biosynthetically interesting sites. An example is afforded by obtusadiol (**109**) which was obtained from the red alga, *Laurencia obtusa*.

A number of other structurally interesting diterpenoids have been isolated from fungi. These are exemplified by the cotylenins (**110**) and the cyathins (**111**).[15] Diterpenoids are also found amongst the defensive secretions of insects. The termites produce an interesting group of trinervitane alcohols exemplified by (**112**) and (**113**).

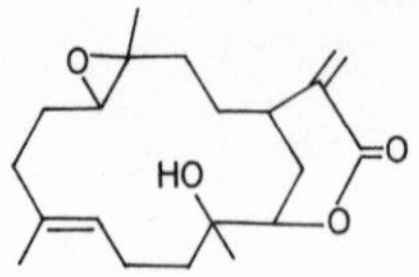

(106)

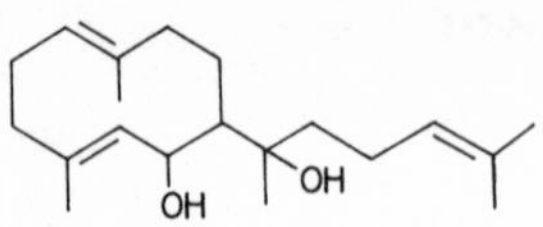

(107)

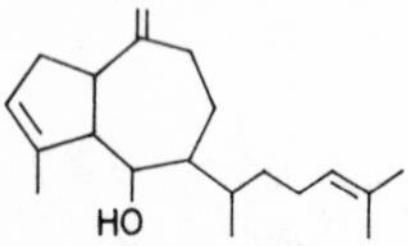

(108)

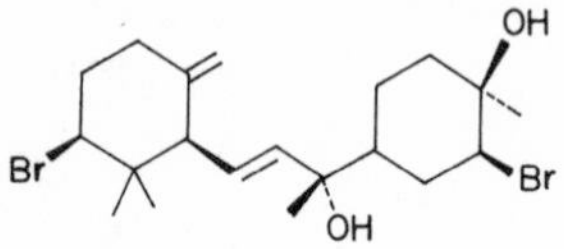

(109)

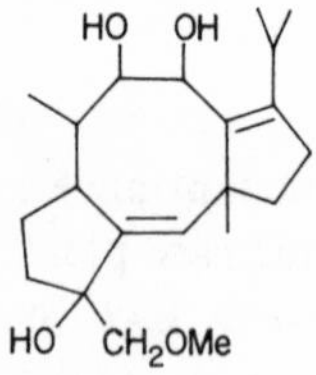

(110)

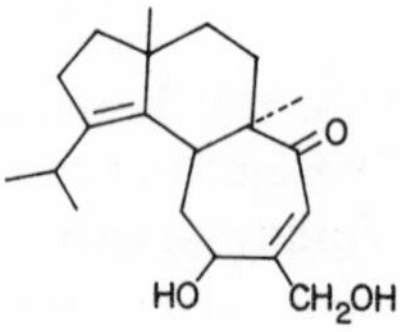

(111)

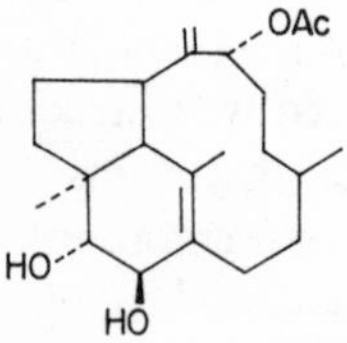

(112)

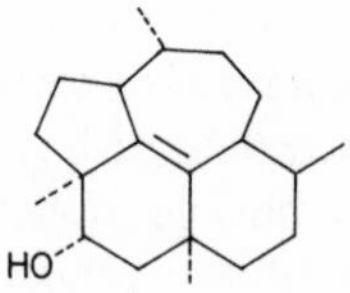

(113)

The biosynthesis of several groups of diterpenoids has been studied in detail. One group of particular interest has been the gibberellin plant hormones. An outline of the pathway leading from ent-kaurene (114) to gibberellic acid (116) is given in Scheme 12. One consequence of these biosynthetic studies have been the realization that some plant growth regulators act by blocking steps in gibberellin biosynthesis such as the cyclization step. The chemical modification of key intermediates in gibberellin biosynthesis has led to several mimics which act as competitive inhibitors

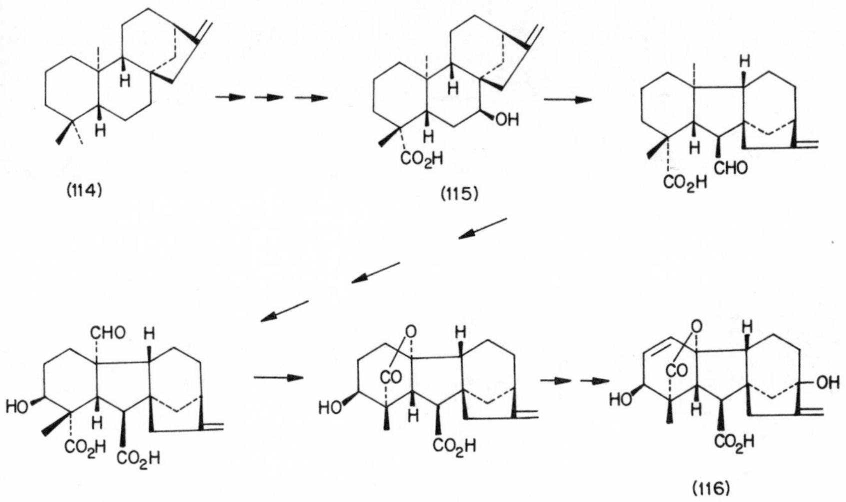

(114) (115)

(116)

Scheme 12

Reagents: i, CF_3CO_2H; ii, Na_2CO_3; iii, $(CH_2OH)_2$, H^+; iv, K-Selectride; v, HNO_2; vi, $h\nu$; vii, hexylborane, H_2O_2; viii, PhSe.SePh, KH; ix, H_2O_2; x, MnO_2; xi, $(CH_2=CH.CH_2)_3Al$; xii, $(EtCO)_2O$, Et_3N, $4\text{-}Me_2N.C_5H_4N$; xiii, KH, DMF; xiv (isopentyl)$_2$BH, H_2O_2; xv, CrO_3—pyr; xvi, K_2CO_3; xvii, H^+; xviii, $Ph_3^+PMeBr^-$, $KOBu^t$, Bu^tOH, THF.

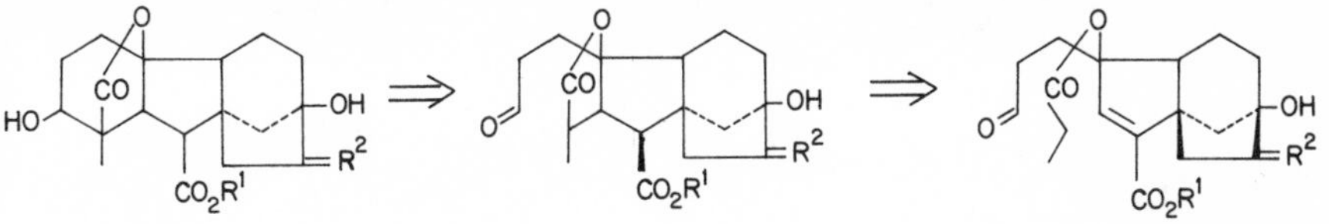

Scheme 13

particularly at the ring contraction step (115) and hence act as novel plant growth regulators.

The total synthesis of the polycyclic diterpenoids has attracted interest over many years. Two recent major successes have been the synthesis of gibberellic acid[26] and aphidicolin. The acid-catalysed degradation of gibberellic acid leads to products with an aromatic ring A and these were the initial targets of synthetic endeavour. This was followed by syntheses of gibberellin A_{15}, gibberellin A_4, and finally in 1978, gibberellic acid (116). A second more economical synthesis of gibberellic acid and gibberellin A_1 (117), first described in 1979 and outlined in Scheme 14, was based on an interesting retrosynthetic analysis (Scheme 13).[27] This envisaged the formation of the C(3)–C(4) bond by an aldol process, the C(4)–C(5) bond by a Michael reaction and the C(1)–C(10) bond through the addition of a nucleophile to a

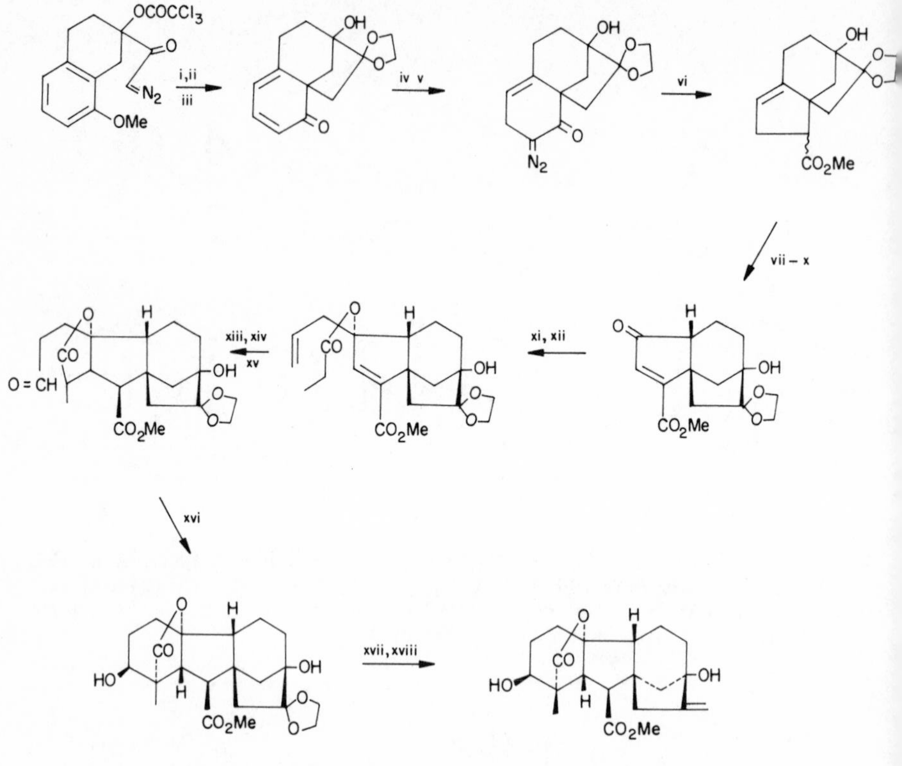

Scheme 14

C(10) ketone. The synthesis was also characterized by the use of diazoketone chemistry in bond formation and ring contraction.

There have been several syntheses of aphidicolin (118) stimulated by its antiviral activity and the requirement to generate a spiran centre at C-9. One synthesis (see Scheme 15) utilizes the mercuric trifluoroacetate catalysed biomimetic cyclization of a triene to generate the A/B ring junction.

4.5   Sesterterpenoids

The sesterterpenoids are a small group of C_{25} compounds which have been isolated[28] from various natural sources such as insect waxes, plant resins, fungi, and marine organisms. Heteronemin (119), a typical marine product, and the lactone (120), a typical fungal product, have recently been isolated.

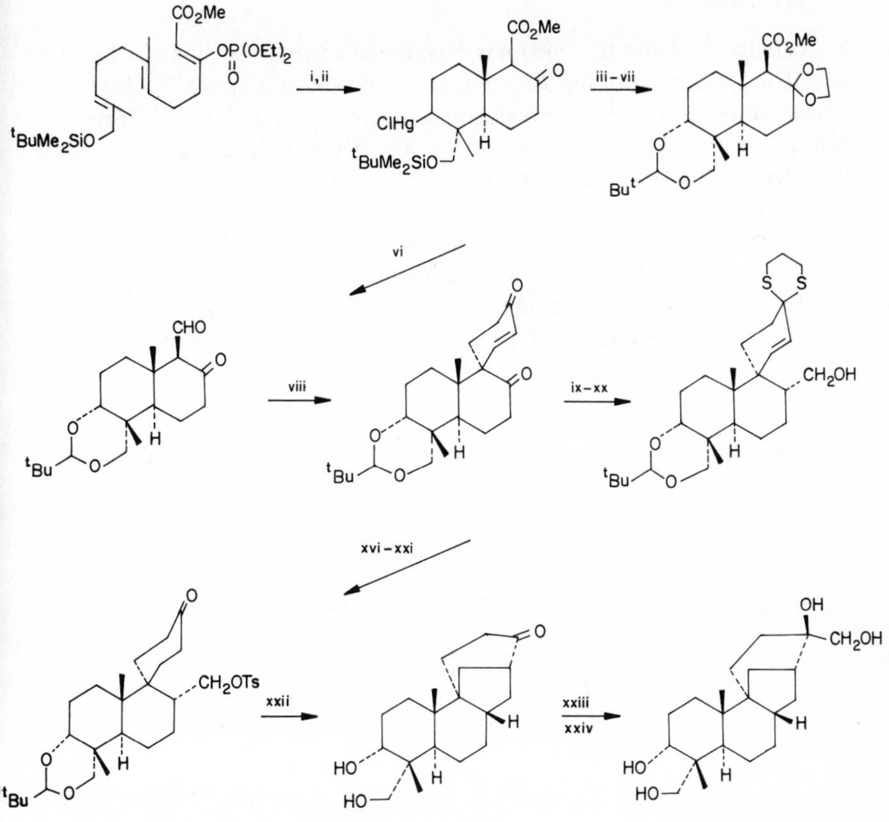

(118)

Scheme 15

Reagents: i, Hg(OCOCF$_3$)$_2$; ii, NaCl; iii, (CH$_2$OH)$_2$, H$^+$; iv, NaBH$_4$, O$_2$; v, CrO$_3$-pyr, ^{n}Bu$_4$N$^+$F$^-$; vi, Li^{s}Bu$_3$BH; vii, ^{t}Bu.CHO, H$^+$; viii, CH$_2$=CH.CO.CH$_3$, K$_2$CO$_3$, DBU; ix, Me$_3$Si.SCH$_2$CH$_2$CH$_2$CH$_2$S.SiMe$_3$, ZnI$_2$; x, Me$_3$SiCN, ZnI$_2$; xi, Bu$^i_2$AlH; xiii, Me$_3$SiLi; xiv, LiN^{i}Pr$_2$; xv, NaBH$_4$; xvi, Bu^{t}Me$_2$SiCl, Et$_3$N, (CH$_3$)$_2$NC$_5$H$_4$N; xvii, 1,3-diiodo-5,5-dimethyl-hydantoin; xix, H$_2$, Pd/C; xx, Bu$_4$N$^+$F$^-$; xxi, TsCl; xxii, LiNBu$^i_2$; xxiii, EtO$_2$CH$_2$CH$_2$OCH$_2$Li; xxiv, CH$_3$CO$_2$H, MeOH.

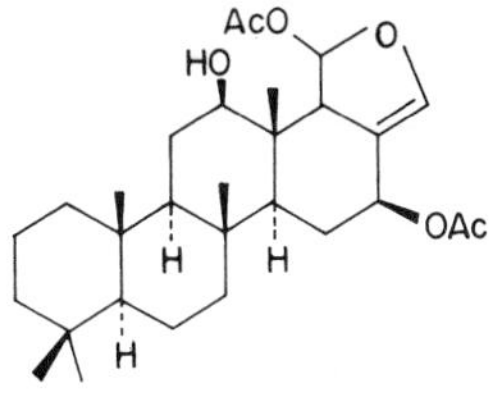

(119)

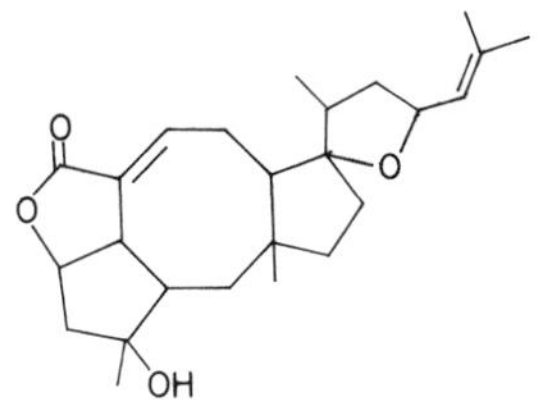

(120)

4.6 Triterpenoids

The formation of the triterpenoids has attracted interest over the last ten years. The cyclization of the 3(S)-isomer of squalene 2,3-epoxide (121) to the triterpenoids (e.g. 122) has been examined in considerable detail.[29] Particular attention has been paid to the substrate specificity of the cyclase. The latter will accept a number of unnatural squalenes and this has led to important generalizations on the stereochemical requirements. These results can be summarized as follows. Firstly the formation of ring A involves a high degree of S_N2-like participation of the Δ^6-bond and an ensuing series of conformationally rigid partially-cyclized carbocationic intermediates. The chiral trisubstituted epoxides Δ^6 and Δ^{10} represent the structural minimum for cyclase

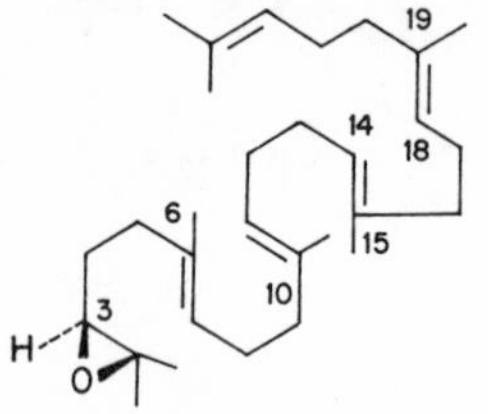

(121) 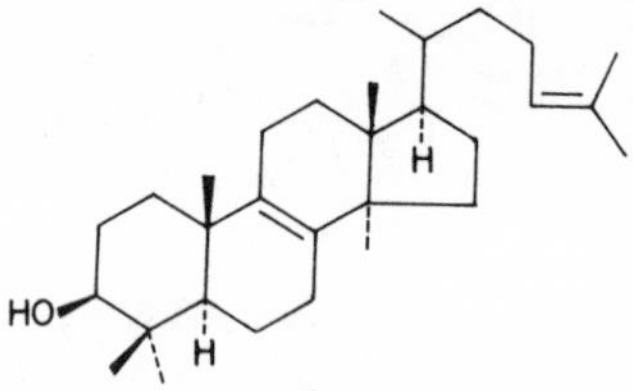   (122)

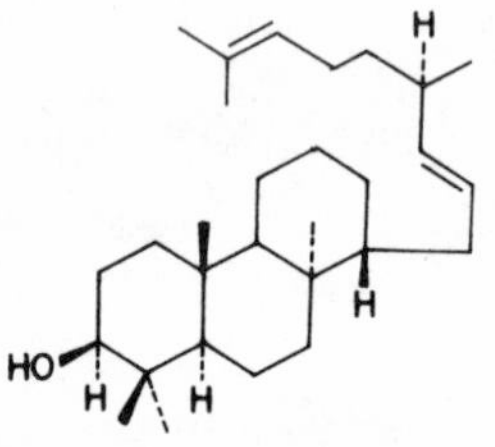

(123)

activity. The Δ^{14} and Δ^{18} double bonds are required for tetracyclization whilst the methyl groups at C-6, C-10 and C-15 are not required. These features are shown in structure 121. The conformational orientations of C-2, C-7, C-6 and Δ^{10} are rather critical. An interesting observation has been made using 15'-nor-18,19-dihydrosqualene-2,3-epoxide as a substrate. This affords the tricyclic product (123) by a hydride transfer from C-18 to C-14 with the implication that these centres are held close together by the cyclase. This would account for the natural substrate affording the non-Markownikoff products at this stage in the cyclization. Considerable attention has been directed at presqualene and several syntheses have been recorded with the material being resolved. Biomimetic polyene cyclizations have also continued to receive attention in the context of triterpenoid formation.[30]

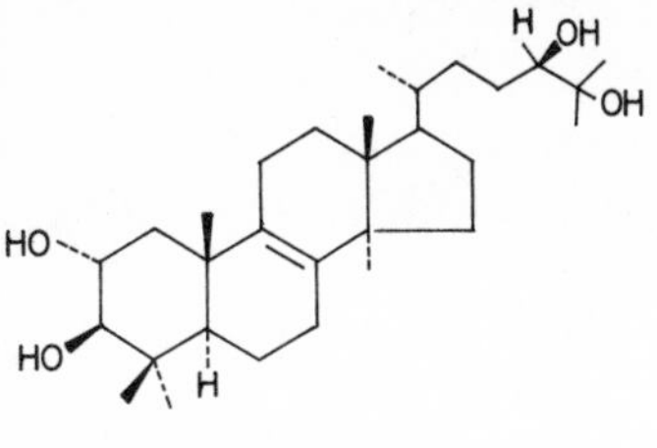

(124)

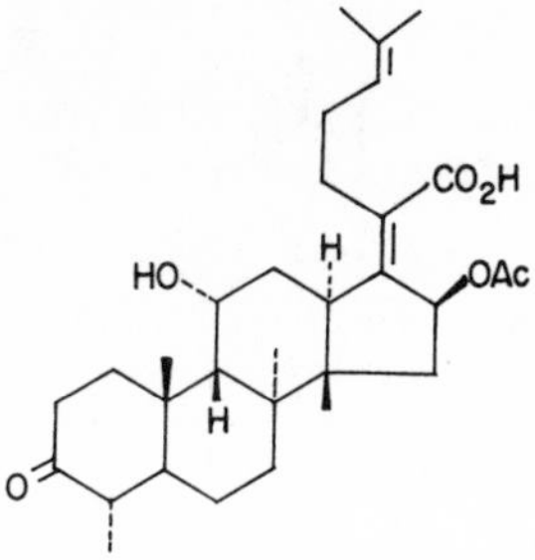

(I25)

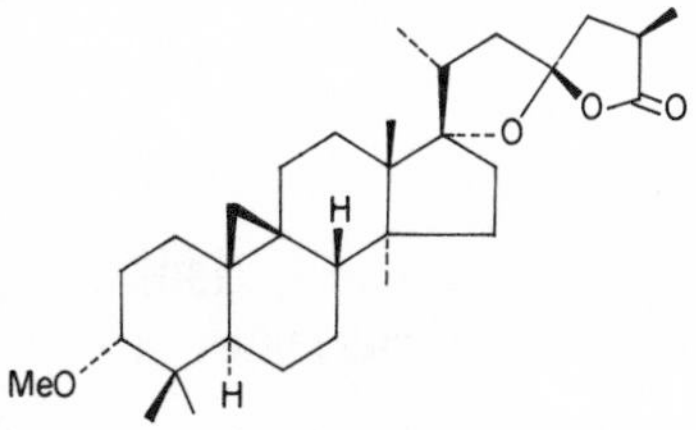

(126)

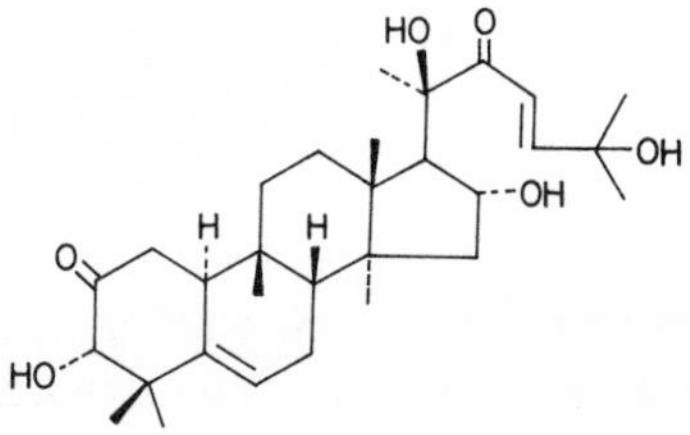

(127)

The search for biologically-active natural products has afforded quite a number of new triterpenoids.[31] Amongst the fungal metabolites, the fasciculols (e.g. **124**), obtained from *Naematoloma fasciculare*, are plant growth regulators, whilst further members of the fusidane antibiotics (e.g. **125**) have been isolated. The biosynthesis of the fungal steroids, unlike those of plants, utilizes the triterpenoid lanosterol rather than cycloartenol. The use of squalene labelled with a chiral methyl group has shown that in the biosynthesis of the latter the proton loss in the formation of the cyclopropane ring involves a retention of configuration. A simple conversion of lanosterol into cycloartenol has been reported. Amongst interesting cycloartane derivatives that have been isolated is abietospiran (**126**) which was obtained in large quantities from the bark of *Abies alba*.

The cucurbitacins have continued to attract attention because of their biological activity. A number of new tumour-inhibitory derivatives exemplified by isocucurbitacin D (**127**) have been described. Other members of this series have been reported to have insect antifeedant activity. Biosynthetic studies suggest that the cucurbitane triterpenoids are formed directly from squalene 2,3-epoxide without the intermediacy of lanostane triterpenoids such as cycloartenol or parkeol. Nevertheless the chemical rearrangement of compounds with the lanostane skeleton into the cucurbitacins has been examined. The boron trifluoride rearrangement of 3β-acetoxy-9β,11β-epoxy-

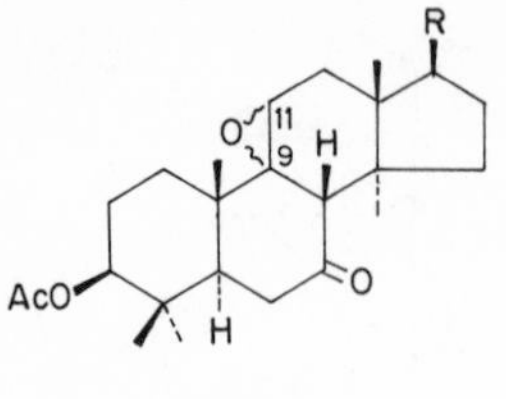

(128)

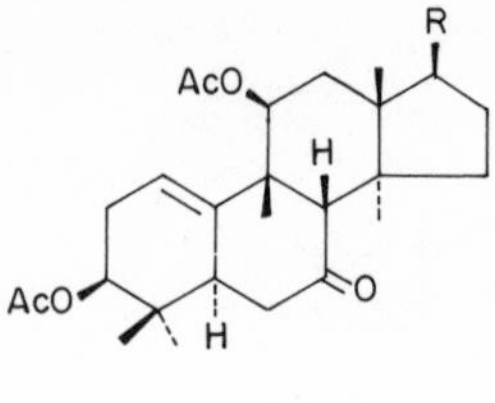

(129)

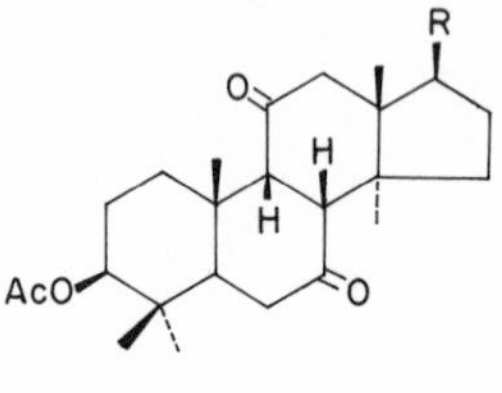

(130)

lanostan-7-one (128) affords (129) whilst its 9α,11α-epimer gives (130). The oxidation of ring C of lanosterol and the rearrangement of ring D to aromatic derivatives has also received attention.

The highly oxidized tetranortriterpenoids which are obtained from timbers of the Meliaceae have attracted a great deal of attention over the past ten years and a number of new compounds have been described. One of the compounds in this series, azadirachtin (131), from *Melia azadarach*, is of interest because of its insect (particularly locust) antifeedant activity. Clausenolide (132), from *Clausena heptaphylla*, is an example in which ring A has been cleaved, whist quite a number of compounds, for example dregeanin (133) have ring B cleaved. An unusual facet of these structures is the presence in a number of cases of ortho-esters bridging polyhydroxylated rings. A group of even more highly oxidized pentanortriterpenoids, the cneorins and tricoccins, have been obtained from the Cneoraceae. The limonoids of *Citrus* species have also received attention and further variants of their structures have been recorded. The bitter principles of the Simaroubaceae have attracted quite a lot of interest because of their anti-leukaemic activity. Bruceantinol (134) is an example. The array of asymmetric centres and variety of functional groups in these quassinoids affords a synthetic challenge which a number of research groups have been investigating.

Amongst the pentacyclic triterpenoids there has been a steady increase in the number of known substances with the lupane, oleanane, ursane and hopane skeleta.[32] An interesting feature is that triterpenoids with the hopane skeleton are widespread components of shale oils.[33] They occur with differing stereochemistry at C-17 and C-21. In a number of cases they have extended side chains as exemplified by (135). Typical of recent isolates amongst the

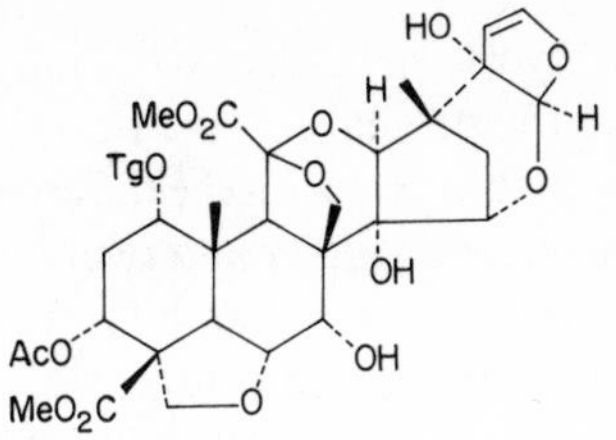

(131)

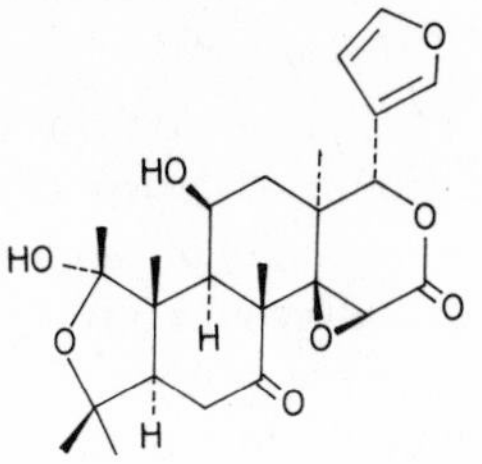

(132)

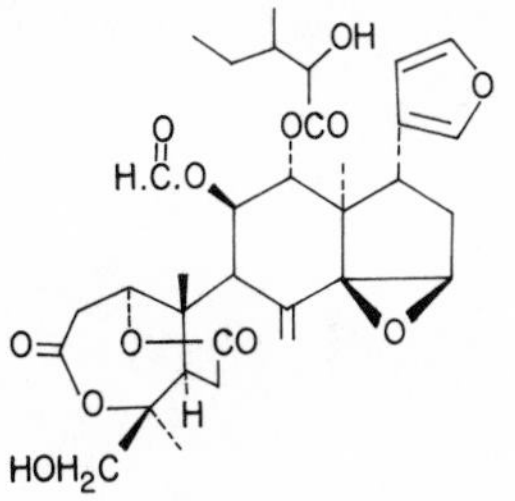

(133)

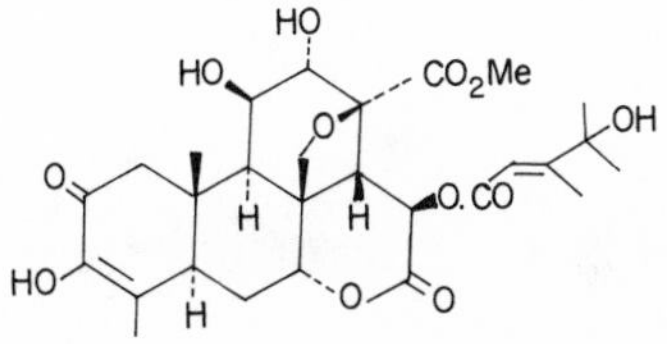

(134)

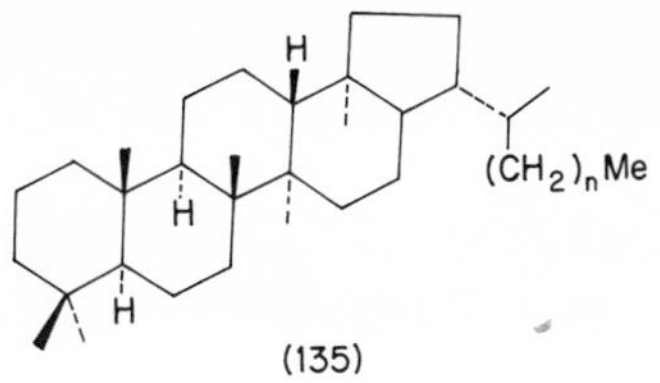

(135)

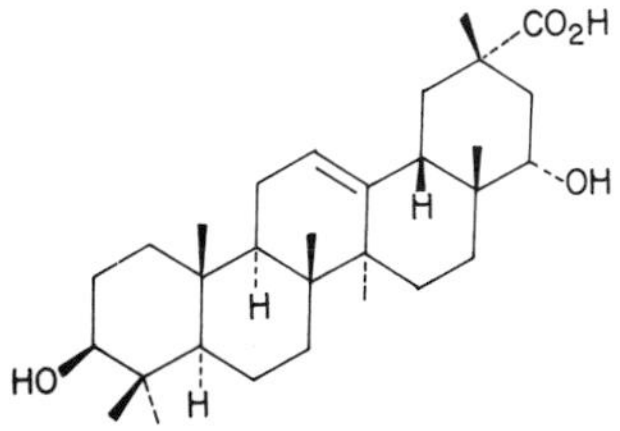

(136)

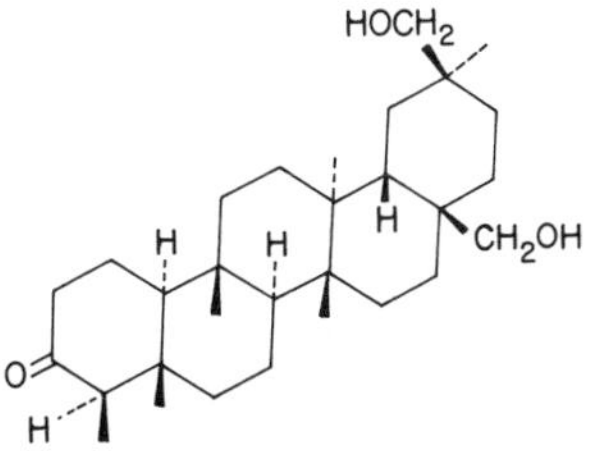

(137)

pentacyclic triterpenes are the β-amyrin derivative, maytenfolic acid (136) and the friedelin derivative, maytenfoliol (137) which were isolated as anti-leukaemic triterpenes from *Maytenus diversifolia*. Their structures were established by X-ray analysis.

4.7 Carotenoids

Although the carotenoids do not have the polycyclic structures associated with the other groups of terpenoids, nevertheless the presence of the polyene system with the possibility of geometrical isomerism together with the different permutations of end-groups introduces a wide structural variety and over 400 carotenoids are now known.[34] The application of more sophisticated separation techniques such as h.p.l.c. has been particularly important, whilst physical methods such as u.v., mass and both ^{1}H and ^{13}C n.m.r. spectrometry have played a major role in structural elucidation.[35,36] Carotenogenesis in *Phycomyces blakesleeanus* has been extensively studied, and a number of new carotenoids have been isolated from mutant systems and cultures in which the dehydrogenation steps are inhibited by diphenylamine.

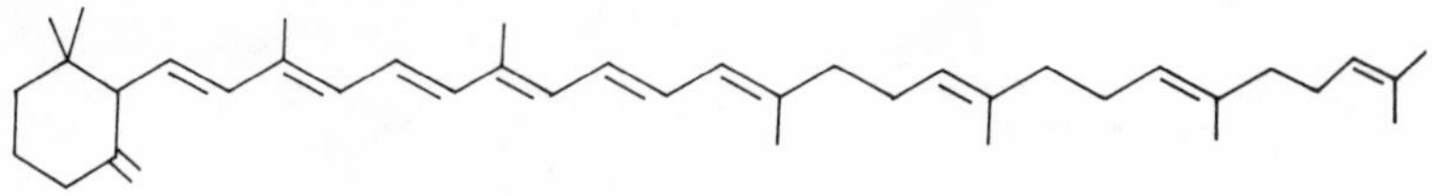

(138)

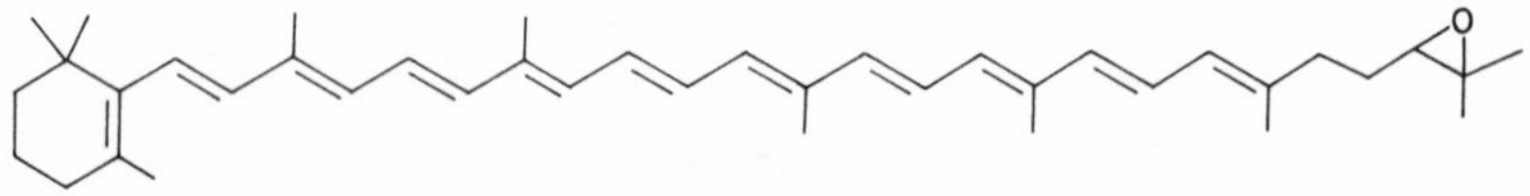

(139)

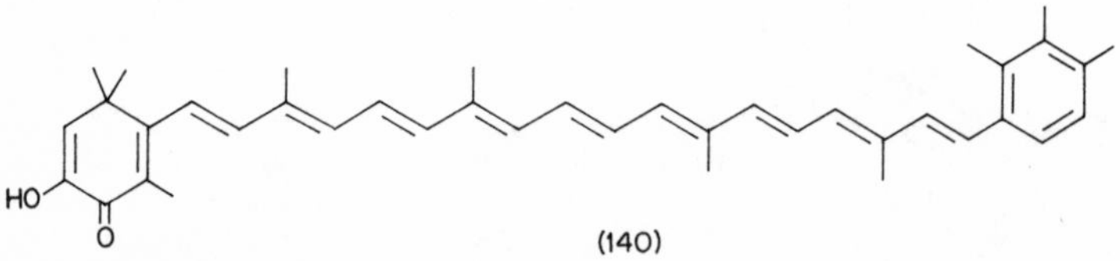

(140)

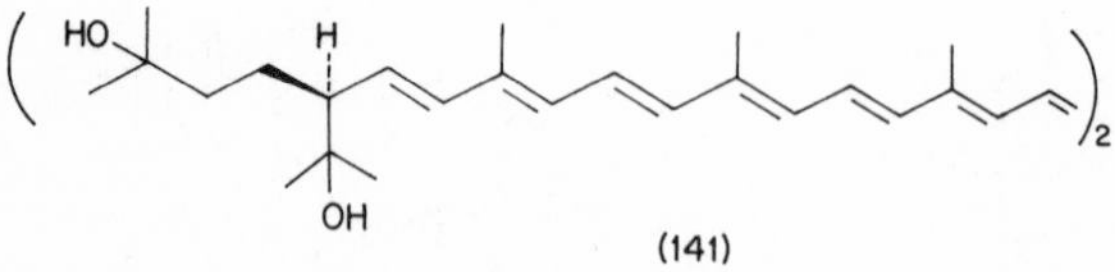

(141)

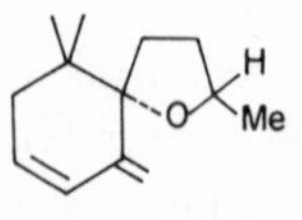

(142)

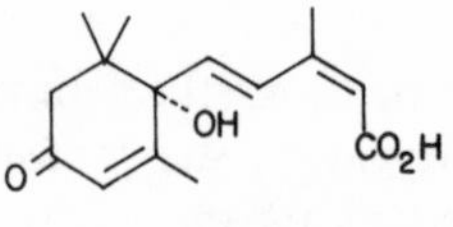

(143)

Some novel carotenoids are exemplified by the carotene (138) isolated from ladybird beetles, the epoxide (139) from a tomato mutant and the diosphenol tedanin (140) obtained from the marine sponge, *Tedania digitata*. The latter is of interest because it possesses an aromatic end group. Although the majority of carotenoids are C_{40} compounds, a number of bacterial C_{30} carotenoids and prenylated carotenoids (e.g. bacterioruberin, 141) have been isolated. Many fragments of carotenoids (apocarotenoids) are found as natural products. Recent examples are the vitispiranes (142) from grapes.

There have been a number of studies on the biological activity and metabolism of the plant hormone, abscisic acid (143)[37] which is probably a degraded carotenoid.

Carotenoids often occur in association with protein, and these complexes, such as that obtained from lobster, have been the subject of several investigations.

The last ten years have seen the determination of the stereochemistry of many carotenoids.[38] Circular dichroism and other spectroscopic measure-

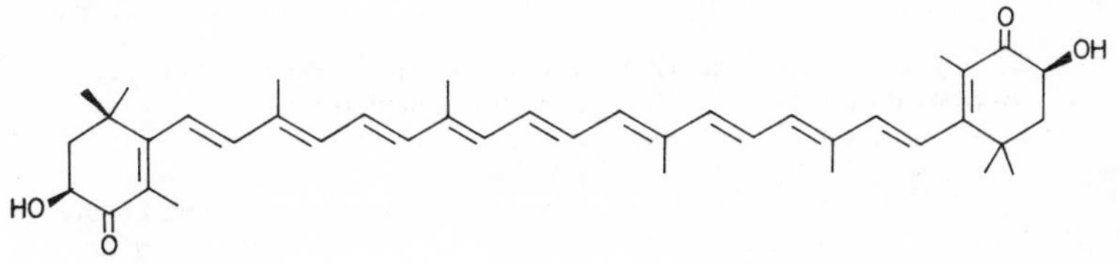

(144)

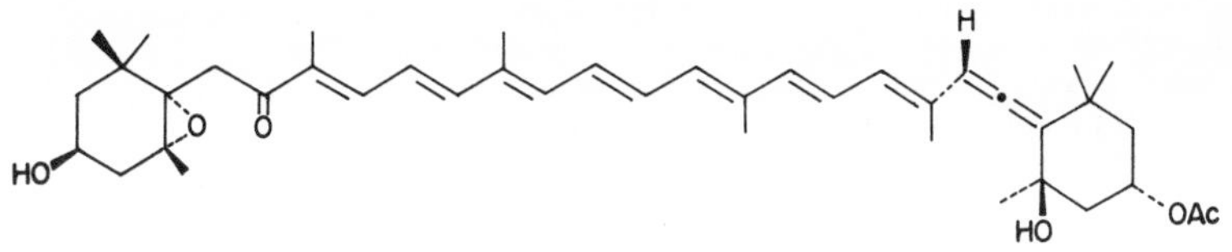

(145)

ments have played a major role in this. Thus astaxanthin is (144) whilst the allene, fucoxanthin is (145).

Recently the study of the biosynthesis of the carotenoids has been concerned not only with the sequence of events from prephytoene pyrophosphate via phytoene but also with the stereochemical course of the carotene cyclizations in the conversion of lycopene to β,β'-carotene.[39,40]

Vitamin A is an important degraded carotenoid which has been the subject of continued chemical investigation in the context of vision.[41] Geometrical isomers of retinal form the prosthetic groups of two classes of biologically important proteins, the rhodopsins and bacteriorhodopsins in which the aldehyde is linked via a Schiff's base to the ε-amino group of a lysine residue. Although the free retinal absorbs light at c. 390 nm and the corresponding Schiff's base at c. 440 nm, both rhodopsins have absorption maxima that are red-shifted to 460–580 nm. A number of models involving protonated Schiff's bases have been constructed to account for the further interaction betweeen the protein and the chromophore. Significant studies have been reported on point group models concerning the role of vitamin A in the visual process.

In this chapter I have endeavoured to portray something of the advances in structure elucidation, synthesis, biosynthesis and chemistry which have taken place in the terpenoids over the last ten years. There remain many structural gaps together with problems of partial and total synthesis and a particular need to confirm stereochemistry. Although many aspects of terpenoid biosynthesis have been elucidated, little is known of the enzymology or control of the various steps whilst the relationship between terpenoid structure and biological activity remains to be explored in many cases.

References

1. For more detailed annual reviews see 'Terpenoids and Steroids', Specialist Periodical Reports, ed. K.H. Overton and J.R. Hanson, Royal Society of Chemistry, London, vols. 1–12.
2. For reviews see E. Guenther, G. Gilbertson and R.T. Koenig, 1977, Anal. Chem., 49, 83R; G. Gilbertson and R.T. Koenig, 1981, Anal. Chem., 53, 61R.
3. R. Baker and J.W.S. Bradshaw, 1979, Ann. Reports (B), Roy. Soc. of Chem., London, 76, 404; J.M. Brand, J.C. Young and R.M. Silverstein, 1979, Prog. Chem. Org. Nat. Prod., 37, 1.
4. D.J. Faulkner, 1977, Tetrahedron, 33, 1421.
5. S. Damtoft, S. Rosendal Jensen and B.J. Nielsen, 1981, Phytochemistry, 20, 2717.
6. 'Biosynthesis of Isoprenoid Compounds', ed. J.W. Porter and S.L. Spurgeon, John Wiley, New York, 1981; D.V. Banthorpe and B.V. Charlwood, in 'Encyclopedia of Plant Physiology' ed. E.A. Bell and B.V. Charlwood, Springer, Berlin, 1980, vol. 8, p. 185.
7. For more detailed reviews also see J.R. Hanson in 'Biosynthesis'. Specialist Periodical Reports, Royal Society of Chemistry, London, vols. 1–7.
8. A.F. Thomas in 'Total Synthesis of Natural Products' ed. J.W. ApSimon, Wiley-Interscience, New York, vol. 2 (1973) and vol. 4 (1980).
9. M. Elliott and N.F. James, 1978, Chem. Soc. Rev., 7, 473.
10. C.A. Henrick, 1977, Tetrahedron, 33, 1845; K. Mori in ref. 8, vol. 4.
11. C.A. Bunton, O. Cori and D. Hachey, 1979, J. Org. Chem., 44, 3238.
12. N.H. Fischer, E.J. Olivier and H.D. Fischer, 1979, Prog. Chem. Org. Nat. Prod., 38, 47.
13. Y. Asakawa, 1982, Prog. Chem. Org. Nat. Prod., 42, 1.
14. Ch. Tamm, 1974, Prog. Chem. Org. Nat. Prod., 31, 63.
15. W.A. Ayer and L.M. Browne, 1981, Tetrahedron, 37, 2199.
16. G.A. Cordell, 1976, Chem. Revs., 76, 425.
17. D.E. Cane, 1980, Tetrahedron, 36, 1109.
18. D. Arigoni, 1975, Pure Appl. Chem., 41, 219.
19. R.M. Coates, 1976, Prog. Chem. Org. Nat. Prod., 33, 73; Sukh Dev, 1981, Prog. Chem. Org. Nat. Prod., 40, 49.
20. E. Fujita, 1981, Bull. Inst. Chem. Res. (Kyoto) 59, 381 and previous reviews in this series.
21. A.J. Weinheimer, C.W.J. Chang and J.A. Matson, 1979, Prog. Chem. Org. Nat. Prod., 36, 285.
22. E. Fujita, Y. Nagao and M. Node, 1976, Heterocycles, 5, 793.
23. F. Piozzi, 1981, Heterocycles, 15, 1489.

24. J.E. Graebe and H.J. Ropers in 'Phytohormones', Elsevier, Amsterdam, 1978, vol. 1, p. 107.
25. 'Gibberellins, Chemistry, Physiology and Use', ed. J.R. Lenton, British Plant Growth Regulator Group Monograph, no. 5, Wantage, 1980.
26. E.J. Corey, R.L. Danheiser, S. Chandrasekaran, G.E. Keck, B. Gopalan, S.D. Larsen, P. Siret, J.L. Gras, 1978, *J. Amer. Chem. Soc.*, 100, 8034.
27. L.N. Mander, 1983, *Acc. Chem. Res.*, 16.
28. G.A. Cordell, 1977, *Progr. Phytochem.*, 4, 209.
29. E. Caspi, 1980, *Accounts Chem. Res.*, 13, 97; E.E. van Tamelen, 1982, *J. Amer. Chem. Soc.*, 104, 6480.
30. W.S. Johnson, 1976, *Bio-organic Chemistry*, 5, 51.
31. P. Pant and R.P. Rastogi, 1979, *Phytochemistry*, 18, 1095.
32. R.F. Chandler and S.N. Hooper, 1979, *Phytochemistry*, 18, 711.
33. G. Ourisson, P. Albrecht and M. Rohmer, 1979, *Pure Appl. Chem.*, 51, 709.
34. 'Carotenoids, Chemistry and Biochemistry', ed. G. Britton and T.W. Goodwin, Pergamon Press, Oxford, 1982.
35. C.H. Eugster, 1979, *Pure Appl. Chem.*, 51, 463.
36. R.F. Taylor and M. Ikawa, 1980, *Meth. Enzymol.*, 67, 233.
37. J.A.D. Zeevaart, 1979, *Amer. Chem. Soc. Symp. Ser.*, 111, 99.
38. S. Liaaen-Jensen, 1980, *Prog. Chem. Org. Nat. Prod.*, 39, 123.
39. T.W. Goodwin, 1979, *Pure Appl. Chem.*, 51, 593.
40. G. Britton in 'Comprehensive Organic Chemistry' vol. 5. ed. E. Haslam, Pergamon Press, Oxford, 1979, p. 1025.
41. D.S. Goodman, 1979, *Fed. Proc.*, 38, 2501.

5 Steroids

B.A. MARPLES

Inevitably, the choice of subjects covered in this short chapter is biased towards the author's own interests. It is hoped that a significant proportion of the work recorded here is of interest to others also.

5.1 General chemistry

Particular consideration is given in this section to protection and deprotection of simple functional groups. The use of new reagents for selected transformations and certain novel mechanistic aspects are also considered.

5.1.1 *Alcohols and their derivatives*

The protection and deprotection of steroidal alcohols allowing selective manipulation of various functional groups has been a subject of considerable interest over many years. In certain cases selective protection or deprotection of polyhydroxy compounds or their derivatives may be achieved directly owing to the varying degree of steric hindrance at different positions in the steroid nucleus. For example, selective hydrolysis of the 3α-acetoxy group in peracetylated bile acids has been reported[1] with methanolic HCl, as has selective base-catalysed hydrolysis of the 3α-formyloxy group in performylated derivatives.[2] The facile reactions at C-3 in the bile acids are easily understood, as the 3α-substituents have the equatorial conformation. However, the well-known slower rate of acetylation of the 12α-hydroxy group versus the 7α-hydroxy group in cholic acid is less easily rationalized. Recent work has ascribed[3] this to intramolecular catalysis by the 3α-acetoxy group and the 12α-hydroxy group, the latter being more effective as it can protonate the N-acetylpyridinium ion, promoting attack by the 7α-hydroxy group (1).

The substituent at C-3 has an effect on the reactivity of the 12α-hydroxy group and the 21-methyl group in the side chain appears to be responsible for some retarding of its reactivity.[4]

Protection of steroid alcohols as their methyl ethers sometimes causes problems owing to the relatively vigorous conditions required for their subsequent deprotection. New methods of deprotection[5] have been applied to steroidal methyl ethers and include the use of $FeCl_3/Ac_2O$,[6] Me_3SiI,[7] $PhSSiMe_3$,[8] $EtSH/BF_3 \cdot Et_2O$[9] and $BBr_3/NaI/15$ crown 5.[10] Benzyl ethers may also be cleaved by the last two reagents and selective cleavage of oestradiol dimethyl ether to give the 3-methyl ether may be achieved with $EtSH/BF_3 \cdot Et_2O$. Selective methylation of oestrogens at the phenolic hydroxy group is not difficult but a new approach, particularly useful for g.l.c. analysis of mixtures, involves extraction from an alkaline solution into dichloromethane containing iodomethane and employs a tetrahexylammonium salt as a phase-transfer catalyst.[11]

The use of steroidal trimethylsilyl ethers for g.l.c. is routine but the t-butyldimethylsilyl ethers are often preferred for preparative work since they are less labile[12,13] and the silylation may be achieved selectively, as exemplified by the preparation of 3-t-butyldimethylsilyloxyandrost-5-en-17β-ol from the 3β,17β-diol.[14] Interestingly, the use of Bu^tMe_2SiI generated in situ from the phenyl selenosilane and iodine allowed the silylation of the 5α-hydroxycholestane (2).[15] Cleavage of t-butyldimethylsilyl ethers which is routinely achieved with aqueous acetic acid or tetrabutylammonium fluoride[14] may also be brought about using NBS-$DMSO$-H_2O[16] and trityl or lithium tetrafluoroborate.[17] As would be expected the less hindered t-butyldimethylsilyl ethers are cleaved more readily than those which are more hindered, and useful selectivity may be observed.[12]

Over the last decade or so, a variety of new oxidizing agents have been developed and have been used to oxidize steroid alcohols. Typically, pyridinium chlorochromate is an efficient and mild example.[18] Reagents which offer some selectivity of reaction are of particular interest and these include a number of oxidants adsorbed on solid supports.[19] Silver carbonate on celite was used selectively to oxidize one hydroxy group in steroids containing two or more oxidizable hydroxy groups. For example, the 3α-hydroxy group of the bile acids was selectively oxidized[20] as was the 3β-hydroxy group of 3,6α- and 3,7-dihydroxy-5α-steroids.[21] The selectivity shows some dependence on solvent. Chloral on alumina was used to selectively oxidize the 17β-hydroxy group of 5α-androstane-3β,17β-diol[22] and bromine or chlorine with $HMPA$[23] and chlorine with pyridine[24] showed selectivity for oxidation of the secondary hydroxy group of primary, secondary diols. Typically, the 3β,23-diol (3) has been converted to the 3-keto-analogue.[23] The transformation of the 3β,19-dihydroxyandrostene (4) to the

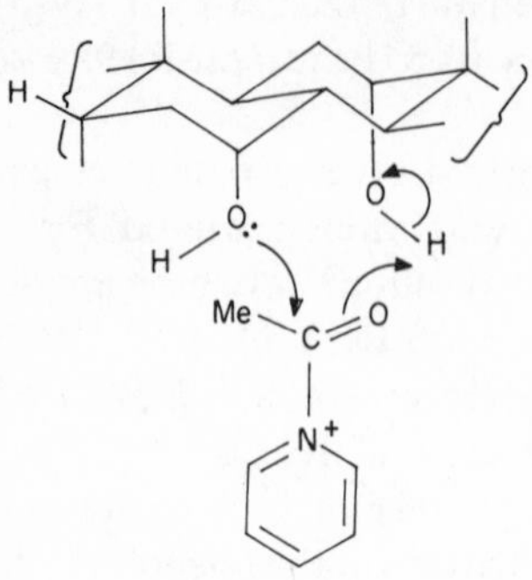

(1)

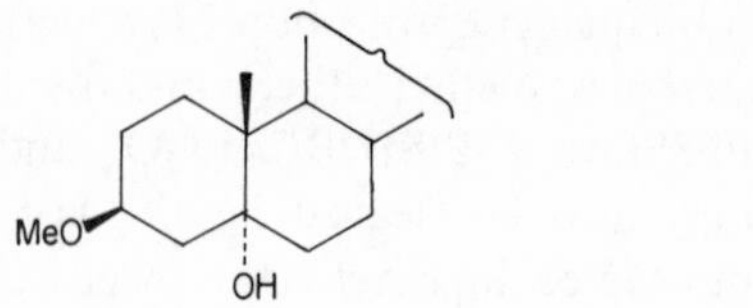

(2)

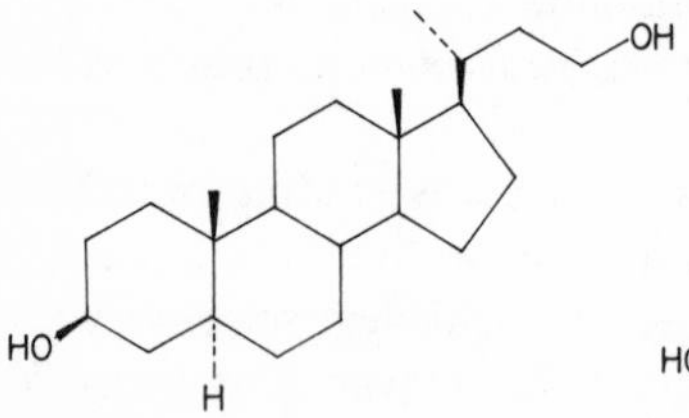

(3)

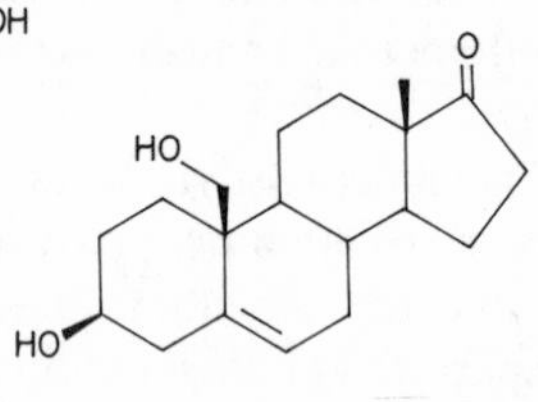

(4)

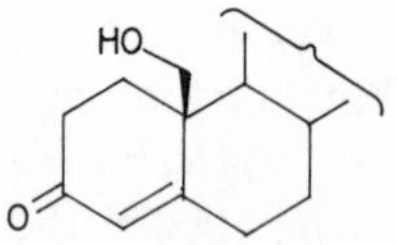

(5)

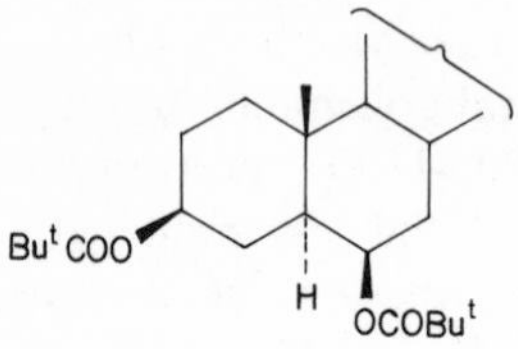

(6)

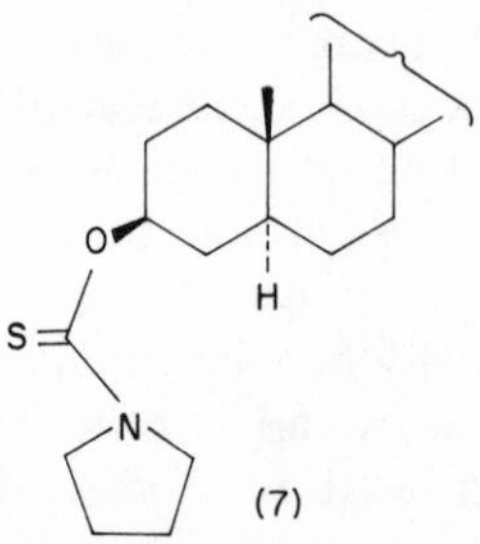

(7)

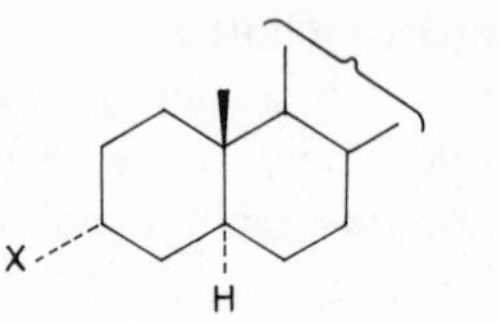

(8) X = RCO$_2$　　(11) X = SCN

(9) X = PhO　　　(12) X = CN

(10) X = PhS　　　(13) X = N$_3$

ketol (5) through the bistrityl ether which was successively treated with Ph$_3$CBF$_4$ and dilute acid is relevant to these studies.[25] Benzeneseleninic anhydride has proved to be a useful oxidizing agent and may be employed to convert the ketones to their α,β-unsaturated derivatives.[26] The oxidation of cholesterol to cholest-5-en-3-one without significant isomerization to cholest-4-en-3-one may be achieved with chromium trioxide-pyridine complexes in methylene chloride[27] and particularly good results have been reported for pyridinium chlorochromate on alumina.[28]

Considerable interest has been shown recently in the deoxygenation of steroidal alcohols via various free radical processes.[29] Typically, the alcohols may be converted into the thiobenzoates, the S-methyl dithiocarbonates, the thiocarbonylimidazolides or the Se-phenyl selenocarbonates all of which may be reduced with tributylstannane. An advantage of these processes is the avoidance of cationic intermediates and their associated rearrangements. Accordingly, high yields of cholest-5-ene were obtained from cholesterol by these procedures. A similar reaction sequence involved conversion of the alcohol to the chloroformate followed by treatment with tripropylsilane in the presence of t-butyl peroxide.

Reductive cleavage of sterically hindered esters may be achieved with lithium in ethylamine or potassium in t-butylamine containing 18-crown-6. The use of the latter reagent was nicely exemplified in the conversion of the dipivaloate (6) to 5α-cholestan-3β-ol. The former reagent may also be used to reduce sterically unhindered S-methyl dithiocarbonates and particularly thiocarbonylamide esters. For example, the 5α-cholestan-3β-ol derivative (7) gave 5α-cholestane in high yield.

The reactions of steroidal alcohols with active hydrogen compounds in the presence of diethyl azodicarboxylate and triphenylphosphine are very useful in providing a variety of derivatives in which the configuration of the reacting carbon is inverted.[30] Thus, 5α-cholestan-3β-ol was smoothly converted into the 5α-cholestane derivatives (8)–(13) by reaction with RCO_2H, PhOH, PhSH, HSCN, HCN and HN_3 respectively[31 – 33] and similarly the vitamins D were converted into their epimers and other derivatives.[32] Whereas cholest-4-en-3β-ol underwent simple S_N2 type substitution,[34] the reactions of cholesterol were dependent on the particular nucleophile used and, with benzoic acid, the products appear to be derived from an intermediate homoallylic cation.[35] In general, the ease of reaction depends on the nucleophilicity of the anion derived from the active hydrogen compound and the steric hindrance suffered by the steroidal hydroxy group. Accordingly, the most striking results were obtained in the 3-hydroxy-steroids. A number of other S_N2 type processes which are also particularly applicable to saturated 3-hydroxy-steroids allow a variety of derivatives to be prepared with clean inversion at the reacting carbon atom. Reaction of the alcohols with N-halogenosuccinimide (Cl, Br, or I) and either triphenylphosphine or triphenyl phosphite in THF gave the halogeno-derivatives.[36] The fluoro-drivatives may be prepared by reaction with diethylaminosulphurtrifluoride,[37, 38] phenyltetrafluorophosphorane or diphenyltrifluorophosphorane.[39] Magnesium iodide is an effective milder alternative to sodium iodide in converting tosylates to the inverted iodides,[40] and tosylates may be converted to the inverted alcohols with tetrabutylammonium hydroxide in DMSO or N-methylpyrrolidone.[41]

5.1.2 Epoxide ring opening reactions

The characteristic preferred diaxial opening of epoxides in acid-catalysed reactions with nucleophiles has long been recognized. Relatively recent studies have demonstrated the influence of neighbouring groups on such reactions. For example, 4β-acetoxy-5,6β-epoxy-5β-cholestane (14) reacted with HBr to give the 6α-bromo-5β-hydroxy-compound (16), the product of diequatorial opening of the epoxide. The absence of diaxial products is ascribed to the strengthening of the C–5–O bond by the electron-withdrawing 4β-acetoxy group. In agreement with this suggestion is the reaction of the hydroxy-epoxide (15) which, under similar conditions, was less specific and

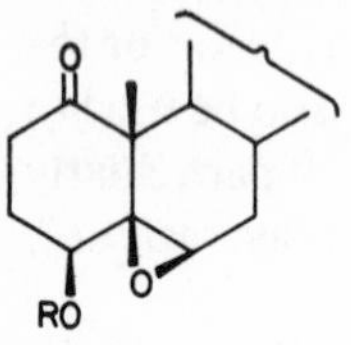

(14) R = Ac   (16) R = Ac   (18)
(15) R = H   (17) R = H

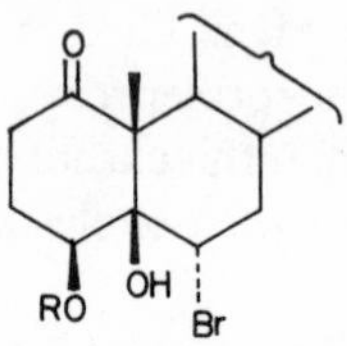

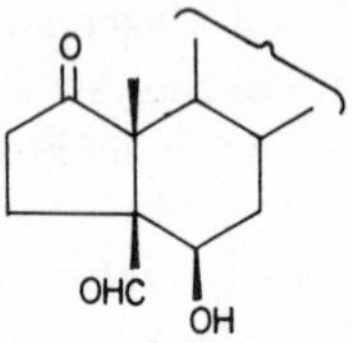

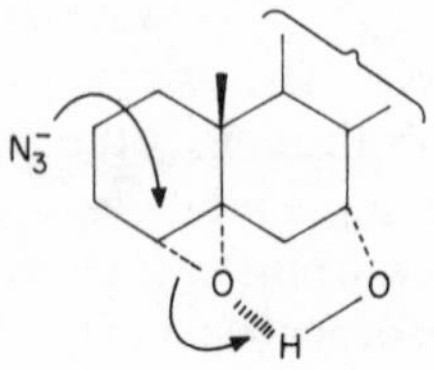

(19)

gave a mixture of the 6α-bromo-4β,5β-dihydroxy-compound (17) and the rearrangement product (18) arising from C–5–O bond cleavage.[42] A neighbouring hydroxy-group may be important in determining the rate of nucleophilic epoxide-opening reactions. Thus, a 7α-hydroxy accelerates the rate of reaction of azide ion with a 4α,5α-epoxide by intramolecular hydrogen bonding (19).[43] A recent study of particular interest has shown that functional groups at C-19 may markedly influence the outcome of reactions of 2,3-, 4,5-, 5,6-, 6,7- and B-homo-6,7-α-epoxides with HX.[44] In general, for 19-hydroxy-, -methoxy- and -acetoxy-compounds the major reaction resulted in the formation of a 5-membered ring by participation of the 19-oxygen atom lone pair of electrons ($5(O)^n$ participation). Not surprising the reactivity decreases across the series OH > OMe > OAc. Typically, 2α,3α-epoxy-19-methoxy-5α-cholestane (20) or its 19-acetoxy-analogue (21) reacted with aqueous perchloric acid to give the tetrahydrofuran (22) and the methoxy-diol (23) or the acetoxy-diol (24) respectively. An efficient competing reaction involves the formation of a 6-membered ring by participation of the carbonyl

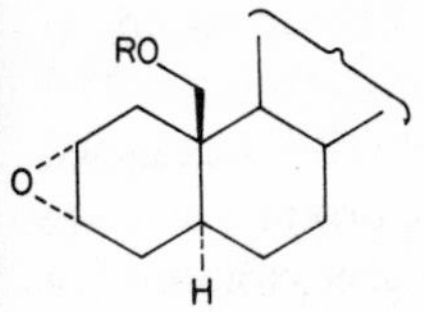

(20) R = Me
(21) R = Ac

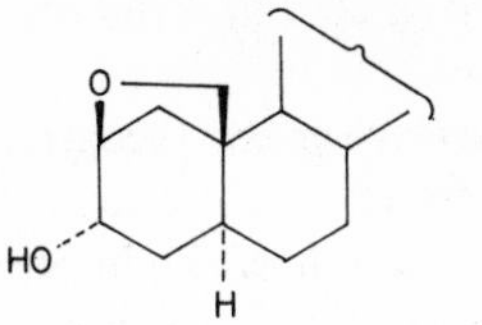

(22)

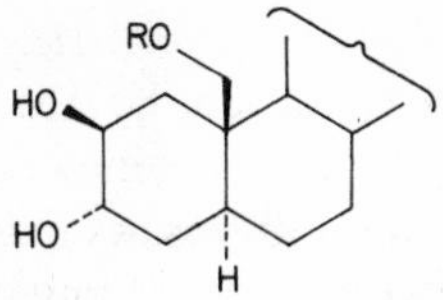

(23) R = Me
(24) R = Ac

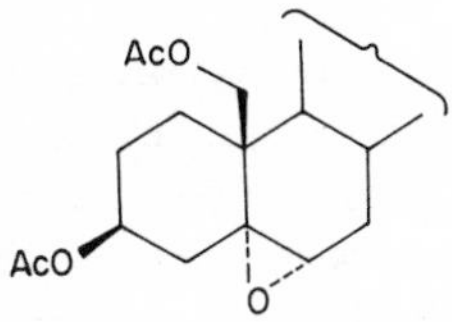

(25)

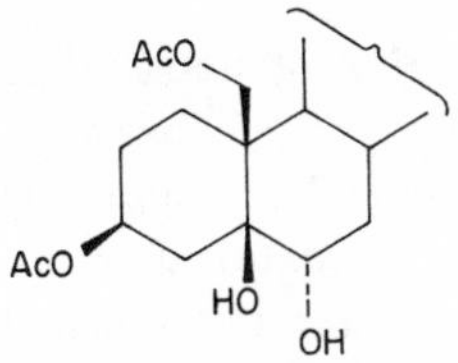

(26)

oxygen atom of the 19-acetoxy-group $(6(O)^{\pi,n}$ participation). Accordingly, $3\beta,19$-diacetoxy-5,6α-epoxy-5α-cholestane **(25)** reacted with aqueous perchloric acid to give the diol **(26)** which is the diequatorial product. Reactions with aqueous HBr do not always follow the same course owing to the presence of the more nucleophilic bromide ion which may compete effectively with the intramolecular nucleophile. Other factors which are significant in determining the reaction course are the angle of approach of the participating group, the

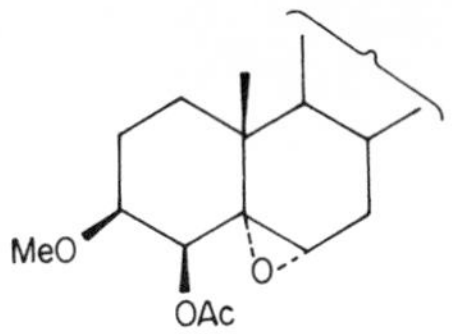

(27)

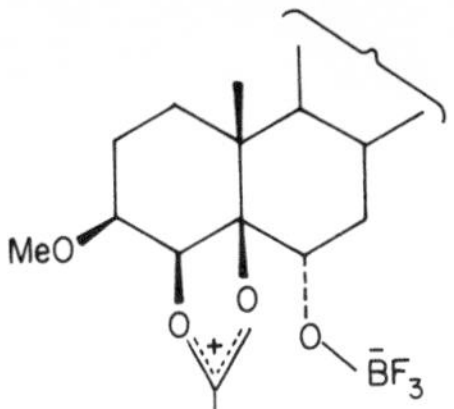

(28)

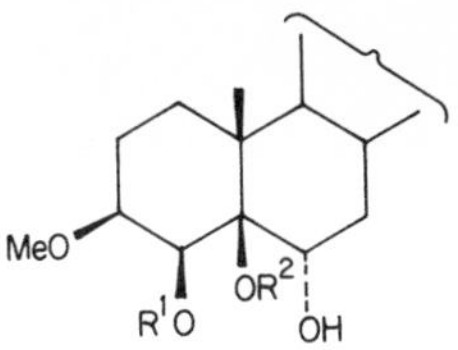

(29) R^1 = H , R^2 = Ac
(30) R^1 = Ac , R^2 = H

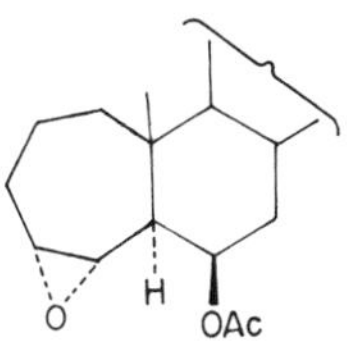

(31)

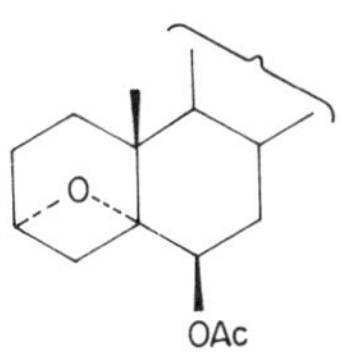

(32)

steric compression at the reaction site, and the presence of polar groups at C-3.[45]

Participation by the 4β-acetoxy-group occurred in the BF$_3$-catalysed diequatorial opening of the 5α,6α-epoxycholestane (27). The product mixture (29) and (30) results from the reaction of the intermediate acetoxonium ion (28) with water.[46] Similar formation of acetoxonium ions has been reported in the opening of the epoxide (31)[47] and the 3α,5α-oxetane (32).[48]

5.1.3　Unsaturated steroids

The well-known reaction of steroidal alkenes with iodine and silver carbo-xylates normally affords the cis- or trans- (diaxial) diols from an intermediate acyloxonium ion dependent upon reaction conditions (Woodward or Prevost). The trans- diequatorial 2α,3β-diol (34) has been prepared (58%) from 5α-cholest-2-ene through the use of iodine with silver salts of sterically demanding carboxylic acids (e.g. mesitoic acid). Steric inhibition of axial approach to C-3 of the 2β,3β-acyloxonium ion (33) is responsible for the slightly preferred diequatorial cleavage.[49] Even greater selectivity, leading to a high yield (77%) of the diequatorial product (34) was achieved through treatment of the benzoyloxonium ion (36) with silver benzoate. The ben-zoyloxonium ion (36), which is subject to severe steric inhibition of axial approach at C-2 was prepared in situ by reaction of the orthoester (35) with boron trifluoride. Reaction of 5α-cholest-2-ene with iodine and potassium iodate in acetic acid provided 2β-acetoxy-3α-iodo-5α-cholestane which was converted into the 2β,3β-diol[50] while a similar reaction in aqueous dioxan

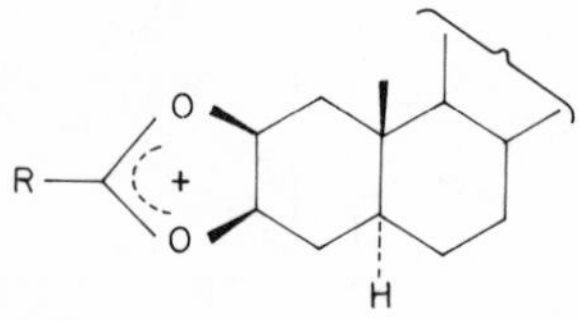

(35) 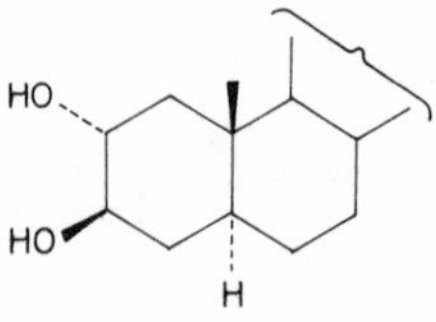　　　　　　　　(34)

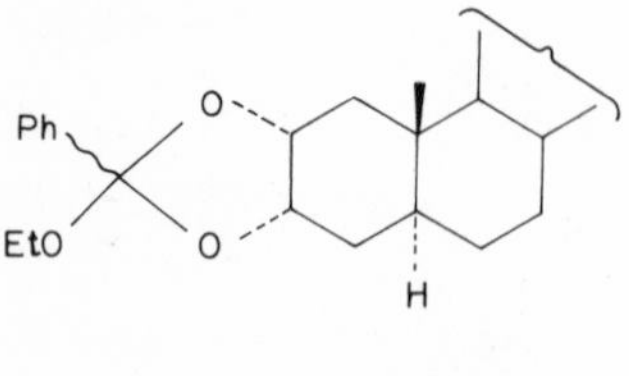

(35) 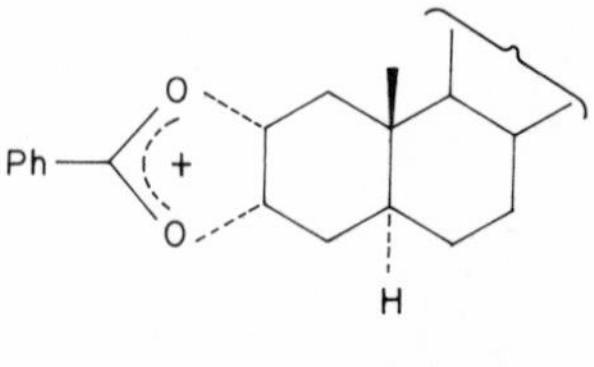　　　　　　　　(36)

containing sulphuric acid gave the $2\beta,3\alpha$-diol.[51] Further iodo-acyloxylation procedures employ the use of iodine triacetate[52] and thallium(I) carboxylates and iodine[53] and thallium(III) acetate is useful for the conversion of Δ^2-compounds to the $2\beta,3\beta$-diols.[54]

Transannular participation of the 5α-hydroxy-group occurs in the reactions of 5-hydroxy-5α-cholest-2-enes (37) with hypobromous acid and with iodine-silver acetate[56] leading to $2\alpha,5\alpha$-epoxides (38) and (39) respectively. Studies on the participation by 19-substituents in the reactions of Δ^2-, Δ^3-, Δ^4-, Δ^5- and Δ^6-compounds with electrophilic reagents complement the related work on the acid-catalysed cleavage reactions of 19-substituted epoxides (see earlier).[44,57] In the Δ^5-series such reactions were useful for the introduction of the 5β-hydroxy-group, a feature of some cardenolides.[58]

Protection of the $\Delta^{5,7}$-diene moiety was usefully carried out by reaction with 4-phenyl-1,2,4-triazoline-3,5-dione to give adducts of the type (40). Regeneration of the $\Delta^{5,7}$-diene was reported by reaction with lithium aluminium hydride,[59] potassium hydroxide,[59] and potassium carbonate in DMSO or DMF,[60] and by heating in tetramethylguanidine or collidine.[61] Useful protection was also achieved in the tricarbonyl iron adduct (41) from which the $\Delta^{5,7}$-diene may be regenerated by treatment with ferric chloride.[62]

The spin-forbidden addition of ground-state triplet oxygen to ergosteryl acetate to give the endoperoxide (42) may be achieved thermally in the presence of Lewis acids or ammoniumyl radicals.[63–65] The endoperoxide (42) is the well-known adduct of ergosteryl acetate and singlet oxygen which is not thought to be involved in these and related photochemical reactions. Rather it is suggested that the radical-cation (43) is a key intermediate.[64,65]

$5E$- and $5Z$-vitamins $D^{66,67}$ and some of their derivatives[68] reacted with sulphur dioxide to give the sulphones (44) and (45). Extrusion of sulphur dioxide from the mixed adducts thermally[66,67] or by the use of methanolic potassium hydroxide[66,68] gave the $5E$-vitamins D or their derivatives although, in the thermal process, isomerization to other products was reported to be a problem.[66] The preparation of 6,19,19-trideuteriated $5E$-vitamins D_3 and some derivatives was achieved by carrying out the extrusion of sulphur dioxide in MeOD-Bu^{t}OK-D$_2$O.[68]

Vitamin D_2 and ergosterol were converted to their palladium π-allyl complexes (46) and (47) respectively.[69] Interest in palladium π-allyl complexes in general[70,71] stems from their potential as synthetic intermediates owing to their ready oxidation to allylic alcohols[72] and enones[73] and their reactivity with nucleophiles.[71,74] The π-allyl complexes are also formed in the palladium(O)-catalysed nucleophilic displacements of allylic acetates.[71] A detailed study of the formation of palladium π-allyl complexes (α-4(6η)-PdCl complexes) (48) from Δ^4-3-oxo-steroids showed that Lewis acid (PdCl$_2$ or PdCl$_4^{2-}$)-catalysed enolization with loss of the 6β-H is followed by in-

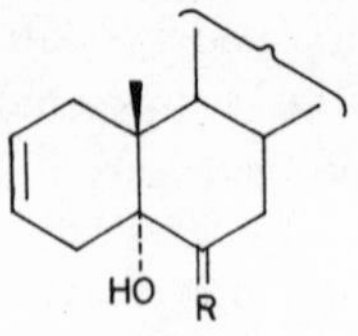

(37)

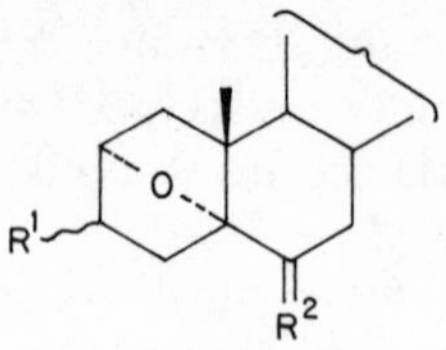

(38) R^1 = β-Br
(39) R^1 = α-OAc

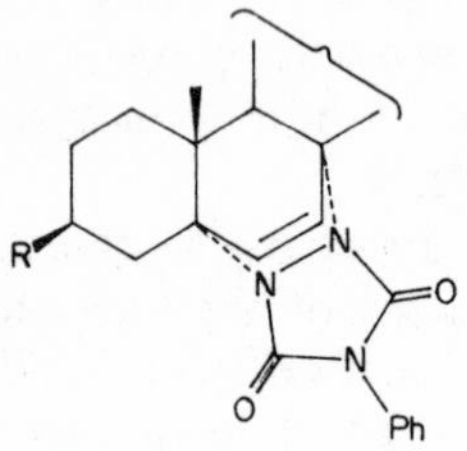

(40)

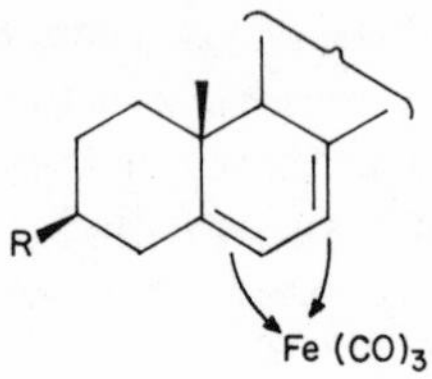

(41)

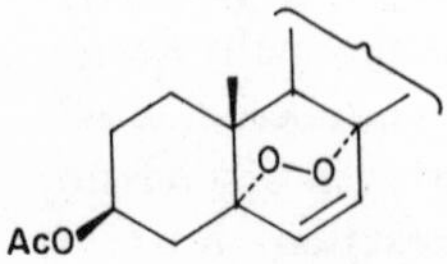

(42)

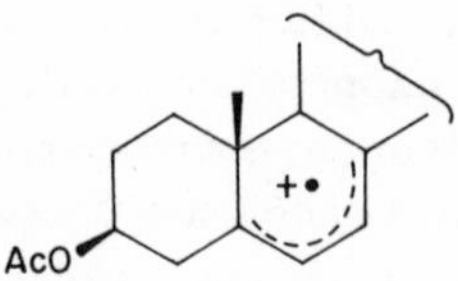

(43)

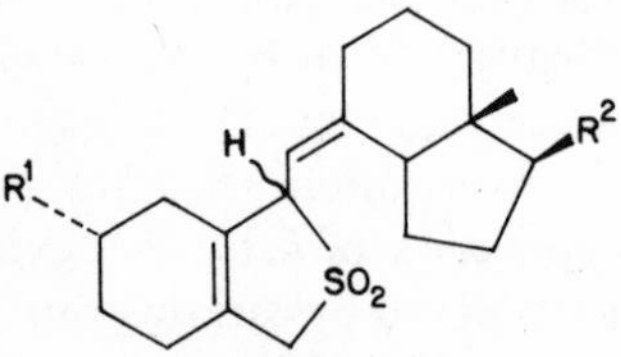

(44) 6α-H
(45) 6β-H

tramolecular α-face transfer of the Pd from oxygen to carbon.[75] Normally, α- and β-(4-6η)-PdCl complexes are formed by *syn* elimination of the 6-H in Δ[4]- compounds containing no 3-oxo-group.[76]

The use of the CrO_3-3,5-dimethylpyrazole complex[77,78] for direct allylic oxidation of alkenes to α,β-enones may in some cases be more efficient than the comparable CrO_3-pyridine complexes.[79]

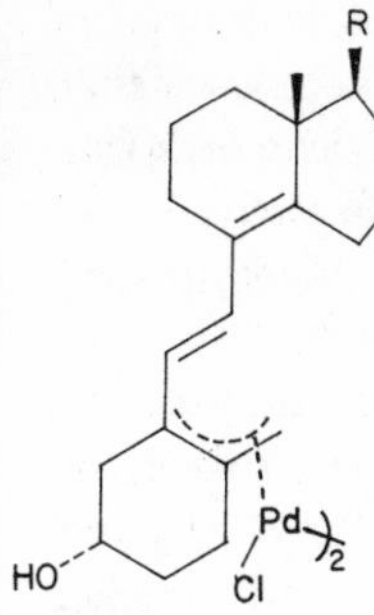

(46) 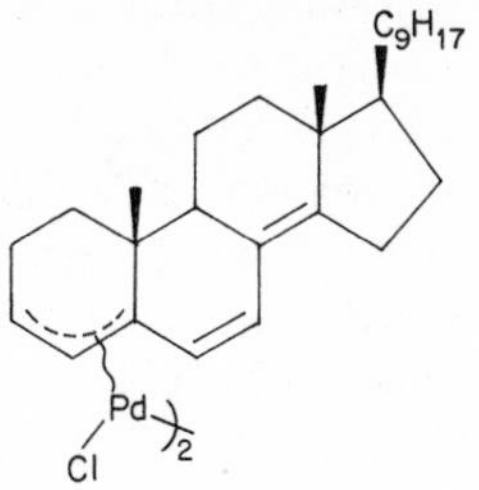   (47) 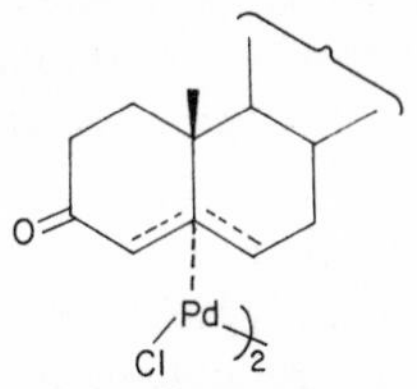   (48)

5.1.4   Carbonyl compounds

As indicated earlier, steroidal ketones are readily dehydrogenated to the α,β-enones directly using benzeneseleninic anhydride[26,80] as an alternative to DDQ and SeO_2. Typically, 5α-cholestan-3-one gave in high yield the 1,4-dien-3-one and cholesta-4,6-dien-3-one which may be oxidized to cholesta-1,4,6-trien-3-one. Interestingly, cholesta-1,4-dien-3-one was not converted to the 1,4,6-trien-3-one. Dehydrogenations were also achieved[81] using $PdCl_2$-HCl-Bu^{t}OH and are exemplified by the conversion in high yield of 5α-cholestan-3-one to the Δ^1-3-oxo-compound and 5β-cholestan-3-one to the Δ^4-3-oxo-compound.

Selective reduction of dicarbonyl compounds is preparatively important and some interesting results have been obtained with poly-(N-isopropyl-iminoalane).[82,83] For example selective reduction of the 6-oxo-group of $3\alpha,5$-cyclo-5α-androstane-6,17-dione and $3\alpha,5$-cyclo-5α-pregnane-6,20-dione was achieved using this reagent[83] as was the selective reduction of the 3-oxo-group of progesterone and androst-4-ene-3,17-dione.[82] Selective reduction of the 3-oxo-group in 3,6-, 3,17-, and 3,20-diketones at position 3 was also achieved by use of dehydrated alumina impregnated with propan-2-ol.[84]

The well-established 'product development control' in hydride reductions of 3-oxo-steroids, whatever the explanation,[85] leads preferentially to the equatorial 3-hydroxy-compound. The use of the Mitsunobo reaction[30] (see earlier) allows the preparation of the axial products indirectly. Direct approaches involve the reduction of the 5α- or 5β-3-oxo-compounds with K- and L-selectride[86,87] or with bornan-2-exo-yloxyaluminium dichloride.[88] The stereochemical outcome of the hydride reductions of steroidal α,β-enones may be predicted using an approach vector analysis which suggests that the *quasi*-axial substituents at C-5 and C-6 in the cyclohexenone are of great importance.[89]

Protection of steroidal ketones and aldehydes as their acetals or thioacetals has found considerable general use. Novel conversions of the ketones to their acetals[90] have included the acid-catalysed reaction with diethylene orthocarbonate[91] and with 2-methylenepropane-1,3-diol.[92] The acetals derived from the latter reaction may be cleaved without the use of acid. Methylthiotrimethylsilane was used to convert aldehydes and ketones to their bis(methylthio)acetals under mild conditions without acid-catalysis and selective reaction at C-3 was achieved in 3,17-diketones.[93] Similar selective conversion of progesterone and androst-4-en-3,17-dione to the 3-ethylenethioacetals was achieved with $Me_3SiSCH_2CH_2SSiMe_3$-ZnI_2,[94] and wet silica gel is a useful selective reagent for the cleavage of acetals of α,β-enones.[95] In steroid chemistry in particular, some cleavages of thioacetals using the traditional mercury(II) salts are not very efficient. A number of new reagents have been developed[96] and have found application in steroid chemistry. Worthy of particular mention are MeI-H_2O-$MeOH$,[97] isoamyl nitrite,[98] periodic acid,[99] and the versatile benzeneseleninic anhydride,[100] which is also of value in deprotection of hydrazones and semicarbazones.[101] Trimethylsilyl iodide is a useful anhydrous and neutral deacetalization agent.[102,103]

1,2-Transposition of carbonyl groups in steroids is often very desirable, typically from C-12 to C-11. A number of recently developed methods for such 1,2-transpositions have been applied to steroids. For example oestrone methyl ether was converted[104] to the isomeric 16-ketone through the 16-phenylthio-16-ene (49) as indicated in Scheme 1. The first step in this sequence involves sulphenylation of the 17-oxo-steroid. In general, sulphenylation-desulphenylation[105,106] and selenylation-deselenylation[107] have found applications in steroid chemistry for the conversion of carbonyl compounds to α,β-enones. 5α-Cholestan-3-one gave a mixture of the 2- and 4-ketones

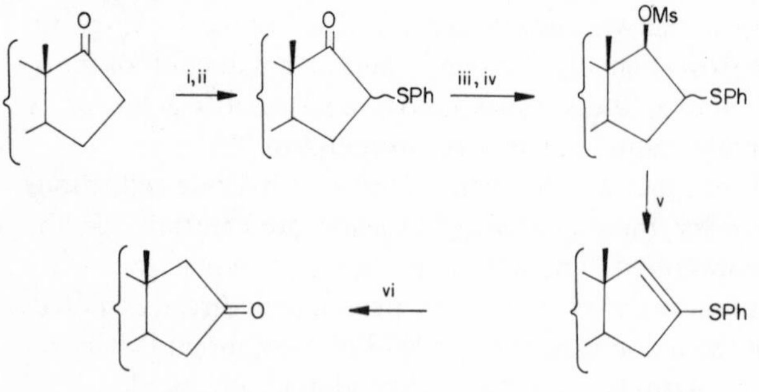

(49)

Scheme 1

Reagents: I, Li N-cyclohexyl-N-isopropylamide; ii, PhSCl; iii, $NaBH_4$; iv, $MeSO_2Cl$—py; v, Bu^tOK—DMSO; vi, $HgCl_2/MeCN/H_2O$

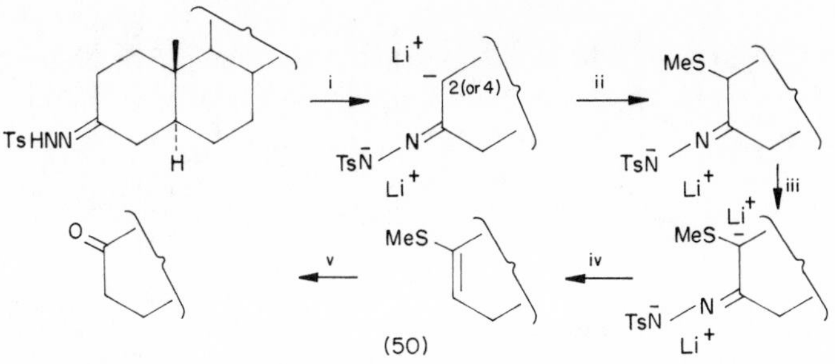

Scheme 2

Reagents: i, BuLi—TMEDA; ii, MeSSMe; iii, BuLi; iv, NH$_4$Cl,Δ; v, HgCl$_2$

through the tosylhydrazone which was transformed via a Shapiro reaction to the enol thioether (**50**) in one pot (Scheme 2).[108] The benzenesulpho-nylhydrazone of 5α-cholestan-3-one has been converted to the vinyl silane (**51**) which gave the epoxide (**52**) with MCPBA. Reduction with LiAlH$_4$ and oxidation of the resultant hydroxy-silanes (**53**) and (**54**) afforded a mixture of the 2- and 3-ketones.[109,110]

α-Hydroxylation of steroidal ketones is very useful from the synthetic and degradative standpoint. Of several new methods of α-hydroxylation of carbonyl compounds, those which have found application in steroids include the MCPBA.[111] Pb(OAc)$_4$,[112] or OsO$_4$–N-methylmorpholine-N-oxide[113] oxidation of enol silyl ethers and the MoO$_5$–pyridine–HMPA oxidation of lithium enolates.[114]

5.2 Rearrangements

5.2.1 Carbocation rearrangements

Wagner-Meerwein rearrangements of steroids have been extensively studied over many years.[115] Backbone rearrangements, which involve multiple Wagner-Meerwein rearrangements of the hydrogen atoms and methyl groups attached to the carbon atoms at ring junctions, have been known for some time,[115] the earliest examples being in the triterpenes.[116]

Such backbone rearrangements are also recognized as being involved in the biosynthetic pathways from squalene 2,3-epoxide to tetracyclic triterpenes such as lanosterol.[117] Recently, attention has focused on the finer details of the mechanisms of *in vitro* backbone rearrangements of steroids with particular reference to their being concerted or not. It has been shown that cholest-5-ene rearranges in refluxing acetic acid containing toluene-*p*-sulphonic acid to give the 5β,14β-dimethyl-18,19-bisnorcholest-13(17)-ene

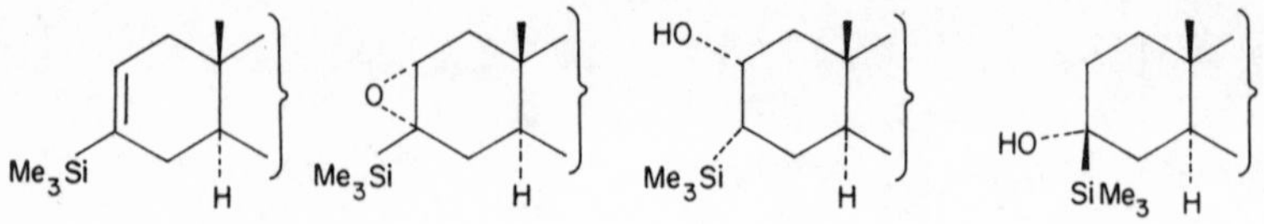

(51)   (52)   (53)   (54)

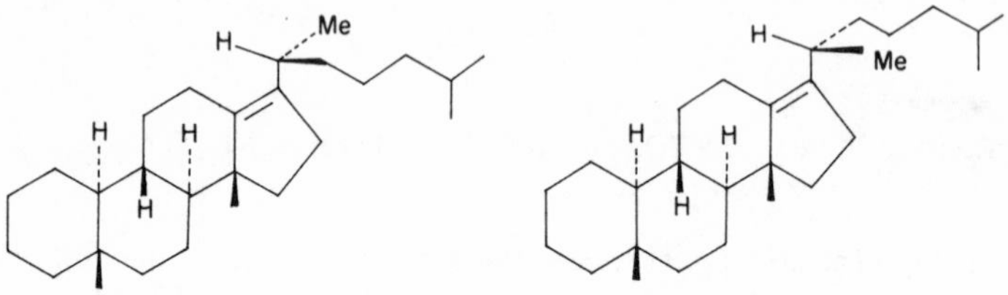

(55)   (56)

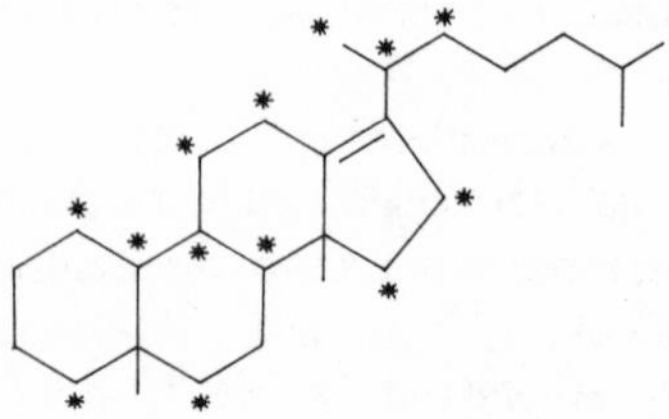

(57)   * deuterium incorporated

(55) which slowly equilibrates to a mixture of (55) and its 20 S-isomer (56).[118] In AcOD the rearrangements proceeded with incorporation of deuterium in all the backbone and immediately adjacent carbon atoms (57) indicating that the individual steps of the backbone rearrangement occur relatively slowly compared with carbocation ⇌ alkene interconversions at each backbone site. The driving force in such backbone rearrangements in cholestenes appears to be the relief of torsional strain suffered by the side chain.[119] In certain cases backbone rearrangements in deuteriated acids or solvents may lead to incorporation of deuterium in the angular methyl groups at C-5[120] or C-14.[116] In such cases it is suggested that protonated cyclopropanes or spiran intermediates may be involved (Scheme 3). However, not all backbone rearrangements necessarily involve alkene or cyclopropane intermediates. For example treatment of 4β-acetoxy-5α-cholestan-5-ol with D_2SO_4-DOAc-Ac_2O at room temperature gave a mixture of Δ^9-, $\Delta^{8(14)}$- and $\Delta^{13(17)}$- products with no incorporation of deuterium.[121] It is suggested that the

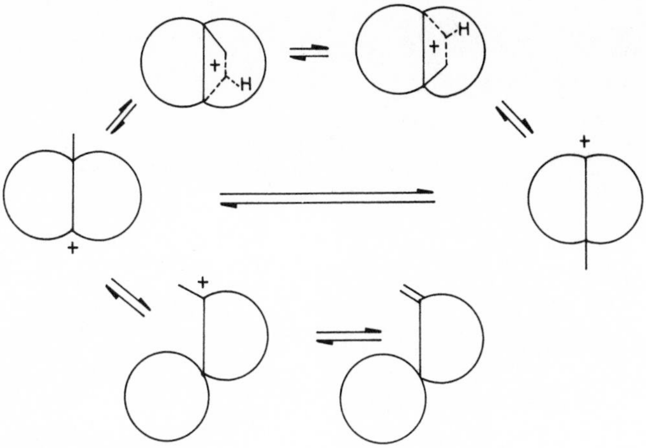

Scheme 3

absence of alkene or cyclopropane intermediates may be a function of the mild conditions employed. It is apparent that a fully concerted series of 1,2-shifts is not involved since products derived from intermediate carbocations were isolated. The general view of backbone rearrangements including those arising from Lewis acid-acatalysed cleavages of 4,5- and 5,6-epoxides[122,123] is that a fully concerted mechanism is rather unlikely.[124] The well known Westphalen rearrangement[125] of 6β-substituted-5α-hydroxy-steroids to the 5β-methyl-Δ^9-compounds [e.g. (58)→(59)] in H_2SO_4-Ac_2O-AcOH is a partial backbone rearrangement and the author favours a mechanism involving a concerted shift of the 10β-methyl group and departure of the 5α-OSO_2OAc group based on a study of 19- and 4-methyl-5α-hydroxy-steroids and other derivatives.[124] The difference in extent of rearrangements observed for hydroxy-steroids under Westphalen-type conditions versus those usually observed for epoxides with Lewis acids in aprotic solvents is almost certainly, in part, a function of the solvents and reagents used.[124] Additionally, the rearrangements in general are influenced by subtle conformational, steric, and electronic effects.[121–124,126]

The rearrangement of epicholesterol with $Tl(OAc)_3$ to the 3α,10α-epoxy-5β-methyl-compound (60) is an interesting variation of the Westphalen rearrangement[127] and involves the sequence shown in Scheme 4. The 3α,10α-epoxy-5β-methyl-compound is one of the rearrangement products derived from BF_3-catalysed cleavage of the oxetane (32).[48] The lability of the oxetane (32) contrasts with that of the 3β,5β-oxetanes (61)–(63) which were reported to be stable toward various acid catalysts.[128] The 3β,5β-oxetane (61) had been reported[128] as the product of solvolysis of 3β-tosyloxy-5β-cholestan-5-ol-6-one (64) but reinvestigation[48,129] indicated that the product should be

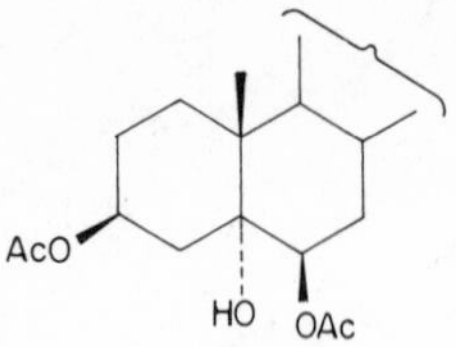

(58) 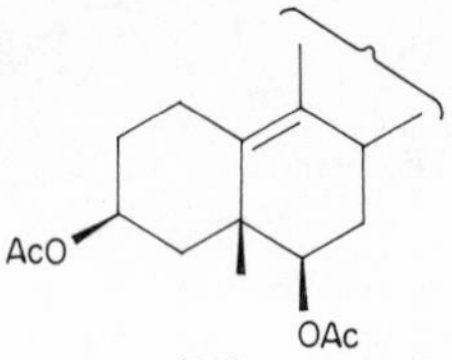 (59)

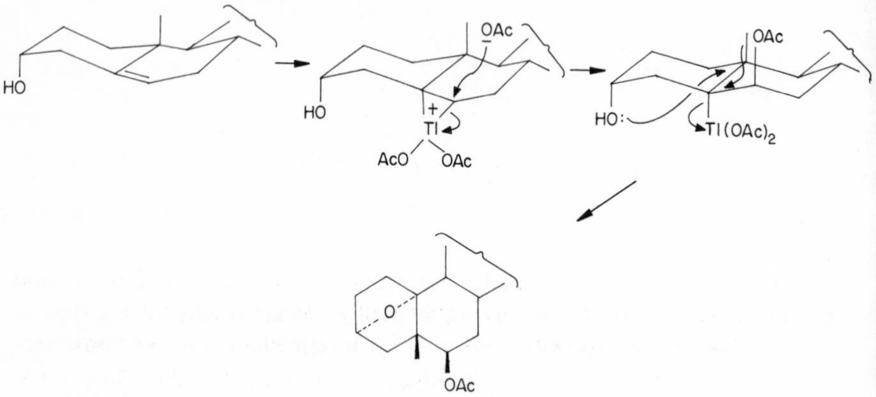

(60)

Scheme 4

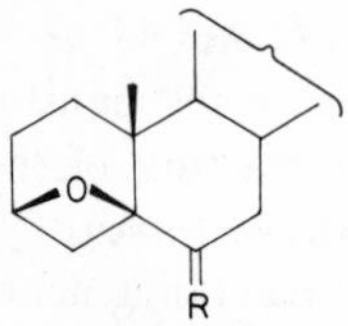

(61) R = O
(62) R = α-OAc, H
(63) R = α-OCOCF$_3$, H

formulated as the rearranged A-homo-B-nor-compound (65). Such a structure, which could be derived by the pathway shown in Scheme 5, would be expected to be considerably more stable than the 3β,5β-oxetane (61) towards acids.

Acetolysis of D-homo-5α-androstan-17aβ-yl tosylate (66) was accompanied by isomerization at C-13 and gave the 17aα-acetoxy-13α-derivative (69).[130] The key intermediate appears to be the C-homo-Δ^{12a}-compound (68) which undergoes extensive protonation and deprotonation since, in AcOD, deuterium is widely incorporated [see (69)]. The product (69) is presumably

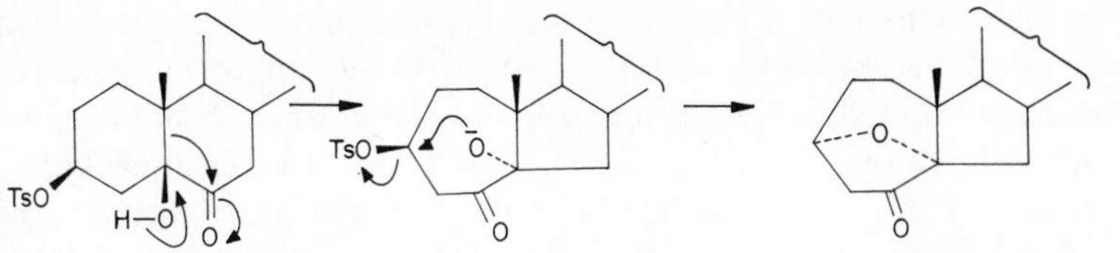

(64)   (65)

Scheme 5

thermodynamically favoured. Similar isomerization at C-13 has been observed as a minor pathway in the solvolyses of the 17-methyl analogues (67)
which gave complex mixtures of products.[131,132] Isomerization at C-14
occurred in the treatment of oestr-4-ene-3,17-dione (70) with HF-SbF$_5$ and
the thermodynamically stable 14β-isomer (71) was formed. It has been
established that both carbonyl oxygen atoms are protonated forming a
hydroxyallyl cation (74) in ring A and, at 0°C, C-4 is additionally protonated
promoting migration of the cationic centre between C-5 and C-14 through a
deprotonation-protonation mechanism.[133] Added cyclohexane or methylcyc-
lopentane trapped the migrating carbocation by a hydride transfer mainly to

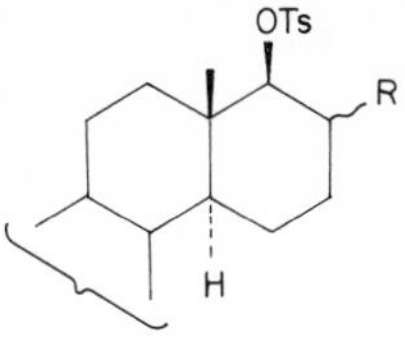

(66) R = H

(67) R = Me   (68)

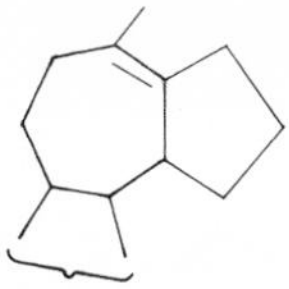

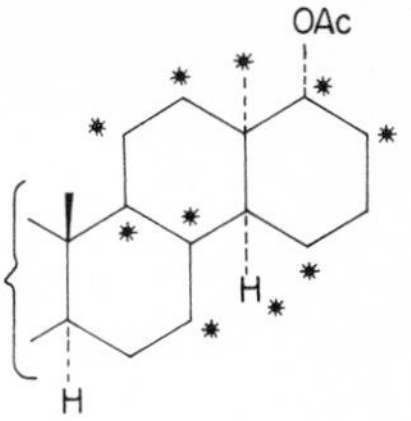

(69) *deuterium incorporated

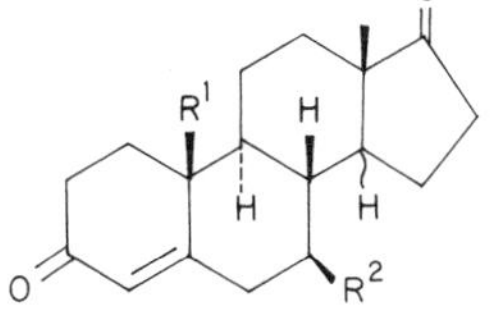

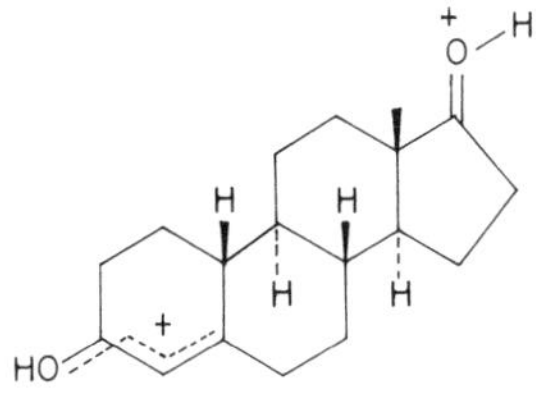

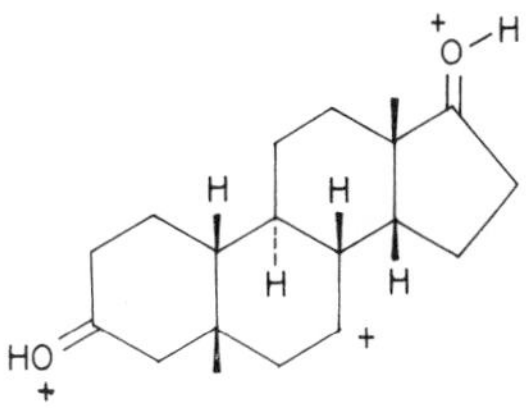

(70) 14α-H, R^1 = R^2 = H   (74)   (75)

(71) 14β-H, R^1 = R^2 = H

(72) 14α-H, R^1 = Me, R^2 = H

(73) 14β-H, R^1 = H, R^2 = Me

the 8β-position. This work forms part of an extensive study of reactions in the hyperacid medium HF-SbF$_5$ which include a variety of other reductive rearrangements and dienone-phenol, phenol-phenol, and phenol-dienone rearrangements all of which may be accompanied by changes in backbone stereochemistry.[134-139] Of particular interest is the rearrangement of androst-4-ene-3,17-dione (72) into 7β-methyl-14β-oestr-4-ene-3,17-dione (73)[139] through the trication (75) which rearranges by successive shifts of the methyl group from C-5 to C-6 and C-7.

Aromatization of steroid rings by other than the traditional dienone-phenol and dienol-benzene rearrangements has been studied in some detail. An extensive study of the acid-catalysed reactions of A/B-trifunctional steroids demonstrated that the major products are A-ring aromatic compounds irrespective of the positions of the functional groups.[140-144] For example the 7-hydroxy-5,6-epoxides (76) each reacted with HBr in acetic acid to give the 4-methyloestra-1,3,5(10)-triene (77) and not the anthrasteroid.[140] Additionally, the $\Delta^{2,5}$-diene-4-oxo-compound (78) rearranged with HBr in acetic

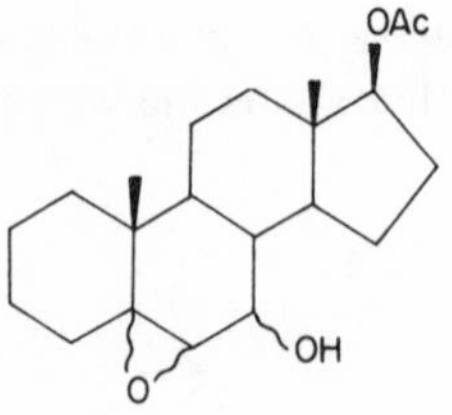

(76) 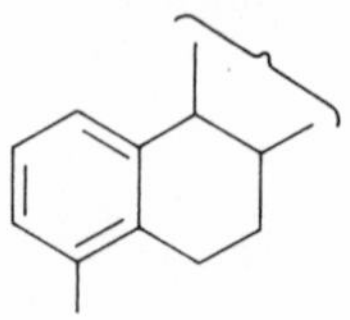　　　　　　(77)

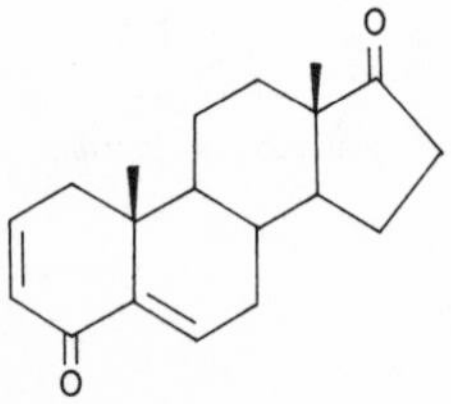

(78) 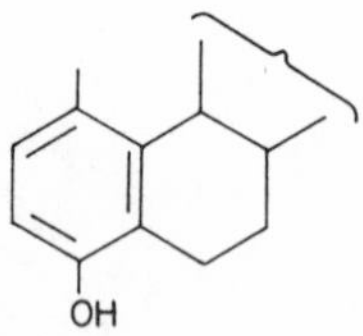　　　　　　(79)

acid to the 1-methyl-4-hydroxy-oestra-1,3,5(10)-triene (79),[141] and the major products from the 3α,5α-cyclo-steroids (80) and (81) were the 4-methyloestra-1,3,5(10)-trienes.[142] Interestingly, the 1α,2α- and 2α,3α-methylene-enones (82) and (83) did not undergo aromatization in HBr in acetic acid but merely cleaved to give the bromomethyl compounds (84) and (85) respectively.[143] In general, in those reactions leading to the 4-methyloestra-1,3,5(10)-trienes from 10-methyl A/B-trifunctional steroids the spirocarbocations (86) are

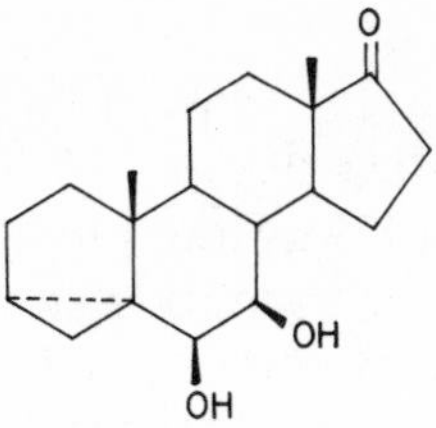

(80)

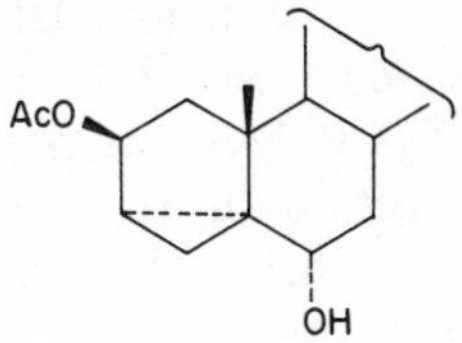

(81)

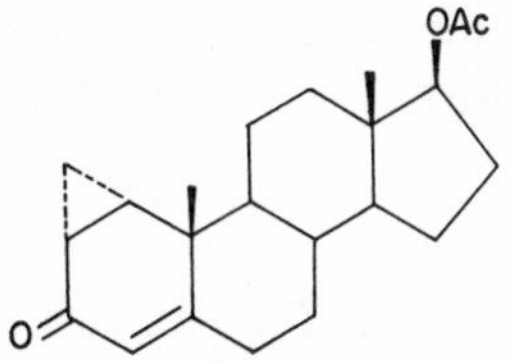

(82)

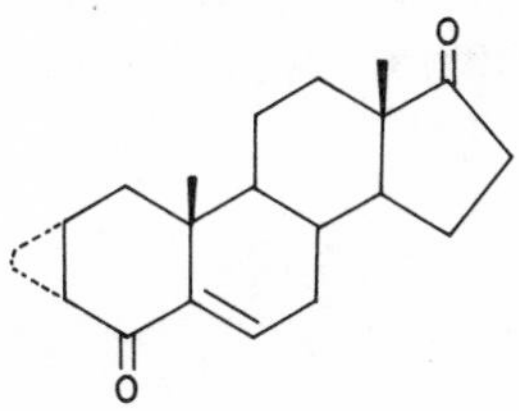

(83)

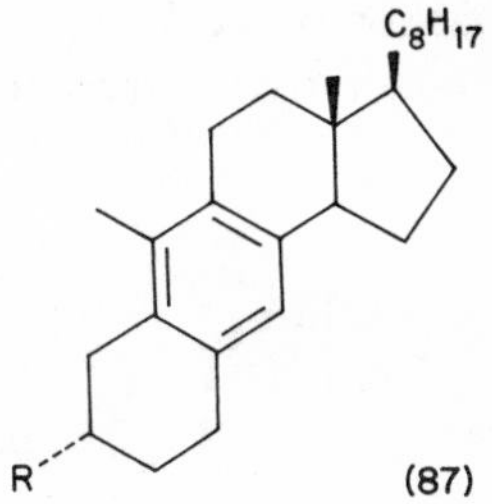

(87)

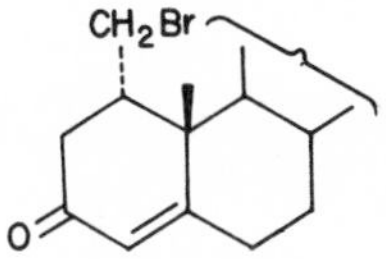

(84)

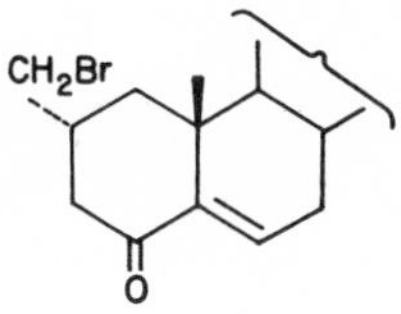

(85)

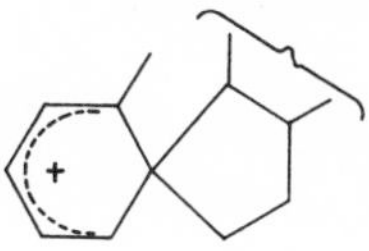

(86)

thought to be involved as in the related dienol-benzene rearrangements.[145]

The adducts (40) of cholesta-5,7-dienes and 4-phenyl-1,2,4-triazoline-3,5,-dione reacted with $BF_3 \cdot Et_2O$ to give the anthrasteroids (87).[146] The rearrangement appears to be initiated by coordination of the BF_3 at NCO group attached to C-5.[147] The ene reaction products (88) of 4,4-dimethyl-$\Delta^{5,7}$-3-oxo-compounds and dimethyl azodicarboxylate similarly rearranged with BF_3 to give the anthrasteroids (90).[148] However the 3β-alcohol (89) under similar conditions led to the A-seco-B-ring aromatic compound (91).

Aromatization of ring C may be achieved with appropriate functionality in rings B, C, and D. Thus, the 17-hydroxy-$\Delta^{4,6,9(11)}$-3-oxo-compound (92) was converted in formic acid to the C-ring aromatic enone (93)[149] as was the 17-hydroxy-$\Delta^{4,8}$-3,11-dione (95) to the phenolic compound (94).[150]

5.2.2   *Miscellaneous rearrangements*

Aspects of the well-known[151] D-homo-annulation of 17-hydroxypregnan-20-ones have been further studied and it is suggested that the anomalous migration of C-16 in the BF_3-catalysed reactions of 17α-hydroxypregnan-20-ones is owed to the involvement of the relatively rigid transition state (96) in which the oxygen atoms are held in a *syn* arrangement by a bridging proton.[152] A similar hydrogen-bridged transition state is thought to be involved in the kinetically-controlled thermal D-homoannulations.

Thermolysis of 5-vinyl-3-oxo-steroids (97) and (98) was observed to give the bicyclo[2.2.1]heptanes (99) and (100) respectively presumably via the ene reaction of the Δ^2-enols.[153] The A-nor-5-propenyl-ketone (102) was a minor product from the reaction of (98) and is believed to arise from the thermolysis

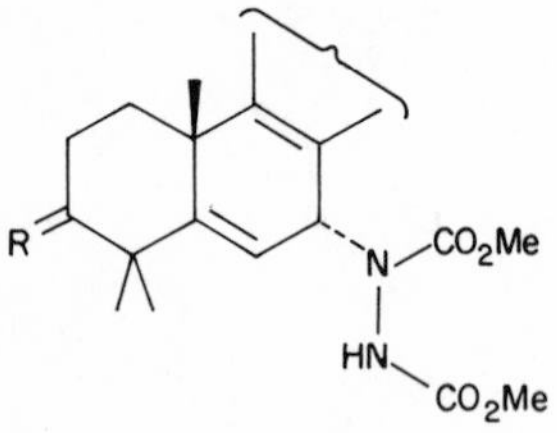

(88) R = O
(89) R = β-OH, H

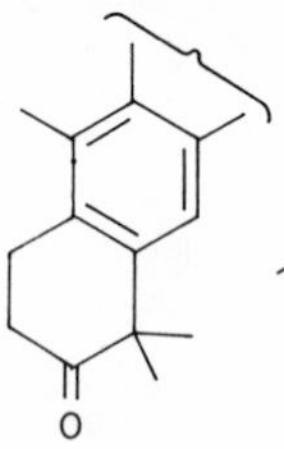

(90)

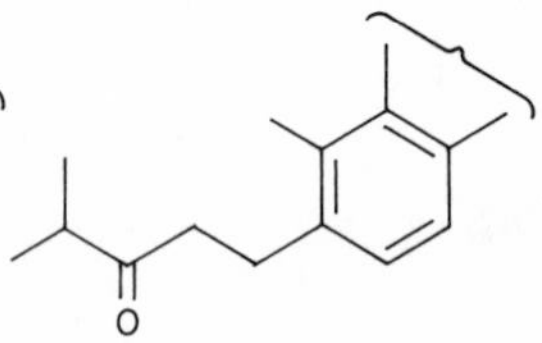

(91)

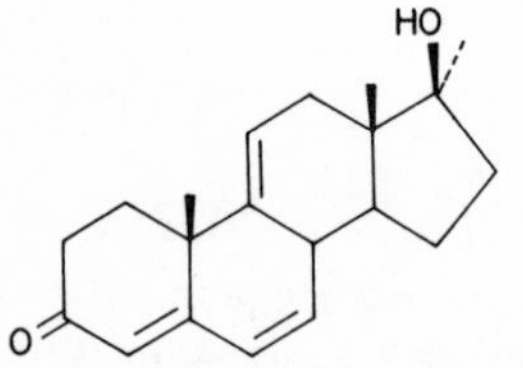

(92)

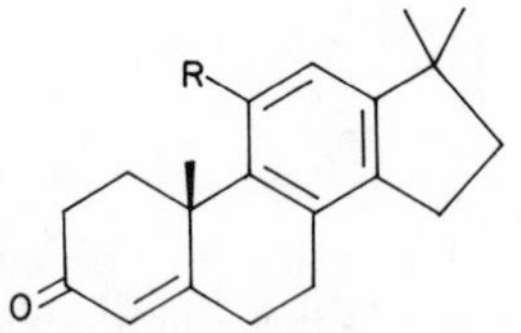

(93) R = H
(94) R = OH

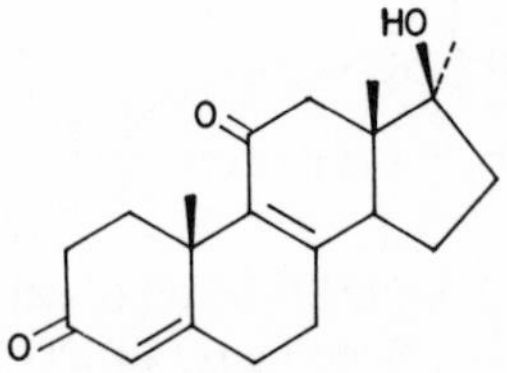

(95)

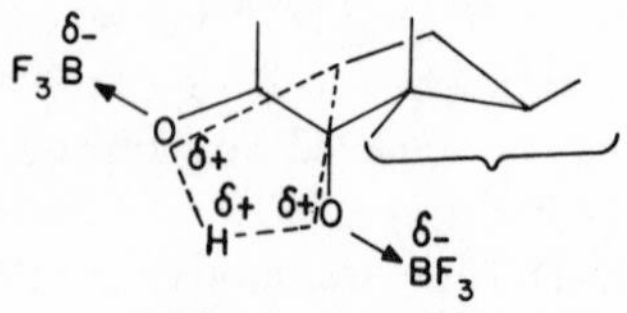

(96)

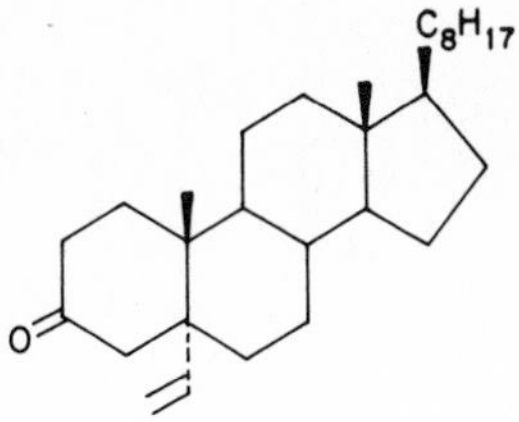

(97) 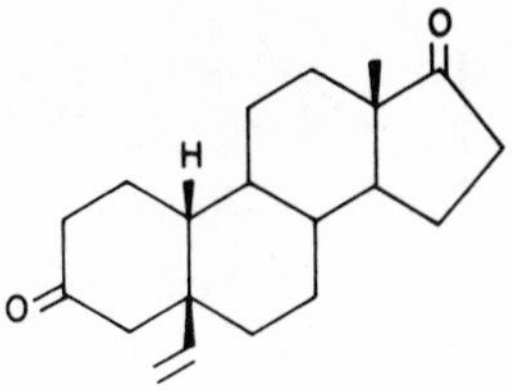 (98)

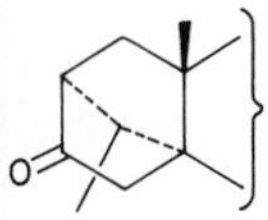

(99) 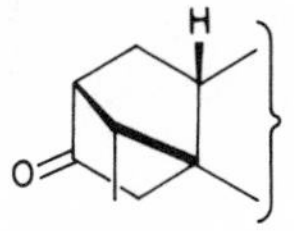 (100) 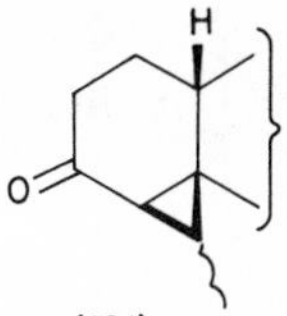 (101) 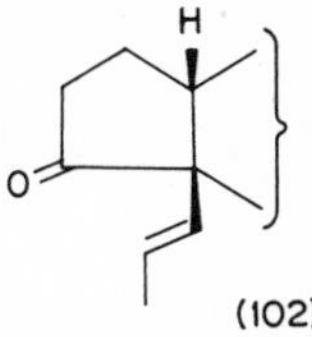 (102)

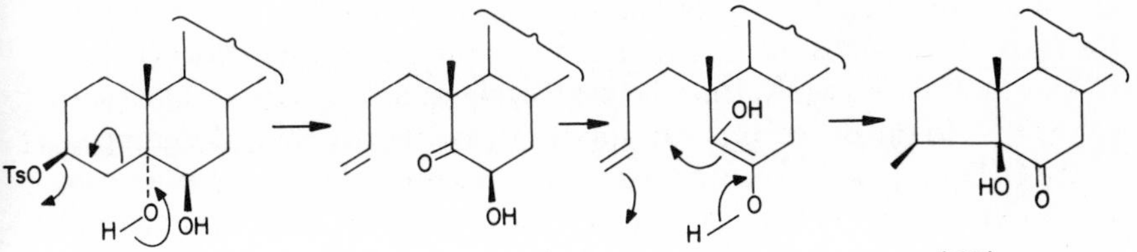

(103) (104) (105)

Scheme 6

of the cyclopropane (101) which could arise by an ene reaction of the Δ^3-enol of (98). The A-nor-5β-hydroxy-6-ketone (105) is a by-product of the reaction of 3β-tosyloxy-5,6β-dihydroxy-5α-cholestane (103) with Bu^{t}OK-Bu^{t}OH at 50°C and appears to arise via a novel mild ene reaction of the enol of the unsaturated acyloin (104) (Scheme 6).[154] An adduct (106) of ergosteryl acetate and nitrosocarbonylbenzene was reported to undergo a novel [3,3]sigmatropic rearrangement to give the dioxazine (107).[155]

A synthetically useful, non-photochemical, epimerization at C-13 of 17-oxo-compounds involved treatment of the derived oximes with boiling acetic anhydride-pyridine. The resultant enamides (108) and enimides (109), which may be formed through 13,17-seco-radicals (110) and (111), are readily hydrolysed to the 13α-17-oxo-compounds.[156]

The well known Beckmann, Schmidt, and Baeyer-Villiger rearrangements in steroids have in some cases been subject to re-examination with particular consideration of the migratory aptitudes of the α-carbon atoms.[157,158] A

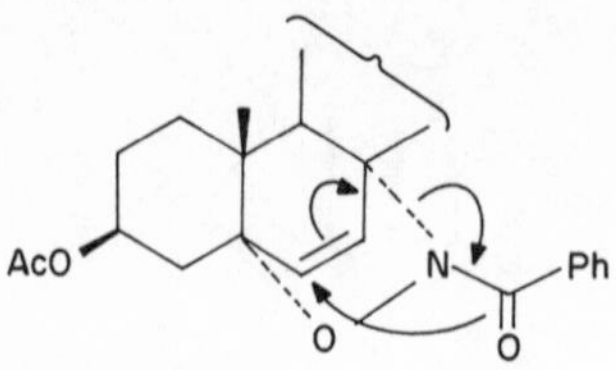

(106) 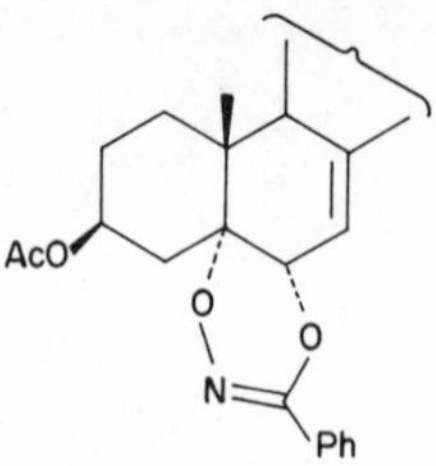   (107)

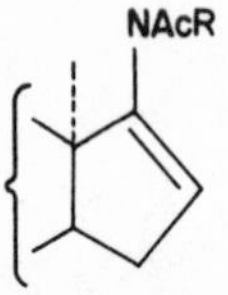

(108) R = H   (110)   (111)
(109) R = Ac

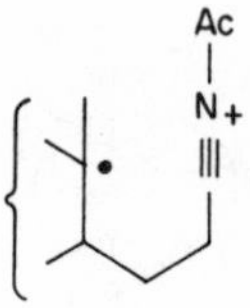

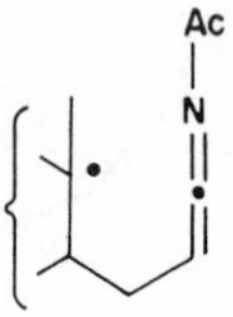

recent investigation employing [13]C n.m.r. spectroscopy has established that for 3-oxo-compounds and their derivatives the migratory aptitudes of C-2 and C-4 are more or less equivalent whether the migration is to carbon, nitrogen or oxygen.[157] In the Beckmann rearrangement, as expected, only the group *anti* to the OH group of the oxime migrates and, accordingly, if the *syn*- and *anti*-oximes are present in unequal proportions and equilibration does not occur then the ratio of products will reflect the proportions of *syn*- and *anti*-oximes. A reinvestigation of the Baeyer-Villiger rearrangement of 6-oxo-5α-cholestanes has established that C-7 competes effectively with C-5 in its migratory aptitude and that electron withdrawing groups at C-3 decrease that of the latter.[158]

5.2.3   *Photochemical rearrangements*

General reviews[159,160] on the photochemistry of steroids include extensive coverage of photo-rearrangements, some of which will be highlighted here with further novel work. An extensive study of the photo-Beckmann rearrangements of steroidal oximes indicated[161] that in general an oxaziridine is formed in the excited singlet state from the oxime which may undergo a

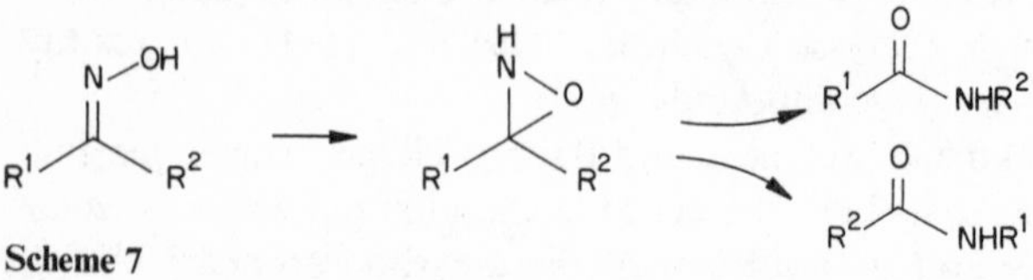

Scheme 7

photochemical *syn* ⇌ *anti* pre-equilibration (Scheme 7). In the main, the rearrangements of the oxaziridines to the lactams are thought to be concerted as the configuration of migrating chiral α-carbon atoms remains unchanged. The regioselectivity of the photo-Beckmann rearrangement is low. There were two notable exceptions to the rule of retention of configuration of the migrating α-carbon in the 17-ketoximes of androstanes and D-norandrostanes where partial inversion occurred at C-13. It is suggested that an open chain mechanism operates in these cases only.

Photolysis of the anions (112) of 17-nitro-steroids gave high yields of the hydroxamic acids (114) through the N-hydroxy-oxaziridine anion (115).[162–164] The anion of the analogous 13α-17-nitro-steroids gave a

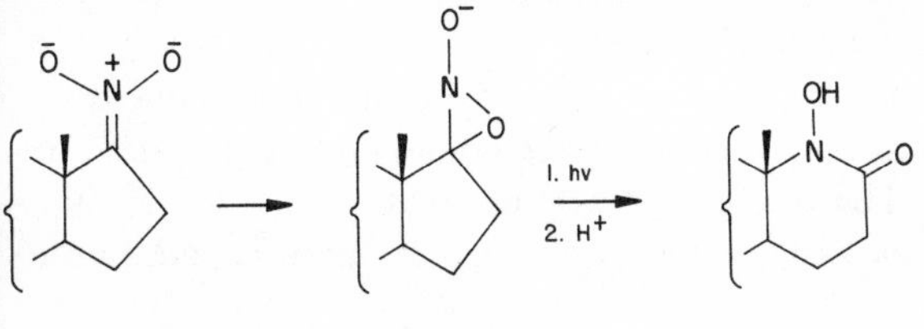

(112)   (113)   (114)

relatively low yield of hydroxamic acid[163] and the anions of 3-, 6- and 20-nitro-steroids[163,164] gave very much reduced yields and the alternative products were alkenes and ketones. In all cases where hydroxamic acid was formed, the more substituted carbon atom preferentially migrated with retention of configuration. The products may be accounted for by the three modes of decomposition of the intermediate N-hydroxyoxaziridine anion indicated in Scheme 8, although the carbene intermediate may arise directly

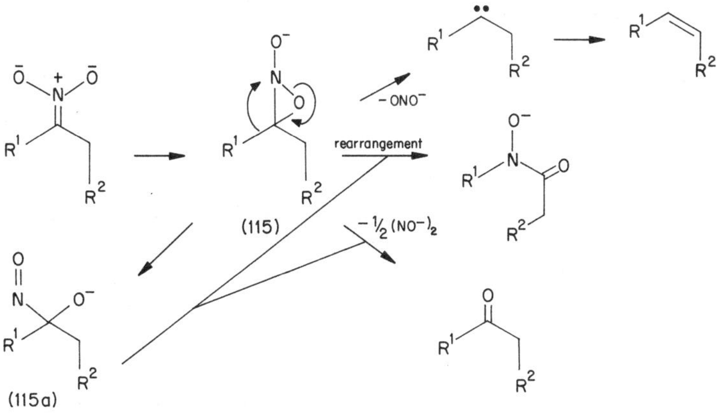

(115a)

Scheme 8

from the nitronate anion. Support for the suggested intermediacy of a carbene leading to alkene is apparent from the isolation of 17,18-cyclo-steroids in photolysis of the anions of the 17-nitrosteroids.[162–164] On the basis of molecular orbital calculations on a model system it is suggested by Yamada and his co-workers that the N-hydroxy-oxaziridine anion (115) is only a transient species and that the hydroxynitroso-anion (115a) is the key intermediate in which stereoelectronic factors determine the regioselectivity of the rearrangements leading to hydroxamic acid.[165] Interestingly, the regioselectivity of the photorearrangements contrasts with that observed for some N-alkyloxaziridines (116).[166] Hydroxamic acids of the type (114) have long been known as photo-products of 17β-yl nitrites, and recent work has established that the cyclization of the intermediate nitroso-aldehydes (117) is a thermal rather than photochemical process.[167]

Photo-deconjugation of 6-nitrocholest-5-enes[168,169] and 4-nitrocholest-4-enes[169] involved formation of the *aci*-nitro-compound [e.g. (118)] by abstraction of the γ-H. The *aci*-form may then tautomerize to the non-conjugated nitro-compound or may react further under suitable conditions to give other products.

The photolyses of $\Delta^{1,4}$-3-oxo-steroids and Δ^4-3-oxo-steroids are well known.[159,160] It has recently been observed that predisone and prednisolone derivatives behave quite typically[170,171] contrary to earlier reports. Thus, lumiprednisone acetate has the structure (119). Interesting related observations are the photoisomerization of the unsaturated lactone (120) to the lumi-lactone (121)[172] and the similar conversion of 4-methoxycholesta-1,4-dien-3-one to the 1-methoxy-1,5-cyclosteroid (122).[173]

The photolyses of the epoxylactone (123)[174] and the epoxylactam (124)[175] resemble the photolyses of simple α,β-epoxy-ketones[160] and gave the keto-lactone (125) and the keto-lactam (126) respectively. Studies on α,β-epoxy-ketones have been extended[176,177] and the photoisomerization of the exocyclic epoxycholestanones (127) to the diketones (128) involves a partial epimerization at C-5 via hydrogen abstraction from C-5 by the exocyclic carbonyl group of (128) followed by hydrogen transfer from C-2 to the resultant C-5 radical.

The photochemistry of β,γ-unsaturated ketones has been an area of intensive study and a substantial number of steroidal examples have been reported. The early work is included in excellent reviews.[178,179] The major pathways observed are the 1,2-acyl shift (the oxa-di-π-methane rearrangement),[180] usually from the excited triplet state, and the 1,3-acyl shift, usually from the excited singlet state. Typically, the 3-oxo-$\Delta^{5(10)}$-steroids (129) may give the cyclopropyl ketones (130) or the cyclobutanones (131).[181,182] Recent studies on the $\Delta^{9(11)}$-analogue (132)[183] and the related A-homo-enone (133)[184] complement the earlier work and give similar results. The dienone

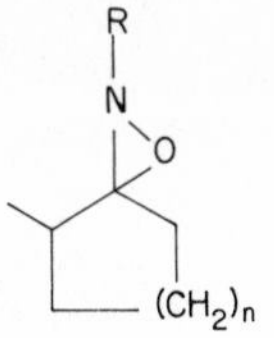

R
N
O

(CH₂)ₙ

(116) n = 1 or 2

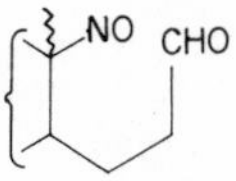

NO CHO

(117)

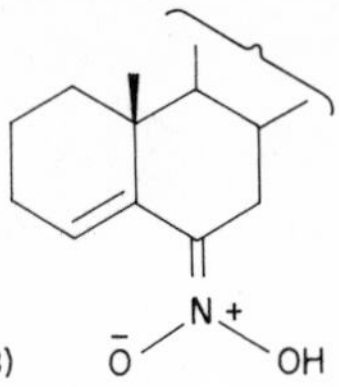

N⁺
O⁻ OH

(118)

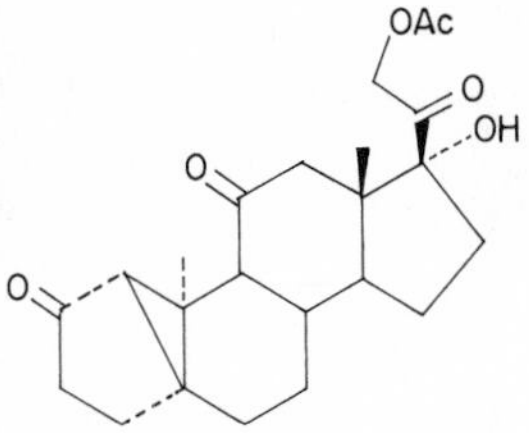

OAc
O
OH
O
O

(119)

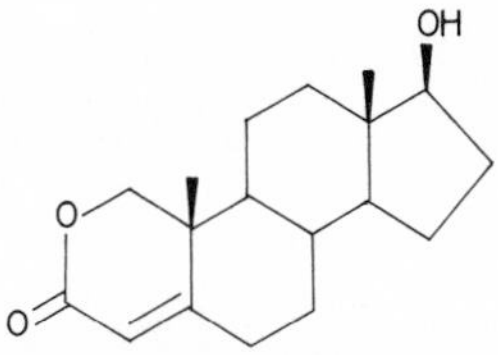

OH
O
O

(120)

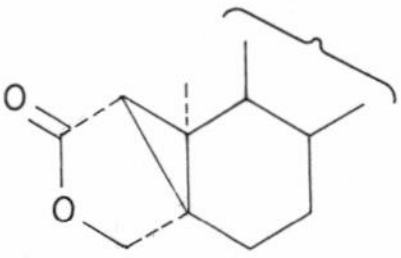

O
O

(121)

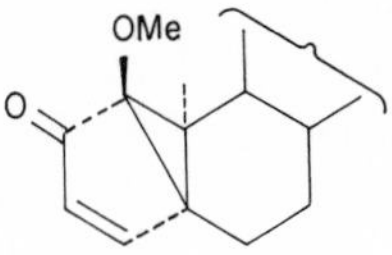

OMe
O

(122)

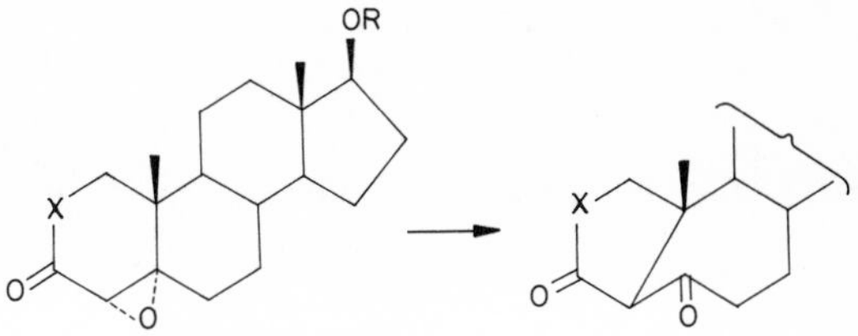

OR
X
O
O

(123) R = Ac, X = O
(124) R = H, X = NH

→

X
O
O

(125) X = O
(126) X = NH

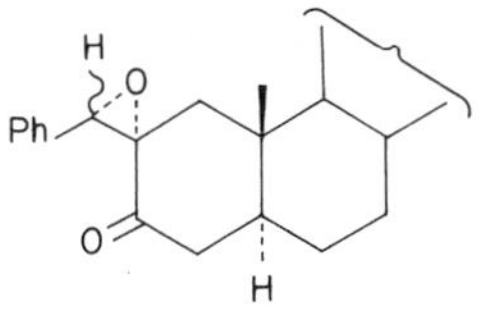

H
O
Ph
O
H

(127) *cis* and *trans*

O
Ph
O
H

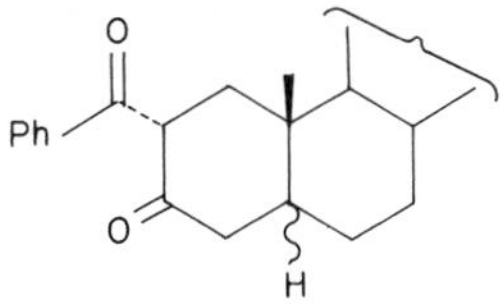

(128)

(**134**) underwent a di-π-methane rearrangement to give the cyclopropyl-enone (**135**)[185] and the diene-lactone (**136**) behaved analogously to give (**137**) (*inter alia*).[186]

Functionalization of unactivated carbon atoms by the use of the Barton, hypoiodite, and related reactions has been studied for some time and has had many important applications.[187,188] Recently, the photolysis of 20-peracetoxy-20-nitriles (**138**) leading to the 18-cyano-20-oxo-steroids (**139**) has been shown to be useful (Scheme 9).[188,189] Photolysis in acetone of a 20-

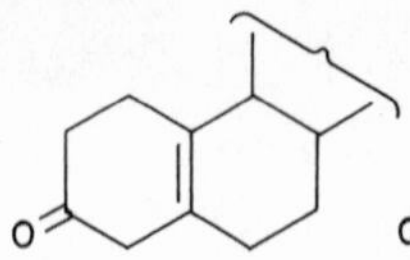

(129) 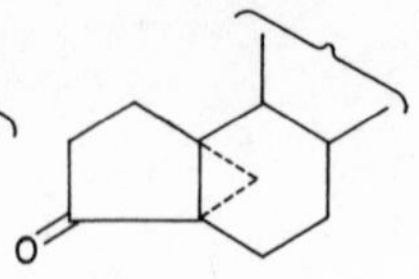   (130) 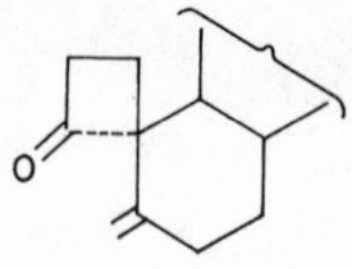   (131)

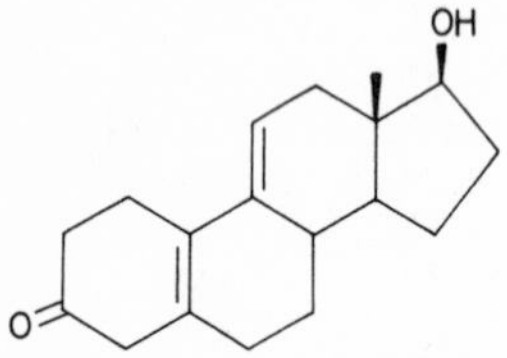

(132) 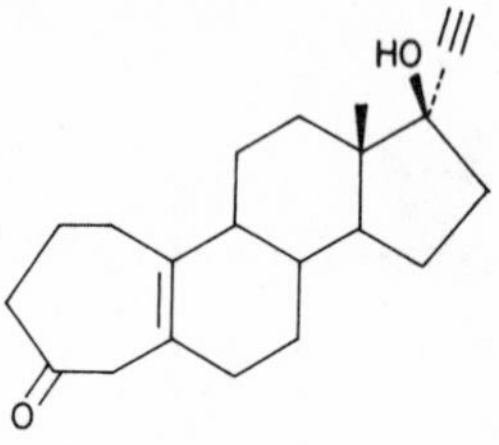   (133)

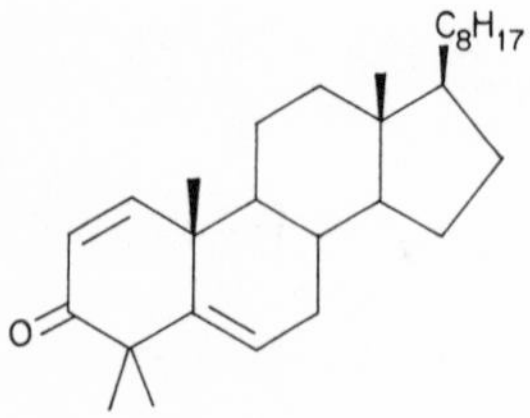

(134) 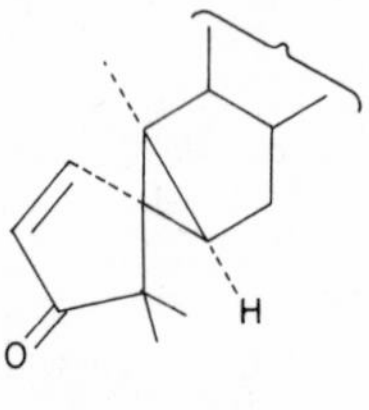   (135)

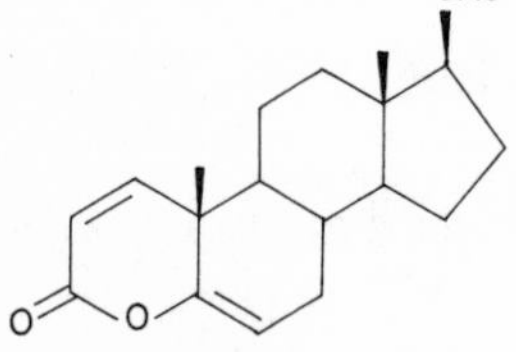

(136) 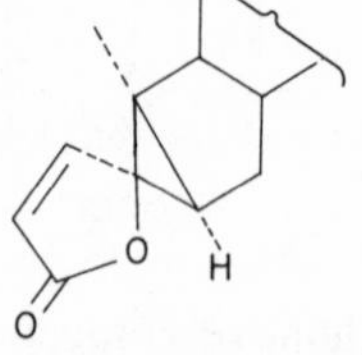   (137)

azido-20-nitrile (140) has been shown to lead also to 18-cyano-20-oxo-steroids,[190] and the photolysis of the 6β-nitroamino-cholestane (141) and the 20-nitroamine (142) with Pb(OAc)$_4$-I$_2$ or HgO-I$_2$ has recently been reported to lead to functionalization of C-19 and C-18 respectively through the generation of nitroaminyl radicals. The primary products from (141) and

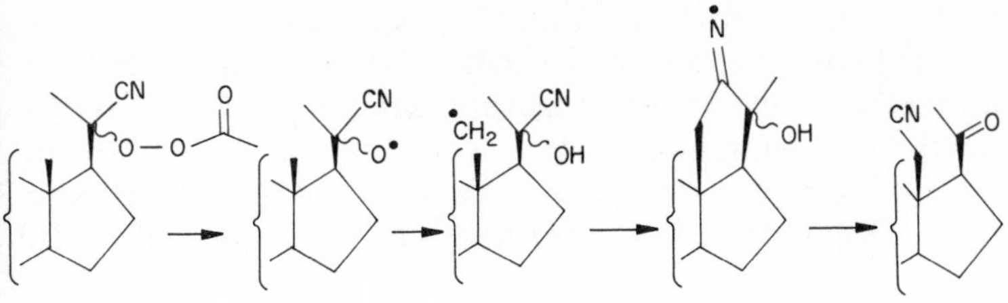

(138)   (139)

Scheme 9

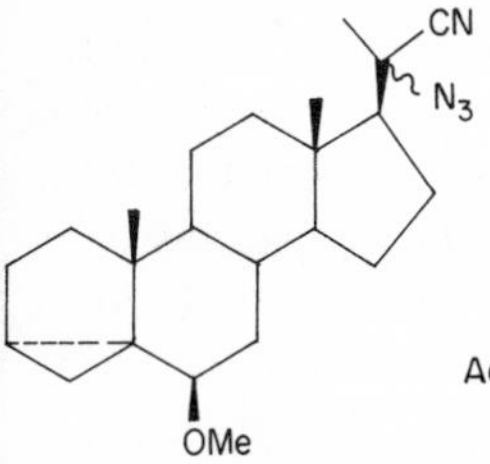

(140) 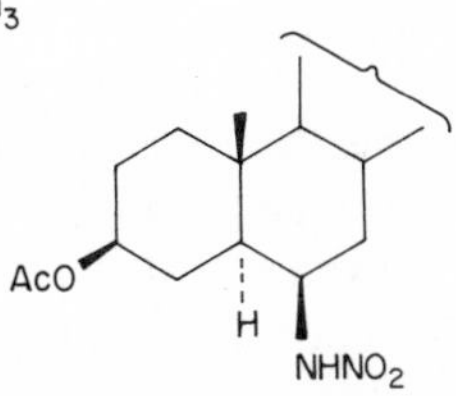   (141) 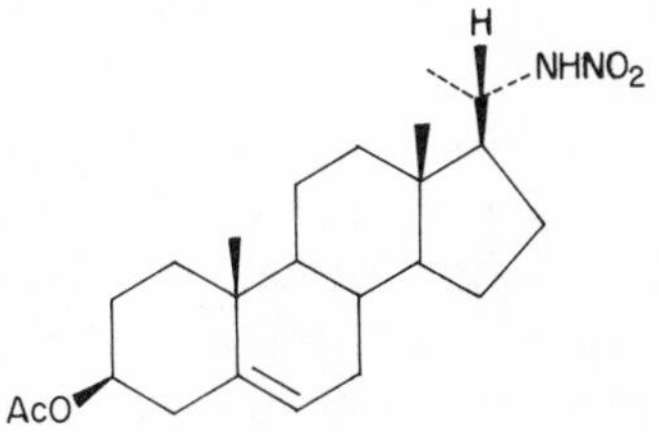   (142)

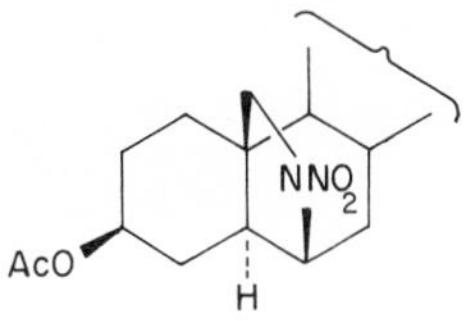

(143) 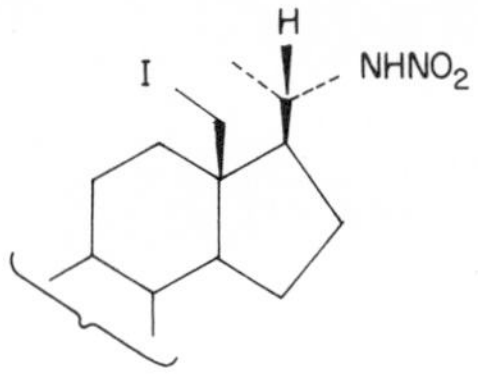   (144)

(**142**) were the N-nitropyrrolidine (**143**) and the 18-iodo-20-nitroamine compound (**144**).[191]

5.3 Biomimetic syntheses

The two broad areas discussed in this section are remote functionalizations and polyene cyclizations. The former have similarities to the Barton and related reactions. They are regarded as biomimetic in that they mimic enzyme-controlled reactions in which it is generally believed there is a close enzyme-substrate interaction by which specific sites are held in appropriate conformations for attack by suitable reactive centres on the enzymes. In remote functionalization reactions, suitable groups appended at various positions in the steroid nucleus may be activated and caused to selectively attack other positions quite remote from their position of attachment. As will be seen, the selectivity depends on the chain length of the group attached and on other factors. Polyene cyclizations are biomimetic in that they mimic the *in-vivo* squalene 2,3-epoxide to tetracyclic triterpene conversions alluded to earlier.

5.3.1 *Remote functionalization*

The major contributions in these template directed reactions were made by Breslow and his co-workers and much of this work has been recently reviewed.[192] The early work involved incorporation of a substituted benzophenone moiety which by photolysis led to an excited triplet capable of hydrogen abstraction. For example, the photolysis of 3α-cholestanyl (p-benzoylphenyl)acetate (**145**) gave the Δ[14]-steroid (**146**) by successive α-hydrogen abstraction from C-14 and C-15 (Scheme 10). The initial attack at C-14 is preferred but some attack at C-12 and C-7 also occurs and the resulting diradicals couple to give hydroxy-lactones. Extension of the carbon chain

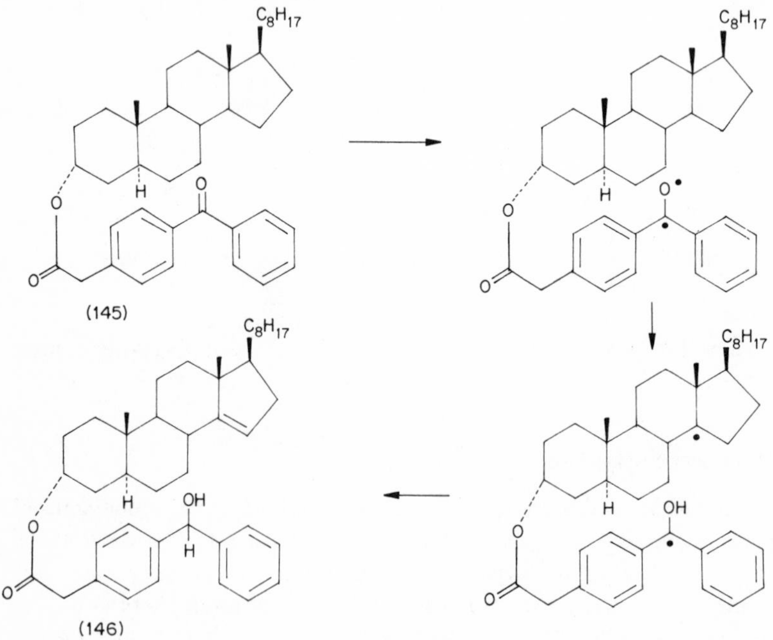

(145)

(146)

Scheme 10

between the ester group and the benzophenone moiety would be expected to modify the position of attack of the excited triplet. This is indeed the case and in the ester (147) the attack was largely at C-17 although some attack at C-14 was also observed. Greater selectivity of attack, principally at tertiary C–H bonds, was achieved by generating the ArICl moiety in an appropriate position. For example, the reaction of the 3α-cholestanyl *m*-iodobenzoate (148) with a chlorine atom from the photolysis of PhICl$_2$ (or other sources) generated the radical (150). Abstraction of a hydrogen atom from C-9 and capture of a chlorine atom led to the 9α-chloro-derivative (149). The radical (150) may also be generated by irradiation directly of the iododichloride (151). Dehydrochlorination of 9α-chloro-compounds gave the $\Delta^{9(11)}$-compounds which may be readily converted to the 11-oxygenated steroids. Accordingly, this process has opened the way to an alternative approach to the synthesis of cortisone and its derivatives.[193,194] Chlorination of C-9 was also achieved by a similar reaction with the 17α-yl *m*-iodobenzoate (152) and chlorination at C-17 was achieved in the iodoaryl ester (153) thus allowing degradation of the

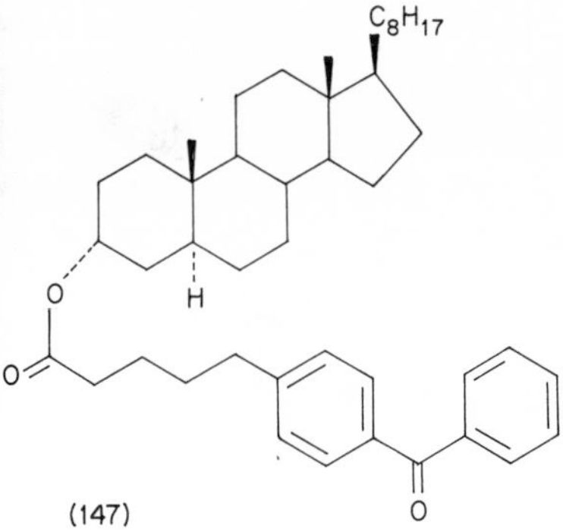

(147)

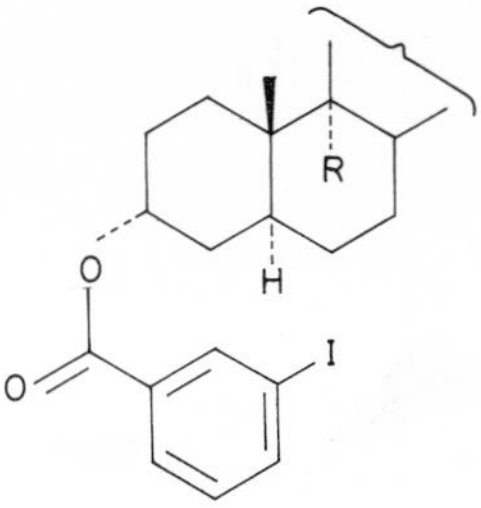

(148) R = H

(149) R = Cl

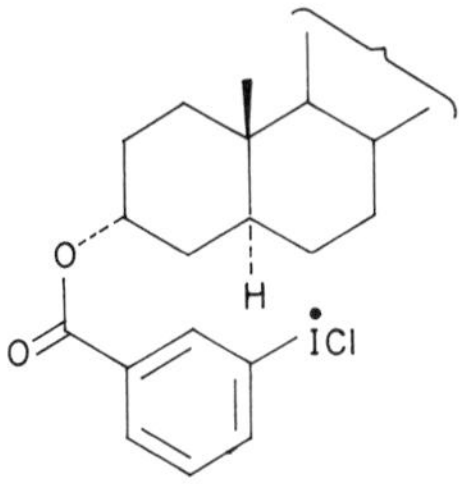

(150)

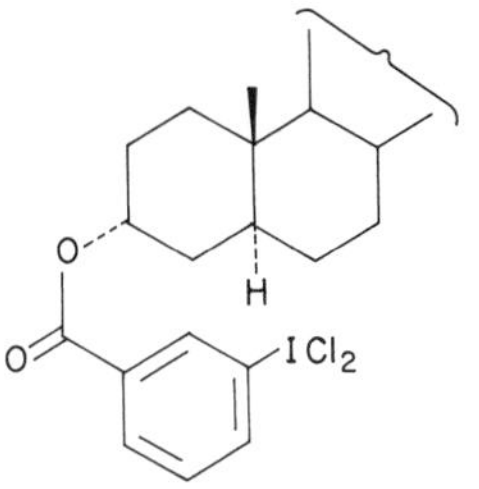

(151)

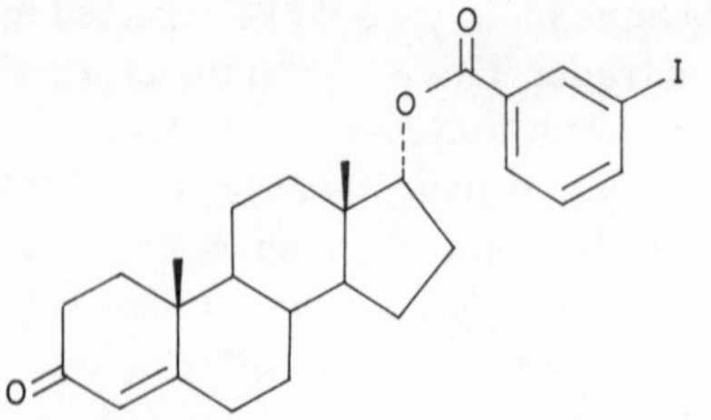

(152)

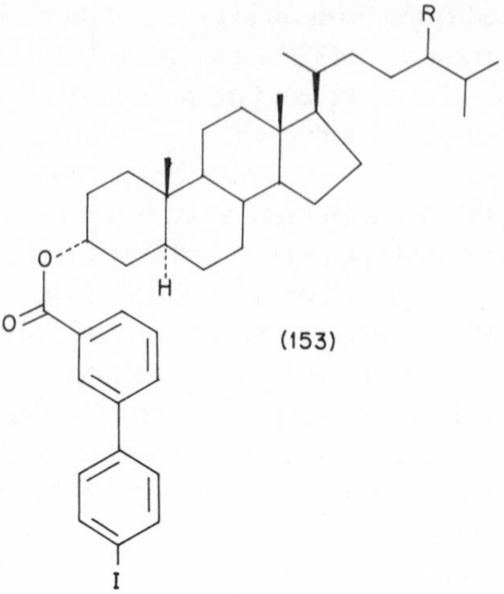

(153)

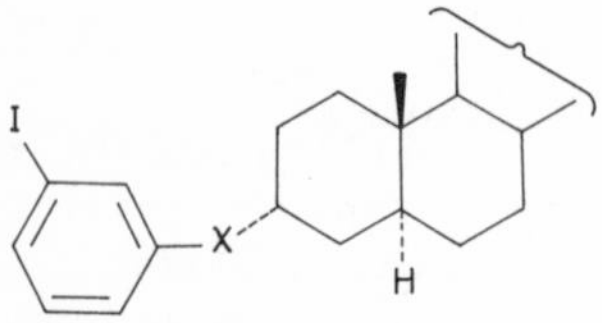

(154) X = -CONH-

(155) X = -CH₂O-

(156) X = -SO₃-

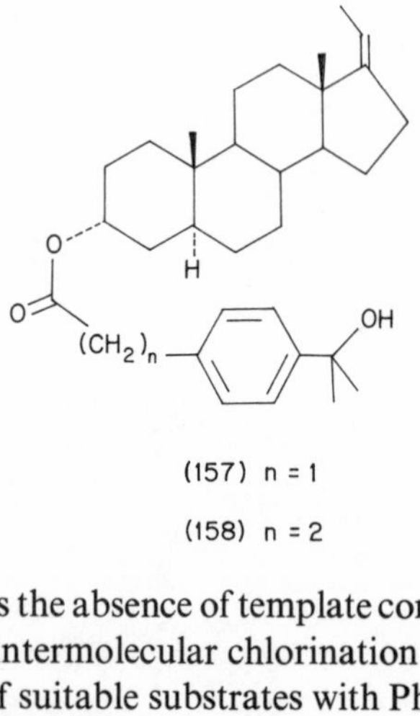

(157) n = 1

(158) n = 2

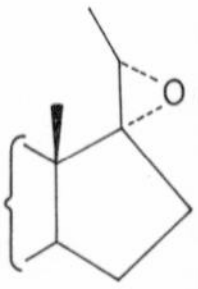

(159)

hydrocarbon side chain. In 5β-steroids the absence of template control may be relatively unimportant and selective intermolecular chlorination at C-14 has proved to be possible by photolysis of suitable substrates with PhICl₂. These reactions have proved useful in the synthesis of cardenolides.[195] Chlorination at C-9 was recently effected by photolyses of the cholestane derivatives (154), (155) and (156) with PhICl₂.[196]

An interesting development is template directed epoxidation of hydroxy-alkenes with BuᵗOOH and Mo(CO)₆. Thus the 3α-pregnenyl ester (157) gave the epoxide (159) whereas the related homologue (158) and the m-derivative (160) failed to react. These reactions demonstrate that, as expected, the catalyst complexes with the hydroxy group and that the geometrical requirements of the template are reasonably strict.[197]

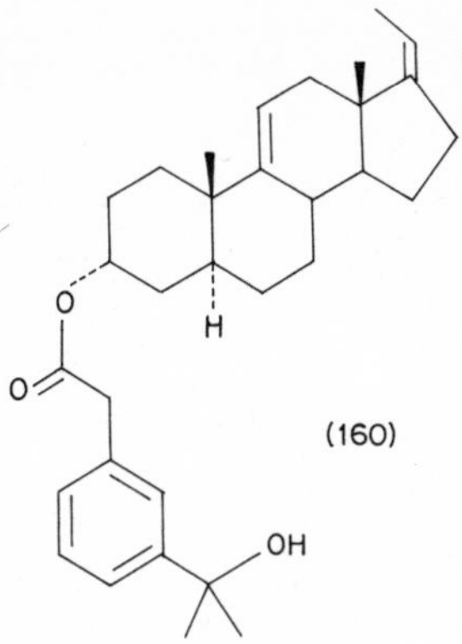

(160)

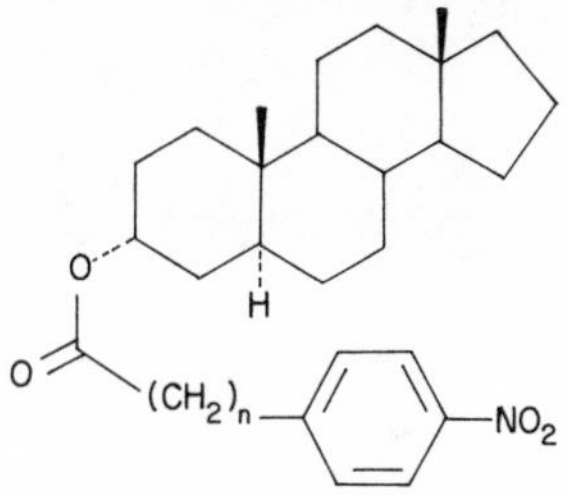

(161) n = 0

(162) n = 1

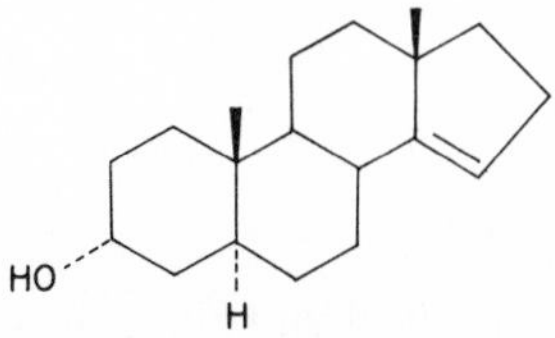

(163)

Further demonstration of the importance of the substrate–template geometry is observed in the photolyses of the nitro-derivatives (161) and (162). No remote functionalization of the former was observed while the latter gave the Δ^{14}-compound (163) after dehydration and saponification albeit in rather low yield.[198]

5.3.2 Polyene cyclizations

In the biosynthesis of tetracyclic triterpenes and steroids the key step is the cyclization of squalene 2,3-epoxide. The cyclization products are determined by the conformation of the squalene 2,3-epoxide which, in turn, is probably determined by the enzymes present. The conformation (164) (chair-boat-chair-boat) of squalene 2,3-epoxide is that believed to be required for eventual formation of lanosterol (165) (Scheme 11) and it may be noted that the cyclization involves a series of trans-electrophilic additions to the trans-double bonds of the squalene-2,3-epoxide. The potential of laboratory processes which could mimic this cyclization and, in one step, thereby stereospecifically generate a series of chiral centres has been recognized for some years and the work of major contributors to this area of study has recently been reviewed.[199,200,201]

The work of van Tamelen[199] and co-workers on the cyclization of polyene epoxides is perhaps closest in concept to the biosynthetic pathways. It was

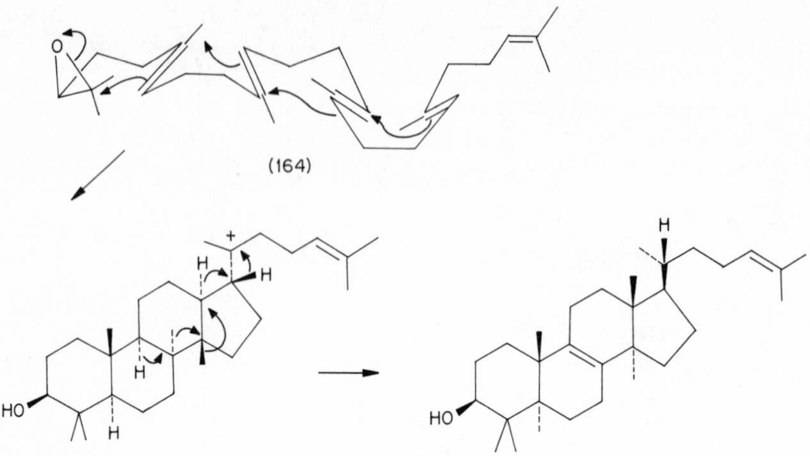

(164)

(165)

Scheme 11

shown that squalene 2,3-epoxide itself did not behave the same way *in vitro* as *in vivo* and none of the expected tetracyclic products were obtained. However, use of substrates containing a preformed D-ring did allow the elaboration of some tetracyclic (and pentacyclic) products. For example, the epoxide (166) could be cyclized with $SnCl_4$ in CH_3NO_2 to give isoeuphenol (167). An

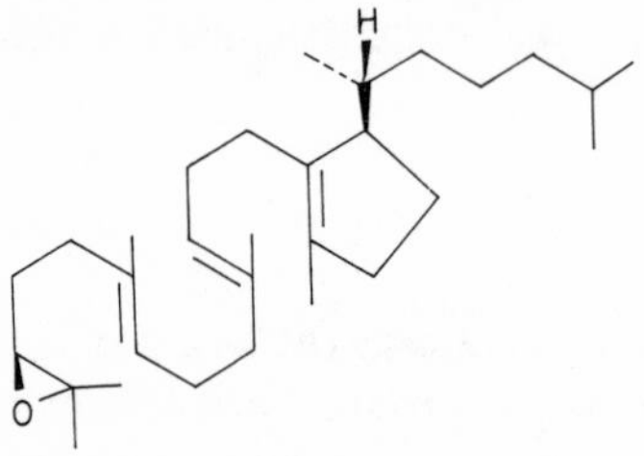

(166)

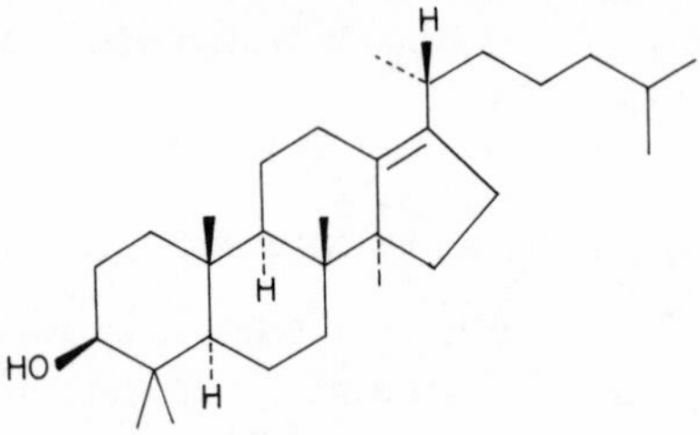

(167)

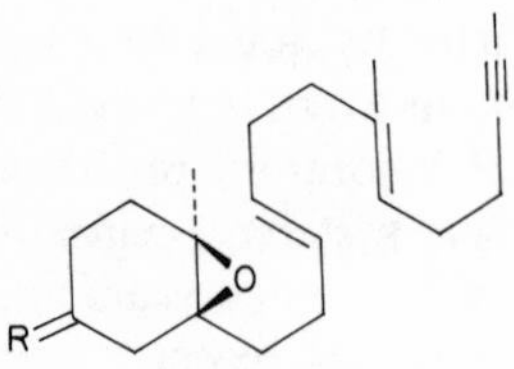

(168)   R = β- OAc , H
(169)   R = (OCH$_2$)$_2$

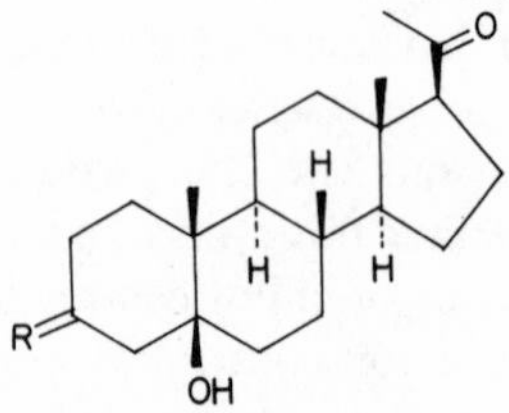

(170)   R = β- OAc, H
(171)   R = (OCH$_2$)$_2$

extension of this methodology has allowed the synthesis of racemic progesterone[199,202] from the tetracyclic products (**170**) and (**171**) which were obtained, respectively, from the acetylenic epoxides (**168**) and (**169**). The use of acetylenic terminator groups in polyene cyclizations had been previously and extensively reported by Johnson and his co-workers. Indeed, syntheses of racemic progesterone (**178**) by cyclization of the acetylenic allylic alcohols

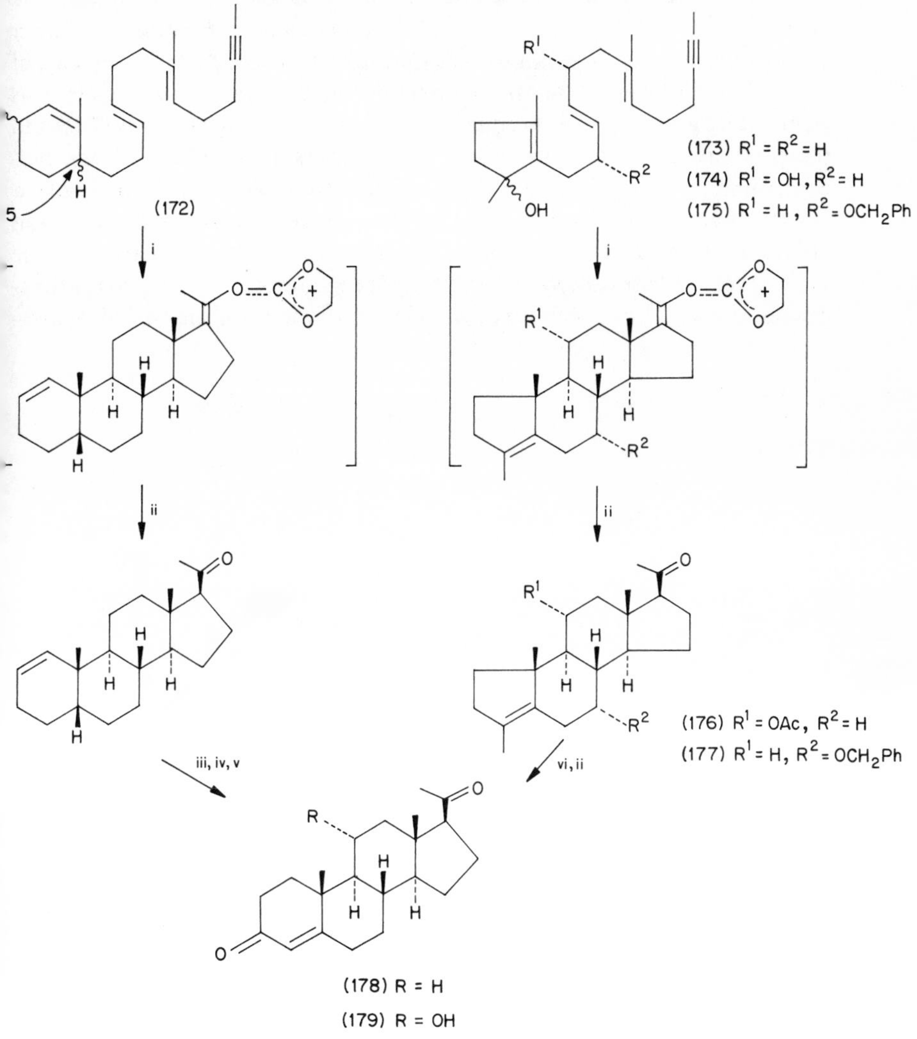
(172)
(173) R¹ = R² = H
(174) R¹ = OH, R² = H
(175) R¹ = H, R² = OCH₂Ph
(176) R¹ = OAc, R² = H
(177) R¹ = H, R² = OCH₂Ph
iii, iv, v
vi, ii
(178) R = H
(179) R = OH

Scheme 12

Reagents: i, $CF_3CO_2H—OC(OCH_2)_2$; ii, $H_2O—\bar{O}H$; iii, t-butyl chromate; iv, DDQ; v, $H_2—Rh(PPh_3)_3I$; vi, O_3

(172) and (173) have been reported (Scheme 12). Recently, the route from (172) has been developed to provide natural progesterone by using the substrate (172) with the R configuration at pro-C-5.[203] The synthesis of natural 11α-hydroxyprogesterone (179) from the hydroxyacetylenic polyene (174) with the R configuration at pro-C-11 represents a very exciting development.[204] 11α-Hydroxyprogesterone (179) is a well-known intermediate in the synthesis of corticosteroids and an attractive alternative approach[205] from the tetracycle (176) is outlined in Scheme 13. Other approaches to corticosteroids include the cyclization $(CF_3CO_2H-CH_2Cl_2)$ of the hydroxypolyene (180) which contains the propargyl silane moiety as terminating group.[206] The product of this cyclization was the allene (182) and there is precedent for the single step conversion of such allenes into compounds containing the corticosteroid side chain (see later). Accordingly, it is anticipated that cyclization of the hydroxy-analogue (181) could lead to an abbreviated and efficient synthesis of corticosteroids. Asymmetric polyene cyclization syntheses additional to those of progesterone[203] and 11α-hydroxy-progesterone[204] have been reported. Thus, cyclization of the benzyloxyacet-

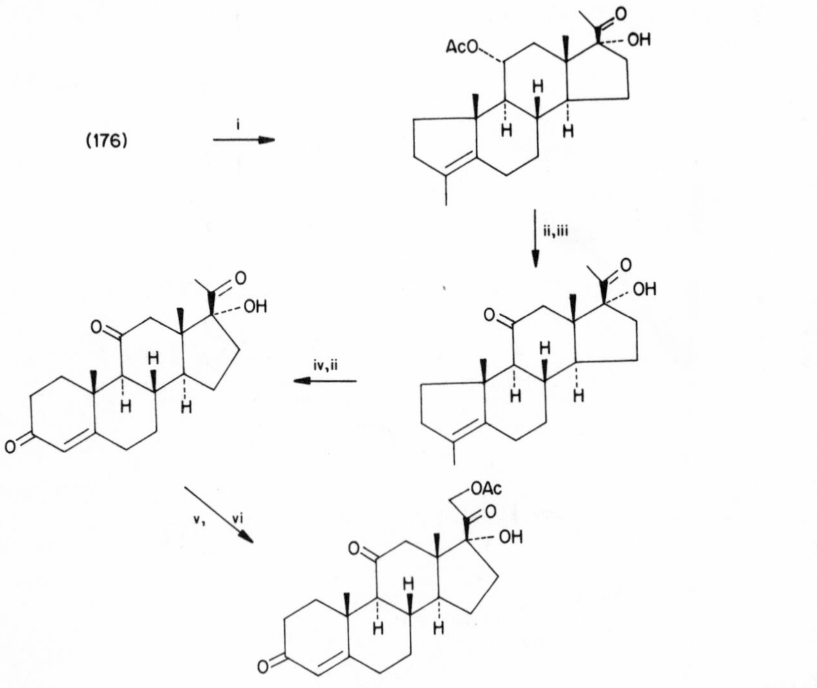

Scheme 13

Reagents: i, $Bu^tOK-O_2-(EtO)_3P$; ii, $\bar{O}H$; iii, CrO_3-H^+; iv, O_3; v, I_2-CoO; vi, KOAc.

ylenic polyene (175) is diastereoselective and gave the racemic tetracycle (177) as a major product and this has been converted[207] into racemic spironolactone (183). Assuming that enantiomerically pure forms of the benzyloxacetylenic polyene (175) may be available it is anticipated that the asymmetric synthesis of spironolactone will be developed. Some aspects of the diastereoselectivity of these polyene cyclizations remain unclear[208] although the absence of 11β-hydroxy-products in the cyclization of the hydroxyacetylenic polyene (174) may be ascribed to the 1,3-diaxial interaction between the methyl group at *pro*-C-10 and the β-hydroxy-group at *pro*-C-11. The remarkable diastereoselectivity of the cyclization of the benzyloxyacetylenic polyene (175) conceivably could arise from the steric interaction between a β-substituent at *pro*-C-7 and the *pro*-C-15 methylene group in a transition state involving extensive coiling of the polyene chain and, by implication, a concerted reaction.[208] Similar diastereoselectivity in the cyclization of the thiophene derivatives (184) and (185) is ascribed to this.[209]

Evidence consistent with a concerted process is available from studies related to a synthesis of oestrone[200] in which the key step was the Lewis acid-catalysed cyclization of the aromatic dienol (186; R^2 = H) to the tetracycle

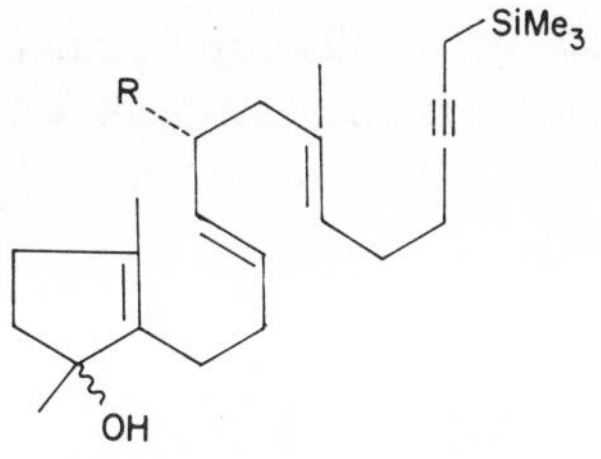

(180) R = H

(181) R = OH

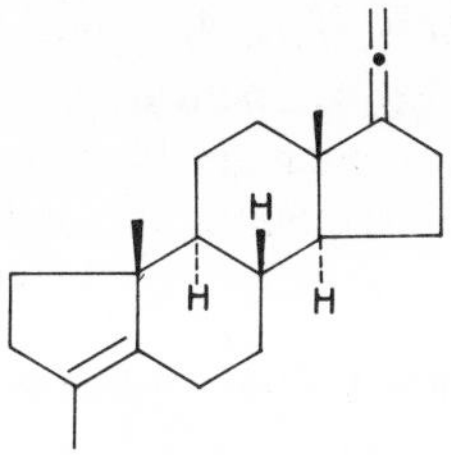

(182)

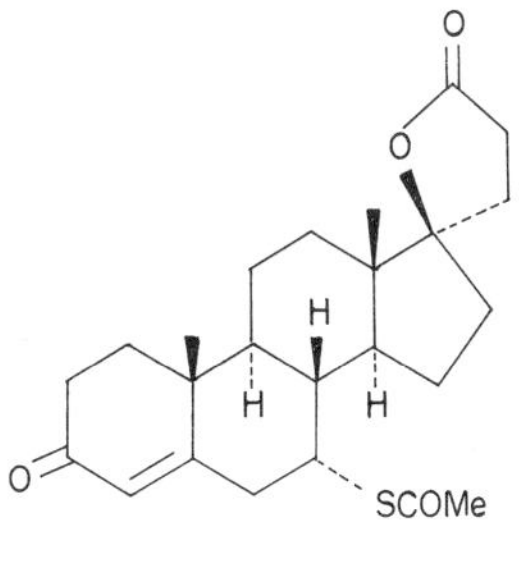

(183)

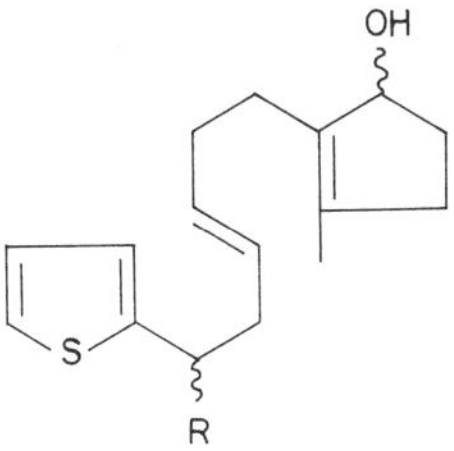

(184) R = Me

(185) R = Bu^t

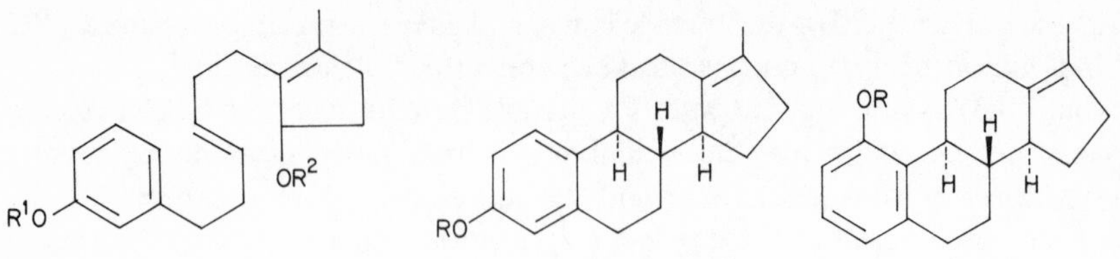

(186)   (187)   (188)

(**187**). A co-product in this cyclization is the *ortho* isomer (**188**) and the *ortho :*
para ratio was found to be dependent on the nature of the allylic group OR^2
(R^2 = H, SiMe$_3$, PhCO) and on the substituent on the phenolic oxygen. Other
kinetic data on a series of other substrates with a variety of substituents in the
A-ring further supported the possibility of a concerted mechanism. It is clearly
understood that although the cyclization of (**186**) and related compounds may
be concerted, it could be dangerous to extrapolate from these systems to other
polyenes. Indeed, in an investigation of epoxypolyenes, van Tamelen and his
co-workers concluded that only the first ring formation occurs in a concerted
process.[210]

In the foregoing discussion the syntheses of the various polyenes have been
neglected owing only to the shortage of space available. It will be appreciated
that the syntheses are by no means trivial but the stereochemical problems are
of course much more manageable than those associated with traditional total
synthesis involving sequential annelation reactions.

5.4 Novel total and partial syntheses

5.4.1 *Intramolecular cycloaddition reactions of o-quinodimethanes*

The generation *in situ* of *o*-quinodimethanes from benzocyclobutenes in
molecules containing a suitable dienophile (Scheme 14) may lead to the
synthesis of tetracyclic compounds and this approach to the total synthesis of
steroids has been studied by principally two groups.[211,212] The approach may
be exemplified by the synthesis of racemic D-homo-oestrone methyl ether
(**190**) from the methoxybenzocyclobutene (**189**) by heating in *o*-

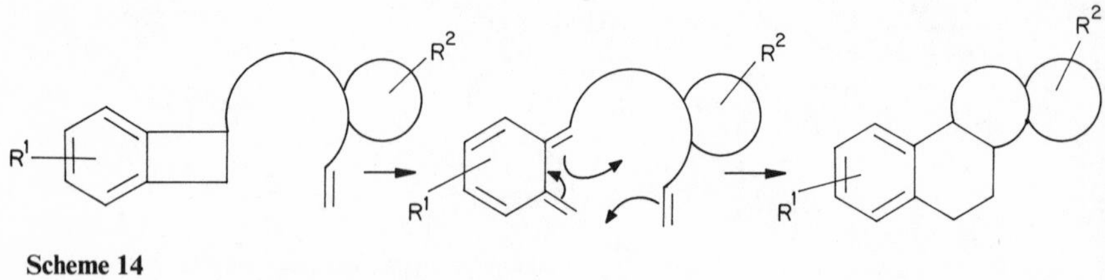

Scheme 14

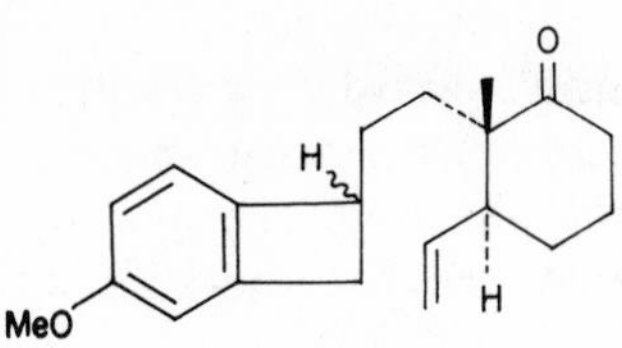

(189)

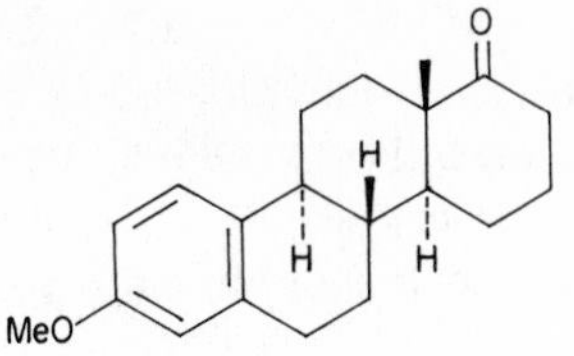

(190)

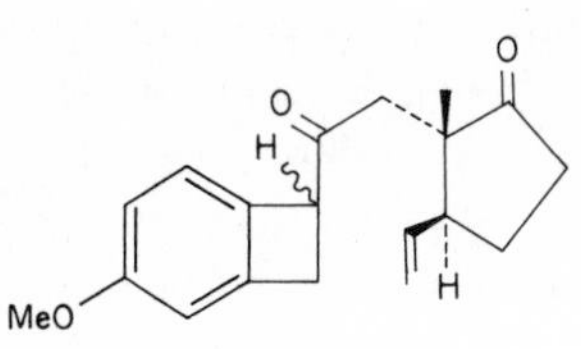

(191)

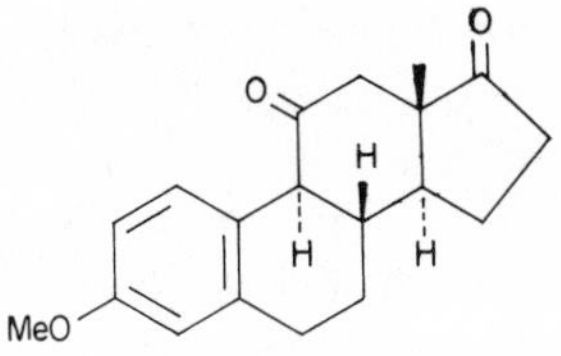

(192)

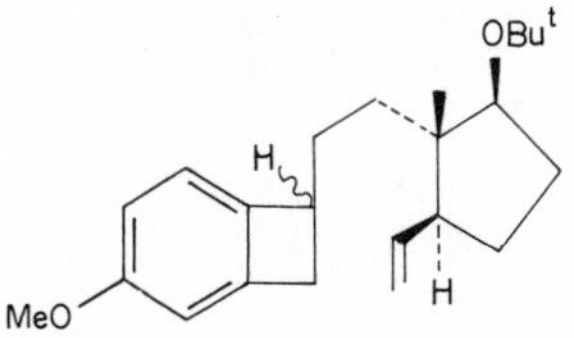

(193)

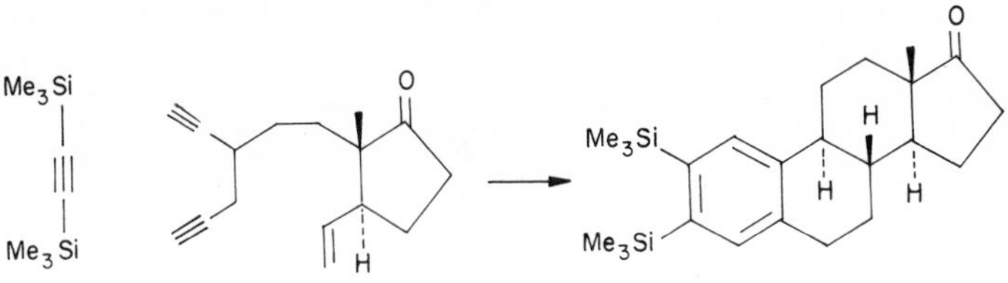

(195)   (194)   (196)

dichlorobenzene. Since (190) may be converted to oestrone this approach constitutes a new total synthesis. The stereoselectivity of the cycloaddition reaction was predicted assuming an *exo*-transition state in which steric interactions are minimal.[211] A similar synthesis of natural (+)-11-oxo-oestrone methyl ether (192) employed the thermolysis of the benzocyclo-butene (191) with the absolute configuration as depicted at *pro*-C-13 and *pro*-C-14.[212] A further enantioselective synthesis is that of oestradiol *via* the pyrolysis of the optically active benzocyclobutene (193).[211] The CpCo(CO)$_2$ catalysed co-oligeromization of the 1,5-hexadiyne (194) and bis (trime-thylsilyl)acetylene (195) gave the 2,3-bis(trimethylsilyl)oestratrienone (196) and this and related reactions proceed through benzocyclobutenes.[213] It has proved possible to selectively remove the trimethylsilyl group at C-2 (CF_3CO_2H-$CDCl_3$-CCl_4 at $-30°$) and to oxidatively cleave the remaining aryl-silicon bond ($Pb(OCOCF_3)_4$) thereby providing racemic oestrone in a very short synthetic sequence. It is anticipated that the use of an optically active 1,5-hexadiyne (194) may lead to an enantioselective synthesis of oestrone.

In the examples selected above only A-ring aromatic steroids have been synthesized. Clearly, the general synthetic strategy would equally allow the construction of D-homo-D-ring aromatic compounds and, typically, the optically active benzocyclobutene (197) afforded the aromatic product (198) which may be converted to (+)-5α-dihydropregnenolone (199), a valuable synthetic intermediate.[214] The similar D-homo-D-ring aromatic substrate (201) which is an intermediate in a synthesis of β-ecdysone (202) is available from the benzocyclobutene (200).[211]

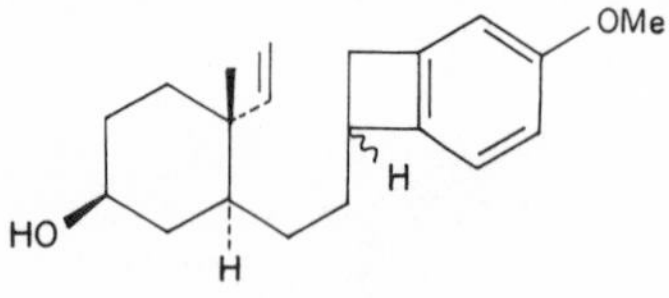

(197) 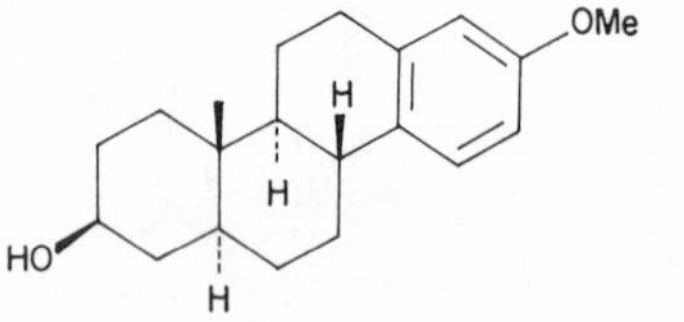   (198)

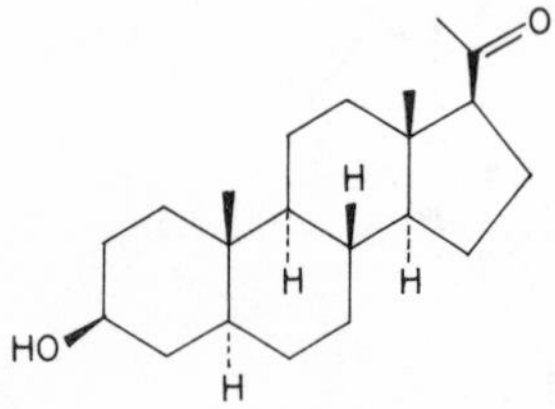

(199) 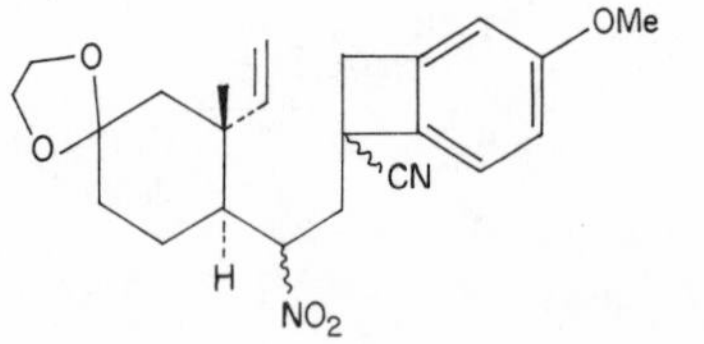   (200)

5.4.2 Novel structures and side-chain syntheses

The novel synthetic approaches discussed thus far have been primarily concerned with the construction of the steroid ring systems. A considerable effort over the last decade or so has also gone into novel approaches to the side-chains of steroids paying particular attention to stereoselectivity. This has, in part, arisen owing to the discovery of a wide variety of steroids with different functionalities in the side-chain.[215] β-Ecdysone (202), which was alluded to earlier, is a typical insect moulting hormone. Other interesting

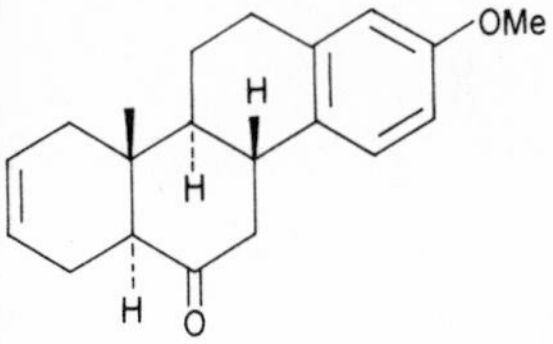

(201)

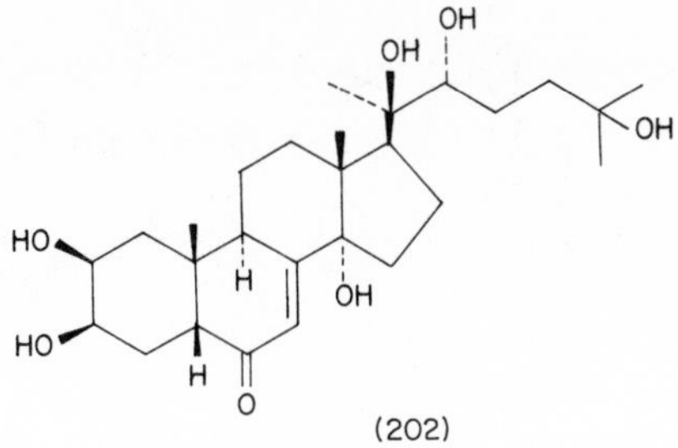

(202)

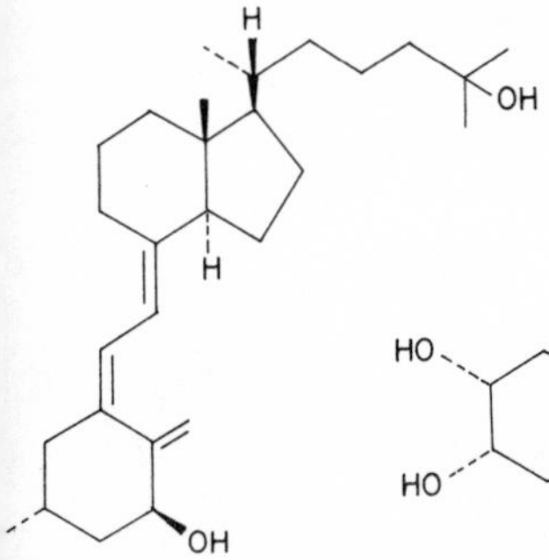

(203) 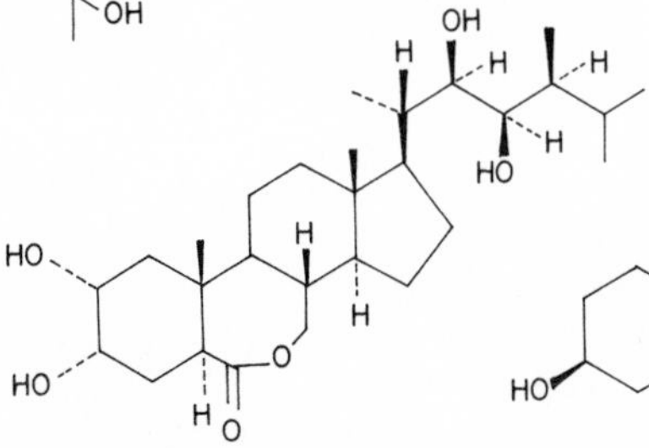   (204) 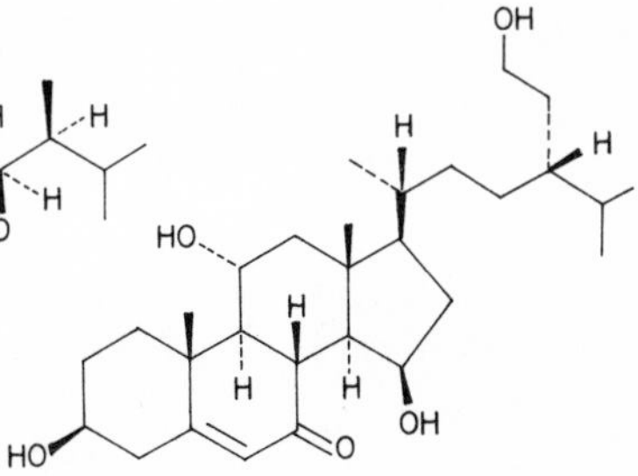   (205)

groups include the hydroxylated metabolites of vitamin D, e.g. 1α,25-
dihydroxyvitamin D_3 (203), the plant growth promoters, e.g. brassinolide
(204), the fungal sex hormones, e.g. oogoniol (205), the marine steroids, e.g.
gorgosterol (206), and the anti-tumour active withanolides, e.g. withaferin A
(207).

Side-chain syntheses have recently been reviewed quite comprehensively.[215,216] Some selected examples are discussed below. The synthetic
utility of palladium-π-allyl complexes has been alluded to earlier[70-74] and a
particular application to the stereoselective development of side-chains
deserves special mention. Reaction of the allylic acetate (208) in a catalytic
palladium reaction employing Ph_3P-$(Ph_3P)_4Pd$-$PhSO_2CHNaCO_2Me$ gave
the ester (209) with the natural C-20 configuration, whereas the stoichio-
metric palladium reaction between the Π-allyl complex (211) and
$PhSO_2CHNaCO_2Me$ in the presence of $Ph_2PCH_2CH_2PPh_2$ afforded the C-
20 epimer (210). The catalytic reaction has been employed in the synthesis of
the side-chains of cholesterol[217] and ecdysone.[218] The Lewis acid-catalysed
ene reaction between the Z-$\Delta^{17(20)}$-compound (212) and methyl propiolate
afforded the 16,22-diene (213) also with the natural configuration at C-20
owing to the well-ordered transition state and the preferred attack on the α-
face of the 17 (20)-double bond.[219,220] Similar stereoselectivity was achieved

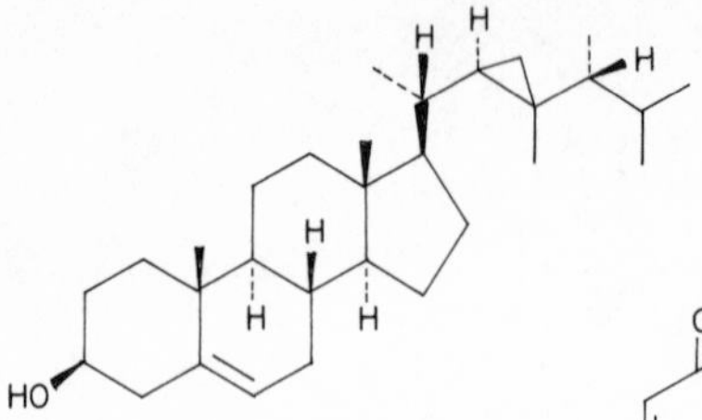

(206)

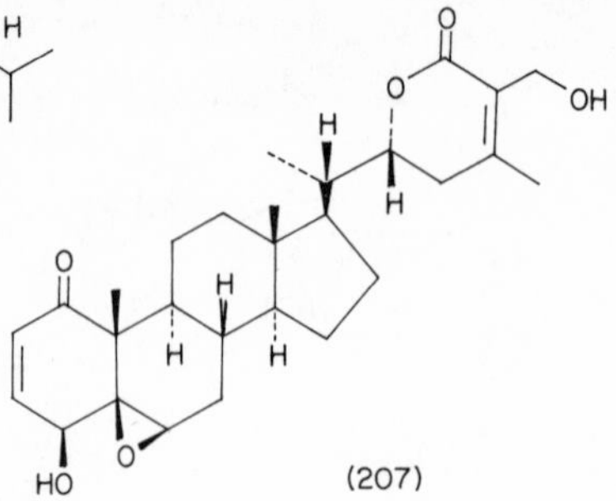

(207)

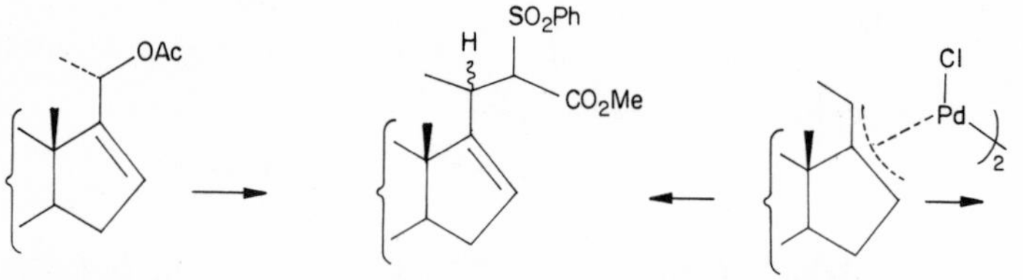

(208)

(209) B-H

(210) α-H

(211)

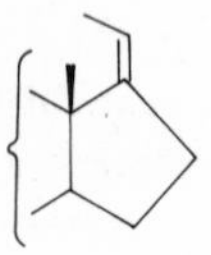

(212)

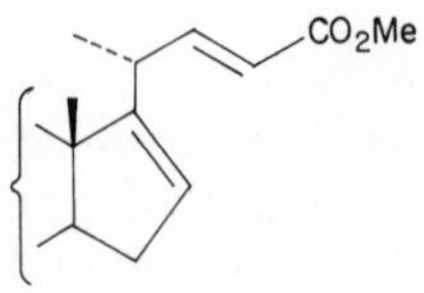

(213)

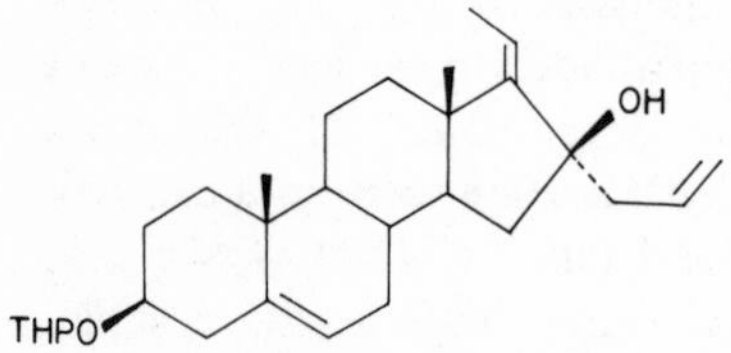

(214)

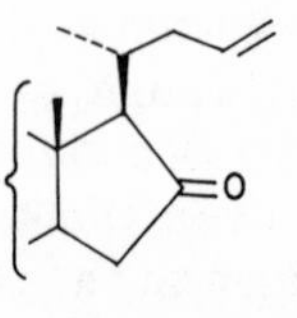

(215)

in a recent synthesis of desmosterol in which the key step was the oxy-Cope rearrangement of the anion of the unsaturated alcohol (214) to the unsaturated ketone (215).[221] The Claisen rearrangement which also has a highly ordered chair-like six-membered transition state has found several applications in stereoselective side-chain and other syntheses.[215] For example, the unsaturated ester (216) underwent the Carroll variation of the Claisen

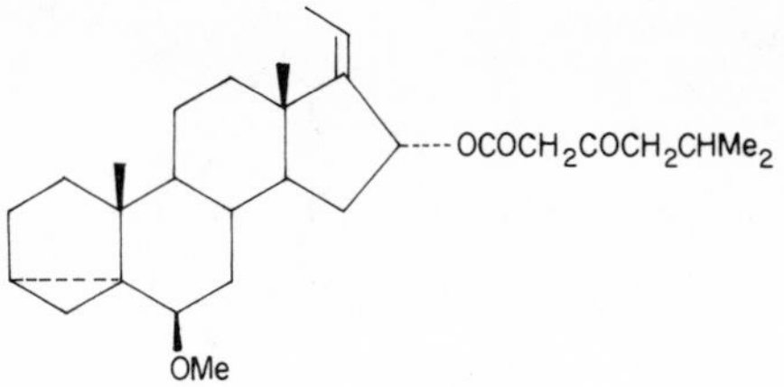

(216)

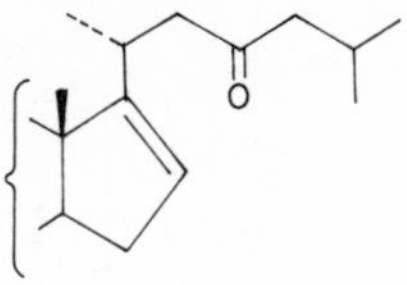

(217)

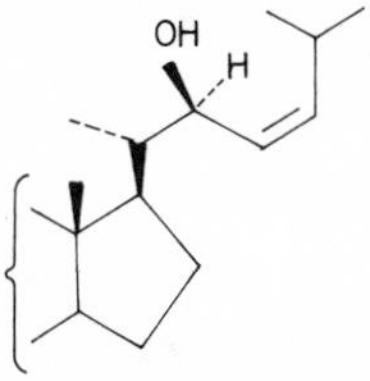

(218)

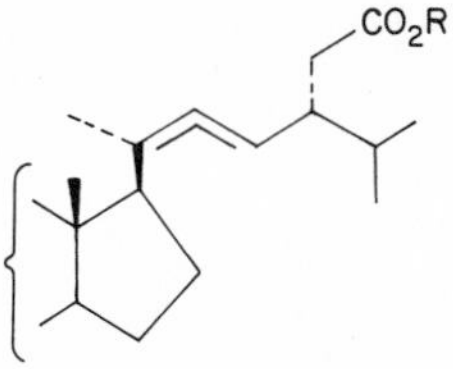

(219)

rearrangement to afford the ketone (217)[222] and a key step in the synthesis of compounds with the oogoniol side-chain was the conversion of the 22-hydroxy-Δ^{23}-compounds (218) to the esters (219) by treatment with trialkyl orthoacetate.[223,224]

Very extensive investigations[225] of marine steroids over the last decade or so have revealed a wide variety of side-chain structures in addition to a group of very interesting A-nor-steroids, e.g. 3β-hydroxy-methyl-5α-cholestane (220), with conventional and unconventional side-chains. The Claisen rearrangement has found application in the construction of some of the unusual side-chains which are encountered. A synthesis of stellasterol (221) which has the 24S-configuration incorporated[226] the Claisen rearrangement of the allylic alcohol (222) with triethyl orthoacetate to the unsaturated ester (223). Similarly, a synthesis of dinosterol (224), which may be a biosynthetic precursor to gorgosterol (226), employed the Claisen rearrangement of the allylic alcohol (225) with triethyl orthopropionate to the unsaturated esters (226) which are epimeric at C-25.[227] Following the discovery of the biological importance of the hydroxylated derivatives of vitamin D there has been considerable renewed interest in this area.[228–231] The Claisen rearrangement of the allylic alcohols (227) and (228) with triethyl orthopropionate was employed, as above, to afford the unsaturated esters (229) which were converted to Windaus and Grundmann's C_{19} ketone (230) which has been converted to vitamin D_2.[232] An alternative approach to (230) employed the reaction of the aldehyde (231) with the lithio derivative of the sulphone (232).

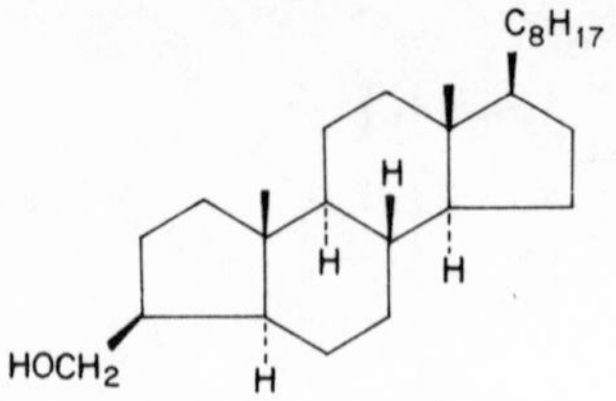

(220)

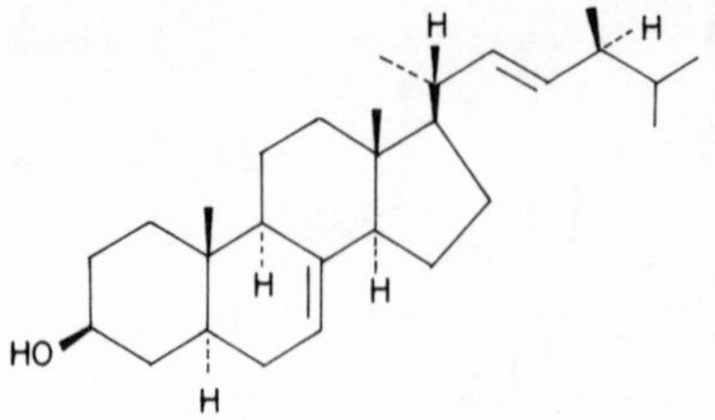

(221)

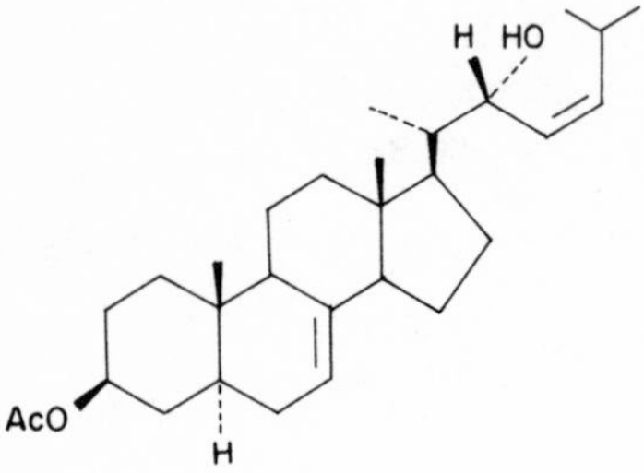

(222)

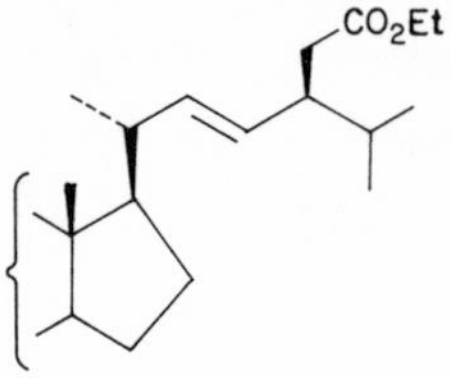

(223)

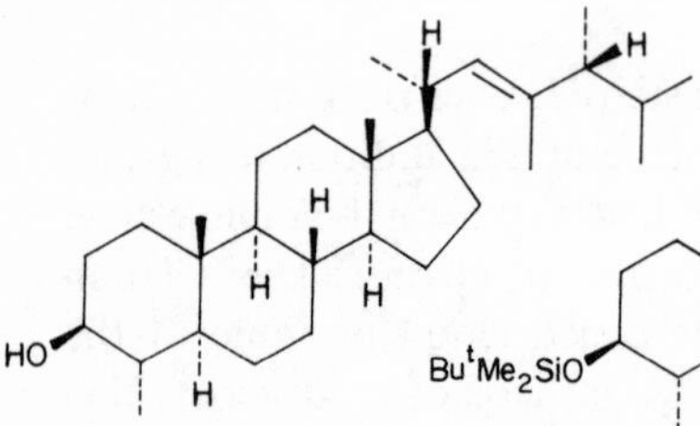

(224)

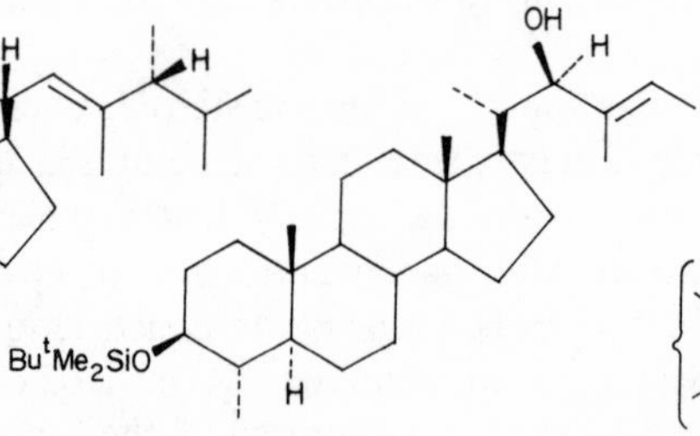

(225)

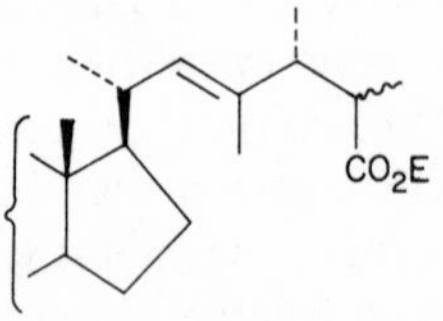

(226)

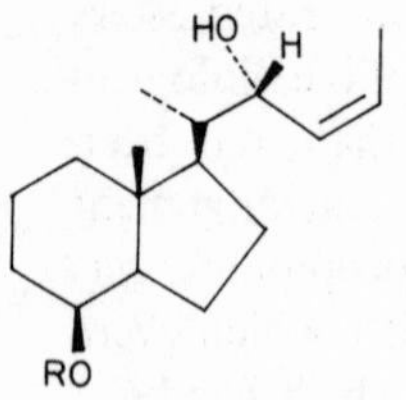

(227)

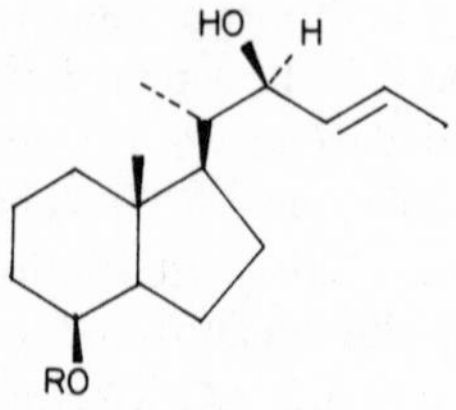

(228)

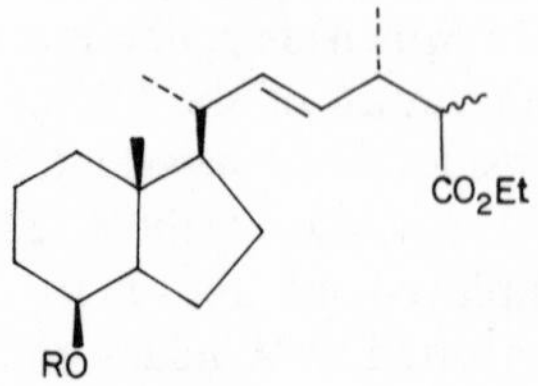

(229)

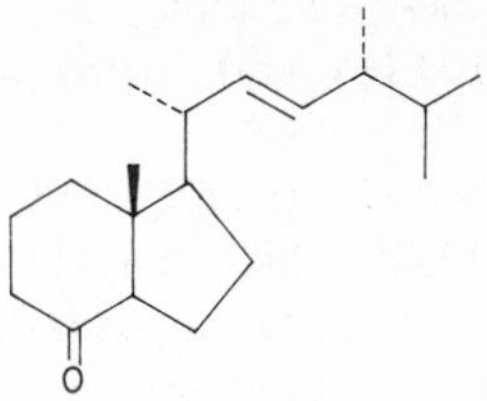

(230)

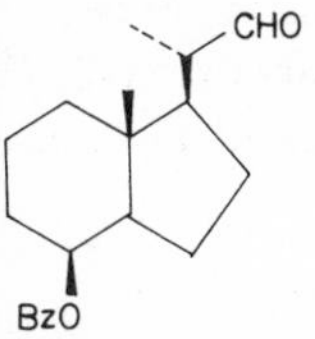

(231)

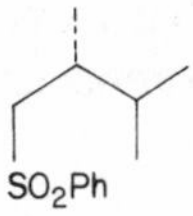

(232)

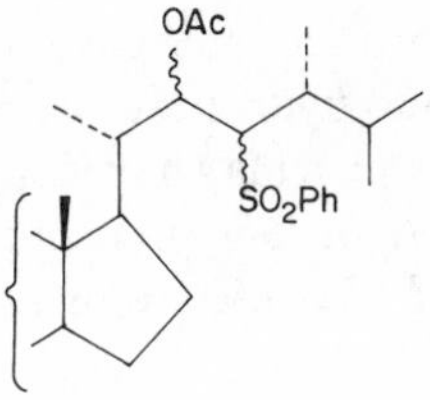

(233)

Acetylation gave the β-acetoxy-sulphone (233) which on sodium amalgam reduction and hydrolysis afforded the alcohol (234). No Z-isomer of (234) was detected. The more obvious Wittig-type approach to the construction of the 22,23-double bond was found to be unsuitable.[233]

Novel approaches to the development of the dihydroxyacetone side-chain of corticosteroids have been investigated.[215,234–237] As mentioned earlier, allenes such as (182) may be directly oxidized with OsO_4 to afford the dihydroxyacetone side-chain.[234] Oxidation of the 17β-acetyl group with PhIO-KOH-MeOH gave the hydroxy-acetal (235) which, after acetylation

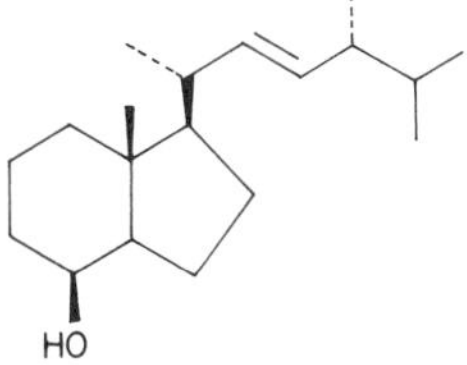

(234)

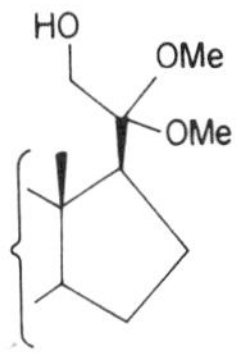

(235)

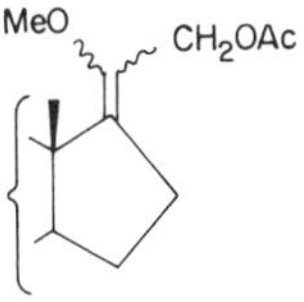

(236)

and heating in xylene with toluene-*p*-sulphonic acid, gave the acetoxy-enol ether (236). Oxidation with MCPBA and hydrolysis gave the required corticosteroid side-chain.[235]

Reaction of 17-oxo-steroids with $(EtO)_2POCH(NC)Me-KH$ afforded the 20-isocyano-$\Delta^{17(20)}$-compound (237) which may be transformed by two procedures to the compounds bearing the dihydroxyacetone side chain.[236] For example, hydrolysis of (237) to the N-formyl-enamine (238) followed by epoxidation (MCPBA) and rearrangement (40°C) gave compound (239). Azeotropic removal of water gave the unstable intermediate (240) which on bromination and hydrolysis gave the 21-bromo-17α-hydroxy-20-oxo-compound (241) which may be readily converted to the 17α,21-dihydroxy-20-oxo-compound. A further procedure was reported from the condensation products (242) of 17-oxo-steroids and ethyl isocyanoacetate.[237]

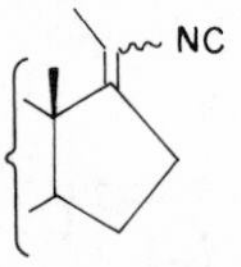

(237) 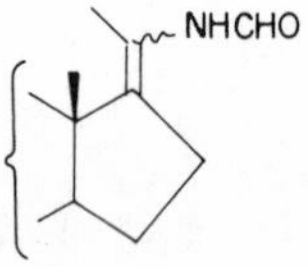   (238) 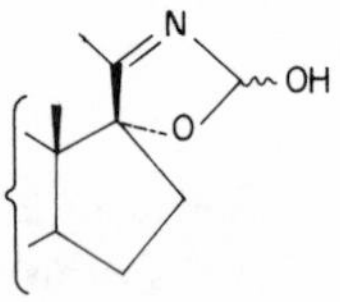   (239) 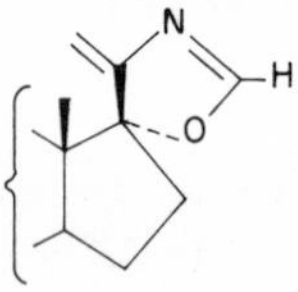   (240) 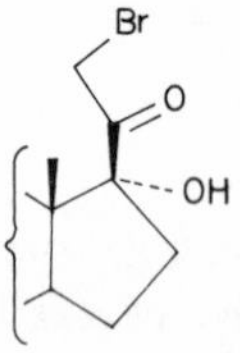   (241)

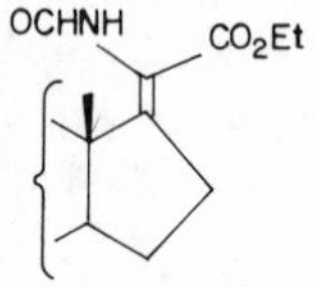

(242)

References

1. J.R. Dias and R. Ramachandra, 1977, *Synth. Comm.*, 7, 293.
2. K.-Y. Tserng and P.D. Klein, 1977, *Steroids*, 29, 635.
3. J.F. Baker and R.T. Blickenstaff, 1975, *J. Org. Chem.*, 40, 1579.
4. K.F. Atkinson and R.T. Blickenstaff, 1974, *Steroids*, 23, 895.
5. M.V. Bhatt and S.U. Kulkarni, 1983, *Synthesis*, 249.
6. B. Ganem and V.R. Small, 1974, *J. Org. Chem.*, 39, 3728.
7. M.E. Jung and M.A. Lyster, 1977, *J. Org. Chem.*, 42, 3761.
8. S. Hanessian and Y. Guindon, 1980, *Tetrahedron Lett.*, 21, 2305.
9. M. Node, H. Hori and E. Fujita, 1976, *J. Chem. Soc. Perkin Trans. I*, 2237.
10. H. Niwa, T. Hida and K. Yamada, 1981, *Tetrahedron Lett.*, 22, 4239.
11. J.D. Daley, J.M. Rosenfeld and E.V. Younglai, 1976, *Steroids*, 27, 481.
12. H. Hosoda, K. Yamashita, H. Sagae and T. Nambara, 1975, *Chem. and Pharm. Bull. Japan*, 23, 2118.
13. G. Phillipou, D.A. Bigham and R.F. Seamark, 1975, *Steroids*, 26, 516.

14. H. Hosoda, D.K. Fukushima and J. Fishman, 1973, *J. Org. Chem.*, 38, 4209.
15. M.R. Detty and M.D. Seidler, 1981, *J. Org. Chem.*, 46, 1283.
16. R.J. Batten, A.J. Dixon, R.J.K. Taylor and R.F. Newton, 1980, *Synthesis*, 234.
17. B.W. Metcalf, J.P. Burkhart and K. Jund, 1980, *Tetrahedron Lett.*, 21, 35.
18. E.J. Corey and J.W. Suggs, 1975, *Tetrahedron Lett.*, 2647.
19. A. McKillop and D.W. Young, 1979, *Synthesis*, 401, 481.
20. K.-Y. Tserng, 1978, *J. Lipid Res.*, 19, 501.
21. F.J. Kakis, M. Fetizon, N. Douchkine, M. Golfier, P. Morgues and T. Prange, 1974, *J. Org. Chem.*, 39, 523.
22. G.H. Posner, R.B. Perfetti and A.W. Runquist, 1976, *Tetrahedron Lett.*, 3499.
23. M. Al Neirabeyeh, J.-C. Ziegler, B. Gross and P. Canbère, 1976, *Synthesis*, 811.
24. J. Wicha and A. Zarecki, 1974, *Tetrahedron Lett.*, 3059.
25. M.E. Jung and L.M. Speltz, 1976, *J. Am. Chem. Soc.*, 98, 7882.
26. D.H.R. Barton, A.G. Brewster, R.A.H.F. Hui, D.J. Lester, S.V. Ley and T.G. Back, 1978, *J. Chem. Soc. Chem. Comm.*, 952.
27. E. Piers and P.M. Worster, 1977, *Can. J. Chem.*, 55, 733.
28. Y.-S. Cheng, W.-L. Liu and S. Chen, 1980, *Synthesis*, 223.
29. D.H.R. Barton and W.B. Motherwell, 1981, *Pure Appl. Chem.*, 53, 1081.
30. O. Mitsunobu, 1981, *Synthesis*, 1, and references cited therein.
31. M.S. Manhas, W.H. Hoffman, B. Lal and A.K. Bose, 1975, *J. Chem. Soc. Perkin Trans. I*, 461.
32. H. Loibner and E. Zbiral, 1976, *Helv. Chim. Acta*, 59, 2100.
33. E. Crochowski, E. Falent and J. Jurczak, 1978, *Pol. J. Chem.*, 335.
34. G. Grynkiewicz and H. Burzynska, 1976, *Tetrahedron*, 32, 2109.
35. R. Aneja, A.P. Davies and J.A. Knaggs, 1975, *Tetrahedron Lett.*, 1033.
36. A.K. Bose and B. Lal, 1973, *Tetrahedron Lett.*, 3937.
37. S. Rozen, Y. Faust and H. Ben-Yakov, 1979, *Tetrahedron Lett.*, 1823.
38. T.G.C. Bird, G. Felsky, P.M. Fredericks, E.R.H. Jones and G.D. Meakins, 1979, *J. Chem. Res. (S)*, 388.
39. Y. Kobayashi, I. Kumadaki, A. Ohsawa, M. Honda and Y. Hanzawa, 1975, *Chem. and Pharm. Bull. Japan*, 23, 196.
40. J. Gore, P. Place and M.L. Roumestant, 1973, *J. Chem. Soc. Chem. Comm.*, 821.
41. D.B. Cowell, A.K. Davis, D.W. Mathieson and P.D. Nicklin, 1974, *J. Chem. Soc. Perkin Trans. I*, 1505.
42. E. Glotter, P. Krinsky, M. Rejtö and M. Weissenberg, 1976, *J. Chem. Soc. Perkins Trans. I*, 1442.
43. D.H.R. Barton and Y. Houminer, 1973, *J. Chem. Soc. Chem. Comm.*, 839.
44. P. Kočovský, L. Kohout and V. Černý, 1980, *Coll. Czech. Chem. Comm.*, 45, 559, and references cited therein.
45. P. Kočovský and V. Černý, 1980, *Coll. Czech. Chem. Comm.*, 45, 3190, 3199.
46. T.H. Campion, G.A. Morrison and J.B. Wilkinson, 1976, *J. Chem. Soc. Perkin Trans. I*, 2509.
47. H. Velgová and V. Černý, 1973, *Coll. Czech. Chem. Comm.*, 38, 575.
48. R.W.G. Foster and B.A. Marples, 1979, *Tetrahedron lett.*, 2071.
49. R. Caputo, L. Mangoni and L. Previtera, 1975, *Steroids*, 25, 619.
50. L. Mangoni, M. Adinolfi, G. Barone and M. Parrilli, 1975, *Gazzetta*, 105, 377.
51. M. Parrilli, M. Adinolfi, G. Barone, G. Laonigri and L. Mangoni, 1975, *Gazzetta*, 105, 1301.
52. R.C. Cambie, D. Chambers, P.S. Rutledge and P.D. Woodgate, 1977, *J. Chem. Soc. Perkin Trans. I*, 2231.
53. R.C. Cambie, R.C. Hayward, J.L. Roberts and P.S. Rutledge, 1974, *J. Chem. Soc. Perkin Trans. I*, 1858.
54. E. Glotter and A. Schwartz, 1976, *J. Chem. Soc. Perkin Trans. I*, 1660.
55. P. Kočovský and V. Černý, 1977, *Coll. Czech. Chem. Comm.*, 42, 353.
56. P. Kočovský and V. Černý, 1977, *Coll. Czech. Chem. Comm.*, 42, 163.

57. P. Kočovský and V. Černý, 1980, *Coll. Czech. Chem. Comm.*, 45, 3023, 3030, and references cited therein.
58. P. Kočovský, 1980, *Tetrahedron Lett.*, 21, 555.
59. D.H.R. Barton, T. Shiori and D.A. Widdowson, 1971, *J. Chem. Soc. (C)*, 1968.
60. M. Tada and A. Oikawa, 1978, *J. Chem. Soc. Chem. Comm.*, 727.
61. M. Anastasia and M. Derossi, 1979, *J. Chem. Soc. Chem. Comm.*, 164.
62. D.H.R. Barton, A.A.L. Gunatilaka, T. Nakanishi, H. Patin, D.A. Widdowson and B.R. Worth, 1976, *J. Chem. Soc. Perkin Trans. I*, 821.
63. D.H.R. Barton, R.K. Haynes, G. Leclerc, P.D. Magnus and I.D. Menzies, 1975, *J. Chem. Soc. Perkin Trans. I*, 2055.
64. R.K. Haynes, 1978, *Aust. J. Chem.*, 31, 121.
65. R. Tang, H.J. Yue, J.F. Wolff and F. Mares, 1978, *J. Amer. Chem. Soc.*, 100, 5248.
66. W. Reischl and E. Zbiral, 1979, *Helv. Chim. Acta*, 62, 1763.
67. S. Yamada and H. Takayama, 1979, *Chem. Lett.*, 583.
68. W. Reischl and E. Zbiral, 1979, *Monatsh. Chem.*, 110, 1463.
69. D.H.R. Barton and H. Patin, 1977, *J. Chem. Soc. Chem. Comm.*, 799.
70. J.Y. Satoh and C.A. Horiuchi, 1979, *Bull. Chem. Soc. Japan*, 52, 2653, and references cited therein.
71. B.M. Trost, 1980, *Acc. Chem. Res.*, 13, 385, and references cited therein.
72. D.N. Jones and S.D. Knox, 1975, *J. Chem. Soc. Chem. Comm.*, 166.
73. J.Y. Satoh and C.A. Horiuchi, 1981, *Bull. Chem. Soc. Japan*, 54, 625.
74. D.J. Collins, W.R. Jackson and R.N. Timms, 1977, *Aust. J. Chem.*, 30, 2167.
75. D.J. Collins, W.R. Jackson and R.N. Timms, 1980, *Aust. J. Chem.*, 33, 2663.
76. J.A.M. Peters, N.P. Van Vliet and F.J. Zeelan, 1979, *Recl. Trav. chim.*, 98, 459.
77. R.J. Chorvat and B.N. Desai, 1979, *J. Org. Chem.*, 44, 3974.
78. W.G. Salmond, M.A. Barta and J.L. Havens, 1978, *J. Org. Chem.*, 43, 2057, and references cited therein.
79. E. Mappus and C.-Y. Cuilleron, 1979, *J. Chem. Res. (S)*, 42.
80. D.H.R. Barton, D.J. Lester and S.V. Ley, 1980, *J. Chem. Soc. Perkin Trans. I*, 2209.
81. E. Mincione, G. Ortaggi and A. Sirna, 1977, *Synthesis*, 773.
82. M.P. Paradisi, G.P. Zecchini and A. Romeo, 1977, *Tetrahedron Lett.*, 2369.
83. M.P. Paradisi and G.P. Zecchini, 1981, *Tetrahedron*, 37, 971.
84. G.H. Posner, A.W. Runquist and M.J. Chapdelaine, 1977, *J. Org. Chem.*, 42, 1202.
85. D.C. Wigfield, 1979, *Tetrahedron*, 35, 449.
86. R. Contreras and L. Mendoza, 1979, *Steroids*, 34, 121.
87. W.G. Dauben and J.W. Ashmore, 1978, *Tetrahedron Lett.*, 4487.
88. D. Nasipuri, P.R. Mukherjee, S.C. Pakrashi, S. Datta and P.P. Ghosh-Dastidar, 1976, *J. Chem. Soc. Perkin Trans. I*, 321.
89. J.E. Baldwin, 1976, *J. Chem. Soc. Chem. Comm.*, 738.
90. F.A.J. Meskens, 1981, *Synthesis*, 501.
91. D.H.R. Barton, C.C. Dawes and P.D. Magnus, 1975, *J. Chem. Soc. Chem. Comm.*, 432.
92. E.J. Corey and J.W. Suggs, 1975, *Tetrahedron Lett.*, 3775.
93. D.A. Evans, K.G. Grimm and L.K. Truesdale, 1975, *J. Amer. Chem. Soc.*, 97, 3229.
94. D.A. Evans, L.K. Truesdale, K.G. Grimm and S.L. Nesbitt, 1977, *J. Am. Chem. Soc.*, 99, 5009.
95. F. Huet, A. Lechevallier, M. Pellet and J.M. Conia, 1978, *Synthesis*, 63.
96. B.-T. Gröbel and D. Seebach, 1977, *Synthesis*, 357.
97. H.L. Wang-Chang, 1972, *Tetrahedron Lett.*, 1989.
98. K. Fuji, K. Ichikawa and E. Fujita, 1978, *Tetrahedron Lett.*, 3561.
99. J. Cairns and R.T. Logan, 1980, *J. Chem. Soc. Chem. Comm.*, 886.
100. N.J. Cussans, S.V. Ley and D.H.R. Barton, 1980, *J. Chem. Soc. Perkin Trans. I*, 1650.
101. D.H.R. Barton, D.J. Lester and S.V. Ley, 1977, *J. Chem. Soc. Chem. Comm.*, 445.
102. M.R. Detty, 1979, *Tetrahedron Lett.*, 4189.
103. M.E. Jung, W.A. Andrus and P.L. Ornstein, 1977, *Tetrahedron Lett.*, 4175.
104. B.M. Trost, K. Hiroi and S. Kurozumi, 1975, *J. Amer. Chem. Soc.*, 97, 438.

105. B.M. Trost, 1978, *Chem. Rev.*, 78, 363.
106. B.M. Trost, 1978, *Acc. Chem. Res.*, 11, 453.
107. K.B. Sharpless, R.F. Lauer and A.Y. Teranishi, 1973, *J. Amer. Chem. Soc.*, 95, 6137.
108. T. Nakai and T. Mimura, 1979, *Tetrahedron Lett.*, 531.
109. W.E. Fristad, T.R. Bailey and L.A. Paquette, 1980, *J. Org. Chem.*, 45, 3028.
110. L.A. Paquette, W.E. Fristad, D.S. Dime and T.R. Bailey, 1980, *J. Org. Chem.*, 45, 3017.
111. G.M. Rubottom and J.M. Gruber, 1978, *J. Org. Chem.*, 43, 1599.
112. P.J. Kocienski, B. Lythgoe and D.A. Roberts, 1980, *J. Chem. Soc. Perkin Trans. I*, 897.
113. J.P. McCormick, W. Tomasik and M.W. Johnson, 1981, *Tetrahedron Lett.*, 22, 607.
114. E. Vedejs, D.A. Engler and J.E. Telschow, 1978, *J. Org. Chem.*, 43, 188.
115. D.N. Kirk and M.P. Hartshorn, 'Steroid Reaction Mechanisms', Elsevier, Amsterdam, 1968, pp. 228, 353.
116. Y. Nakatani, G. Ponsinet, G. Wolff, J.L. Zundel and G. Ourisson, 1972, *Tetrahedron*, 28, 4249, and references cited therein.
117. K.E. Suckling and C.J. Suckling, 'Biological Chemistry', Cambridge University Press, 1980, p. 114.
118. D.N. Akponaye, R.D. Farrant and D.N. Kirk, 1981, *J. Chem. Res. (S)*, 210.
119. D.N. Kirk and P.M. Shaw, 1975, *J. Chem. Soc. Perkin Trans. I*, 2284.
120. M.M. Janot, F. Frappier, J. Thierry, G. Lukacs, F.X. Jarreau and R. Goutarel, 1972, *Tetrahedron Lett.*, 3499.
121. E.T.J. Bathurst, J.M. Coxon and M.P. Hartshorn, 1974, *Aust. J. Chem.*, 1505 and references cited therein.
122. J.W. Blunt, M.P. Hartshorn and D.N. Kirk, 1966, *Tetrahedron*, 22, 3195.
123. B.N. Blackett, J.M. Coxon, M.P. Hartshorn and K.E. Richards, 1969, *Tetrahedron*, 25, 4999 and other papers in this series.
124. J.G. Ll. Jones and B.A. Marples, 1972, *J. Chem. Soc. Perkin Trans. I*, 792, and references cited therein.
125. Ref. 115, p. 257, and references cited therein.
126. I.G. Guest and B.A. Marples, 1973, *J. Chem. Soc. Perkin I*, 900, and references cited therein.
127. A. Schwartz and E. Glotter, 1977, *J. Chem. Soc. Perkin Trans. I*, 2470.
128. A.T. Rowland, 1966, *Steroids*, 7, 527.
129. V. Dave and E.W. Warnhoff, 1978, *J. Org. Chem.*, 43, 4622.
130. I. Khattak, D.N. Kirk, C.M. Peach and M.A. Wilson, 1975, *J. Chem. Soc. Perkin Trans. I*, 916.
131. Y. Gopichand and H. Hirschmann, 1979, *J. Org. Chem.*, 44, 185.
132. H. Hirschmann, F.B. Hirschmann and Y. Gopichand, 1979, *J. Org. Chem.*, 44, 180.
133. J.-C. Jacquesy, R. Jacquesy and G. Joly, 1975, *Bull. Soc. chim. France*, 2281.
134. J.-C. Jacquesy, R. Jacquesy and G. Joly, 1975, *Bull. Soc. chim. France*, 2289.
135. R. Jacquesy and H.L. Ung, 1976, *Bull. Soc. chim. France*, 1889.
136. R. Jacquesy and H.L. Ung, 1977, *Tetrahedron*, 33, 2543.
137. R. Jacquesy, C. Narbonne and H.L. Ung, 1979, *J. Chem. Res. (S)*, 288.
138. R. Jacquesy and C. Narbonne, 1979, *J. Chem. Soc. Chem. Comm.*, 765.
139. R. Jacquesy and C. Narbonne, 1978, *Bull. Soc. chim. France*, 163.
140. D. Baldwin and J.R. Hanson, 1975, *J. Chem. Soc. Perkin Trans. I*, 1941, and references cited therein.
141. J.R. Hanson, D. Raines and S.G. Knights, 1980, *J. Chem. Soc. Perkin Trans. I*, 1311, and references cited therein.
142. J.R. Hanson and S.G. Knights, 1980, *J. Chem. Soc. Perkin Trans. I*, 1306, and references cited therein.
143. J.R. Hanson and S.G. Knights, 1981, *J. Chem. Soc. Perkin Trans. I*, 25, and references cited therein.
144. J.J. Jagodzinski, J. Gumulka and W.J. Szczepek, 1981, *Tetrahedron*, 37, 1015.
145. See J.R. Hanson and H.J. Wilkins, 1974, *J. Chem. Soc. Perkin Trans. I*, 1388, for exceptions.
146. N. Bosworth, A. Emke, J.M. Midgley, C.J. Moore, W.B. Whalley, G. Ferguson and W.C. Marsh, 1977, *J. Chem. Soc. Perkin Trans. I*, 805.
147. A. Emke, J.M. Midgley and W.B. Whalley, 1980, *J. Chem. Soc. Perkin Trans. I*, 1779.

148. H. de Nijs and W.N. Speckamp, 1973, *Tetrahedron Lett.*, 813.
149. A.B. Turner, 1979, *J. Chem. Soc. Perkin Trans. I*, 1333.
150. C.L. Hewett, S.G. Gibson, I.M. Gilbert, J. Redpath and D.S. Savage, 1973, *J. Chem. Soc. Perkin Trans. I*, 1967.
151. Ref. 115, p. 294.
152. D.N. Kirk and C.R. McHugh, 1978, *J. Chem. Soc. Perkin Trans. I*, 173.
153. P. Yates and F.M. Winnik, 1981, *Can. J. Chem.*, 59, 1641.
154. D.S. Brown, R.W.G. Foster, B.A. Marples and K.G. Mason, 1980, *Tetrahedron Lett.*, 21, 5057.
155. G.W. Kirby and J.W.M. Mackinnon, 1977, *J. Chem. Soc. Chem. Comm.*, 23.
156. R.B. Boar, F.K. Jetuah, J.F. McGhie, M.S. Robinson and D.H.R. Barton, 1977, *J. Chem. Soc. Perkin Trans. I*, 2163.
157. V. Dave, J.B. Stothers and E.W. Warnhoff, 1980, *Can. J. Chem*, 58, 2666, and references cited therein.
158. M.S. Ahmad, G. Moinuddin and I.A. Khan, 1978, *J. Org. Chem.*, 43, 163, and references cited therein.
159. J.A. Waters, Y. Kondo and B. Witkop, 1972, *J. Pharm. Sci.*, 61, 321.
160. O. Jeger and K. Schaffner, 1970, *Pure and Appl. Chem.*, 21, 247.
161. H. Suginome, N. Maeda, Y. Takahashi and N. Miyata, 1981, *Bull. Chem. Soc. Japan*, 54, 846, and references cited therein.
162. S.H. Imam and B.A. Marples, 1977, *Tetrahedron Lett.*, 2613.
163. G.J. Edge, S.H. Imam and B.A. Marples, 1981, *J. Photochem.*, 17, 97.
164. K. Yamada, S. Tanaka, K. Naruchi and M. Yamamoto, 1982, *J. Org. Chem.*, 47, 5283.
165. O. Kikuchi, T. Kanekiyo, S. Tanaka, K. Naruchi and K. Yamada, 1982, *Bull. Chem. Soc. Japan*, 1509.
166. E. Oliveros, M. Riviere, J.P. Malrieu and Ch. Teichteil, 1979, *J. Am. Chem. Soc.*, 101, 318, and references cited therein.
167. H. Suginome, N. Yonekura, T. Mizuguchi and T. Masamune, 1977, *Bull. Chem. Soc. Japan*, 50, 3010.
168. J.T. Pinhey, E. Rizzardo and G.C. Smith, 1978, *Aust. J. Chem.*, 31, 97, 113.
169. J.S. Cridland, P.J. Moles, S.T. Reid and K.T. Taylor, 1976, *Tetrahedron Lett.*, 4497.
170. J.R. Williams, R.H. Moore, R. Li and J.F. Blount, 1979, *J. Am. Chem. Soc.*, 101, 5019.
171. J.R. Williams, R.H. Moore, R. Li and C.M. Weeks, 1980, *J. Org. Chem.*, 45, 2324.
172. V. Ferrer, J. Gomez and J.-J. Bonet, 1977, *Helv. Chim. Acta*, 60, 1357.
173. A. Enger, A. Feigenbaum, J.-P. Pete and J.L. Wolfhugel, 1978, *Tetrahedron*, 34, 1509.
174. M.J. Caus, A. Cánovas and J.-J. Bonet, 1980, *Helv. Chim. Acta*, 63, 473.
175. A. Cánovas and J.-J. Bonet, 1980, *Helv. Chim. Acta*, 63, 486.
176. J. Muzart and J.-P. Pete, 1978, *Tetrahedron*, 34, 1179.
177. A. Murai, N. Iwasa and T. Masamune, 1980, *Bull. Chem. Soc. Japan*, 53, 259.
178. K.N. Houk, 1976, *Chem. Rev.*, 76, 1.
179. W.G. Dauben, G. Lodder and J. Ipaktschi, 1975, *Fortschr. Chem. Forsch.*, 54, 73.
180. S.S. Hixon, P.S. Mariano and H.E. Zimmerman, 1973, *Chem. Rev.*, 73, 531.
181. J. Pusset, M.-T. Le Goff and R. Beugelmans, 1975, *Tetrahedron*, 31, 643, and references cited therein.
182. J.R. Williams and A. Abdel-Magid, 1981, *Tetrahedron*, 37, 1675, and references cited therein.
183. J.R. Williams, H. Salama and J.D. Leber, 1977, *J. Org. Chem.*, 42, 102.
184. J.R. Williams and G.M. Sarkisian, 1980, *J. Org. Chem.*, 45, 5088.
185. L.J. Dolby and M. Tuttle, 1975, *J. Org. Chem.*, 40, 3786.
186. J.A. Vallet, J. Boix, J.-J. Bonet, M.C. Briansó, C. Miravittles and J.L. Briansó, 1978, *Helv. Chim. Acta*, 61, 1158.
187. D.H.R. Barton, N.K. Basu, M.J. Day, R.H. Hesse, M.M. Pechet and A.N. Starrat, 1975, *J. Chem. Soc. Perkin Trans. I*, 2243.
188. R.W. Freerksen, W.E. Pabst, M.L. Raggio, S.A. Sherman, R.R. Wroble and D.S. Watt, 1977, *J. Amer. Chem. Soc.*, 99, 1536, and references cited therein.
189. G. Neef, U. Eder, G. Haffer, G. Sauer and R. Wiechert, 1980, *Chem. Ber.*, 113, 1106.

190. A.D. Barone and D.S. Watt, 1978, *Tetrahedron Lett.*, 3673.
191. R. Hernández, A. Rivera, J.A. Salazar and E. Suárez, 1980, *J.C.S. Chem. Comm.*, 958.
192. R. Breslow, 1980, *Acc. Chem. Res.*, 13, 170.
193. R. Breslow, B.B. Snider and R.J. Corcoran, 1974, *J. Am. Chem. Soc.*, 96, 6792.
194. R. Breslow, R.J. Corcoran, B.B. Snider, R.J. Doll, P.L. Khanna and R. Kaleya, 1977, *J. Amer. Chem. Soc.*, 99. 905.
195. S.F. Donovan, M.A. Avery and J.E. McMurry, 1979, *Tetrahedron Lett.*, 3287, and references cited therein.
196. D. Wolner, 1979, *Tetrahedron Lett.*, 4613.
197. R. Breslow and L.M. Maresca, 1977, *Tetrahedron Lett.*, 623.
198. P.C. Scholl and M.R. van de Mark, 1973, *J. Org. Chem.*, 38, 2376.
199. E.E. van Tamelen, 1981, *Pure Appl. Chem.*, 53, 1259.
200. W.S. Johnson, 1976, *Bioorg. Chem.*, 5, 51.
201. W.S. Johnson, 1976, *Angew. Chem. (Int. Edn.)*, 15, 9.
202. E.E. van Tamelen and J.R. Hwu, 1983, *J. Amer. Chem. Soc.*, 105, 2490.
203. W.S. Johnson, B.E. McCarry, R.L. Markezich and S.G. Boots, 1980, *J. Amer. Chem. Soc.*, 102, 352.
204. W.S. Johnson, R.S. Brinkmeyer, V.M. Kapoor and T.M. Yarnell, 1977, *J. Amer. Chem. Soc.*, 99, 8341.
205. W.S. Johnson, C.R. Fagundo and A.G. Ravelo, private communication of unpublished observations.
206. R. Schmid, P.L. Huesmann and W.S. Johnson, 1980, *J. Amer. Chem. Soc.*, 102, 5122.
207. W.S. Johnson, D.J. Dumas and D. Berner, 1982, *J. Amer. Chem. Soc.*, 104, 3510.
208. W.S. Johnson, D. Berner, D.J. Dumas, P.J.R. Nederlof and J. Welch, 1982, *J. Amer. Chem. Soc.*, 104, 3508.
209. A.A. Macco and H.M. Buck, 1981, *J. Org. Chem.*, 46, 2655, and references cited therein.
210. E.E. van Tamelen and D.R. James, 1977, *J. Amer. Chem. Soc.*, 99, 950.
211. T. Kametani and H. Nemoto, 1981, *Tetrahedron*, 37, 3, and references cited therein.
212. W. Oppolzer, 1978, *Synthesis*, 793, and references cited therein.
213. R.L. Funk and K.P.C. Volhardt, 1979, *J. Amer. Chem. Soc.*, 101, 215.
214. T. Kametani, K. Suzuki and H. Nemoto, 1980, *J. Org. Chem.*, 45, 2204.
215. J. Redpath and F.J. Zeelan, 1983, *Chem. Soc. Rev.*, 12, 75.
216. D.M. Piatak and J. Wicha, 1978, *Chem. Rev.*, 78, 199.
217. B.M. Trost and T.R. Verhoeven, 1978, *J. Amer. Chem. Soc.*, 100, 3435.
218. B.M. Trost and Y. Matsumura, 1977, *J. Org. Chem.*, 42, 2036.
219. W.G. Dauben and T. Brookhart, 1981, *J. Am. Chem. Soc.*, 103, 237.
220. A.D. Batcho, D.E. Berger, M.R. Uskoković and B.B. Snider, 1981, *J. Amer. Chem. Soc.*, 103, 1292.
221. M. Koreeda, Y. Tanaka and A. Schwartz, 1980, *J. Org. Chem.*, 45, 1173.
222. M. Tanabe and K. Hayashi, 1980, *J. Amer. Chem. Soc.*, 102, 862.
223. J. Wiersig, N. Waespe-Sarcevic and C. Djerassi, 1979, *J. Org. Chem.*, 44, 3374.
224. M. Anastasia, A. Fiecchi and A. Scala, 1979, *J. Chem. Soc. Chem. Comm.*, 858.
225. C. Djerassi, 1981, *Pure Appl. Chem.*, 53, 873.
226. M. Anastasia and A. Fiecchi, 1981, *J. Org. Chem.*, 46, 1726.
227. A.Y.L. Shu and C. Djerassi, 1981, *Tetrahedron Lett.*, 4627.
228. P.E. Georghiou, 1977, *Chem. Soc. Rev.*, 6, 83.
229. H. Jones and G.H. Rasmusson, 1980, *Fortschr. Chem. Org. Naturst.*, 39, 63.
230. R.I. Yakhimovich, 1980, *Russ. Chem. Rev.*, 49, 371.
231. B. Lythgoe, 1980, *Chem. Soc. Rev.*, 9, 449.
232. B. Lythgoe, D.A. Roberts and I. Waterhouse, 1977, *J. Chem. Soc. Perkin Trans. I*, 2608.
233. P.J. Kocienski, B. Lythgoe and D.A. Roberts, 1978, *J. Chem. Soc. Perkin Trans. I*, 834.
234. M. Biollaz, W. Haefliger, E. Velarde, P. Crabbé and J.H. Fried, 1971, *Chem. Comm.*, 1322.
235. R.M. Moriarty, L.S. John and P.C. Du, 1981, *J. Chem. Soc. Chem. Comm.*, 641.
236. D.H.R. Barton, W.B. Motherwell and S.Z. Zard, 1981, *J. Chem. Soc. Chem. Comm.*, 774.
237. L. Nédélec, V. Torelli and M. Hardy, 1981, *J. Chem. Soc. Chem. Comm.*, 775.

6 Amino acids, peptides and proteins

B.W. BYCROFT and A.A. HIGTON

For the purposes of this review, those amino acids which are directly coded for in ribosomal protein biosynthesis will be referred to as typical amino acids and the same terminology will be used when referring to ribosomally synthesized peptides. Atypical amino acids are found in a wide range of organisms, both as free entities or within peptides containing a variety of structural features not usually found in protein. Generally these peptides are constructed by non-ribosomal processes. However, atypical amino acid features can result from the post-translational modification of a typical amino acid residue within a peptide or protein molecule and be observed as an anomalous amino acid on total hydrolysis.

This very general classification of amino acids and peptides based on biochemical origin is convenient and this chapter will follow this broad format. But it is not intended, nor would it be possible, to cover the biochemical aspects of amino acid, peptide or protein biosynthesis and the reader is referred to excellent reviews in these areas.[1-5] The chemistry related to this broad group of natural products has expanded enormously over the past decade and it will only be possible to highlight significant developments in the various areas. Since most novel structural features occur in atypical amino acids and peptides, the emphasis here has been placed on primary structure and synthesis; whereas for typical peptides and proteins the focus has been centred on methodology relating to gross structure, modification and synthesis.

There is increasing biochemical evidence, and indeed simple structural analysis would support the view, that the β-lactam antibiotics can be considered as atypical peptides or derivatives of atypical amino acids. Consequently a review of this type would be incomplete without brief reference to this important group of compounds. The clinical, biochemical, and pharmacological importance of the β-lactams is self-evident and the selection of material in the other areas, has, to some extent, reflected their biological significance. More detailed treatment of all the subjects covered can be found in the excellent annual review series produced by the Royal Society of Chemistry.[6]

6.1 Amino acids

Until recently atypical amino acids present in peptides and proteins could be divided into three main groups; those formed by post-translational C-hydroxylation, N-methylation or aromatic substitution,[7] those formed by cross-chain interaction of two or more amino acid[8] residues, or the interaction of an amino acid side-chain with a non-amino acid structural unit in complex biomacromolecules. An example of this latter group is the fluorescent Y-base (1)* derived from the yeast, wheat germ and rat liver phenylalanine t-RNA.[9] The structure of (1) was determined on only 300 µg of sample using high resolution n.m.r. and mass spectrometry and confirmed by synthesis.[10]

An important recent development has been the identification of γ-carboxyglutamic acid (2)[11] and γ-carboxyaspartic acid (3)[12] as components of various proteins. Acid hydrolysis results in the decarboxylation of (2) and (3) to give glutamic and aspartic acids, respectively, and they can only be isolated after basic hydrolysis. Acid (2) was first identified as a component of the modified blood coagulating protein prothrombin, which undergoes vitamin K-mediated post-translational γ-carboxylation of a large number, if not all of the glutamic acid residues.[13] The resulting γ-carboxy groups appear to be essential for calcium binding, and (3) has been identified in bone and other tissues.[14] It is also apparently universally distributed in ribosomal proteins.[15] Amino acid analysis of alkaline hydrolysates of E. coli ribosomal protein indicated the presence of (3). The distribution and biological significance of (3) are as yet unknown but a systematic study of proteins known to bind divalent metals may lead to its detection in other biological systems. γ-Carboxyglutamic acid may be synthesized simply from pyroglutamic acid.[16]

Atypical peptides, on hydrolysis, yield amino acids of much greater structural diversity, and microbial peptide antibiotics are a particularly rich source. Actinoidinic acid (4), a bis-phenylglycine derivative, is present in hydrolysates of the complex antibiotics actinoidin, ristomycin and vancomycin.[17] The structures of these important antibiotics will be described later, but the anomalous amino acid can be considered to be derived by cross-chain phenolic coupling. Capreomycidine (5) and its (S) epimer are derived from the

---

Footnote Fischer nomenclature is used throughout for the chirality at the α-centre of amino acids. The L-configuration can be assumed unless otherwise indicated.

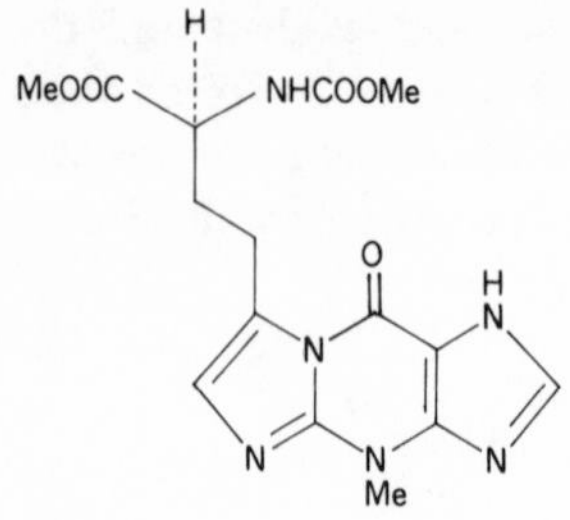

(1)

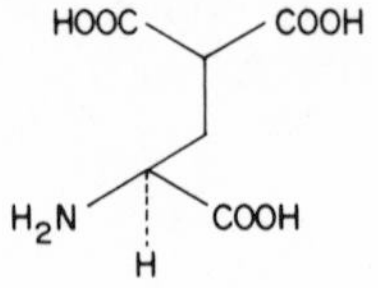

(2)

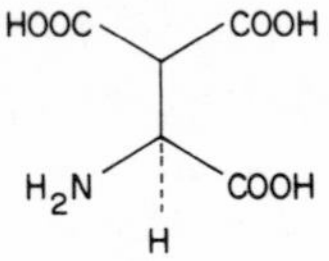

(3)

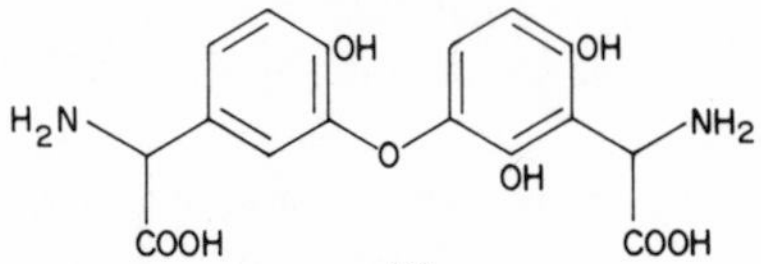

(4)

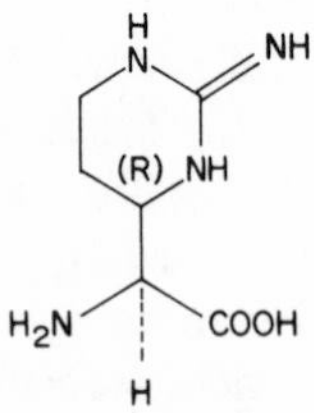

(5)

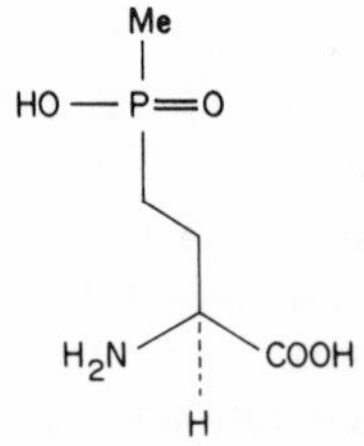

(6)

cyclic tuberculostatic antibiotic capreomycin,[18] and the linear atypical peptide elastatinal,[19] respectively. The immediate precursor is likely to be arginine, but whether or not the cyclization occurs at the amino acid or peptide stage is not established. The unique phosphorus-containing amino acid phosphinothricin (6)[20] identified as a component of the tripeptide herbicide bialaphos (antibiotic SF 1293), itself has antibiotic properties and is readily available as a synthetic compound.[21] Phosphinothricin is a novel natural product containing a C–P–C bond sequence. Preliminary studies on the biosynthesis[22] of bialaphos indicated that the C–P–C bond is formed after reduction of phosphoenol pyruvate to give a phosphite ester which then rearranges to give a phosphinic acid intermediate. The P-demethyl derivative of (6) has been isolated from culture broths of mutant strains of the producing organism. Thus, in this case, the modification of the amino acid occurs before it is incorporated into a peptide. There are many examples, some of which will be described later, where the structural features of the modified amino acid residue within an atypical peptide are so unstable to hydrolytic conditions that they cannot be isolated intact. The sulfoximine amino acid (7) is labile to both

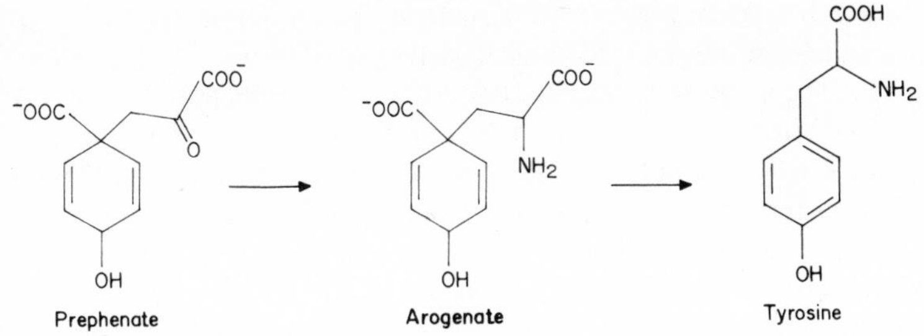

Scheme 1 Alternative pathway to tyrosine

acid and base but can be obtained from the parent tripeptide by enzymatic hydrolysis.[23]

The recent developments in chromatographic techniques coupled with more detailed analysis of plant, animal and microbial materials have led to the characterization of a large number of free atypical amino acids. For many of these, comparatively little is known of their biological significance, biosynthesis, and fate within the organisms in which they occur. However this is not true of arogenic acid (8) or pretyrosine[24] which is now established as an important intermediate in an alternative biosynthetic pathway to tyrosine (Scheme 1). In a number of micro-organisms it appears to be the only pathway which operates,[25] whereas plants seem to possess both routes.[26]

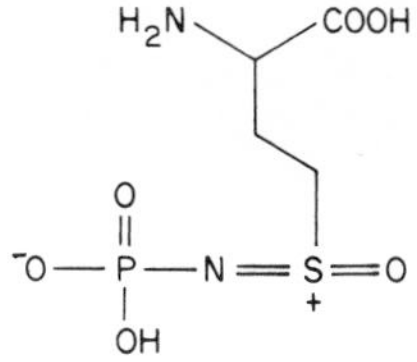

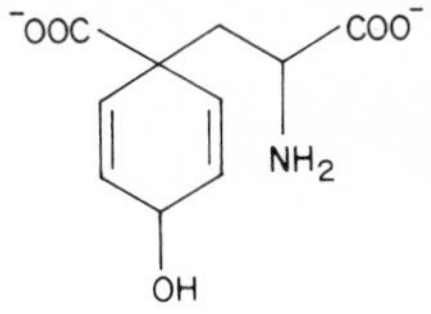

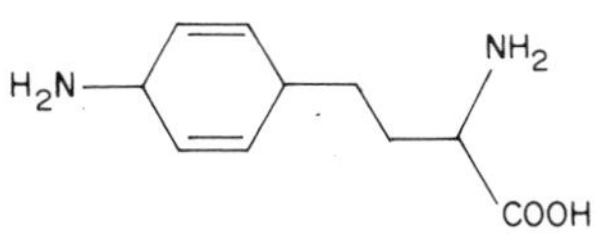

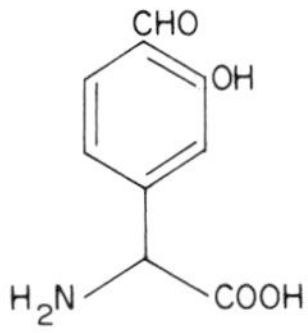

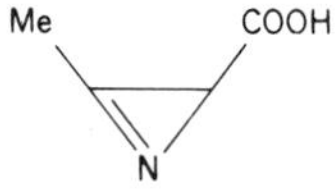

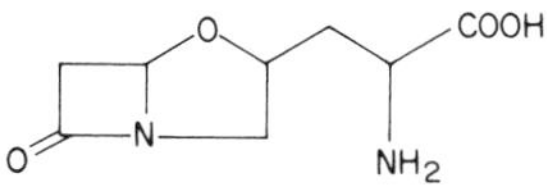

Excellent reviews are available on microbial[27] and plant[28] amino acids and here only a selection of the more recent structures, which outline the variety of novel features and noteworthy biological activity which continue to emerge, are presented. The actinomycetes have proved a rich source, revealed mainly as the result of antibiotic screening. Amiclenomycin (9) from *Streptomyces lavendulae* is a further example of an unstable dihydroaromatic system[29] with some inhibitory activity against mycobacteria. Naturally-occurring phenyl-glycine derivatives are relatively rare but the formyl derivative, forphenicine (10) has recently been isolated from *Actinomyces fulvoviridis* and shown to inhibit alkaline phosphatase and increase the number of antibody-forming cells.[30] Unusual small ring systems are also evident. Azirinomycin (11),[31] although highly unstable, demonstrates broad spectrum antibacterial activity. The bicyclic system of 3-alanylclavam (12) is considerably more stable and exhibits both antibacterial and antifungal properties.[32] This molecule is related to the β-lactam antibiotics of the clavulanic acid family (see later).

Relatively simple amino acids turn up as a result of antibiotic screens. One such example is 2-amino-3-butynoic acid (13)[33] which is active against some Gram positive organisms. This compound, the related vinylglycine (14) and

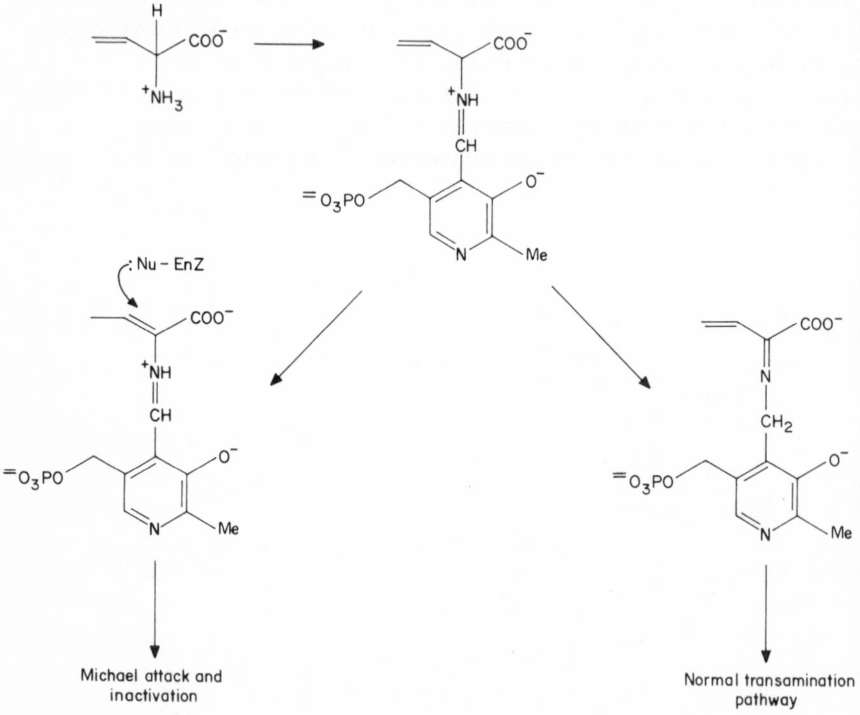

Michael attack and inactivation

Normal transamination pathway

Scheme 2 Schematic representation of suicide-substrate inactivation of pyridoxal-dependent transaminase

the corresponding vinyl ether derivative rhizobitoxin (15)[34] have attracted considerable attention as irreversible inhibitors of a number of pyridoxal-dependent transaminase enzyme systems. This specific type of enzyme inactivation which involves the enzyme catalysing the formation of its own active-site inactivator (Scheme 2) has been referred to as mechanism-based inactivation, K_{cat} inhibition or suicide-substrate inactivation.[35] Rhizobitoxin inactivates bacterial and plant β-cystathionases and inhibits the conversion of methionine into ethylene. The related compound (16)[36] possesses similar properties. Both have been synthesized.[37,38]

Interesting molecules containing unusual heterocyclic systems attached to an α-amino acid framework still continue to be isolated. The structures of the insecticidal compound ibotenic acid (17) isolated from the mushroom *Amanita muscaria*,[39] in addition to its dihydro derivative, tricholomic acid[40] (18) are well established. But more recently the related isoxazole acivicin (19), derived from a *Streptomyces*,[41] has been shown to possess notable antitumour properties. (17) and (18) have found significant application in neuro-

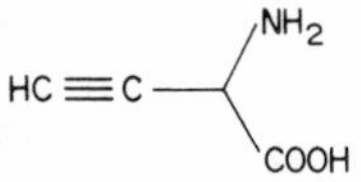

(13) 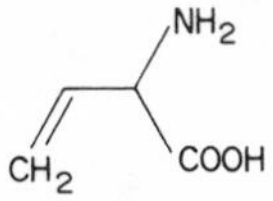   (14)

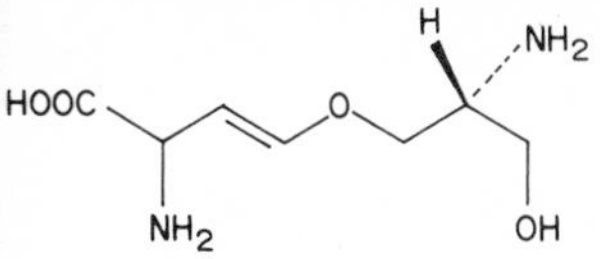

(15) 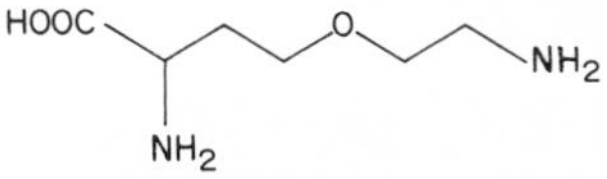   (16)

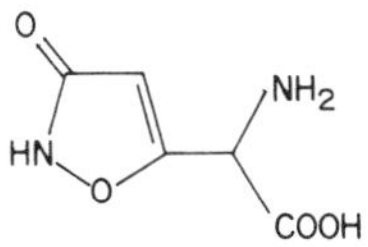

(17) 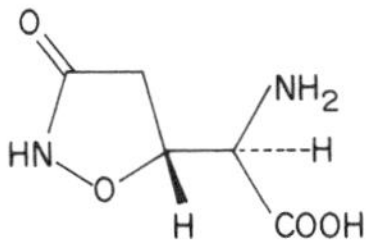   (18)

physiology as agonists of glutamate, which is now considered to be a neurotransmitter in both vertebrates and invertebrates. The acidity of the NH attached to the heterocyclic systems in (17) and (18) as well as in the unique oxadiazolidindione system found in quisqualic acid (20) is approximately that of the γ-carboxyl group of glutamate. Quisqualic acid is found in the seeds of the plant *Quisqualis indica*[42] and is one of the most potent analogues of glutamate yet observed.

The field of neuroexcitatory amino acids is one of increasing interest.[43] Kainic (21)[44] and domoic (22)[45] acids, which are isolated from marine algae, have been studied extensively in recent years. They are potent and long-acting excitants of neurons in amphibian and mammalian tissues. It has been suggested that the pyrrolidine ring acts as a lock in these molecules orientating the two carboxylic acid functions in the supposed conformationally active

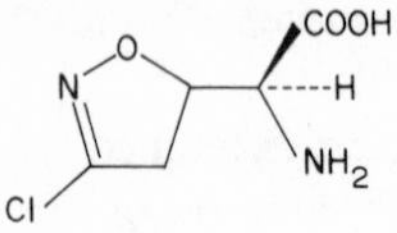

(19) 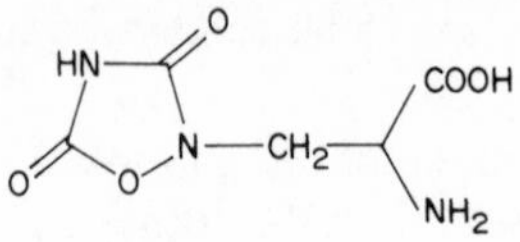   (20)

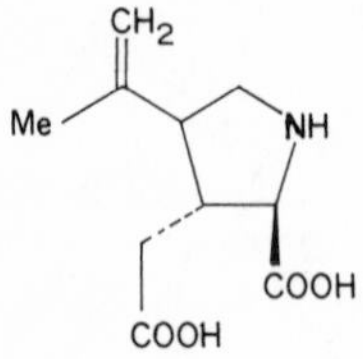

(21) 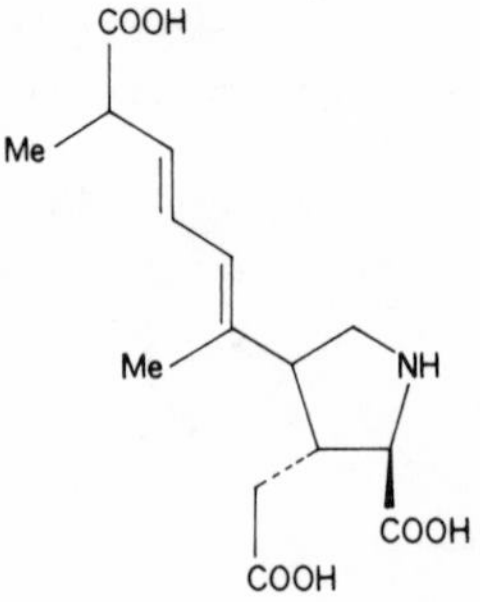   (22)

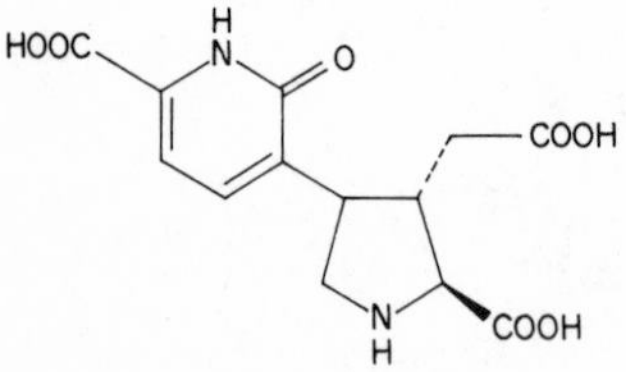

(23) 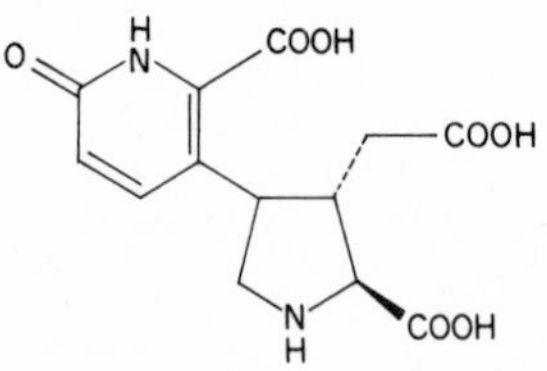   (24)

form of glutamate.[46] The kainate-like pyridone derivatives (23) and (24), termed acromelic acids A and B, are fungal products, possibly with similar neuro-excitatory properties.[47] The dicarboxylic acid derivative of proline (25)[48] from the seeds of *Pentaclethra macrophylla* does not appear to possess neuro-excitatory activity.

It has been known for some time that the lower homologue of proline, azetidine-2-carboxylic acid, is widely distributed in various species of higher plants. Mugineic acid (26)[49] and nicotianamine (27)[50] are complex amino acid derivatives which are iron-chelating agents and have been implicated in the role of iron uptake and transport in higher plants.

Free amino acids in animals tend to be relatively simple derivatives of the typical amino acids formed by methylation[51] or the interaction with primary metabolic products. The arginine-related amino acid octopine (28)[52] is a component of scallop and octopus muscle; whereas hypusine (29)[53] which is

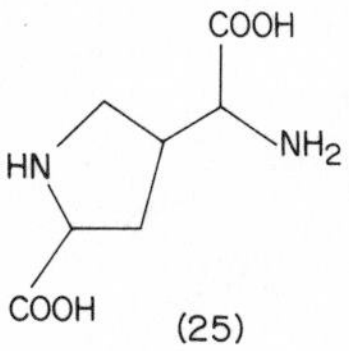

(25)

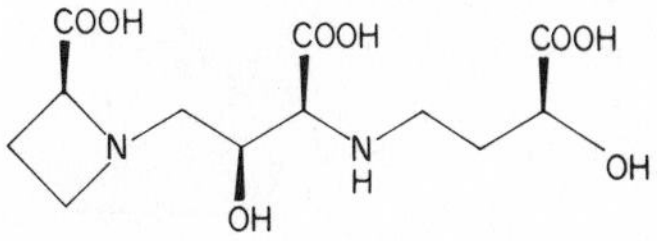

(26)

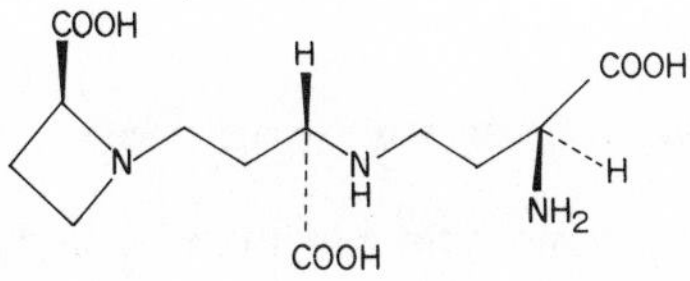

(27)

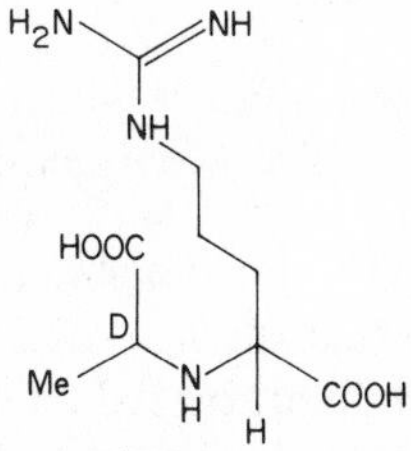

(28)

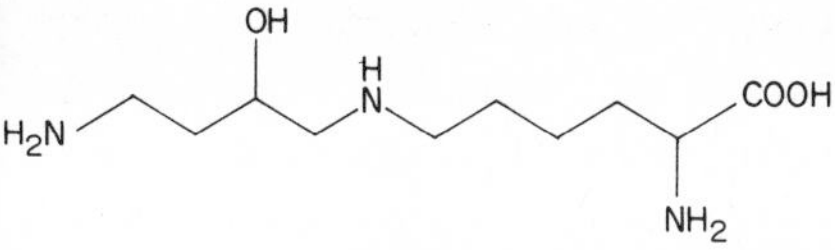

(29)

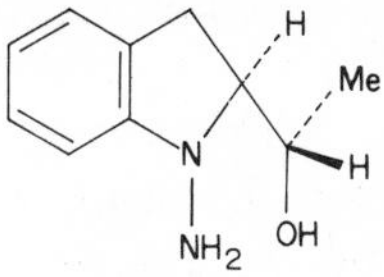

(30)

widely distributed in the liver, kidney, muscle, blood and brain of mammals, is lysine-derived.

There are many standard methods available for the synthesis of protein and simple atypical amino acids which have been well tried and tested over the years. Since these are well known and have been adequately covered elsewhere,[54] attention is focused on the advances which have taken place in asymmetric synthesis. In general the standard methods give rise to racemic products, the subsequent resolution of which by enzymatic[55] and chemical methods[56] is often long and tedious. Until relatively recent times, attempts to develop a synthetic routes which were satisfactory both in yield and chiral efficiency had met with little success.[57]

The chiral induction method developed by Corey (Scheme 3) was the first to combine both.[58] It overcame the major problem of the poor optical efficiency

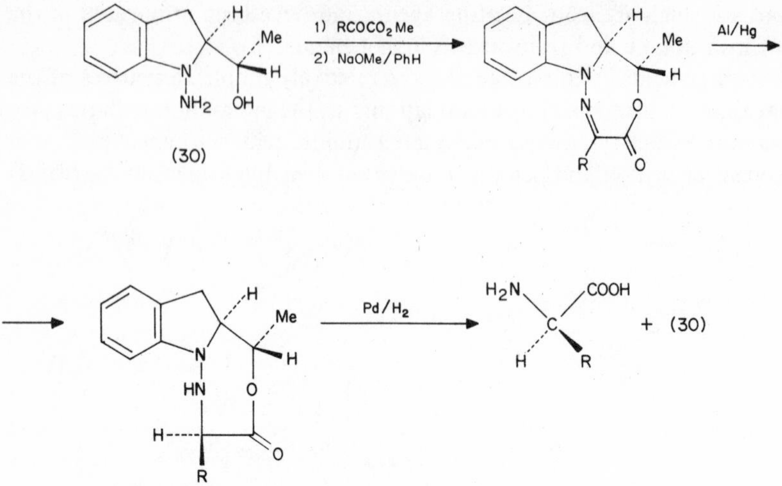

Scheme 3 Corey asymmetric synthesis of amino acids (illustrated for D-enantiomers)

by locking the chiral agent (**30**) and the amino acid precursor in a ring structure so that the heterogeneous reduction of the imine occurred stereo-selectively. Subsequent reduction and hydrolysis of this intermediate afforded the amino acid in high chemical yield and with an enantiomeric excess greater than 90%. In addition the chiral reagent could be regenerated at the end of the reaction sequence, which represented a further significant feature of the route. Since this important development, similar methods which employ amino acids[59,60] as the chiral synthon rather than the less readily available (**30**) have been developed. These allow the efficient synthesis of optically active amino acids from the corresponding α-keto acid and ammonia.

The major drawback of this type of chiral induction approach is that one molecule of chiral reagent is required for the synthesis of each molecule of amino acid. The chiral reagent can be employed much more efficiently if it is incorporated in a catalyst so that a relatively small number of chiral molecules can catalyse the formation of a large number of product molecules. This principle has been elegantly exploited in the reduction of acyldehydroamino acid derivatives (Scheme 4).[61] Homogeneous catalytic hydrogenation is

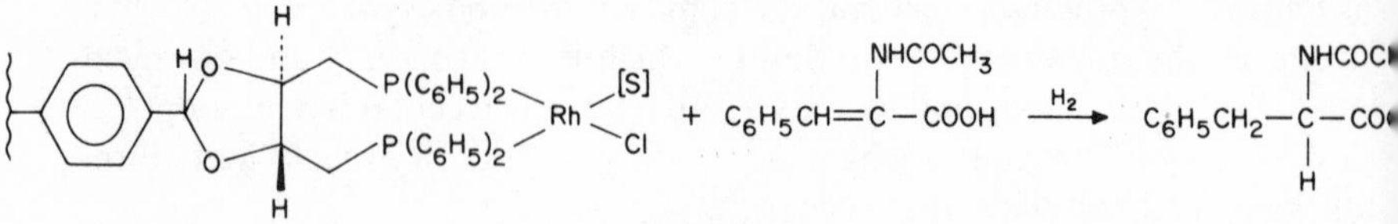

Scheme 4 Chiral hydrogenation of N-acetyldehydroamino-acids, using polystyrene-supported 2,3-O-isopropylidene-2,3-dihydroxy-1,4-bis(diphenylphosphino)butane (DIOP).

effected with chiral rhodium catalysts of the type (31). The interaction of the catalyst with the substrate is almost totally stereoselective and provides either the D- or the L- enantiomer with greater than 90% optical efficiency depending on the chirality of the catalyst ligand.[62] The method has been used commercially for the synthesis of the therapeutically significant agent L-dihydroxyphenylalanine (L-dopa).[63] An extensive review which covers asymmetric synthesis in general provides more detailed information concerning amino acids with respect to both protein and atypical amino acids.[64]

Depending on their complexity, atypical amino acids pose the usual problems of organic synthesis, but within the amino acid framework and consequently all the constraints that that implies. There are numerous examples of elegant synthesis from which one can draw for illustration. These are cited in the various review articles[6,54,65] and inevitably any selection is bound to be subjective. However the syntheses of acivicin[66] (19) and kainic acid[67] (21) do reflect some salient aspects of chemistry and stereochemical control. These are outlined in Schemes 5 and 6 respectively.

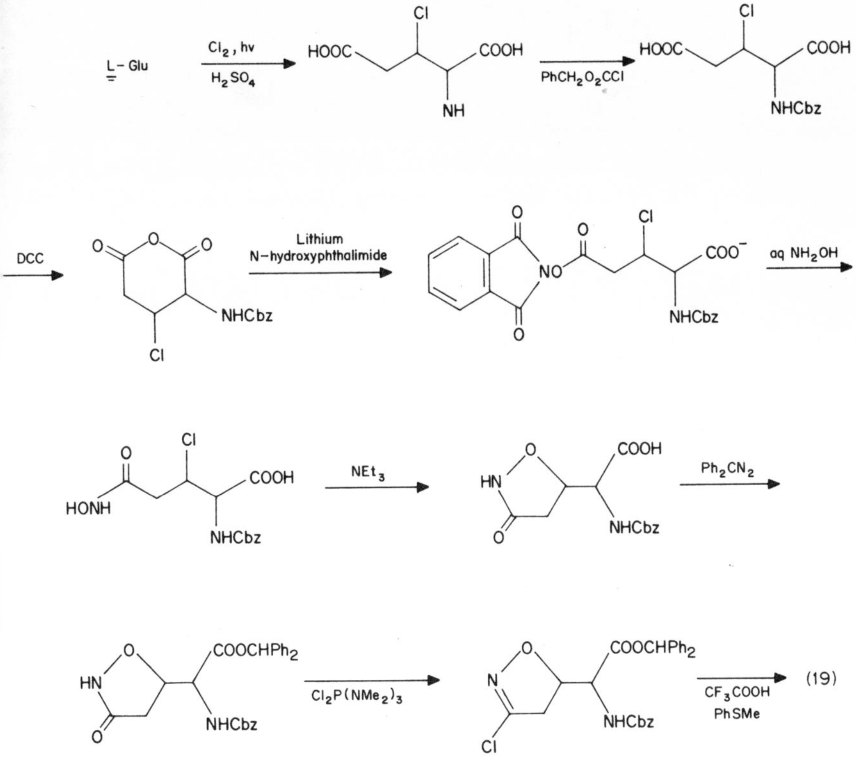

Scheme 5   Synthesis of acivicin

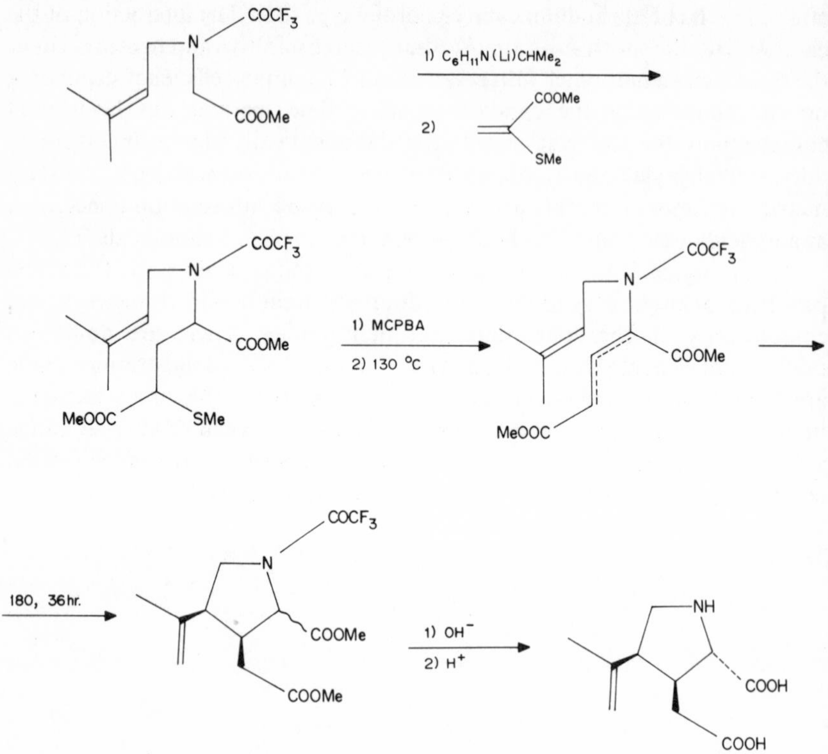

Scheme 6 Synthesis of kainic acid

Finally, no review of amino acids would be complete without reference to the sophistication of techniques and methodology for detection and analysis of amino acids which have advanced substantially over the past decade.[68−70]

6.2 Atypical peptides

Atypical peptides are structurally very diverse and encompass all compounds containing two or more amino acid residues linked by an amide bond but which possess some structural feature not characteristic of protein. This may be simply the presence of an atypical amino acid residue which could include protein amino acids with the D-configuration or at a higher oxidation level. The anomalous feature can also be an unusual amide linkage, e.g. a cyclic or γ-glutamyl peptide, or alternatively an ester linkage within a linear or cyclic structure (depsipeptide or cyclic depsipeptide, respectively). Frequently atypical peptides contain a combination of the above structural features and, in more complex structures, these may be masked by further extensive interaction between adjacent amino acid residues or across ring structures.

The latter modifications lead to an array of novel peptide-related structures which have attracted considerable attention from natural product chemists. Some have been detected due to their toxicity to plants or animals,[71] while others have found application as antibacterial, antitumour and antiviral agents, or as immunostimulants and immunosuppressants.[72] There is no doubt that these molecules have provided a testing challenge to the battery of sophisticated physical, spectroscopic and chemical techniques now available to the chemist for structural elucidation, both with respect to their size and complexity. In many cases it has been necessary to resort to X-ray

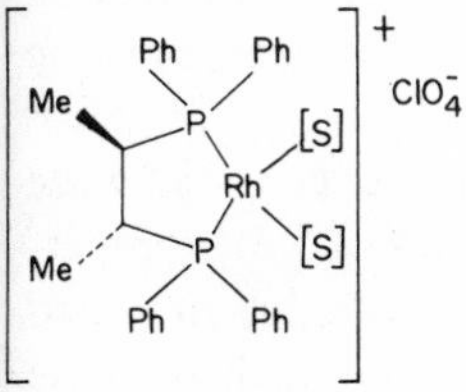

(31)

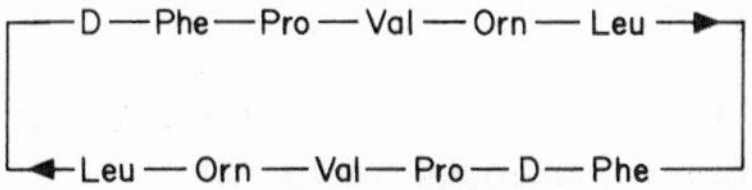

(32)

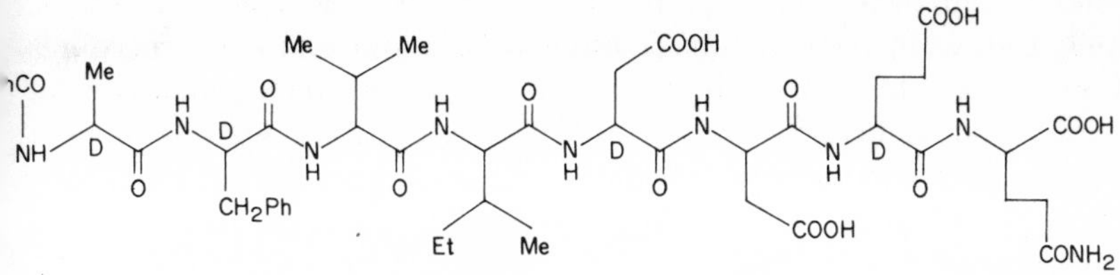

(33)

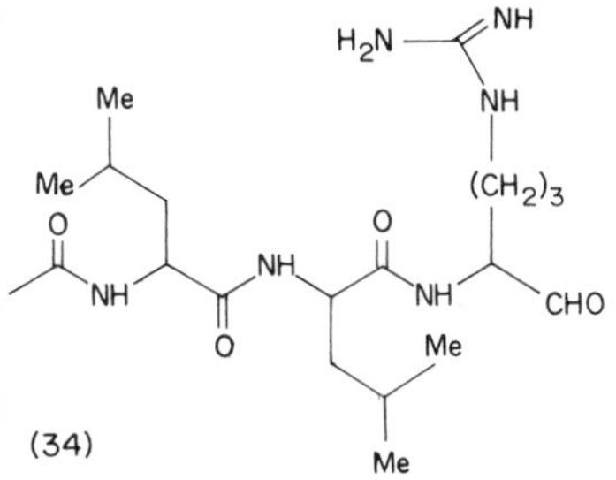

(34)

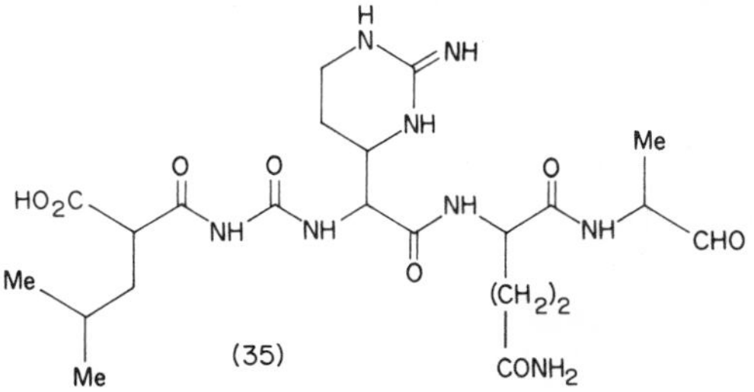

(35)

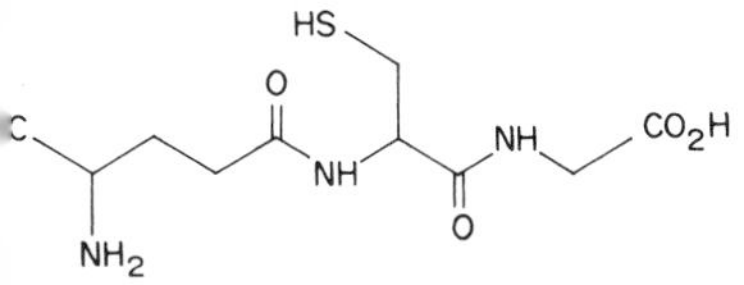

(36)

crystallographic analysis to confirm or adjust certain structures.[73] Mention must also be made here of the recent development of fast atom bombardment mass spectrometry, FABS, which is revolutionizing structural determination and sequencing of both atypical and normal peptide molecules.[74-76] Molecular and informative sequence ions can be observed from an underivatized sample and it would appear that molecular size is limited only by the resolving power of the instrument.

So far there seems to be an upper limit of c. 2000 daltons for the molecular size of atypical peptides[77] and this may reflect the fact that they are biosynthetically derived by nonribomosal pathways. For relatively simple atypical peptides it is probable that the multienzyme thiotemplate mechanism of the type now well authenticated for gramicidin S (32) and related compounds operates,[78-80] but the generality of this process remains to be established.

For convenience atypical peptides have been divided and covered under the following main headings; linear peptides, cyclic dipeptides (2,5-dioxopiperazines), cyclic peptides (homodetic and heterodetic), large, highly modified peptides, peptide alkaloids and β-lactam antibiotics. Within each of these groups, selective examples of the various types of anomalous features are illustrated and, where appropriate, reference has been made to specific techniques used in the structure determination, novel synthetic procedures and significant chemical or biological properties. Synthetic analogues of hormones and related peptides containing atypical features are becoming of increasing interest in medicinal chemistry. Although they are not specifically covered, their significance for drug design[72,81] and pharmacology[82] is noteworthy.

6.2.1   Linear peptides

The sawfly toxin (33) is composed only of protein related amino acids, four of which possess the D-configuration. Free D-amino acids are known in insects, but this peptide would appear to be unique in animal systems. Both the structure and sequence were established by FAB mass spectrometry and the amino acid configurations using a chiral g.l.c. system.[83] The leupeptins (34) are a group of simple microbially-derived acyl tripeptides in which the only structural modification is represented by the reduction of the carboxyl terminal group to an aldehyde.[84] These compounds together with elastatinal (35) are inhibitors of various proteases and the aldehyde function is essential for inhibitory activity. Elastatinal also contains an oxidized and cyclized arginine residue and a non-protein amino acid unit.[85]

Unusual peptide linkages, as typified by γ-glutamyl peptides, are frequently encountered in all living systems. Glutathione (36) is the most widely distributed small peptide and there is evidence to suggest it is present in all living cells. It was isolated and characterized sixty years ago, and a considerable number of biological functions have been ascribed to it.[86,87] One of its major functions is the mediation of amino acid transport across mammalian cell membranes by a process which is referred to as the γ-glutamyl cycle.[88] Related γ-glutamyl cycles operate in micro-organisms and plants and, in addition to glutathione, plants and fungi produce other γ-glutamyl peptides. Large numbers of these peptides have been reported,[89,90] but very little is known about their function. In edible fungi it has been suggested they make a significant contribution to the taste, e.g. lentinic acid (37).[91]

The tripeptide (38), isolated more than twenty years ago, has come to prominence recently, as the common precursor of all the penicillin and cephalosporin antibiotics;[92] thus establishing that they represent highly modified atypical peptides. Because of their importance, the β-lactam antibiotics are described later under a separate heading. However there are other naturally-occurring peptide β-lactams without antibacterial activity, as illustrated by (39) and (40). The former is the so-called wildfire toxin[93] and the latter an antimetabolite which inhibits glutamine synthetase.[94]

It is beyond the scope of this review to describe in any detail the atypical

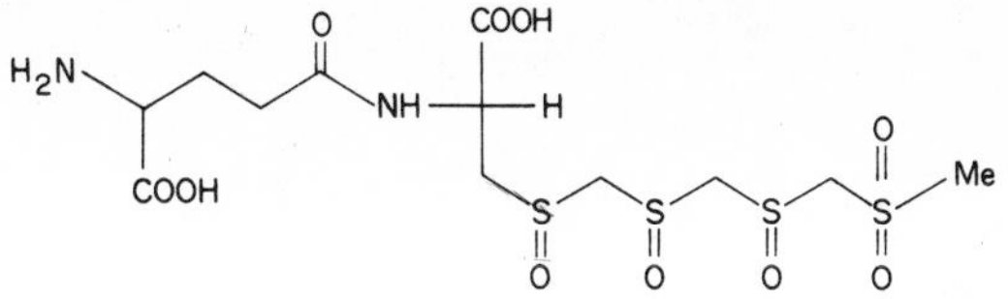

(37)

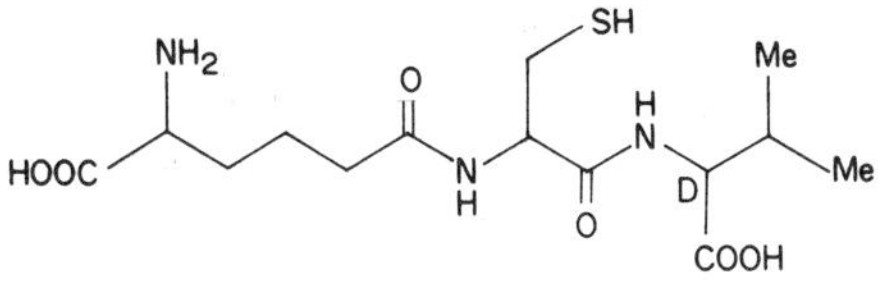

(38)

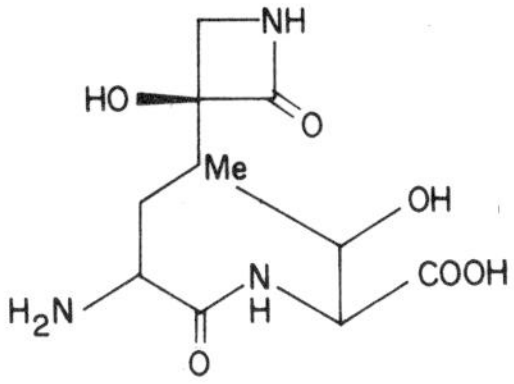

(39)

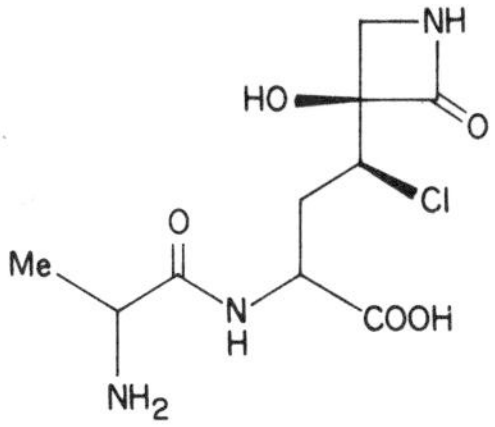

(40)

peptide which is an important constituent of the complex macromolecules associated with bacterial cell walls. It will suffice to indicate that the macromolecular structure consists of a network of peptidoglycan in which linear polysaccharide chains are cross-linked by short peptides.[95] The amino acids within these peptides vary for different bacteria and only the basic structural unit (41) of the peptidoglycan of *Staphylococcus aureus* is shown.[96] Certain fragments of bacterial cell wall are known to act as immuno-stimulants.[97] Such activity has been reported for the novel tetrapeptide (42) which has been isolated from a *Streptomyces* species.[98] The structural similarities to the cross-linking peptides are self-evident.

The antibiotic malonomycin (43) which contains a tetramic acid unit is one of the only two natural products to incorporate an aminomalonic acid residue.[99] The other is the fish attractant arcamine (44).[100] Although marine natural product chemistry has turned up a wide variety of novel halogenated

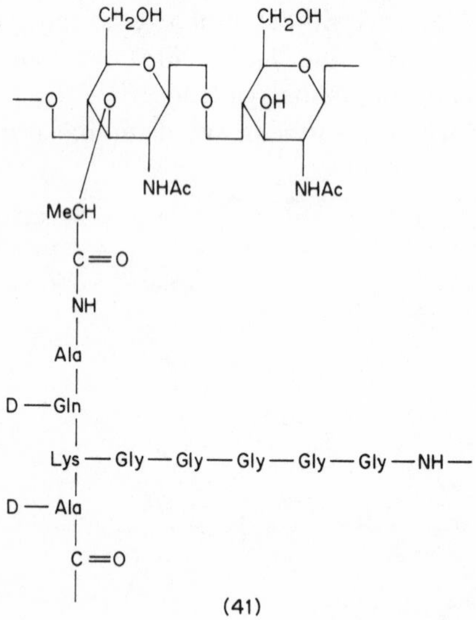

(41)

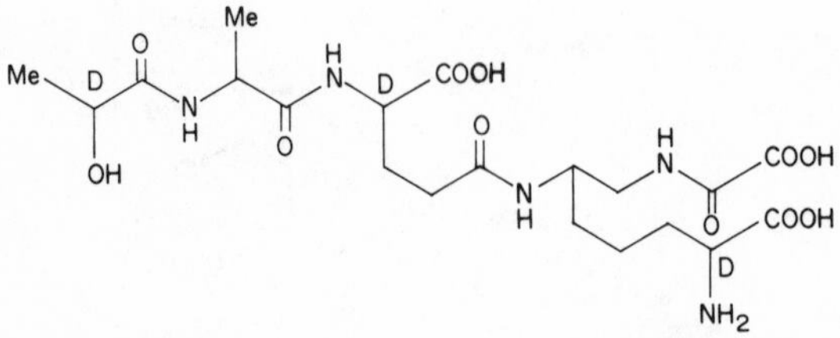

(42)

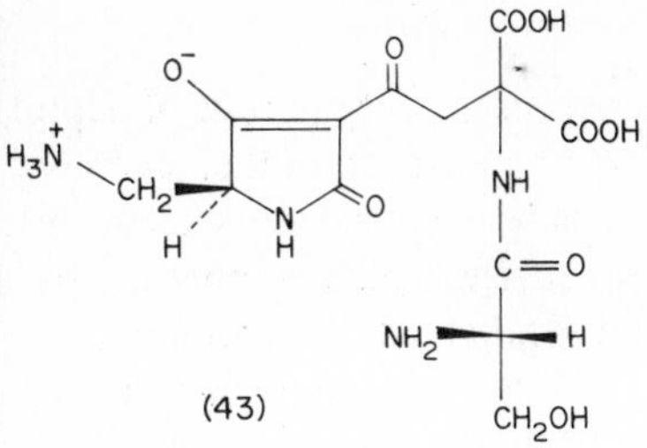

(43)

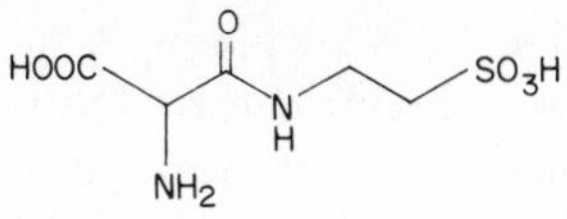

(44)

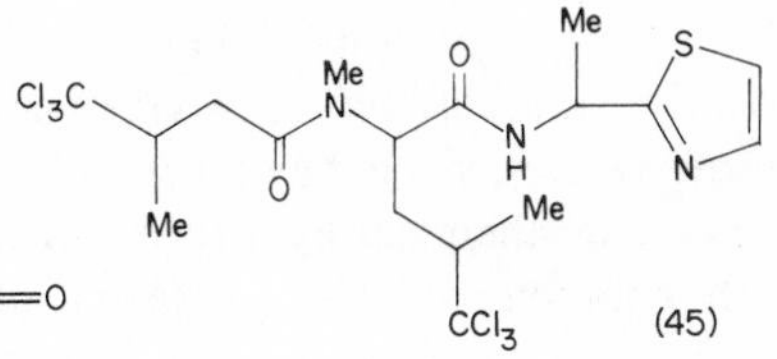

(45)

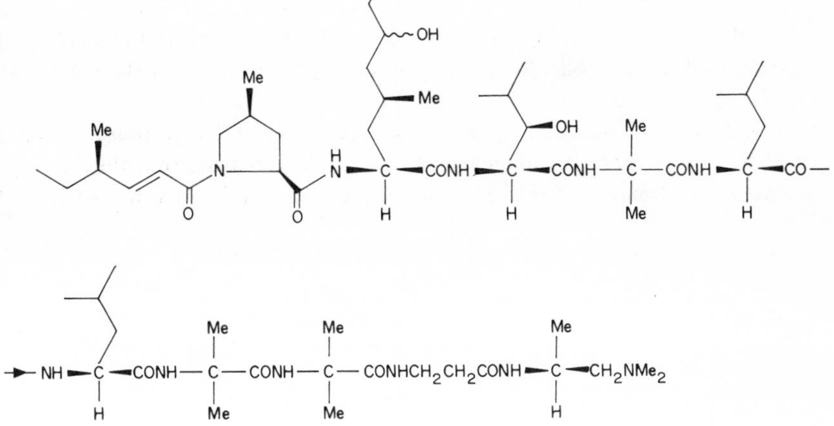

(46)

compounds, the number of peptides among them is relatively small. The isolation and characterization of (45) from the sponge *Dysidea herbacea* is an encouraging development,[101] and also noteworthy is the presence of a thiazole unit which occurs frequently in large modified peptides.

The problems of structure determination become more demanding as the molecular size increases, and until relatively recently the upper limit for reliable structures was c. 1000 daltons. It was for this reason that those peptides, whether linear or cyclic, above this limit were allocated to a special category. Leucinostatin A[102] (46) approaches this rather arbitrary limit, but demonstrates admirably the complexity of the structural units which make up the large atypical peptides. α-Aminoisobutyric acid (Aib) is common to many of the large linear peptides and its presence is claimed to impart particular conformational properties (see later).

6.2.2   Cyclic dipeptides (2,5-dioxopiperazines)

Cyclic dipeptides are amongst the most numerous and ubiquitous of peptide derivatives found in nature. They also differ in some aspects of their chemistry from other cyclic peptides, and therefore it has been considered expedient to treat them as a separate group. An excellent review which comprehensively deals with all aspects relating to natural and synthetic 2,5-dioxopiperazines, and covers the literature up to the end of 1973, offers an important primary source of information on these compounds.[103] Briefer general reviews[104,105] and one on biosynthesis[106] have appeared since then.

Considerable care has to be taken in the isolation of these compounds and their derivatives, since they are readily formed from peptides and proteins on enzymatic or chemical hydrolysis, as well as thermolysis. Indeed it appears that the characteristic bitter taste of a number of aged alcoholic beverages and roasted cocoa beans are due, in part, to *cyclo*-dipeptides formed by one or other of these processes. However, their careful isolation, particularly from extracts and culture filtrates of micro-organisms, testifies to their genuine metabolic character.

A selection of microbial cyclic dipeptides derived from primary L-amino acids for which there is sufficient evidence to demonstrate that they are not artefacts are shown in Table 6.1. They have been isolated from a variety of

Table 6.1   Some naturally occurring 2,5-dioxopiperazines derived from primary L-amino acids

cyclo-L-Pro-L-Leu	*Aspergillus fumigatus*
cyclo-L-Pro-L-Val	*Aspergillus ochraceus*
cyclo-L-Pro-L-Phe	*Rosellinia necatrix*
cyclo-L-Pro-Gly	Yeast extract
cyclo-L-Pro-L-Tyr	Yeast extract
cyclo-L-Ala-L-Leu	*Aspergillus niger*
cyclo-L-Phe-L-Phe	*Streptomyces noursei*
cyclo-L-His-L-Pro	Human blood

other sources, e.g. *cyclo*-Gly-L-Pro has recently been identified in the extracts of the starfish *Luidia clatharata*.

It is interesting to note the relatively large number of proline derivatives in this group of natural products. This is perhaps not too surprising when it is considered that, in order to form dioxopiperazines, the dipeptide precursors have to adopt a *cis* geometry about the peptide linkage, and it is well established that the activation energy for the *cis-trans* isomerism of *N*-acyl proline derivatives is considerably lower than for other peptide bonds. Unlike most other cyclic peptides, the two peptide bonds in dioxopiperazines possess the *cis* configuration and this does influence the topology and chemistry of these molecules.

Table 6.2

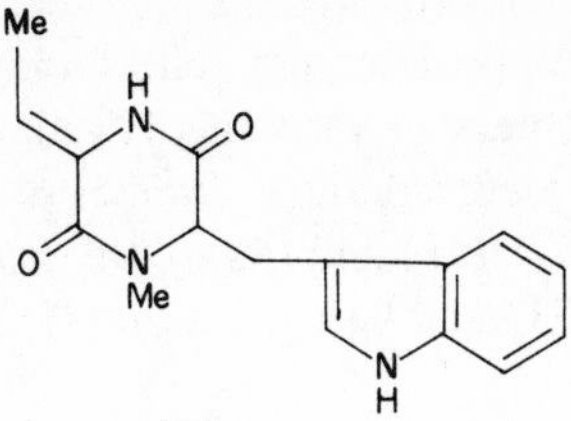

(47)[107]

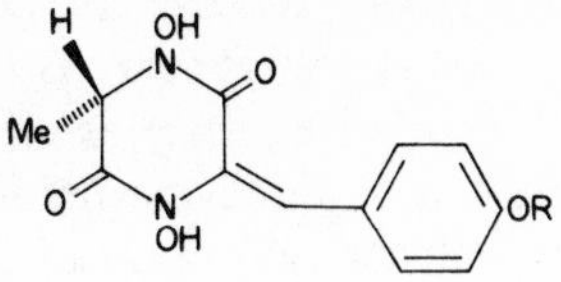

(48)[108]

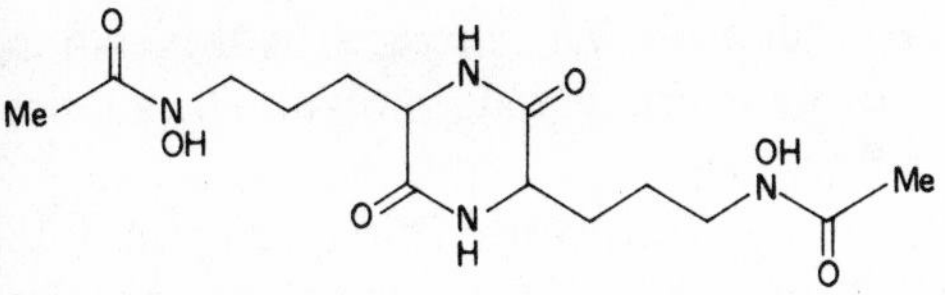

(49)[109]

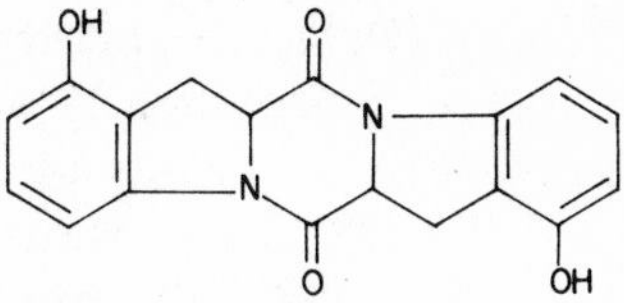

(50)[110]

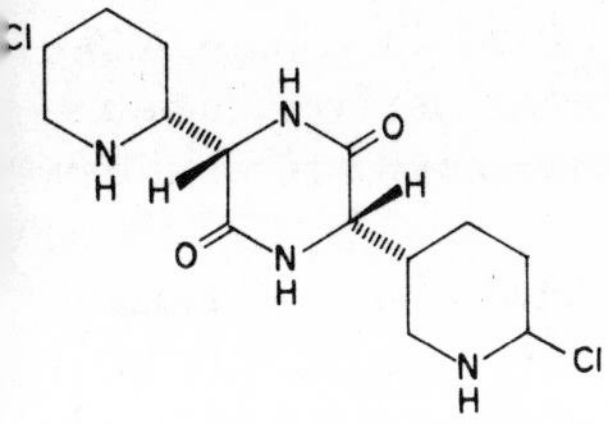

(51)[111]

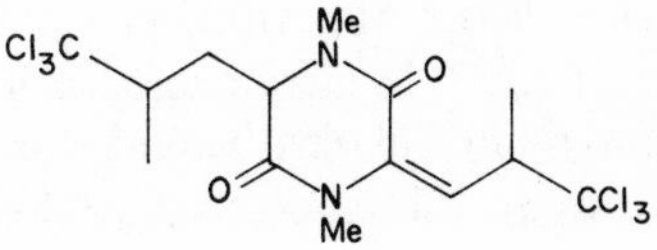

(52)[112]

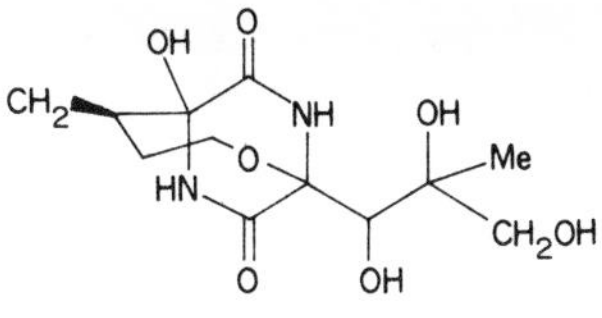

(53)[113]

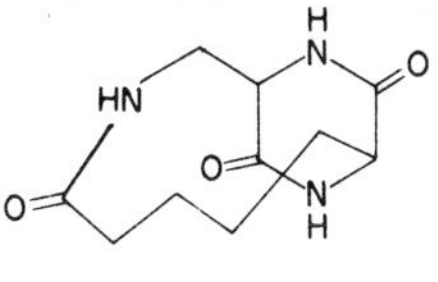

(54)[114]

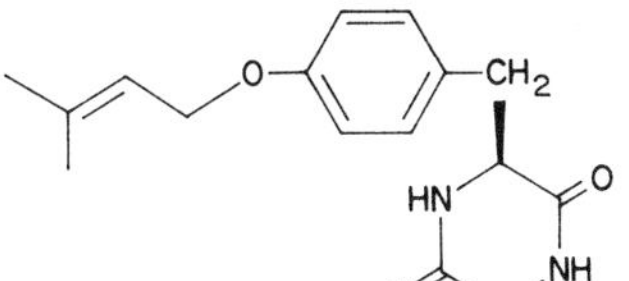

(55)[115]

The collection of compounds in Table 6.2 illustrates the range of structural modifications that have been observed on the basic dioxopiperazine skeleton. N-methylation is relatively common and, as with proline, may reflect a lower activation energy for the *cis-trans* isomerization about an N-methyl-peptide bond. Oxidative processes are responsible for a considerable number of the features i.e. α,β-dehydroamino acid residues (47), (48) and (52), N-hydroxylation (48) and (49), cyclization (50), (51) and (53), and halogenation (51) and (52).

Whether or not these modifications occur before or after the formation of the dioxopiperazine remains an open question. In the case of rhodotorulic acid (49), there is evidence to suggest that ornithine undergoes hydroxylation before dimerization;[116] whereas for bipolarimide (50), biosynthetic studies point to the intermediacy of *cyclo*-Phe-Phe.

It is perhaps not surprising that common structural feature arise in both synthetic and natural products when they are focused toward a particular medicinal application. The chlorine containing microbial metabolite (51) has significant antitumour properties, and it is interesting to note the β-chloramine function, characteristic of the nitrogen mustards, which has been incorporated in a number of clinically used synthetic anticancer drugs. Bicyclomycin (53) has also found limited clinical application for the treatment of certain Gram-negative infections. However, to date this represents a novel structural type not matched by synthesis.

In (55) the p-hydroxyl group of tyrosine is alkylated with a mevalonate-derived C-5 unit. Isoprenylation of cyclic dipeptides is more commonly associated with tryptophan and, over the past two decades, an array of mainly fungal metabolites of this type have been characterized (Table 6.3). Lysergic acid is an isoprenylated tryptophan derivative, albeit substantially modified, and is present in all the ergot peptides (56).[124–126] Ergot, the sclerotia of the fungus *Claviceps purpurea* which is parasitic to rye and related grain crops, is the toxin involved in the medieval condition referred to as St. Anthony's fire. It caused severe disturbance of mental and circulatory functions leading ultimately to death.

Clinically, ergot peptides find application as vasoconstrictors in the control

(56)*

*R^1, R^2: see p. 297.

Table 6.3

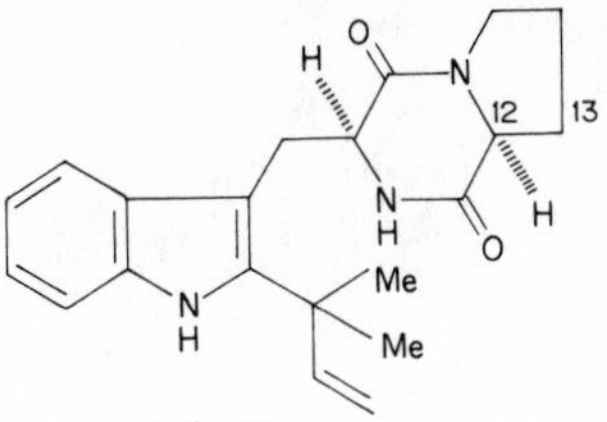

Ref. 117

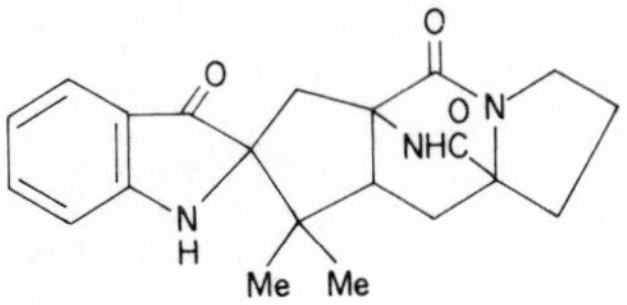

Ref. 120

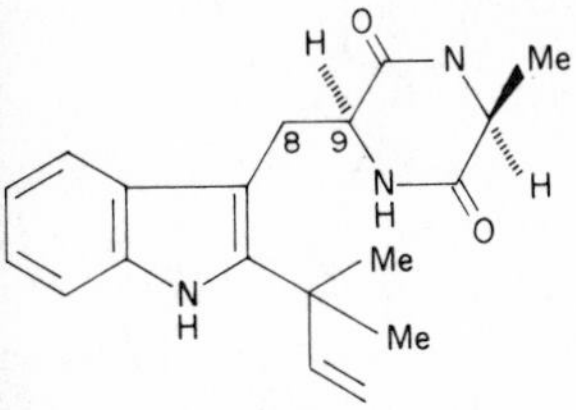

Ref. 118

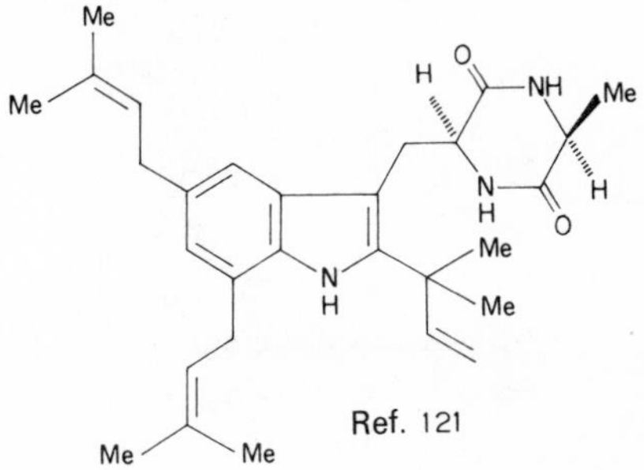

Ref. 121

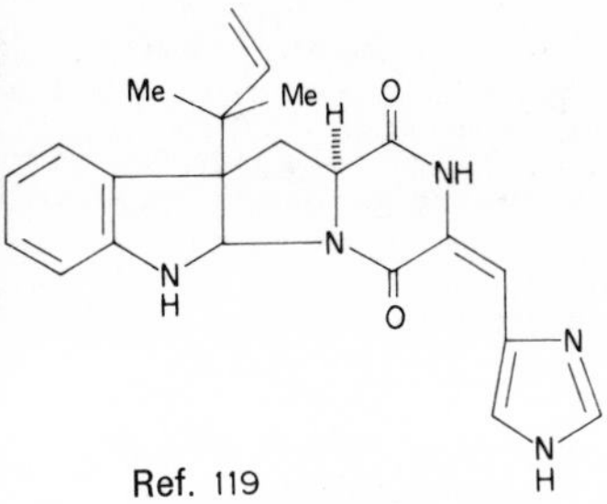

Ref. 119

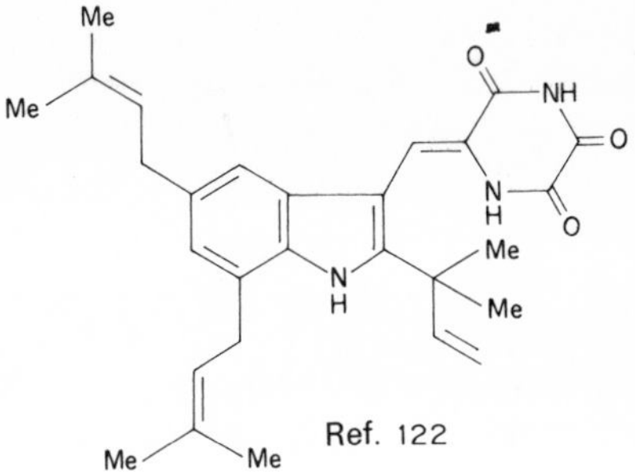

Ref. 122

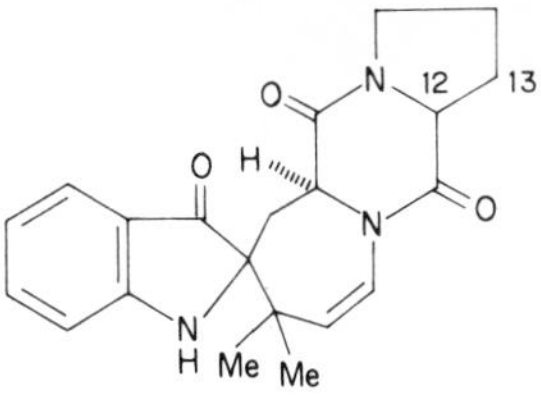

Ref. 117

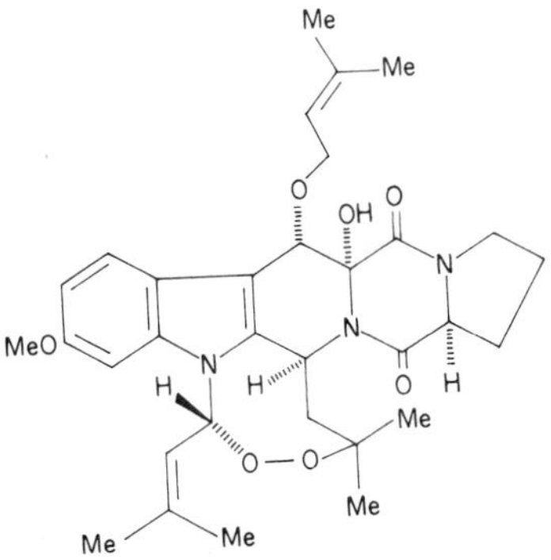

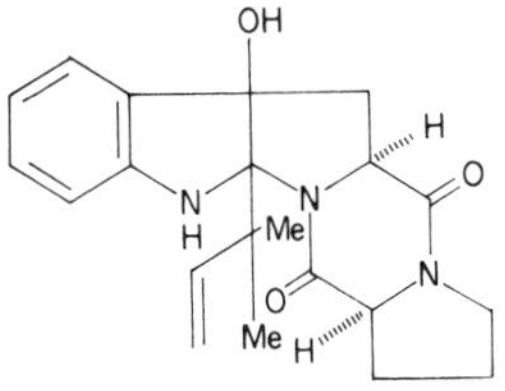

Ref. 120

Ref. 123

Table 6.4

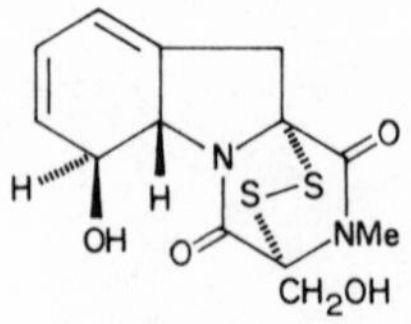

Ref. 130

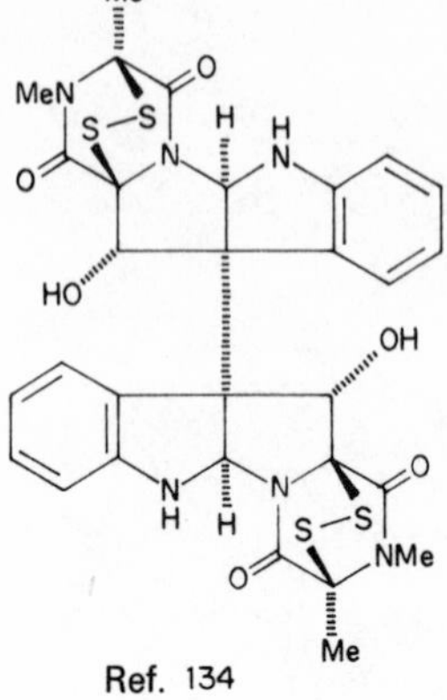

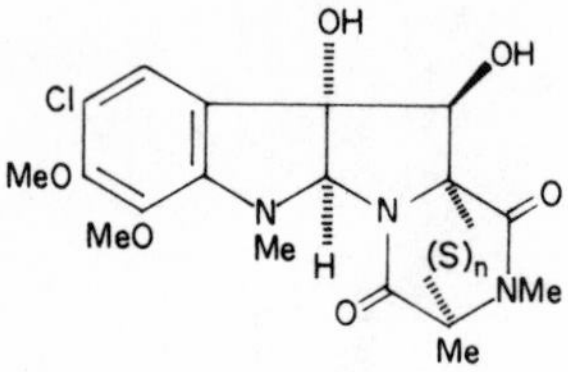

n = 2 **Ref. 131**
n = 4 **Ref. 132**

Ref. 134

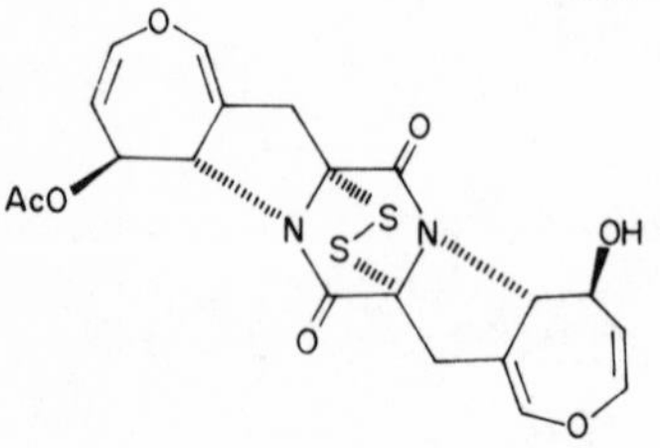

Ref. 133

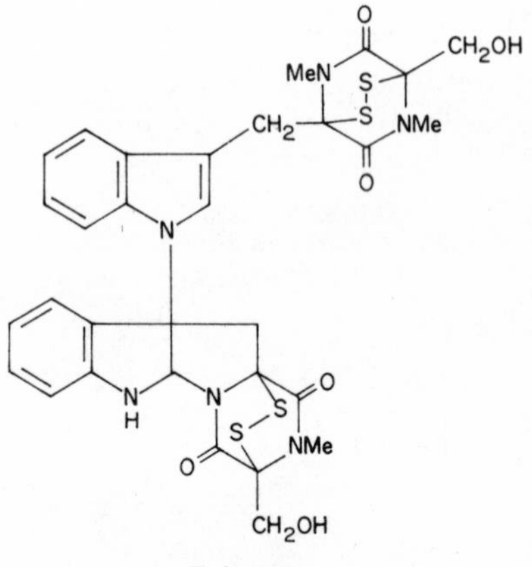

Ref. 135

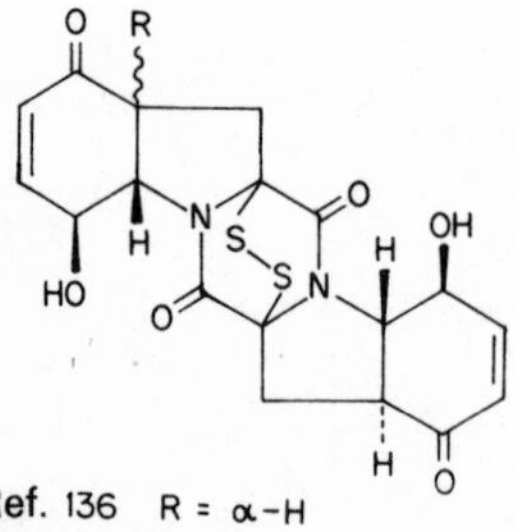

Ref. 136 R = α–H
Ref. 137 R = β–H

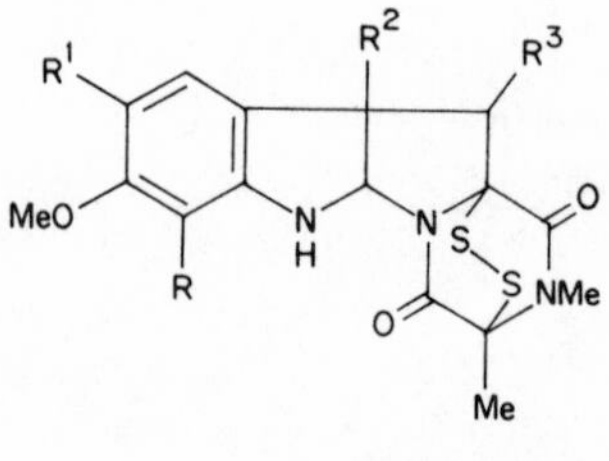

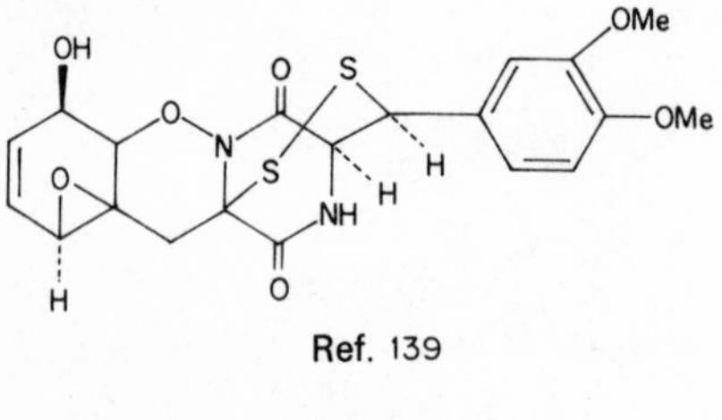

Ref. 139

R = H, R^1 = Cl, R^2 = R^3 = OH
R = OMe, R^1 = R^3 = H, R^2 = Cl **Ref. 138**

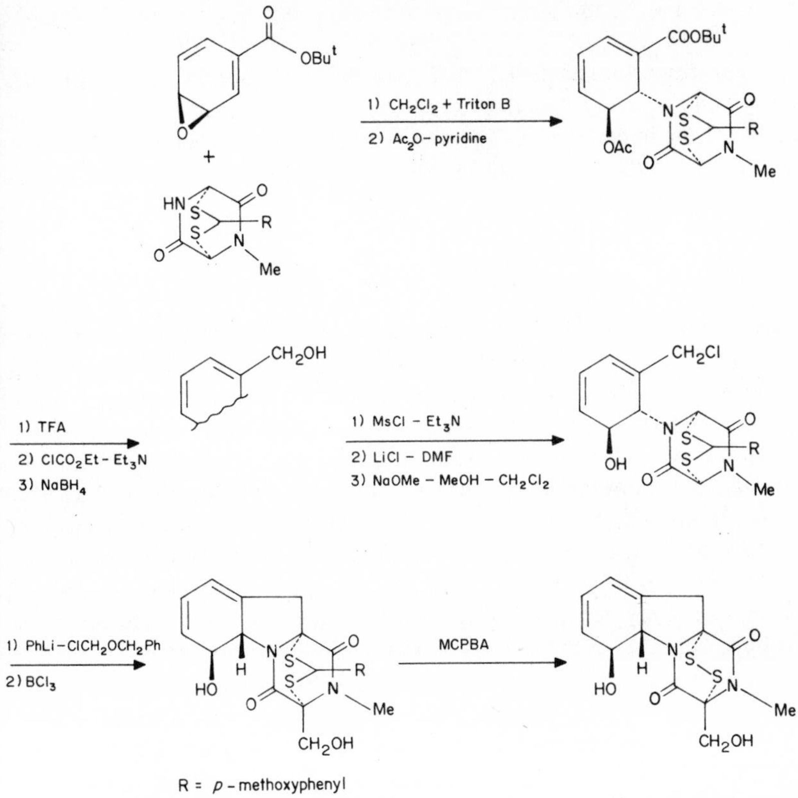

Scheme 7   Synthesis of gliotoxin.

of haemorrhage after childbirth, and in the treatment of migraine. These peptides are clearly related to dioxopiperazines and occur as pairs of compounds diastereomeric at C-8 within the lysergic acid unit. To date a series of some twenty or more members which differ in the component amino acids at the positions indicated by the R groups have been characterized from various fungi. However, it appears that the proline residue is an essential feature in the natural series. An elegant synthetic route to the ergot peptides has been developed and analogues of proline introduced into synthetic ergot peptides.[127] The biosynthesis of the peptide portion of the molecule is not understood, but an N-acyl acyclic tripeptide is implicated.[128 – 129]

The dioxopiperazines shown in Table 6.4 belong to the sulphur-bridged gliotoxin and sporidesmin groups. The mechanism by which the sulphur is inserted into these molecules is not understood; but probably occurs on an intact cyclic dipeptide via an acylimine intermediate. Many of the metabolites listed possess potent antibacterial and antiviral properties, but are also highly toxic.

The natural products listed in Tables 6.2–6.4 contain many complex chemical and stereochemical features and have been one of the testing grounds for the impressive advances in synthetic organic chemistry. Many of these are cited in the review articles, but that of gliotoxin is worthy of special note[140] (Scheme 7).

6.2.3   *Cyclic homodetic peptides*

The terms homodetic and heterodetic are commonly used to classify and describe cyclic peptides. The first of these groups contains cyclic compounds derived from amino acids through the formation of amide linkages in the usual way. Heterodetic peptides are compounds with ring systems involving both amide and other hetero-atom linkages, and are dealt with in the following section. Many of the larger homodetic peptides are well known because of their biological importance as antibiotics, toxins, and ion transport regulators, and the decade has seen substantial advances in the structure, synthetic, and functional aspects of *cyclo*-peptide chemistry. The presence of the same sort of atypical features within the amino acid residues, as observed in the linear and cyclic dipeptides, is manifest, and these, as well as the aspects outlined above, have been reviewed.[141] Cyclopeptides have a greater rigidity than their linear counterparts, due not only to the conformational constraints of the ring, but also to strong transannular hydrogen bonding. The increasing application of physical methods, principally X-ray analysis, n.m.r., and optical (ORD and CD) spectroscopy, is permitting detailed consideration of the overall topology and its relationship to biological function.[142–144] Clearly, detailed discussion of conformation is beyond the bounds of this review, but the salient points which relate to the chemical or biological properties of the molecules will be discussed where appropriate.

A considerable number of naturally-occurring cyclic peptides containing from four to eleven amino acid residues are now known and it is only possible to highlight the various ring sizes and structural variations that have so far been encountered. In some of the larger molecules, the peptide ring is often obscured by transannular modifications or structural interaction between adjacent amino acid residues. No cyclic homodetic tripeptides have been isolated to date, and only a few naturally occurring tetrapeptides have been fully characterized. The amino acid sequence of tentoxin (57), a fungal phytotoxin, was the subject of some controversy until it was resolved by an X-ray analysis of the dihydro derivative.[145] The reduction of the didehydro-phenylalanine unit results in the stereospecific formation of the D-isomer of N-methyl phenylalanine, and both the N-methyl amino acid units in the crystal

(57)

(58)

lattice conformation of dihydrotentoxin adopt *cis* peptide bonds. There is increasing crystallographic and solution evidence that proline and *N*-methyl amino acids in cyclic peptides often posses *cis* amide bonds, and there is speculation that this may be one of their functions in microbial peptides. The advantage of *cis* amides, particularly in small peptides, is the considerable release of ring strain which they allow. It was therefore surprising that the X-ray analysis of the dihydro derivative of chlamydocin (58), a highly cytostatic cyclic tetrapeptide, revealed an all-*trans* conformation for the peptide bonds.[146] This structure represents the first naturally-occurring or synthetic *cyclo*-tetrapeptide for which this has been observed, although each of the peptide units is significantly distorted from planarity. Other points of interest concerning this molecule are the previously unknown L-α-amino decanoic acid and the α-methylalanine which is rarely encountered in small peptides.

The malformins, a family of cyclic pentapeptide mycotoxins, derive their name from the malformations that they produce in germinating bean plants. The toxins were isolated some fifteen years ago and extensively investigated over the intervening period. The structure originally proposed for the main component, malformin A, has recently been re-examined and the alternative structure (59) formulated and subsequently confirmed by synthesis.[147]

The members of the viomycin, tuberactinomycin, and capreomycin families[148-152] have found limited clinical use for the treatment of tuberculosis, and are a particularly intriguing group, in that they are all cyclic-penta-peptides with notable structural, conformational, and biosynthetic features (60). The families have a common pentapeptide skeleton which contains variants of the modified arginine residue, a unique didehydroserine ureide unit and two molecules of L-diaminopropionic acid. The ring system is homo-detic, but includes an amide linkage involving a β-amino group of one of the diaminopropionic acids. They differ in the point of attachment of the β-lysine side-chain; in the viomycin and tuberactinomycin group it is to the α-amino group of one of the diaminopropionic acids, and to the β-amino of the other diaminopropionic for the capreomycins.

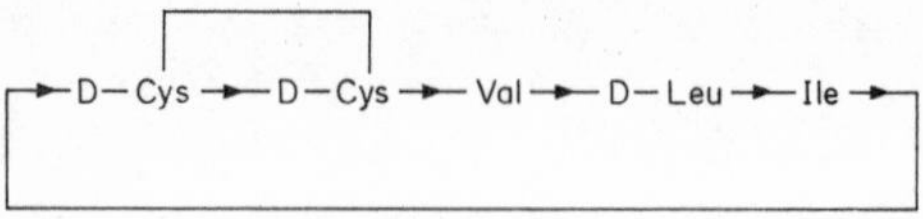

(59)

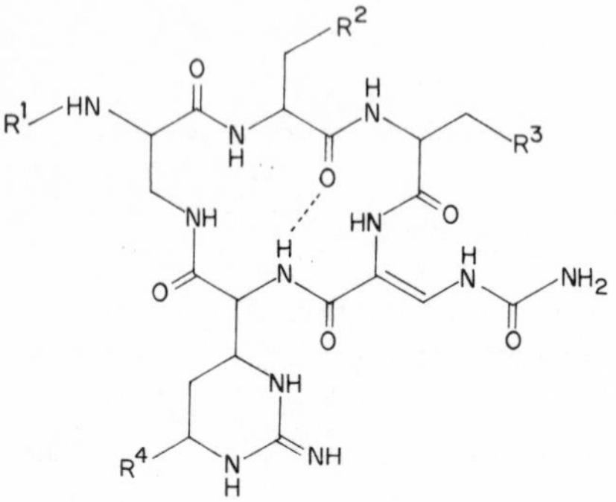

(60)*

   The peptides contain only L-amino acids and crystallographic analysis of
viomycin and tuberactinomycin O demonstrate that the amides all have the
trans geometry. A further significant feature revealed by X-ray data and
subsequently corroborated by solution studies, is the hydrogen bonded
chelate ring, the conformation of which is similar to the β-turn structures
common to many other cyclic peptides (see later). In the viomycin family, the
corner positions of the β-turns are occupied by L-serine and the didehydroserine ureide unit.

   The most fully authenticated series of cyclic hexapeptides are the iron-
containing metabolites of the siderochrome class. The subject has been
reviewed[153-156] and here only the general structural features and properties
are outlined. Ferrichrome (**61**) was first isolated twenty years ago and shown
to be a potent microbial growth factor. Related compounds were subsequently isolated and it was discovered that the metal-free peptides were
produced in high yield when the organisms were grown in media lacking iron.
This observation together with other evidence has led to the suggestion that
these compounds act as cellular transport agents for iron in aerobic micro-
organisms. The iron-binding site is furnished by three acylated N-hydroxy-
ornithine units incorporated in a cyclic hexapeptide.

   The cyclic heptapeptide rhizonin A (**62**), a fungal mycotoxin, has very
recently been characterized and is an excellent example of the application of
the full armory of spectral and chemical techniques for structural determination. The N-methylfurylalanine residue is unique and the two units

*R¹, R², R³, R⁴; see p. 297.

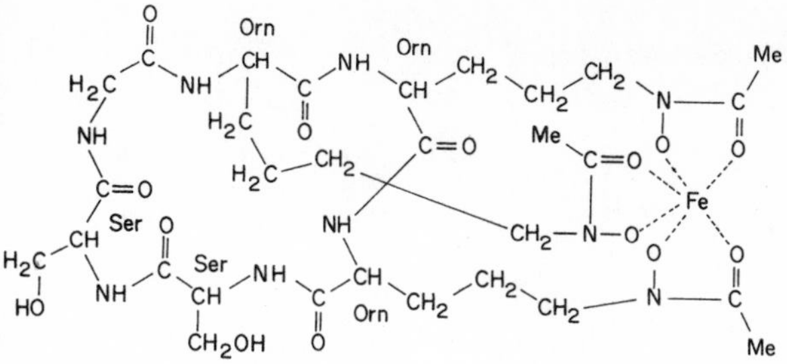

(61)

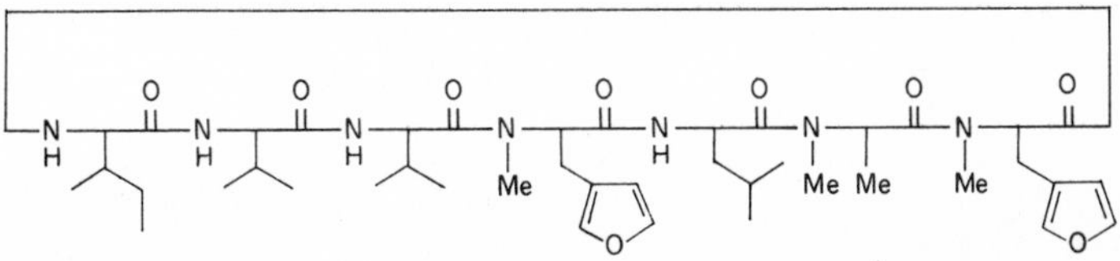

(62)

within the molecule possess opposite absolute stereochemistry; the same applies to the two valine, leucine and *allo*-isoleucine residues.[157]

The poisonous green mushroom *Amanita phalloides* is sometimes called the 'deathcap'; it resembles the edible mushroom in appearance and flavour and is a common cause of poisoning. The chemistry and biological properties have been investigated almost continuously over the past forty years, and these efforts have resulted in the isolation and structural elucidation of the major components.[158,159] These can be divided into two groups, namely phallotoxins and amatoxins.

Phallotoxins are cyclic heptapeptides bridged by the side-chains of the tryptophan and cysteine residues, whose common skeleton is shown (63). With the exception of the residue indicated, all the amino acids possess the L-configuration. The more recently identified cyclic hexapeptides of the virotoxin family (64) are products of the related fungus *A. virosa* which lack the bridging unit.[160] The amatoxins (65) have a larger ring (octapeptides) with the bridging sulphur atom oxidized to a sulphoxide and a hydroxyl substituent on the 6-position of the indole nucleus. All the *Amanita* toxins contain a γ-hydroxylated amino acid which is a prerequisite of toxicity. Despite the similar structures, the groups of toxins have different modes of action. Phalloidin damages the membranes of the liver cells, releasing potassium ions and enzymes. The amanitins have a much more specific action, binding strongly to RNA-polymerases, and for this reason they are widely used as powerful biochemical tools.[161]

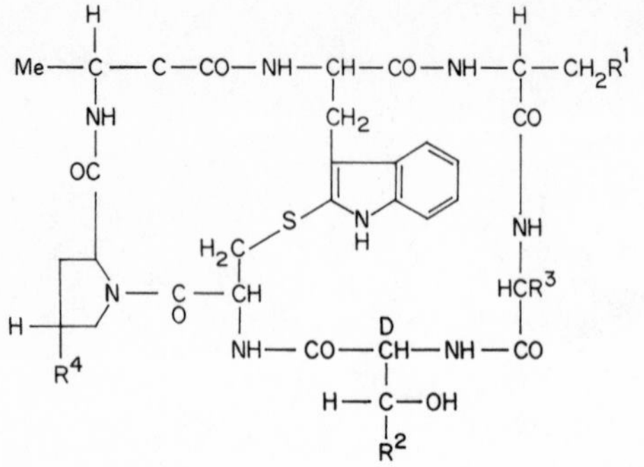

(63)

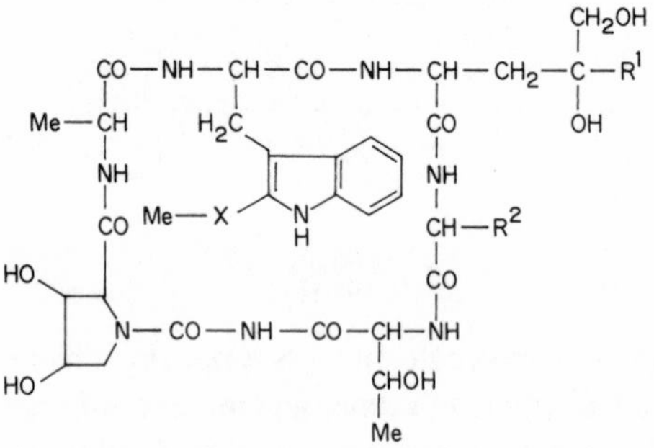

(64)

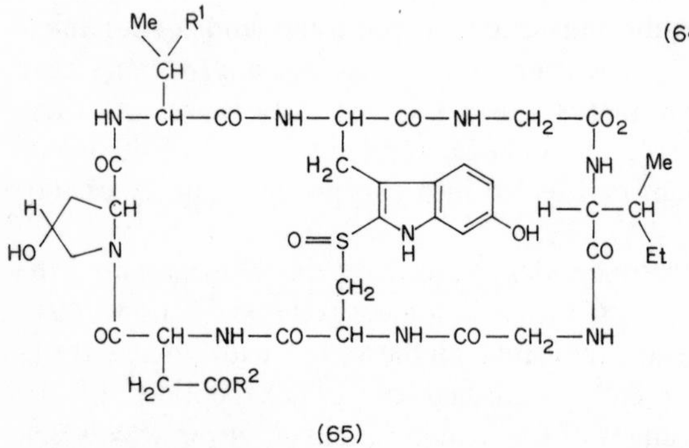

(65)

The patellamides (**66**), isolated from a Caribbean tunicate, are a family of closely related cyclic octapeptides with antitumour properties. Four of the peptide bonds are masked by two thiazole and two oxazoline units which are presumably derived from cysteine and serine or threonine, respectively, by appropriate transformation.[162] The cyclic decapeptide antamanide (**67**) is produced in small amounts by *Amanita phalloides*[163] and surprisingly counteracts the lethal effects of its toxins. The antitoxin is believed to act by preventing the accumulation of the toxins in liver cells. The amino acids all

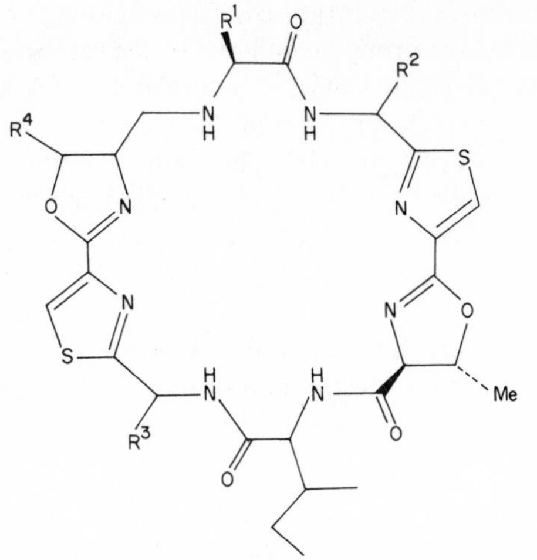

(66)

	R^1	R^2	R^3	R^4
Patellamide A	Et MeCH	Me$_2$CH	Me$_2$CH	H
B	Me$_2$CHCH$_2$	PhCH$_2$	Me	Me
C	Me$_2$CH	PhCH$_2$	Me	Me

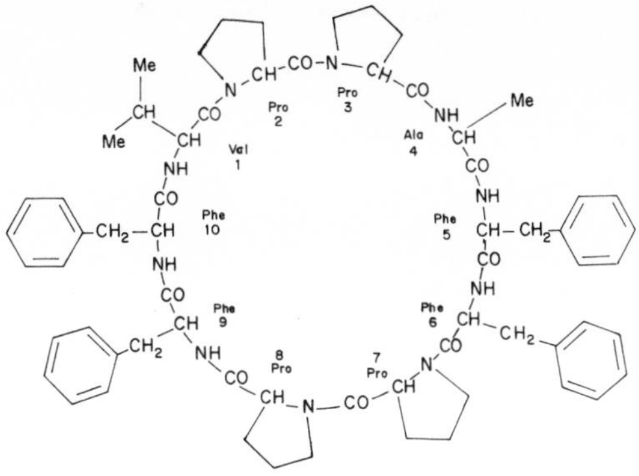

(67)

possess the L-configuration and there is considerable conformational equilibration of at least four forms.[164-165] Antamanide is an ionophore and forms complexes with alkali metal ions, the most stable complexes being with Na$^+$ and Ca^{2+}.[166] Gramicidin S and tyrocidine antibiotics have been studied extensively over the past two decades and provide the important model for the biosynthesis of atypical peptides. They have also been models for many of the innovations in the conformational study of peptides in recent years.[164,167] Their relatively simple structures and high biological activity have presented an important area for structure-activity investigations. Although gramicidin S exists in solution and the crystal lattice as one predominant conformer, several different topological structures have been proposed. The weight of crystallo-graphic and spectral evidence is now firmly behind one of the earlier models which is usually referred to as 'the pleated sheet structure' (68).[168] A further significant conformational feature is the presence of the two β-turns or loops which are characteristics of many cyclic peptides. The corner positions of the β-turns are occupied by the D-phenylalanine and L-proline residues and conformational energy considerations accord with the established stability of this type of system. Similar conclusions have been drawn for the tyrocidine family.

Finally, the largest series of fully characterized homodetic peptides are represented by the cyclosporins (69). Their immunosuppressant properties

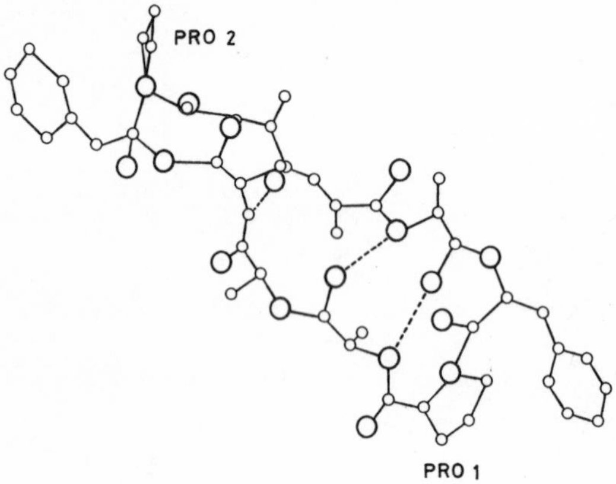

(68)

which have found clinical application in transplant surgery have sustained[169] a strong interest in their chemistry.

6.2.4   Cyclic heterodetic peptides (depsipeptides)

Cyclic peptides in which one or more of the amide bonds is replaced by an ester linkage are referred to as depsipeptides or peptide lactones and they represent the most numerous of the heterodetic peptides. Cyclic depsipeptides may be broadly divided into two main groups, namely those possessing a regular alternating array of peptide and ester linkages, and those with irregular insertion of ester bonds. Valinomycin (70), the enniatins (71), and beauvericin (72), most of which were characterized twenty years ago, all belong in the

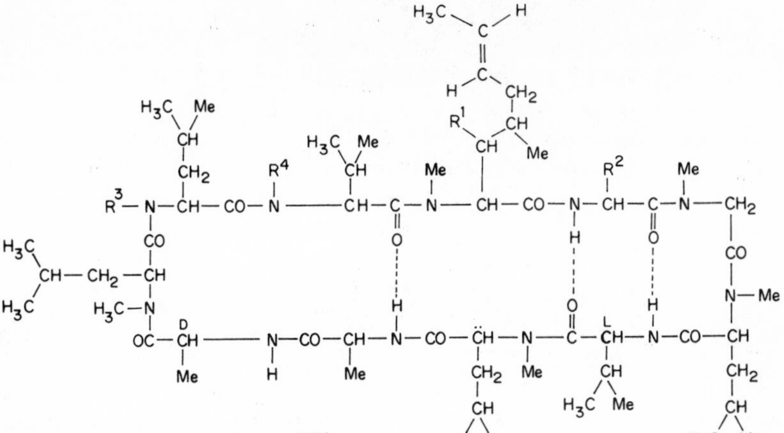

(69)

		R^1	R^2	R^3	R^4
Cyclosporin	A	OH	Et	Me	Me
	B	OH	Me	Me	Me
	C	OH	CH(OH)Me	Me	Me
	D	OH	CHMe$_2$	Me	Me
	E	OH	Et	Me	H
	F	H	Et	Me	Me
	G	OH	Pr	Me	Me
	H	OH	Et	Me	Me
	I	OH	CHMe$_2$	H	Me

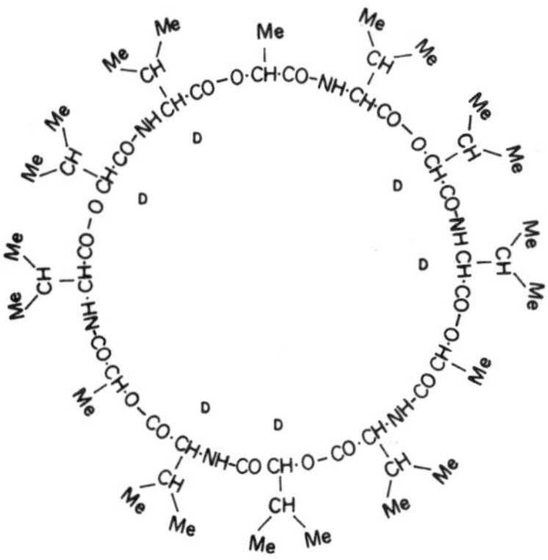

(70)

(71) Enniatin A, R = CHMeEt
B, R = CHMe$_2$
C, R = CH$_2$CHMe$_2$

(72) Beauvericin, R = CH$_2$Ph

former category.[170,171] The observations in the mid-60s that valinomycin and related compounds possessed unique selective ion-transporting properties led to a resurgence of interest in these compounds which were subsequently referred to as ionophores. These peptides form lipid-soluble complexes with polar cations of which K$^+$, Na$^+$, Ca^{2+}, Mg^{2+}, and the biogenic amines are the most significant biologically. A variety of physical studies indicate that the complexation-decomplexation kinetics and diffusion rates of ionophores and their complexes across lipid barriers are so favourable that their transport turnover numbers through biological and artificial membranes attain values in some cases exceeding the turnover of some enzyme systems. The biological applications of general ionophores which also include the polyethers and synthetic compounds, have been reviewed.[172,173]

The crystal structures of uncomplexed valinomycin,[174] and its complexes with several ions[175] have been reported. Uncomplexed valinomycin exists as a mixture of conformers in solution with the proportions depending on temperature and solvent.[176] The molecule possesses a hydrophilic and a hydrophobic face. On complex formation, the ion enters the molecule and the ester groups turn inwards towards the centre, forming a lipid-soluble rigid sphere which easily accommodates K$^+$, Rb$^+$, but is too large for Na$^+$ and Li$^+$ (see Fig. 6.1). The K/Na complexing selectivity at 10,000/1 is higher than for any other ionophore. The complex formation does not involve any *cis-trans* isomerization of the peptide bonds. Equally extensive investigations on the enniatins and beauvericin and their complexes are documented in the review articles already cited.

The most comprehensively studied cyclic depsipeptides belonging to the second category are undoubtedly the actinomycins.[177-179] The general structure of the group (73) is well established and a considerable number have now been characterized, principally from *Streptomyces* species. The most commonly isolated compound is actinomycin D, which possesses two

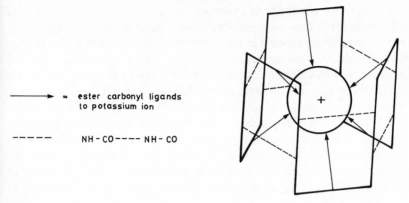

Figure 6.1 Schematic representation of the potassium ion complex of valinomycin showing the ester carbonyl ligands to the potassium ion and the intramolecular hydrogen bonds.

(73) Actinomycin D, A = X = Sar, B = Y = Pro, C = Z = D-Val

" C_2, A = X = Sar, B = Y = Pro, C = D-Val,

                                    Z = D-alloisoleu

(74) Echinomycin, R = --CH$_2$-S-CH-

                                SMe

(75) Triostin A, R = -CH$_2$-S-S-CH$_2$-

structurally identical peptide rings. Metabolites of this series are referred to as the *iso*-series, and those with differing peptide rings as the *aniso*-series.

The considerable variation in the amino acids within the peptide residues reflects the relatively low specificity in the biosynthesis. This can be exploited to produce higher yields of specific antibiotic by adding the appropriate amino acids to the organism. In some cases, this can be used to produce new antibiotics. For example, the actinomycins containing azetidine-2-carboxylic acid instead of proline or pipecolic acid are produced when the organism is supplied with an extraneous source of azetidine-2-carboxylic acid.[180]

The actinomycins are potent inhibitors of DNA-dependent RNA synthesis; i.e. the transcription in protein biosynthesis. Actinomycin D has found limited application in the clinical treatment of certain types of tumours. Its

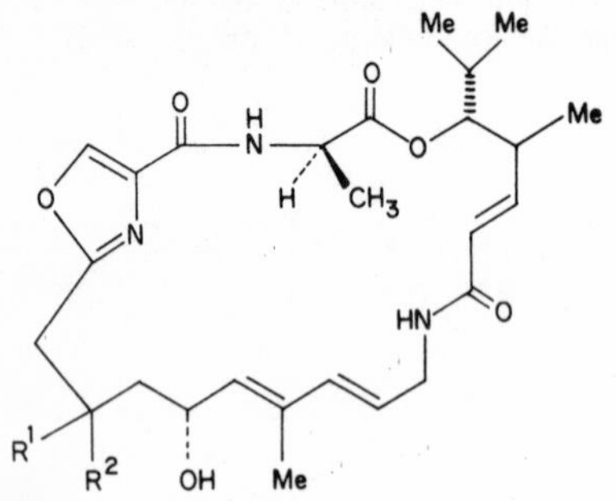

(76)

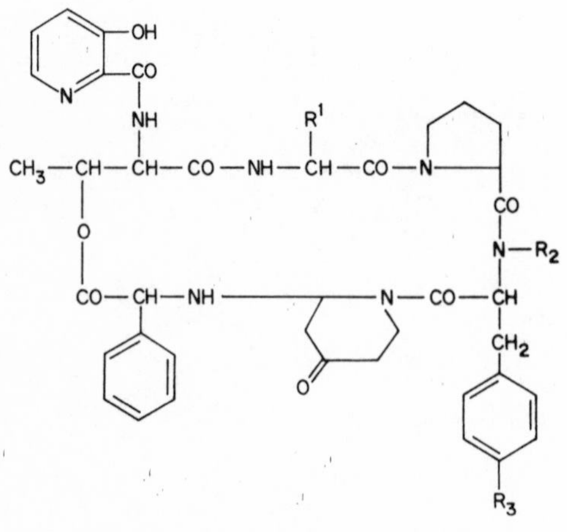

(77)

e.g. Virginiamycin S_1 R^1 = Et , R^2 = Me, R^3 = H,

mode of action involves the formation of highly stable complexes with DNA, which precludes the latter from fulfilling its biological function. The current model for the interaction of the DNA double helix with actinomycin is based on the X-ray data obtained from a crystalline complex containing actinomycin and deoxyguanosine.[181] In this model the phenoxazone chromophore is intercalated between adjacent G-C base pairs of DNA, where the guanine residues are on opposite DNA strands and the two amino groups of the guanines form specific hydrogen bonds with both cyclic peptides which fit into a narrow groove. Similar intercalating properties have been ascribed to members of the echinomycin[182] group which are also potent antibiotics. Echinomycin (74)[183] and triostin A[184] (75) are the principal members and are probably related biosynthetically by a process involving methylation of the disulphide bridge followed by rearrangement.[185] Although the peptide ring is locked by the disulphide bridge, triostin A exists in solution as a pair of interconverting conformers which are a consequence of the readily reversed chirality of the disulphide bridge.[186]

The virginiamycin family of antibiotics are represented by two distinct groups of cyclic depsipeptides[187] which co-occur and exhibit remarkable synergism with respect to their antibacterial activity. The group A compounds, represented by madumycin II (76), are highly modified; whereas the group B (77) possess a more characteristic depsipeptide structure.

6.2.5 Large modified peptides

This somewhat arbitrary grouping of atypical peptides which fall in the 1–3K dalton molecular weight range serves to illustrate the difficulties encountered with molecules in the border area between classical organic natural products and biomacromolecules. Passing reference has already been made to the advances in methodology which have allowed these complex molecules to come realistically within the compass of the natural product chemist. However they still present formidable problems and many structures have required minor adjustments. In general, these metabolites contain the same

Dha = dehydroalanine
Dhb = dehydrobutyrine
Ala—S—Ala = lanthionine
Abu—S—Ala = methyllanthionine   (78)

types of structural modifications as observed in the peptides described in the preceding sections.

The antibiotic subtilin from *Bacillus subtilis* has been assigned structure (78) on the basis of an extensive investigation.[188] The molecule is composed of 32 amino acid residues, eight of which, i.e. four 3-methyl-lanthionines, two dehydroalanines, and one each of dehydro-α-aminobutyric acid and lanthionine, are not found in protein. The alanine moieties of lanthionine and all four aminobutyric acid moieties of the 3-methyl-lanthionine residues possess the D-configuration. The novel thio-ether bridges probably result from intramolecular addition of cysteine thiol groups to dehydroalanine or dehydrobutyrine units which in turn could be derived by the dehydration of serine and threonine, respectively. This leads to the interesting speculation that the precursor of subtilin could well be a linear peptide derived solely from primary amino acids and constructed under ribosomal control. There is evidence that the biosynthesis of a closely related peptide, nisin,[189] may well be formed on a ribosomal template.[190]

Ac-Aib-Pro-Aib-Ala-Aib-Ala-Gln-Aib-Val-Aib-Gly-Leu-Aib-Pro-Val-Aib-Aib-Glu(OH)-Gln-Phol

(79)

The linear ion-transporting peptide alamethicin (79) is smaller than subtilin and is noteworthy because of the presence of a considerable number of α-aminoisobutyric acid (Aib) residues and a terminal β-phenylalaninol (Phol). Although the molecule contains relatively few modified amino acid residues, their sequencing has posed difficulties and it has been necessary to revise the structure a number of times.[191] The formulation (79) is now generally accepted and similar revisions have been made to the structures of other members of a growing family which are now referred to as peptaibophols for obvious reasons.[192,193]

The complexity of thiostrepton (80) could only be unravelled by an X-ray analysis which revealed a variety of modified units previously observed in smaller metabolites.[194] This particular structure has served as a reference for investigations which have led to a whole series of related compounds.[195-200]

The absence of X-ray crystallographic data for a reference compound of the bleomycin family (81) has hampered the structural studies, and again it has been necessary to adjust some of the earlier structural proposals.[201] However the current formulations are now well-founded, since two independent total syntheses of bleomycin A$_2$ have been achieved.[202,203]

The possible clinical potential of these potent antitumour agents has stimulated interest in their mode of action. On the basis of extensive studies, a

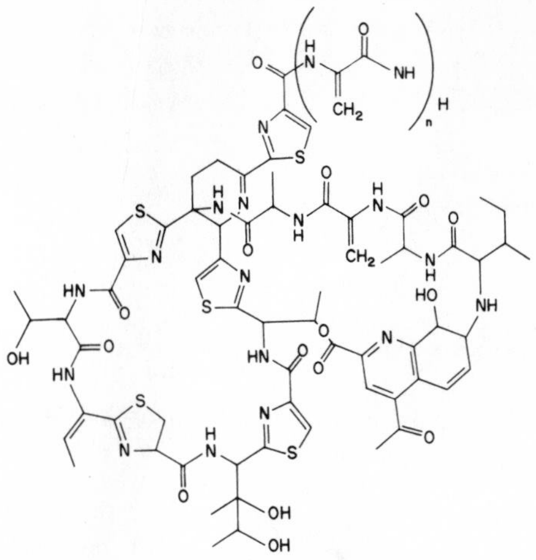

(80)

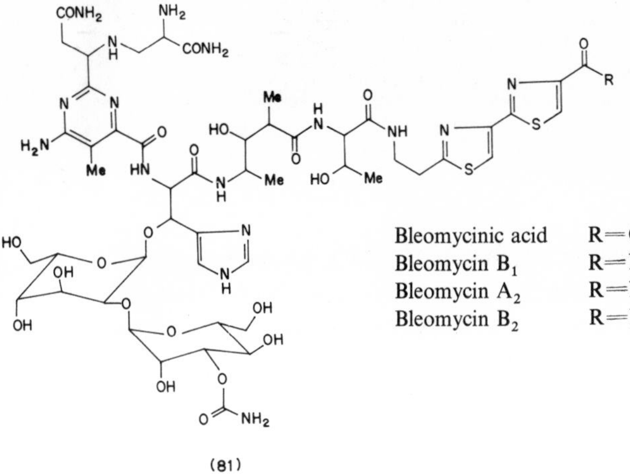

Bleomycinic acid	R=OH
Bleomycin B_1	R=NH$_2$
Bleomycin A_2	R=NH(CH$_2$)$_3$·S$^+$Me$_2$
Bleomycin B_2	R=NH(CH$_2$)$_4$NH·C—NH$_2$
	‖
	NH

(81)

working hypothesis has evolved which proposes a complex and subtle series of steps. These involve the formation of a ferrous complex which then binds molecular oxygen and the interaction of the 'tail' of the complex with double-stranded DNA. This positions the oxygen complex in such a way that oxidative cleavage of one of the DNA strands occurs at a deoxyribose site. The general scheme is illustrated in Fig. 6.2. Thus the type of reaction of bleomycins envisaged with DNA is that of a quasi-enzyme system.[204]

The vancomycin group of antibiotics are a group of glycopeptides

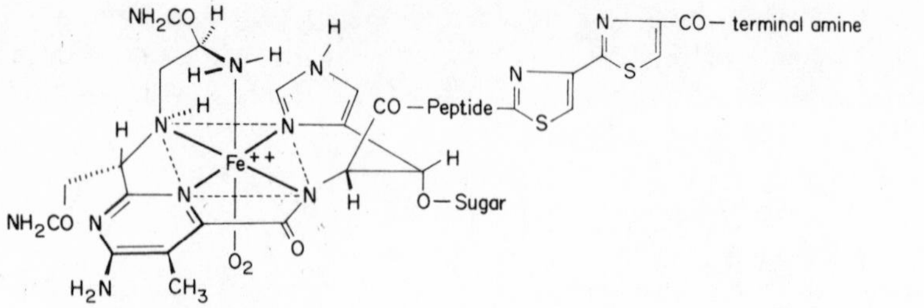

Figure 6.2 Schematic representation of the proposed interaction of bleomycin with double-stranded DNA.

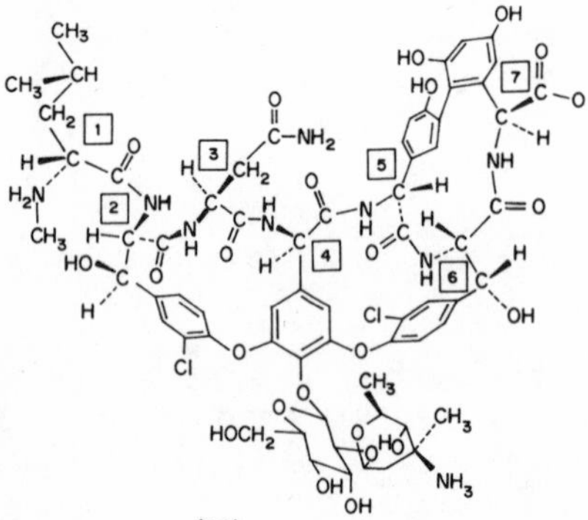

(82)

composed of a central and biologically active modified peptide core and peripheral carbohydrate units. They are broad spectrum antibiotics that have aroused particular interest because bacteria do not really develop resistance to them. An excellent review which covers all aspects of the chemistry and biology, including mode of action has appeared recently.[205] Unfortunately minor revisions have been made to the structures of the main members of the group, vancomycin[206] (82) and ristocetin,[207] since its publication, but this does not detract from its value. The network of biphenyl and biphenyl ether linkages are modifications which can formally be derived from the oxidative interactive of adjacent hydroxyphenylglycine residues. Although unusual, this type of phenolic coupling is not unique to the vancomycins, and has been observed in plant-derived cyclic hexapeptides.[207]

6.2.6 Cyclopeptide alkaloids

Although strictly heterodetic, this group of closely related peptides are invariably referred to as peptide alkaloids, and they are conveniently discussed under this heading. Isolated mainly from plants of the Rhamnaceae family, the numbers with established structures are increasing rapidly and within the structures of eighty or so now known, definite chemotaxonomic relationships are beginning to emerge.

Structurally they can be divided into three distinct classes based either on the number of atoms in the ring; i.e. those containing 13, 14, and 15 atoms or alternatively the way in which the common hydroxystyrylamino group is incorporated into the cyclic system. The largest group, containing a 14-membered ring and with a p-hydroxystyrylamino group, is further subdivided into the frangalanine (83), integerrine (84), and amphibine-B (85) types. The smaller classes containing 13- and 15-membered rings are characterized by zizyphine-A (86) and mucronine-A (87). The chemistry, structural elucidation, and distribution of these plant products have been admirably reviewed[209] and regularly updated in the specialist reports.

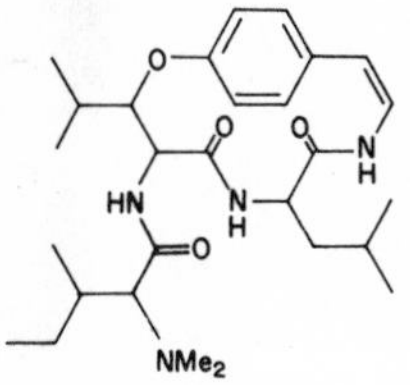

(83)

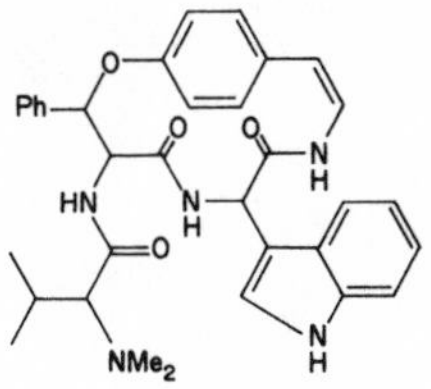

(84)

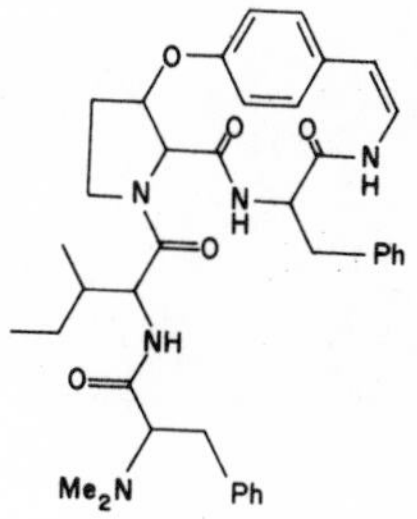

(85)

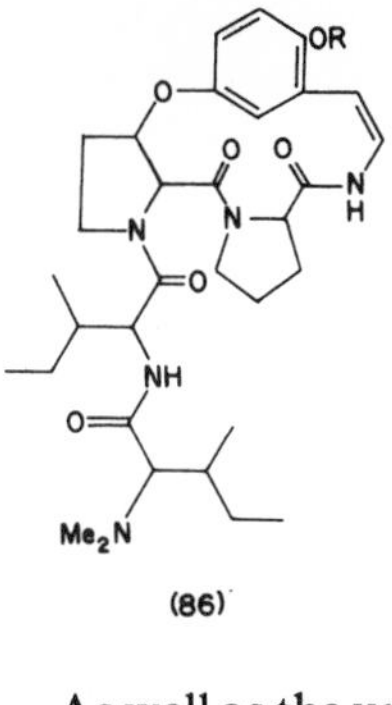

(86)

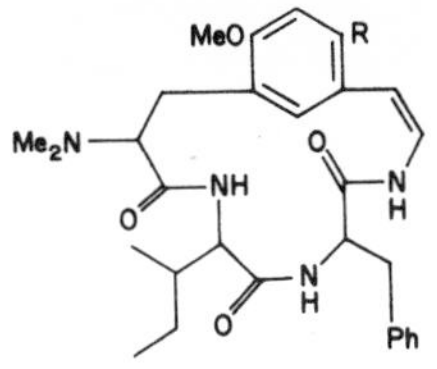

(87)

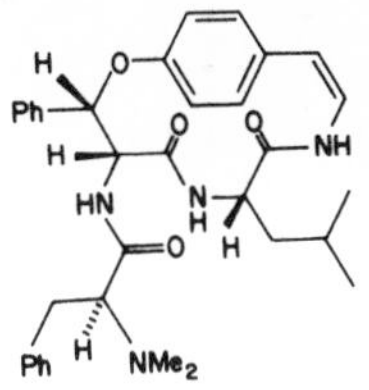

(88)

As well as the unique hydroxystyrylamino group, all the alkaloids possess a terminal N-dimethylamino acid residue which is rarely observed in microbial peptides. In the main, the amino acids of the cyclopeptides have the L-configuration. Scrutianine-E (88) appears to be an exception in that it contains the novel D-amino acid threo-β-phenylserine. In view of the considerable structural diversity of the microbial peptides, it is surprising that the cyclopeptide alkaloids represent the only major group of plant peptides. Since these compounds are basic and were originally isolated as alkaloids, it is possible that this represents a distorted picture of the distribution of atypical peptides in the plant kingdom, and that more careful investigation of plant material may lead to the isolation of other types of peptide.

6.2.7 *β-Lactam antibiotics*

The β-lactams find the widest clinical use of all the antibiotics and as a consequence their chemistry and biology occupies the efforts of countless researchers throughout the world, as is evident from even the most cursory

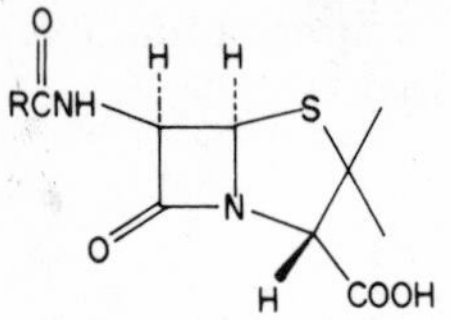
O
H H
RCNH
S
N
O
H
COOH

(89)

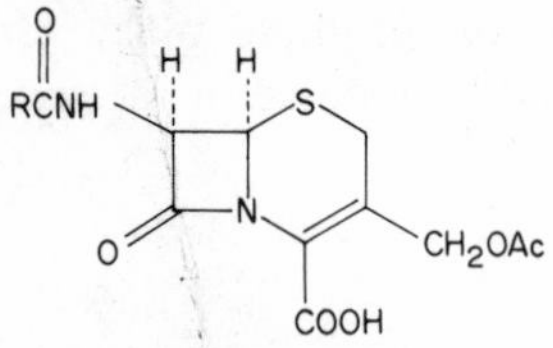
O
H H
RCNH
S
N
O
CH2OAc
COOH

(90)

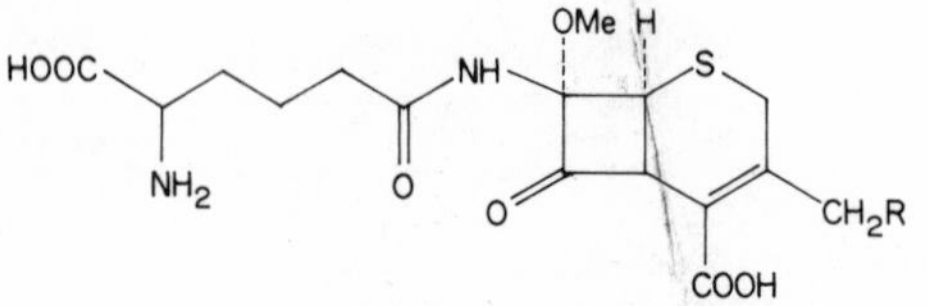
HOOC
NH2
O
NH
OMe H
S
N
O
CH2R
COOH

(91)

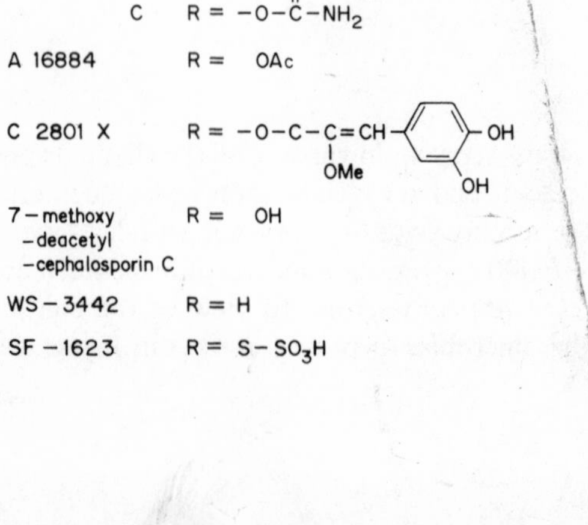
Cephamycin A R = –O–C(=O)–C(OMe)=CH–C6H4–O–SO3H
B R = –O–C(=O)–C(OMe)=CH–C6H4–OH
C R = –O–C(=O)–NH2
A 16884 R = OAc
C 2801 X R = –O–C–C(OMe)=CH–C6H3(OH)–OH
7–methoxy
–deacetyl
–cephalosporin C R = OH
WS–3442 R = H
SF –1623 R = S–SO3H

inspection of the major journals. Those readers wishing to familiarize themselves with what seems to be an overwhelming field should look no further than a recent excellent trilogy[210] which covers all aspects of chemistry and biology. The sole purpose of attempting to include them in a review on peptides is to emphasize their relationship to atypical peptides and amino acid derivatives.

The penicillins (89) and cephalosporins (90) should be familiar to almost everyone, as they are by far the oldest and most important groups.[211,212] It was noted in an earlier section that the linear peptide (38) is the common precursor to both sets of compounds. The sequential relationship in general terms is outlined in Scheme 8. The detailed biochemical pathways are currently under active investigation[210] and beyond the scope of this review. However their peptide origin is beyond doubt.

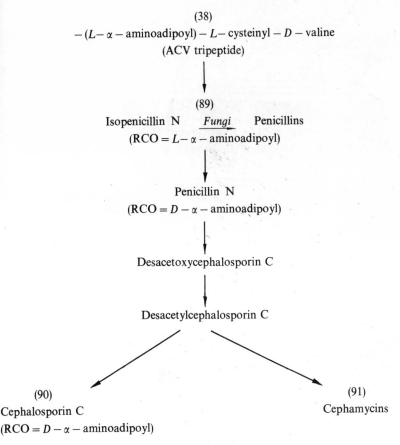

(38)

$- (L-\alpha-\text{aminoadipoyl}) - L - \text{cysteinyl} - D - \text{valine}$

(ACV tripeptide)

(89)

Isopenicillin N   *Fungi*   Penicillins

$(RCO = L-\alpha-\text{aminoadipoyl})$

Penicillin N

$(RCO = D-\alpha-\text{aminoadipoyl})$

Desacetoxycephalosporin C

Desacetylcephalosporin C

(90)

Cephalosporin C

$(RCO = D-\alpha-\text{aminoadipoyl})$

(91)

Cephamycins

Scheme 8   Relationship of ACV tripeptide to the penicillins and cephalosporins.

In the early seventies after more than twenty-five years of intense activity, it looked as though interest in the β-lactams was at last beginning to wane. It then underwent a remarkable rejuvenation, stimulated by the isolation of several new families of microbial antibiotics, all having in common a β-lactam functionality but with widely differing gross structures. The first of these, the 7-methoxycephalosporins[213] (cephamycins) (91) were clearly structurally related to the cephalosporins; this was later confirmed by biosynthetic studies[214] (see Scheme 8). These were closely followed by clavulanic acid, the nocardicins, thienamycin, and the olivanic acids.[210,215]

Clavulanic acid (92) is produced by the same organism, *Streptomyces clavuligerus*, which produces the cephamycins. However the scant evidence currently available suggests that it is biosynthesized by an independent process, possibly from glutamic acid;[216] (92) is a relatively poor antibiotic in

(92)   (93)   R = CH$_2$OH, COOH

its own right, but by virtue of its ability to inhibit the β-lactamase enzymes, it can potentiate the activity of certain penicillins against resistant bacteria.[217] Several related metabolites (93) have been isolated but none possess such potent β-lactamase inhibitory activity.

The nocardicins (94) are obviously atypical peptides unrelated to other β-lactam families. The oxime function at the amino terminal position is unusual but not unique, as is the D-*p*-hydroxyphenylglycine residue. Although their antibacterial activity does not match that of many other β-lactams, there is still some interest in their novel immunostimulating activity.[215]

Thienamycin and the olivanic acids were the first members to be reported of what has now become a substantial group of metabolites (see Table 6.5). These compounds, all of which are extraordinarily potent broad spectrum antibiotics and β-lactamase inhibitors, are generally termed the carbapenems.[210,215] The olivanic acids differ from the rest of the series in that they are epimeric at C-6 and are often referred to as epithienamycins. The only evidence pertaining to their biochemical origin would suggest that, as with clavulanic acid, glutamic acid may well be involved. Their clinical application to date has been limited by their instability, particularly toward mammalian dehydropeptidases; an observation which serves to highlight their relationship to atypical peptides.

This new impetus continues with the recent isolation of the novel series of

Table 6.5

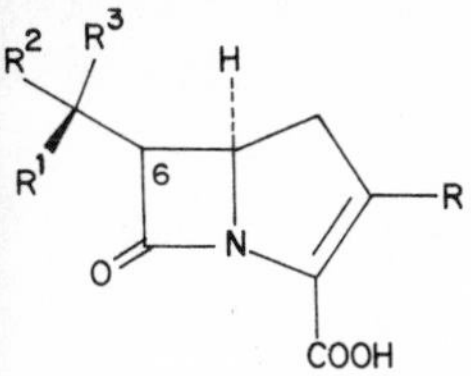

Name	Configuration at C-6	R	R^1	R^2	R^3
NS5	R	$-SCH_2CH_2NH_2$	H	H	Me
PS5	R	$-SCH_2CH_2NHAc$	H	H	Me
PS7	R	$-SCH=CHNHAc$	H	H	Me
Epithienamycin A	R	$-SCH_2CH_2NHAc$	OH	H	Me
Epithienamycin B	R	$-SCH=CHNHAc$	OH	H	Me
Epithienamycin C	S	$-SCH_2CH_2NHAc$	OH	H	Me
Epithienamycin D	S	$-SCH=CHNAc$	OH	H	Me
MM 27696	R	$-SCH=CHNHCOEt$	OH	H	Me
C-19393E$_5$	R	$-S(O)CH=CHNHAc$	OH	H	Me
8U-207	R	$-SCH_2CH_2NH_2$	OSO_3H	H	Me
MM 17880	R	$-SCH_2CH_2NHAc$	OSO_3H	H	Me
MM 13902	R	$-SCH=CHNHAc$	OSO_3H	H	Me
MM 4550	R	$-S(O)CH=CHNHAc$	OSO_3H	H	Me
Pluracidomycin B	R	$-S(O)CH_2COOH$	OSO_3H	H	Me
Pluracidomycin C	R	$-S(O)CH(OH)_2$	OSO_3H	H	Me
SF 2103A	R	$-SO_3H$	OSO_3H	H	Me
Thienamycin	S	$-SCH_2CH_2NH_2$	Me	H	OH
N-Acetylthienamycin	S	$-SCH_2CH_2NHAc$	Me	H	OH
N-acetyldehydrothienanycin	S	$-SCH=CHNHAc$	Me	H	OH
PS6	R	$-SCH_2CH_2NHAc$	H	Me	Me
PS8	R	$-SCH=CHNHAc$	H	Me	Me
Carpetimycin A	R	$-S(O)CH=CHNHAc$	OH	Me	Me
KA 6643 D	R	$-SCH_2CH_2NHAc$	OSO_3H	Me	Me
KA 6643 F	R	$-SCH=CHNHAc$	OSO_3H	Me	Me
C-19393 S$_2$	R	$-S(O)CH=CHNHAc$	OSO_3H	Me	Me

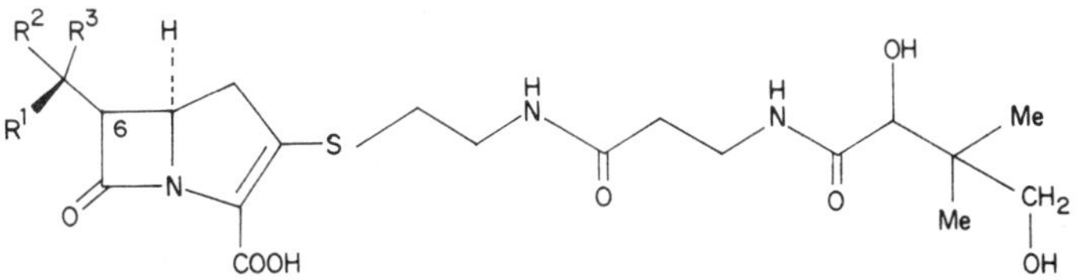

Name	Configuration at C-6	R^1	R^2	R^3
OA-6129 A	R	H	H	Me
OA-6129 B$_1$	S	H	H	Me
OA-6129 B$_2$	R	OH	H	Me
OA-6129 C	R	OSO_3H	H	Me

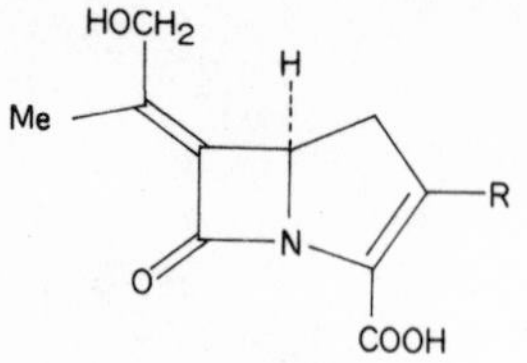

Asparenomycin A	R = —S(O)CH=CHNHAc
B	R = —S(O)CH₂CH₂NHAc
C	R = —SCH=CHNHAc

Asparenomycin A $\quad$ R = —S(O)CH=CHNHAc
B $\quad$ R = —S(O)CH$_2$CH$_2$NHAc
C $\quad$ R = —SCH=CHNHAc

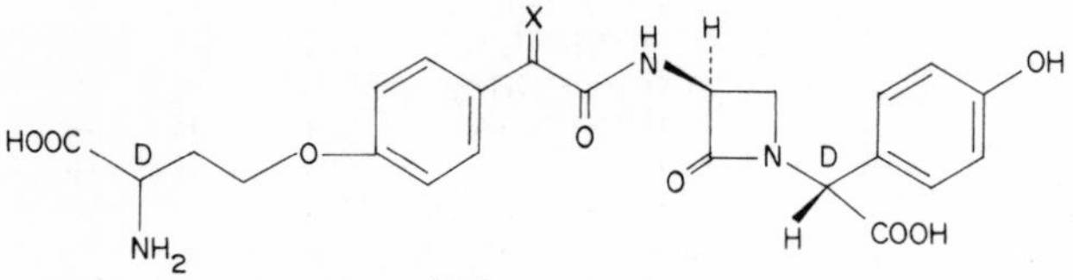

(94)

Nocardicin $\quad$ AX = NOH (*syn* with respect to acyl imine group)
$\qquad\qquad$ BX = NOH (*anti* with respect to acyl imine group)
$\qquad\qquad$ CX = CONH$_2$ Nocardicin DX = 0

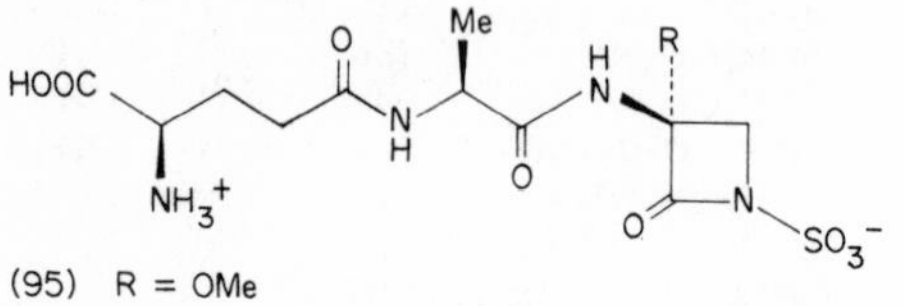

(95) $\quad$ R = OMe
(96) $\quad$ R = H

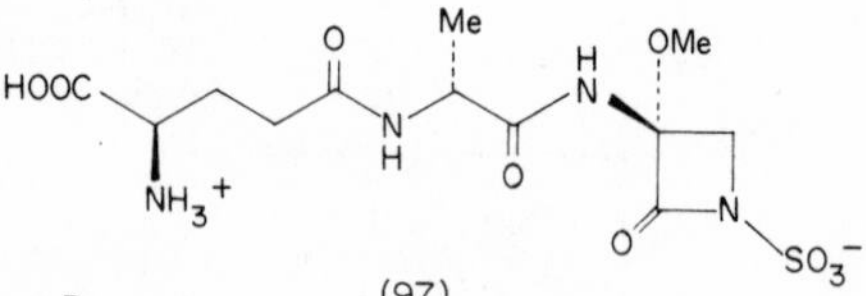

(97)

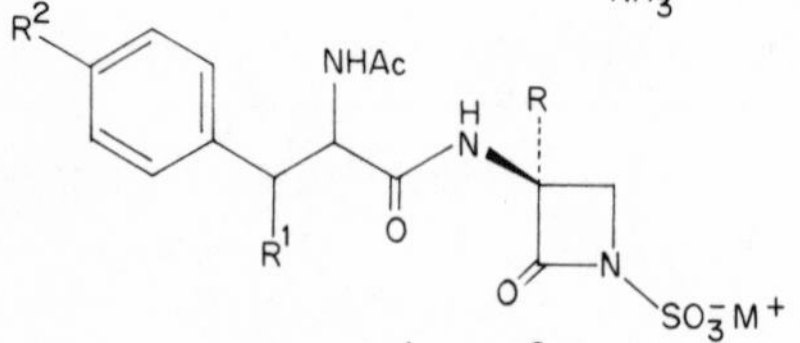

	R	R¹	R²	M
(98)	OMe	H	H	Na
(99)	OMe	H	OH	K
(100)	H	H	OH	K
(101)	OMe	OH	OSO₃Na	Na
(102)	OMe	OSO₃Na	OSO₃Na	Na

OMe

AcNH

(103)

monocyclic β-lactams (**94–102**) from *Pseudomonas*[218] strains and *Agrobacterium radiobacter*.[219] These are now referred to as the monobactams; the simpler compound (**103**) which contains the basic skeleton has been obtained from *Chromobacterium violaceum*.[220]

The β-lactams are so effective as antibiotics because they act as inhibitors of bacterial peptidoglycan synthesis (cell wall inhibitors). The wall peptidoglycan is both indispensable and unique to prokaryotes so that antibiotics have no corresponding targets in eukaryotes, which explains their high selective toxicity. The advances which are occurring[210] in the understanding of the basic molecular biology in this area may ultimately allow the medicinal chemist to compete on equal terms with micro-organisms.

6.3 Typical peptides and proteins

It is obviously impossible to do justice to the enormous strides that have been made over the past decade with respect to the understanding of the chemistry and biochemistry of peptides and proteins. There is also no doubt that some of the significant developments in molecular biology have altered and will continue to alter radically the chemist's view of this area, and in constructing this brief overview their impact upon the interests of natural product chemists was borne very much in mind. Those requiring a broader perspective are referred to the excellent review series[221–224] and annual reports.

The distinction between peptides and proteins is arbitrary, and historically has been based on molecular size. The ability to pass through a natural dialysis membrane usually defines the upper limit for a peptide and this corresponds to a molecular weight of *c.* 10 000 or approximately a hundred amino acid residues. The majority of peptides are derived by enzymatic cleavage of larger protein molecules, the biosynthesis of which takes place at the ribosomes and is under rigid genetic control. The main outline of protein biosynthesis is now well documented and a general understanding of the process is necessary for an appreciation of the application of the advances in molecular biology to protein studies.[225]

6.3.1 *Separation and isolation of peptides and proteins*

The majority of peptides and proteins are normally isolated by techniques reasonably familiar to the organic chemist interested in water-soluble natural products. The well-tried and tested procedures of selective precipitation by salts or organic solvents, ion exchange chromatography, adsorption chromatography and gel filtration are regular tools which have been developed to a high level of sophistication.[226]

However, the tendency of many proteins to denature even under the mildest of conditions has led to the development of very sensitive and highly specific techniques. Ligand affinity chromatography takes advantage of the high affinity of many proteins for specific chemical groups. The general principle is outlined in Scheme 9.[227] A protein that recognizes the group X can be purified by passing a crude extract through a column containing X or a derivative covalently bound to an insoluble support. For example, the plant protein concanavalin A can readily be purified by passing through a column that

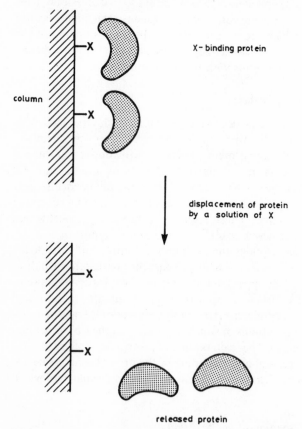

Scheme 9   Principles of affinity chromatography illustrated by the purification of the protein concanavalin A on a column with covalently attached glucose residues (X).

contains covalently attached glucose, whereas other proteins are not adsorbed. The bound concanavalin A can subsequently be released by washing with a concentrated solution of glucose.

Immunoaffinity chromatography is an extension of this basic principle which offers to revolutionize the isolation and purification of peptides and proteins. Antibodies raised to a particular peptide or protein (antigen) are highly specific to that antigen and, if bound to a support, can be used to effect selective separation from other proteins. Until relatively recently, the application of this potentially invaluable method was severely restricted because of the considerable technical difficulties of raising homogeneous antibodies. The advances with monoclonal antibodies have resolved many of these problems and using partially purified protein,[228] the procedure required to produce and isolate specific hybridomas has become routine. The immunoglobulins from these are chemically homogeneous and directed against the specific antigenic determinant. Due to the strong interaction of the antigen with antibodies, the release of the protein can sometimes be troublesome, but methods to overcome this problem are slowly evolving.

6.3.2 *Primary structure determination of peptides and proteins*

In general, proteins are composed of the twenty principal amino acids which are coded for by the triad of bases in the nucleic acids. When other amino acids are present they are formed by post-translational events which have resulted in the modification of the principal amino acids.

With the exception of an interesting new method to be described later, the general strategy and tactics associated with protein sequencing have not changed dramatically,[229-231] although they have become more instrumented and sophisticated. Amino acid analysis and automated sequencing using modified Edman procedures are cases in point.[232] A greater range of enzymes suitable for the preparation of fragments convenient for sequence analysis, as well as enzymes for *N*-terminal analysis are now available. No significant advances have been made with chemical cleavage and cyanogen bromide remains the only notable method. The application of mass spectrometric techniques for sequencing is well established[232,234] and likely to increase in importance with the greater application of the fast atom bombardment (FABS) technique.[235,236]

The lack of significant developments contrasts sharply with the major breakthrough in DNA sequencing using the enzyme replication techniques.[234,238] As the amino acid sequence of a protein is directly related to the base sequence in the corresponding DNA molecule, it follows that the determination of the latter can be translated into a protein sequence. Where such a protein/gene relationship is established, there is no doubt that DNA sequencing is the more rapid and convenient procedure, and is likely to become more important in the future.

6.3.3   *Conformations and interactions of peptides and proteins*

The critical determinant of the biological function of a protein is its conformation which is prescribed by its amino acid sequence. To date our understanding of protein conformation has stemmed predominantly from crystallographic studies, theoretical considerations, and the data from a wide range of other physical and spectroscopic methods which have been used to explore the dynamics of solution conformation.[239] These have led, not only to an appreciation of the overall three-dimensional arrangement of the amino acid residues within molecules but also an understanding of the inherent factors within the amino acid sequence which determine the folding and interactions within the chain.[240-242]

Nuclear magnetic resonance spectroscopy is an excellent probe for recognizing molecular conformation, because it responds to individual atoms in their specific environment. Since it can be used favourably to study molecules in solution it fulfils the requirement for the investigation of intra- and intermolecular interactions. In addition it also provides considerable information on molecular dynamics.

Considerable advances in resolution and sensitivity in n.m.r. spectroscopy have been brought about by the use of superconducting magnets and the development of many instrumental techniques. These have been particularly useful for conformational assignments to small peptides, including peptide hormones as well as peptide antibiotics.[243-245] Larger molecules yield n.m.r. data of corresponding complexity, but information, especially the detection of conformational changes accompanying interactions with other molecules or changes in physical conditions can be deduced.[246-248] Data from infra-red, Raman and fluorescence[249] spectroscopy, as well as circular dichroism[250] and other spectroscopic methods, have made significant contributions to the knowledge of the solution conformation of peptides and proteins.

Much effort has centred on theoretical studies aimed at predicting the secondary structure from the amino acid sequence.[251,252] These are essentially based on energy calculations from van der Waals interactions between atoms making up an amino acid residue and between adjacent amino acid residues, as well as taking into account peptide–water interactions. Although *ab initio* prediction of conformation for large molecules is still well out of reach, good correlations with known solution conformers have been obtained for some small peptides.[253] However, the most likely advances will probably come from semi-empirical approaches which take into consideration data from crystallography, other physical methods, simplified energy calculations, and computer-assisted model building.[254]

6.3.4 *Post-translational modification of peptides and proteins*

Proteins are assembled from the twenty principal amino acids specified by the genetic code. However, over 140 amino acids or derivatives have been identified as constituents of proteins of different organisms.[255] This diversity is brought about by the post-translational modification of the principal amino acids by a variety of processes.

Many of these modifications, such as esterification of the carboxylic acid groups of aspartic and glutamic acid residues, acetylation[256] and phosphorylation[257] of the hydroxyl groups of serine and threonine, and the carboxylation of glutamic acid, are lost on hydrolysis of the protein, and their presence poses considerable difficulties with regard to their location within the structure. On the other hand, many proteins from mammalian systems contain *N*-methylated derivatives of the basic amino acids lysine, arginine and histidine,[258] also several (anomalous) amino acids including 4-hydroxyproline and 5-hydroxylysine have been observed in structural protein. All these amino acids are stable to hydrolysis and are readily detected and identified on amino acid analysis.

The complexity and significance of post-translational modification is best illustrated with reference to collagen[259] which is the major fibrous component of skin, bone, tendon and cartilage. It is unusually rich in glycine and proline and also contains 4-hydroxyproline and 5-hydroxylysine along with other modified amino acid residues. The basic triple-stranded helical structural unit is termed tropocollagen (Fig. 6.3), and its conformational stability is de-

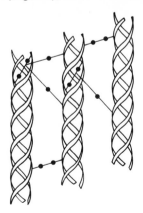

Figure 6.3 Schematic representation of triple stranded collagen demonstrating intrahelical and interhelical crosslinks.

pendent on the locking effect of the proline and hydroxyproline, as well as on the interstrand hydrogen bonding provided by the hydroxy groups of the hydroxyproline. The chain of collagen fibres is further cross-linked by interactions of specific amino acid side chains, as indicated in Schemes 10 and 11.

Elastin, which relates to collagen, is an insoluble protein in the elastic fibres of connective tissue. Like collagen, elastin is rich in proline, but contains little hydroxyproline or hydroxylysine. Elastin is synthesized as a soluble precursor which then becomes cross-linked in a variety of ways. Desmosine (**104**), one of these cross-links, is derived from the interaction of four lysine chains.[260]

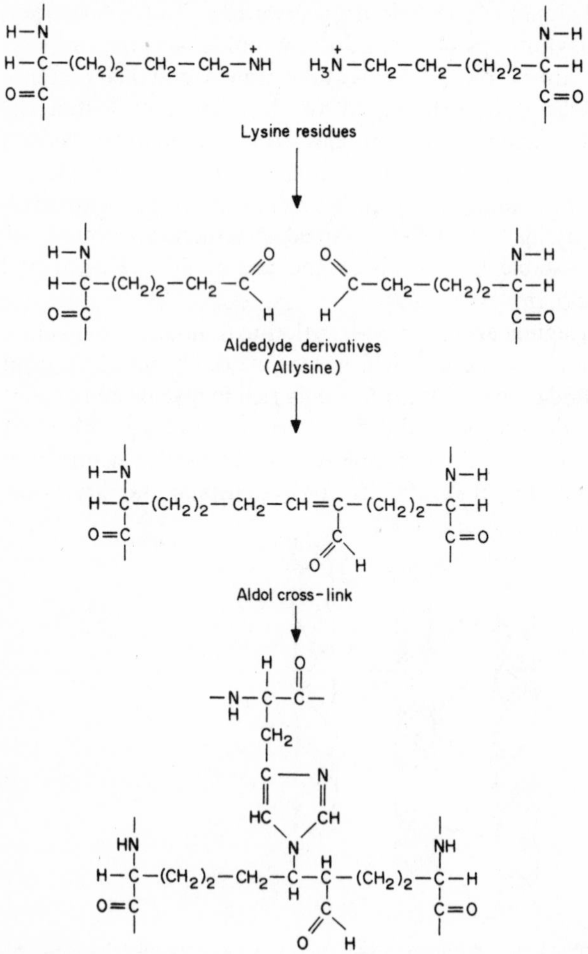

$$
\begin{array}{ll}
\text{H-N} & \text{N-H} \\
\text{H-C}-(CH_2)_2-CH_2-CH_2-\overset{+}{N}H \qquad H_3\overset{+}{N}-CH_2-CH_2-(CH_2)_2-\text{C-H} \\
\text{O=C} & \text{C=O}
\end{array}
$$

Lysine residues

$$
\begin{array}{ll}
\text{H-N} & \text{N-H} \\
\text{H-C}-(CH_2)_2-CH_2-C\overset{\diagup O}{\diagdown H} \qquad O\overset{\diagdown}{\diagup}C-CH_2-(CH_2)_2-\text{C-H} \\
\text{O=C} & \text{C=O}
\end{array}
$$

Aldedyde derivatives
(Allysine)

$$
\begin{array}{ll}
\text{H-N} & \text{N-H} \\
\text{H-C}-(CH_2)_2-CH_2-CH=C-(CH_2)_2-\text{C-H} \\
\text{O=C} \qquad\qquad\qquad\quad C \qquad\qquad\quad \text{C=O} \\
\qquad\qquad\qquad\qquad\quad O \quad H
\end{array}
$$

Aldol cross-link

$$
\begin{array}{c}
\text{H} \quad \text{O} \\
-N-C-C- \\
\text{H} \\
CH_2 \\
C - N \\
\parallel \qquad \parallel \\
HC \qquad CH \\
N \\
HN \qquad\qquad H \qquad\qquad NH \\
\text{H-C}-(CH_2)_2-CH_2-C-C-(CH_2)_2-\text{C-H} \\
\text{O=C} \qquad\qquad\quad H \quad C \qquad\qquad \text{C=O} \\
\qquad\qquad\qquad\qquad O \quad H
\end{array}
$$

Scheme 10

unique to itself. Considerable difficulties, in terms of functional group protection, racemization, and coupling procedures, to mention but a few, have had to be overcome and these endeavours have found expression in notable syntheses of innumerable peptide hormones, biologically active peptides and countless of their analogues. In addition, the formidable problems associated with the total synthesis of simple enzymes have been faced and tackled with some success.[267,268]

However, these have all been achieved within the framework of the general strategies and methodology developed in the sixties and the advances have centred on the refinement and extension of the individual steps involved in both solution and solid phase synthesis. In the latter case, considerably improved supports are now available that overcome many of the constraints associated with the polystyrene resins.

The futility of attempting to survey either the advances in methodology or the individual syntheses is highlighted by the recent publication of a major series on peptides, the first volume of which was solely occupied by methods available for peptide bond formation; other individual volumes cover solid phase synthesis, problems associated with racemization and protection of functional groups.[264] The authors here are therefore opting out and pointing those readers interested in this direction, while for those completely unfamiliar with the field, an earlier monograph[270] and a review article[271] are still probably the best introductions.

However, it does seem appropriate at this stage to view the current status of chemical synthesis against the background of the remarkable developments in molecular biology. Already recombinant DNA technology has yielded the commercial production of human insulin[272] and it is not unreasonable to suppose that within the next decade cloning techniques will improve dramatically, so that they are likely to be the methods of choice for producing peptides with twenty or more amino acid residues. These will not be limited to peptides corresponding to natural fragments of DNA, since the synthesis of polynucleotides of defined base sequence appears to be fraught with fewer difficulties than the synthesis of the peptide.[273,274] Furthermore, because the micro-organism will replicate innumerable copies it will only be necessary to undertake the synthesis once!

These prospects do not forecast the demise of peptide synthesis, far from it, but there will inevitably be a shift in direction and emphasis. The value of chemical methods for the total synthesis of small peptides[267] such as oxytocin, vasopressin, etc., will probably remain unchallenged. Also the growing application of semi-synthesis[268] for which once again human insulin is a prominent example. But it is likely that the real potential will lie in the development of syntheses of atypical peptides which are likely to remain out of the reach of biotechnology for some time.

References

1. H.R.V. Arnstein (ed.), 1975, M.T.P. Int. Rev. Science, Biochemistry Series/Vol. 7, 'Synthesis of Amino Acids and Proteins', Butterworths, London.
2. B.J. Miflin, 1980, in 'Biochemistry of Plants', ed. P.K. Stumpf, Academic Press, New York, Vol. 5.
3. P.J. Lea, 1978, *Int. Rev. Biochem.*, 18, 79.
4. O. Ciferri, 1975, in 'Chemistry and Biochemistry of Plant Proteins', ed. J.B. Harborne and C.F. van Sumere, Academic Press, London, p. 113.
5. B.V. Milborrow (ed.), 1973, 'Biosynthesis and its Control in Plants', Academic Press, London, p. 21, 49, 323.
6. Specialist Periodical Report, 'Amino-Acids, Peptides and Proteins', series editor R.C. Sheppard, The Royal Society of Chemistry, London.
7. L. Beevers, 1982, in 'Nucleic Acids and Proteins in Plants', Vol. 1, 'Structure, Biochemistry and Physiology of Proteins', ed. D. Boulter and B. Parthier, Springer, Berlin, p. 136.
8. M. Friedman, 1977. 'Advances in Experimental Medicine and Biology', Vols. 86A, 86B, Plenum Press, New York.
9. K. Nakanishi, N. Furutachi, M. Funamizu, D. Grunberger and I.B. Weinstein, 1970, *J. Amer. Chem. Soc.*, 92, 7617.
10. M. Funamizu, A. Terahara, A.M. Feinberg and K. Nakanishi, 1971, *J. Amer. Chem. Soc.*, 93, 6706.
11. J. Stenflo, P. Fernlund, W. Egan and P. Roepstorff, 1974, *Proc. Natl. Acad. Sci., USA*, 71, 2730.
12. M.R. Christy, R.M. Barkley, T.H. Koch, J.J. van Buskirk and W.M. Kirsch, 1981, *J. Amer. Chem. Soc.*, 103, 3935.
13. P. Fernlund, J. Stenflo, P. Roepstorff and J. Thomson, 1975, *J. Biol. Chem.*, 250, 6125.
14. P.V. Hauschka, P.A. Friedman, H.P. Traverso and P.M. Gallop, 1976, *Biochem. Biophys. Res. Comm.*, 71, 1207.
15. J.J. van Buskirk and W.M. Kirsch, 1978, *Biochem. Biophys. Res. Comm.*, 82, 1329.
16. S. Danishefsky, E. Berman, L.A. Clizbe and M. Hirama, 1979, *J. Amer. Chem. Soc.*, 101, 4385.
17. G.A. Smith, K.A. Smith and D.H. Williams, 1975, *J. Chem. Soc. Perkin I*, 2108.
18. B.W. Bycroft, D. Cameron and A.W. Johnson, 1971, *J. Chem. Soc. C.*, 3040.
19. A. Okura, H. Morishima, T. Takita, T. Aoyagi, T. Takeuchi and H. Umezama, 1975, *J. Antibiot.*, 28, 337.
20. E. Bayer, K.H. Gugel, K. Hagale, H. Hagenmaier, S. Jessipow, W.A. Konig and H. Zaher, 1972, *Helv. Chim. Acta*, 55, 224.
21. C. Wasielewski and K. Antczak, 1981, *Synthesis*, 540.
22. H. Seto, T. Sasaki, S. Imai, T. Tsuruoka, A. Satoh, M. Kojima, S. Inouye, T. Niida and N. Otake, 1983, *J. Antibiot.*, 36, 96.
23. J.P. Scannell, D.L. Pruess, T.C. Demny, H.A. Ax, F. Weiss, T. Williams and A. Stempel, 1972, in 'Chemistry and Biology of Peptides', Proceedings of the 3rd American Peptide Symposium, ed. J. Meienhofer, Ann Arbor Science Publishers, Michigan, p. 415.
24. L.O. Zamir, R.A. Jensen, B.H. Arison, A.W. Douglas, G. Albers-Schonberg and J.R. Bowen, 1980, *J. Amer. Chem. Soc.*, 102, 4499.
25. A.M. Fazel and R.A. Jensen, 1979, *J. Bacteriol.*, 140, 580.
26. J.L. Rubin and R.A. Jensen, 1979, *Plant Physiol.*, 64, 727.
27. J.P. Scannell and D.L. Preuss, 1974, in 'Chemistry and Biochemistry of Amino Acids, Peptides and Proteins', Vol. 3, ed. B. Weinstein, Dekker, New York, p. 189.
28. E.A. Bell, 1980, in 'Encyclopaedia of Plant Physiology', Vol. 9 (Secondary Plant Products), New Series, eds. E.A. Bell and B.V. Charlwood, Springer, Berlin, p. 403.
29. Y. Okami, T. Kitahara, M. Hamada, H. Naganawa, S. Kondo, K. Maeda, T. Takeuchi and H. Umezawa, 1974, *J. Antibiot.*, 27, 656.
30. T. Aoyagi, T. Yamamoto, K. Kojiri, F. Kojima, M. Hamada, T. Takeuchi and H. Umezawa, 1978, *J. Antibiot.*, 31, 244.

31. E.O. Stapley, D. Hendlin, M. Jackson and A.K. Miller, 1971, *J. Antibiot.*, 24, 42.
32. United States Patent, 1980, 4202819; *Chem. Abs.* 1980, 93, 130567.
33. Y. Kuroda, M. Okuhara, T. Goto, M. Kohsaka, H. Aoki and H. Imanaka, 1980, *J. Antibiot.*, 33, 132.
34. L.D. Owens, J.F. Thompson, R.G. Pitcher and T. Williams, 1972, *J. Chem. Soc. Chem. Comm.*, 714.
35. C. Walsh, 1982, *Tetrahedron*, 38, 871.
36. D.L. Pruess, J.P. Scannell, M. Kellett, H.A. Ax, J. Janacek, T.H. Williams, A. Stempel and J. Berger, 1972, *J. Antibiot.*, 27, 229.
37. D.D. Keith, J.A. Tortora, K. Ineichen and W. Leimgruber, 1975, *Tetrahedron*, 31, 2633.
38. D.D. Keith, R. Yang, J. Tortora and M. Weigele, (1978), *J. Org. Chem.*, 43, 3711.
39. T. Wieland, 1968, *Science*, 159, 946.
40. H. Iwasaki, T. Kamiya, O. Oka and J. Ueyanagi, 1965, *Chem. Pharm. Bull.*, 13, 753.
41. D.G. Martin, D.J. Duchamp and C.G. Chidester, 1973, *Tetrahedron Lett.*, 2549.
42. Pan Pei-Chuan, Fang Sheng-Din and Tsai Chun-chao, 1976, *Scientia Sinica*, 19, 691.
43. J.C. Watkins, 1978, in 'Kainic acid as a Tool in Neurobiology', ed. E.G. McGeer, Raven Press, New York, p. 37.
44. Y. Ueno, H. Nawa, J. Ueganagi, H. Morimoto, R. Nakamori and T. Matsuoka, 1955, *J. Pharmacol. Sci. Japan*, 75, 807.
45. T. Takemoto and K. Draigo, 1960, *Arch. Pharmacol.*, 293, 627.
46. H. Shinozaki and I. Shibuya, 1976, *Neuropharmacol.*, 15, 145.
47. K. Konno, H. Shirahama and T. Matsumoto, 1983, *Tetrahedron Lett.*, 24, 939.
48. A. Weller, J. Jadot, G. Dardenne, M. Marlier and J. Cashmir, 1975, *Phytochemistry*, 14, 1347; E.I. Mbadiwe, 1975, *Phytochemistry*, 14, 1351.
49. T. Takemoto, K. Nomoto, S. Fushiya, R. Ouchi, K. Kusano, H. Hikino, S. Takegi, Y. Matsuura and M. Kakuda, 1978, *Proc. Japan Acad. Sci., Ser. B*, 54, 469.
50. M. Noguchi, H. Sakuma and E. Tamaki, 1968, *Phytochemistry*, 7, 1861.
51. T. Nakajima, Y. Matsuoka and Y. Kakimoto, 1971, *Biochim. Biophys. Acta*, 230, 212.
52. J.L. Irvine and D.W. Wilson, 1939, *J. Biol. Chem.*, 127, 555.
53. T. Shiba, H. Mizote, T. Kaneko, T. Nakajima, Y. Kakimoto and I. Sano, 1971, *Biochem. Biophys. Acta*, 244, 523.
54. P.M. Hardy, 1979, 'Comprehensive Organic Chemistry', Vol. 5, eds. D. Barton and W.D. Ollis, Pergamon, London, p. 193.
55. J.P. Greenstein, M. Winitz, 1961, 'Chemistry of the Amino Acids', Vol. 1, Wiley, New York, p. 728, T. Barth and H. Maskova, 1971, *Coll. Czech. Chem. Comm.*, 36, 2398.
56. A. Collet, M.-J. Brienne and J. Jaques, 1980, *Chem. Rev.*, 80, 215; G. Glaschke, 1980, *Angew. Chem. (Int. Ed. Engl.)*, 19, 13.
57. J.D. Morrison and H.S. Mosher, 'Asymmetric Organic Reactions', Prentice Hall, New Jersey, 1971, p. 297.
58. E.J. Corey, R.J. McCaully and H.S. Sachdev, 1970, *J. Amer. Chem. Soc.*, 92, 2476.
59. B.W. Bycroft and G.R. Lee, 1975, *J. Chem. Soc. Chem. Comm.*, 988.
60. N. Izumiya, S. Lee, T. Kanmera and H. Aoyagi, 1977, *J. Amer. Chem. Soc.*, 99, 8346.
61. S. Alexandratos, 1981, in 'Optical Resolution Procedures for Chemical Compounds', ed. P. Newman, Vol. 2, p. 1085.
62. M.D. Fryzuk and B. Bosnich, 1978, *J. Amer. Chem. Soc.*, 100, 5491.
63. German Patent 1972, 2, 210, 938; *Chem. Abs.*, 77, 165073.
64. J.W. ApSimon and R.P. Sanguin, 1979, *Tetrahedron*, 35, 2797.
65. D. Valentine and J.W. Scott, 1978, *Synthesis*, 329.
66. R.B. Silverman and M.W. Hollowday, 1981, *J. Amer. Chem. Soc.*, 103, 7357.
67. W. Oppolzer and H. Andres, 1979, *Helv. Chim. Acta*, 62, 2282.
68. A. Niederweiser and G. Pataki, 1971, 'New Techniques in Amino Acid, Peptide and Protein Analysis', Ann Arbor Science Publ., Michigan.
69. J.W. Payne and J.M. Tuffnell, 1980, in 'Microorganisms and Nitrogen Sources', ed. J.W. Payne, Wiley, England, p. 727.

70. J.M. Rattenbury (ed.) 'Amino Acid Analysis (Symposium) 1979', Publ. 1981, Horwood, England.
71. R.J. Cole and R.H. Cox, eds., 1981, 'Handbook of Toxic Fungal Metabolites', Academic Press, New York.
72. P.S. Ringrose, 1980, in 'Microorganisms and Nitrogen Sources', ed. J.W. Payne, Wiley, England, p. 641.
73. P.M. Hardy, 1982, in 'Amino Acids, Peptides and Proteins', R.S.C. London, Vol. 13, p. 293, and earlier volumes.
74. M. Barber, R.S. Bordeli, R.D. Sedgwick and A.N. Taylor, 1981, *J. Chem. Soc. Chem. Comm.*, 325.
75. M. Barber, R.S. Bordoli, G.J. Elliott, R.D. Sedgwick, A.N. Taylor and B.N. Green, 1982, *J. Chem. Soc. Chem. Comm.*, 936.
76. J.L. Gower, 1983, *Int. J. Mass Spectrometry Ion Physics*, 46, 431.
77. M. Bodansky and D. Perlman, 1964, *Nature*, 204, 840.
78. F. Lipman, 1980, *Advances in Microbial Physiol.*, 21, 227.
79. K. Kurahashi, S. Komura, K. Akashi and C. Nishio, 1982, in 'Peptide Antibiotics, Biosynthesis and Function', eds. H. Kleinkauf and H. von Dohren, Walter de Gruyter, Berlin, p. 275.
80. H. Kleinkauf, 1979, *Planta Medica*, 35, 1.
81. P.S. Farmer, 1980, in 'Drug Design', E.J. Ariens (ed.), Academic Press, New York, Vol. 10, p. 119.
82. J.S. Morley, 1980, *Trends in Pharmacol. Sci.*, 1, 463.
83. D.H. Williams, S. Santikarn, P.B. Delrichs, F. de Angelis, J.K. McLeod and R.J. Smith, 1982, *J. Chem. Soc. Chem. Comm.*, 1394.
84. K. Suzukake, T. Fujiyama, H. Hayashi and M. Hori, 1979, *J. Antibiot.*, 32, 523.
85. H. Umezawa, M. Ishizuka, T. Aoyagi and T. Takeuchi, 1976, *J. Antibiot.*, 29, 857.
86. A. Meister, 1975, in 'Metabolic Pathways', ed. D.M. Greenburg, 3rd edition, Vol. 7, Academic Press, London, p. 101.
87. A. Meister and S.S. Tate, 1976, *Ann. Rev. Biochem.*, 45, 559.
88. A. Meister, 1973, *Science*, 180, 33.
89. T. Kasei and P.O. Larsen, 1980, *Fortschr. Chem. Org. Naturst.*, 39, 173.
90. Y. Aoyagi, T. Suguhara, T. Hasegawa and T. Suzuki, 1982, *Agric. Biol. Chem.*, 46, 987, 1939.
91. G. Höfle, R. Gmelin, H.-H. Luxa, M. N'Galamulume-Treves and S.I. Hatanaka, *Tetrahedron Lett.*, 1976, 3129.
92. H.R.V. Arnstein, M. Artman, D. Morris and E.J. Toms, 1960, *Biochem. J.*, 76, 353.
93. W.W. Stewart, 1971, *Nature*, 229, 174.
94. J.P. Scannell, D.L. Pruess, J.F. Blount, H.A. Ax, M. Kellett, F. Weiss, T.C. Demny, T.H. Williams and A. Stempel, 1975, *J. Antibiot.*, 21, 1.
95. L. Leive (ed.), 1973, 'Bacterial Membranes and Walls', Marcel Dekker, New York.
96. D.J. Tipper and A. Wright, 1978, in 'The Bacteria', ed. J. Sokatch, Academic Press, New York, Vol. 7.
97. P. Lefrancier, J. Choay, M. Derrien and I. Lederman, 1977, *Int. J. Peptide Protein Res.*, 9, 249.
98. K. Hemmi, H. Takeno, S. Okada, S. Nakaguchi, Y. Kitaura and M. Hashimoto, 1981, *J. Amer. Chem. Soc.*, 103, 7026.
99. J.L. van der Baan, J.W.F.K. Barnick and F. Bickelhaupt, 1978, *Tetrahedron*, 34, 223.
100. D. Schipper, J.L. van der Baan and F. Bickelhaupt, 1979, *J. Chem. Soc., Perkin I*, 2017.
101. R. Kazlauskas, R.O. Lidgard, R.J. Wells and W. Vetter, 1977, *Tetrahedron Lett.*, 3183.
102. Y. Mori, M. Tsuboi, M. Suzuki, K. Fukushima and T. Arai, 1982, *J. Chem. Soc. Chem. Comm.*, 94.
103. P.G. Sammes, 1975, *Fortschr. Chem. Org. Naturst.*, 32, 51.
104. B.W. Bycroft, 1979, in 'Comprehensive Organic Chemistry', eds. D. Barton and W.D. Ollis, Pergamon, London, Vol. 5, p. 259.
105. M.J.O. Anteunis, 1978, *Bull. Soc. Chem. Belg.*, 87, 627.

106. L.C. Vining and J.L.C. Wright, 1977, in The Chemical Society Specialist Periodical Report; Biosynthesis 5 ed. J.D. Bu'Lock, p. 246.
107. K. Katinuma and K.L. Rinehart, 1974, *J. Antibiot.*, 27, 733.
108. A.J. Birch, R.A. Massey-Westropp and R.W. Richards, 1956, *J. Chem. Soc.*, 3717.
109. C.L. Atkin and J.B. Neilands, 1968, *Biochemistry*, 7, 3734.
110. C.M. Maes, P.S. Steyn, P.H. van Rooyen and C.J. Rabie, 1982, *J. Chem. Soc., Chem. Comm.*, 350.
111. B.H. Arison and J.L. Beck, 1973, *Tetrahedron*, 29, 2743.
112. R. Kazlauskas, P.T. Murphy and R.J. Wells, 1978, *Tetrahedron Lett.*, 4945.
113. T. Kamiya, S. Maeno, M. Hashimoto and Y. Milne, 1972, *J. Antibiot.*, 25, 576.
114. I.R. Shimi, N. Abdalla and S. Fathey, 1977, *Antimicrob. Agent Chemother.*, 11, 373.
115. J.P. Ferezou, A. Quesneau, M. Barbier, A. Kollmann and J.F. Bousquet, 1980, *J. Chem. Soc. Perkin I*, 113.
116. H.A. Akers, M. L'inas and J.B. Neilands, 1972, *Biochemistry*, 11, 2283.
117. P.S. Steyn, 1973, *Tetrahedron*, 29, 107.
118. A. Dossena, R. Marchelli and A. Pochini, 1974, *J. Chem. Soc. Chem. Comm.*, 771.
119. P.M. Scott, M.A. Merrien and J. Polonski, 1976, *Experientia*, 32, 140.
120. A.J. Birch and J.J. Wright, 1970, *Tetrahedron*, 26, 2329.
121. A.J. Birch, G.E. Blance, S. David and H. Smith, 1961, *J. Chem. Soc.*, 3128.
122. G. Casnati, A. Pochini and R. Ungaro, 1973, *Gazzeta*, 103, 141.
123. N. Eickmann, J. Clardy, R.J. Cole and J.W. Kirksey, 1975, *Tetrahedron Lett.*, 105.
124. A. Stoll and A. Hoffman, 1965, in 'The Alkaloids', R.H.F. Manske (ed.), Academic Press, New York, Vol. VIII, p. 725.
125. M.L. Bianchi, N.C. Perellino, B. Gioia and A. Minghetti, 1982, *J. Nat. Prod.*, 45, 191.
126. R. Brunner, P.L. Stutz, H. Tscherter and P.A. Stadler, 1979, *Can. J. Chem.*, 57, 1638.
127. D.C. Horwell, 1980, *Tetrahedron*, 36, 3123.
128. P. Stutz, R. Brunner and P.A. Stadler, 1973, *Experientia*, 29, 936.
129. J. Stuchlik, A. Krojicek, L. Cvak, J. Spacil, P. Sedmera, M. Flieger, J. Vakoun and Z. Rehacek, 1982, *Coll. Czech. Chem. Comm.*, 47, 3312.
130. M.R. Bell, J.R. Johnson, B.S. Wildi and R.B. Woodward, 1958, *J. Amer. Chem. Soc.*, 80, 1001.
131. J. Fridrichsons and A. McL. Mathieson, 1965, *Acta Cryst.*, 18, 1043.
132. E. Francis, R. Rahman, S. Safe and A. Taylor, 1972, *J. Chem. Soc. Perkin I*, 470.
133. R. Nagarajan, L.L. Huckstep, D.H. Lively, D.C. DeLong, M.M. Marsh and N. Nuess, 1968, *J. Amer. Chem. Soc.*, 90, 2980.
134. H. Minato, M. Matsumoto and T. Katayama, 1973, *J. Chem. Soc., Perkin I*, 1819.
135. A.G. McInnes, A. Taylor and J.A. Walter, 1976, *J. Amer. Chem. Soc.*, 98, 6741.
136. G. Deffieux, M. Gadret, J.M. Leger and A. Carpy, 1979, *Acta Cryst.*, B35, 2358.
137. G. Deffieux, M.J. Filleau and R. Baute, 1978, *J. Antibiot.*, 31, 1106.
138. R. Rahman, S. Safe and A. Taylor, 1978, *J. Chem. Soc. Perkin I*, 1476.
139. R.D. Stipanovic and C.R. Howell, 1982, *J. Antibiot.*, 35, 1326.
140. T. Fukuyama and Y. Kishi, 1976, *J. Amer. Chem. Soc.*, 98, 6723.
141. T. Wieland and C. Birr, 1976, in 'International Review of Science', Organic Chemistry, Series Two, ed. D.H. Hey, Butterworths, London, Vol. 6, p. 183.
142. F.A. Bovey, A.I. Brewster, D.J. Patel, A.E. Tonelli and D.A. Torchia, 1972, *Acc. Chem. Res.*, 5, 193.
143. Yu. A. Ovchinnikov and V.T. Ivanov, 1974, *Tetrahedron*, 30, 1871.
144. C.M. Deber, V. Madison and E.R. Blont, 1976, *Acc. Chem. Res.*, 9, 106.
145. W.L. Meyer, G.E. Templeton, C.I. Grable, R. Jones, L.F. Kuyper, R.B. Lewis, C.W. Sigel and S.H. Woodhead, 1975, *J. Amer. Chem. Soc.*, 97, 3802.
146. J.L. Flippen and I.L. Karle, 1976, *Biopolymers*, 15, 1081.
147. M. Bodansky and G.L. Stahl, 1974, *Proc. Natl. Acad. Sci. U.S.A.*, 71, 2791.
148. B.W. Bycroft, 1972, *J. Chem. Soc., Chem. Comm.*, 660.
149. S. Nomoto, T. Teshima, T. Wakamiya and T. Shiba, 1978, *Tetrahedron*, 34, 921.
150. S. Nomoto and T. Shiba, 1979, *Bull. Chem. Soc. Japan*, 52, 540.

151. H. Yoshioka, T. Aoki, H. Goko, K. Nakatsu, T. Noda, H. Sakakibara, T. Take, A. Nagata, J. Abe, T. Nakamiya, T. Shiba and T. Kaneko, 1971, *Tetrahedron Lett.*, 2043.
152. R.M. Stroud, 1973, *Acta Cryst.*, B29, 677.
153. W. Keller-Schierlein, V. Prelog and H. Zahner, 1964, *Fortschr. Chem. Org. Naturst.*, 22, 279.
154. J.B. Neilands, 1973, in 'Inorganic Biochemistry', G.L. Eichorn, ed., Elsevier, New York, Vol. 1, p. 167.
155. M.C. Stephenson and C. Ratledge, 1980, in 'Microorganisms and Nitrogen Sources', J.W. Payne, ed., Wiley, London, p. 693.
156. J.B. Neilands and C. Ratledge, 1980, in 'C.R.C. Handbook of Microbiology', 2nd ed., A.I. Laskin and H.A. Lechavalier, eds., C.R.C. Press, Vol. 2.
157. P.S. Steyn, A.A. Tuinman, F.R. van Heerden, P.H. van Rooyen, P.L. Wessels and C.J. Rabie, 1983, *J. Chem. Soc., Chem. Comm.*, 47.
158. T. Wieland and O.P. Wieland, 1972, in 'Microbial Toxins', A. Ciegler, S.J. Ayl and S.K. Adis, eds., Academic Press, New York, Vol. VIII, p. 248.
159. T. Wieland and H. Faulstich, 1978, *Crit. Rev. Biochem.*, 5, 185.
160. A. Hadjiolov, M.D. Dabeva and V.V. Mackedonski, 1974, *Biochem. J.*, 138, 321.
161. P.D. Thut and T.J. Lindell, 1974, *Mol. Pharmacol.*, 10, 146; L. Fiume and G. Barbanti, 1974, *Experientia*, 30, 76.
162. C.M. Ireland, A.R. Dursa, R.A. Newman and M.P. Hacker, 1982, *J. Org. Chem.*, 47, 1807.
163. T. Wieland, J. Faesel and W. Konz, 1969, *Annalen*, 722, 197.
164. Yu. A. Ovchinnikov, V.T. Ivanov, V.F. Bystrov and A.I. Miroshnikov, 1972, in 'The Chemistry and Biology of Peptides', ed. J. Meienhofer, Ann Arbor Science, Ann Arbor, p. 111.
165. H. Faulstich and T.H. Wieland, 1973, in 'The Peptides', ed. H. Hanson and H.D. Jakubke, Elsevier, Amsterdam, p. 312.
166. T. Wieland, H. Faulstich and W. Burgermeister, 1972, *Biochem. Biophys. Res. Comm.*, 47, 984.
167. L.K. Ramachandran, 1975, *Biochem. Rev.*, 46, 1.
168. D.C. Hodgkin and B.M. Oughton, 1957, *Biochem. J.*, 65, 752.
169. S.E. Hull, R. Karlsson, P. Main, M.M. Woolfson and E.J. Dodson, 1978, *Nature*, 275, 206.
170. M.M. Shemyakin, N.A. Aldanova, E.I. Vinogradova and M. Yu. Feigina, 1963, *Tetrahedron Lett.*, 1921.
171. R.L. Hamil, C.E. Higgins, H.E. Broz and M. Gorman, 1969, *Tetrahedron Lett.*, 4255.
172. E.P. Bakker, 1979, in 'Antibiotics', F.E. Hahn, ed., Springer, Berlin, Vol. V (I), p. 67.
173. E.F. Gale, E. Cundliffe, P.E. Reynolds, M.H. Richmond and M.J. Waring, 1981, 'Molecular Bases of Antibiotic Action', 2nd edn., Wiley, London, p. 226.
174. G.D. Smith, W.L. Duax, D.A. Langs, G.T. DeTitta, J.W. Edmonds, D.C. Rohrer and C.M. Weeks, 1975, *J. Amer. Chem. Soc.*, 97, 7242.
175. K. Neupert-Laves and M. Dobler, 1975, *Helv. Chim. Acta*, 58, 432.
176. D.J. Patel and A.E. Tonnelli, 1973, *Biochemistry*, 12, 486.
177. J. Meienhofer and E. Atherton, 1977, in 'Structure Activity Relations among Semisynthetic Antibiotics', D. Perlmann, ed., Academic Press, New York, p. 427.
178. H. Brockmann, 1960, *Fortschr. Chem. Org. Naturstoff*, 18, 1.
179. U. Hollstein, 1974, *Chem. Rev.*, 74, 625.
180. J.V. Formica and M.A. Apple, 1976, *Antimicrob. Agent Chemother.*, 9, 214.
181. H.M. Sobell, S.C. Jain, T.D. Sakore, C.E. Nordman, 1972, *J. Mol. Biol.*, 68, 1.
182. M.J. Waring, 1979, in 'Antibiotics', ed. F.E. Hahn, Springer, Berlin, Vol. V, 2, p. 173.
183. G.D. Martin, S.A. Mizsak, C. Biles, J.C. Stewart, L. Baczynskyj and P.A. Meulmann, 1975, *J. Antibiot.*, 28, 332.
184. H. Otsuka, J. Shoji, K. Kawano and Y. Kyogoko, 1976, *J. Antibiot.*, 29, 107.
185. A. Dell, D.H. Williams, H.R. Morris, G.A. Smith, J. Feeney and G.C.K. Roberts, 1975, *J. Amer. Chem. Soc.*, 97, 2497.
186. J.R. Kalman, T.J. Blake, D.H. Williams, J. Feeney and G.C.K. Roberts, 1979, *J. Chem. Soc., Perkin I*, 1313.
187. C. Cocito, 1979, *Microbiol. Rev.*, 43, 145.

188. E. Gross, H.H. Kiltz and E. Nebellin, 1973, *Z. Physiol. Chem.*, 354, 810.
189. E. Gross and J.L. Morell, 1971, *J. Amer. Chem. Soc.*, 93, 810.
190. L. Ingram, 1970, *Biochem. Biophys. Acta*, 224, 263.
191. R.C. Pandey, J.C. Cook and K.L. Rinehart, 1977, *J. Amer. Chem. Soc.*, 99, 8469.
192. R.C. Pandey, J.C. Cook and K.L. Rinehart, 1977, *J. Amer. Chem. Soc.*, 99, 5203.
193. R.C. Pandey, J.C. Cook and K.L. Rinehart, 1977, *J. Antibiot.*, 31, 241.
194. B. Anderson, D.C. Hodgkin and M.A. Viswamitra, 1970, *Nature*, 225, 233.
195. A. Olesker, L. Valente, L. Barata, G. Lukacs, W.E. Hull, K. Tori, K. Tokura, K. Okabe, M. Ebata and H. Otsuka, 1978, *J. Chem. Soc. Chem. Comm.*, 577.
196. K.L. Tori, K. Tokura, Y. Yoshimura, K. Okabe, H. Otsuka, F. Inagaki and T. Miyazawa, 1979, *J. Antibiot.*, 32, 1072.
197. B.W. Bycroft and M.S. Gowland, 1978, *J. Chem. Soc. Chem. Comm.*, 256.
198. O.D. Hensens and G. Albers-Schonberg, 1978, *Tetrahedron Lett.*, 3648.
199. T. Endo and H. Yonehara, 1978, *J. Antibiot.*, 31, 623.
200. L.J. Pearce and K.L. Rinehart, 1979, *J. Amer. Chem. Soc.*, 101, 5069.
201. T. Takita, Y. Muraoka, T. Nakatani, A. Fuji, Y. Umezawa, H. Naganawa and H. Umezawa, 1978, *J. Antibiot.*, 31, 801.
202. T. Takita, Y. Umezawa, S. Aito, H. Morishima, H. Naganawa, M. Umezawa, T. Tsuchiya, T. Miyake, S. Kageyama, S. Umezawa, Y. Muraoka, M. Suzuki, M. Otsuka, M. Narita, S. Kobayashi and M. Ohno, 1982, *Tetrahedron Lett.*, 23, 521.
203. Y. Aoyagi, K. Katano, H. Suguna, J. Primeau, L.-H. Chang and S.M. Hecht, 1982, *J. Amer. Chem. Soc.*, 104, 5537.
204. S.K. Carter, S.T. Crooke and H. Umezawa, eds., 1978, 'Bleomycin, Current Status and New Developments', Academic Press, New York.
205. D.H. Williams, V. Rajananda, M.P. Williamson and G. Bojeson, 1980, in 'Topics in Antibiotic Chemistry', ed. P. Sammes, Ellis Horwood, England, Vol. 5, p. 119.
206. C.M. Harris and T.M. Harris, 1982, *J. Amer. Chem. Soc.*, 104, 293.
207. C.M. Harris and T.M. Harris, 1982, *J. Amer. Chem. Soc.*, 104, 363.
208. S.D. Jolad, J.J. Hoffmann, S.J. Torrance, R.M. Wiedhopf, J.R. Cole, S.K. Arora, R.B. Bates, R.L. Gargiulo and G.R. Kriek, 1977, *J. Amer. Chem. Soc.*, 99, 8040.
209. R. Tschesche and E.U. Kaussmann, 1975, in 'The Alkaloids', R.H.F. Manske (ed.), Academic Press, New York, Vol. 15, p. 165.
210. R.B. Morin and M. Gorman, eds., 1982, 'Chemistry and Biology of β-Lactam Antibiotics', Vols. 1–3, Academic Press, New York.
211. E.H. Flynn, ed., 1972, 'Cephalosporins and Penicillins', Academic Press, New York.
212. P. Sammes, ed., 1980, 'Topics in Antibiotic Chemistry', Ellis Horwood, England, Vol. 4.
213. E.O. Stapley, 1981, in 'β-Lactam Antibiotics', M. Salton and G.D. Shockman, eds., Academic Press, New York, p. 327.
214. J. O'Sullivan and E.P. Abraham, 1980, *Biochem. J.*, 186, 613.
215. R.D.G. Cooper, 1979, in 'Topics in Antibiotic Chemistry', P.G. Sammes, ed., Ellis Horwood, Vol. 3, p. 39.
216. S.W. Elson, R.S. Oliver, B.W. Bycroft and E.A. Faruk, 1982, *J. Antibiot.*, 35, 81.
217. M. Cole, 1980, *Phil. Trans., Royal Soc., London*, B289, 207.
218. A. Imada, K. Kitano, K. Kintaka, M. Muroi and M. Asai, 1981, *Nature*, 289, 590.
219. J.S. Wells, W.H. Trejo, P.A. Principe, K. Bush, N. Georgopapadakou, D.P. Bonner and R.P. Sykes, 1982, *J. Antibiot.*, 35, 295 and 300.
220. W.L. Parker, W.H. Koster, C.M. Cimarusti, D.M. Floyd, W.-C. Liu and M.L. Rathnum, 1982, *J. Antibiot.*, 35, 189.
221. 'The Proteins', 1976 onwards, eds. H. Neurath and R.L. Hill, Academic Press, New York.
222. 'Enzyme Structure', 1970 onwards, eds. C.H. Hirs and S.N. Timasheff, Academic Press, New York.
223. *Annual Review of Biochemistry*, Annual Reviews Inc., Palo Alto.
224. *Advances in Protein Chemistry*, Academic Press, New York.
225. L. Stryer, 1981, 'Biochemistry', W.H. Freeman and Co., San Francisco.
226. T.G. Cooper. 1977, 'The Tools of Biochemistry', Wiley, London.

227. J. Turkova, 1978, 'Affinity Chromatography', Elsevier, Amsterdam.
228. C. Milstein, 1980, *Sci. Amer.*, 243, 66.
229. W.H. Konigsberg and H.M. Steinman, 1977, in 'The Proteins', eds. H. Neurath and R.L. Hill, Academic Press, New York, Vol. 3, p. 1.
230. 'Advanced Methods in Protein Sequence Determination', 1977, ed. S.B. Needleman, Springer-Verlag, Heidelberg.
231. L.R. Croft, 1980, 'Introduction to Protein Sequence Analysis', Wiley, London.
232. H. Niall, 1977, in 'The Proteins', eds. H. Neurath and R.L. Hill, Academic Press, New York.
233. H.G. Khorana, G.E. Gerber, W.C. Herlihy, C.P. Gray, R.J. Anderegg, K. Nihei and K. Biemann, 1976, *Proc. Nat. Acad. Sci. U.S.A.*, 73, 5046.
234. A.D. Auffret, T.J. Blake and D.H. Williams, 1981, *Eur. J. Biochem.*, 113, 333.
235. M. Barber, R.S. Bordoli, R.D. Sedgwick and A.N. Tyler, 1981, *Nature*, 293, 270.
236. D.H. Williams, C. Bradley, G. Bojesen, S. Santikarn and L.C.E. Taylor, 1981, *J. Am. Chem. Soc.*, 103, 5700.
237. A.M. Maxam and W. Gilbert, 1977, *Proc. Nat. Acad. Sci.*, 74, 560.
238. F. Sanger, S. Nicklen and A.R. Coulson, 1977, *Proc. Nat. Acad. Sci.*, 74, 5463.
239. G.E. Schulz and R.H. Schirmer, 1979, 'Principles of Protein Structure,' Springer-Verlag, Heidelberg.
240. R.L. Baldwin, 1975, *Ann. Rev. Biochem.*, 44, 453.
241. M. Karplus and D.L. Weaver, 1976, *Nature*, 260, 404.
242. P.Y. Chou and C.D. Fasman, 1978, *Ann. Rev. Biochem.*, 47, 251.
243. Yu. A. Ovchinnikov and V.T. Ivanov, 1975, *Tetrahedron*, 31, 2177.
244. C.M. Deber, V. Madison and E.R. Blout, 1976, *Acc. Chem. Res.*, 9, 106.
245. H. Kessler, 1982, *Angew. Chem.*, 21, 512.
246. W. Egan, H. Shindo and J.S. Cohen, 1977, *Ann. Rev. Biophys. Bioeng.*, 6, 383.
247. 'Biological Magnetic Resonance', eds. I.J. Berliner and J. Reuben, Plenum Press, London, 1978.
248. 'Biological Applications of Magnetic Resonance', ed. R.G. Shulman, Academic Press, New York, 1979.
249. W.C. Galley and J.G. Milton, 1979, *Photochem. Photobiol.*, 29, 179.
250. P.C. Kahn, 1979, *Meth. Enzymol.*, 61, 339.
251. G.D. Rose, 1978, *Nature*, 272, 586.
252. J.D. Kuntz, G.M. Crippen and P.A. Kollman, 1979, *Biopolymers*, 18, 939.
253. D.W. Weatherford and F.R. Salemme, 1979, *Proc. Nat. Acad. Sci. U.S.A.*, 76, 19.
254. A.C.T. North, A.K. Denson, A.C. Evans, L.O. Ford and T.V. Willoughby, 1980, in 'Biomolecular Structure, Conformation, Function and Evolution', ed. R. Srinivasan, Pergamon, Vol. 1, 59.
255. R. Uy and F. Wold, 1977, *Science*, 198, 891.
256. I. Isenberg, 1979, *Ann. Rev. Biochem.*, 48, 159.
257. C.S. Ruben and C.M. Rosen, 1975, *Ann. Rev. Biochem.*, 44, 831.
258. W.K. Paik and S. Kim, 1975, *Adv. Enzymol.*, 42, 227.
259. D.R. Eyre, 1980, *Science*, 207, 1315.
260. L.B. Sandberg, 1976, *Int. Rev. Connect. Res.*, 7, 159.
261. S.J. Chan and D.F. Steiner, 1977, *Trends in Biochem. Sci.*, 2, 254.
262. R. Burgus, M. Butcher, M. Amoss, N. Ling, M. Monahan, J. Rivier, R. Fellows, R. Blackwell, W. Vale and R. Guillemin, 1972, *Proc. Nat. Acad. Sci. U.S.A.*, 69, 278.
263. *See* D.M. Moran and A.J. Garman, 1981, in 'Amino Acids, Peptides and Proteins', Vol. 12, 150, and the corresponding section in earlier volumes, ed. R.C. Sheppard, R.C.S., London.
264. B.S. Cooperman, 1978, in 'Bioorganic Chemistry', ed. E.E. van Tamelen, Academic Press, New York.
265. V. Chowdry and S.H. Westheimer, 1979, *Ann. Rev. Biochem.*, 48, 293.
266. M. Das and C.F. Fox, 1979, *Ann. Rev. Biophys. Bioeng.*, 8, 165.
267. G.R. Pettit, 1976, 'Synthetic Peptides', Vols. 1–5, Elsevier, Amsterdam.
268. 'Semisynthetic Peptides and Proteins', 1978, eds. R.E. Offord and C. di Bello, Academic Press, London.

269. 'The Peptides', 1981, eds. E. Gross and J. Meienhofer, Vols. 1–4, Academic Press, New York.
270. 'Peptide Synthesis', 1976, eds. M. Bodanszky, Y.S. Krausner and M.A. Ondetti, John Wiley, London.
271. R.C. Sheppard, 1978, in 'Comprehensive Organic Chemistry', eds. D.H. Barton and W.D. Ollis, Pergamon Press, Vol. 5, 321.
272. L. Villa-Komaroff, A. Efstratiadis, S. Broome, P. Lomedico, R. Tizard, S.P. Naber, W.L. Chick and W. Gilbert, 1978, *Proc. Nat. Acad. Sci.*, 73, 3727.
273. R. Crea, A. Kraszewski, T. Hirose and K. Itakura, 1978, *Proc. Nat. Acad. Sci.*, 75, 5765.
274. H.G. Khorana, 1979, *Science*, 203, 614.

*Structure 56

	R	R^1	R^2
Ergotamine	Me	CH_2Ph	α-H
Ergotaminine	Me	CH_2Ph	β-H
Ergosine	Me	CH_2CHMe_2	α-H
Ergosinine	Me	CH_2CHMe_2	β-H
Ergostine	Et	CH_2Ph	α-H
Ergostinine	Et	CH_2Ph	β-H
Ergocristine	$CHMe_2$	CH_2Ph	α-H
Ergocristinine	$CHMe_2$	CH_2Ph	β-H
Ergocryptine	$CHMe_2$	CH_2CHMe_2	α-H
Ergocryptinine	$CHMe_2$	CH_2CHMe_2	β-H
Ergocornine	$CHMe_2$	$CHMe_2$	α-H
Ergocorninine	$CHMe_2$	$CHMe_2$	β-H
Ergobutyrine	Et	Et	α-H
Ergobutine	$CHMe_2$	Et	α-H
Ergovaline	Me	$CHMe_2$	α-H
Ergoptine	Et	CH_2CHMe_2	α-H
Ergonine	Et	$CHMe_2$	α-H

*Structure 60

	R^1	R^2	R^3	R^4	Ref.
Viomycin	$H_2N(CH_2)_3CH(NH_2)CH_2CO$	OH	OH	OH	148
Capreomycin IA	H	OH	X	H	149
IB	H	H	X	H	
IIA	H	OH	X	NH_2	150
IIB	H	H	X	NH_2	
Tuberactinomycin A	$H_2N(CH_2)_2CH(OH)CH(NH_2)CH_2CO$	OH	OH	OH	151
B	$H_2N(CH_2)_3CH(NH_2)CH_2CO$	OH	OH	OH	
N	$H_2N(CH_2)_2CH(OH)CH(NH_2)CH_2CO$	OH	OH	H	
O	$H_2N(CH_2)_3CH(NH_2)CH_2CO$	OH	OH	H	152
X	$= NHCOCH_2CH(NH_2)(CH_2)_3NH_2$				

7 Alkaloids

I.R.C. BICK

The term alkaloid was formerly restricted to bases of plant origin, but more recently it has been widened to include substances that occur in animals and micro-organisms, and it now comprises many compounds with only feeble basic properties. A number of textbooks[1,2,3] are available which give a general background to modern alkaloid chemistry, and the developments during the last decade have been well surveyed in two comprehensive series of reviews.[4,5] There has been a great deal of activity in the alkaloid field during this period, particularly as regards the application of increasingly sophisticated methods for structural determination, and in the areas of synthesis and biosynthesis. Although the latter aspect will not be dealt with directly in this survey, biogenetic concepts nevertheless provide a convenient framework for classifying, under a comparatively few simple headings, the daunting array of alkaloid structures now known. These headings correspond to the units, for the most part simple amino acids, from which alkaloids are considered to be derived in nature; in some cases the derivation has been proved experimentally, in others it is still uncertain.

7.1 Alkaloids derived from ornithine

These alkaloids generally have a pyrrolidine ring incorporated into their structures, and in some recent examples this ring is attached to a nucleus derived from some different source, as in the case of the flavonoid base phyllospadine (1).[6] Other examples of this sort, like brevicolline (76) and macrostomine (34), appear under subsequent headings.

Until recently the Proteaceae were not known to produce alkaloids, but over forty bases have now been isolated from Australian and New Caledonian members of this family.[7] Apart from simple pyrrolidine derivatives such as

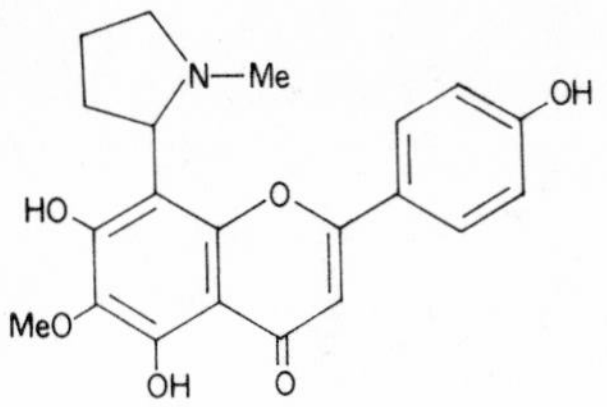

(1) Phyllospadine

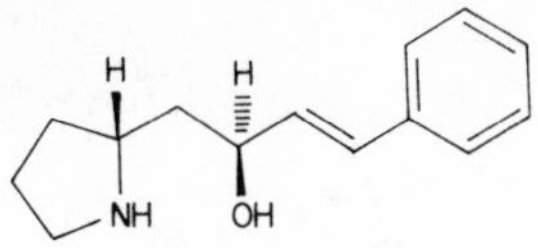

(2) Darlinine

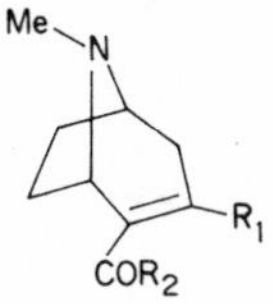

(3) Ferruginine, R_1=H, R_2=Me

(4) Chalcostrobamine, R_1 = OH,
     R_2 = CH = CHPh

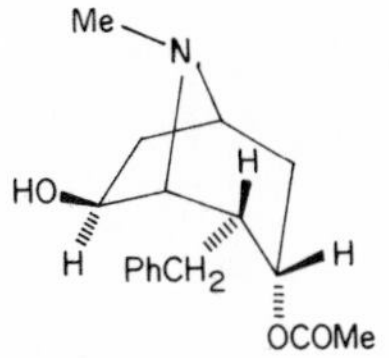

(5) Knightoline

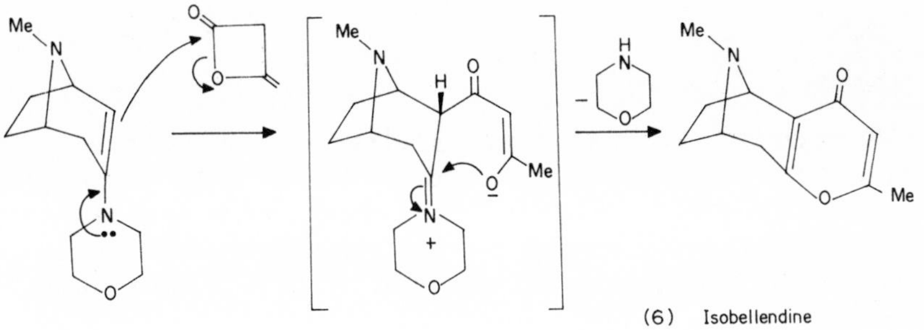

(6) Isobellendine

Scheme 1

darlinine (2),[8] these plants produce a wide range of tropanes, mostly with unusual substituents, such as ferruginine (3),[9] chalcostrobamine (4)[10] and knightoline (5).[10] The most interesting, however, have a fused γ-pyrone ring like isobellendine (6).[11] The structures, deduced mainly from spectroscopy, have in some cases been confirmed by X-ray crystallography and by synthesis. An effective route to isobellendine starts from an enamine of tropinone (Scheme 1).[12]

The tropane alkaloids in general continue to attract attention by reason of their interesting chemical and pharmacological properties. An intriguing hydride shift involved in the reaction of 3α-tropanol (7), benzoyl chloride and alkali has been studied[13] with 3α-deuteriotropan-3β-ol and 3β-deuteriotropan-3α-ol. Only the latter gives the reaction which involves deuterium transfer (Scheme 2).

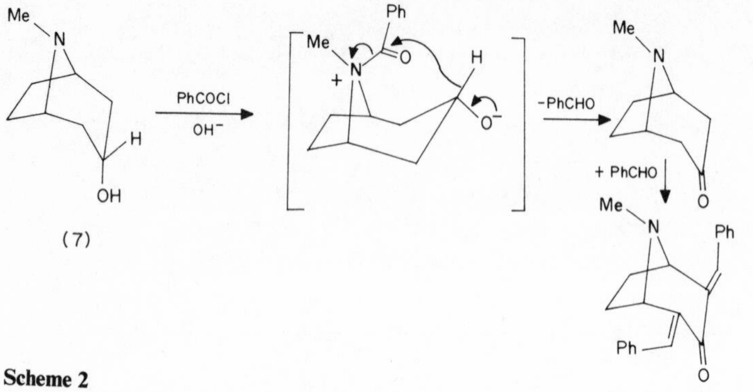

(7)

Scheme 2

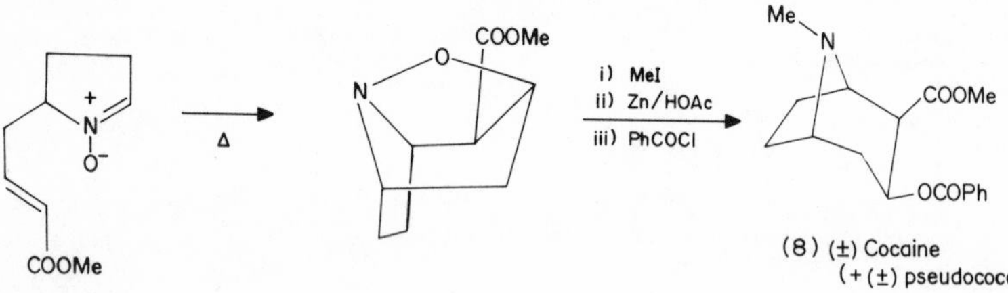

(8) (±) Cocaine
(+ (±) pseudococaine)

Scheme 3

Some novel approaches to tropane synthesis have appeared recently: the nitrone-based route, for instance, has been used in a stereospecific synthesis of racemic cocaine (**8**)[14] (Scheme 3). Another strategy involves a condensation, promoted by transition-metal carbonyls, of pyrroles with tetrabromoacetone (Scheme 4).[15] The tropenol (**9**) so formed can be reduced to tropine, or epoxidized to give scopine.

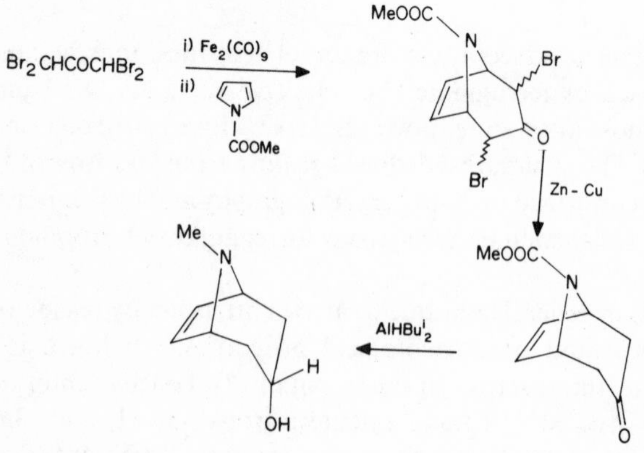

(9)

Scheme 4

Several bases have been reported with structures analogous to the tropanes, such as the carbinolamine physoperuvine (10),[16] and the homotropane anatoxin-a (11),[17] a neurotoxin isomeric with ferruginine (3) that is produced by certain algae. Their structures were deduced by spectroscopy and by X-ray crystallography, and a partial synthesis of anatoxin-a starting from (−)cocaine has confirmed its structure and stereochemistry.[17]

Interest in the pharmacology, particularly the hepatoxicity, of the pyrrolizidine alkaloids has stimulated a good deal of activity in the isolation of new alkaloids, in their structural determination—often quite involved because

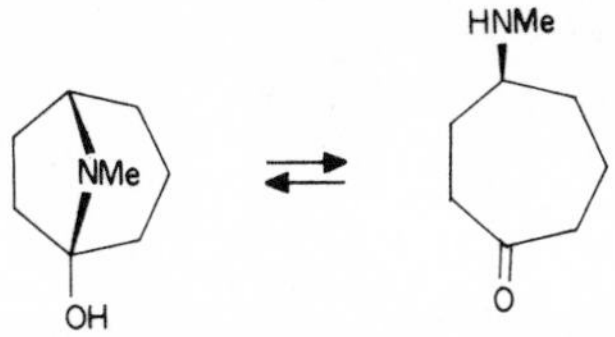

(10) Physoperuvine

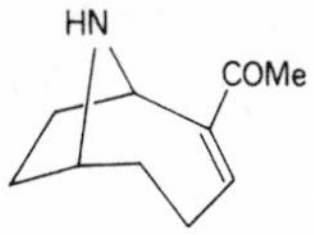

(11) Anatoxin-a

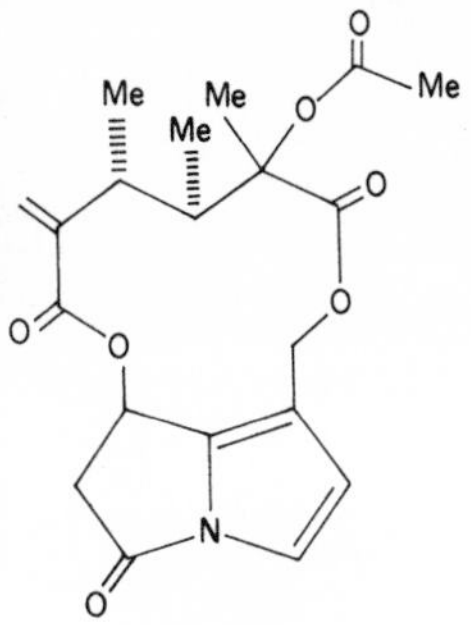

(12) Pterophorine

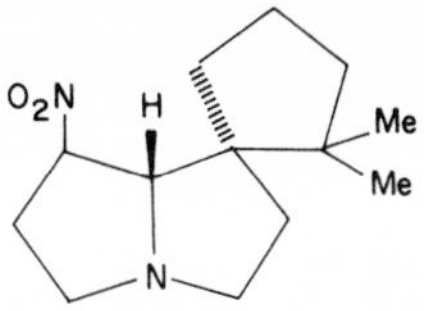

(13) Nitropolizonamine

of the complex necic acids present—and in their synthesis. Some of the examples reported recently, such as pterophorine (12),[18] have unusual structural features; another novel pyrrolizidine, nitropolizonamine (13), has been found in the defensive secretions of millipedes.[19]

The use of pyrrolizidines of plant origin by certain butterflies as sex pheromones is now well-established, and a subsidiary use may be for defence against predators.[20] A brief and efficient synthesis[21] of the butterfly pheromone (14) (Scheme 5) can be modified to give a range of pyrrolizidine derivatives. Another ingenious synthetic approach is shown in Scheme 6: a mixture of diastereomeric esters is obtained, which can be separated and reduced with LAH to the corresponding pyrrolizidine alcohols isoretronecanol and trachelanthamidine.[22]

Many alkaloids, like nicotine (15), have some extra amino acid besides ornithine in their make-up. A remarkably simple synthesis of nicotine along

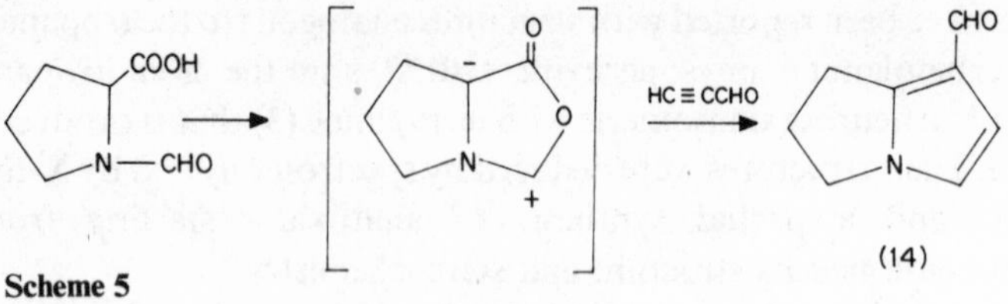

Scheme 5

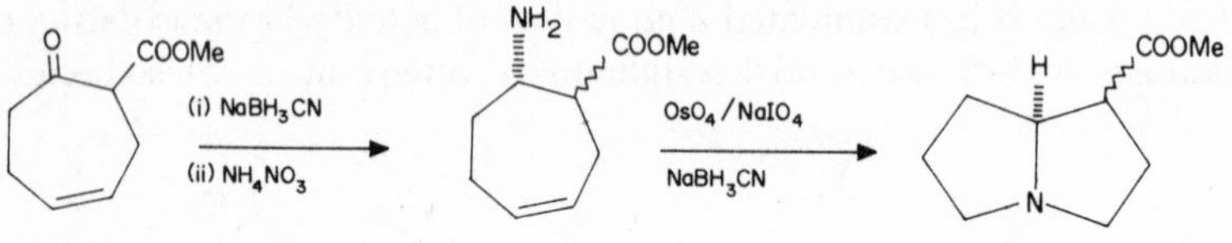

(14)

Scheme 6

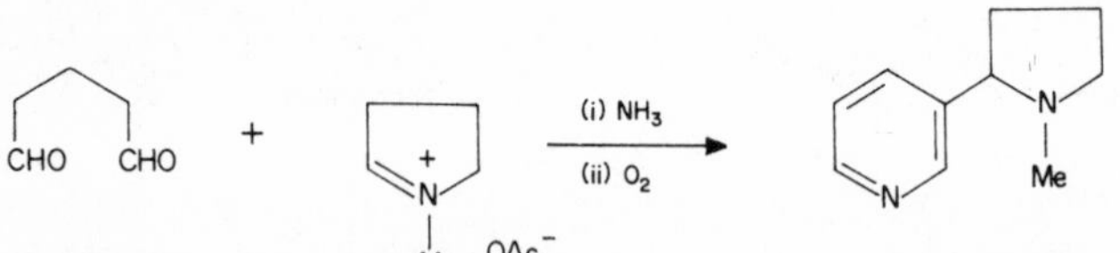

(15) Nicotine

Scheme 7

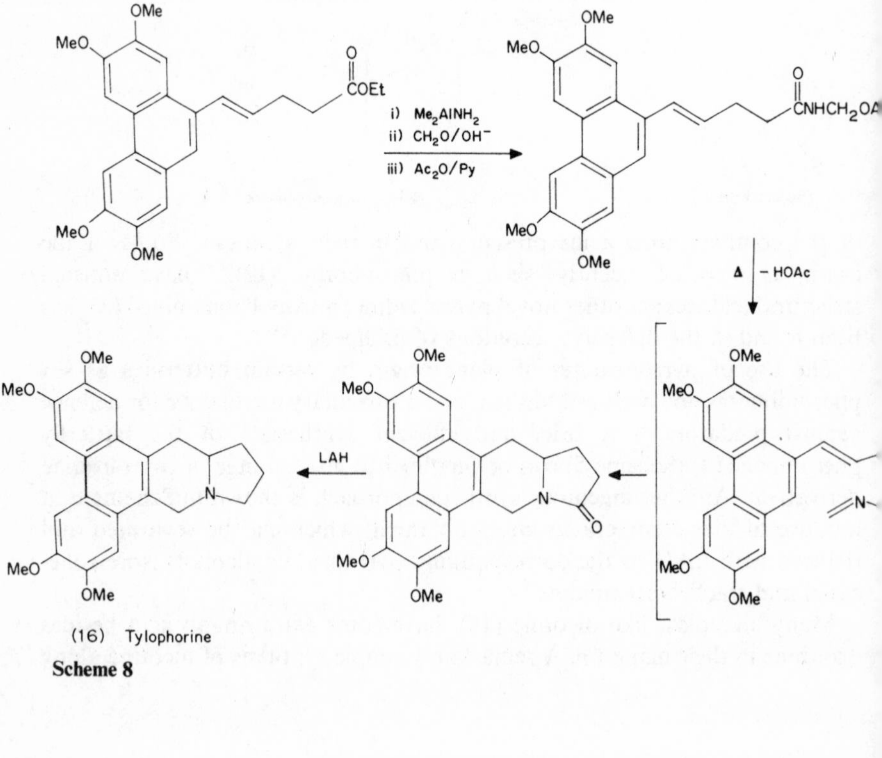

(16) Tylophorine

Scheme 8

biomimetic lines has been achieved in fair yield (Scheme 7).[23] The structures of the phenanthroindolizidine alkaloids incorporate units derived from tyrosine and phenylalanine as well as ornithine. In spite of their limited numbers, they have attracted attention because of their highly toxic and vesicant properties; some of them in addition have anti-tumour activity. The synthesis of tylophorine (16) shown in Scheme 8 employs an imino Diels-Alder reaction to form both fused heterocyclic rings simultaneously.[24] Alkaloids related to the phenanthroindolizidines include vincetene (17),[25] which occurs in the same plant as tylophorine.

7.2 Alkaloids derived from lysine

Alkaloids in this category commonly have a piperidine or quinolizidine ring, and some of the simpler ones are counterparts of the pyrrolidines or pyrrolizidines derived from ornithine, like anabasine, the analogue and isomer of nicotine. This well-known alkaloid has attracted attention lately as an anti-smoking agent, and also because of its stimulating effect on respiratory muscle. The alkaloid hoveine (18), which has marked hypotensive activity, has two tetrahydroanabasine units incorporated into its structure.[26]

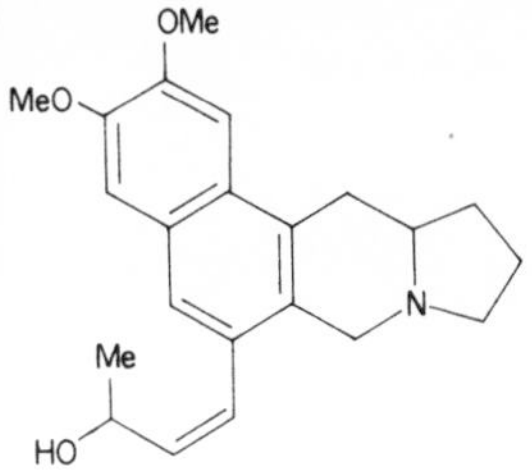

(17) Vincetene

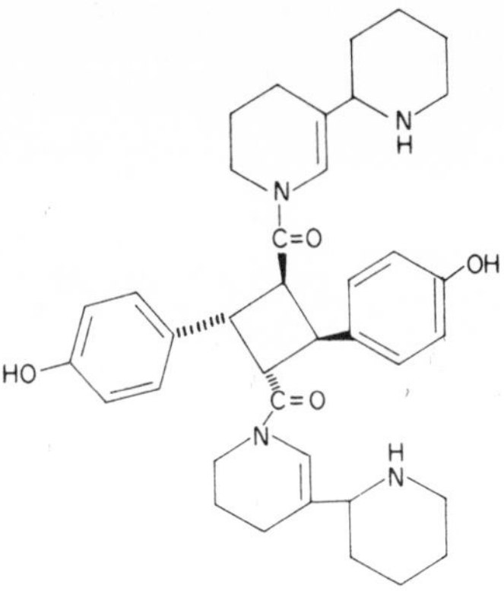

(18) Hoveine

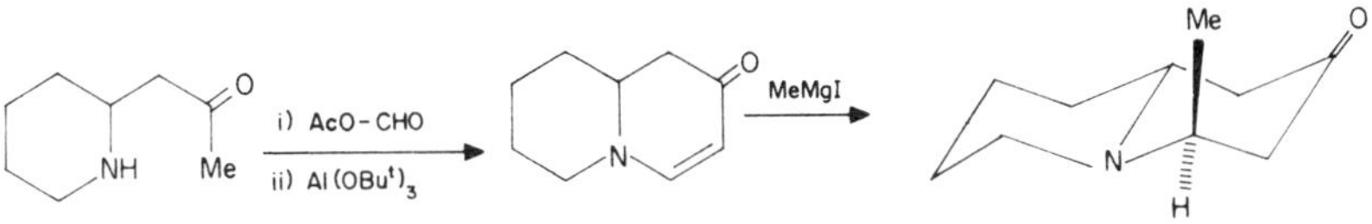

(20) Isopelletierine

(19) Myrtine

Scheme 9

Myrtine (**19**) is an example of a simple quinolizidine alkaloid from the recent literature. Its structure was deduced by spectroscopy, and confirmed by a synthesis[27] that starts from the well-known pomegranate alkaloid isopelletierine (**20**) (Scheme 9). The more complex quinolizidine alkaloids have two or more lysine units in their constitution, such as the *Lycopodium* alkaloids, which have been studied particularly by Canadian chemists. A number of new structural types, both simple and complex, have been found recently: the simplest is phlegmarine (**21**), whose structure was deduced largely from mass and MIKE spectroscopy.[28] It is considered to be a likely stage in the biosynthesis of the more complex *Lycopodium* alkaloids from lysine via

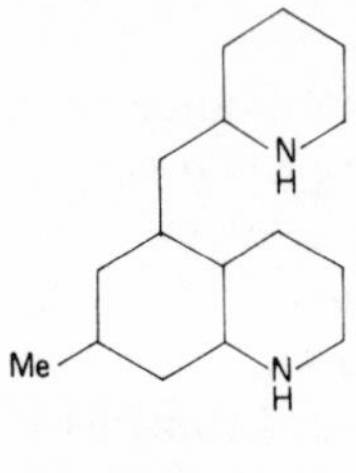

(21) Phlegmarine

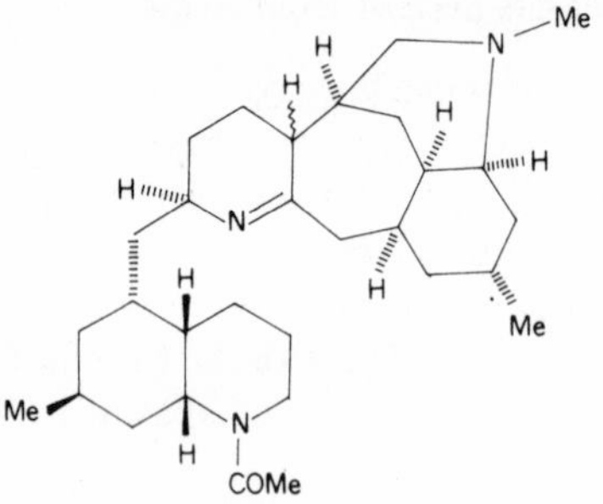

(22) Lucidine B

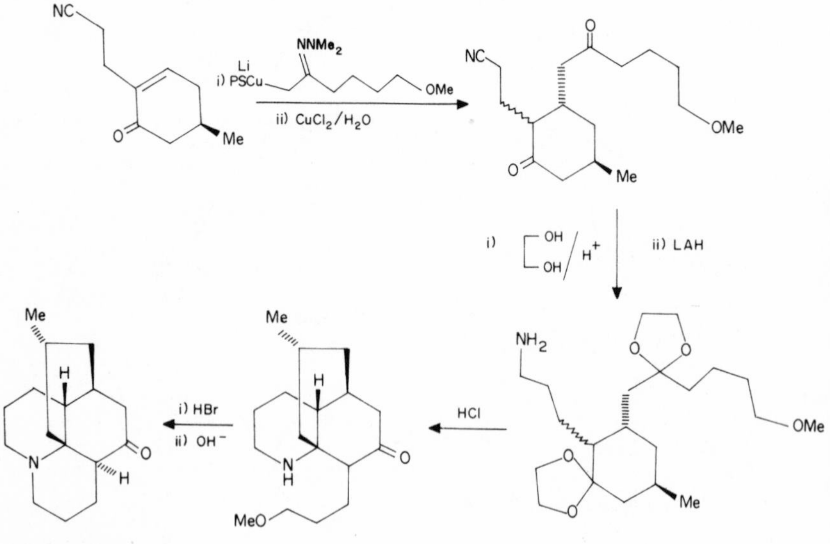

(23) Lycopodine

Scheme 10

isopelletierine (**20**). Some of these, such as lucidine B (**22**), appear to have an extra unit of the isopelletierine type incorporated into their molecules; the structure and stereochemistry of (**22**) have been determined by X-ray crystallography on a lucidine B derivative.[29]

Some elegant syntheses of *Lycopodium* alkaloids have been devised recently. The method shown in Scheme 10 for the principal alkaloid, lycopodine (**23**), can be modified to produce various others.[30] In the synthesis of luciduline (**24**) (Scheme 11), an alkaloid that appears to have only one lysine

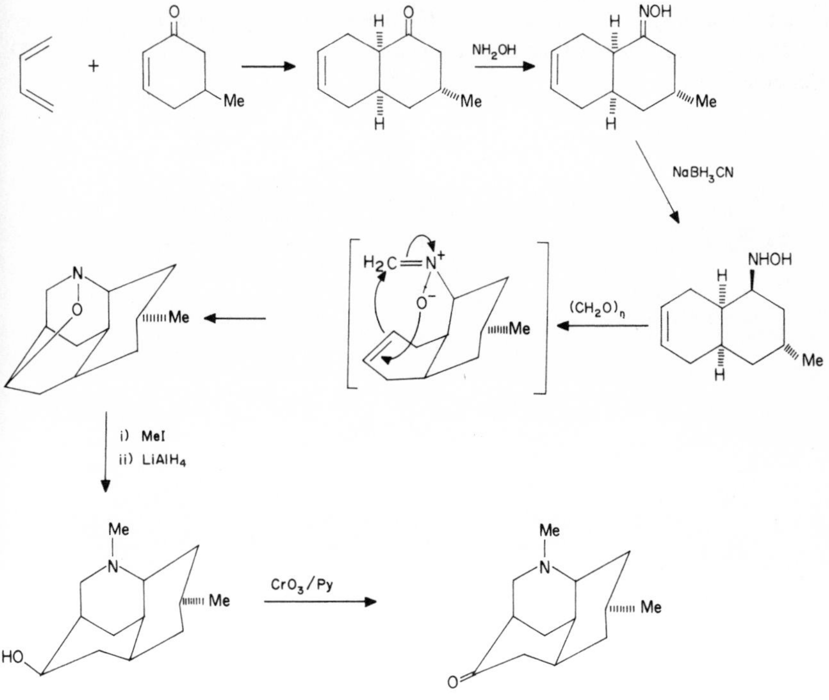

(24) Luciduline

Scheme 11

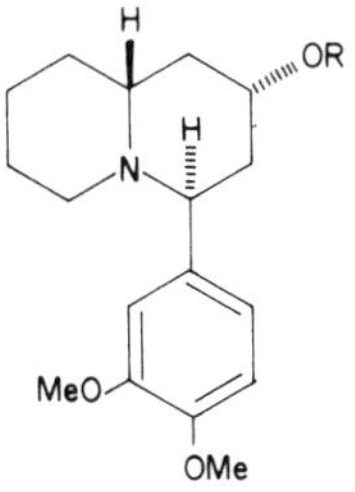

(25) Lasubine – I (R = H)

(26) Subcosine – I (R = COCH $\stackrel{E}{=}$ CH— with OMe, OMe substituents)

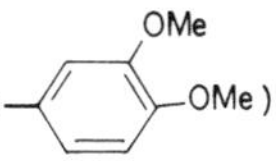

unit in its make-up, an intramolecular addition of a nitrone to a double bond is employed.[31]

Among the numerous alkaloids with a quinolizidine nucleus are those produced by the plant family Lythraceae. Recently some simple examples with a phenylquinolizidine structure have been reported from this source, such as lasubine-I **(25)**;[32] it is accompanied by subcosine-I, in which the hydroxyl is esterified by a substituted cinnamic acid residue **(26)**.[32] The majority of lythraceous alkaloids also have an extra unit of this kind, which is, however, usually attached to the benzene ring by a diphenyl or a diphenyl ether link so as to form a macro ring. There has lately been considerable synthetic activity directed towards the preparation of certain of these bases such as decaline **(27)** (Scheme 12).[33]

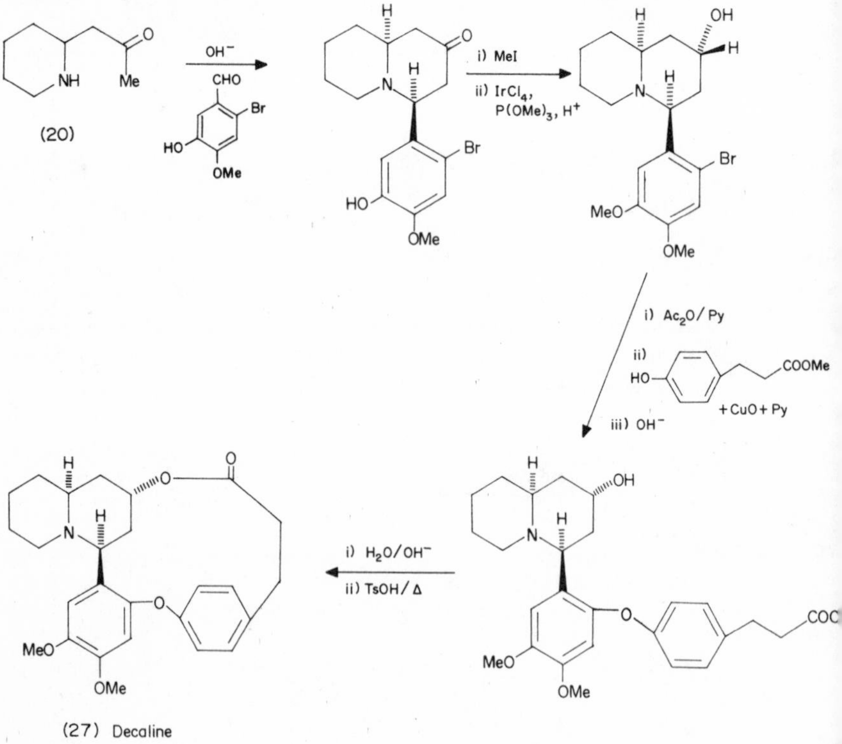

(27) Decaline

Scheme 12

The family Leguminosae produces a variety of quinolizidine alkaloids, the nuclei of which appear to be composed entirely of lysine units. The number of known alkaloids, some of which have turned up in other plant families, has increased considerably during the last ten years, and bases with new types of

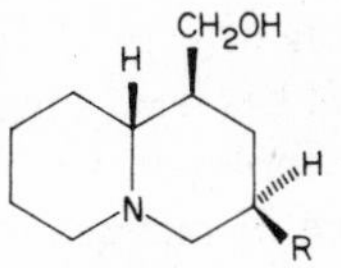

(29) Epilupinine (R = H)
(28) Pohakuline

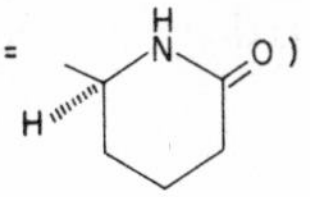

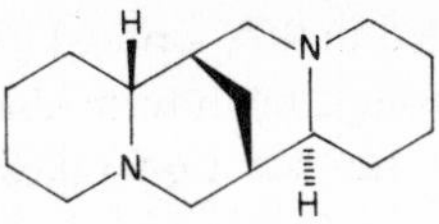

(30) Sparteine

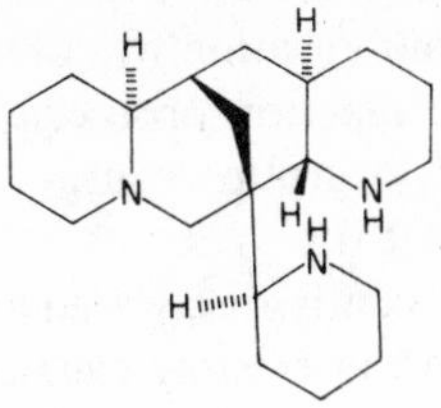

(31) Ormosanine

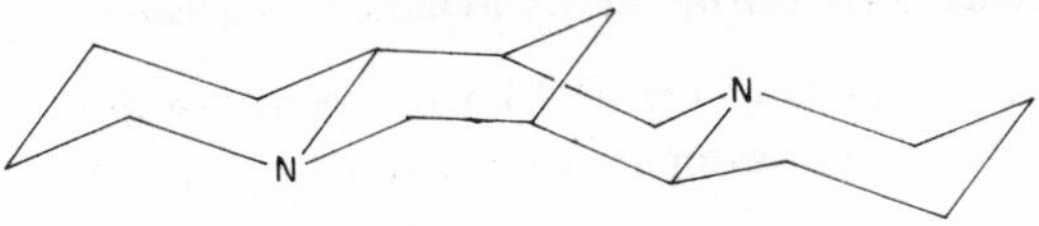

(R =)

(32) Sparteine, preferred conformation

structures have been found, including pohakuline (28).[34] This alkaloid bears an obvious structural relationship not only to the simple lupin bases like epilupinine (29), but also to the more complex sparteine (30) and ormosanine (31) types composed of three and four lysine units respectively. These latter alkaloids have received a good deal of attention from synthetic chemists, but apart from this, sparteine, although it is one of the most widely distributed and best known of leguminous alkaloids, still attracts interest by reason of its pharmacology and other aspects of its chemistry, especially its stereochemistry. A careful series of spectroscopic and X-ray crystallographic studies have been made on sparteine and on its analogues and derivatives, particularly its N-oxides, of which four are known.[35] It appears that in solution, the preferred conformation of (−)sparteine is as shown in (32) with one ring in the boat form;[36] however, the presence of other substances can alter this: in particular, sparteine is an excellent bidentate ligand for metals such as magnesium and zinc, and it can induce partial asymmetric synthesis in the Grignard and Reformatsky reactions.[37] Thus when benzaldehyde, zinc and

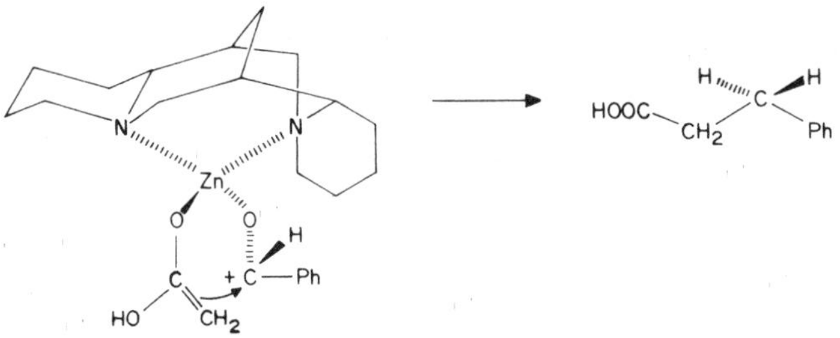

Scheme 13

ethyl bromoacetate are reacted in the presence of (−)sparteine, a product with the S-configuration was obtained in 95% optical yield. It has been suggested that the reaction proceeds through an intermediate complex in which the bulky aryl group ensures that the reaction takes place in one sense only (Scheme 13).[37]

The recently-discovered alkaloid tsukushinamine-A (33) proved to have a novel and interesting cage-type structure, which was deduced spectroscopically and confirmed by X-ray crystallography.[38]

7.3 Alkaloids derived from phenylalanine or tyrosine

A number of variants of the well-known benzylisoquinoline structure have been found recently; examples are provided by macrostomine (34),[39] which

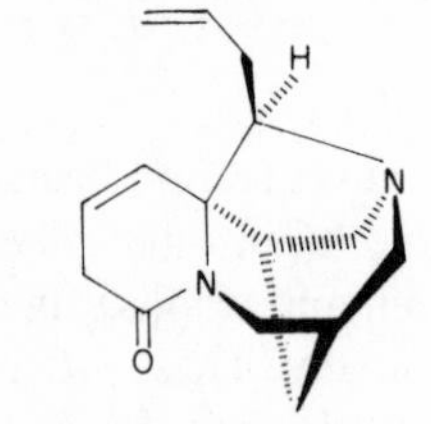

(33) Tsukushinamine A

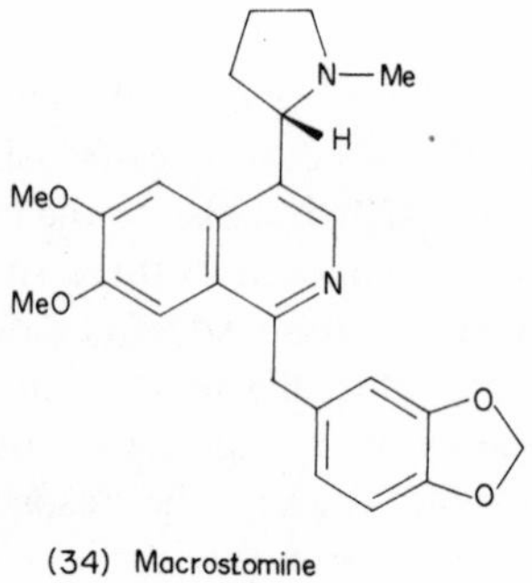

(34) Macrostomine

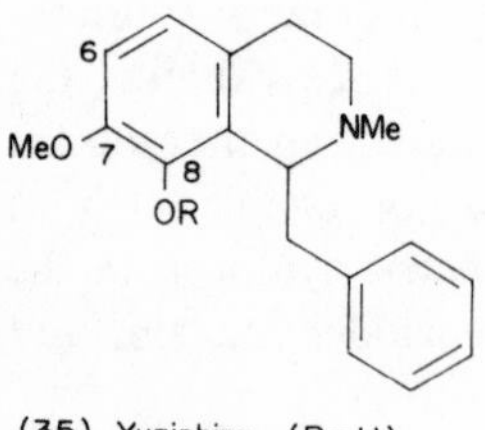

(35) Yuziphine (R = H)
(36) Longifolidine (R = Me)

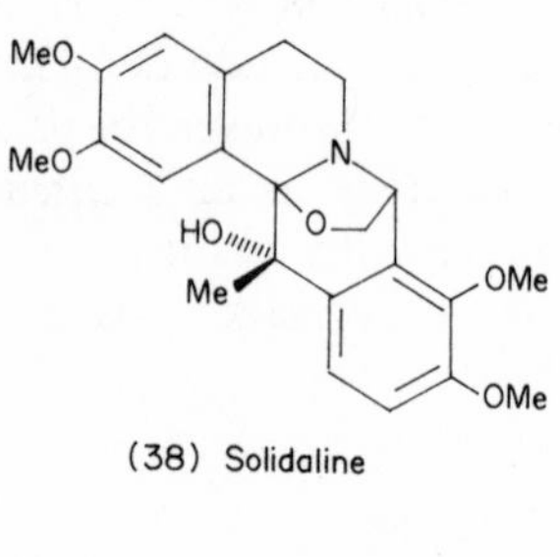

(38) Solidaline

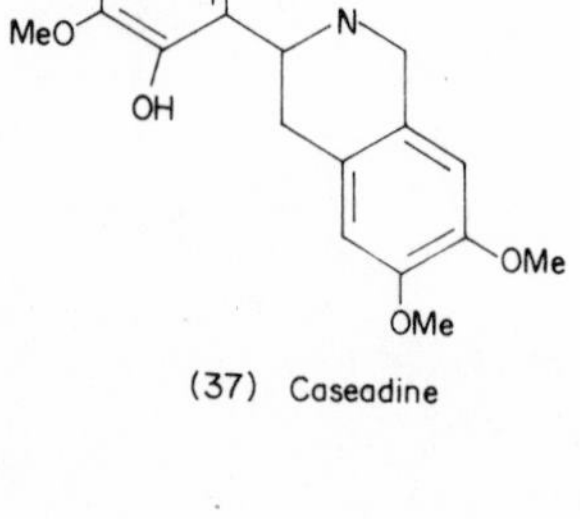

(37) Caseadine

has an extra pyrrolidine unit, and the alkaloids yuziphine **(35)**[40] and longifolidine **(36)**[41] with their unusual 7,8-substitution pattern of oxy groups. The same pattern also occurs in the protoberberine series with the alkaloid caseadine **(37)**, whose structure has been finally proved by synthesis.[42] Other unusual alkaloids in this series include solidaline **(38)**,[43] puntarenine **(39)**,[44] and chilenine **(40)**:[45] ring B has been expanded in the last two, at the expense of ring C in the case of **(40)**. A recent general synthesis of protoberberines may be illustrated by the preparation of scoulerine **(41)** (Scheme 14).[46]

The structures of the secoberberines, an interesting group of which corydalisol **(42)**[47] is an example, have been determined by correlation with the corresponding protoberberines or phthalideisoquinolines. Hypecorine **(43)**[48] is closely related, and mecambridine **(44)**,[49] into which a new 'berberine

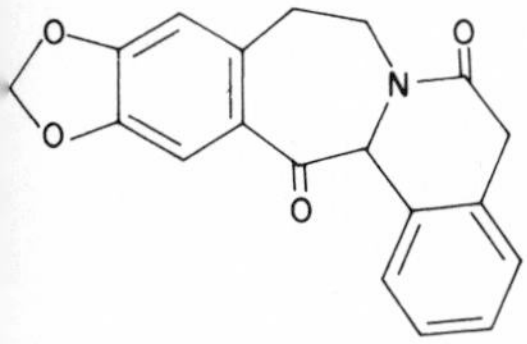

(39) Puntarenine

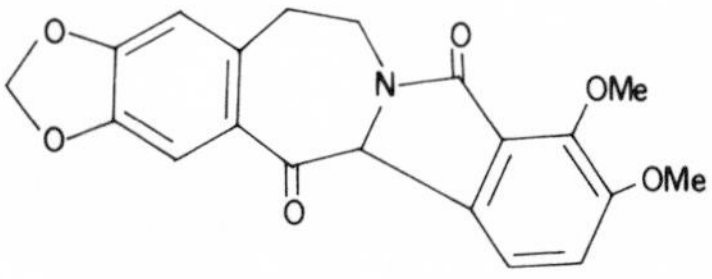

(40) Chilenine

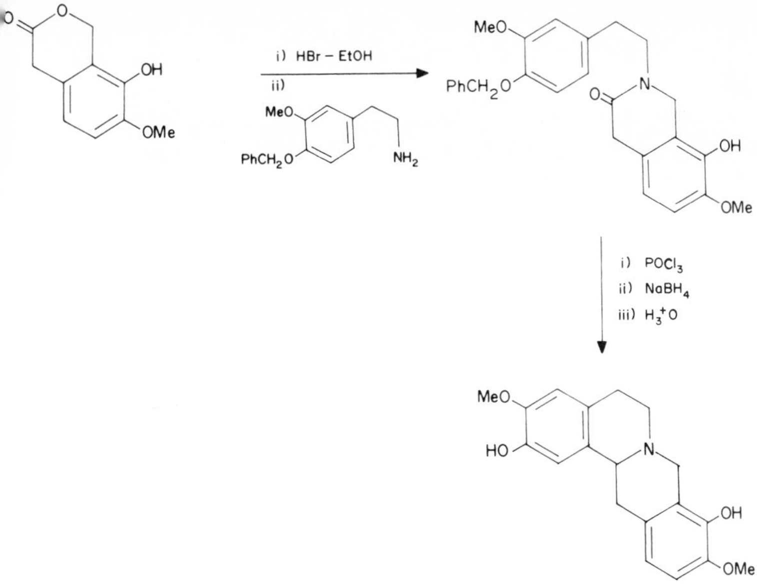

(41) Scoulerine

Scheme 14

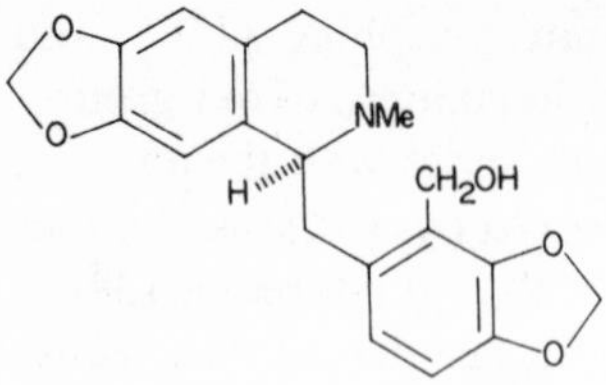

(42) Corydalisol

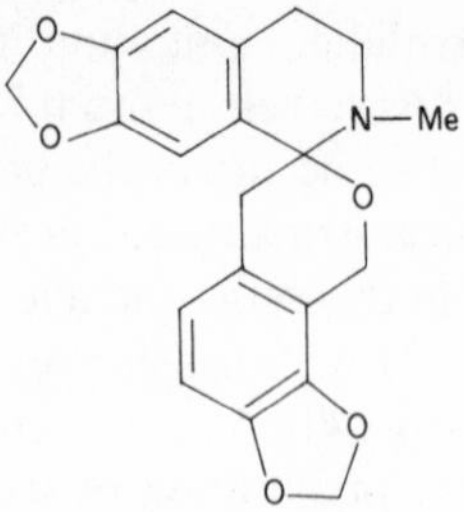

(43) Hypecorine

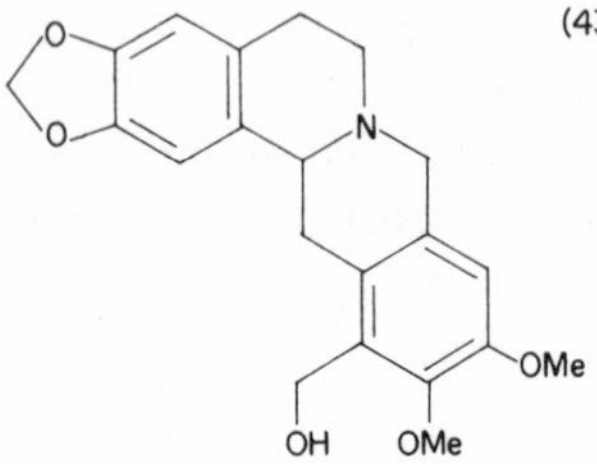

(44) Mecambridine

bridge' appears to have been inserted, may be allied biogenetically as well.

Since the discovery that apomorphine is useful in treating Parkinson's disease, there has been a good deal of interest taken in the aporphine group in general. They have a range of other pharmacological activities, and many new examples have been discovered lately. Their properties, in particular their spectra, have been carefully examined, and attention has been devoted to improved synthetic methods. Photocyclization of the 8-bromo derivatives or diazonium salts of benzylisoquinolines gives proaporphines,[50] which can then be rearranged under acid conditions, or by further irradiation,[51] to aporphines. One of the most efficient methods of aporphine synthesis, which may be illustrated by the preparation of glaucine (45) starting from laudanosine (46), was developed by Kupchan and his associates (Scheme 15).[52]

A number of interesting new aporphinoids have been described, including pontevedrine (47)[53] and imerubine (48),[54] whose structures were confirmed by synthesis and by X-ray crystallography respectively. Perhaps the most remarkable is the iso-oxoaporphine menisporphine (49).[55]

It is a striking fact that all 27 of the known spirobenzylisoquinoline alkaloids, of which O-methylcorpaine (50)[56] is a recent example, have a methylenedioxy group in ring D, and evidence has been adduced[57] that steric compression is a factor in the biosynthesis of this group.

The bisbenzylisoquinoline series and its analogues now form one of the largest single groups of alkaloids, and interest has been focused on them particularly since the report of antitumour properties associated with certain members, including tetrandrine (51) and thalicarpine (58). An X-ray study[58]

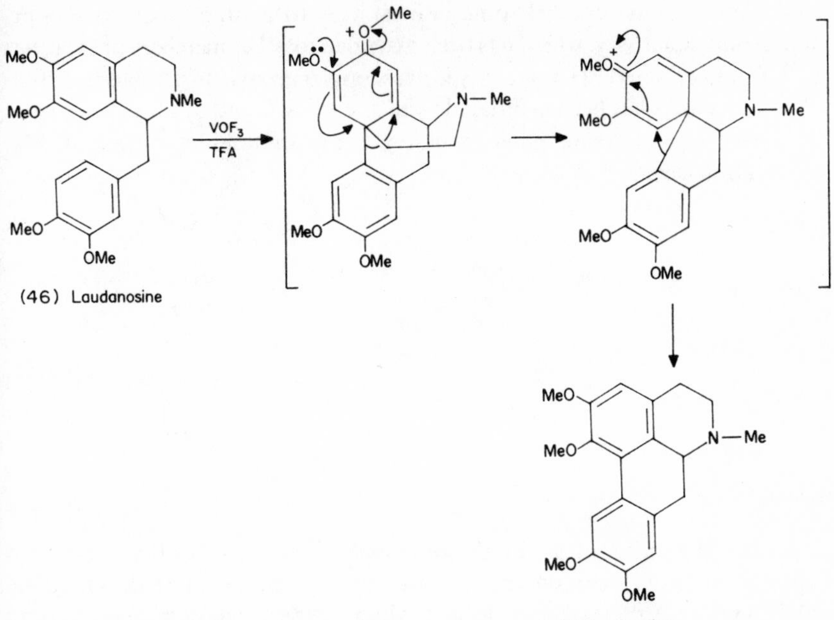

(46) Laudanosine

(45) Glaucine

Scheme 15

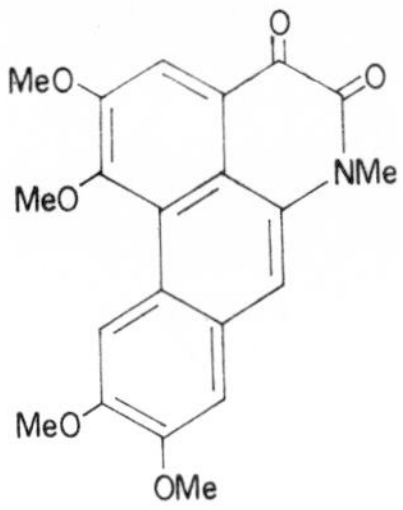

(47) Pontevedrine

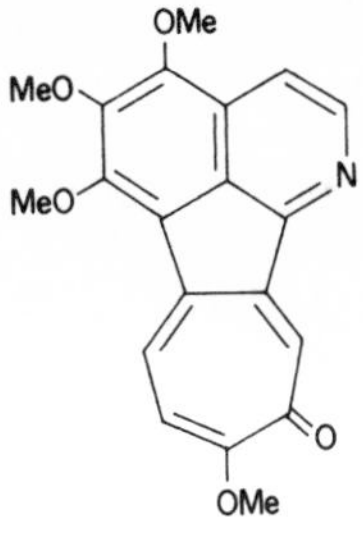

(48) Imerubine

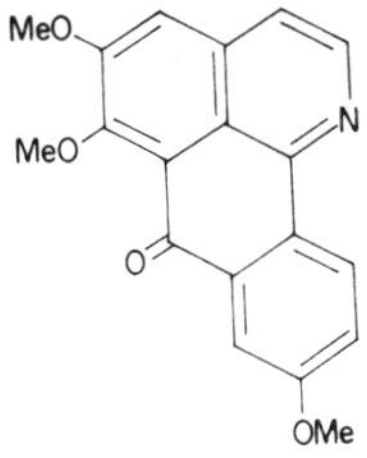

(49) Menisporphine

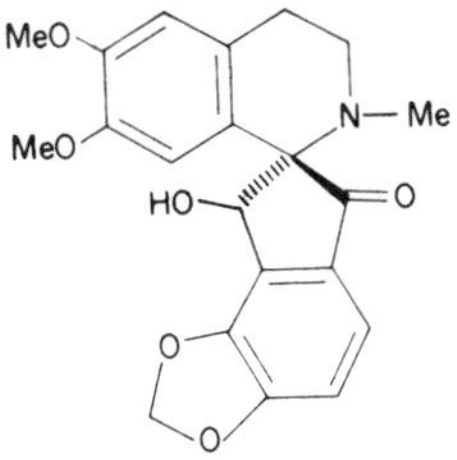

(50) O-Methylcorpaine

of the former has revealed some interesting information about its conformation and reactivity, which has been extended to other members of the series by ^{13}C n.m.r.[59] and ^{1}H n.m.r.[60] spectroscopy. The structural determination of new members has been facilitated by m.s. studies,[61] and by the development of new degradative methods including ceric ammonium nitrate (CAN) oxidation (Scheme 16).[62]

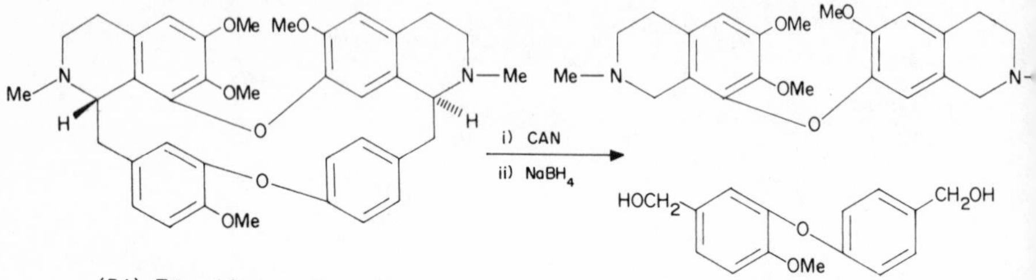

(51) Tetrandrine

Scheme 16

Among the more unusual types of bisbenzylisoquinolines that have been reported recently, stepinonine (52)[63] has a benzazepine unit, pakistanamine (53)[64] a proaporphine unit, while pennsylpavine (54)[65] has one aporphine and one pavine unit, and beccapoline (55)[66] has two aporphine units; the novel alkaloid coyhaiquine (56)[67] probably represents a biological oxidation

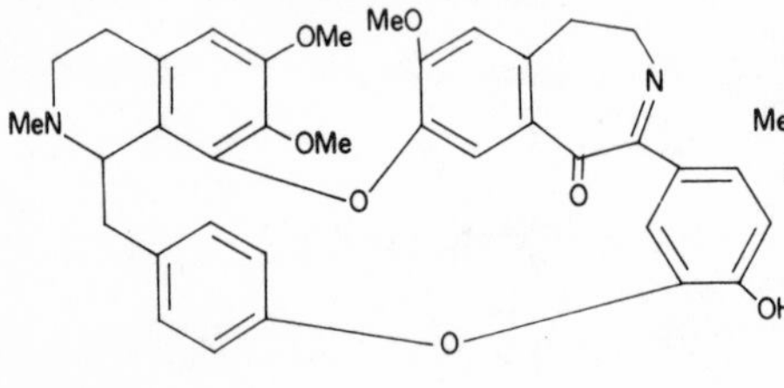

(52) Stepinonine

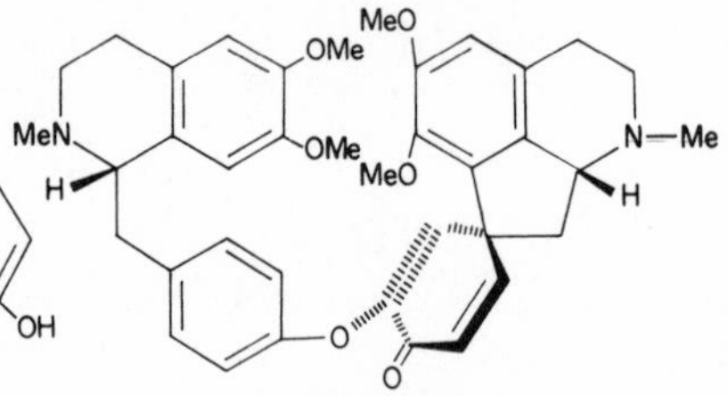

(53) Pakistanamine

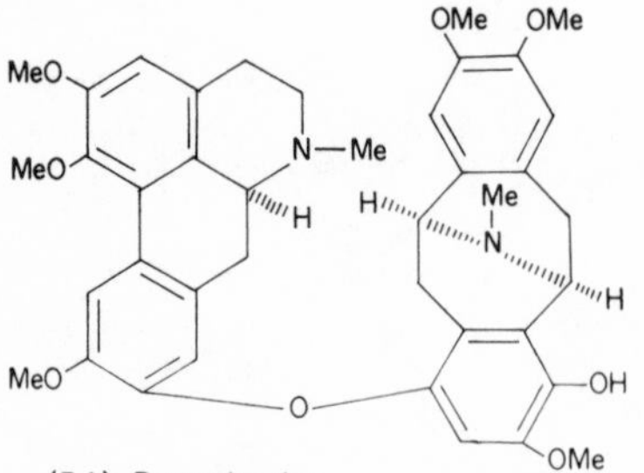

(54) Pennsylpavine

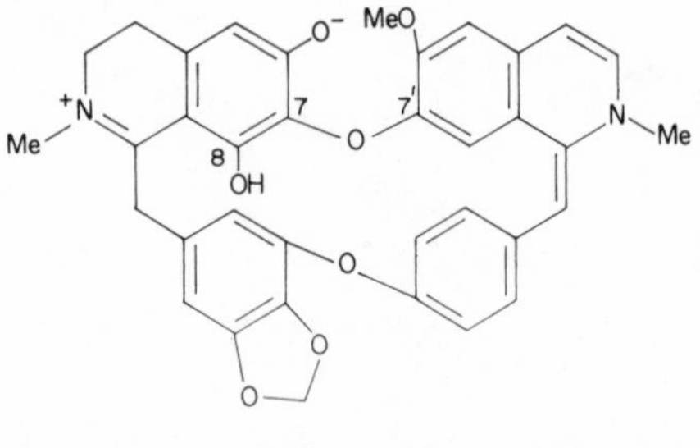

(57) Daphnine

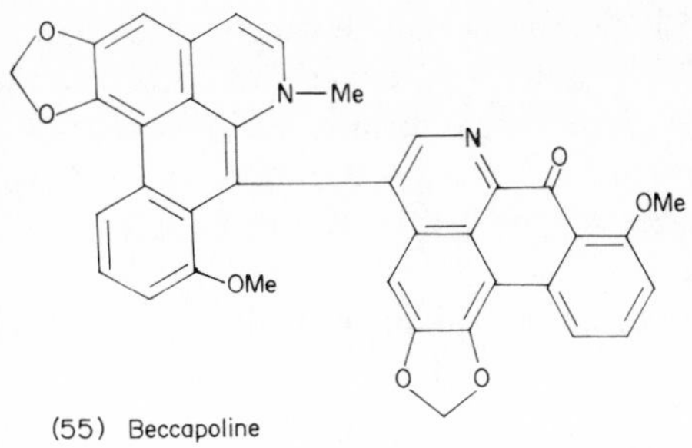

(55) Beccapoline

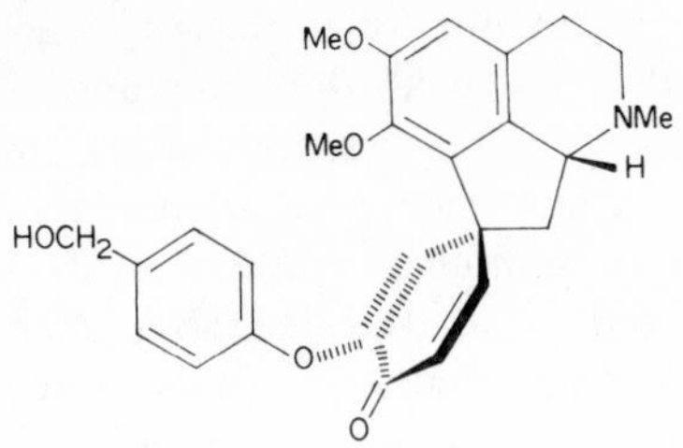

(56) Coyhaiquine

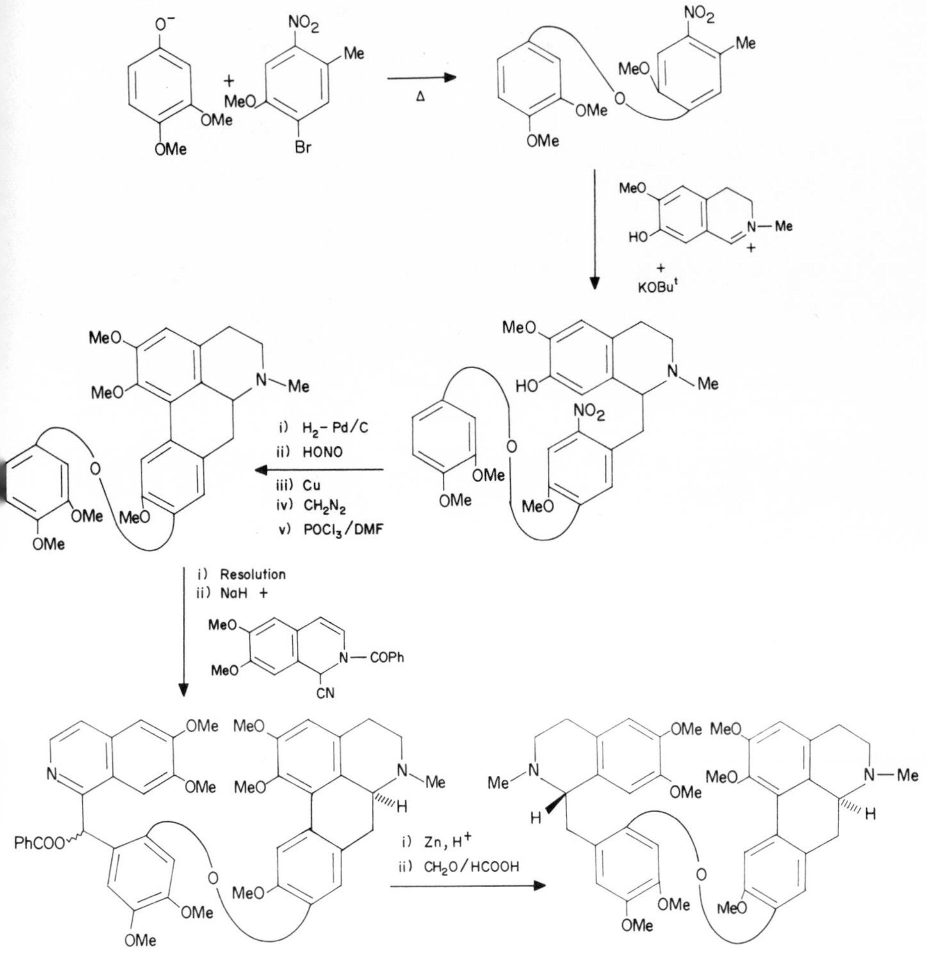

Scheme 17

(58) Thalicarpine

product of a dimeric base such as (53). One of the unique structural features of the yellow alkaloid daphnine (57)[68] is a 7–7¹ either link between the isoquinoline residues, which may have been formed by a Smiles-type rearrangement of the commonly-found 8–7¹ type link during its biosynthesis. An example of synthesis in the series is provided by Kupchan and Liepa's route to thalicarpine (58) (Scheme 17).[69]

A new synthesis[70] of the more important morphinane alkaloids starts from a benzylisoquinoline derivative (Scheme 18). The dihydrothebainone (59)

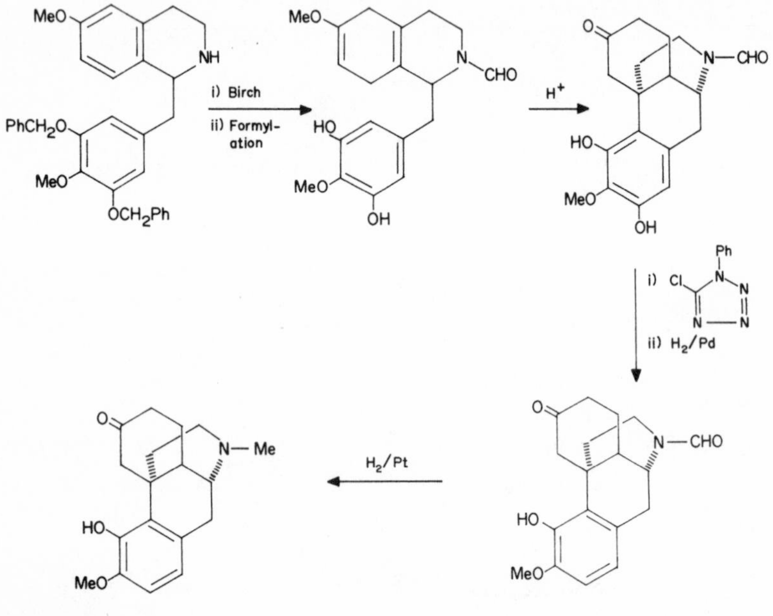

(59) Dihydrothebainone

Scheme 18

which is formed can be converted in good yield into thebaine (60) (Scheme 19)[71], codeine (61), or morphine (62). Alternatively, the penultimate stage can be transformed into codeine through codeinone. The conversion of thebaine into codeine and morphine is currently of considerable interest, since a process is under study that is designed to supplement the supply of the latter alkaloids starting from *Papaver bracteatum*, a plant which unlike *P. somniferum*, produces very largely thebaine. A high-yield procedure[72] for this conversion is sketched in Scheme 20.

The alkaloids of the *Erythrina* spp. continue to attract interest, in part because of their physiological effects which include paralysis of smooth muscle, and a good deal of ingenuity has been displayed in the development of

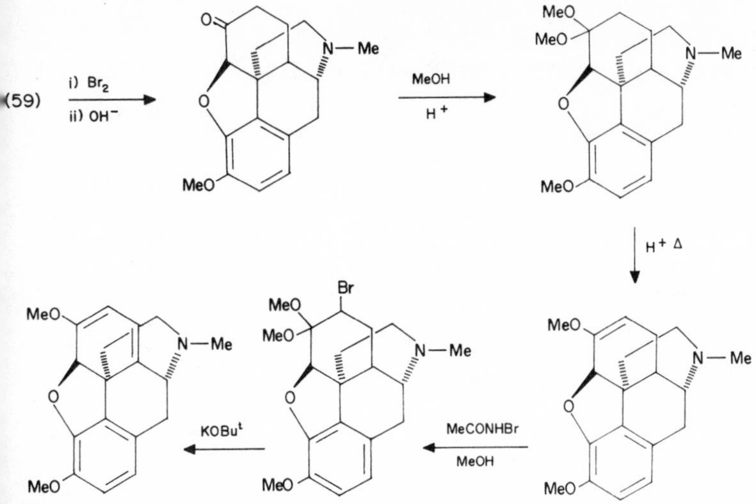

(60) Thebaine

Scheme 19

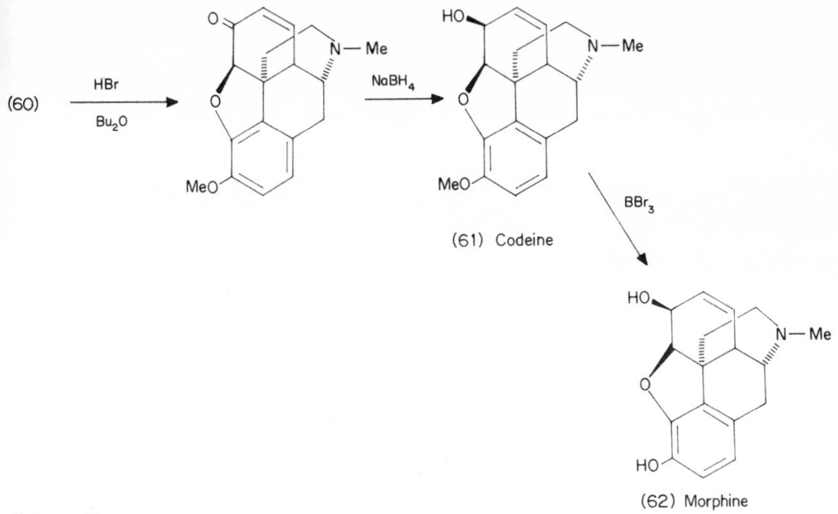

(61) Codeine

(62) Morphine

Scheme 20

a satisfactory synthesis for these bases. A stereo-controlled synthesis of erysostrine (**63**) has recently been reported by Japanese workers,[73] who used a Diels-Alder reaction to construct ring D (Scheme 21).

The family Liliaceae and its close allies provide a series of alkaloids whose members are higher homologues of the mono- and bisbenzylisoquinolines,

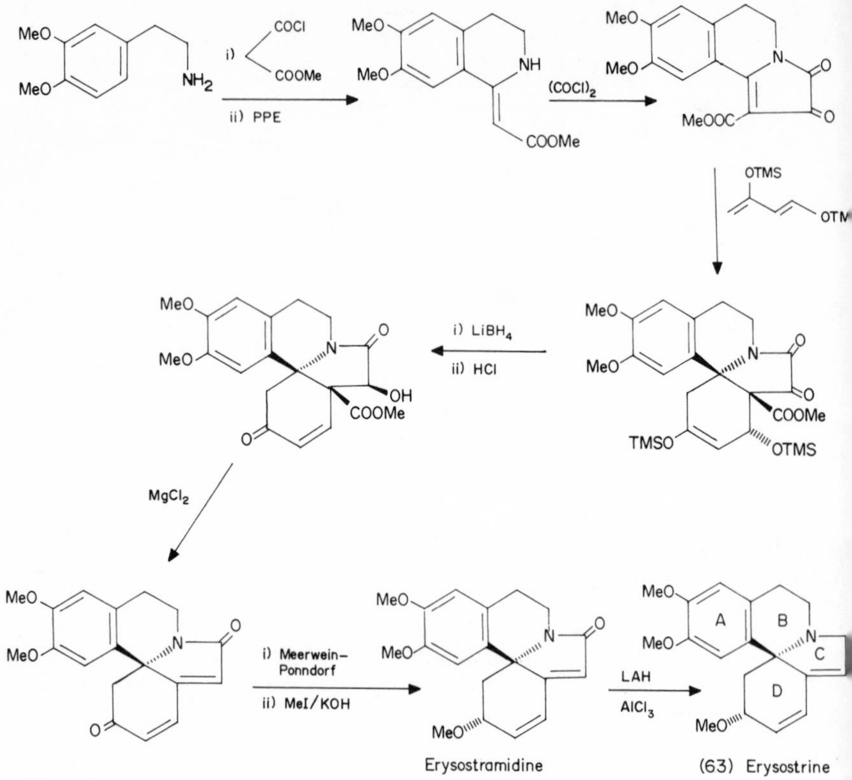

Scheme 21

aporphinoids, and morphinanes. The most important of these alkaloids, col-chicine, is biosynthetically related to the homomorphinanes, and although it has been synthesized several times, new strategies are still being explored. The homoerythrina alkaloids are somewhat exceptional in that they occur not only in plants belonging to the Liliaceae, but also in others quite unrelated to the latter family. One such plant, *Cephalotaxus harringtonia*, in addition to various homoerythrina alkaloids such as 3-epischelhammericine (**64**), also produces a group of alkaloids with the analogous cephalotaxine (**65**) structure; a molecular rearrangement involving a loss of a carbon atom has evidently taken place during their formation. Certain of these *Cephalotaxus* alkaloids, like harringtonine (**66**), an ester of cephalotaxine, show potent antileukemic activity; they have been the subject of several ingenious syntheses such as that shown in Scheme 22,[74] which are designed to supplement the limited supply of the natural alkaloids.

Apart from this group of alkaloids that have one more carbon than the

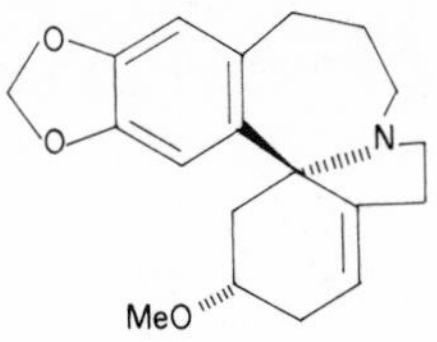

(64) 3-Epischelhammericine

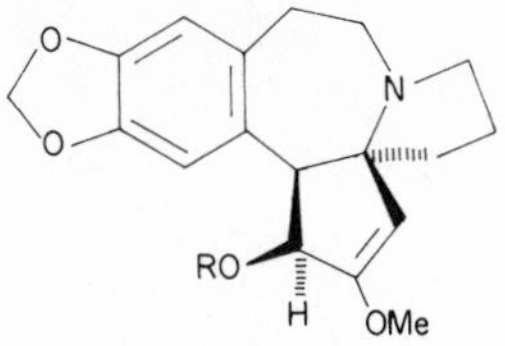

(65) Cephalotaxine (R = H)
(66) Harringtonine

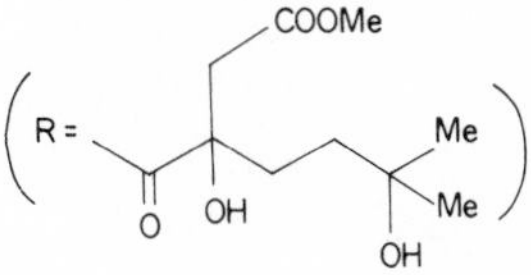

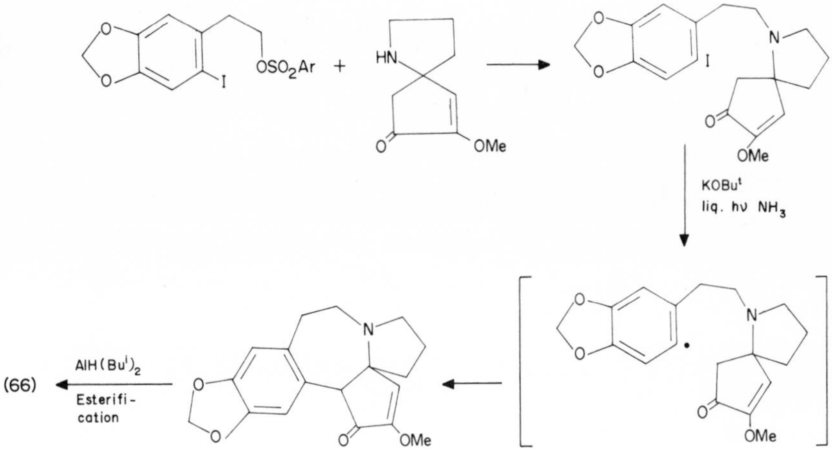

Scheme 22

benzylisoquinolines and their relatives, there is also a series with one less carbon, of which the arylisoquinolinium derivative (67)[75] and rufescine (68)[76] are recent examples. More complex types occur in the plant family Amaryllidaceae, but these show little or no structural analogy with the benzylisoquinoline alkaloids. There are over a hundred known amaryllidaceous alkaloids, divided into several structural types, with a wide range of pharmacological properties. As a result they have been the focus of a good deal of synthetic effort. Galanthamine (69) amongst other properties exhibits analgesic activity comparable to that of morphine, and the asymmetric synthesis[77] shown in Scheme 23 can be modified to give either the natural (−) galanthamine or its enantiomer.

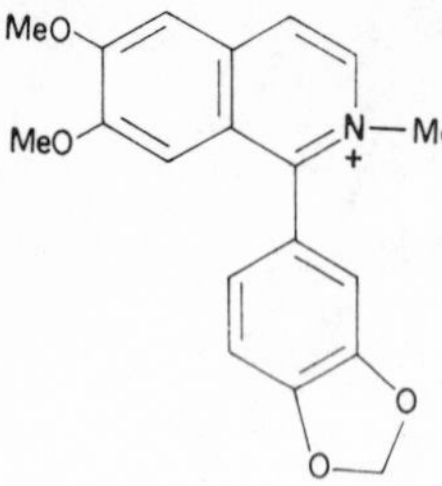
MeO
MeO
N—Me
+
O
O

(67)

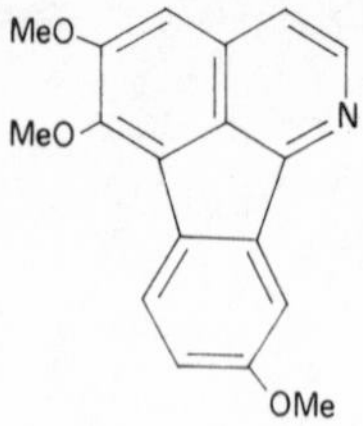
MeO
MeO
N
OMe

(68) Rufescine

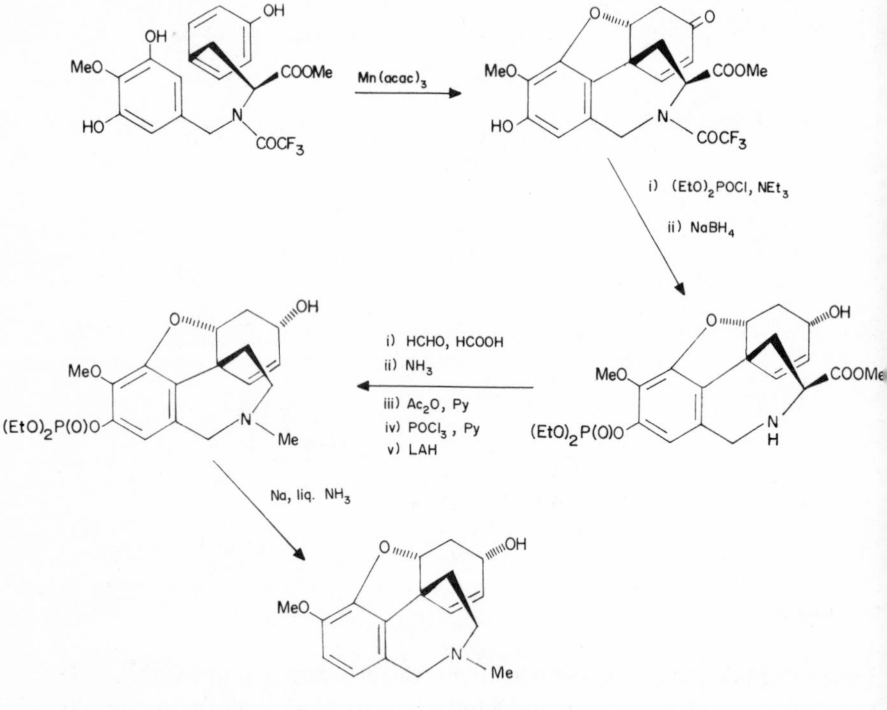
OH
OH
MeO
HO
COOMe
N
COCF3
Mn(acac)3
O
O
MeO
HO
COOMe
N
COCF3
i) (EtO)2POCl, NEt3
ii) NaBH4
O
OH
MeO
(EtO)2P(O)O
COOMe
N
H
i) HCHO, HCOOH
ii) NH3
iii) Ac2O, Py
iv) POCl3, Py
v) LAH
O
OH
MeO
(EtO)2P(O)O
N
Me
Na, liq. NH3
O
OH
MeO
N
Me

(69) (+) Galanthamine

Scheme 23

7.4 Alkaloids derived from tryptophan

Alkaloids with an indole nucleus constitute the largest group of all with about 1500 members, but the indole residue may in some cases be derived from units other than tryptophan, as in the case of 3-methylcarbazole (**70**)[78] and greenwayodendrin-3-one (**71**).[79] In the structure of most indole alkaloids,

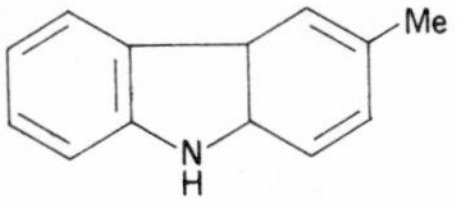

(70) 3-Methylcarbazole 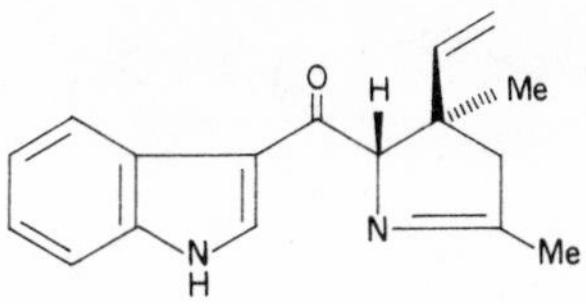   (72) Borreline 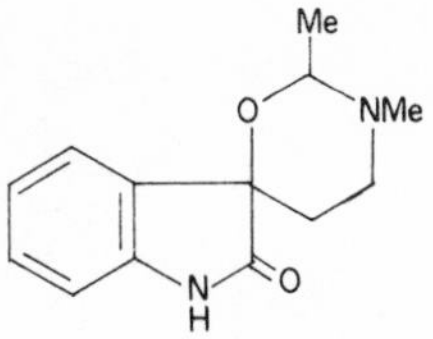   (73) Donaxarine

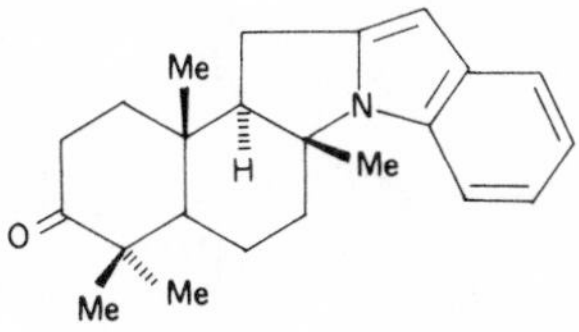

(71) Greenwayodendrin-3-one

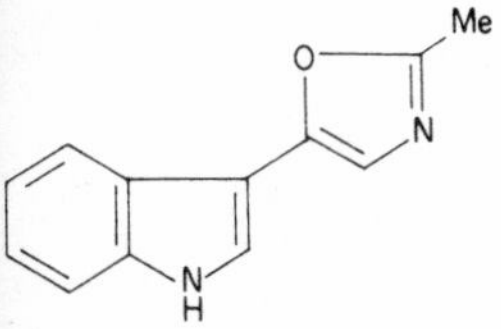

(74) Pimprinine 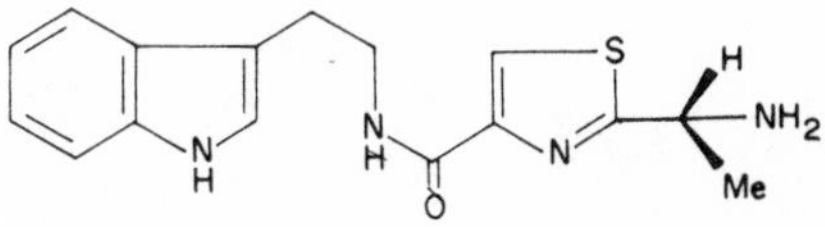   (75)

however, the ethanamine chain of the original tryptophan can still be discerned; some recent simple examples include borreline (72),[80] donaxarine (73),[81] pimprinine (74),[82] and the tryptamide (75).[83] The structures of these compounds were deduced from spectroscopic evidence, and have been confirmed by synthesis in the case of the latter two examples which are elaborated by micro-organisms. Another unusual indole alkaloid, brevicolline (76),[84] has a pyrrolidine unit attached to a carboline nucleus; it has recently been synthesized biomimetrically (Scheme 24).

Some β-carbolines with unusual features include (77),[85] which has an N-methoxyl group, cecilin (78),[86] with a hydroxybenzyl substituent on C-1, and (79),[87] in which not only the ethanamine chain, but also the carboxyl group of the original tryptophan has been retained. More complex examples in which the carboxyl carbon is still present include the so-called tremorigenic alkaloids isolated from various moulds. A good deal of interest has been taken in these recently because of their effect in producing tremors in experimental animals. Their structures have been determined mainly by spectroscopy, and that of tryptoquivaline L (80) has now been confirmed by synthesis.[88] Another group of cytotoxic mould products, the cytochalasans, cause a series of remarkable effects in animal cells. Their structures have been determined by X-ray

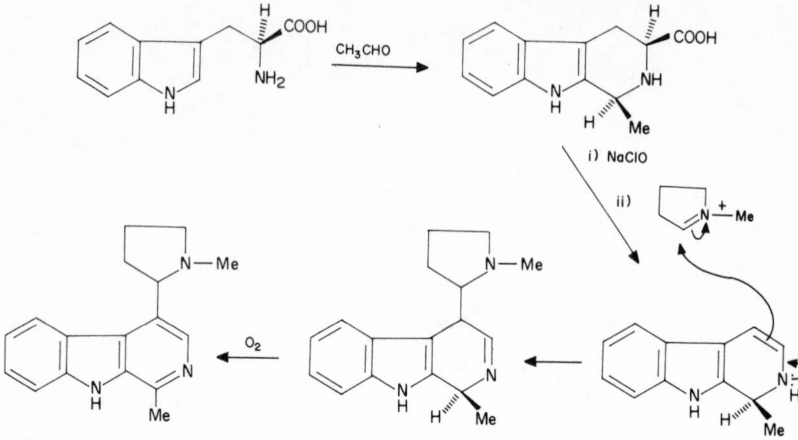

Scheme 24

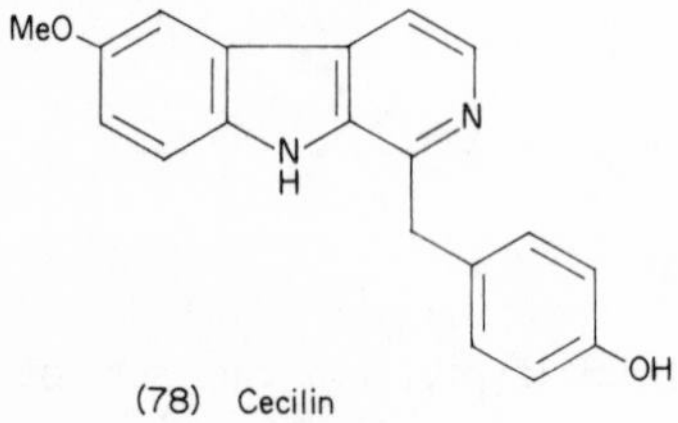

(78) Cecilin

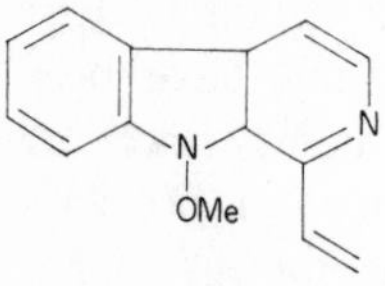

(77)

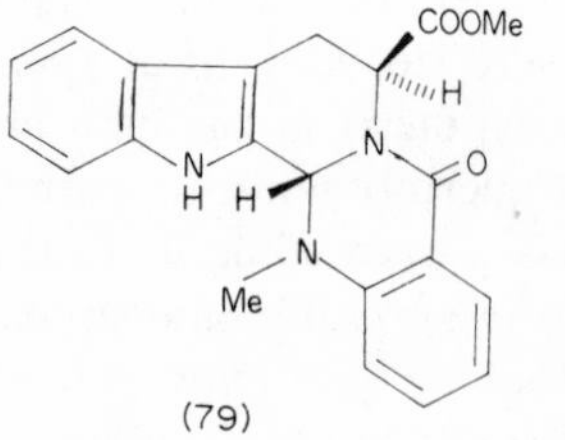

(79)

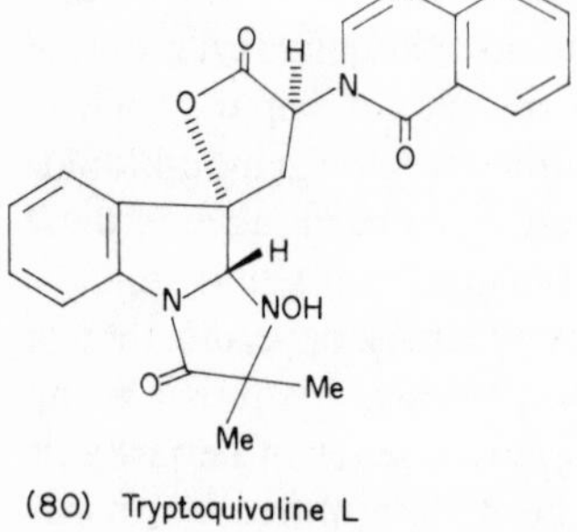

(80) Tryptoquivaline L

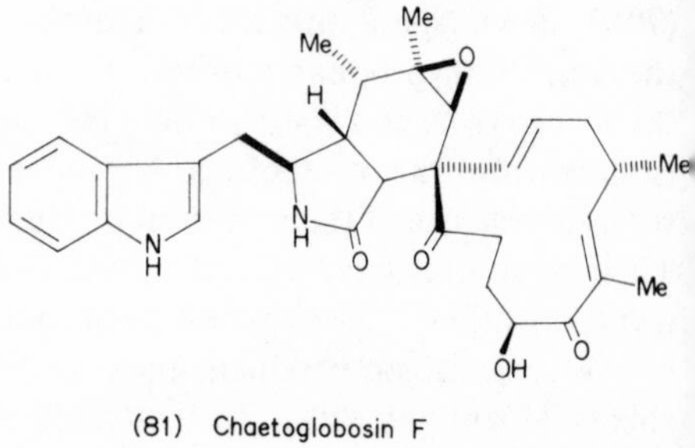

(81) Chaetoglobosin F

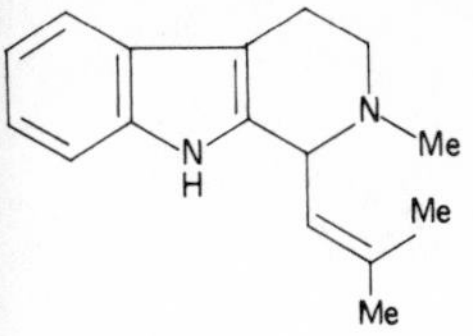

(82) Borrerine

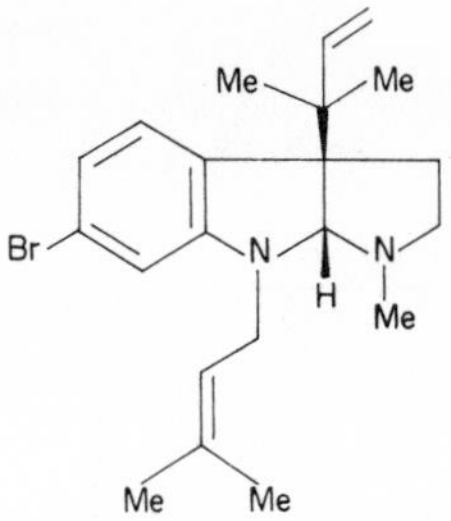

(83) Flustramine A

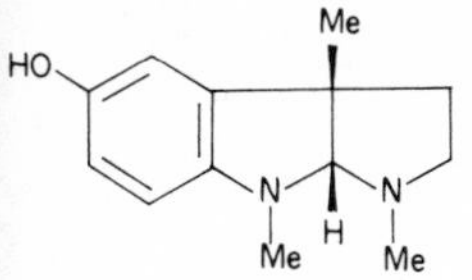

(84) Eseroline

crystallography, and by chemical and spectroscopic correlation with one another. A feature of their structure is a macro ring, and an example is furnished by chaetoglobosin F (81).[89]

The indole alkaloids mentioned so far all contain in addition to a tryptamine residue some extra unit in their structures which may be another amino acid, one or more acetate units, or any one of a number of other metabolites. The unit most frequently found in indole alkaloids, however, is isoprenoid in origin. Simple examples are provided by borrerine (82),[90] which occurs in a *Borreria* sp. from West Africa, and flustramine A (83).[91] The latter, which was isolated from a marine bryozoan, resembles the well-known alkaloid eseroline (84) in structure, but it has two prenyl units, one of which is reversed, in place of two methyl groups.

The ergoline alkaloids, well known as toxic products of ergot and other fungi, have now been found in small quantities in certain higher plants.[92] They continue to attract attention largely because of their intriguing pharmacological properties which include prolactin release. A recent synthesis[93] of lysergic acid (85) based on a suggestion made by Woodward is shown in Scheme 25. Among the moulds that produce ergoline-type alkaloids is *Penicillium roquefortii*, used in the production of roquefort cheese; in addition, however, it elaborates some interesting metabolites such a roquefortine (86)[94] and marcfortine A (87).[95] The structures of these unique mould products, which appear to be constructed from simple amino acids and isoprene units, have been deduced by spectroscopy and by X-ray crystallography. Another intriguing mould metabolite, cryptoechinulin G (88),[96] has three isoprene units in its structure.

There is an extensive range of indole alkaloid structures in which the non-

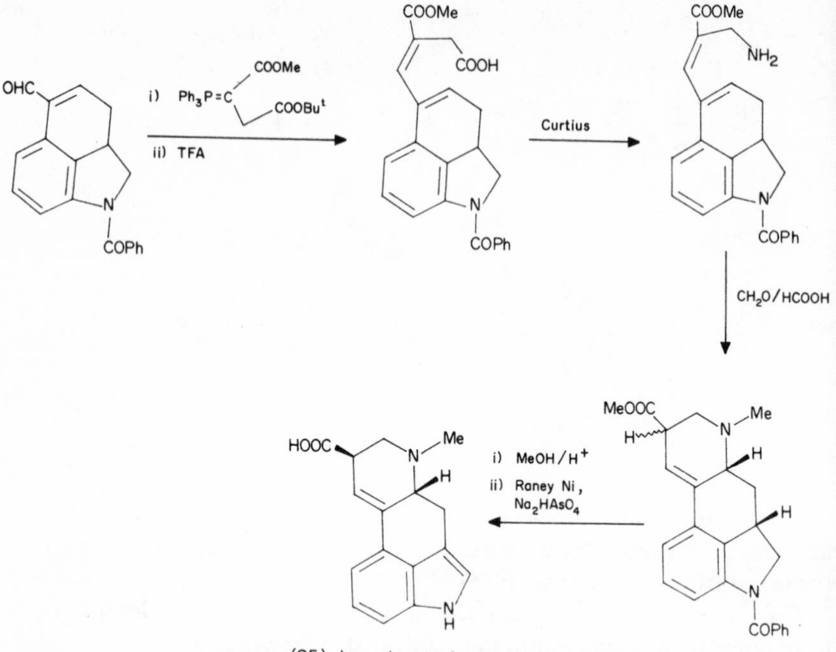

(85) Lysergic acid (+ epimer)

Scheme 25

tryptamine unit is terpenoid in origin; however, in nearly all cases the terpenoid moiety has undergone a rearrangement to an iridoid before its incorporation into the alkaloid structure. An exception is borrecapine (89),[97] which occurs along with borreline (72) in a *Borreria* sp. from South America, and which has an unrearranged open-chain terpene unit attached to the tryptamine moiety. Plants of the genus *Aristotelia*, which grow in countries bordering the South Pacific, are likewise able to incorporate an unchanged terpene unit into the structures of the alkaloids they produce, although there may be some subsequent rearrangement. Examples are fruticosonine (90),[98]

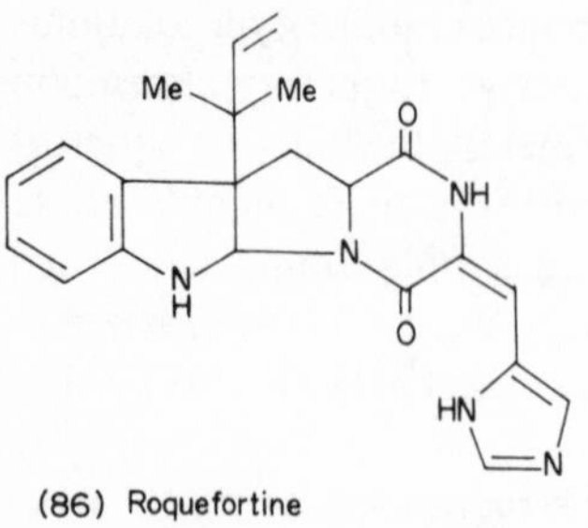

(86) Roquefortine

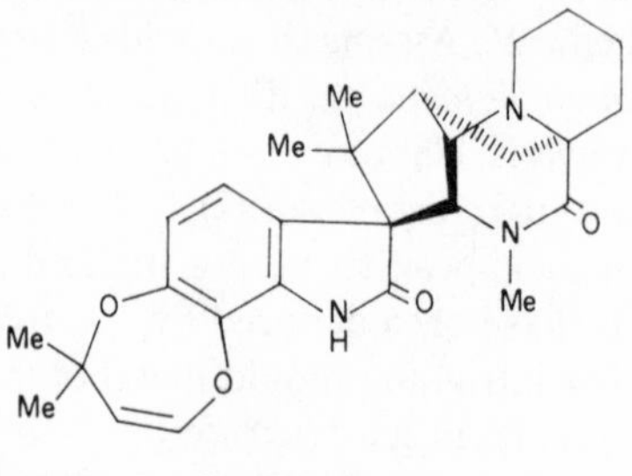

(87) Marcfortine A

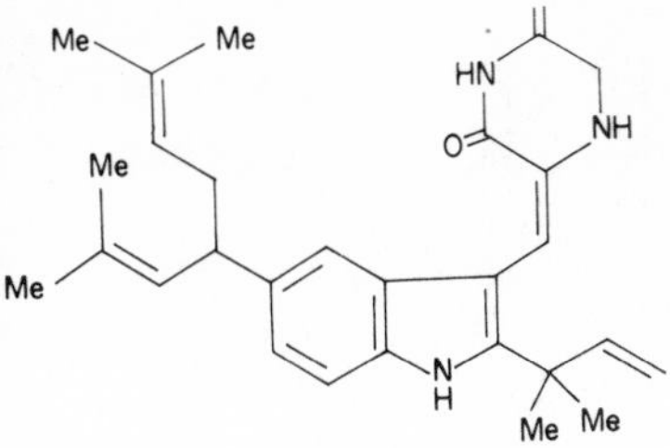

(88) Cryptoechinulin G

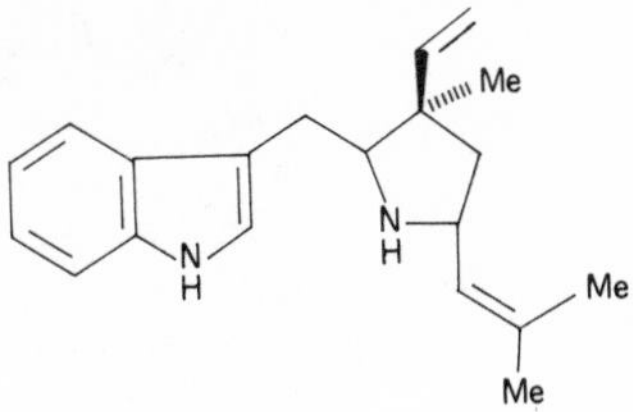

(89) Borrecapine

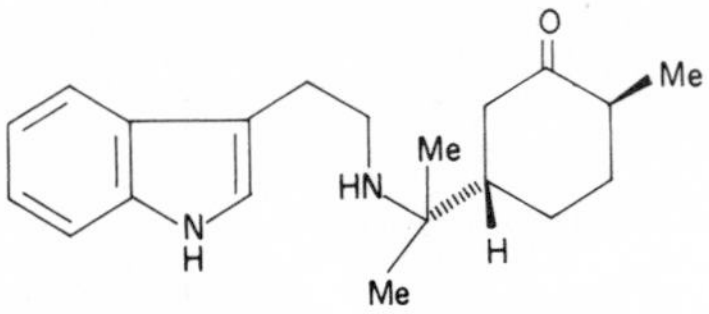

(90) Fruticosonine

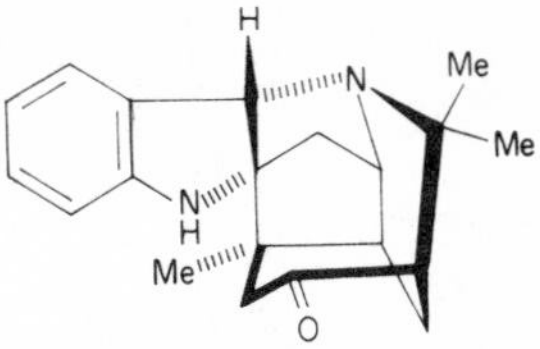

(91) Aristone

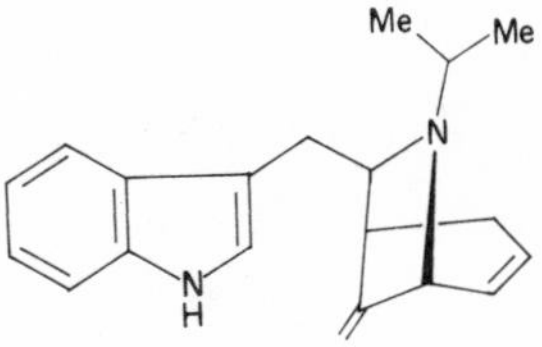

(92) Peduncularine

with a virtually intact terpene unit, and aristone (91),[99] where a rearrangement involving the tryptamine residue has evidently taken place. The structure of the former was determined spectroscopically and confirmed by synthesis; the structures and relative configurations of both have been established by X-ray crystallography. In the formation of peduncularine (92),[100] whose structure was deduced by degradation and spectroscopy, the terpene portion appears to have suffered rearrangement so as to produce the unique N-isopropyl group.

The most widespread of these alkaloids is aristoteline (93),[101] whose structure and absolute stereochemistry were determined by X-ray crystallography. A very neat synthesis of (93) and of another *Aristotelia* alkaloid, makomakine (94), starting from (−)β-pinene, has recently appeared (Scheme 26).[102]

The majority of known indole alkaloids have a unit of the iridoid secologanin (96) incorporated into their structures, and the key intermediate formed from this and tryptamine has been identified as the glucosidic alkaloid strictosidine (95).[103] In the course of a great deal of synthetic activity in this

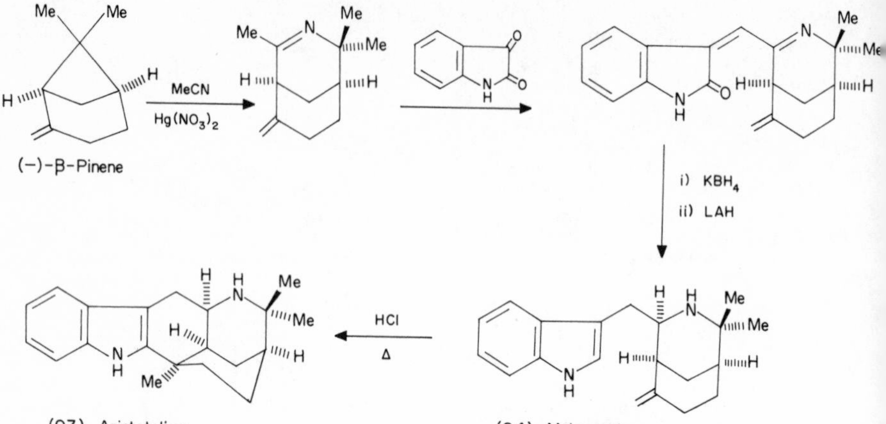

(−)-β-Pinene

MeCN
Hg(NO₃)₂

i) KBH₄
ii) LAH

(93) Aristoteline   HCl   (94) Makomakine
                            Δ

Scheme 26

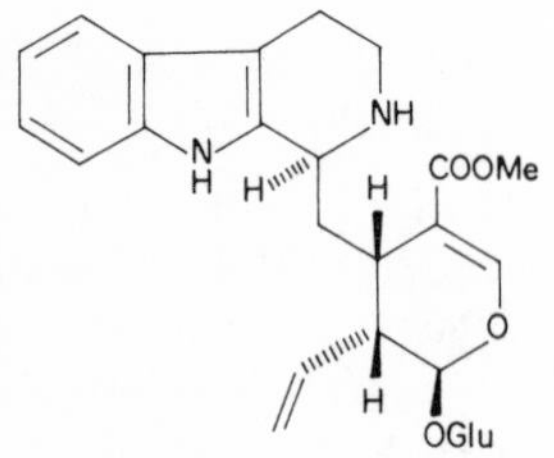

(95) Strictosidine

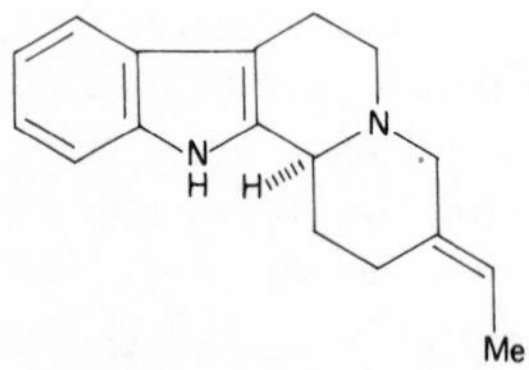

(98) Deplancheine

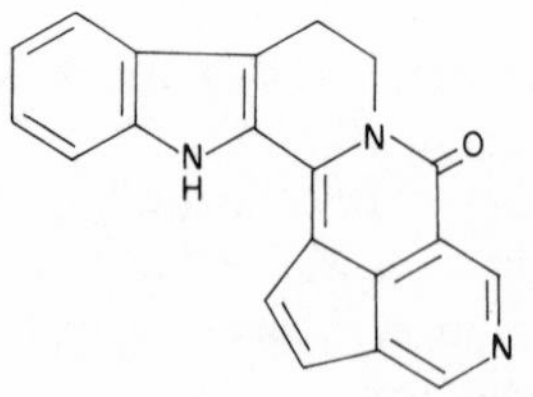

(99) Naulafine

area, some interesting biomimetic syntheses have been developed such as that shown in Scheme 27 for yohimbine (97).[104]

The ajmalicine group has interesting and useful pharmacological properties, and an extremely skilful asymmetric synthesis[105] for certain of these bases has been devised, one example of which is shown in Scheme 28. Among the recently-discovered alkaloids that seem to be related biosynthetically to the corynantheine-ajmalicine group are deplancheine (98),[106] which has three carbons less, and naulafine (99),[107] into which an extra nitrogen has been

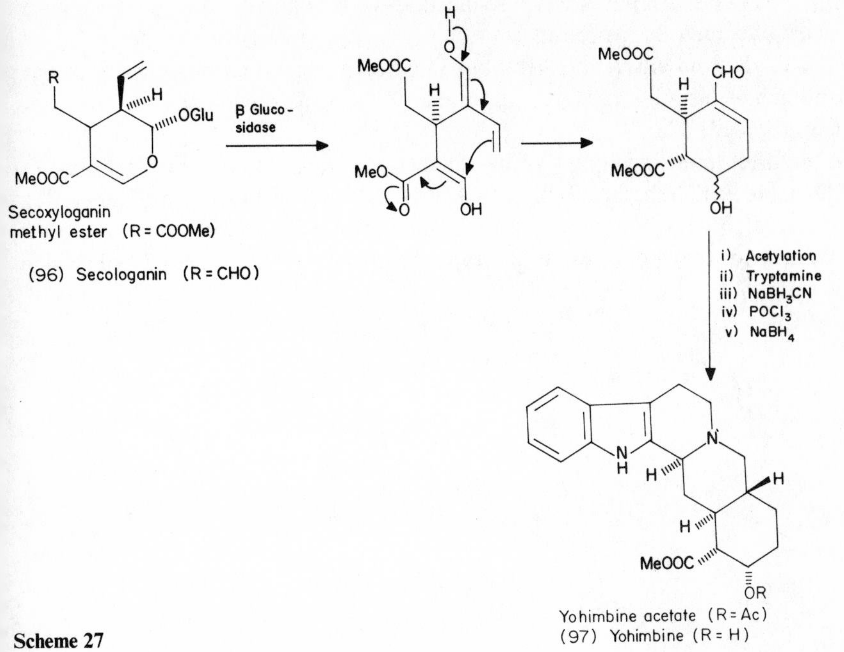
R
OGlu
MeOOC
Secoxyloganin methyl ester (R = COOMe)
(96) Secologanin (R = CHO)
β Glucosidase
MeOOC
MeO
OH
CHO
i) Acetylation
ii) Tryptamine
iii) NaBH$_3$CN
iv) POCl$_3$
v) NaBH$_4$
Yohimbine acetate (R = Ac)
(97) Yohimbine (R = H)

Scheme 27

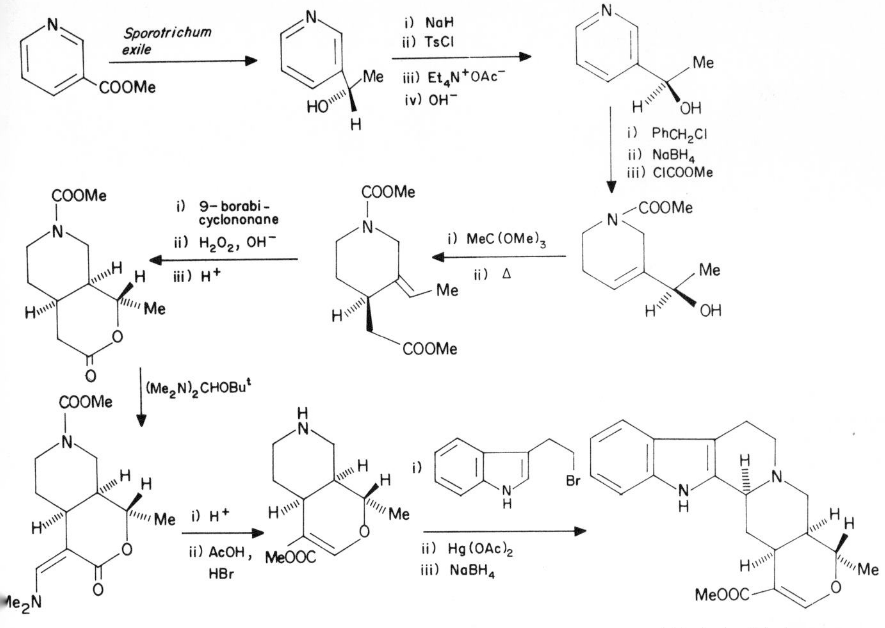
COOMe
Sporotrichum exile
i) NaH
ii) TsCl
iii) Et$_4$N$^+$OAc$^-$
iv) OH$^-$
i) PhCH$_2$Cl
ii) NaBH$_4$
iii) ClCOOMe
i) MeC(OMe)$_3$
ii) Δ
i) 9-borabicyclononane
ii) H$_2$O$_2$, OH$^-$
iii) H$^+$
(Me$_2$N)$_2$CHOBu^t
i) H$^+$
ii) AcOH, HBr
MeOOC
Br
ii) Hg(OAc)$_2$
iii) NaBH$_4$
Tetrahydroalstonine

Scheme 28

inserted. The structures were deduced spectroscopically, and in the former case confirmed by synthesis.

The alkaloid ellipticine (100) has inspired a good deal of synthetic work, and in a beautifully conceived approach[108] the nucleus was constructed in one step (Scheme 29).

A brief and ingenious route to the aspidospermine-type alkaloids is illustrated in Scheme 30 by the synthesis[109] of aspidospermidine (101), starting from tryptamine oxindole. A small group of alkaloids with structures related to the aspidospermine type has recently been found, of which

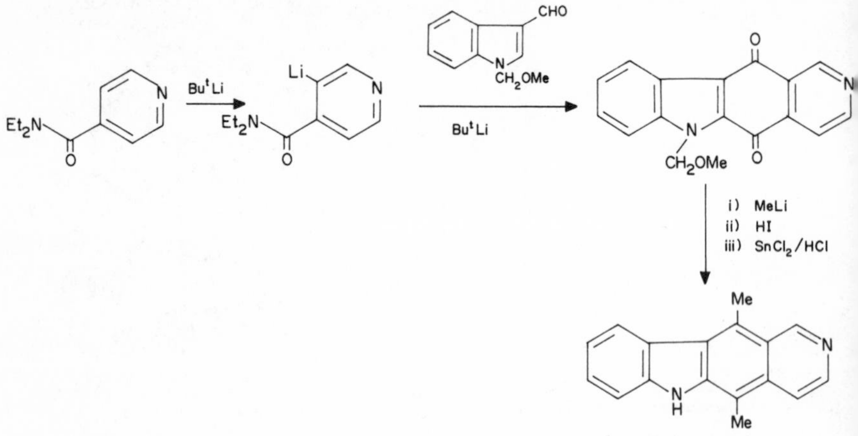

Scheme 29

(100) Ellipticine

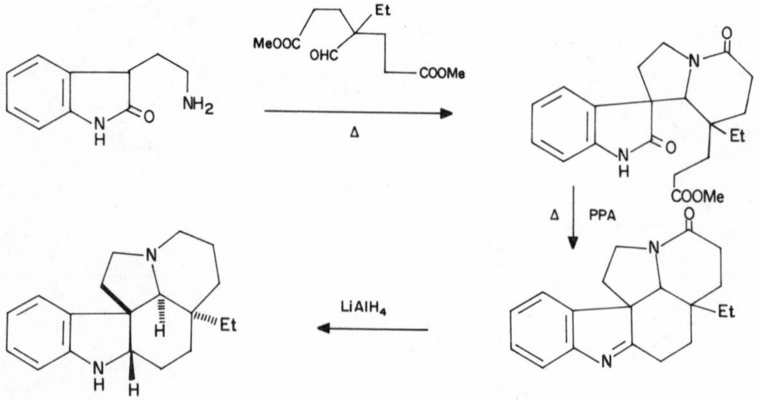

(101) ± Aspidospermidine

Scheme 30

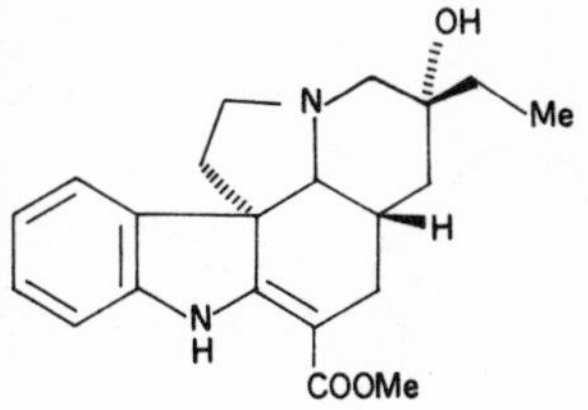

(102) Pandoline

pandoline (102)[110] is an example; its structure and stereochemistry were determined largely be spectroscopy.

A spectacularly successful asymmetric synthesis[111] of ibogamine (103) has only four steps and an overall yield of 17%. Two of the steps comprise palladium-catalysed cyclizations, the second being mediated also by silver (Scheme 31). The method has been extended to include the synthesis of (±) catharanthine (110) as well.

The alkaloids tacamine (104),[112] koumine (105),[113] and ngouniensine (106)[114] have all been isolated during the last few years, and the structures deduced for them do not correspond to any previously known types of indole bases. Ervitsine,[115] from a Madagascan plant, also has a novel type of structure (107), which was established by X-ray crystallography and by correlation with the known alkaloid methuenine. The technique of interrelating alkaloids belonging to different structural types by skilfully-devised transformations has proved valuable in many other instances in clarifying points of structure and stereochemistry: a good example is provided by the conversion[116] of tabernaemontanine (108) into eravatamine (109) using a

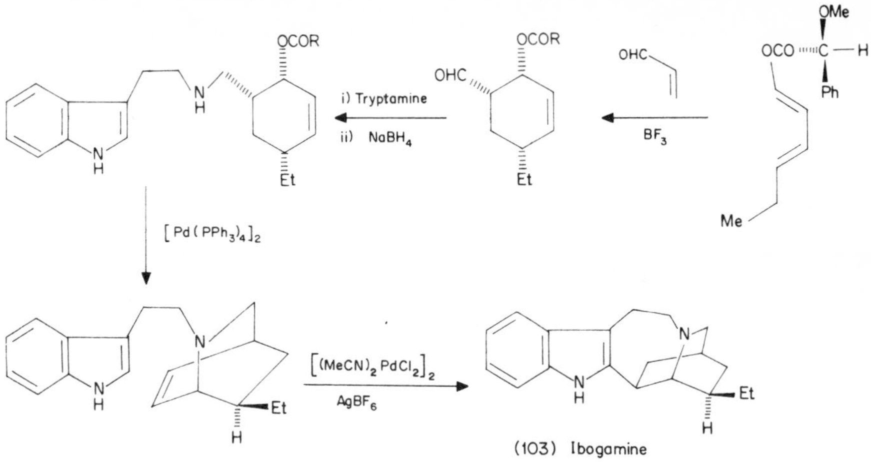

(103) Ibogamine

Scheme 31

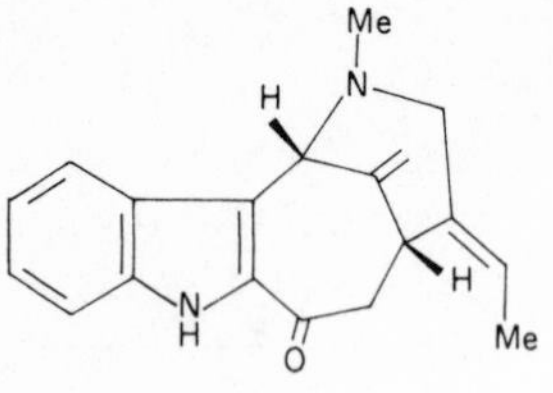

(107) Ervitsine

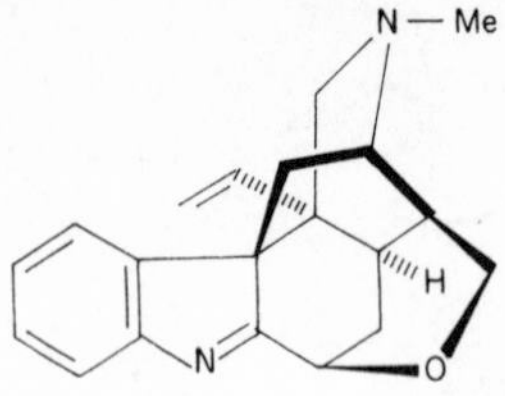

(105) Koumine

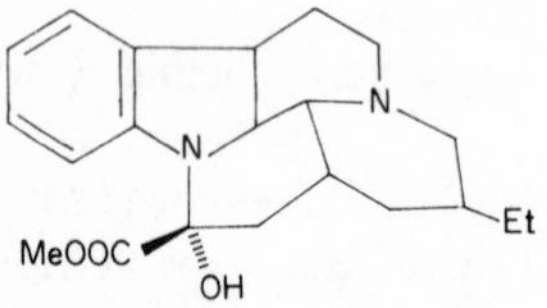

(104) Tacamine

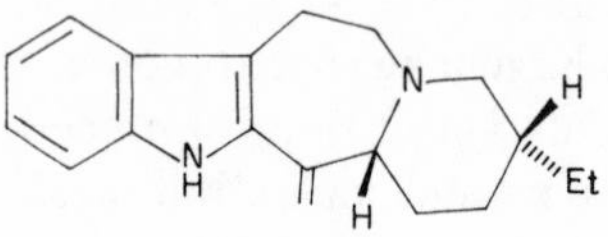

(106) Ngouniensine

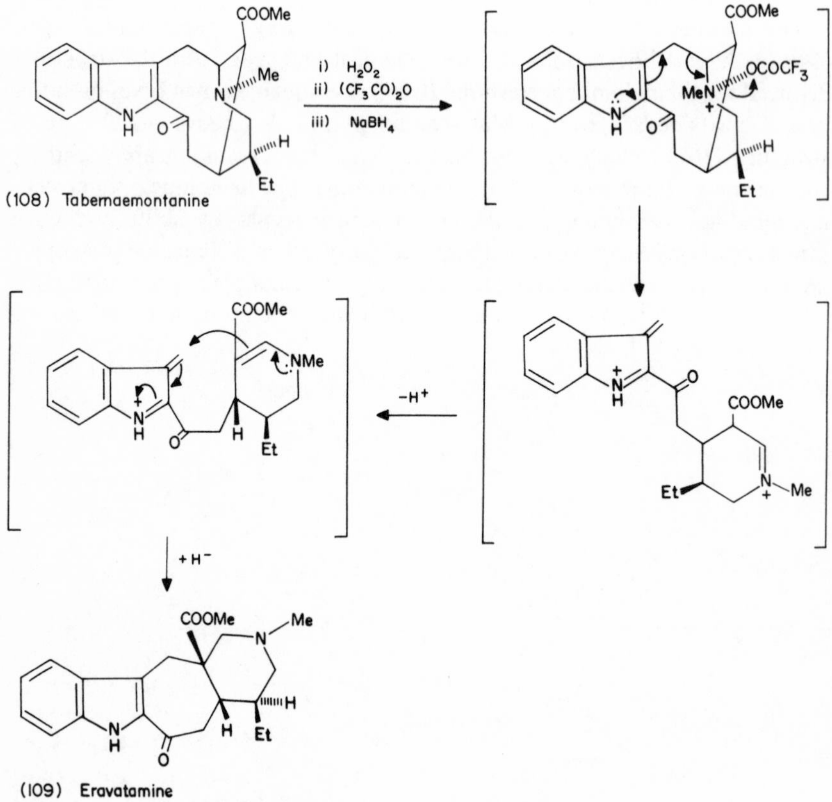

(108) Tabernaemontanine

i) H₂O₂
ii) (CF₃CO)₂O
iii) NaBH₄

−H⁺

+ H⁻

(109) Eravatamine

Scheme 32

Polonovski reaction on the N-oxide. The reaction probably proceeds as shown in Scheme 32.

The Polonovski reaction has also proved to be of crucial importance in linking together two indole alkaloid units to form a bisindole: many alkaloids of the latter type are now known, and they include the valuable antitumour and antileukaemic agents vinblastine (VBL, **112**) and vincristine (VCR), both of which are present in tiny quantities only in the plant *Catharanthus roseus*. The first synthesis[117] of VBL was accomplished as shown in Scheme 33.

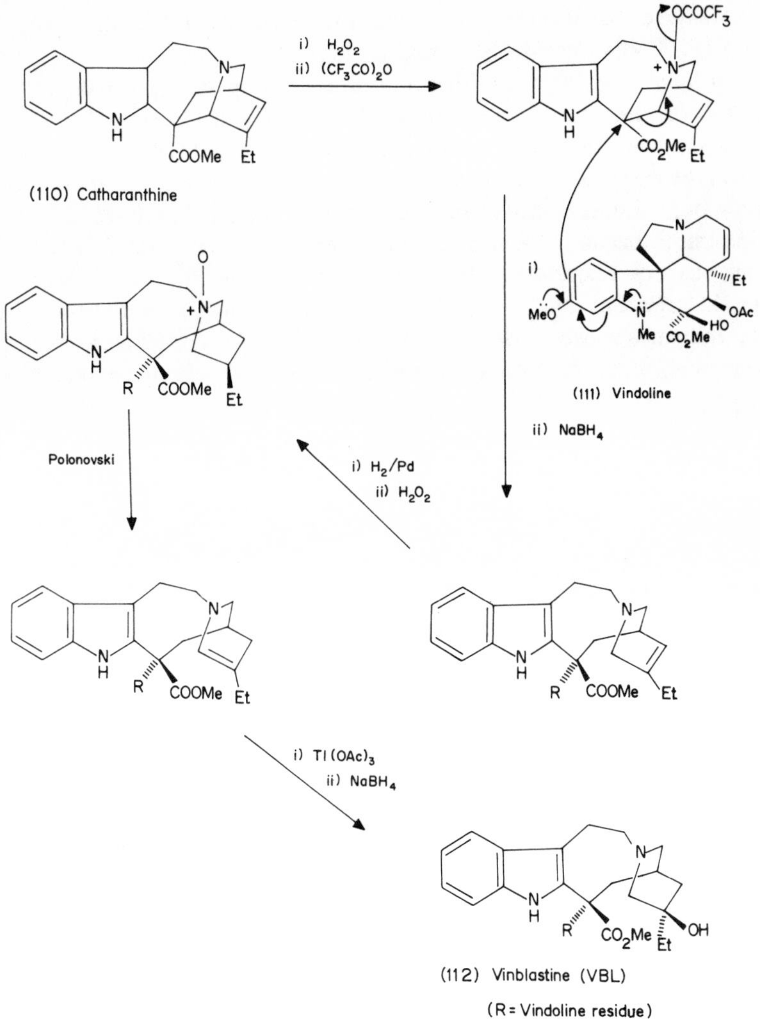
(110) Catharanthine
i) H_2O_2
ii) $(CF_3CO)_2O$
$OCOCF_3$
CO_2Me
i)
(111) Vindoline
ii) $NaBH_4$
Polonovski
i) H_2/Pd
ii) H_2O_2
i) $Tl(OAc)_3$
ii) $NaBH_4$
(112) Vinblastine (VBL)
(R = Vindoline residue)

Scheme 33 (R = vindoline residue)

Vindoline (111), an alkaloid of the aspidospermine type, had been synthesized separately; to achieve the correct configuration at the point of junction between this and the catharanthine (110) unit, it was found necessary to carry out the reaction at as low a temperature as possible so that the concerted reaction shown would be favoured; moreover, to attain the right stereochemistry around the tertiary alcohol group, a careful choice of oxidizing agent had to be made.

VCR is the more valuable of the two agents, but it is present in lesser amount. Among the various methods for transforming VBL into VCR is a low-temperature oxidation with chromium trioxide to convert the methylimino group of VBL into an N-formyl group.[118]

The *Cinchona* alkaloids, another important group derived from tryptophan, have a quinoline instead of an indole nucleus. Until recently they were known to occur only in the rubiaceous genera *Cinchona* and *Remija* and in a single *Strychnos* species, but cinchonine and some other known analogues have now been found in leaves of the olive tree and of the privet.[119] The alkaloid camptothecine (113) also has a quinoline ring system, but its skeleton bears an obvious relationship to that of strictosidine (95) with the two nitrogen heterocycles interchanged. Camptothecine has high antitumour activity, and since it occurs only in small amounts in a rare Chinese tree, a good deal of effort has been directed to its synthesis. A short route is sketched in Scheme 34.[120]

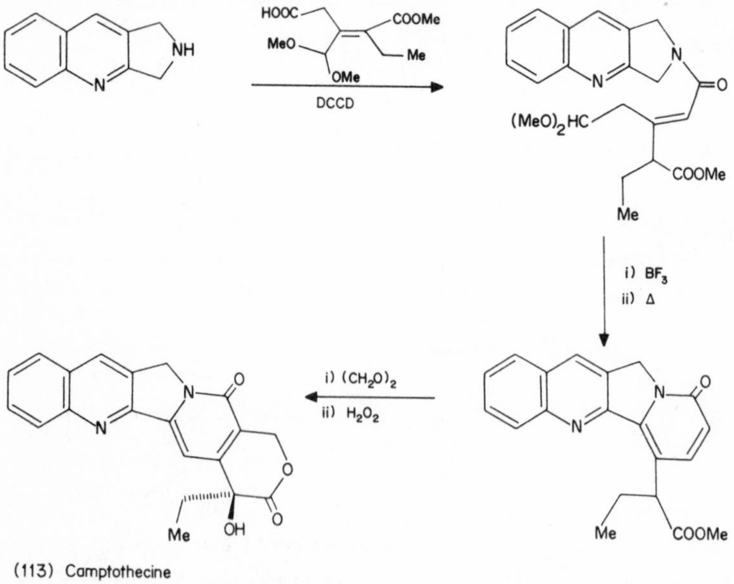

(113) Camptothecine

Scheme 34

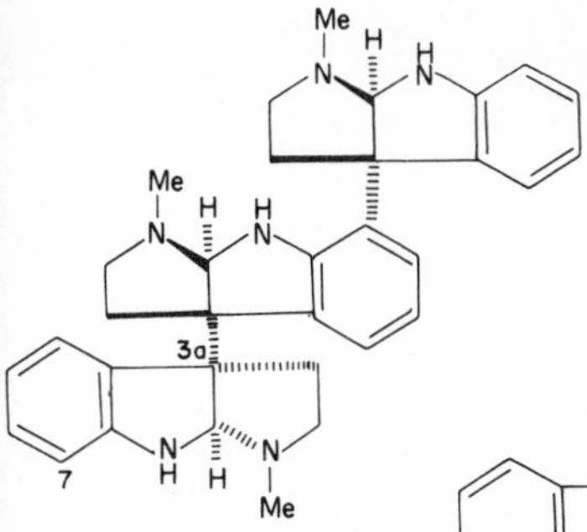

(116) Spinaceamine

(115) Hodgkinsine

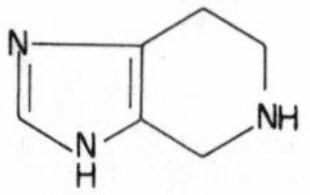

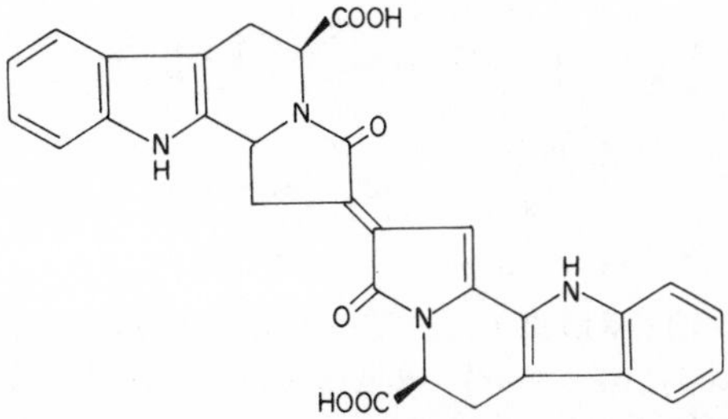

(114) Trichotomine

Apart from bisindoles like VBL and VCR, and the fascinating blue alkaloid trichotomine (114),[121] a series of oligoindoles are now known. Hodgkinsine (115)[122] is a trimer made up of units with the eseroline (84) skeleton joined together by 3a-3′a′ and 7′-3″a″ links. Others have been isolated recently with four and five units of the same kind linked in one or other of these positions.[123]

7.5 Alkaloids which originate in other amino acids

Some well-known alkaloids such as pilocarpine are derived from histidine, and the alkaloid spinaceamine (116),[124] which was recently isolated from an amphibian venom, also appears to belong to this group; anatanine (117),[125] from a leguminous plant, may have a similar origin. There is also a considerable group of alkaloids derived from anthranilic acid, and two recent examples that most likely belong to this series are the quinazoline (118)[126] and the quinolone alkaloid melochinone (119),[127] from the plant families Rutaceae and Sterculiaceae respectively. The structure of (118) was confirmed by synthesis, and that of melochinone by X-ray crystallography.

Nicotinic acid is another amino acid which serves as a starting point for a number of important and well-known alkaloids; recent examples are provided by the simple pyridine derivative navenone A (120),[128] a sea-slug alarm pheromone, and by duckein (121),[129] from a lauraceous plant native to the Amazon region. The cytotoxic alkaloid sesbanine has the unusual

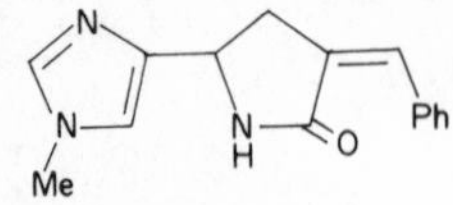
N
N
Me
N
H
O
Ph

(117) Anatanine

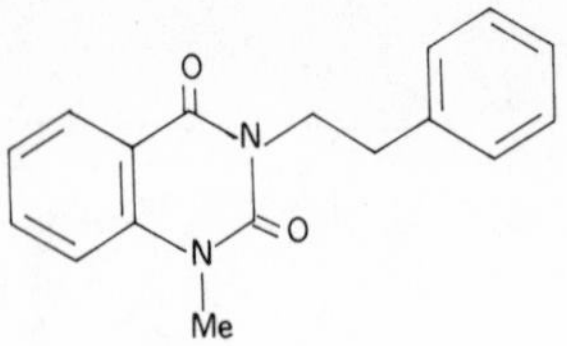
O
N
N
O
Me

(118)

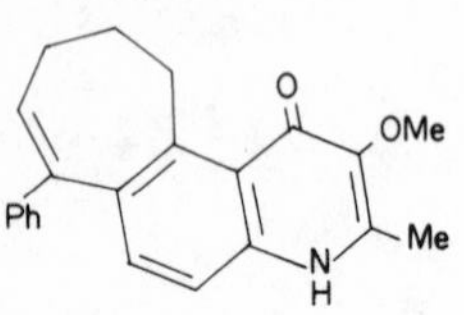
O
OMe
Ph
Me
N
H

(119) Melochinone

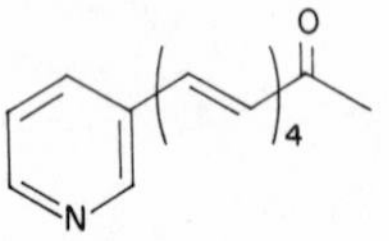
O
4
N

(120) Navenone A

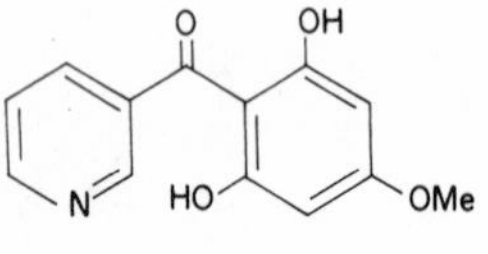
O
OH
N
HO
OMe

(121) Duckein

structure (**122**) which was established by X-ray crystallography and confirmed by several syntheses, one of which[130] is shown in Scheme 35.

The plant family Celastraceae produces a range of complex alkaloids derived from nicotinic acid and highly oxygenated sesquiterpene units. The best known of these alkaloids is evonine (**123**),[131] whose structure and stereochemistry were finally established after a long series of degradative and spectroscopic studies; they have now been confirmed by synthesis.[132] Among the interesting members of this family is khat, a drug plant acceptable to Islam and well-known in the folk-lore of Arabia and East Africa, where it plays a role similar to wine in western society. The main physiological effects seem to be due to the presence of norpseudoephedrine (**125**), but several alkaloids

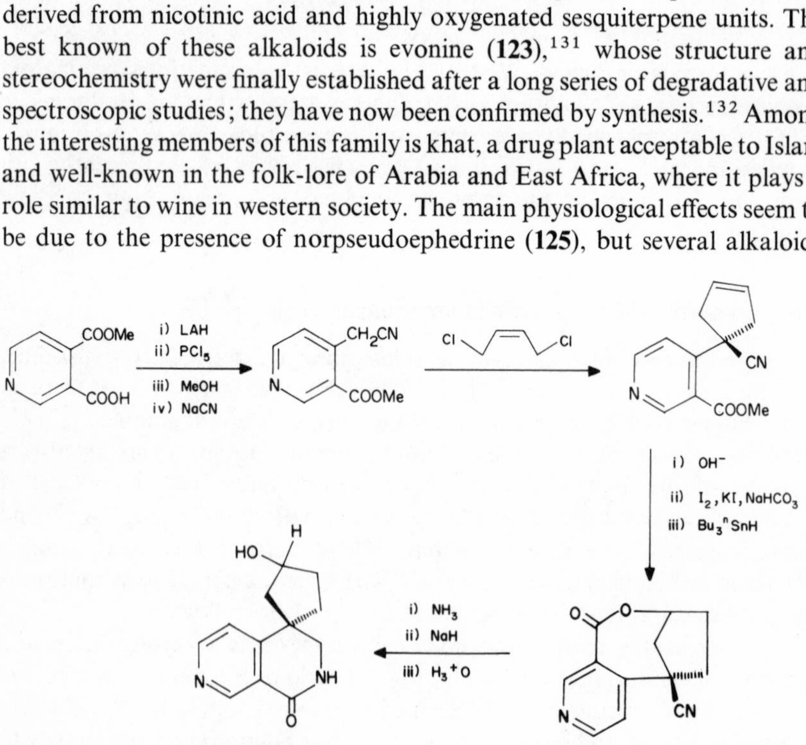
COOMe
N
COOH
i) LAH
ii) PCl5
iii) MeOH
iv) NaCN
CH2CN
N
COOMe
Cl
Cl
CN
N
COOMe
i) OH⁻
ii) I2, KI, NaHCO3
iii) Bu3ⁿSnH
O
O
N
CN
i) NH3
ii) NaH
iii) H3⁺O
H
HO
N
NH
O

(122) Sesbanine

Scheme 35

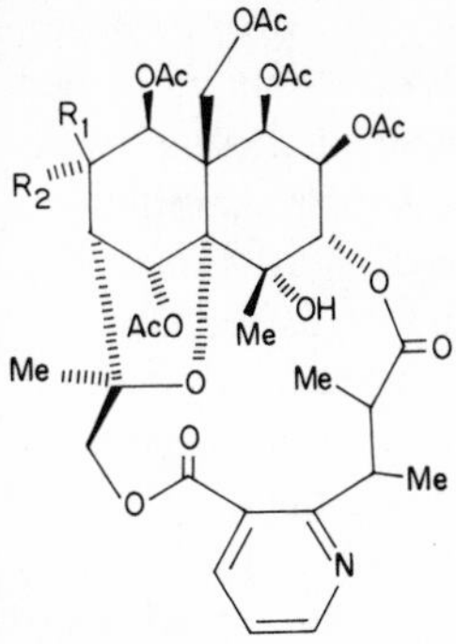

(123) Evonine (R₁,R₂=0)

(124) (R₁ = OAc, R₂= H)

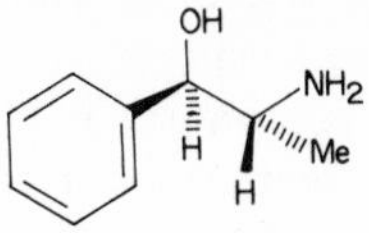

(125) (+)-Norpseudoephedrine

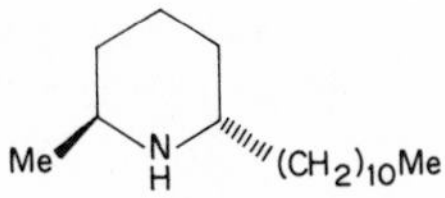

(126) Solenopsin A

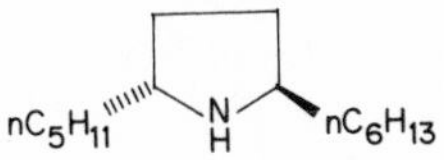

(127)

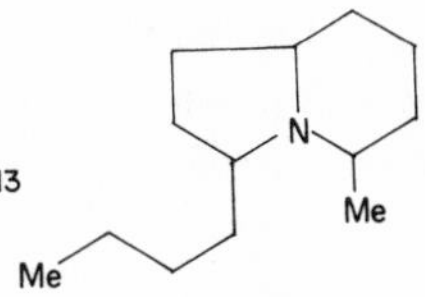

(128)

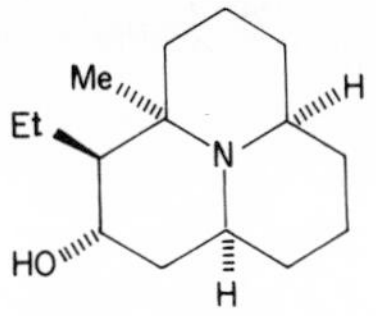

(130) Poranthericine

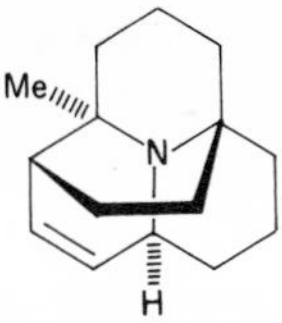

(131) Porantherine

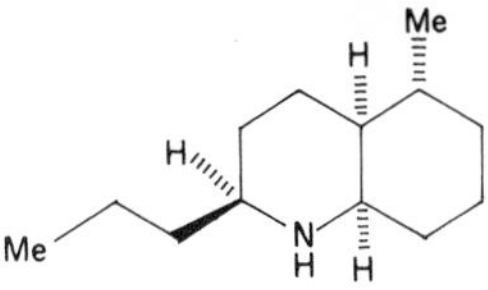

(132) Pumiliotoxin C

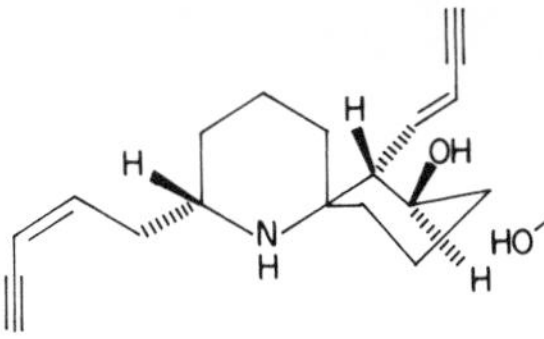

(133) Histrionicotoxin

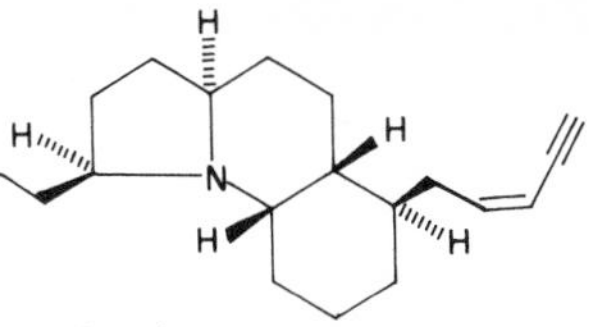

(134) Gephyrotoxin

analogous to evonine have also been isolated,[133] with a general structure of the type (124) in which some of the acetate groups are replaced by hydroxyls, or by other ester units such as nicotinate.

7.6 Alkaloids derived from polyketides

Although the great majority of alkaloids have one or more amino acids incorporated into their structures, there are some that seem to be constructed almost entirely from other building blocks except for the nitrogen atom. Many of these bases have a skeleton composed of a polyketide chain formed from acetate units, and they usually have an odd number of carbon atoms because of decarboxylation at the end of the chain. Some simple examples are found among the insect venoms, defence substances and pheromones. The piperidine derivative solenopsin A (126)[134] has been isolated from fire-ant venom, and its structure was proved by synthesis; on the other hand thief-ant venom contains a pyrrolidine derivative (127),[135] while a trail pheromone (128)[136] of the Pharaoh ant has these two nuclei fused together.

From a structural point of view, the most interesting of this group of bases are the perhydro [9b] azaphenylenes elaborated by ladybird beetles for defence against predators. An example is precoccinelline (129),[137] whose structure has been confirmed by several syntheses including that shown in Scheme 36. A group of alkaloids with remarkably similar structures to (129) have been isolated from a euphorbiaceous shrub growing in Australia. They include poranthericine (130) and porantherine (131), whose structures were

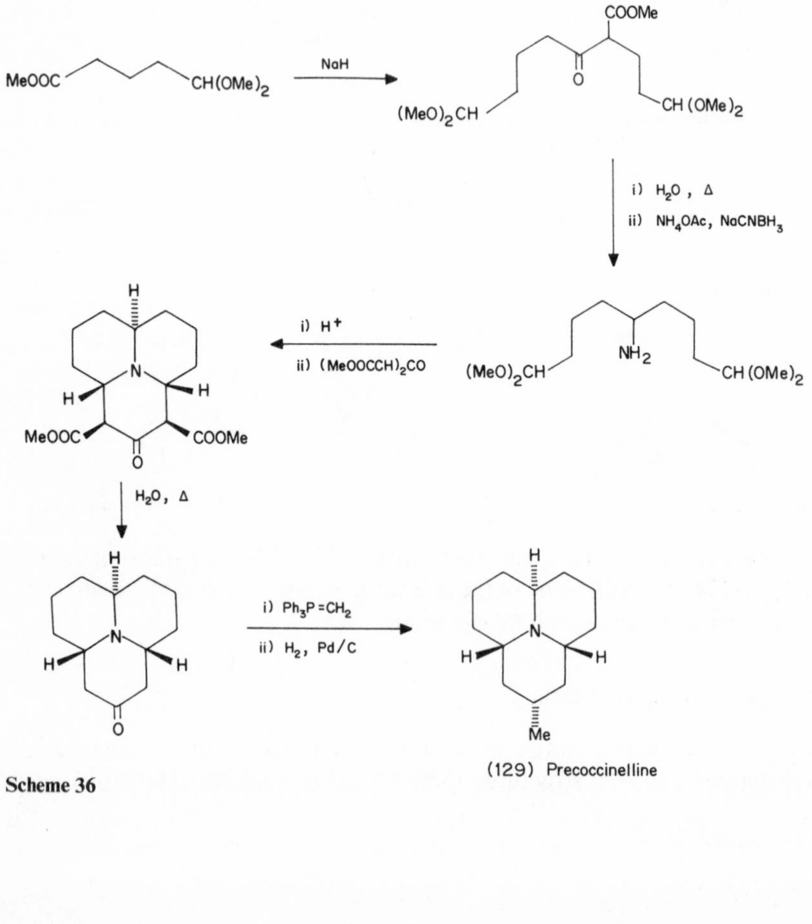

Scheme 36

(129) Precoccinelline

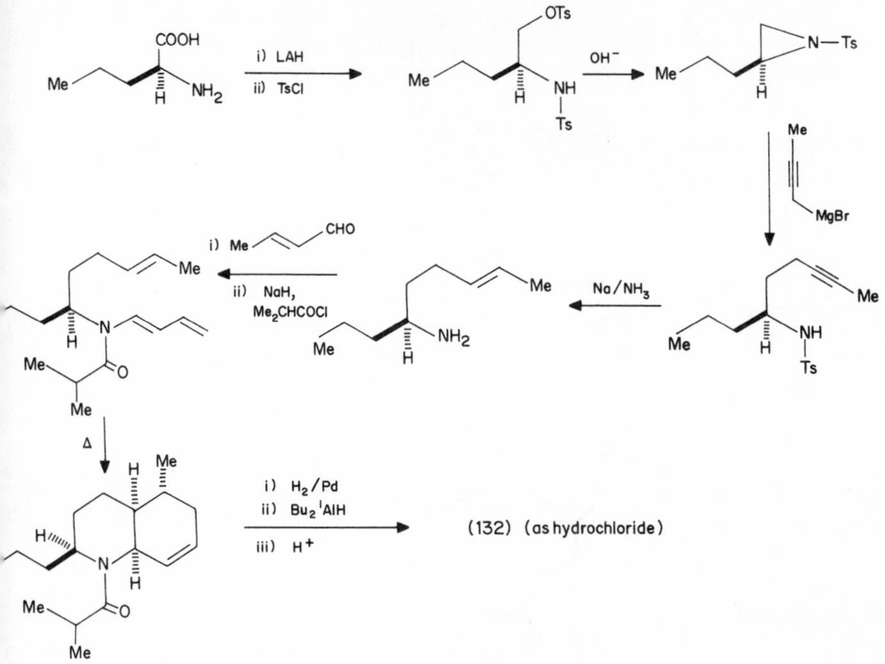

Scheme 37

established by X-ray crystallography;[138] the latter has also been synthesized.[139]

A series of toxins that have some structural relationship with certain of the above-mentioned insect alkaloids has been isolated[140] from the skins of tropical frogs of the genus *Dendrobates* found in Colombia and neighbouring countries, where they are used as blow-dart poisons. Some of the toxins, such as pumiliotoxin C (132),[141,143] histrionicotoxin (133),[141] and gephyrotoxin (134),[141,142] possess extremely potent biological activity, and have given rise to considerable interest and synthetic activity. An enantioselective synthesis[143] of natural (−)pumiliotoxin-C hydrochloride is shown in Scheme 37.

7.7 Alkaloids of terpenoid and steroidal origin

Apart from the alkaloids that are built up from tryptophan or some other amino acid plus a terpenoid, there are others which appear to be constructed from isoprenoid units only, such as polyzonimine (135),[144] a millipede defence agent whose structure was deduced by spectroscopy and confirmed by synthesis and by X-ray crystallography, and the sesquiterpene tropolone manicoline A (136).[145] Other interesting sesquiterpene alkaloids studied in

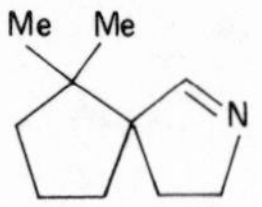

(135) Polyzonimine

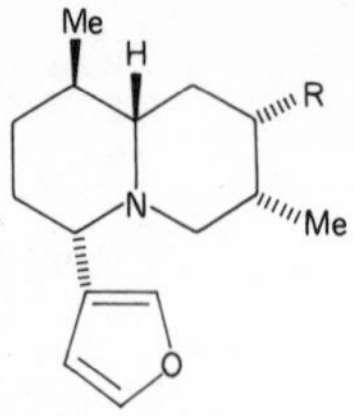

(136) Manicoline A

(137) Isocastoramine (R = OH)
(138) Deoxynupharidine (R = H)

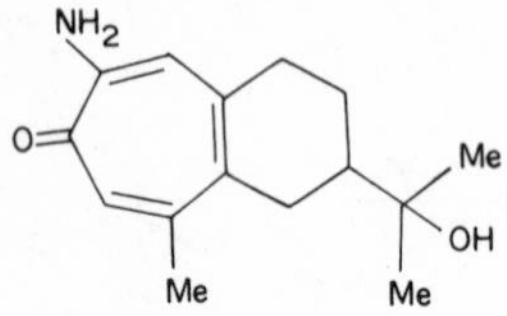

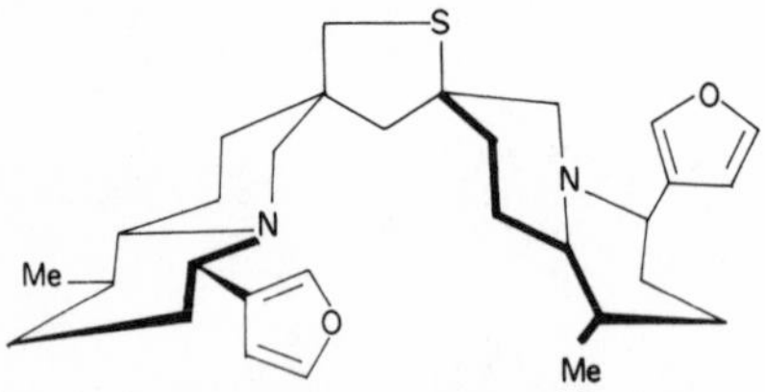

(139) Thiobinupharidine

recent years include the chief constituent of the scent gland of the Canadian beaver, isocastoramine (137),[146] which is accompanied by various analogues such as deoxynupharidine (138).[147] This and some further analogues of (137) also occur in the rhizomes of water-lilies belonging to the genus *Nuphar*. The structure and stereochemistry of isocastoramine were determined by spectroscopy and by correlation with deoxynupharidine, whose constitution had been established by several syntheses. Other types of *Nuphar* alkaloids include a number of unique sulphur-containing dimeric bases such as thiobinupharidine (139),[148] whose structure has been determined by X-ray crystallography. Several sulphoxides derived from these dimeric bases have also been isolated and their stereochemistry studied,[149] mainly by spectroscopic methods.

To the three main groups of diterpene alkaloids, a further group with a novel type of skeleton has now been added; the new bases have been isolated from *Anopteris* spp. native to Australia, extracts of which show anticancer activity. The basic structure of these alkaloids was established[150] by X-ray crystallography, and the details for the individual alkaloids, of which the principal one is anopterine (140), were determined by ¹H n.m.r. spectroscopy.[150]

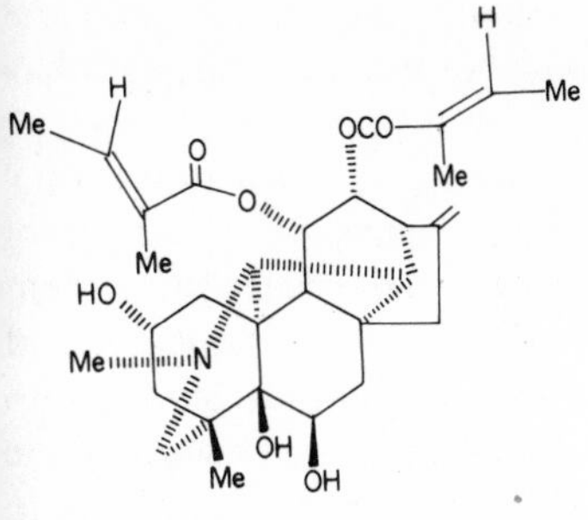

(140) Anopterine

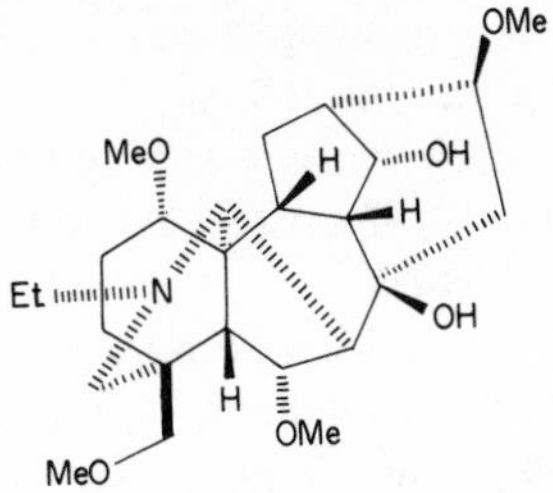

(141) Chasmanine

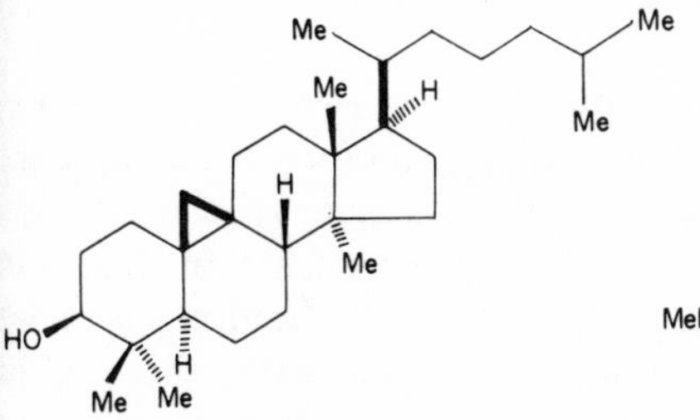

(142) Cycloartanol

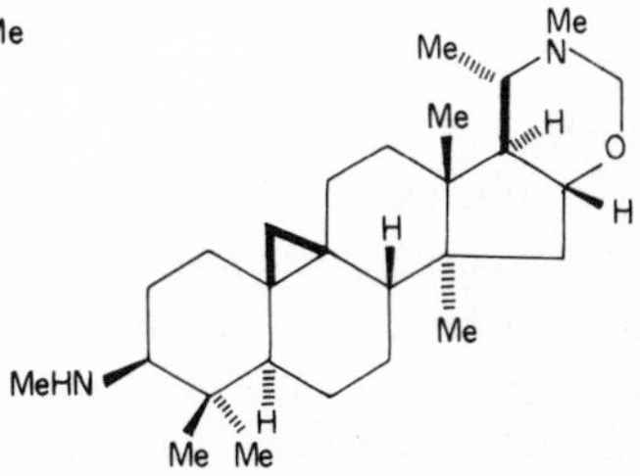

(143) Buxozine C

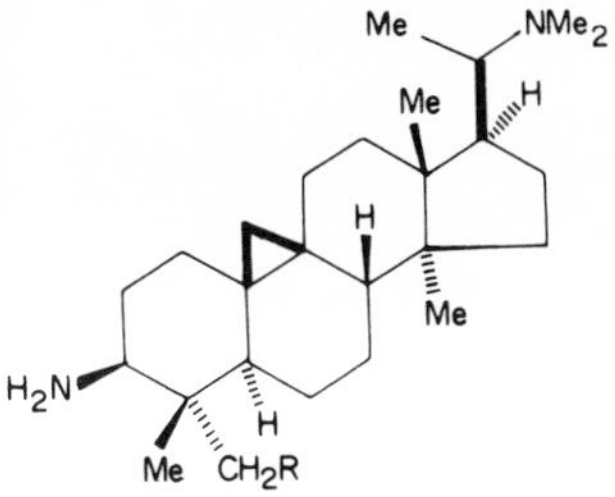

(144) Cyclobuxidine F, R = OH
(145) Cycloprotobuxine F, R = H

The field of diterpene alkaloids has been dominated during the last decade by the massive synthetic programme of Wiesner and his collaborators, who have succeeded in developing stereo- and regiospecific methods for the construction[151] of such complex examples as chasmanine (141), with seven interlocking ring systems and thirteen chiral centres. Reviews[152] are available of these monumental achievements in molecular engineering, which have extended over many years and have involved the preparation and study of a large number of model compounds.

Several dimeric diterpene alkaloids have recently been found, and the complex task of deducing their structures from ^{13}C and ^1H n.m.r. data has been successfully accomplished.[153]

The *Buxus* alkaloids are usually classified as steroidal, but structurally they are perhaps more closely related to the triterpenes, in particular to lanosterol and cycloartanol (142). A recent example with an unusual oxazine ring system is buxozine C (143), whose structure was deduced by spectroscopy and degradation.[154] Cyclobuxidine F (144) has a methyl and a carbinol group attached to C-4, and there has been some conflicting evidence regarding the relative orientation of these two groups. The question has finally been resolved[155] by a careful comparison of the ^{13}C n.m.r. spectra of (142), (144), and cycloprotobuxine F (145), from which it is clear that the carbinol group has an equatorial α configuration, and in consequence the methyl carbon attached to C-4 in (144) shows a distinct γ-effect.

A good deal of interest has been focused recently on one of the best-known steroidal alkaloids, solasodine (146), as a starting-point for the manufacture of biologically-active steroidal hormones. Extensive surveys[156] in various parts of the world have been made to determine the most suitable *Solanum* species for its extraction, and new methods for isolating it[157] and converting it into progesterone derivatives have been described, such as that sketched in Scheme 38.[158] A number of related bases have been reported, including solaverbascine (147), isolated recently from a *Solanum* sp. grown in Vietnam;

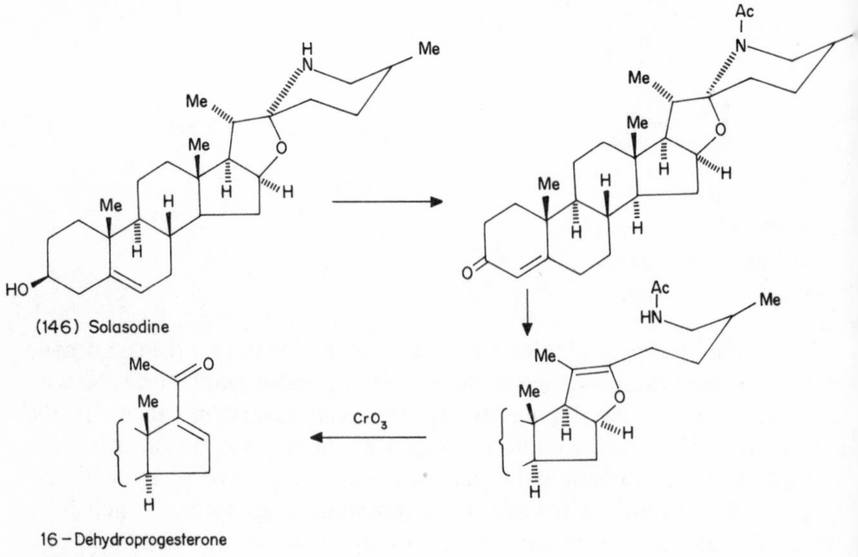

(146) Solasodine

16 – Dehydroprogesterone

Scheme 38

(147) Solaverbascine

(148) Tomatillidine

its structure was proved by oxidation with manganese dioxide, which converted it, via the imine, into a mixture of (146) and its C_{22} epimer, tomatidine.[159] The structure of tomatillidine has been revised to (148) as a result of a spectroscopic study of the alkaloid and its derivatives, and the new structure has been confirmed by a conversion of solasodine (146) into (148), the final stage of which involves ring contraction, catalysed by silica gel, of a piperid-1-en-3-one ring to the unusual pyrrolone ring occurring in (148).[160]

Some interesting new steroidal alkaloids of fungal origin have a nitrogen atom inserted into ring D.[161] They are antibiotics that are active against a wide variety of moulds and bacteria; an example is (149). The *Salamandra* alkaloids on the other hand have ring A expanded to an azepine with the inclusion of an amino group. It is considered likely that these bases, which have antibiotic properties and are secreted by the skin glands of various amphibia, have the function of protecting the animal against infection. The original structure ascribed to cycloneosamandaridine has now been revised to (150) in the light of further synthetic and spectroscopic evidence.[162]

7.8 Miscellaneous alkaloids

Apart from the peptide group of alkaloids which is treated in chapter 6, there remain a number of types that are difficult to classify under any of the preceding headings. Some of these alkaloids are guanidine derivatives, such as alchorneine (151),[163] zoanthoxanthin (152),[164] and saxitoxin (153).[165] Alchorneine, which has an unusual N-methoxyl group, occurs in a euphorbiaceous plant, and its structure was deduced by a combination of degradation and spectroscopy. The other two bases are of marine origin, and their structures were established by X-ray crystallography. Saxitoxin, one of the most potent neurotoxins known, is produced by various micro-organisms that proliferate off the coasts of north America, forming the so-called red tides which infest various types of shellfish and render them extremely toxic when eaten. The structure of saxitoxin has been confirmed by a skilful synthesis[166] that takes advantage of subtle differences in reactivity between carbonyl and

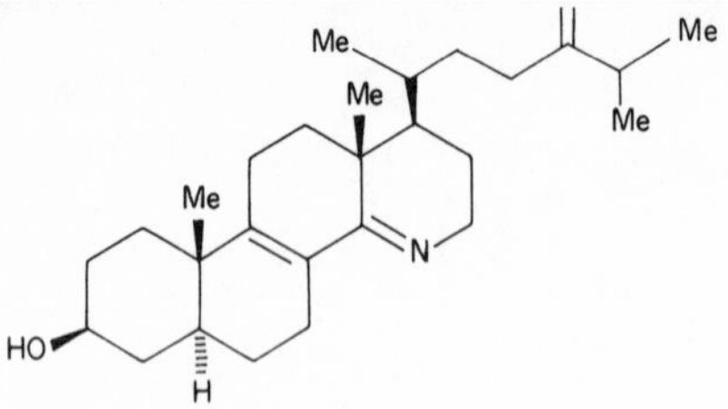

(149)

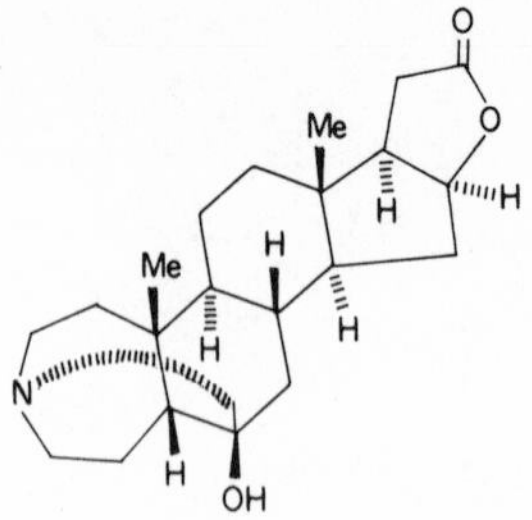

(150) Cycloneosamandaridine

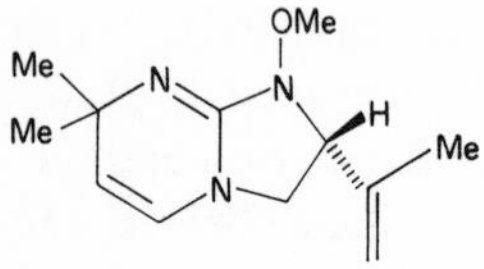

(151) Alchorneine

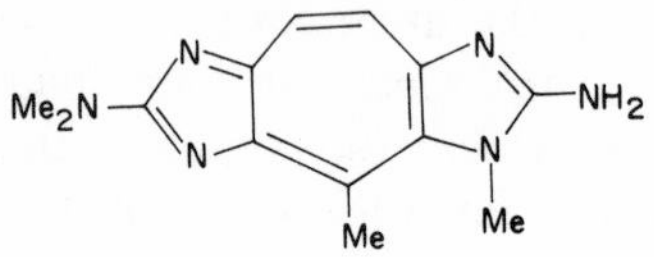

(152) Zoanthoxanthin

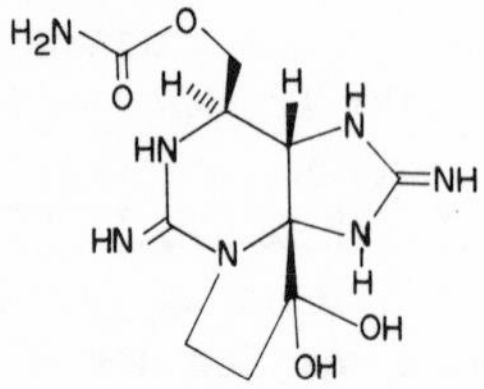

(153) Saxitoxin

acetal groups on the one hand, and their sulphur analogues on the other (Scheme 39).

Another type of alkaloid not included in the previous sections is that related to spermidine (154), a base which in turn is known to be formed from putrescine and methionine. Palustrine (155)[167] and maytenine (156)[168] are examples of spermidine-derived alkaloids; the former has a macro ring and occurs in a plant known as marsh horsetail, while maytenine was obtained from a *Maytenus* sp. (Celastraceae). This genus, however, is now better known for its production of alkaloids of quite a different type which have attracted widespread interest because of their unique macrocyclic structure and their potent antileukemic properties. The structure and stereochemistry of the principal alkaloid, maytansine (157),[169] were deduced largely by spectroscopy and confirmed by X-ray crystallography. There has been a good deal of competition to achieve a total synthesis of maytansine between at least six groups, two of which have reported success quite recently and almost simultaneously.[170]

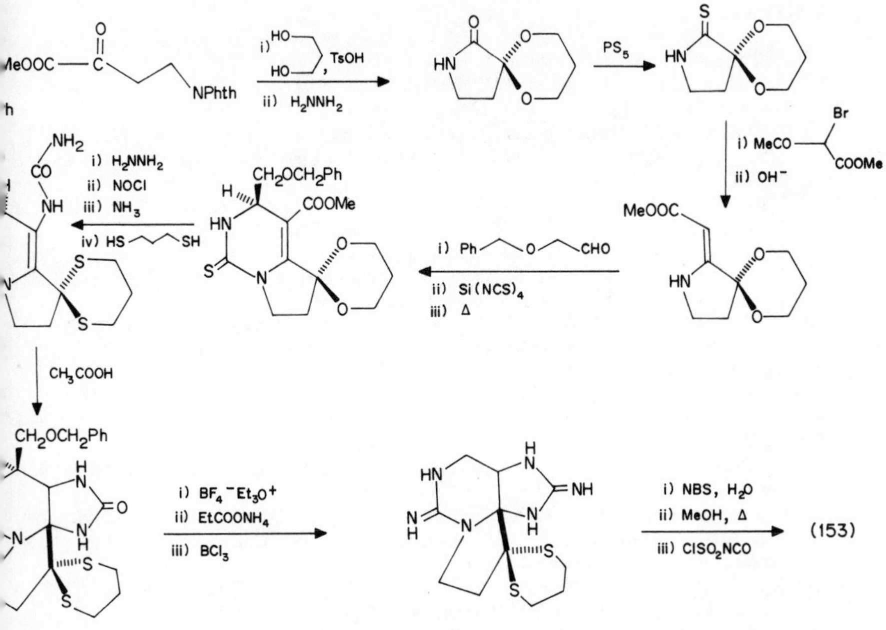

Scheme 39

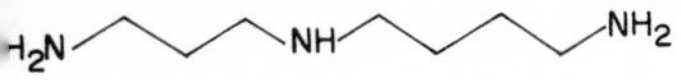

(154) Spermidine

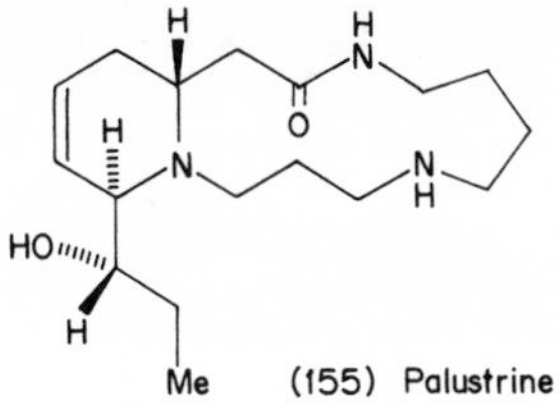

(155) Palustrine

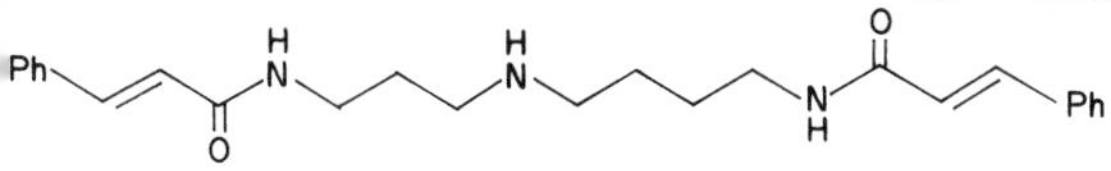

(156) Maytenine

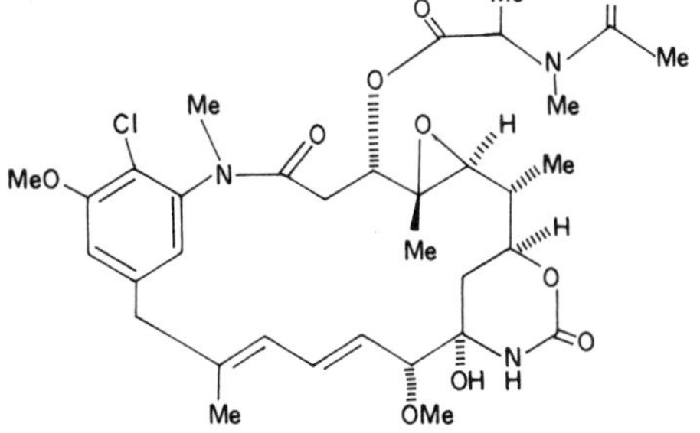

(157) Maytansine

References

1. 'Alkaloid Chemistry'; M. Hesse, 1981, Wiley, New York.
2. 'The Alkaloids', D.R. Dalton, 1979, Dekker, New York.
3. 'Introduction to Alkaloids', G.A. Cordell, 1981, Wiley, New York.
4. Specialist Periodical Reports: 'Alkaloids', ed. J.E. Saxton, vols. 3–5, 1973–1975; ed. M.F. Grundon, vols. 6–12, 1976–1982, The Chemical Society/Royal Society of Chemistry, London.
5. 'The Alkaloids', ed. R.H.F. Manske, vols. 14–16, 1973–1977; ed. R.H.F. Manske and R.G.A. Rodrigo, vols. 17–20, 1979–1981, Academic Press, New York.
6. M. Takagi, S. Funahashi, K. Ohta, and T. Nakahayashi, 1980, *Agr. Biol. Chem.*, 44, 3019.
7. I.R.C. Bick, J.W. Gillard, H.-M. Leow, M. Lounasmaa, J. Pusset, and T. Sévenet, 1981, *Planta Medica*, 41, 379, and references therein.
8. I.R.C. Bick, J.W. Gillard, and H.-M. Leow, 1979, *Aust. J. Chem.*, 32, 2523.
9. I.R.C. Bick, J.W. Gillard, and H.-M. Leow, 1979, *Aust. J. Chem.*, 32, 2071.
10. M. Lounasmaa, J. Pusset, and T. Sévenet, 1980, *Phytochemistry*, 19, 949, 953.
11. I.R.C. Bick, J.W. Gillard, and H.-M. Leow, 1979, *Aust. J. Chem.*, 32, 1827.
12. M. Lounasmaa, T. Langenskiold, and C. Holmberg, 1981, *Tetrahedron Lett.*, 22, 5179.
13. P.J. KocziensKi and M. Kirkup, 1975, *J. Org. Chem.*, 40, 2998.
14. J.J. Tufariello, G.B. Mullen, J.J. Tegeler, E.J. Trybulski, S.C. Wong, and Sk. A. Ali, 1979, *J. Amer. Chem. Soc.*, 101, 2435.
15. Y. Hayakawa, Y. Baba, S. Kalino, and R. Noyosi, 1978, *J. Amer. Chem. Soc.*, 100, 1786.
16. A.B. Ray, Y. Oshima, H. Hikino, and C. Kabuto, 1982, *Heterocycles*, 19, 1233; A.R. Pinder, 1982, *J. Org. Chem.*, 47, 3607.
17. H.F. Campbell, O.E. Edwards, J.W. Elder, and R.J. Kolt, 1979, *Pol. J. Chem.*, 53, 27.
18. F. Bohlmann, C. Zdero, and M. Grenz, 1977, *Chem. Ber.*, 110, 474.
19. J. Meinwald, T. Smolanoff, A.T. McPhail, R.W. Miller, T. Eisner, and K. Hicks, 1975, *Tetrahedron Lett.*, 2367.
20. J.A. Edgar, M. Boppré, and D. Schneider, 1979, *Experientia*, 35, 1447.
21. M.T. Pizzorno and S.M. Albonico, 1978, *Chem. and Ind.*, 349.
22. R.F. Borch and B.C. Ho, 1977, *J. Org. Chem.*, 42, 1225.
23. E. Leete, 1972, *J. Chem. Soc. Chem. Comm.*, 1091.
24. S.M. Weinreb, N.A. Khatri, and J. Shringarpure, 1979, *J. Amer. Chem. Soc.*, 101, 5073.
25. H. Budzikiewicz, L. Faber, E.-G. Herrmann, F.F. Perrollaz, U.P. Schlunegger, and W. Wiegrebe, 1979, *Annalen*, 1212.
26. J.S. Fitzgerald, S.R. Johns, J.A. Lamberton, A.H. Redcliffe, A.A. Sioumis, and H. Suares, 1972, *Anal. Quim.*, 68, 737.
27. P. Slosse and G. Hootelé, 1979, *Tetrahedron Lett.*, 4587.
28. L. Nyembo, A. Goffin, C. Hootelé, and J.C. Braekman, 1978, *Can. J. Chem.*, 56, 851.
29. W.A. Ayer, L.M. Browne, Y. Nakahara, M. Tori, and L.T.J. Delbaere, 1979, *Can. J. Chem.*, 57, 1105.
30. E. Kleinman and C.H. Heathcock, 1979, *Tetrahedron Lett.*, 4125.
31. W. Oppolzer and M. Petrzilka, 1976, *J. Amer. Chem. Soc.*, 98, 6722.
32. K. Fuji, T. Yamada, E. Fujita, and H. Murata, 1978, *Chem. Pharm. Bull.*, 26, 2513.
33. M. Hanaoka, N. Ogawa, and Y. Arata, 1973, *Tetrahedron Lett.*, 2355.
34. M.M. Kadooka, M.Y. Chang, H. Fukami, P.J. Scheuer, J. Clardy, B.A. Solheim, and J.P. Springer, 1976, *Tetrahedron*, 32, 919.
35. W. Boczoń, G. Pieczonka, and M. Wiewiórowski, 1977, *Tetrahedron*, 33, 2565.
36. A.S. Sadykov, F.G. Kamayev, V.A. Korenevsky, V.B. Leontev, and Yu. A. Ustynyuk, 1972, *Org. Magn. Resonance*, 4, 837.
37. M. Guetté, J. Capillon, and J.P. Guetté, 1973, *Tetrahedron*, 29, 3659.
38. S. Ohmiya, K. Higashiyama, H. Otomasu, J. Haginiwa, and I. Murakoshi, 1979, *Chem. Pharm. Bull.*, 27, 1055; J. Bordner, S. Ohmiya, H. Otomasu, and J. Haginiwa, 1980, *Chem. Pharm. Bull.*, 28, 1965.
39. V.A. Mnatsakanyan, V. Preiniger, V. Šimánek, J. Juřina, A. Klásek, L. Dolejš, and F. Šantavý, 1977, *Coll. Czech. Chem. Comm.*, 42, 1421.

40. R. Ziyaev, T. Irgashev, I.A. Israilov, N.D. Abdullaev, M.S. Yunusov, and S. Yu. Yunusov, 1977, *Khim. prirod. Soedinenii*, 239.
41. I.R.C. Bick, T. Sévenet, W. Sinchai, B.W. Skelton, and A.H. White, 1981, *Aust. J. Chem.*, 34, 195.
42. T.R. Govindachari, B.R. Pai, H. Suguna, and M.S. Premila, 1977, *Heterocycles*, 6, 1811.
43. R.H.F. Manske, R. Rodrigo, H.L. Holland, D.W. Hughes, D.B. MacLean, and J.K. Saunders, 1978, *Can. J. Chem.*, 56, 383.
44. V. Fajardo, V. Elango, S. Chattopadhyay, L.M. Jackson, and M. Shamma, 1983, *Tetrahedron Lett.*, 24, 155.
45. V. Fajardo, V. Elango, B.K. Cassels, and M. Shamma, 1982, *Tetrahedron Lett.*, 23, 39.
46. G.D. Pandey and K.P. Tiwari, 1979, *Heterocycles*, 12, 1327.
47. G. Nonaka and I. Nishioka, 1975, *Chem. Pharm. Bull.*, 23, 294.
48. L.D. Yakhontova, M.N. Komarova, M.E. Perel'son, K.F. Blinova, and O.N. Tolkachev, 1972, *Khim prirod. Soedin*, 624.
49. Z. Horii, C. Iwata, and Y. Nakashita, 1978, *Chem. Pharm. Bull.*, 26, 481.
50. C. Casagrande and L. Canonica, 1975, *J. Chem. Soc. Perkin I*, 1647.
51. S.F. Hussain, M.T. Siddiqui, G. Manikumar, and M. Shamma, 1980, *Tetrahedron Lett.*, 21, 723.
52. S.M. Kupchan and C.-K. Kim, 1975, *J. Amer. Chem. Soc.*, 97, 5623.
53. L. Castedo, R. Estévez, J.M. Saá, and R. Suau, 1978, *Tetrahedron Lett.*, 2179.
54. J.V. Silverton, C. Kabuto, K.T. Buck, and M.P. Cava, 1977, *J. Amer. Chem. Soc.*, 99, 6708.
55. J. Kunitomo and M. Satoh, 1982, *Chem. Pharm. Bull.*, 30, 2659.
56. N.N. Margvelashvili, O.F. Lasskaya, A.T. Kir'yanova, and O.N. Tolkachev, 1976, *Khim. prirod. Soedin*, 123.
57. G. Blaskó, S.F. Hassain, and M. Shamma, 1982, *J. Amer. Chem. Soc.*, 104, 1599.
58. C.J. Gilmore, R.F. Bryan, and S.M. Kupchan, 1976, *J. Amer. Chem. Soc.*, 98, 1947.
59. L. Koike, A.J. Marsaioli, F. de A.M. Reis, and I.R.C. Bick, 1982, *J. Org. Chem.*, 47, 4351.
60. F. Scheinmann, E.F.V. Scriven, and O.N. Ogbeide 1980, *Phytochemistry*, 19, 1837.
61. J. Baldas, I.R.C. Bick, T. Ibuka, R.S. Kapil, and Q.N. Porter, 1972, *J. Chem. Soc. Perkin I*, 592; J. Baldas, I.R.C. Bick, M.R. Falco, J.X. de Vries, and Q.N. Porter, 1972, *J. Chem. Soc. Perkin I*, 597; J. Baldas, I.R.C. Bick, T. Ibuka, R.S. Kapil, and Q.N. Porter, 1972, *J. Chem. Soc. Perkin I*, 599.
62. I.R.C. Bick, J.B. Bremner, M.P. Cava, and P. Wiriyachitra, 1978, *Aust. J. Chem.*, 31, 321.
63. T. Ibuka, T. Konoshima, and Y. Inubushi, 1972, *Tetrahedron Lett.*, 4001.
64. M. Shamma, J.L. Moniot, S.Y. Yao, G.A. Miawa, and M. Ikram, 1972, *J. Amer. Chem. Soc.*, 94, 1381; H. Guinaudeau, V. Elango, M. Shamma, and V. Fajardo, 1982, *J. Chem. Soc. Chem. Comm.*, 1122.
65. M. Shamma and J.L. Moniot, 1974, *J. Amer. Chem. Soc.*, 96, 3338.
66. A. Jössang, M. Leboeuf, and A. Cavé, 1982, *Tetrahedron Lett.*, 5147.
67. V. Fajardo, H. Guinaudeau, V. Elango, and M. Shamma, 1982, *J. Chem. Soc. Chem. Comm.*, 1350.
68. J. Guilhem and I.R.C. Bick, 1981, *J. Chem. Soc. Chem. Comm.*, 1007.
69. S.M. Kupchan, A.J. Liepa, V. Kameswaran, and K. Sempuku, 1973, *J. Amer. Chem. Soc.*, 95, 2995.
70. H.C. Beyerman, T.S. Lie, L. Maat, H.H. Bosman, E. Buurman, E.J.M. Bijsterveld, and H.J.M. Sinnige, 1976, *Rec. Trav. Chim. Pays-Bas*, 95, 24.
71. D.D. Weller and H. Rappoport, 1975, *J. Med. Chem.*, 19, 1171.
72. F. Calvo Mondelo, Swiss Pat. 619227, 1980 (1981, *Chem. Abstr.*, 94, 30984); K.C. Rice, 1977, *J. Med. Chem.*, 20, 164.
73. T. Sano, J. Toda, N. Kashiwaba, Y. Tsuda, and Y. Iitaka, 1981, *Heterocycles*, 16, 1151; T. Sano, J. Toda, and Y. Tsuda, 1982, *Heterocycles*, 18, 229.
74. M.F. Semmelhack, B.P. Chong, R.D. Stauffer, T.D. Rogerson, A. Chong, and L.D. Jones, 1975, *J. Amer. Chem. Soc.*, 97, 2507.
75. S. Agurell, I. Granelli, K. Leander, B. Luning, and J. Rosenblom, 1974, *Acta Chem. Scand. (B)*, 28, 239.
76. M.P. Cava, K.T. Buck, and A.I. da Rocha, 1972, *J. Amer. Chem. Soc.*, 94, 5931.

77. K. Shimizu, K. Tomioka, S.-I. Kamada, and K. Koga, 1977, *Heterocycles*, 8, 277; 1978, *Chem. Pharm. Bull.*, 26, 3765.
78. S. Roy, P. Bhattacharyya, and D.P. Chakaborty, 1974, *Phytochemistry*, 13, 1017.
79. C.M. Hasan, T.M. Healey, P.G. Waterman, and C.H. Schwalbe, 1982, *J. Chem. Soc. Perkin Trans. I*, 2807.
80. A. Jössang, H. Jacquemin, J.-L. Pousset, A. Cavé, M. Damak, and C. Riche, 1977, *Tetrahedron Lett.*, 1219.
81. K.A. Ubaidullaev, R. Shakirov, and S. Yu. Yunusov, 1976, *Khim. prirod. Soedinenii*, 553.
82. Y. Oikawa, T. Yoshioka, K. Mohri, and O. Yonemitsu, 1979, *Heterocycles*, 12, 1457.
83. M. Onda and Y. Konda, 1978, *Chem. Pharm. Bull.*, 26, 2167.
84. E. Leete, 1979, *J. Chem. Soc. Chem. Comm.*, 821.
85. H. Wagner and T. Nestler, 1978, *Tetrahedron Lett.*, 2777.
86. L.M.G. Aguiar, R. Braz Filho, O.R. Gottlieb, J.G.S. Maia, S.L.V. Pinho, and J.R. de Sousa, 1980, *Phytochemistry*, 19, 1859.
87. B. Danieli, G. Lesma, and G. Palmisano, 1979, *Experientia*, 35, 156.
88. G. Büchi, P.R. DeShong, S. Katsumura, and Y. Sugimura, 1979, *J. Amer. Chem. Soc.*, 101, 5084.
89. S. Sekita, K. Yoshihira, S. Natori, and H. Kuwano, 1977, *Tetrahedron Lett.*, 2771.
90. J.-L. Pousset, J. Kerharo, G. Maynart, X. Monseur, A. Cavé, and R. Goutarel, 1973, *Phytochemistry*, 12, 2308.
91. J.S. Carlé and C. Christophersen, 1979, *J. Amer. Chem. Soc.*, 101, 4012; 1980, *J. Org. Chem.*, 45, 1586.
92. T.-C. Choong and H.R. Shough, 1977, *Tetrahedron Lett.*, 3137.
93. V.W. Armstrong, S. Coulton, and R. Ramage, 1976, *Tetrahedron Lett.*, 4311; R. Ramage, V.W. Armstrong, and S. Coulton, 1981, *Tetrahedron*, 37, Suppl. 1, 157.
94. P.M. Scott, M.A. Merrien, and J. Polonsky, 1976, *Experientia*, 32, 140.
95. J. Polonsky, M.A. Merrien, T. Prangé, C. Pascard, and S. Moreau, 1980, *J. Chem. Soc. Chem. Comm.*, 601.
96. G. Garli, R. Cardillo, and C. Fuganti, 1978, *Tetrahedron Lett.*, 2605.
97. A. Jössang, J.-L. Pousset, H. Jacquemin, and A. Cavé, 1977, *Tetrahedron Lett.*, 4317.
98. N. Chaichit, B.M. Gatehouse, I.R.C. Bick, M.A. Hai, and N.W. Preston, 1979, *J. Chem. Soc. Chem. Comm.*, 874.
99. M. Bittner, M. Silva, E.M. Gopalakrishna, W.H. Watson, V. Zabel, S.A. Matlin, and P.G. Sammes, 1978, *J. Chem. Soc. Chem. Comm.*, 79.
100. H.P. Ros, R. Kyburz, N.W. Preston, R.T. Gallagher, I.R.C. Bick, and M. Hesse, 1979, *Helv. Chim. Acta*, 62, 481.
101. B.E. Anderson, G.B. Robertson, H.P. Avey, W.F. Donovan, I.R.C. Bick, J.B. Bremner, A.J.T. Finney, N.W. Preston, R.T. Gallagher, and G.B. Russell, 1975, *J. Chem. Soc. Chem. Comm.*, 511.
102. C. Mirand, G. Massiot, and J. Lévy, 1982, *J. Org. Chem.*, 47, 4169.
103. C.R. Hutchinson, A.H. Heckendorf, P.E. Daddona, E. Hagaman, and E. Wenkert, 1974, *J. Amer. Chem. Soc.*, 96, 5609.
104. R.T. Brown and S.B. Pratt, 1980, *J. Chem. Soc. Chem. Comm.*, 165.
105. M. Uskoković, R.L. Lewis, J.J. Partridge, C.W. Despreaux, and D.L. Pruess, 1979, *J. Amer. Chem. Soc.*, 101, 6742, and references therein.
106. R. Besselièvre, J.P. Cossom, B.C., Das, and H.-P. Husson, 1980, *Tetrahedron Lett.*, 21, 63.
107. F. Hotellier, P. Delaveau, and J.-L. Pousset, 1979, *Planta Med.*, 35, 242.
108. M. Watanabe and V.A. Snieckus, 1980, *J. Amer. Chem. Soc.*, 102, 1457.
109. J.-Y. Laronze, J. Laronze-Fontaine, J. Lévy, and J. Le Men, 1974, *Tetrahedron Lett.*, 491.
110. J. Bruneton, A. Cavé, E.W. Hagaman, N. Kunesch, and E. Wenkert, 1976, *Tetrahedron Lett.*, 3567, and references therein.
111. B.M. Trost, S.A. Godleski, and J.P. Genêt, 1978, *J. Amer. Chem. Soc.*, 100, 3930; B.M. Trost, S.A. Godleski, and J.L. Belletire, 1979, *J. Org. Chem.*, 44, 2052.
112. T.A. van Beek, P.P. Lankhors, R. Verpoorte, and A.B. Svendsen, 1982, *Tetrahedron Lett.*, 23, 4827.
113. F. Khuong-Huu, A. Chiaroni, and C. Riche, 1981, *Tetrahedron Lett.*, 22, 733.

114. G. Massiot, M. Zèches, P. Thépenier, M.-J. Jacquier, L. Le Men-Olivier, and C. Delaude, 1982, *J. Chem. Soc. Chem. Comm.*, 768.
115. M. Andriantsiferana, R. Besselièvre, C. Riche, and H.-P. Husson, 1977, *Tetrahedron Lett.*, 2587.
116. A. Husson, Y. Langlois, C. Riche, H.-P. Husson, and P. Potier, 1973, *Tetrahedron*, 3095.
117. P. Mangeney, R.Z. Andriamialisoa, N. Langlois, Y. Langlois, and P. Potier, 1979, *J. Amer. Chem. Soc.*, 101, 2243.
118. J.P. Kutney, J. Balsevich, T. Honda. P.H. Liao, H.P.M. Thiellier, and B.R. Worth, 1978, *Heterocycles*, 9, 201, and references therein.
119. G. Schneider and W. Kleinert, 1972, *Planta Med*, 22, 109.
120. J.C. Bradley and G. Büchi, 1976, *J. Org. Chem.*, 41, 699.
121. S. Iwadare, Y. Shizuri, K. Yamada, and Y. Hirata, 1974, *Tetrahedron Lett.*, 1177.
122. J.P. Fridrichsons, M.F. Mackay, and A. McL. Mathieson, 1974, *Tetrahedron*, 30, 85.
123. K.P. Parry and G.F. Smith, 1978, *J. Chem. Soc. Perkin I*, 1678; N.K. Hart, S.R. Johns, J.A. Lamberton, and R.E. Simmons, 1974, *Aust. J. Chem.*, 27, 639.
124. H.J. Preusser, G. Habermehl, M. Sablofski, and D. Schmall-Haury, 1975, *Toxicon*, 13, 285.
125. F. Khuong Huu, H. Monseur, G. Ratle, G. Lukacs, and R. Goutarel, 1973, *Tetrahedron Lett.*, 1757.
126. D.L. Dreyer and R.C. Brenner, 1980, *Phytochemistry*, 19, 935.
127. G.J. Kapadia, B.D. Paul, J.V. Silverton, M.H. Fales, and E.A. Sokoloski, 1975, *J. Amer. Chem. Soc.*, 97, 6814.
128. H.L. Sleeper and W. Fenical, 1977, *J. Amer. Chem. Soc.*, 99, 2367.
129. D. de B. Corrêa and O.R. Gottlieb, 1975, *Phytochemistry*, 14, 271.
130. J.C. Bottaro and G.A. Berchtold, 1980, *J. Org. Chem.*, 45, 1176.
131. Y. Shizuri, H. Wada, K. Sugiura, K. Yamada, and Y. Hirata, 1973, *Tetrahedron*, 29, 1773.
132. K. Sugiura, K. Yamada, and Y. Hirata, 1975, *Chem. Lett.*, 579.
133. L. Crombie, W.M.L. Crombie, D.A. Whiting, O.J. Braenden, and K. Szendrei, 1978, *J. Chem. Soc. Chem. Comm.*, 107.
134. K. Fuji, K. Ichikawa, and E. Fujita, 1979, *Chem. Pharm. Bull.*, 27, 3183, and references therein.
135. T.H. Jones, M.S. Blum, and H.M. Fales, 1979, *Tetrahedron Lett.*, 1031.
136. T.H. Jones and P.J. Kropp, 1974, *Tetrahedron Lett.*, 3503.
137. R.V. Stevens and A.W.M. Lee, 1979, *J. Amer. Chem. Soc.*, 101, 7032, and references therein.
138. W.A. Denne and A. McL. Mathieson, 1973, *J. Cryst. Mol. Struct.*, 3, 79, 87, 139.
139. E.J. Corey and R.D. Balanson, 1974, *J. Amer. Chem. Soc.*, 96, 6516.
140. J.W. Daly, G.B. Brown, M. Mensah-Dwumah, and C.H. Myers, 1978, *Toxicon*, 16, 163.
141. J.W. Daly, D. Witkop, T. Tokuyama, T. Nishikawa, and I.L. Karle, 1977, *Helv. Chim. Acta*, 60, 1128, and references therein.
142. R. Fujimoto and Y. Kishi, 1981, *Tetrahedron Lett.*, 22, 4197.
143. W. Oppolzer and E. Flaskamp, 1977, *Helv. Chim. Acta*, 60, 204.
144. J. Smolanoff, A.F. Kluge, J. Meinwald, A. McPhail, R.W. Miller, K. Hicks, and T. Eisner, 1975, *Science*, 188, 736.
145. J. Polonsky, J. Varenne, T. Prangé, C. Pascard, H. Jacquemin, and A. Fournet, 1981, *J. Chem. Soc. Chem. Comm.*, 731.
146. B. Maurer and G. Ohloff, 1976, *Helv. Chim. Acta*, 59, 1169.
147. J. Szychowski, A. Leniewski, and J.T. Wróbel, 1978, *Chem. and Ind.*, 273.
148. R.T. LaLonde and C.F. Wong, 1977, *Pure Appl. Chem.*, 49, 169.
149. R.T. LaLonde and C.F. Wong, 1978, *Can. J. Chem.*, 56, 56.
150. W.A. Denne, S.R. Johns, J.A. Lamberton, A. McL. Mathieson, and H. Suares, 1972, *Tetrahedron Lett.*, 2727.
151. K. Wiesner, 1979, *Pure Appl. Chem.*, 51, 689.
152. S.W. Pelletier and S.W. Page, in Specialist Periodical Reports: Alkaloids, ed. M.F. Grundon, The Chemical Society/Royal Society of Chemistry, London, 1982, vol. 12, p. 248; 1981, vol. 11, p. 218; 1979, vol. 9, p. 233; 1978, vol. 8, p. 229; S.W. Pelletier and N.V. Mody, in 'The Alkaloids', ed. R.H.F. Manske and R.G.A. Rodrigo, Academic Press, New York, 1981, vol. 18, p. 99; 1979, vol. 17, p. 1.

153. S.W. Pelletier, N.V. Mody, Z. Djarmati, and S.D. Lajsić, 1976, *J. Org. Chem.*, 41, 3042.
154. Z. Votický, L. Dolejš, O. Bauerová, and V. Paulík, 1977, *Coll. Czech. Chem. Comm.*, 42, 2549; *Phytochemistry*, 16, 1860.
155. M. Sangaré, F. Khuong-Huu, D. Herlem, A. Milliet, B. Septe, G. Berenger, and G. Lukacs, 1975, *Tetrahedron Lett.*, 1791.
156. D.M. Harrison, in Special Periodical Reports: Alkaloids, ed. M.F. Grundon, The Chemical Society/Royal Society of Chemistry, London, 1976, vol. 6, p. 289; 1978, vol. 8, pp. 256–6; 1980, vol. 10, p. 235; 1981, vol. 11, p. 232; 1982, vol. 12, p. 284.
157. J.D. Mann, 1978, *Adv. Agron.*, 30, 207 (1979, *Chem. Abstr.*, 91, 44430); J. Rodriguez, R. Segovia, E. Guerreiro, F. Ferretti, G.Z. de Sosa, and R. Ertola, 1979, *J. Chem. Technol. Biotechnol.*, 29, 525 (1980, *Chem. Abstr.*, 92, 59105).
158. G. Adam, H.Th. Huong, and M. Lischewski, Ger. (East) Pat. 132437 (1979, *Chem. Abstr.*, 91, 91840).
159. G. Adam, H.Th. Huong, and N.H. Khoi, 1980, *Phytochemistry*, 19, 1002.
160. G. Kusano, T. Takemoto, Y. Sato, and D.F. Johnson, 1976, *Chem. Pharm. Bull.*, 24, 611.
161. C.D. Jones, U.S. Pat. 4001246 (1977, *Chem. Abstr.*, 87, 53493); U.S. Pat. 4008238 (1977, *Chem. Abstr.*, 86, 171737); U.S. Pat. 4001246 (1977, *Chem. Abstr.*, 87, 53493); J.W. Chamberlin, U.S. Pat. 4039547 (1978, *Chem. Abstr.*, 88, 38080).
162. K. Oka and S. Hara, 1977, *J. Amer. Chem. Soc.*, 99, 3859; 1978, *J. Org. Chem.*, 43, 4408.
163. F. Khuong-Huu, J.P. Le Forestier, and R. Goutarel, 1972, *Tetrahedron*, 28, 5207.
164. L. Cariello, S. Crescenzi, G. Prota, S. Capasso, F. Giordano, and L. Mazzarella, 1974, *Tetrahedron*, 30, 3281.
165. E.J. Schantz, V.E. Ghazarossian, H.K. Schnoes, F.M. Strong, J.P. Springer, J.P. Pezzanite, and J. Clardy, 1975, *J. Amer. Chem. Soc.*, 97, 1238.
166. H. Tanino, T. Nakata, T. Kaneko, and Y. Kishi, 1977, *J. Amer. Chem. Soc.*, 99, 2818.
167. P. Ruedi and C.H. Eugster, 1978, *Helv. Chim. Acta*, 61, 899.
168. G. Engbert, K. Klinga, Raymond-Hamet, E. Schlittler, and W. Vetter, 1973, *Helv. Chim. Acta*, 56, 474.
169. S.M. Kupchan, Y. Komoda, W.A. Court, G.J. Thomas, R.M. Smith, A. Karim, C.J. Gilmore, R.C. Haltiwanger, and R.F. Bryan, 1972, *J. Amer. Chem. Soc.*, 94, 1354.
170. A.I. Mayers, P.J. Reider, and A.L. Campbell, 1980, *J. Amer. Chem. Soc.*, 102, 6597; E.J. Corey, L.O. Weigl, A.R. Chamberlin, H. Cho, and D.H. Hua, 1980, *J. Amer. Chem. Soc.*, 102, 6613.

8 Nucleosides, nucleotides and nucleic acids

J.B. HOBBS

8.1 Introduction

The last decade has witnessed great advances in research on the nucleic acids and their components, both in synthetic organic chemistry and in all fields contiguous to it. The potential value of nucleosides as antibiotic, antineoplastic and antiviral agents[1] has stimulated the search for more effective methods of nucleoside synthesis and, particularly, the synthesis of nucleoside analogues.[2,3] Much progress has also been made in the synthesis of nucleotides, with particular elegance in the syntheses of nucleotides containing chiral phosphate groups. Also, the development of rapid sequencing methods for the nucleic acids has represented a huge stride forward towards understanding gene structure and function, and this, with the inception of genetic engineering techniques, has resulted in a great demand for synthetic oligonucleotides of defined sequence, and spurred the development of new methods of oligonucleotide synthesis of remarkable speed and efficiency.

The huge amount of material available means that the topics described in this chapter are necessarily selective, and synthetic aspects of the chemistry have been emphasized. Only representative (rather than exhaustive) literature references are cited. While the reviews in nucleoside chemistry listed will be found invaluable for extra reading, a useful annual update in nucleotide and polynucleotide chemistry is available in a Specialist Periodical Report.[4]

8.2 Nucleosides

8.2.1 *Nucleoside synthesis: general methods*

8.2.1.1 *Nucleoside synthesis by base glycosylation.* Earlier base glycosylation procedures, whereby a heavy metal derivative of a nucleic acid base, or the base itself, was fused[5] with a sugar halide or acetate, have largely been superseded by Friedel-Crafts-type reactions in which the silylated base is condensed with a fully acylated sugar derivative in the presence of a Lewis acid, in 1,2-dichloroethane or acetonitrile as solvent.[6] While stannic chloride was employed in most earlier reactions of this type, Vorbrüggen[7] and his colleagues have studied base glycosylation intensively, and described tri-methylsilyl (TMS) perchlorate or (more safely) triflate or nonaflate as highly selective and efficient catalysts for the reaction. Vorbrüggen distinguishes three reversible processes which occur simultaneously during the 'Silyl-Hilbert-Johnson' reaction:

(a) Formation of the electrophilic sugar cation (1).
(b) Formation of a σ-complex (2) between the silylated base (3) and the Lewis acid (Friedel-Crafts catalyst) (4).
(c) Electrophilic reaction of (1) with (3).

By using ^{13}C n.m.r. to investigate the mode and rate of formation of (2) it was found that strong Lewis acids, such as stannic chloride, drive reaction (b) to the right, especially if the base moiety is electron rich, as in the case of silylated 5-methoxyuracil. Then in (2) (or in the ion-pair formed as (2) dissociates, in which the Lewis acid still blocks attack at N-1), N-3 is available for glycosylation, and thus a high proportion of the kinetic product, the N-3

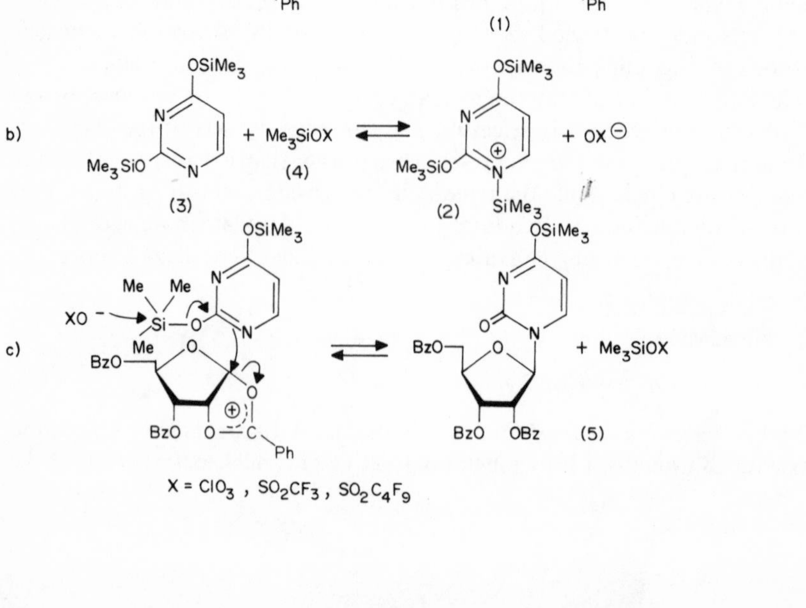

$$X = ClO_3 \, , \, SO_2CF_3 \, , \, SO_2C_4F_9$$

nucleoside, is formed. Using the weaker Lewis acids (4), little of the σ-complex (2) is formed, and N-1 is thus more available for glycosylation. A nucleophilic solvent (acetonitrile) also tends to suppress formation of (2) by competing for the Lewis acid. Thus, by using species (4) as catalysts in nucleophilic solvents, very high yields of the acylated nucleosides (5) can be obtained. The reaction works well for base analogues, and the β-anomers are formed, in accordance with Baker's Rule. With silylated purine bases (e.g. (6)), ^{13}C n.m.r. evidence indicates reaction occurring at N-1 to afford (7), which may then become-desilylated at N-9, giving (8). Since (6) could become glycosylated at N-1 or N-3, and (8) at N-3, N-7 or N-9, it is likely that several kinetic products are formed initially, which rearrange rapidly to form the thermodynamic product, the N-9 nucleoside.

If β-nucleosides containing arabinose are to be formed by direct condensation, reaction of the base with a 2,3,5-tri-O-benzylarabinofuranosyl halide is the method of choice.[8] A phase transfer reaction method has been found rather effective:[9] the base (9) and the arabinosyl halide (10) are dissolved in a two-phase system (dichloromethane or 1,2-dimethoxyethane: 50% aqueous sodium hydroxide) in the presence of tetra-n-butylammonium bisulphate or

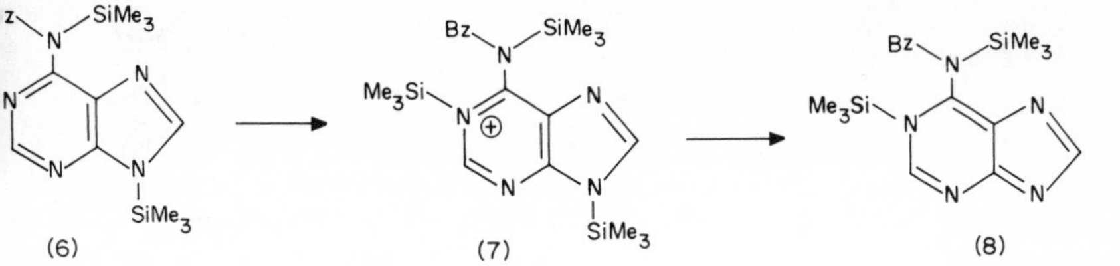

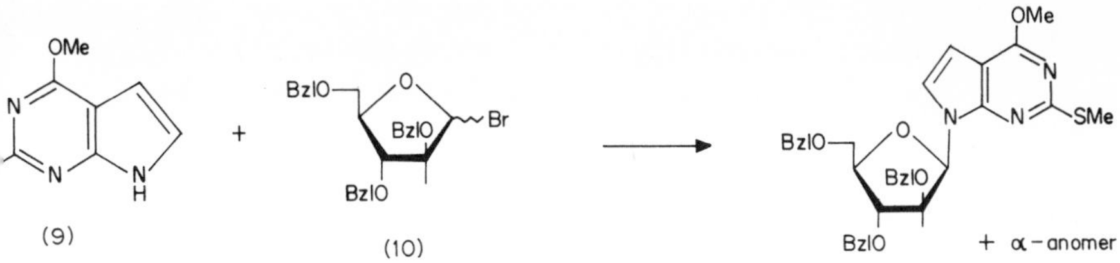

benzyltriethylammonium chloride as phase-transfer catalyst, and shaken hard, to afford a mixture of (11) and its α-anomer. The proportions of the anomers depend on the nature and the quantity of catalyst added, and it is thought that the catalyst alters the proportions of the anomers of halide (10) prior to condensation. Compound (11) may subsequently be converted to ara-7-deazaguanosine.

8.2.1.2 *Nucleoside synthesis by transglycosylation.* In some instances, conditions similar to those employed for base glycosylation may be used to exchange the sugar attached to the base in a pre-formed nucleoside for another sugar residue. Thus, treatment of $2',3'$-O-isopropylideneinosine with acetobromoglucose and mercuric cyanide in nitromethane affords 9-(2,3,4,6-tetra-O-acetyl-β-D-glucopyranosyl) hypoxanthine in moderate yield, together with the corresponding N^7-isomer.[10] It is thought that an intermediate of type (12) is formed, which fragments to form the (isolated) 1,5-anhydroribose derivative and N^7-glucosylated hypoxanthine as the initial product, followed by an $N^7 \rightarrow N^9$ glucosyl shift.

More commonly, the base attached to the sugar in a preformed nucleoside may be replaced by another base, and replacement of a pyrimidine, usually

(13) B = Ura - 1; R = CF$_3$CO
(14) B = Gua - 9; R = H
(15) B = Ura - 1; R = H
(16) B = N^9 - (2-chloro)-
      hypoxanthinyl; R = H

(17) B = Purin - 9 - yl
(18) B = Ura - I

(12)

(19)

uracil, by a purine base may afford a preparatively advantageous route to nucleoside analogues. Thus, silylation of $2'$-trifluoracetamido-$2'$-deoxyuridine (13) followed by treatment with N^2-palmitoylguanine and TMS-triflate in acetonitrile affords, after removal of the protecting groups, a good yield of the antibiotic $2'$-amino-$2'$-deoxyguanosine (14).[11] While exceptions have been described,[12] in general an acylated oxygen or nitrogen substituent at the 2-position of the sugar ring is required for transglycosylation in this way, presumably to stabilize a cationic sugar intermediate.[13]

Switching the base residue via bacterial transglycosylation has also proven synthetically useful. Thus, incubation of (15) with 2-chlorohypoxanthine in phosphate buffer in the presence of an appropriate strain of *Erwinia* or *Enterobacter* affords (16), which may then be converted to (14) with ammonia.[14] Purine D-arabinonucleosides (17), which are of value for their antitumour properties, may be prepared similarly by transfer of the sugar moiety from *ara*-uridine (18) to adenine or 2-chlorohypoxanthine.[15,16] The process involves phosphorolysis of the uridine analogue by uridine phosphorylase to afford the corresponding 1-phosphorylated sugar, which then condenses with the purine base via the agency of purine nucleoside phosphorylase.

8.2.1.3 *Nucleoside synthesis via sugars bearing nitrogen substituents at C-1.* Other novel nucleoside syntheses have formed the base moiety *in situ* by condensation involving an appropriately substituted sugar derivative bearing a nitrogen function at C-1. Thus, amination of D-ribose with methanolic ammonia, followed by condensation with 2,2-dimethoxypropane, affords (19), which condenses with appropriately substituted N-ethoxycarbonylacryl-amides to form 5-substituted uridine derivatives, and with N-acylated formimidates to give imidazole nucleosides.[17] The ribosylhydrazone (20) condenses with the thioamide of monoethyl oxalate to afford (21) which is cyclized with triethyl orthoformate to give (22). Treatment with ammonia then affords (23), and reductive debenzylation gives the 1*H*-1,2,4-triazole nucleoside virazole (24), a powerful antiviral agent.[18] If the ribosyl azide (25) is treated firstly with triphenyl phosphine, and subsequently with an isocyanate, ribosyl carbodiimides of type (26) are formed, which undergo cycloaddition with hydrazoic acid to afford tetrazole nucleosides (27).[19] Alternatively, (26) on treatment with triphenyl phosphine oxide affords the

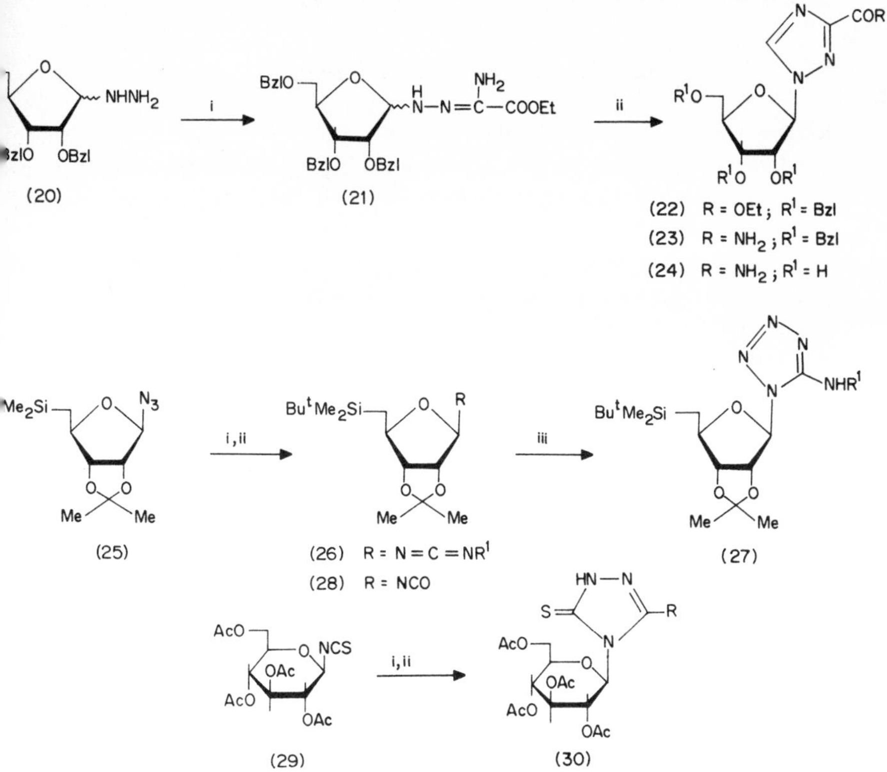

(20)

(21)

(22) R = OEt; R^1 = Bzl
(23) R = NH_2; R^1 = Bzl
(24) R = NH_2; R^1 = H

(25)

(26) R = N=C=NR^1
(28) R = NCO

(27)

(29)

(30)

Reagents: i, K, CS_2; ii, NaOH; iii, NH_3–MeOH

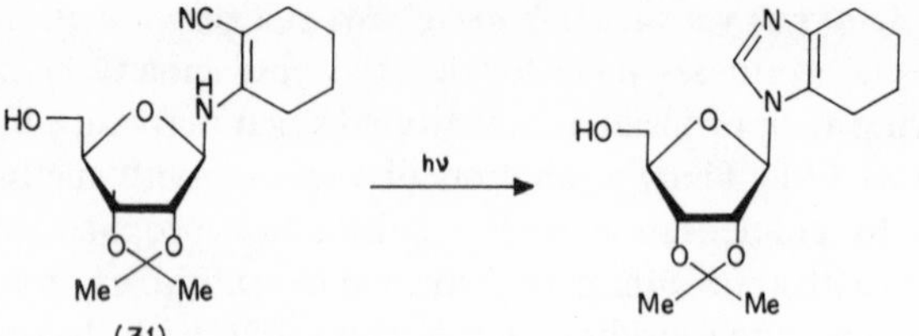

(31)　　　　　　　　　　　　(32)

isocyanate (28), another useful synthon. The glucopyranosyl isothiocyanate (29) readily condenses with acylhydrazines to form the corresponding N-acylthiosemicarbazides which are cyclized in alkali affording triazole nucleosides of type (30).[20]

An unusual nucleoside synthesis involving a photorearrangement has been described.[21] 2-Cyanocyclohexanone condenses with (19) to form the ribofuranosylenaminonitrile (31), together with its α-anomer, and irradiation in acetonitrile then affords the tetrahydrobenzimidazole (32), the α-anomer of (31) similarly giving rise to the α-anomer of (32).

8.2.2 Novel nucleoside reactions of general utility

8.2.2.1 *Introduction of chirality at C-5'.* (5'R)-[²H₁] adenosine has been synthesized, in effect, by Pfitzner-Moffatt oxidation of the 5'-dideutero-adenosine derivative (33) to afford the aldehyde (34), which is then reduced with the adduct formed from α-pinene and 9-borabicyclo[3.3.1]nonane (a species known to reduce asymmetrically) to afford (35) in c. 60% optical purity. After deblocking, the chiral adenosine has been converted to its 5'-triphosphate and used to elucidate the stereochemistry of S-adenosyl-methionine formation.[23]

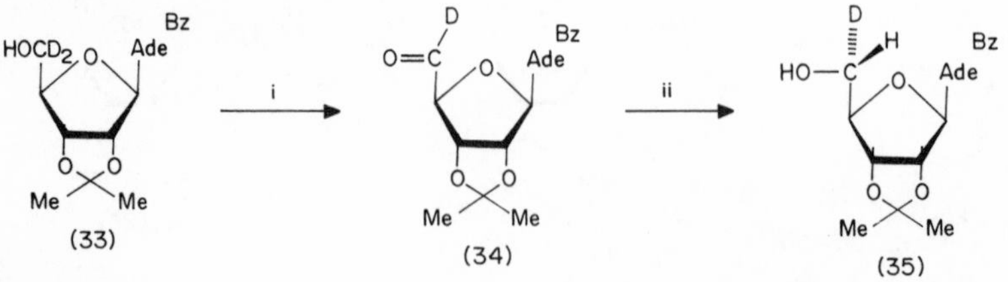

(33)　　　　　　　　　(34)　　　　　　　　　(35)

8.2.2.2 *Silylation.* Silyl groups have been introduced as protecting groups for the sugar hydroxy functions of nucleosides with spectacular success, notably by Ogilvie and his co-workers.[24] The nucleoside is treated with *tert*-butyldimethylsilyl (TBDMS) chloride (the reagent of choice) and excess imidazole in DMF. Attack occurs initially at the 5'-OH group, then at the 2'-

(if present) and 3'-OH groups. The 2',5'- and 3',5'-bis (TBDMS) nucleosides are separated cleanly on TLC plates, and treatment of these with 80% acetic acid at 100°C results in selective desilylation at the 5'-position, affording the pure 2'- or 3'-silylated compounds. These have found wide use in oligonucleotide synthesis. Silylation as described above but using 1,4-diaza[2.2.2]bicyclooctane and silver salts in place of imidazole confers selective silylation at the 3'-position, and the 3',5'-bis (TBDMS) nucleoside is formed almost exclusively.[25] Alternatively, treatment of a ribonucleoside with 1,3-dichloro-1,1,3,3-tetraisopropyldisiloxane selectively silylates the 3'- and 5'-OH functions, leaving the 2'-OH free.[26] All silyl protecting groups are removed rapidly by fluoride ion, usually in the form of tetra-*n*-butylammonium fluoride.

8.2.2.3 *Synthesis of deoxynucleosides.* The reduction of ribonucleosides to 2'- and 3'-deoxynucleosides has previously offered experimental difficulty. Reduction of 3'-bromo-3'-deoxy-*xylo*-adenosine (blocked at the 2'- and 5'-positions, and formed by treating adenosine with 2-acetoxyisobutyryl bromide) with palladium and hydrogen certainly affords some 3'-deoxyadenosine (the antibiotic cordycepin), but also gives some 2',3'-dideoxyadenosine.[27] Electrochemical reduction of an analogous iodo-derivative affords the 2',3'-unsaturated 2',3'-dideoxy product.[28] However, 3',5'-di-O-acetyl-2'-bromo- and 2'-chloro-2'-deoxyuridine are efficiently reduced to 3',5'-di-O-acetyl-2'-deoxyuridine by tri-*n*-butyltin hydride in the present of azobisisobutyronitrile (AIBN)[29] and this tin radical reduction has now been used to reduce 2'-[30] or 3'-[31] thiocarbonyl esters of nucleosides to the corresponding 2'- or 3'-deoxynucleosides with reasonable efficiency. These esters are formed by treating the target OH-group with thiocarbonyldiimidazole, phenoxythiocarbonyl chloride or a similar reagent, the other OH-groups having previously been blocked.

8.2.2.4 *Alkylation at oxygen and nitrogen.* 2'-O-Methylated nucleosides occur in tRNA and in 'cap' structures found at the 5'-termini of some viral RNA molecules (see 8.3.7.2), and their role has therefore aroused interest. While modest yields of 2'- and 3'-O-methyl adenosine and -cytidine can be obtained via treatment of the 5'-blocked nucleosides with diazomethane in 1,2-dimethoxyethane, nucleosides with an acidic proton present on the base (uridine, guanosine, inosine) become alkylated on the base. However, all ribonucleosides can be monomethylated giving a mixture of the 2'- and 3'-O-methyl isomers by adding dilute diazomethane slowly to the unprotected nucleosides in methanol in the presence of stannous chloride dihydrate.[32] This salt reacts rapidly with diazomethane affording stannous methoxide, which undergoes ligand exchange with Brønsted acids, such as the *cis*-glycol system

(36)   (37)

of ribonucleosides, to give (36), which is in equilibrium with ring-opened forms (37), each of which contains (formally, at least) a highly nucleophilic alkoxide group at which methylation takes place.[33] With purine nucleosides the 3'-O-methylated product predominates, while for pyrimidine nucleosides the 2'-O-methylated isomer is favoured.

Alkylation of nucleosides using alkyl halides, trimethyl phosphate, dialkyl sulphates and alkyl methanesulphonates results in alkylation of the base residue, and this process is strongly catalysed by fluoride ions.[34] The position and degree of alkylation are dependent on the agent used and the base present, with N-3 of pyrimidine nucleosides being particularly susceptible. No alkylation of the sugar OH-groups occurs under these conditions.

8.2.2.5 *Alkylation at carbon.* New methods for the introduction of an alkyl group at C-6 of purine nucleosides have been described. Thus, 2',3',5'-tri-O-acetylinosine may be treated with tosyl chloride and triethylamine to give the 6-O-tosylated purine riboside, which reacts with acetoacetic ester in THF in the presence of sodium hydride to give (38).[35] Alternatively, the same reagents may be used to displace a 6-methanesulphonyl substituent, affording the same product.[36] Note that a concomitant retro-Claisen condensation has occurred. Treatment of (38) with base, and then with acid, affords 6-methylpurine riboside. A more versatile route to 6-alkyl and 6-aryl purine ribosides involves coupling 9-(2,3,5-tris-(O-TBDMS)-β-D-ribofuranosyl)-6-chloropurine with alkyl and aryl Grignard reagents in the presence of dichloro[1,3-bis(diphenyl-phosphine)propane]nickel(II).[37]

Organometallic intermediates have been much used recently in the synthesis of nucleoside analogues, and an up-to-date review is recommended.[38] 5-Chloromercuriuracil and -cytosine-N^1-ribosides react with alkenes, alkyl acrylates, allyl chloride, allyl alcohols and allyl acetates in the presence of

(38)   (39)   (40)

lithium tetrachloropalladate to form C-5-substituted pyrimidine nucleosides via the agency of organopalladium intermediates.[39] Allyl chlorides react predominantly to form species of type (39), while alkyl acrylates afford (E) alkyl 3-(5-uridyl) propenoates (40). Allylic alcohols and acetates afford the same products as the chlorides, but couple more slowly and less cleanly. Hydrolysis of (40:R = Et, sugar = 2-deoxyribofuranosyl), followed by bromination with NBS, has been used to prepare (E)-5-(2-bromovinyl)-2'-deoxyuridine, a powerful antiviral agent.[40] The vinyl group has also been introduced by direct vinylation of 2'-deoxy-5-iodo-uridine and -cytidine with vinyl acetate in DMF in the presence of diacetato-bis(triphenylphosphine)palladium(II).[41] An alternative method of forming carbon-carbon bonds at C-5 of the uracil ring consists in condensing 5-hydroxyuridine (which presumably reacts as the 5-ketopyrimidine) with Wittig reagents such as methoxycarbonylmethylidene-triphenylphosphorane. 5-Methoxycarbonylmethyluridine and 5-carbamoyl-methyluridine, two minor constituents of tRNA, have been made in this way.[42]

Substitution of uridine at the 6-position may be performed by lithiation, which results in 'Umpolung' of the normal reactivity at this position. Treatment of 2',3'-O-isopropylideneuridine or its 5'-methoxymethyl derivative with lithium diisopropylamide in THF at − 78°C affords (41) or (42), of which the former reacts with alkyl halides to give 6-alkyluridines,[43] and the latter with aldehydes, ketones, acyl and aroyl halides, and esters to introduce a variety of groupings at C-6.[44]

(41) R = Li
(42) R = CH₃OCH₂

(43)

(44) R¹ = H or Me

(45)

(46) B = Gua – 9 ; R = H

(47)

(48)

8.2.2.6 *Conversion of ketopyrimidines to aminopyrimidines.* Convenient new methods for converting uridine and thymidine to cytidine and 5-methyl-2'-deoxycytidine have been described. Persilylation of uridine with HMDS and TMS chloride affords an O⁴-trimethylsilylpyrimidin-2-one of type (43) which gives cytidine on treatment with ammonia, or the corresponding 4-alkylamino-

356   THE CHEMISTRY OF NATURAL PRODUCTS

pyrimidin-2-ones on treatment with primary or secondary amines.[45] Alternatively, treatment of a sugar-protected uridine or thymidine residue with 4-chlorophenylphosphorodichloridate and 1,2,4-triazole affords a 4-(1,2.4-triazol-1-yl)pyrimidin-2-one of type (44) in which the triazole ring is also readily replaced by ammonia or primary or secondary amines.[46]

8.2.2.7   *Halogenation and related studies.* 5-Halogenopyrimidine nucleosides have aroused much interest as potential antiviral and antitumour agents. Fluorine is readily introduced at the 5-position of sugar-protected uridine species by treatment with trifluoromethyl hypofluorite in chlorotrifluoro-methane at − 78°C,[47] or by treatment with elemental fluorine in acetic acid.[48] An addition-elimination mechanism occurs in each case. The pseudohalogen thiocyanogen chloride reacts readily with free or sugar-protected uridine in acetic acid to afford 5-thiocyanatouridine, which is easily reduced to 5-mercaptouridine.[49] A number of novel methods have been developed for halogenation of the sugar ring in nucleosides, and for details of these (and transformations of the carbohydrate moiety in general) a superlative review is recommended.[50]

Treatment of 2′,3′,5′-tri-*O*-acetyladenosine with *n*-pentyl nitrite in DMF under nitrogen, followed by irradiation at reflux temperature affords 2′,3′,5′-tri-*O*-acetyl-nebularine (45).[51] Presumably an intermediate 6-diazo species is photolysed to afford a purinyl radial which abstracts hydrogen from the solvent. If the same reaction is performed in carbon tetrachloride, bromoform, or methylene iodide, the corresponding 6-halopurine species is formed efficiently in each case.[52]

8.2.2.8   *Other studies.* 'Open-ring' riboside and deoxyriboside analogues lacking C-3′ or the C-3′-C-4′ bond have aroused interest since some species of this type (e.g. (46)) exhibit strong antiviral activity. They may be prepared by treating a 1,3-dioxolane (47) with TMS iodide in cyclohexene at − 78° to afford (48) which is then used to alkylate a suitable purine base at N-9.[53]

Ara-adenosine is a valuable antitumour drug, and some effort has been expended in devising new methods for its synthesis. 8,2′-*O*-Cycloadenosine (49)

(49)   (50)   (51)

undergoes ring-opening with hydrogen sulphide in pyridine,[54] or with ethanolic hydrazine,[55] to afford 8-mercapto- and 8-hydrazo-*ara*-adenosine, respectively, which on treatment with Raney nickel or mercuric oxide, respectively, yield *ara*-adenosine. Alternatively, cyclization of 8-carboxamido-2′-*O*-tosyladenosine (**50**) in aqueous pyridine affords the same compound in one step: compound (**51**) is believed to occur as an intermediate.[56]

8.2.3   C-Nucleosides

In C-nucleosides, the familiar C-N glycosidic bond is replaced by a C-C bond. The natural occurrence of nucleosides of this type in tRNA, and particularly as antibiotics,[3] has spurred much effort towards their synthesis. Useful recent reviews[57] may be recommended. Most C-nucleoside syntheses start by construction of a functionalized one to four carbon moiety which is attached to C-1 of the sugar ring, and the base is subsequently formed by condensation and cyclization reactions. It is instructive to consider some of the ways in which these derivatized sugars are constructed and employed.

8.2.3.1   *C-Nucleoside synthesis via sugar derivatives*. Ribosyl cyanides such as (**52**) are frequently employed as starting material in C-nucleoside syntheses, being easily prepared from (**53**) by treatment with TMS-cyanide and stannic chloride in acetonitrile.[58] Hydrolysis of (**52**) with aqueous acid and subsequent treatment with thionyl chloride affords (**54**), which on condensation with *tert*-

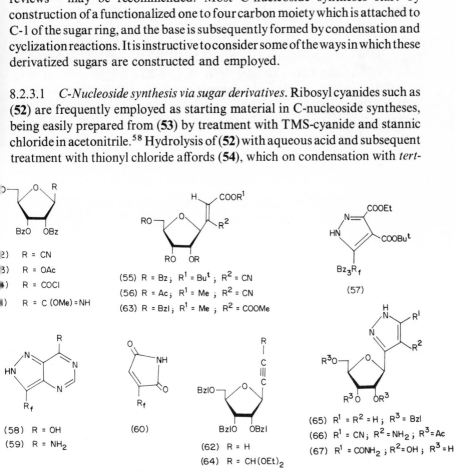

(52) R = CN
(53) R = OAc
(54) R = COCl
( ) R = C (OMe) = NH

(55) R = Bz ; R¹ = Buᵗ ; R² = CN
(56) R = Ac ; R¹ = Me ; R² = CN
(63) R = Bzl ; R¹ = Me ; R² = COOMe

(57)

(58) R = OH
(59) R = NH₂

(60)

(62) R = H
(64) R = CH(OEt)₂

(65) R¹ = R² = H ; R³ = Bzl
(66) R¹ = CN ; R² = NH₂ ; R³ = Ac
(67) R¹ = CONH₂ ; R² = OH ; R³ = H

butyloxycarbonylmethylenetriphenylphosphorane in the presence of hydrogen cyanide yields predominantly (55). A minor variation on this route affords (56). 1,3-Dipolar cycloaddition between (55) and ethyl diazoacetate yields (57), and, following removal of the *tert*-butyl group, a Curtius reaction, and subsequent condensation of the product with formamidine acetate, yields formycin B (58) after deacylation.[59] If diazoacetonitrile is used in place of ethyl diazoacetate, formycin (59) is obtained.[59] Hydrolysis and cyclization of (56) using sulphuric acid and acetic anhydride, followed by deacylation, affords showdomycin (60).[60] Treatment of (52) with sodium methoxide gives the corresponding carboximidic acid methyl ester (61) which condenses with acyl hydrazines and, following cyclization, affords C-nucleoside analogues of (24), bredinin, and related compounds.[61]

Another much-used stratagem in C-nucleoside synthesis involves introduction of an alkyne moiety at C-1 of the sugar. Treatment of 2,3,5-tri-*O*-benzylribose with ethynyl magnesium bromide in THF, and subsequently with tosyl chloride in pyridine, gives predominantly (62), which upon further treatment with carbon monoxide in methanol in the presence of mercuric chloride and palladium chloride yields (63). Alkaline hydrolysis of (63), followed by cyclization with trifluoroacetic anhydride, affords the correspondingly ribosylated maleic anhydride, which on cleavage with ammonia, recyclization with acetyl chloride and debenzylation gives (60).[62] Another synthesis of formycin (59) begins with the condensation of (64) (prepared analogously to (62)) with hydrazine to afford (65),[63] and a later intermediate in this synthesis, (66), serves as starting material for the synthesis of pyrazofurin (67).[64]

In an extensive series of papers on C-nucleoside synthesis, Fox and his colleagues have used 5-*O*-trityl-2,3-*O*-isopropylideneribose (68) as starting

(68) R = OH
(69) R = Cl
(70) R = C≡C. COOEt
(71) R = CH (COOEt)$_2$
(72) R = CH$_2$COOEt
(73) R = CH$_2$COOMe
(74) R = CH$_2$CN
(75) R = CH$_2$CHO
(76) R = CHBrCHO

(77) R^1 = OMe ; R^2 = COOEt
(78) R^1 = OMe ; R^2 = CN
(79) R^1 = NMe ; R^2 = CN
(80) R^1 = OH ; R^2 = CONH$_2$
(82) R^1 = OH ; R^2 = CN

(81)

material, converting it to the ribofuranosyl chloride (**69**) with triphenyl-phosphine and carbon tetrachloride. Then, treatment with the silver salt of ethyl propynoate furnishes (**70**), of which the β-anomer condenses with guanidine to afford, after deprotection, 2-amino-4-oxo-6-(β-D-ribofuranosyl)pyrimidine.[65] Alternatively, cycloaddition of various 1,3-dipoles to the triple bond of (**70**) gives a number of other C-nucleosides. Treatment of (**69**) with diethyl sodiomalonate gives (**71**), providing an alternative substrate for condensation and cyclization.[66]

Reaction of Wittig reagents with (**68**) has also been used to generate key precursors for C-nucleoside synthesis. Treatment with the appropriate alkoxycarbonylmethylidenetriphenylphosphoranes gives (**72**)[67] and (**73**),[68] while use of cyanomethylidenetriphenylphosphorane (or sodium diethyl cyanomethylphosphonate) affords (**74**).[69] Reduction of (**73**) with diisobutyl aluminium hydride affords (**75**), which may be brominated via its enol acetate to give (**76**), which condenses with thioamides to afford a number of ribofuranosylthiazole C-nucleosides.[68] Treatment of (**71**) with sodium hydride and ethyl formate and subsequent methylation with methyl iodide gives (**77**), which condenses with guanidine to afford the corresponding derivative of pseudoisocytidine, and may also be used to prepare pseudouridine.[67] Similar formylation and methylation of (**74**) affords (**78**), which also condenses with urea, thiourea and guanidine to form protected pseudocytidines, etc.[70] Condensation of (**74**) with bis(dimethylamino)-*tert*-butoxymethane in DMF affords a key synthon (**79**) which may be converted in three steps to (**80**), which then condenses with carbonyldiimidazole to afford oxazinomycin (**81**).[69] Mild acidic hydrolysis of (**79**) gives (**82**), which has been utilized in construction of the C-nucleoside isosteres of inosine and adenosine.[71]

(83)   (84)   (85)

In some cases a complete ring system may be attached to C-1 of the sugar at the inception of C-nucleoside synthesis. Thus, treatment of (**53**) with 1,2-bis(trimethylsilyloxy)cyclobut-1-ene and stannic chloride in methylene chloride gives (**83**), which on conversion to the silylated enol and treatment with nitrosyl chloride affords oxime (**84**) which in turn forms nitrile (**85**) on standing in methylene chloride. Successive treatments of (**85**) with ammonia and trifluoracetic anhydride then afford showdomycin (**60**).[72] Treatment of 2,4-

(86)   (87)   (88)   (89)

bis(*tert*-butyloxy)-5-bromopyrimidine with lithium butyl affords the 5-lithiated base, which adds to 2,3,5-tri-*O*-benzylribose in THF at $-78°$ to form (86) and its C-1′ epimer. Cyclization with ethanolic HCl followed by debenzylation with boron trichloride then affords pseudouridine (87), and the α-anomer is prepared similarly.[73] Condensation of 2-methylthio-4-thiopyrimidine with 1,2,3,5-tetra-*O*- acetyl-D-ribofuranose using boron trifluoride etherate affords the pyrimidine S-nucleoside (88), which is rearranged upon irradiation to form the C-nucleoside (89).[74]

8.2.3.2   *C-Nucleoside synthesis via other cyclic precursors.* Some elegant and versatile routes to C-nucleosides starting from non-carbohydrate precursors have been described. Noyori and his colleagues have coupled α,α,α′,α′-tetrabromoacetone to substituted furans in the presence of diiron nonacarbonyl or a zinc-silver couple, and reduced the product with a zinc-copper couple to afford bicyclic ketones of general structure (90). *Cis*-hydroxylation with osmium tetroxide, followed by acetonide formation and Baeyer-Villiger oxidation leads, after resolution, to the lactone (91) which on condensation with bis(dimethylamino)-*tert*-butoxymethane afford (92). The functional relationship of (92) to species (77)–(80), and thus its utility in C-nucleoside synthesis, will readily be appreciated.[75] Moreover, appropriate choice of a substituted furan (i.e. R^1, $R^2 \neq H$ in (90)–(92)) allows C-4′-alkylated,[76] C-4′-hydroxymethylated,[77] and C-2′-methylated[78] pyrimidine C-nucleosides to be prepared. Flexibility of choice of the bromoketone component used in preparing (90) (i.e. $R^3 \neq H$ in (90)–(92)) allows C-5′-modified C-nucleosides to be constructed;[79] and hydrolysis of (91) followed by homologation and functionalization of the carboxymethylene moiety affords a general route to pyrimidine homo-C-nucleosides.[80]

An alternative route to (91) involves reduction of 2,3,4,4-tetrachloro-8-oxabicyclo[3.2.1]octa-2,6-diene (93) with lithium hydride and lithium aluminium hydride to give (94) which is *cis*-hydroxylated and converted to the acetonide (as in (90)→(91)), ozonolysed in methanol, reduced with sodium

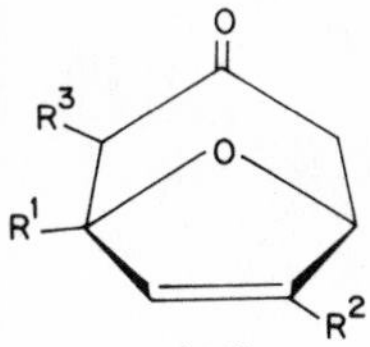

(90)

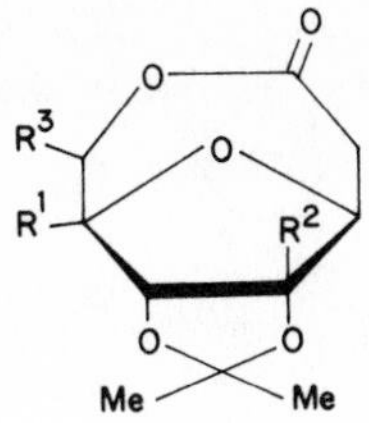

(91)

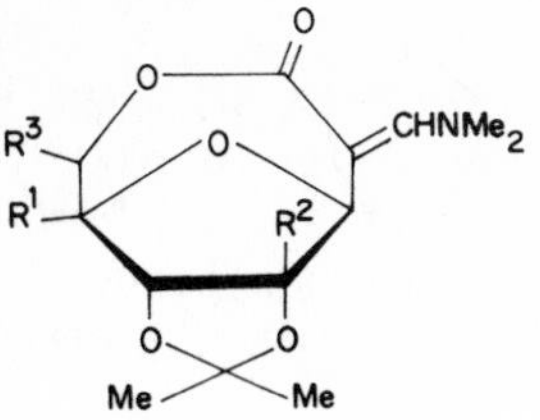

(92)

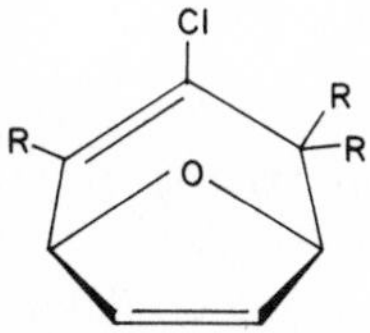

(93) R = Cl
(94) R = H

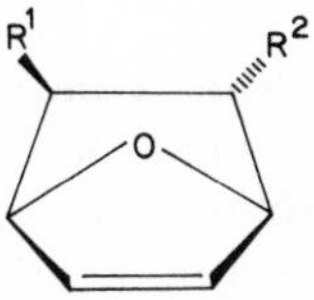

(95) R¹ = NO₂(or COOMe)
    R² = COOMe (or NO₂)

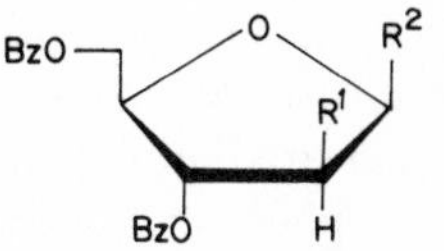

(96) R¹ = OH; R² = CH₂OH
(97) R¹ = H; R² = COOH

borohydride, and cyclized to (91) for use in the synthesis of triazole homo-C-nucleosides.[81]

Just and his colleagues have utilized the products of cycloaddition of furan and 3-nitromethacrylate, (95), for the synthesis of D,L-showdomycin and its 2'-epimer.[82]

8.2.3.3 *Other studies.* The method described above (section 8.2.2.3) for reduction of ribonucleosides to the 2'- and 3'-deoxyribonucleosides via thiocarbonyl esters[30,31] has been used for the conversion of C-ribonucleosides to C-2'-deoxyribonucleosides with similar success.[83] It has also been used to convert (96) to the corresponding 2'-deoxy species for use in a synthesis of β-2'-deoxyshowdomycin[84] which, following further oxidation to (97) with pyridinium dichromate, essentially follows Kalvoda's route.[60]

N-Methylation of pseudouridine (87) using DMF-dimethylacetal affords 1,3-dimethylpseudouridine, in which the N(1)-C(2)-N(3) unit may be replaced by guanidine, methylguanidine or thiourea, in the presence of ethanolic ethoxide, to afford a new C-nucleoside.[85] Thus, thiourea affords 2-thiopseudouridine as the product, and guanidine gives pseudoisocytidine.

8.2.4 *Synthetic routes to some unusual nucleosides*

A useful synthetic route to nucleosides in which the ribofuranose ring is replaced by the corresponding carbocycle starts from (98), the hydrolysis product of the adduct formed between tosyl cyanide and cyclopentadiene.[86] *Cis*-hydroxylation of (98), followed by methanolysis, protection of the amino group by

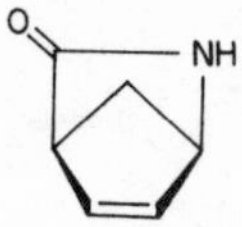

(98)

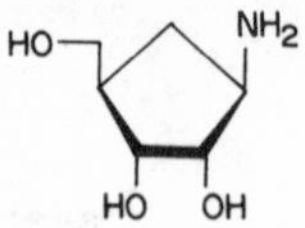

(99)

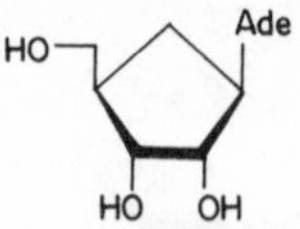

(100)

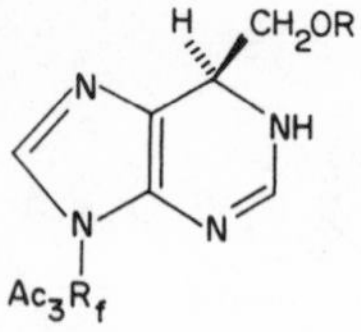

(101) R = H

(102) R = MeSO$_2$

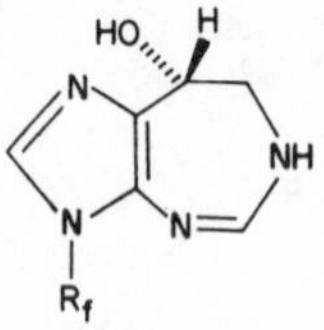

(103)

acetylation, reduction with calcium borohydride and deacetylation affords the carbocyclic 'ribofuranosylamine' (99)[87] (and its 'lyxo'-counterpart), from which aristeromycin (100) may be elaborated by standard methods.[88]

Upon irradiation in methanol, nebularine triacetate (45) undergoes stereospecific photoaddition to afford (101), which upon mesylation gives (102). Treatment with potassium tert-butoxide results in ring-expansion, yielding coformycin (103) after deacetylation.[89] Some difficulty has been encountered in attempting to reproduce this reaction.

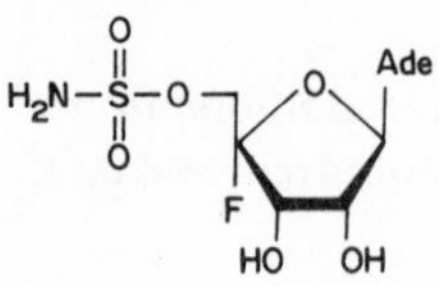

(104)

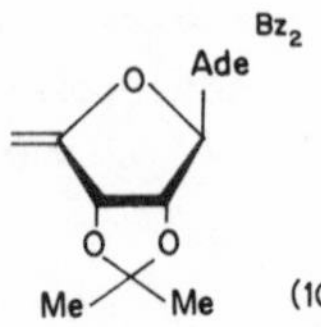

(105)

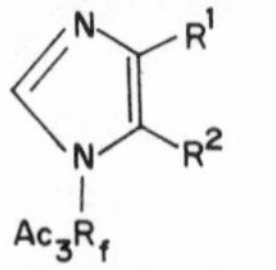

(106) R^1 = CN ; R^2 = NH$_2$

(107) R^1 = CONH$_2$; R^2 = NHMe

(108) R^1 = CONH$_2$; R^2 = N(CN)Me

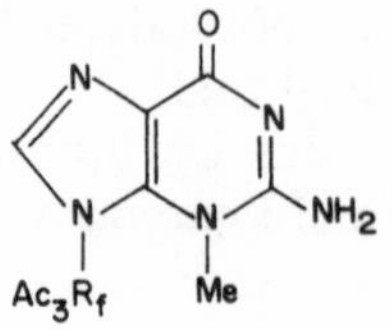

(109)

Nucleocidin (104) has been prepared via the addition of iodine fluoride across the exocyclic double bond of (105). The iodine atom at C-5' in the desired isomer of the product is displaced by azide, and the new product photolysed in acid to afford the 5'-aldehyde, reduced to the alcohol, and finally treated with sulphamoyl chloride to give (104) after deprotection.[90]

The rare nucleoside wyosine, found in tRNA, has been prepared by methylation and hydrolysis of the aminocyanoimidazole nucleoside (106) to afford (107),[91] which is treated with cyanogen bromide to give (108), and subsequent ring-closure with ethoxide yields the acetylated N^3-methylguanosine (109). Treatment of (109) with bromoacetone and subsequent deacylation gives wyosine (110) which is extremely acid-labile, the $t_{\frac{1}{2}}$ in 0.1M hydrochloric acid being 45 seconds.[92]

A more formidable synthetic task, the total synthesis of nucleoside Q from tRNA, has also been realized.[93] The critical stage of the synthesis involves allylic bromination of the 7-methyl-7-deazaguanosine derivative (111) with NBS, followed by condensation with the acetonide of (3S, 4R, 5S)-4,5-dihydroxy-

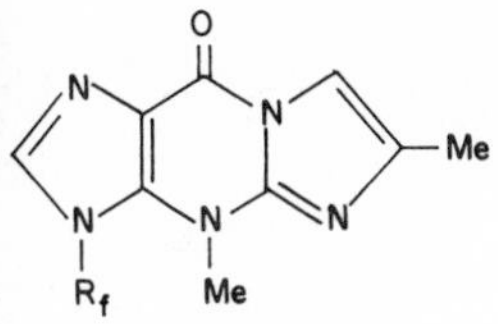

(110) 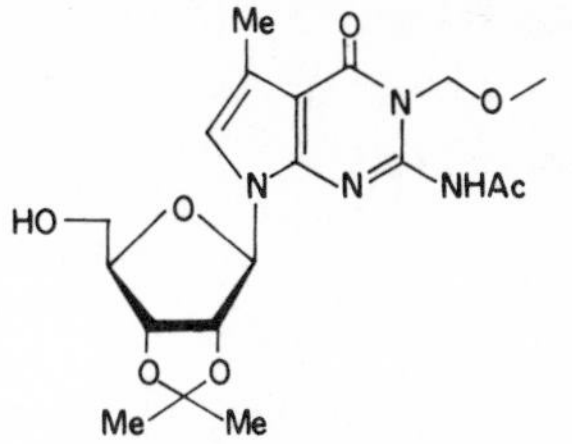 (111) 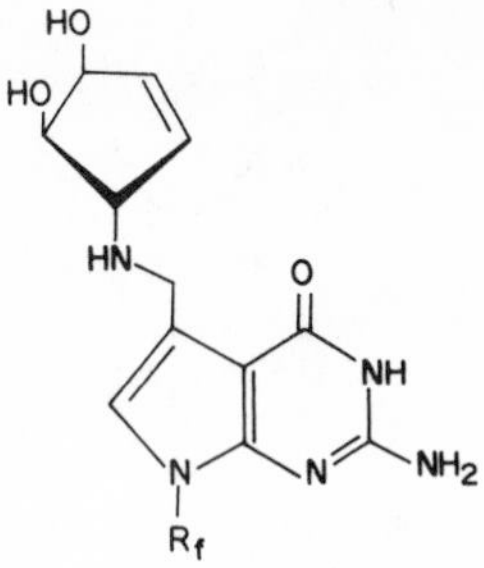 (112)

cyclopent-1-en-3-ylamine to afford, after deblocking, nucleoside Q (112), identical with the natural product.

8.3 Nucleotides

8.3.1 *Synthesis of nucleoside monophosphates*

The method most widely applied for phosphorylation of unprotected nucleosides utilizes phosphoryl chloride in (usually) trimethyl or triethyl phosphate at around 0 °C.[94] Phosphorylation occurs predominantly (and often exclusively) at the 5'-position, affording phosphorochloridates of type (113) as initial products, which are hydrolysed by water to nucleoside 5'-phosphates. The phosphorylating agent has been suggested to be (114). Ribonucleosides, deoxyribonucleosides, and a wide range of nucleoside analogues may be phosphorylated in this way. If thiophosphoryl chloride is used instead, the corresponding 5'-phosphorothioates are obtained.[95] A technique of similar efficiency which is highly selective for the 5'-OH group involves treating the nucleoside with phosphoryl chloride in the presence of water and pyridine (or, better, with pyrophosphoryl chloride and pyridinium chloride) in acetonitrile.[96] The phosphorylating agent is thought to be the species (115), and it has been hypothesized that the protons derived from pyridinium chloride increase the conductivity of the solution and suppress the reactivity of secondary alcoholic functions. 2-(N,N-Dimethylamino)-4-nitrophenyl phosphate (116),[97] bis(2-*tert*-butylphenyl)phosphorochloridate (117),[98] and bis(2,2,2-tri-chloro-1,1-dimethylethyl)phosphorochloridate (118)[99] have all been described as reagents specific for phosphorylation of unprotected nucleosides at the 5'-position. Phosphorylation by (116) is acid-catalysed, presumably by protonation of the dimethylamino moiety, while steric hindrance probably determines the positional selectivity of (117) and (118). Alternatively, enzymic phosphorylation of the 5'-position of nucleosides may be advantageous, using wheat

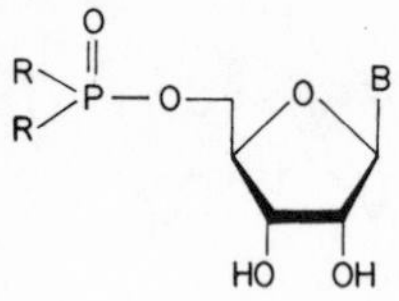

(113) R = Cl
(121) R = NH₂

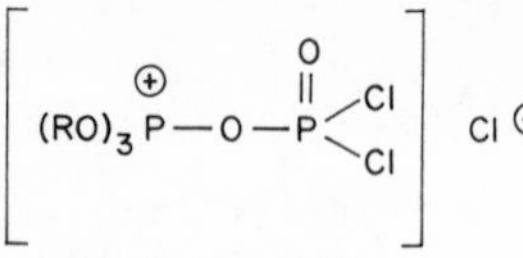

(114) R = Me or Et

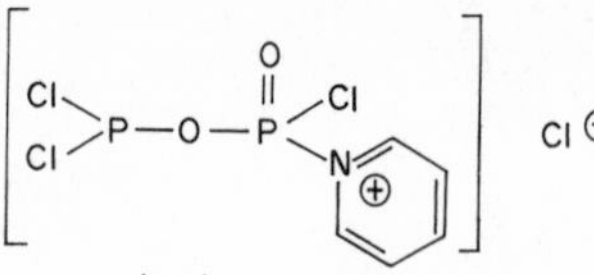

(115)

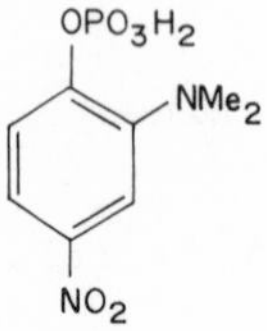

(116)

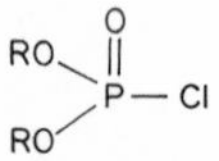

(117) R = (2-Buᵗ)C₆H₄

(118) R = Cl₃C.CMe₂

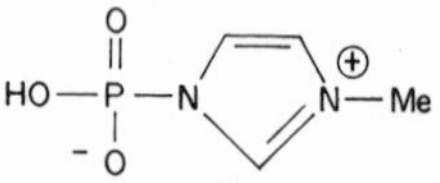

(119)

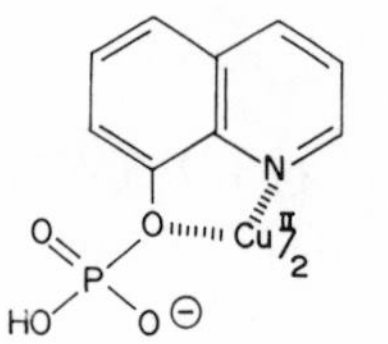

(120)

shoot (or a similar) phosphotransferase and 4-nitrophenyl phosphate as phosphate donor.[100]

Phosphorylation of 2',3'-O-isopropylidene nucleosides at the 5'-position may be performed using phosphorous acid and mercuric chloride in N-methylimidazole in which the phosphorylating species is thought to be (119),[101] or 8-quinolyl phosphate in the presence of cupric chloride in pyridine, where (120) is probably the reactive agent. Ribonucleoside 5'-phosphites may be prepared via phosphorylation of the 2',3'-O-isopropylidene nucleosides with phosphorus trichloride in trimethyl phosphate,[102] or by condensation with phosphorous acid using 2,4,6-triisopropylbenzenesulphonyl chloride (TPS-Cl). Silylation of these compounds following by oxidation with sulphur[103] affords the corresponding 5'-phosphorothioates, while replacement of sulphur by alkyl or arylthiols or selenols gives the corresponding S-(or Se-) alkyl or aryl phosphorothioates or phosphoroselenoates.[104] Nucleosides may also be thiophosphorylated by heating with thiophosphate monoanion in DMF.[105] The phosphorylating agent is probably monothiopyrophosphate which appears to be formed under these conditions.

If the phosphorodichloridate (113) initially formed on phosphorylation of a nucleoside by phosphoryl chloride is treated with ammonia, the corresponding phosphordiamidate (121) is formed.[106] Controlled hydrolysis of (121) in weak acid affords the phosphoramidate, which may be treated with phosphate or pyrophosphate in DMF to give the nucleoside di-or triphosphate.

Phosphorylation of a nucleoside at the 2'-or 3'-positions is best carried out by masking all OH-groups other than the one to be phosphorylated, and then utilizing one of the phosphorylating agents developed for oligonucleotide synthesis. While specific reagents for phosphorylating these positions are available,[107] none enjoys wide usage. A review giving a comprehensive account of nucleoside phosphorylation is available.[108]

8.3.2   Conversion of nucleoside monophosphates to polyphosphates

Three methods are predominantly favoured for converting a nucleoside 5'-monophosphate to the 5'-di or triphosphate. In the first, treatment with diphenylphosphorochloridate affords a P^1,P^1-diphenyl-P^2-(5'-nucleosidyl) pyrophosphate, and subsequent attack by inorganic phosphate or pyrophosphate displaces diphenylphosphate to give the desired product. This method has also been used to prepare adenosine 5'-(α-thiotriphosphate) (ATPαS) and adenosine 5'-(α-thiodiphosphate) (ADPαS) from adenosine-5'-phosphorothioate.[109] In the second, the nucleoside-5'-phosphate is coupled with morpholine using DCC to afford the 5'-phosphoromorpholidate, which is treated with phosphate or pyrophosphate, displacing morpholine to give the corresponding 5'-polyphosphate. An alternative route to the phosphoromorpholidate consists in treating the base-protected nucleoside with 2,2,2-tribromoethyl phosphoromorpholinochloridate (122), which selectively phosphorylates the 5'-position, followed by removal of the tribromoethyl group with a zinc-copper couple.[110] In the third, the nucleoside 5'-phosphate is treated with carbonyldiimidazole to afford the 5'-phosphorimidazolidate, and imidazole is in turn displaced by phosphate or pyrophosphate.[111] If excess reagent is used, the 2',3'-cyclic carbonate of the phosphorimidazolidate may be formed, but following phosphorylation the carbonate ring is easily destroyed on hydrolysis at pH 10.5.[112]

(122)

(123)   R = H or OH

A recent interesting method of performing this conversion requires treatment of a nucleoside -5'-phosphate with bis(n-butyl)phosphinothioyl bromide in pyridine to form a compound of type (123), which is stable in aqueous pyridine, but which reacts rapidly with phosphate or pyrophosphate in dry pyridine in the presence of silver salts to give the corresponding nucleoside di- or triphosphate.[113]

Treatment of ATP with DCC affords adenosine 5'-trimetaphosphate (124). The ring is readily opened by nucleophilic attack to give a γ-substituted ATP derivative, providing a high-yield general method for preparing analogues of type (125).[114]

The reactions described above permit positional isotopic labelling in the polyphosphate chain to be carried out. For instance, activation of ADP with carbonyldiimidazole, followed by treatment with $[^{18}O_4]$ orthophosphate affords $[\beta\gamma\text{-}^{18}O, \gamma\text{-}^{18}O_3]$ ATP (126), while similar activation of AMP, treatment with $[^{18}O_4]$ orthophosphate, reactivation of the $[\alpha\beta\text{-}^{18}O, \beta\text{-}^{18}O_3]$ ADP formed, and condensation with $[^{16}O_4]$ orthophosphate gives $[\alpha\beta\text{-}^{18}O, \beta\text{-}^{18}O_2]$ ATP (127).[115] These species have been utilized in an elegant study of positional exchange of isotopes in nucleoside triphosphates occurring at the active sites of some enzymes. Alternatively, enzymic synthesis may be utilized to effect such positional labelling, and many such $[^{18}O]$-and $[^{17}O]$-labelled nucleotides have been prepared in order to study the effect of the isotopic substitution on the ^{31}P n.m.r. resonances.[116,117].

8.3.3 Nucleoside polyphosphate analogues

A number of nucleotides containing analogues of the polyphosphate chain have been prepared, generally in order to study their interaction with enzymes. The nucleoside phosphorothioates containing sulphur at α, β and γ-phosphorus atoms of the polyphosphate chain have been invaluable in this respect.[118] Generally they are synthesized using the methods described in 8.3.1 and 8.3.2, with thiophosphate used in place of phosphate. Note that the presence of a sulphur atom at P_α or P_β of a triphosphate or P_β of a diphosphate in a non-bridging position confers chirality at these phosphorus atoms.

The analogues (128) and (129), containing imidobisphosphate and methylene-bisphosphonate units, respectively, in place of P_β-O-P_γ of ATP, have long been known and, together with similar derivatives of nucleosides other than adenosine, have found wide use in enzymology, the presence of the imino or methylene bridge conferring general resistance to hydrolysis by nucleoside triphosphate-utilizing enzymes. Treatment of adenosine 5'-phosphoro-morpholidate with the bis(tributylammonium) salts of dichloro-methylenebisphosphonic acid, fluoromethylenebisphosphonic acid and difluoromethylenebisphosphonic acid has recently afforded (130), (131) and (132), respectively.[119] The introduction of electronegative groups at the methylene bridge shifts the pK_a of the fourth ionization of the polyphosphate chain from the value of 8.4 measured for (129), to 7.0, 7.4 and 6.7 in (130), (131) and (132), and thus to values more comparable with that of 7.1 measured for ATP. In terms of dissociation constant and charge distribution, then, the new species are 'better' analogues of ATP. The analogue (133), containing two methylene bridges, has been prepared via the action of 2',3'-O-isopropylidene adenosine on the bismethylene analogue of trimetaphosphate (134), and

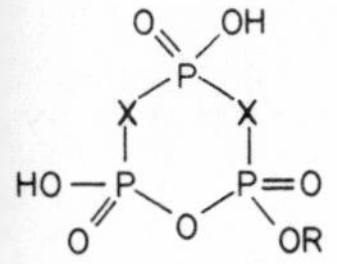

(124) X = O; R = 5' – Ado
(134) X = CH$_2$; R = H
(136) X = O; R = H

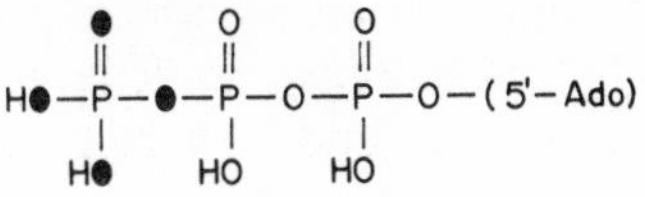

HO—P—•—P—O—P—O—(5'–Ado)

(126) • = ^{18}O

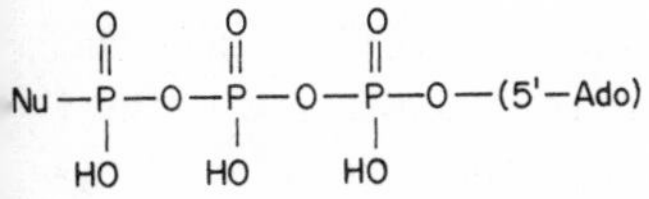

Nu—P—O—P—O—P—O—(5'—Ado)

(125) Nu = Nucleophile

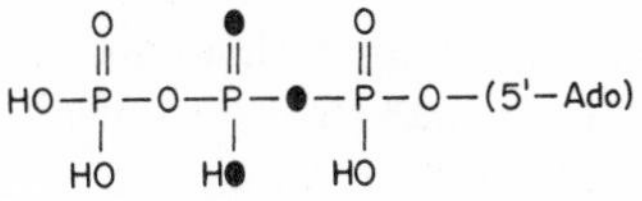

HO—P—O—P—•—P—O—(5'—Ado)

(127)

HO—P—X—P—Y—P—Z—(5'-Ado)

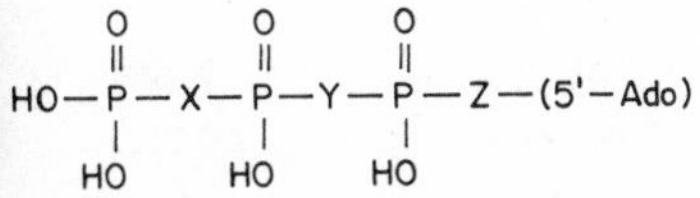

(128) X = NH ; Y = O; Z = O
(129) X = CH$_2$; Y = O; Z = O
(130) X = CCl$_2$; Y = O; Z = O
(131) X = CHF ; Y = O; Z = O
(132) X = CF$_2$; Y = O; Z = O
(133) X = CH$_2$; Y = CH$_2$; Z = O
(135) X = O ; Y = O; Z = NH
(137) X = O-O; Y = O; Z = O

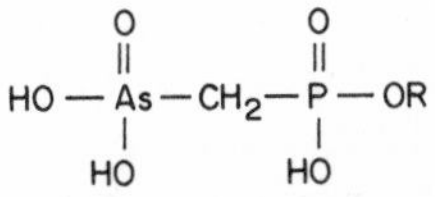

HO — As — CH$_2$ — P — OR

(138) R = H
(139) R = 5' — Ado

HO —P—O—P—O—P—O—(5'-d Thd)

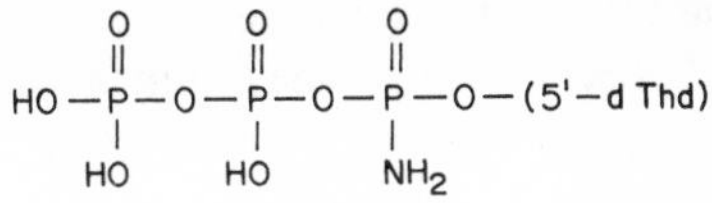

(140)

analogues such as 5'-amino-5'-deoxy ATP (135) via attack of 5'-amino-5'-deoxyadenosine upon trimetaphosphate itself (136).[120] Analogue (137), with a peroxy link between P_β and P_γ, has been formed by treating adenosine 5'-phosphorimidazolidate with peroxodiphosphate.[121] It is generally a poorer substrate than ATP in reactions involving phosphoryl or adenylyl transfer, possibly because the oxygen functions at P_β and P_γ are less readily co-ordinated by metal ions to form a $\beta\gamma$ chelate of the type found in Mg.ATP.$^{2-}$

Treatment of (chloromethyl)phosphonic acid with alkali and arsenious oxide affords (arsonomethyl)phosphonic acid (138) which, if treated with 2',3'-O-isopropylideneadenosine and DCC, yields the arsenical analogue (139) of ADP, after deprotection.[122] The compound is a poor replacement for ADP in enzyme-catalysed reactions, however. When 2'-deoxythymidine-5'-phosphorodiamidate (cf. (121)) is treated with pyrophosphate in DMF, P^1-(2'-deoxythymidine-5')-P^1-aminotriphosphate (140) is formed.[123] In acid, (140) is converted quantitatively into 5'-dTTP, while in ammonia it yields 2'-deoxythymidine-5'-phosphoramidate and pyrophosphate.

8.3.4   Cyclic nucleotides

Interest in the biological role of adenosine 3',5'-monophosphate (141) (cyclic AMP, cAMP) and other cyclic nucleotides as 'second messengers', effecting the intracellular mediation of an extracellular hormonal signal, has led to the synthesis, literally, of hundreds of analogues of cAMP. Many of these have been prepared by performing standard nucleoside chemistry on cAMP itself: this section will consider the ways in which the cyclophosphate ring can be formed.

The most common way of forming the 3',5'-cyclophosphate ring is to treat a nucleoside 5'-phosphate bearing a free 3'-OH group with DCC.[124] In essence, though, any compound in which a good leaving group is attached to the 5'-phosphate of a nucleoside monophosphate can be cyclized in this way upon attack by a free 3'-OH group. For instance, 2'-deoxythymidine-5'-phosphorofluoridate (prepared from 5'-dTMP by treatment with 2,4-dinitro-fluorobenzene in DMF; the 2,4-dinitrophenyl ester of the nucleotide is first formed, and 2,4-dinitrophenate then expelled by fluoride ion[125]) may be cyclized by treatment with potassium *tert*-butoxide to afford 2'-deoxythymidine-3',5'-monophosphate in high yield.[126] Condensation of 8-quinolyl phosphate with the 5'-OH group of a nucleoside using a standard condensing agent (e.g. DCC or TPS-Cl) affords the corresponding 8-quinolyl nucleoside-5'-phosphate (cf. 120), which is cyclized by cupric chloride in pyridine to give the nucleoside-3',5'-monophosphate.[127] In a slightly more unusual preparation, the unprotected ribonucleoside is treated with trichloromethylphosphonyl chloride in triethyl phosphate to afford (142), and the trichloromethyl anion is lost on ring closure using *tert*-butoxide, to give the desired product.[128] Phosphotriester methods (see 8.4.1) have also been utilized to prepare nucleoside 3',5'-monophosphates.[129]

(141) X = O ; R = H
(143) X = S ; R = H
(144) X = O ; R = Alkyl

(142)

(145)

Adenosine 3',5'-monophosphorothioate (143) has been prepared by treating 2',3'-di-O-acetyl adenosine with bis(4-nitrophenyl)thiophosphoryl chloride, deacetylating the product, and cyclization using *tert*-butoxide.[130] This procedure affords a mixture of (*Rp*) and (*Sp*) diastereoisomers; the preparation of the individual isomers will be described below. Analogues of cAMP containing phosphosphoramidate or phosphorothioate groups by virtue of having thiol or amino substituents at the 3'- or 5'-positions of the sugar ring have also been prepared via carbodiimide-mediated cyclization.[131]

Alkyl phosphotriesters of cAMP (**144**) may be prepared by treating cAMP with an arenesulphonyl chloride and the appropriate alcohol.[132] The ethyl ester, in particular, is a powerful anti-tumour agent. The corresponding triesters of cUMP have been formed via direct alkylation with diazoalkanes.[133]

Ribonucleoside 2',3'-monophosphates are easily formed by cyclization of a ribonucleoside 2'-(or 3'-) monophosphate with a condensing agent, as described above for cAMP. It is noteworthy that ribonucleosides may be converted to their 2',3'-monophosphate-5'-phosphate derivatives in a single step by solution in pyrophosphoryl chloride at low temperature followed by neutral buffered hydrolysis of the presumed intermediate (**145**).[134]

8.3.5   Nucleotides containing chiral phosphorus atoms

Some of the most elegant nucleotide chemistry of the last decade has been directed to the synthesis of nucleotides containing stereochemically-defined chiral phosphate and thiophosphate groups. The development of these two classes of compounds has taken place indivisibly, and they are dealt with together here. These nucleotides have been utilized to define the stereochemical course of various enzyme-catalysed reactions, and, while space does not permit coverage of this topic, the reader may be directed to several useful reviews.[118,135,136].

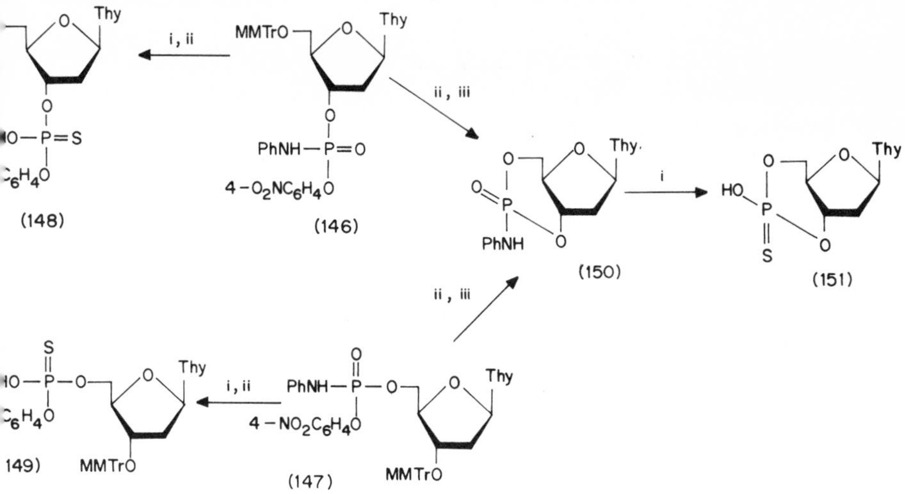

8.3.5.1   Cyclic nucleotides containing chiral phosphorus. Treatment of 4-nitrophenyl phosphorodichloridate with aniline in benzene affords O-(4-

nitrophenyl) phosphoranilidochloridate, which reacts with 5′-[137,138] and with 3′-monomethoxytrityl-2′-deoxythymidine in pyridine to give (146) and (147), respectively, as diastereoisomeric pairs separable on silica gel. On treatment of these compounds with sodium hydride and carbon disulphide or carbon dioxide in benzene, the anilido group is replaced by sulphur, or oxygen, respectively, with retention of configuration. Thus, using carbon disulphide, the chiral thiophosphates (148) and (149) are formed. Further, on removal of the monomethoxytrityl groups and cyclization using *tert*-butoxide in DMF, (146) and (147) afford the stereoisomers of (150). These may be differentiated by [31]P n.m.r. spectroscopy (see below) while the stereoisomers of (149) may be differentiated by the fact that the (*Sp*) isomer is cleaved preferentially by snake venom phosphodiesterase. Knowing this fact, and since configuration is retained in the reaction (147)→(149), the (*Rp*) isomer of (147) has been positively identified, and, since it in turn affords exclusively the (*Sp*) isomer of (150), the ring-closure reaction has been shown to proceed stereospecifically with inversion of configuration. Using this fact allows the (*Sp*) isomer of (146) (which affords only the (*Sp*) isomer of (150)) to be identified, and also the (*Rp*) isomer of (148) (formed exclusively from the (*Sp*) isomer of (146)). Moreover, since treatment of (150) with sodium hydride and carbon disulphide affords the corresponding cyclic phosphorothioate (151) with retention of configuration, the (*Rp*) and (*Sp*) isomer of (151) have been obtained.[137] Stereospecific synthesis of adenosine 3′,5′-(*Sp*)-and-(*Rp*) phosphorothioates has been performed by a related method.[139] $N^6,N^6,O^{2'}$-Tribenzoyl-cAMP, treated with triphenylphosphine and carbon tetrachloride, and then with aniline, affords the cyclic phosphoranilidates (152) and (153) as a chromatographically separable mixture, which on thiation using carbon disulphide and potassium and deblocking afford (154) and (155). Stereoisomeric pairs such as (152) and (153) may be assigned absolute stereochemistry by virtue of the fact that the (*Rp*) isomer (152), in which the anilino group is in the axial position, exhibits a [31]P n.m.r. resonance at higher field than does (153). Moreover, treatment of (152) and (153) with sodium hydride and [^{18}O] benzaldehyde followed by deblocking affords (*Sp*)- and (*Rp*)-adenosine 3′,5′-[^{18}O] monophosphates (156) and (157).[140] A similar procedure has been utilized to prepare the corresponding stereoisomers of [^{18}O]-2′-deoxy-cAMP.[141] The differences in [31]P n.m.r. resonance shifts resulting from the substitution of ^{18}O for ^{16}O in diastereoisomeric pairs such as (156) and (157) are very small, but unequivocal assignment may be made following esterification with diazoethane.

Access to chiral 2′-deoxythymidine-3′,5′-monophosphate analogues has been realized using a different route. Treatment of 2′-deoxythymidine with tris(dimethylamino) phosphine affords (158), which with methanol gives the methyl phosphite (159). Oxidation of (159) with dimethylchloramine gives (160) as stereoisomers separable by chromatography, while treatment of (159)

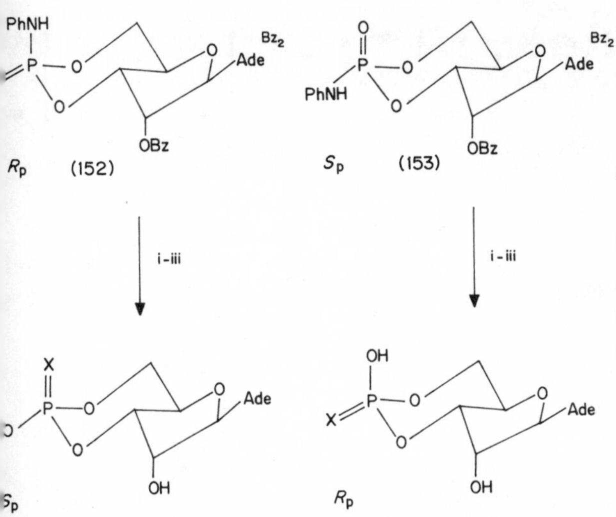

R_p (152) S_p (153)

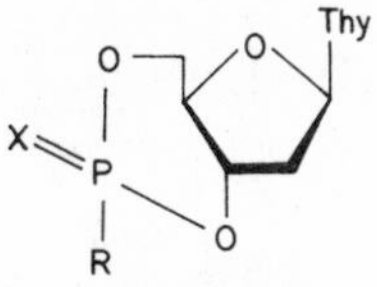

(158) R = Me$_2$N ; X absent
(159) R = MeO ; X absent
(160) R = Me$_2$N ; X = O
(161) R = Me ; X = O
(162) R = MeO ; X = S
(163) R = MeO ; X = ^{18}O

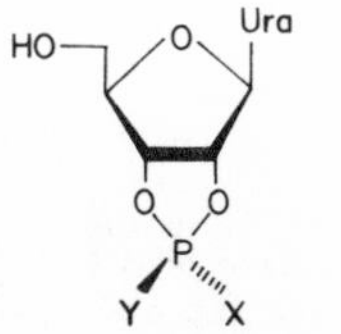

(154) X = S
(156) X = ^{18}O

(155) X = S
(157) X = ^{18}O

(164) X = S ; Y = OH
(165) X = OH ; Y = S

with methyl iodide gives methyl phosphonate (161),[142] and irradiation of (159) with bis(*tert*-butyl) sulphide affords the cyclic phosphorothioate methyl ester (162) as separable stereoisomers which are demethylated by *tert*-butylamine with retention of configuration to form the crystalline cyclic phosphorothioates.[143] Oxidation of (159) using AIBN and ^{18}O$_2$ in benzene affords (163) as separable stereoisomers which may also be demethylated as above. Thus, if (159) is prepared using isotopically-labelled methanol, a route to 2'-deoxythymidine-5'-[^{16}O, ^{17}O, ^{18}O] phosphates of known chirality is established.[144]

Treatment of 5'-acetyluridine with thiophosphoryl tri-imidazolide in the presence of 2,6-lutidine, followed by hydrolysis of the product with water, affords 96% of the *endo* isomer (164) of uridine 2',3'-monophosphorothioate and 4% of the *exo* isomer (165).[145] These compounds have been valuable in determining the stereochemical course of hydrolysis catalysed by ribonucleases.[118]

8.3.5.2 *Nucleoside monophosphates containing chiral phosphorus.* [^{18}O]-Labelled *meso*-hydrobenzoin (167) may be prepared from (S)-mandelic acid (166) in four steps, as indicated. Treatment of (167) with thiophosphoryl bromide in pyridine, and then with 2',3'-di-O-acetyladenosine, affords (168) or (169), depending on the quantity of pyridine used, and these, on reduction with

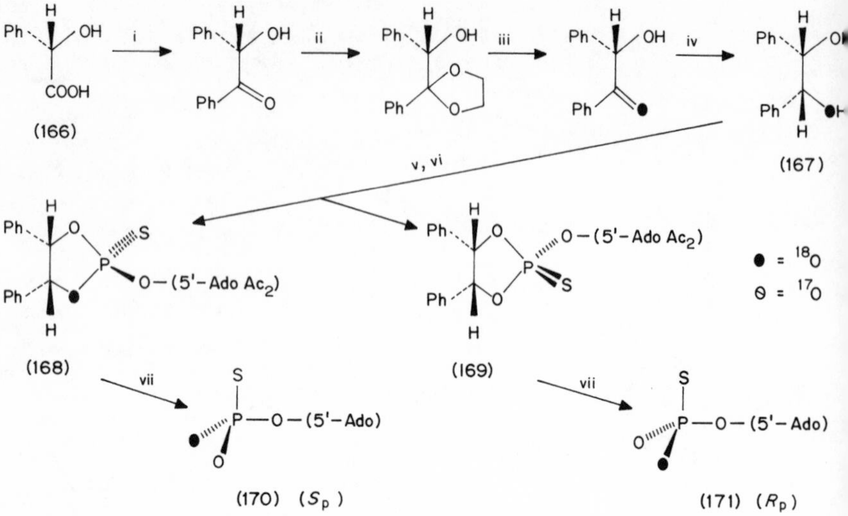

Reagents: i, PhLi; ii, (CH$_2$OH)$_2$, p-TosOH; iii, H$_2^{18}$O; iv, LiAlH$_4$; v, PSBr$_3$, pyr; vi, 2′,3′-di-O-acetyladenosine; vii, Na, liq. NH$_3$.

sodium and liquid ammonia, afford the adenosine (S)-and (R)-5′-[^{18}O] phosphorothioates (170) and (171), respectively.[146]

This useful procedure is similarly applicable to preparation of the corresponding chiral phosphates. If [^{17}O] phosphoryl chloride is employed instead of thiophosphoryl bromide, a single diastereoisomer (172), of the (known) absolute configuration indicated, is obtained, and generates adenosine 5′-[(S)-^{16}O, ^{17}O, ^{18}O] phosphate (173) upon deblocking. The corresponding (Rp) isomer may be obtained either by using [^{17}O]$meso$-hydrobenzoin and [^{18}O] phosphoryl chloride or by starting from (R)-mandelic acid.[147]

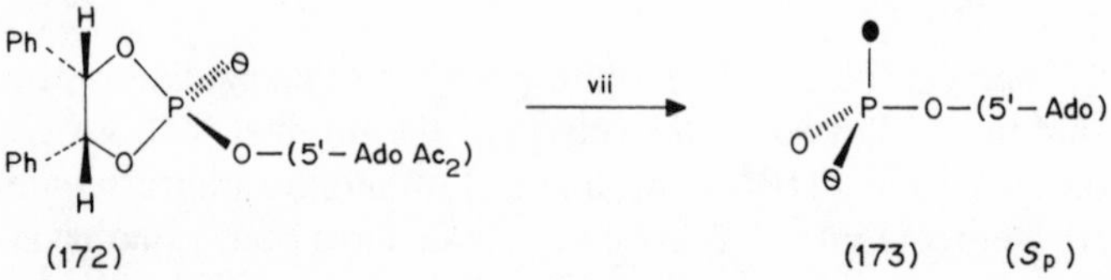

A particularly useful property of the ^{17}O nucleus is its nuclear spin of 5/2, since the resultant nuclear quadrupole causes line-broadening and virtual disappearance from the ^{31}P n.m.r. spectrum of any phosphorus nuclei to which it is attached. Thus, for instance, if (173) is treated with diphenyl phosphorochloridate, and then $tert$-butoxide, cyclization to cAMP occurs, and only the isomer (174) from which ^{17}O has been lost during elimination affords a strong ^{31}P n.m.r. signal.[147] Since the absolute configuration of (174) may be deduced as indicated previously,[141] cyclization with inversion of configuration at phosphorus is seen to have taken place, as shown previously using phosphorothioates.[137,138] In practice the situation is slightly more complex, since ^{17}O is currently only available in c. 50% enrichment, but this does not hinder the successful application of the technique.

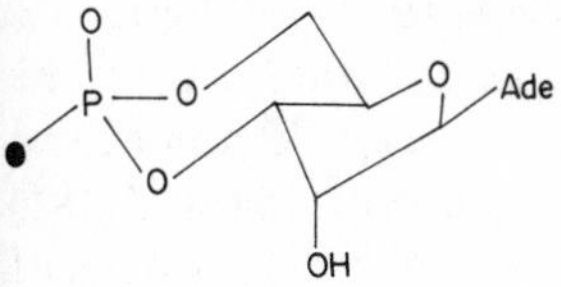

(174)

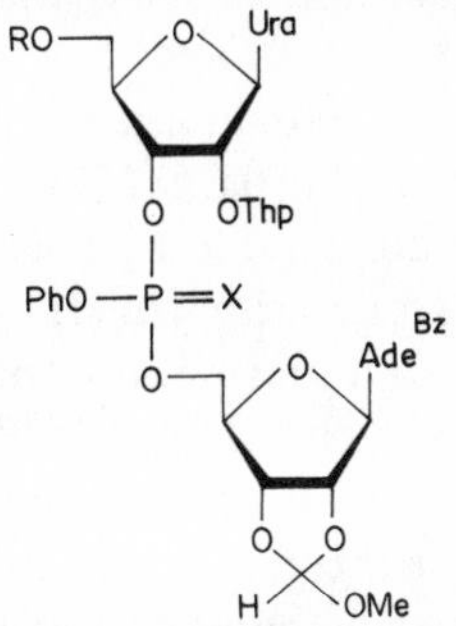

(175) R = 4 - ClC_6H_4OCH$_2$CO; X absent
(176) R = 4 - ClC_6H_4OCH$_2$CO; X = S

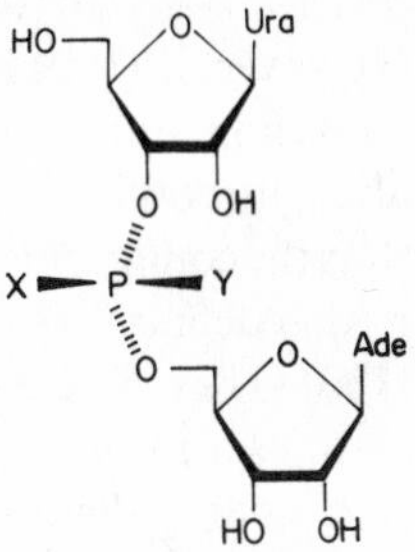

(177) X = O ; Y = S
(178) X = S ; Y = O

Treatment of 5'-O-(4-chlorophenoxy)acetyl-2'-O-tetrahydropyranyluridine with phenylphosphorodichloridite in THF at $-78°C$, followed by N^6-benzoyl-2',3'-O-methoxymethylidene adenosine affords phosphite (175) which upon oxidation with sulphur in pyridine affords the stereoisomers of the phosphorothioate (176) as a separable mixture.[148] Deblocking of the separated isomers affords (Rp)Up(S)A, (177), and (Sp)Up(S)A, (178) thus giving a convenient route to chiral dinucleoside monophosphorothioates of definable absolute configuration. An alternative route via the phosphoranilidate has been described.[149]

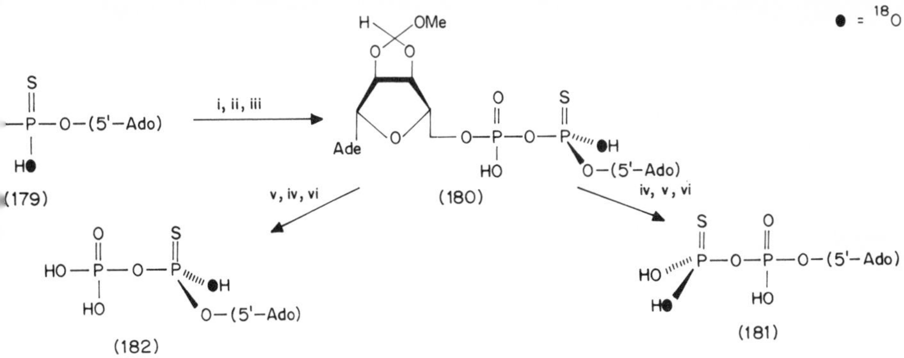

Reagents: i, (PhO)$_2$POCl; ii, 2',3'-O-methoxymethylidene-AMP; iii, DEAE-Sephadex; iv, 1 equiv. NaIO$_4$; v, H$^+$; vi, OH$^-$ *B' = protected base; R' = group protecting 5'-OH; R^2 = H (in 2'-deoxy species) or protected 2'-OH group; R^3 = group protecting phosphate; R^4 = group protecting phosphate oxygen.

8.3.5.3 *Nucleoside polyphosphates containing chiral phosphorus.* If adenosine is treated with thiophosphoryl bromide, and the product hydrolysed in $H_2{}^{18}O$, (179) is obtained, which upon coupling to 2',3'-*O*-methoxyethylidene-AMP using diphenyl phosphorochloridate affords (180) and its diastereoisomer as a separable mixture. Oxidation of (180) with periodate, deblocking of the *cis*-glycol, and elimination of the oxidized residue then yield (181). However, if (180) is first deblocked, and then subjected to limited periodate oxidation and elimination of the oxidized residue, (182) is also formed. The absolute configuration at phosphorus in (182) is assignable by ^{31}P n.m.r., and thus the configuration of the corresponding diastereoisomers of (180) and (181) is also defined. The same is, of course, true for the other diastereoisomers of (180)-(182). Nucleotides with a phosphorothioate group of known absolute configuration at P_β may thus be constructed.[150] This method is easily extended to prepare adenosine 5'-*O*-($\gamma[^{18}O]$-γ-thiotriphosphate) of known chirality at P_γ^{151}, and other methods for these compounds have also been described.[152]

If the 2',3'-*O*-acetyladenosine employed in the reaction to form (172) is replaced by ADP, adenosine 5'-[(S)-^{16}O, ^{17}O, ^{18}O] triphosphate (183) is obtained,[147] and, as indicated earlier, the corresponding (*Rp*) isomer is obtained by varying the reagents in the synthetic sequence. In a not dissimilar synthesis of these compounds, (−) ephedrine is phosphorylated with [^{17}O] phosphoryl chloride and the resultant phosphoramidochloridate is hydrolysed with [^{18}O] lithium hydroxide to afford stereospecifically (184). Protonation of a phosphoramidate potentiates it as a phosphorylating species, of course, so treatment of (184) with ADP as the free acid in DMSO, followed by reductive debenzylation, affords (183).[153] As before, variation in the isotopically labelled reagents, or the stereoisomer of ephedrine used as starting material, will generate the (*Rp*) isomer.

If adenosine 5'-phosphorothioate is condensed with β-cyanoethylphosphate

(183)

(184)

(185)

(186)

$\bullet = {}^{18}O$; $\ominus = {}^{17}O$

using diphenyl phosphorochloridate, the (Rp) and (Sp) isomers of adenosine 5'-[α-thio-β-cyanoethyl] diphosphate, separable using reversed-phase h.p.l.c., are obtained. If, for instance, the (Rp) isomer (185) is treated with cyanogen bromide in [^{18}O] water, the transiently-formed thiocyanate group is displaced with inversion of configuration at P_α, to give (186), which on deblocking with alkali affords (Sp)-[α-$^{18}O_1$] ADP. Thus stereospecific replacement of sulphur by an oxygen isotope with inversion may be performed, and by starting from chiral [^{17}O] adenosine-5'-phosphorothioate, the method can be used to prepare chiral [^{16}O, ^{17}O, ^{18}O] phosphates of known configuration.[154] When adenosine 5'-[α-thiodiphosphate] is oxidized using bromine or NBS in [^{17}O]- or [^{18}O]-labelled water, the sulphur atom is lost and the oxygen isotope introduced with predominant-but not exclusive-inversion at P_α[155,156] Using bromine and [^{17}O] water, 93% inversion is observed, but oxidation of the (Rp) isomer of (151) with NBS in [^{18}O] water proceeds with only 77% inversion.[156]

8.3.6 Reactions for preparing modified nucleotides of general utility

Several reactions of nucleotides with chemical agents lead to modified nucleotide analogues of such widespread utility that they demand inclusion in this chapter. Purine nucleotides are readily brominated at the 8-position, and pyrimidine nucleotides at the 5-position, using bromine water. The 8-bromo function of the purine nucleotides is easily replaced on treatment with nucleophiles, and use of a 1,ω-diaminoalkane affords the corresponding 8-(ω-aminoalkyl)aminopurine nucleotides, which may be coupled to, e.g. CNBr-Sepharose, for use in affinity chromatography.[157] Alternatively a diaminoalkane may be used to displace halide ion from a 6-halogenopurine nucleotide, or else linked to the polyphosphate chain in phosphoramidate linkage via condensation using carbonyldiimidazole or mesitoyl chloride,[158] in order to attach a 'handle' to the nucleotide to permit immobilization. The 8-bromo function is also readily displaced by azide ion, and 8-azidoadenosine-and-guanine nucleotides have been much used for the photoaffinity labelling of enzymes.[159]

Treatment of 5-bromouridylic acid with sodium cyanide in DMSO affords 6-cyanouridylic acid via an addition-elimination sequence, and alkaline hydrolysis of this product gives first the carboxamide, and then orotidylic acid (187),[160] affording a facile route to this important metabolite.

Upon treatment with chloroacetaldehyde, adenosine nucleotides form 1,N^6-etheno-derivatives (188) which are strongly fluorescent, and which have been widely employed as analogues of the corresponding adenosine nucleotides in enzymology. The reaction at 50° in weak acid appears to proceed via attack of the 6-amino group of adenine on the aldehydic group, followed by cyclization and loss of water.[161] Cytidylic acid reacts similarly with chloroacetone to give

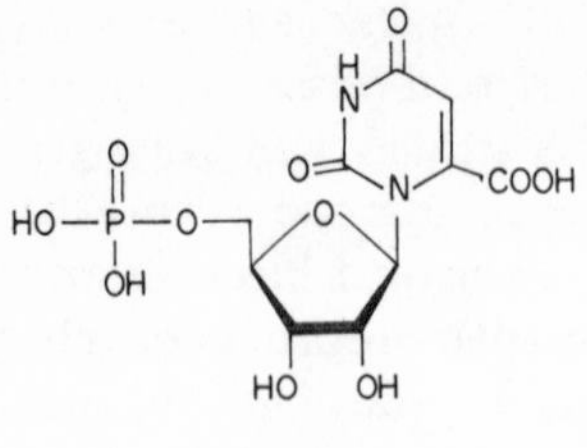

(187)

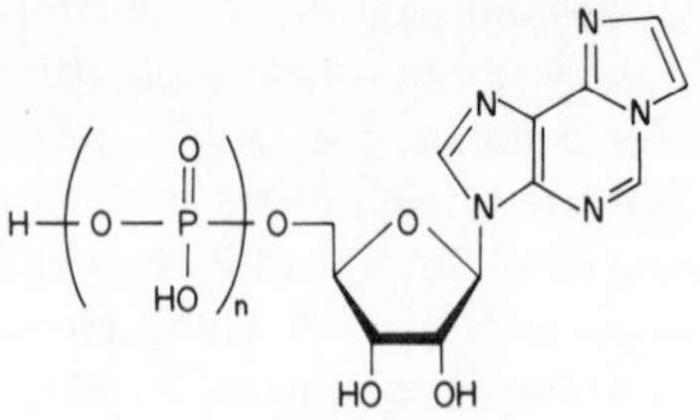

(188)   n = 1—3

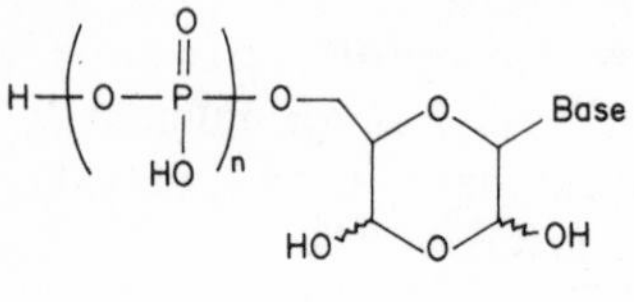

(189)   n = 1—3

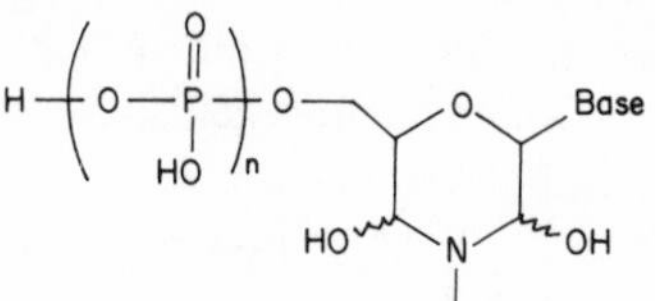

(190)

the methylated $3,N^4$-ethenocytidylic acid which fluoresces strongly in acid solution.[162]

Ribonucleotides are oxidized by periodate to species which, while formally dialdehydes, are hydrated in aqueous solution and condense to form, for the most part, 1,4-dioxan derivatives (189).[163] These species tend to react as if they possess aldehydic functions, however, and have found much use in affinity labelling, during which the ε-amino groups of enzymic active-site lysine residues condense with these compounds forming Schiff bases. This reversible condensation is rendered irreversible by reduction using sodium borohydride or cyanohydridoborate. Moreover, (189) condenses with hydrazides to form (190), and the use of adipic dihydrazide, for instance, allows an alternative method of immobilizing nucleotides for affinity chromatography.[164]

Uridine and cytidine nucleotides are rapidly and quantitatively mercuriated at the 5-position by treatment with mercuric acetate in sodium acetate buffer at pH5, and the same procedure may be used to mercuriate the residues in polynucleotides.[165] The mercuriated species are very strongly bound to sulphydryl-cellulose and other thiol reagents. Moreover, upon treatment with sodium [^{3}H] borohydride, NBS, or iodine, the mercury is replaced by tritium, bromine or iodine respectively, affording a mild and efficient way of introducing a halogen or a tritium label at the mononucleotide level, and in the case of halogenation, at the polynucleotide level.[166] Tritiation is less efficient in polynucleotides, where the polymers undergo incomplete demercuriation. Also, as described earlier for the nucleosides, the mercuriated uridine nucleotides may be styrylated at the 5-position using a styrene derivative and lithium tetrachloropalladate,[167] and this reaction can also be performed at the polynucleotide level.[168]

DNA, covalently attached to cellulose papers, is much used for the detection of electrophoretically separated RNA by blot transfer and hybridization. Essentially the cellulose paper is derivatized so that an arylamino group is covalently attached to the cellulose. Diazotization is then performed, and treatment of the diazo-paper with DNA leads to covalent attachment of the nucleic acid, presumably via attack at C-5 of the pyrimidine bases and C-8 of the purine bases.[169]. A useful review on methods of immobilizing nucleic acids on solid supports is available.[170]

8.3.7 Recently discovered nucleotides of unusual structure

8.3.7.1 *'Magic spot' nucleotides.* When *Escherichia coli* cells are starved of amino acids, they exhibit a 'stringent response' during which protein synthesis is reduced, and two unusual nucleotides, guanosine 3′,5′-bis(diphosphate) (191) and guanosine-3′-diphosphate-5′-triphosphate (192), which have become known as 'magic spot' nucleotides, are produced. Structurally similar adenosine polyphosphates accumulate in *Bacillus subtilis* deprived of a carbon source, during the onset of sporulation. It has been thought that these nucleotides have a role in controlling gene expression during the 'stringent response', although recent evidence suggests that this may not be the case. Compound (191) has been prepared by treating 2′-*O*-(1-methoxyethyl)guanosine with β-cyanoethyl phosphate and DCC, followed by alkali, to give (193), which is bisphosphorylated with, for instance, carbonyldiimidazole and orthophosphate, and then deblocked with weak acid, to give the required compound.[171]

(191) $n_1 = n_2 = 2$; $R = H$
(192) $n_1 = 3$; $n_2 = 2$; $R = H$
(193) $n_1 = n_2 = 1$; $R = CHMe(OMe)$

(194) $R^1 = H$; $R^2 = 5' - (7-MeGuo)$
(195) $R^1 = Me$; $R^2 = 5' - (7-MeGuo)$

8.3.7.2 *'Cap' structures in mRNA.* Sequence analysis of the 5′-ends of viral and nuclear mRNA molecules shows that they are frequently 'capped' with a structure in which 7-methylguanosine is joined by a (5′ → 5′) triphosphate link to a 2′-*O*-methylated nucleoside which is the first residue in the (3′ → 5′)-linked mRNA sequence. The presence of the 'cap' structure is needed for binding and translation of the mRNA to occur. A novel reagent has been devised for synthesis of such structures:[172] treatment of methyl phosphorodichloridate with thiophenol and 7-methylguanylic acid affords (194) in a single step, presumably via formation of a pyrophosphate triester intermediate (195) which is demethylated by thiophenol.

Then, oxidation of (194) using iodine or silver salts in pyridine in the presence of AMP or GMP affords $m^7G(5')ppp(5')A$ or $m^7G(5')ppp(5')G$. Otherwise, more conventional methods have been used for 'cap' structure syntheses, either by treating a nucleoside 5'-phosphorimidazolidate with the 5'-diphosphate of another nucleoside,[173] or by using bis(n-butyl)phosphinothioyl bromide to effect condensation.[113]

8.3.7.3 P', P^1-Bis(5'-nucleosidyl)tetraphosphates.

P^1, P^4-Bis(adenosine-5') tetraphosphate has been found in mammalian cells in concentrations which vary inversely with the doubling time, and it has been speculated that it may behave as a positive growth regulator. It is readily prepared by treating adenosine 5'-phosphoromorpholidate with ATP,[174] a general method of synthesis of compounds of this type. The corresponding bis(guanosine)tetraphosphate is the major purine nucleotide in brine shrimp platelets. Compounds of general formula $G(5')pppp(5')N$ (N = A,C,U,G) are formed during in-vitro transcription of cytoplasmic polyhedrosis virus.

8.3.7.4 '2-5A'.

When mouse L-cells are treated with interferon or double-stranded RNA, a compound is formed which inhibits protein synthesis at subnanomolar concentrations by potentiating a ribonuclease (among other effects). Isolation and degradation of the material has shown it to have the structure pppA2'p5'A2'p5'A, containing the 'unnatural' 2'-5'-phosphodiester link.[175] This compound is usually referred to as '2-5A', and if the terminal 5'-triphosphate chain is absent, it is called 'core trimer'. In fact, a series of compounds of this type, having general formula $pppA2'p(5'A2'p)_n5'A$, has been isolated, and all members of the series with $n > 1$ inhibit protein synthesis. A number of syntheses of 'core trimer', and of '2-5A' itself, have been described. Most of these have used the 'phosphotriester' method of oligonucleotide synthesis (see 8.4.1) although the oligomerization of adenosine-5'-phosphorimidazolidates[176] (see 8.4.2) has also been utilized.

8.4   Nucleic acids

8.4.1   Synthesis of oligonucleotides by stepwise chain elongation

Oligonucleotide synthesis via stepwise chain elongation consists in the formation of a phosphate bridge between two nucleoside residues, repeated so that a chain of nucleoside residues connected by phosphodiester bonds is formed. In general, these internucleotidic links connect the 3′-OH group of one nucleoside unit to the 5′-OH of the next, since oligonucleotide synthesis is usually directed to synthesizing structures found in the nucleic acids. Since the reagents used to phosphorylate the nucleoside units must be directed to phosphorylating a specific OH-group, and these, and the condensing agents used to form the phosphodiester link, may give rise to unwanted side reactions at other sugar OH-groups, or on the base residues, unless these susceptible parts of the molecule are blocked by appropriate protecting groups, the use of protecting groups and their manipulation is paramount importance.

A general scheme of oligonucleotide synthesis is shown. The starting nucleoside (196) is protected on the base residue, and at the 5′-OH and (if present) the 2′-OH groups. It is then phosphorylated by a suitable agent, to form the nucleotide (197), in which R^3 and R^4 may be blocking groups or (in the case of 'phosphodiester' synthesis) R^4 may be a hydrogen atom. In either case, any protecting group R^3 must be removed to unmask a phosphate dissociation, forming (198), while in a separate reaction R^1 is removed to unmask the 5′-OH group, forming (199). Then, (198) is treated with a condensing agent, usually a species which forms a mixed anhydride with the phosphate group, and addition of (199) effects the formation of the new phosphate ester bond in (200), the internucleotidic link being thus made. Then, either any blocking group R^3 is removed from (200) and the product condensed with another molecule of (199), or (more commonly) R^1 is removed from (200) and the product condensed with another molecule of (198), to lengthen the chain. And so on.

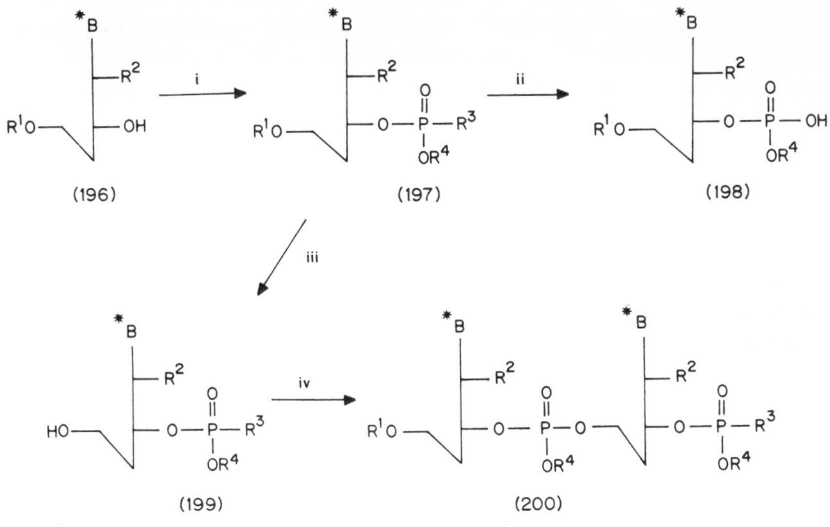

Reagents: i, phosphorylating agent, e.g. (201)–(208); ii, phosphate-deblocking agents; iii, sugar-deblocking agent; iv, (198), condensing agent, e.g. (217)–(221)

It is important to realize that if R^4 is a hydrogen atom, then (200), and all similar products formed after subsequent elongation cycles, will contain only phosphodiester links, and will therefore be charged in aqueous solution, and generally require ion exchange chromatographic techniques to effect purification, rendering the synthetic process lengthy and limiting the quantity of material which can be handled easily. If, however, R^4 is a group blocking the oxygen function, then (200) and all the products of elongation cycles will be uncharged phosphotriesters, relatively soluble in organic solvents, capable of purification by silica gel chromatography (and, particularly, using h.p.l.c., in recent years) and offering far greater ease of handling quantities of material. This, in brief, is the essential difference between the 'phosphodiester' and 'phosphotriester' techniques of oligonucleotide synthesis.

While the 'phosphodiester' method was predominant during the greater part of the 1970s, most notably in Khorana's monumental syntheses of tRNA genes,[177] improvements in methodology and the considerations noted above have now made the 'phosphotriester' technique the method of choice.[178] This chapter will therefore concentrate on aspects of 'phosphotriester' oligonucleotide synthesis. Some useful reviews on polynucleotide synthesis in general[179] and the 'phosphotriester' synthesis in particular[180] are recommended.

8.4.1.1 *Protecting groups for base residues.* A standard set of protecting groups has long been in use to protect the primary amino functions of the nucleic acid bases as amide linkages; for N^6 of adenine, the benzoyl group; for N^4 of cytosine, the anisoyl group; and for N^2 of guanine, the isobutyryl group. All may be removed using concentrated ammonia. In addition, recently the O^6 function of guanine and the O^4 function of uracil have been blocked by arylation,[181] or by the use of the phosphinothioyl group,[182] since these otherwise tend to react with the arylsulphonyl species employed as condensing agents (see 8.4.1.4).

8.4.1.2 *Protecting groups for sugar residues.* The 5'-OH group is most commonly protected by the acid-labile trityl functions, and in particular the monomethoxytrityl and dimethoxytrityl groups, which are introduced via the corresponding trityl chlorides. The use of acidic reagents (80% acetic acid, or 2% benzenesulphonic acid, or trihaloacetic acids) to remove the trityl groups may give rise to depurination, and has been supplanted by the use of zinc bromide in methylene chloride-isopropanol[183] or dialkylaluminium chlorides in hydrocarbon solvents.[184] Base-labile groups introduced to protect 5'-OH include aryloxyacetyl groups,[185] and the laevulinyl group, introduced by condensation with laevulinic acid using 2-chloro-1-methylpyridinium iodide as condensing agent, and removed specifically by hydrazine.[186] The 2'-OH of ribonucleosides is frequently protected by acid-labile functions such as the tetrahydropyranyl or

4-methoxytetrahydropyranyl groups, although the use of the silyl group (particularly TBDMS) at this position[187] has also become popular. It is removed using fluoride ion. A photolabile group, 2-nitrobenzyl, has also been used to protect 2'-OH. It is introduced using 2-nitrobenzyl bromide and sodium hydride, and removed by irradiation at wavelengths > 300 nm.[188]

8.4.1.3 *Phosphorylation of the sugar OH-function* (196)→(197).[189] Many
new agents have been developed for the purpose of phosphorylating the 3'-OH function for the step (196) → (197) indicated in the general scheme. The agents are prepared by successive displacements of chloride from phosphoryl chloride, and may be either bifunctional or monofunctional. The monofunctional reagents (201)–(208) are used, in the presence of a base such as N-methylimidazole, to afford directly species corresponding to (197). The bifunctional reagents (209)–(214) react with (196) in the presence of a base to form the 3'-phosphate ester by replacement of one reactive function, and the second function is usually then replaced by reaction with the alcohol or amine corresponding to one of the R^2 groups of (201)–(206) to afford the corresponding (197) species. An alternative route to these species consists in condensing the nucleoside 3'-OH group of (196) with a phosphodiester (e.g. (215), (216)) using a condensing agent of the type described below.

It should be noted that groups R^1 of (201)–(216) are to be identified with R^4O of (197), and groups R^2 of (201)–(208) with R^3 of (197), and that specific

(201) $R^1 = 2-ClC_6H_4O$; $R^2 = CCl_3CH_2O$

(202) $R^1 = 2-ClC_6H_4O$; $R^2 = PhNH$

(203) $R^1 = 2-Cl-4-Bu^t-C_6H_3O$; $R^2 = CBr_3CH_2O$

(204) $R^1 = 2-ClC_6H_4O$; $R^2 = 4-MeC_6H_4NH$

(205) $R^1 = 4-ClC_6H_4O$; $R^2 = PhNH$

(206) $R^1 = 4-ClC_6H_4O$; $R^2 = CNCH_2CH_2O$

(207) $R^1 = 4-ClC_6H_4O$; $R^2 = 5-Cl-8-$ Quinolyloxy

(208) $R^1 = C_6H_5O$ $R^2 = 4-O_2NC_6H_4O$

(209) $R^1 = 2-ClC_6H_4O$; $R^2 = Cl$

(210) $R^1 = 4-ClC_6H_4O$; $R^2 = Cl$

(211) $R^1 = 2,5-Cl_2C_6H_3O$; $R^2 = Cl$

(212) $R^1 = 2-ClC_6H_4O$; $R^2 = 1,2,4-$ Triazolyl

(213) $R^1 = 4-ClC_6H_4O$; $R^2 = 1,2,4-$ Triazolyl

(214) $R^1 = 2-ClC_6H_4O$; $R^2 = N^1-$oxybenzotriazolyl

(215) $R^1 = 4-ClC_6H_4O$; $R^2 = CNCH_2CH_2O$

(216) $R^1 = 4-ClC_6H_4O$; $R^2 = 5-Cl-8-$ Quinolyloxy

reagents permit the removal of the oxygen-blocking groups of the R^2's (or the anilidates) to afford the phosphate OH group of (198) without affecting groups R^1, or the base- and sugar-protecting groups. Thus, the 2,2,2-trihaloethyl groups are removed with a zinc-copper couple, or using zinc in acetylacetone; the cyanoethyl group is removed with triethylamine in pyridine; the 5-chloro-8-quinolyl group requires zinc chloride in aqueous pyridine for removal; the 4-nitrophenyl group is removed using toluene-p-thiol and triethylamine in acetonitrile; and the anilido-groups are removed using isopentyl nitrite or sodium hydride and carbon dioxide. A number of other phosphate-protecting groups have been described recently, and may in due course become reagents of choice, but all the above-named have proven their worth in complex syntheses.

8.4.1.4 *Condensing agents*.[189] Many new agents have been used to effect condensation of (198) with (199) to give (200). The arenesulphonyl chlorides (217) still find widespread use, but have largely been supplanted by the use of arenesulphonyl-1,2,4-triazolides (218), -3-nitro-1,2,4-triazolides (219), -tetra-zolides (220) and -4-nitroimidazolides (221). Generally these reagents perform the coupling reaction faster than the arenesulphonyl chlorides, giving higher yields and fewer side-reactions (such as sulphonating the 5'-OH component, (199), of the reaction mixture, or sulphonation of the base oxygen functions, such as O^6 of guanine and O^4 of uracil). The tetrazolides (220) decompose on storage and must be prepared fresh. Reagents (218)–(221) are all prepared by treatment of (217) with the appropriate heterocyclic base.

(217)   (218) $R^1 = H$ (220)   (221)
        (219) $R^2 = NO$

$R = 2,4,6-Me_3C_6H_2$; $2,4,6-Pr^i_3C_6H_2$; $4-NO_2C_6H_4$; 8-Quinolyl

The coupling reactions are catalysed by bases, such as 4-(N,N-dimethyl-amino-)pyridine, N-methylimidazole, and tetrazole. It is thought that upon reaction of these condensing agents with (198), a mixed sulphonic-phosphoric anhydride is first formed, which may either react with more (198) to form the corresponding tetrasubstituted pyrophosphate,[190] or with a base to form the appropriate phosphoramidate, which is a more efficient coupling agent than the pyrophosphate when treated with (199).[191] Indeed, even the efficacy of (217) as coupling agent is reportedly much improved by the addition of N-methyl-imidazole to the reaction.[192]

8.4.1.5   *The phosphite method.* In an alternative method for olig‹ synthesis, the internucleotidic link is initially formed as a phosphite subsequently oxidized to the phosphate.[193] Thus, treatment of ‹ phosphorodichloridite (222)–(226)[194] in THF at −70° gives the phosphoromonochloridite (227), which is further treated with the 5'-OH reagent (228) and allowed to warm up, affording (229). The oxidation of (229) to (230) is usually performed with aqueous iodine,[193] although 3-chloroperbenzoic acid in pyridine offers an alternative method.[195] Then, selective removal of the sugar-protecting group R^5 affords a fresh substrate for elongation. Species of type (227) are highly unstable to hydrolysis and aerial oxidation, but if the phosphorodichloridites (222)–(226) are firstly treated with excess $1H$-1,2,4-triazole or $1H$-tetrazole in THF-pyridine at −20°C, and then cooled to −70° and the resultant phosphorodiamidites used, the coupling reaction is similarly efficient and the quantity of by-products is reduced.[196]

This versatile approach also allows the construction of modified internucleotidic links, since oxidation of (229) with iodine and a primary amine, or selenium in DMF, or sulphur in pyridine, results in the formation of phosphoramidate, phosphoroselenoate, or phosphorothioate links respectively.[197]

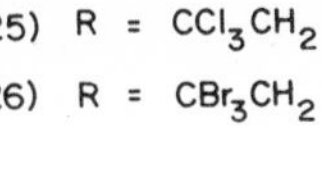

(222) R = Me
(223) R = $2-ClC_6H_4$
(224) R = $4-ClC_6H_4$
(225) R = CCl_3CH_2
(226) R = CBr_3CH_2

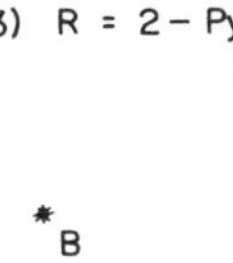

(231) R = $4-O_2NC_6H_4$
(232) R = $2-O_2NC_6H_4$
(233) R = 2- Pyridyl

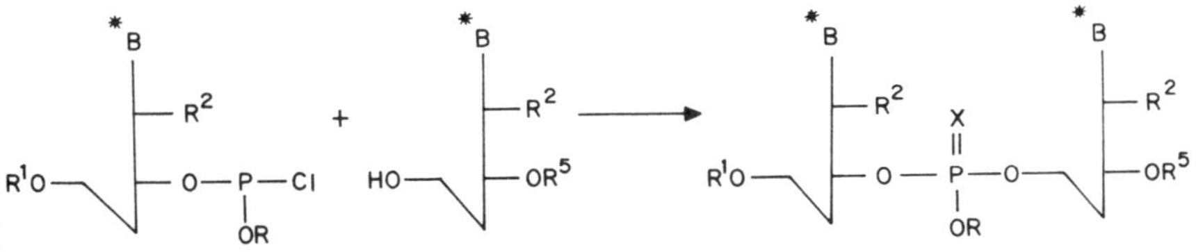

(227)   (228)   (229) X is absent
                              (230) X = O

In (227)–(230), R^1, R^2 and *B have the same significance as in (196)–(200), and R the same as in (222)–(226). R^5 is a sugar -OH protecting group.

8.4.1.6   *Insertion of terminal phosphorus.* It may be desired to synthesize an oligonucleotide with a 5'-phosphorylated terminus, and several reagents have been devised to permit this as part of a 'phosphotriester' synthesis. Selective deblocking of the 5′-OH group, followed by reaction with bis(anilido)phosphorochloridate affords a generally useful method for phosphorylating shorter fully-protected oligonucleotides.[198] More sophisticated reagents such as morpholinophosphorobis-(3-nitro-1,2,4-triazolidate)[199] offer an added advantage in that the phosphoromorpholidate group introduced at the 5′-terminus may subsequently be converted to the 5′-di-or -triphosphate.

8.4.1.7 Final deblocking of the synthesized oligonucleotide.

Following the completion of synthesis and the purification of the fully-protected oligonucleotide, it must be deblocked. Some care must attend the procedure used in this process, since a blocked 5′-OH group or 2′-OH group which is unblocked while the adjacent phosphotriester group is still intact may attack it, with resultant isomerization or chain cleavage.[200] As a general rule, it is best to remove the phosphate-protecting groups first. Aryl groups protecting phosphate (R^1 in (201)–(216), equating to R^4 in (197)–(200) and R in (227), (229) and (230)) are removed rapidly and efficiently by the N^1,N^1,N^3,N^3-tetramethylguanidinium salts of oximates, particularly syn-4-nitrobenzaldoxime (231), syn-2-nitrobenzaldoxime (232) and syn-pyridine-2-aldoxime (233),[201] although these reagents also cause desilylation.[194] Methyl (and 4-nitrophenyl) groups protecting phosphate may be removed with benzene thiol or toluene-4-thiol, although this procedure may give rise to a small amount of internucleotidic cleavage.[202] Alkyl groups may alternatively be removed using tert-butylamine.[203] Any other phosphate-protecting groups may be removed as in section 8.4.1.3. The bases are then deprotected with ammonia, and sugar OH-groups deblocked as indicated in section 8.4.1.2 to give the oligonucleotide, freed of all protecting groups.

(234)

(235) R^1 = DMTr
(236) R^1 = H

(237) R^1 = DMTr ; X is absent
(238) R^1 = DMTr ; X = O
(239) R^1 = H ; X = O

Reagents: for (234)–(235): (196), DCC; for (235)–(236):CCl_3COOH or $ZnBr_2$; for (236)–(237):(227), THF; for (237)–(238):I_2,H_2O.
R, R^2 and *B have the same significance as in (227)–(230).

8.4.1.8 Solid phase synthesis of oligonucleotides.

Solid phase synthesis of oligonucleotides, especially using the phosphotriester methodology, has evolved rapidly during the last few years, and synthetic feats which would have been daunting a decade ago have become commonplace. The phosphite method has proved easily adaptable to solid phase synthesis, and is usually performed on a silica gel support.[204,205] H.p.l.c. grade silica is silylated with (3-aminopropyl) triethoxysilane to give aminopropyl silica which is succinylated with succinic anhydride to give a support phase (234). After blocking unreacted OH-groups on the silica with TMSCl, a base-protected 5'-O-dimethoxytritylnucleoside is coupled to (234), using DCC, to give (235), and then detritylated to give (236). Next, treatment with (227) (or the corresponding phosphorotetrazolidite) in THF affords (237), which is oxidized to (238) with aqueous iodine. After each coupling reaction the product is treated with excess acetic anhydride or diethoxyphosphorotetrazolidite, in order to block any unreacted 5'-OH groups on (236) or its elongated analogues (239, etc.), since these might otherwise react in the next and subsequent rounds of elongation to give, finally, products shorter than the target molecule and having the wrong sequence. This is called 'Failure Sequence Elimination'. Then, detritylation of (238) gives (239) as substrate for the next coupling reaction. This process is conveniently automated,[195,205] and has been used principally for oligodeoxyribonucleotide synthesis. Flushing is necessary between the stages of the addition cycle in order to change solvents or to remove traces of moisture after the oxidation step. The time required for a single addition cycle can be as little as eight minutes! The completed oligonucleotide is deblocked at the phosphate group (usually with benzene thiol, since methyl has mostly been employed as the protecting group), cleaved from the silica support with ammonia, and generally purified using reverse-phase h.p.l.c. before final detritylation at the 5'-terminus. While the claims for product yields vary, the synthesis of a decamer in *overall isolated yield of 30%* is typical.

Solid phase synthesis using the more conventional phosphotriester approach, which has also been automated, has tended to use a polyacrylamide or polyacrylmorpholide(Enzacryl K2) resin,[206] or cross-linked polystyrene,[207] in each case derivatized to afford an aminoalkyl 'handle', which is condensed with a 5'-blocked-3'-O-succinyl nucleoside to give an amide link. The resultant species is thus entirely comparable to (235) with silica replaced by a different polymer. After detritylation, a species such as (240) is used in the

(240)

(241) B = Ade , Ura , Gua , Hyp

addition stage with a suitable condensing agent, e.g. of type (219) or (220). Failure sequence elimination is performed using phenyl isocyanate or acetic anhydride. At the conclusion of synthesis, deblocking and removal from the resin use the procedures described above.

8.4.2 Nucleotide oligomerization

Some interesting studies on the oligomerization of ribonucleoside 5'-phosphorimidazolidates (241) have appeared.[208] To generalize, these condense in neutral buffers in the presence of divalent metal ions to form oligonucleotides which are predominantly 2'-5'-linked if lead salts are used, but contain a higher proportion of 3'-5'-links if zinc salts are employed. Small yields of oligomers up to the pentamer are isolable, but in the presence of the complementary oligoribonucleotide as template, yields are increased dramatically. These reactions are of interest for their potential role in 'prebiotic' oligonucleotide formation. While of rather limited use for oligonucleotide synthesis in general, the method can be used as a rapid route to modest yields of 2'-5'-linked homooligomers.[176]

8.4.3 Nucleic acid sequencing

Two conceptually similar techniques of sequencing nucleic acids have revolutionized this area of molecular biology. Sanger's technique[209] is principally dependent on enzymes for its application, but the other ('Maxam-Gilbert sequencing'[210]) is largely 'chemical'.

Alkylation of DNA using dimethyl sulphate forms 3-methyladenine and 7-methylguanine residues, and labilizes the glycosidic bonds attached to these bases. If the DNA is then heated at pH7, the alkylated bases are removed. In dilute acid, 3-methyladenine is removed preferentially. Treatment with alkali then cleaves the DNA chain at the point of depurination. Hydrazine attacks thymine and cytosine residues at C-6, fragmenting the bases to leave ultimately ribosylhydrazones. In 2M NaCl, however, cytosine is attacked preferentially. In either case, subsequent treatment with base leads to cleavage of the chain. These reactions are applied as follows: a solution containing a fragment of DNA with a 5'-OH terminus, up to several hundred nucleotides in length, is treated with $[\gamma^{-32}P]$ ATP and polynucleotide kinase, which introduces a radiolabel by phosphorylating the 5'-OH terminus. The mixture is then divided into four portions, and each portion is treated with dimethyl sulphate, or hydrazine with and without 2M NaCl, at a concentration calculated to give *one statistical hit per chain*. Then, applying the hydrolytic conditions described above gives:

(1) a solution containing *all possible chain lengths of the DNA* which terminate where A or G was originally present.

(2) as (1), but terminating only where A was originally present.
(3) as (1), but terminating where T or C was originally present.
(4) as (1), but terminating only where C was originally present.

The oligonucleotides in four solutions are then separated electrophoretically side by side on denaturing polyacrylamide gels, which separate the components according to charge (and, thus, to chain length), with the longer chains running further, and the position of the fragments derived from the original 5'-terminus is revealed by autoradiography as a series of dark bands, for each solution (the 'ladder'). The sequence of the DNA fragment is then read simply by inspection: where a dark band is seen in gels (1) and (2), A was present; if seen in (1) but not in (2), G was present; and so on.

This is the 'Maxam-Gilbert' technique as originally described. In the meantime, the use of dimethyl sulphate to effect depurination has been superseded by the use of acid, which is less specific but more easily controlled.[211] Inevitably, analogous sequencing methods for RNA, including a 'chemical' method,[212] have also been devised.

The impact of Sanger's and Gilbert's techniques on molecular biology can hardly be overstated. Just decade ago, Gilbert and his colleagues took three years to sequence a 20-residue fragment of DNA. Now, a competent operator can sequence possibly 1000 nucleotides in a day!

8.4.4 Interaction of nucleic acids with alkylating agents and carcinogens

The interaction of DNA with alkylating agents and other carcinogens has attracted much interest from those seeking to understand the molecular mechanisms of carcinogenesis. Much of the work has been carried out at the nucleotide level, and useful summaries are available.[213] Studies on the alkylation of DNA, homopolynucleotides, and poly[d(A-T)] by dimethyl sulphate, alkyl methanesulphonates and N-methyl- and N-ethyl-N-nitrosourea (MNU and ENU) indicate that MNU and ENU alkylate the phosphodiester links to a large extent, particularly at high or low pH, while the other reagents do so very little or not at all.[214] While the resultant phosphotriesters are hydrolysed in alkali, leading to chain cleavage, they seem to persist in vivo with little hydrolysis,[215] and no evidence has been educed to indicate phosphodiester alkylation as a promutagenic event. Alkylation of the bases leading to point mutations is indicated as the more significant mechanism,[216] and in one study, the formation of 6-O-alkylguanine residues was positively correlated with carcinogenicity.[217] The interaction of DNA and its components with carcinogens such as aflatoxin B, benzanthracene and benzpyrene derivatives, which are potentiated as epoxides prior to covalent interaction with DNA, has also received much attention, and recent reviews may be recommended.[218] Alkylation of guanine residues at N-7 or at N^2 has been implicated as the initial point of attack in several cases.[219]

Certain antibiotics (bleomycin, chartreusin), antitumour agents (neocarzinostatin) and cuprous complexes of 1,10-phenanthroline, usually in the presence of oxygen and a reducing agent (ferrous ion, ascorbate or thiols), lead to strand scission in DNA. Mechanistic studies,[220,221] and identification of the products of the cleavage process,[222] point to hydroxyl radical as the agent mediating strand scission, and an interesting mechanism has been suggested in which it abstracts a hydrogen atom from C-4' of the sugar ring, after which addition of oxygen and a Fenton reaction lead to the peroxide (242). A 1,2-shift followed by hydration of the intermediate cation generates the unstable (243) which decomposes to generate the 3-(base)-2-propenal, glycollic acid, and 3'- and 5'-phosphoryl termini which are the observed products. It has been suggested[221] that hydroxyl radical may be the simplest mutagen and the principal cause of intrinsic mutagenesis.

(242)   (243)

8.5   Supplementary reading

This chapter has concentrated on synthetic chemical aspects of the nucleic acids and their components and neglected the enormous progress made in physical and other studies. Much valuable ^{1}H n.m.r. work is summarized in two reviews,[223] and the results of many X-ray studies are compounded in a book.[224] It has been found that DNA duplexes consisting of d(G-C) pairs assume a left-hand helical arrangement ('Z'-DNA)[225] and a recent article discussing single crystal X-ray analyses of double-stranded DNA molecules of 'A', 'B' and 'Z'-helical structures provides useful supplementary reading.[226]

For general supplementary background in nucleic acid chemistry, two major works[227] contain a huge amount of information on studies prior to c. 1973, and a recent book on nucleotide analogues[228] is essential reading for anyone attempting to come to grips with this burgeoning area of research.

References

1. R.J. Suhadolnik, 'Nucleoside Antibiotics', Wiley-Interscience, N.Y., 1970; *idem*, 'Nucleosides as Biological Probes', Wiley-Interscience, N.Y., 1979.
2. 'Nucleoside Analogues: Chemistry, Biology and Medical Application' (eds. R.T. Walker, E. de Clercq and F. Eckstein), Plenum Press, N.Y., 1979.
3. J.G. Buchanan and R.H. Wightman, 1982, *Top. Antibiot. Chem.*, 6, 229.

4. 'Nucleotides and Nucleic Acids' in *Organophosphorus Chemistry* (Specialist Periodical Reports, The Chemical Society/The Royal Society of Chemistry, London) Vols. 1–14.
5. M. Sekiya, T. Yoshino, H. Tanaka and Y. Ishido, 1973, *Bull. Chem. Soc. Japan*, 46, 556.
6. F.W. Lichtenthaler, P. Voss and A. Heerd, 1974, *Tetrahedron Lett.*, 2141; M. Saneyoshi and E. Satoh, 1979, *Chem. Pharm. Bull.*, 27, 2518.
7. H. Vorbrüggen, K. Krolikiewicz and B. Bennua, 1981, *Chem. Ber.*, 114, 1234; H. Vorbrüggen and G. Höfle, 1981, *ibid.*, 114, 1256.
8. K. Kadir, G. Mackenzie and G. Shaw, 1980, *J. Chem. Soc., Perkin Trans. I*, 2304.
9. F. Seela and H.-D. Winkeler, 1982, *Liebig's Ann. Chem.*, 1634; *idem*, 1982, *J. Org. Chem.*, 47, 226.
10. F.W. Lichtenthaler and K. Kitahara, 1975, *Angew. Chem. (Int. Edn. Engl.)*, 14, 815.
11. M. Imazawa and F. Eckstein, 1979, *J. Org. Chem.*, 44, 2039.
12. M. Imazawa and F. Eckstein, 1978, *J. Org. Chem.*, 43, 3044.
13. J. White and J.B. Hobbs, unpublished results.
14. H. Morisawa, T. Utagawa, S. Yamanaka and A. Yamazaki, 1981, *Chem. Pharm. Bull.*, 29, 3191.
15. H. Morisawa, T. Utagawa, T. Miyoshi, F. Yoshinaga, A. Yamazaki and K. Mitsugi, 1980, *Tetrahedron Lett.*, 479.
16. T.A. Krenitsky, G.W. Koszalska, J.V. Tuttle, J.L. Rideout and G.B. Elion, 1981, *Carbohydr. Res.*, 97, 139.
17. N.J. Cusack, B.J. Hildick, D.H. Robinson, P.W. Rugg and G. Shaw, 1973, *J. Chem. Soc., Perkin Trans. I*, 1720.
18. R.R. Schmidt and D. Heermann, 1981, *Chem. Ber.*, 114, 2825; T. Ito, Y. Nii, S. Kobayashi and M. Ohno, 1979, *Tetrahedron Lett.*, 2521.
19. E. Zbiral and U. Schorkhüber, 1982, *Liebig's Ann. Chem.*, 1870.
20. M. Valentiny, A. Martvon and P. Kovac, 1981, *Collect. Czech. Chem. Comm.*, 46, 2197.
21. J.P. Ferris, V.R. Rao and T.A. Newton, 1979, *J. Org. Chem.*, 44, 4378.
22. R.J. Parry, 1978, *J. Chem. Soc., Chem. Comm.*, 294.
23. R.J. Parry and A. Minta, 1982, *J. Amer. Chem. Soc.*, 104, 871.
24. K.K. Ogilvie, J.L. Sadana, E.A. Thompson, M.A. Quilliam and J.B. Westmore, 1974, *Tetrahedron Lett.*, 2861; K.K. Ogilvie, E.A. Thompson, M.A. Quilliam and J.B. Westmore, 1974, *ibid.*, 2865.
25. G.H. Hakimelahi, Z.A. Proba and K.K. Ogilvie, 1981, *Tetrahedron Lett.*, 22, 5243.
26. W.T. Markiewicz, 1979, *J. Chem. Res. (S)*, 24.
27. A.F. Russell, S. Greenberg and J.G. Moffatt, 1973, *J. Amer. Chem. Soc.*, 95, 4025.
28. R. Mengel and J.-M. Seifert, 1977, *Tetrahedron Lett.*, 4203.
29. J. Brokes, H. Hrebabecky and J. Beranek, 1979, *Collect. Czech. Chem. Comm.*, 44, 439.
30. M.J. Robins and J.S. Wilson, 1981, *J. Amer. Chem. Soc.*, 103, 932; R.A. Lessor and N.J. Leonard, 1981, *J. Org. Chem.*, 46, 4300; J.R. Rasmussen, C.J. Slinger, R.J. Kordish and D.D. Newman-Evans, 1981, *J. Org. Chem.*, 46, 4843.
31. K.K. Ogilvie, G.H. Hakimelahi, Z.A. Proba and N. Usman, 1983, *Tetrahedron Lett.*, 24, 865.
32. M.J. Robins, S.R. Naik and A.S.K. Lee, 1974, *J. Org. Chem.*, 39, 1891.
33. L. Dudycz, A. Kotlicki and D. Shugar, 1981, *Carbohydrate Res.*, 91, 31.
34. K.K. Ogilvie, S.L. Beaucage, M.F. Gillen, D. Entwistle and M. Quilliam, 1979, *Nucleic Acids Res.*, 6, 1695; K.K. Ogilvie, S.L. Beaucage, M.F. Gillen and D.W. Entwistle, 1979, *ibid.*, 2261.
35. A. Yamane, A. Matsuda and T. Ueda, 1980, *Chem. Pharm. Bull.*, 28, 150.
36. S. Sakata, S. Yonei and H. Yoshino, 1982, *Chem. Pharm. Bull.*, 30, 2583.
37. D.E. Bergstrom and P.A. Reddy, 1982, *Tetrahedron Lett.*, 23, 4191.
38. D.E. Bergstrom, 1982, *Nucleosides and Nucleotides*, 1, 1.
39. D.E. Bergstrom and J.L. Ruth, 1976, *J. Amer. Chem. Soc.*, 98, 1587; D.E. Bergstrom and M.K. Ogawa, 1978, *ibid.*, 100, 8106; D.E. Bergstrom, J.L. Ruth and P. Warwick, 1981, *J. Org. Chem.*, 46, 1432.
40. A.S. Jones, G. Verhelst and R.T. Walker, 1979, *Tetrahedron Lett.*, 4415.

41. S.G. Rahim, M.J.H. Duggan, R.T. Walker, A.S. Jones, R.L. Dyer, J. Balzarini and E. de Clercq, 1982, *Nucleic Acids Res.*, 10, 5285.
42. K. Hirota, M. Suematsu, Y. Kuwabara, T. Asao and S. Senda, 1981, *J. Chem. Soc., Chem. Comm.*, 623.
43. H. Tanaka, I. Nasu and T. Miyasaka, 1979, *Tetrahedron Lett.*, 4755.
44. H. Tanaka, H. Hayakawa and T. Miyasaka, 1982, *Tetrahedron*, 38, 2635.
45. H. Vorbrüggen, 1972, *Angew. Chem. (Int. Edn. Engl.)*, 10, 657.
46. W.L. Sung, 1981, *J. Chem. Soc., Chem. Comm.*, 1089; K.J. Divakar and C.B. Reese, 1982, *J. Chem. Soc., Perkin Trans. I*, 1171.
47. M.J. Robins and S.R. Naik, 1971, *J. Amer. Chem. Soc.*, 93, 5277.
48. D. Cech and A. Holy, 1976, *Collect. Czech. Chem. Comm.*, 41, 3335.
49. T. Nagamachi, P.F. Torrence, J.A. Waters and B. Witkop, 1972, *J. Chem. Soc., Chem. Comm.*, 1025.
50. J.G. Moffatt, in Ref. 4.
51. V. Nair and S.G. Richardson, 1979, *Tetrahedron Lett.*, 1181.
52. V. Nair and S.G. Richardson, 1980, *J. Org. Chem.*, 45, 3969.
53. J.D. Bryant, G.E. Keyser and J.R. Barrio, 1979, *J. Org. Chem.*, 44, 3733.
54. M. Ikehara and Y. Ogiso, 1972, *Tetrahedron*, 28, 3695.
55. J.B. Chattopadhyaya and C.B. Reese, 1977, *J. Chem. Soc., Chem. Comm.*, 414.
56. K.J. Divakar and C.B. Reese, 1980, *J. Chem. Soc., Chem. Comm.*, 1191.
57. S.R. James, 1979, *J. Carbohydr., Nucleosides, Nucleotides*, 6, 417; S. Hanessian and A.G. Pernet, 1976, *Adv. Carbohydr. Chem. Biochem.*, 33, 111.
58. F.G. de las Heras and P. Fernandez-Resa, 1982, *J. Chem. Soc., Perkin Trans. I*, 903; K. Utimoto and T. Horiie, 1982, *Tetrahedron Lett.*, 23, 237.
59. L. Kalvoda, 1978, *Collect. Czech. Chem. Comm.*, 43, 1437.
60. L. Kalvoda, 1976, *J. Carbohydr., Nucleosides, Nucleotides*, 3, 47.
61. M.S. Poonian and E.F. Nowoswiat, 1980, *J. Org. Chem.*, 45, 203.
62. J.G. Buchanan, A.R. Edgar, M.J. Power and C.T. Shanks, 1979, *J. Chem. Soc., Perkin Trans. I*, 225.
63. J.G. Buchanan, A.R. Edgar, R.J. Hutchison, A. Stobie and R.H. Wightman, 1980, *J. Chem. Soc., Chem. Comm.*, 237.
64. J.G. Buchanan, A. Stobie and R.H. Wightman, 1980, *J. Chem. Soc., Chem. Comm.*, 916.
65. S.Y.-K. Tam, R.S. Klein, F.G. de las Heras and J.J. Fox, 1979, *J. Org. Chem.*, 44, 4854.
66. H. Ohrui and J.J. Fox, 1973, *Tetrahedron Lett.*, 1951.
67. C.K. Chu, I. Wempen, K.A. Watanabe and J.J. Fox, 1976, *J. Org. Chem.*, 41, 2793.
68. T.J. Cousineau and J.A. Secrist III, 1979, *J. Org. Chem.*, 44, 4351.
69. S. De Bernardi and M. Weigele, 1977, *J. Org. Chem.*, 42, 109.
70. C.K. Chu, U. Reichman, K.A. Watanabe and J.J. Fox, 1977, *J. Org. Chem.*, 42, 711.
71. M.-I. Lim, R.S. Klein and J.J. Fox, 1980, *Tetrahedron Lett.*, 1013; M.-I. Lim and R.S. Klein, 1981, *ibid.*, 22, 25.
72. T. Inoue and I. Kuwajima, 1980, *J. Chem. Soc., Chem. Comm.*, 251.
73. D.M. Brown and R.C. Ogden, 1981, *J. Chem. Soc., Perkin Trans. I*, 723.
74. J.-L. Fourrey, G. Henry and P. Jouin, 1977, *J. Amer. Chem. Soc.*, 99, 6753.
75. R. Noyori, T. Sato and Y. Hayakawa, 1978, *J. Amer. Chem. Soc.*, 100, 2561.
76. T. Sato, M. Watanabe and R. Noyori, 1979, *Tetrahedron Lett.*, 2897.
77. T. Sato and R. Noyori, 1980, *Tetrahedron Lett.*, 2535.
78. T. Sato, H. Kobayashi and R. Noyori, 1980, *Tetrahedron Lett.*, 1971.
79. T. Sato, M. Watanabe and R. Noyori, 1978, *Tetrahedron Lett.*, 4403.
80. T. Sato, K. Marunouchi and R. Noyori, 1979, *Tetrahedron Lett.*, 3669.
81. W.J. Gensler, S. Chan and D.B. Ball, 1975, *J. Amer. Chem. Soc.*, 97, 436.
82. G. Just, T.J. Liak, M.-I. Lim, P. Potvin and Y.S. Tsantrizos, 1980, *Can. J. Chem.*, 58, 2024.
83. K. Pankiewicz, A. Matsuda and K.A. Watanabe, 1982, *J. Org. Chem.*, 47, 485.
84. A.M. Mubarak and D.M. Brown, 1981, *Tetrahedron Lett.*, 22, 683.
85. K. Hirota, K.A. Watanabe and J.J. Fox, 1978, *J. Org. Chem.*, 43, 1193.
86. S. Daluge and R. Vince, 1978, *J. Org. Chem.*, 43, 2311.

87. R.C. Cermak and R. Vince, 1981, *Tetrahedron Lett.*, 22, 2331.
88. R. Vince and S. Daluge, 1980, *J. Org. Chem.*, 45, 531.
89. M. Ohno, N. Yagisawa, S. Shibahara, S. Kondo, K. Maeda and H. Umezawa, 1974, *J. Amer. Chem. Soc.*, 96, 4326.
90. I.D. Jenkins, J.P.H. Verheyden and J.G. Moffatt, 1976, *J. Amer. Chem. Soc.*, 98, 3346.
91. S. Nakatsuka, T. Ohgi and T. Goto, 1978, *Tetrahedron Lett.*, 2579.
92. T. Itaya, T. Watanabe and H. Matsumoto, 1980, *J. Chem. Soc., Chem. Comm.*, 1158.
93. T. Ohgi, T. Kondo and T. Goto, 1979, *J. Amer. Chem. Soc.*, 101, 3629.
94. M. Yoshikawa, T. Kato and T. Takenishi, 1969, *Bull. Chem. Soc. Japan*, 42, 3505.
95. See, e.g., G.R. Gough, D.M. Nobbs, J.C. Middleton, F. Penglis-Caredes and M.H. Maguire, 1978, *J. Med. Chem.*, 21, 520.
96. T. Sowa and S. Ouchi, 1975, *Bull. Chem. Soc. Japan*, 48, 2084.
97. Y. Taguchi and Y. Mushika, 1975, *Chem. Pharm. Bull.*, 23, 1586.
98. J. Hes and M.P. Mertes, 1974, *J. Org. Chem.*, 39, 3767.
99. H.A. Kellner, R.G.K. Schneiderwind, H. Eckert and I.K. Ugi, 1981, *Angew. Chem. (Int. Edn. Engl.)*, 20, 577.
100. J. Giziewicz and D. Shugar, 1975, *Acta Biochim. Pol.*, 22, 87.
101. H. Takaku, Y. Shimada and H. Oka, 1973, *Chem. Pharm. Bull.*, 21, 1844.
102. M. Yoshikawa, M. Sakuraba and K. Kurashio, 1970, *Bull. Chem. Soc. Japan*, 43, 456.
103. T. Hata and M. Sekine, 1974, *J. Amer. Chem. Soc.*, 96, 7363.
104. M. Sekine and T. Hata, 1979, *Chem. Lett.*, 801.
105. D. Dunaway-Mariano, 1976, *Tetrahedron*, 32, 2991.
106. J. Tomasz, 1981, *J. Carbohydr., Nucleosides, Nucleotides*, 8. 557.
107. A. Holy, 1972, *Tetrahedron Lett.*, 157.
108. L.A. Slotin, 1977, *Synthesis*, 737.
109. F. Eckstein and R.S. Goody, 1976, *Biochemistry*, 15, 1685.
110. J.H. van Boom, R. Crea, W.C. Luyten and A.B. Vink, 1975, *Tetrahedron Lett.*, 2779.
111. J.W. Kozarich, A.C. Chinault and S.M. Hecht, 1973, *Biochemsitry*, 12, 4458.
112. M. Maeda, A.D. Patel and A. Hampton, 1977, *Nucleic Acids Res.*, 4, 2843.
113. K. Furusawa, M. Sekine and T. Hata, 1976, *J. Chem. Soc., Perkin Trans. I*, 1711.
114. D.G. Knorre, V.A. Kurbatov and V.V. Samukov, 1976, *FEBS Lett.*, 70, 105.
115. M.R. Webb, 1980, *Biochemistry*, 19, 4744.
116. M. Cohn and A. Hu, 1980, *J. Amer. Chem. Soc.*, 102, 913.
117. J.A. Gerlt, P.C. Demou and S. Mehdi, 1982, *J. Amer. Chem. Soc.*, 104, 2848; S.L. Huang and M.-D. Tsai, 1982, *Biochemistry*, 21, 951.
118. F. Eckstein, 1979, *Acc. Chem. Res.*, 12, 204.
119. G.M. Blackburn, D.E. Kent and F. Kolkmann, 1981, *J. Chem. Soc., Chem. Comm.*, 1188.
120. D.B. Trowbridge, D.M. Yamamoto and G.L. Kenyon, 1972, *J. Amer. Chem. Soc.*, 94, 3816.
121. M.S. Rosendahl and N.J. Leonard, 1982, *Science*, 215, 81.
122. D. Webster, M.J. Sparkes and H.B.F. Dixon, 1978, *Biochem. J.*, 169, 239.
123. A. Simoncsits and J. Tomasz, 1976, *Tetrahedron Lett.*, 3995.
124. See, e.g. W. Wieringa and J.A. Waltersom, 1977, *J. Carbohydr., Nucleosides, Nucleotides*, 4, 189.
125. P.W. Johnson, R. von Tigerstrom and M. Smith, 1975, *Nucleic Acids Res.*, 2, 1945.
126. R. von Tigerstrom, P. Jahnke and M. Smith, 1975, *Nucleic Acids Res.*, 2, 1727.
127. H. Takaku, M. Kato and T. Hata, 1978, *Chem. Lett.*, 681.
128. R. Marumoto, T. Nishimura and M. Honjo, 1975, *Chem. Pharm. Bull.*, 23, 2295.
129. J.H. van Boom, P.M.J. Burgers, P. van Deursen and C.B. Reese, 1974, *J. Chem. Soc., Chem. Comm.*, 618.
130. F. Eckstein, L.P. Simonson and H.P. Bär, 1974, *Biochemistry*, 13, 3806.
131. M. Morr, M.-R. Kula and L. Ernst, 1975, *Tetrahedron*, 31, 1619; M. Morr, 1976, *Tetrahedron Lett.*, 2127.
132. R.N. Gohil, R.G. Gillen and J. Nagyvary, 1974, *Nucleic Acids Res.*, 1, 1691.
133. J. Engels and W. Pfleiderer, 1975, *Tetrahedron Lett.*, 1661.
134. A. Simoncsits and J. Tomasz, 1975, *Biochim. Biophys. Acta*, 395, 74.

135. J.R. Knowles, 1980, *Ann. Rev. Biochem.*, 49, 877.
136. P.A. Frey, 1982, *Tetrahedron*, 38, 1541.
137. W. Niewiarowski, W.J. Stec and W.S. Zielinksi, 1980, *J. Chem. Soc., Chem. Comm.*, 524. W.S. Zielinski and W.J. Stec, 1977, *J. Amer. Chem. Soc.*, 99, 8365.
138. J.A. Gerlt, S. Mehdi, J.A. Coderre and W.O. Rogers, 1980, *Tetrahedron Lett.*, 21, 2385.
139. J. Baraniak, R.W. Kinas, K. Lesiak and W.J. Stec, 1979, *J. Chem. Soc., Chem. Comm.*, 940.
140. J. Baraniak, K. Lesiak, M. Sochacki and W.J. Stec, 1980, *J. Amer. Chem. Soc.*, 102, 4533.
141. J.A. Gerlt and J.A. Coderre, 1980, *J. Amer. Chem. Soc.*, 102, 4531.
142. G.S. Bajwa and W.G. Bentrude, 1978, *Tetrahedron Lett.*, 421; idem, 1980, ibid, 21, 4683; A.E. Sopchik and W.G. Bentrude, 1980, ibid, 21, 4679.
143. A.E. Sopchik and W.G. Bentrude, 1981, *Tetrahedron Lett.*, 22, 307.
144. T.M. Gajda, A.E. Sopchik and W.G. Bentrude, 1981, *Tetrahedron Lett.*, 22, 4167.
145. J.A. Gerlt and W.H.Y. Wan, 1979, *Biochemistry*, 18, 4630.
146. R.L. Jarvest and G. Lowe, 1979, *J. Chem. Soc., Chem. Comm.*, 364.
147. P.M. Cullis and G. Lowe, 1981, *J. Chem. Soc., Perkin Trans. I*, 2317.
148. P.M.J. Burgers and F. Eckstein, 1978, *Tetrahedron Lett.*, 3835; cf. idem, 1979, *Biochemistry*, 18, 592.
149. Z.J. Lesnikowski, J. Smrt, W.J. Stec and W.S. Zielinski, 1978, *Bull. Acad. Pol. Sci., Ser. Sci. Chim.*, 26, 661.
150. J.P. Richard, H.-T. Ho and P.A. Frey, 1978, *J. Amer. Chem. Soc.*, 100, 7756; J.P. Richard and P.A. Frey, 1982, *J. Amer. Chem. Soc.*, 104, 3476.
151. J.P. Richard and P.A. Frey, 1978, *J. Amer. Chem. Soc.*, 100, 7757.
152. G.A. Orr, J. Simon, S.R. Jones, G.J. Chin and J.R. Knowles, 1978, *Proc. Natl. Acad. Sci. U.S.A.*, 75, 2230.
153. W.A. Blättler and J.R. Knowles, 1979, *J. Amer. Chem. Soc.*, 101, 510.
154. R.D. Sammons and P.A. Frey, 1982, *J. Biol. Chem.*, 257, 1138.
155. B.A. Connolly, F. Eckstein and H.H. Füldner, 1982, *J. Biol. Chem.*, 257, 3382.
156. G. Lowe, G. Tansley and P.M. Cullis, 1982, *J. Chem. Soc., Chem. Comm.*, 595.
157. W.L. Dills Jr., J.A. Beavo, P.J. Bechtel, K.R. Myers, L.J. Sakai and E.G. Krebs, 1976, *Biochemistry*, 15, 3724.
158. V.V. Shumyantzeva, N.I. Sokolova and Z.A. Shabarova, 1976, *Nucleic Acids Res.*, 3, 903.
159. H. Bayley and J.R. Knowles, 1977, in 'Methods in Enzymology' (ed. W.B. Jakoby and M. Wilchek), Academic Press, N.Y., Vol. 46, p. 69.
160. T. Ueda, M. Yamamoto, A. Yamane, M. Imazawa and H. Inoue, 1978, *J. Carbohydr. Nucleosides, Nucleotides*, 5, 261.
161. N.K. Kochetkov, V.N. Shibaev, A.A. Kost, A.P. Razzhivin and S.V. Ermolin, 1977, *Dokl. Akad. Nauk S.S.S.R.*, 234, 1339.
162. N.K. Kochetkov, V.N. Shibaev, A.A. Kost, A.P. Razzhivin and A.Y. Borisov, 1976, *Nucleic Acids Res.*, 3, 1341.
163. P.N. Lowe and R.B. Beechey, 1982, *Bioorg. Chem.*, 11, 55.
164. F. Hansske, M. Sprinzl and F. Cramer, 1974, *Bioorg. Chem.*, 3, 367.
165. R.M.K. Dale, E. Martin, D.C. Livingston and D.C. Ward, 1975, *Biochemistry*, 14, 2447.
166. R.M.K. Dale, D.C. Ward, D.C. Livingston and E. Martin, 1975, *Nucleic Acids Res.*, 2, 915.
167. C.F. Bigge, P. Kalaritis and M.P. Mertes, 1979, *Tetrahedron Lett.*, 1653.
168. C.F. Bigge, K.E. Lizotte, J.S. Panek and M.P. Mertes, 1981, *J. Carbohydr., Nucleosides, Nucleotides*, 8, 295.
169. G.R. Stark and J.G. Williams, 1979, *Nucleic Acids Res.*, 6, 195; B. Seed, 1982, *Nucleic Acids Res.*, 10, 1799.
170. H. Potuzak and P.D.G. Dean, 1978, *FEBS Lett.*, 88, 161.
171. J.W. Kozarich, A.C. Chinault and S.M. Hecht, 1975, *Biochemistry*, 14, 981; G.N. Bennett, G.R. Gough and P.T. Gilham, 1976, *Biochemistry*, 15, 4623.
172. I. Nakagawa, S. Konya, S. Ohtani and T. Hata, 1980, *Synthesis*, 556.
173. S. Bornemann and E. Schlimme, 1981, *Z. Naturforsch.*, 36C, 135.
174. E. Rapoport and P.C. Zamecnik, 1976, *Proc. Natl. Acad. Sci. U.S.A.*, 73, 3984.
175. I.M. Kerr and R.E. Brown, 1978, *Proc. Natl. Acad. Sci. U.S.A.*, 75, 256.

176. H. Sawai, T. Shibata and M. Ohno, 1979, *Tetrahedron Lett.*, 4573; *idem*, 1981, *Tetrahedron*, 37, 481.
177. H.G. Khorana, 1979, *Science*, 203, 614; *idem*, 1978, *Bioorg. Chem.*, 7, 351 and references therein.
178. K. Itakura, N. Katagiri, S.A. Narang, C.P. Bahl, K.J. Marians and R. Wu, 1975, *J. Biol. Chem.*, 250, 4592.
179. V. Amarnath and A.D. Broom, 1977, *Chem. Rev.*, 77, 183; M. Ikehara, E. Ohtsuka and A.F. Markham, 1979, *Adv. Carbohydr. Chem. Biochem.*, 36, 135.
180. C.B. Reese, 1976, *Phosphorus and Sulphur*, 1, 245; *idem*, 1978, *Tetrahedron*, 34, 3143; J.H. van Boom, 1977, *Heterocycles*, 7, 1197.
181. S.S. Jones, C.B. Reese, S. Sibanda and A. Ubasawa, 1981, *Tetrahedron Lett.*, 22, 4755.
182. M. Sekine, J.-i. Matsuzaki, M. Satoh and T. Hata, 1982, *J. Org. Chem.*, 47, 571.
183. R. Kierzek, H. Ito, R. Bhatt and K. Itakura, 1981, *Tetrahedron Lett.*, 22, 3761.
184. H. Köster and N.D. Sinha, 1982, *Tetrahedron Lett.*, 23, 2641.
185. E.S. Werstiuk and T. Neilson, 1976, *Can. J. Chem.*, 54, 2689; R.W. Adamiak, R. Arentzen and C.B. Reese, 1977, *Tetrahedron Lett.*, 1431.
186. J.A.J. den Hartog, G. Wille and J.H. van Boom, 1981, *Rec. Trav. Chim. Pays-Bas*, 100, 320.
187. K.K. Ogilvie and R.T. Pon, 1980, *Nucleic Acids Res.*, 8, 2105.
188. E. Ohtsuka, S. Tanaka and M. Ikehara, 1977, *Chem. Pharm. Bull.*, 25, 949.
189. For the original references to the large number of phosphorylating and condensing agents now described, see ref. 4, Vols. 8–13.
190. L.I. Drozdova, V.F. Zarytova and L.M. Khalimskaya, 1981, *Izv. Sib. Otd. Akad. Nauk SSSR, Ser. Khim. Nauk*, 125.
191. A.K. Seth and E. Jay, 1980, *Nucleic Acids Res.*, 8, 5445.
192. V.A. Efimov, S.V. Reverdatto and O.G. Chakhmakhcheva, 1982, *Tetrahedron Lett.*, 23, 961.
193. R.L. Letsinger, J.L. Finnan, G.A. Heavner and W.B. Lunsford, 1975, *J. Amer. Chem. Soc.*, 97, 3278; G.W. Taub and E.E. van Tamelen, 1977, *J. Amer. Chem. Soc.*, 99, 3526.
194. K.K. Ogilvie, N.Y. Theriault, J.-M. Seifert, R.T. Pon and M.J. Nemer, 1980, *Can. J. Chem.*, 58, 2686.
195. T. Tanaka and R.L. Letsinger, 1982, *Nucleic Acids Res.*, 10, 3249.
196. S.L. Beaucage and M.H. Caruthers, 1981, *Tetrahedron Lett.*, 22, 1859.
197. M.J. Nemer and K.K. Ogilvie, 1980, *Tetrahedron Lett.*, 21, 4149.
198. E. Ohtsuka, T. Tanaka and M. Ikehara, 1980, *Chem. Pharm. Bull.*, 28, 120.
199. G. van der Marel, G. Veeneman and J.H. van Boom, 1981, *Tetrahedron Lett.*, 22, 1463.
200. A. Myles, W. Hutzenlaub, G. Reitz and W. Pfleiderer, 1975, *Chem. Ber.*, 108, 2857; J.F.M. de Rooij, G. Wille-Hazeleger, P.M.J. Burgers and J.H. van Boom, 1979, *Nucleic Acids Res.*, 6, 2237.
201. C.B. Reese, R.C. Titmas and L. Yau, 1978, *Tetrahedron Lett.*, 2727; C.B. Reese and L. Zard, 1981, *Nucleic Acids Res.*, 9, 4611.
202. C.B. Reese, R.C. Titmas and L. Valente, 1981, *J. Chem. Soc., Perkin Trans.* 1, 2451.
203. D.J.H. Smith, K.K. Ogilvie and M.F. Gillen, 1980, *Tetrahedron Lett.*, 21, 861.
204. M.D. Matteucci and M.H. Caruthers, 1980, *Tetrahedron Lett.*, 21, 719; K.K. Ogilvie and M.J. Nemer, 1980, *Tetrahedron Lett.*, 21, 4159.
205. M.D. Matteucci and M.H. Caruthers, 1981, *J. Amer. Chem. Soc.*, 103, 3185; G. Alvarado-Urbina, G.M. Sathe, W.C. Liu, M.F. Gillen, P.D. Duck, R. Bender and K.K. Ogilvie, 1981, *Science*, 214, 270.
206. A.F. Markham, M.D. Edge, T.C. Atkinson, A.R. Greene, G.R. Heathcliffe, C.R. Newton and D. Scanlon, 1980, *Nucleic Acids Res.*, 8, 5193; M.L. Duckworth, M.J. Gait, P. Goelet, G.F. Hong, M. Singh and R.C. Titmas, 1981, *ibid.*, 9, 1691; P. Dembek, K. Miyoshi and K. Itakura, 1981, *J. Amer. Chem. Soc.*, 103, 706.; M.J. Gait, H.W.D. Matthes, M. Singh and R.C. Titmas, 1982, *J. Chem. Soc., Chem. Comm.*, 37.
207. K. Miyoshi, R. Arentzen, T. Huang and K. Itakura, 1980, *Nucleic Acids Res.*, 8, 5507; H. Ito, Y. Ike, S. Ikuta and K. Itakura, 1982, *ibid.*, 10, 1755.
208. H. Sawai, 1976, *J. Amer. Chem. Soc.*, 98, 7037; H. Sawai and M. Ohno, 1981, *Bull. Chem.*

Soc. Japan, 54, 2759; *idem*, 1981, *Chem. Pharm. Bull.*, 29, 2237; R. Lohrmann, P.K. Bridson and L.E. Orgel, 1980, *Science*, 208, 1464.

209. F. Sanger, S. Nicklen and A.R. Coulson, 1977, *Proc. Natl. Acad. Sci. U.S.A.*, 74, 5463; F. Sanger, 1981, *Science*, 214, 1205.

210. A.M. Maxam and W. Gilbert, 1977, *Proc. Natl. Acad. Sci. U.S.A.*, 74, 560; W. Gilbert, 1981, *Science*, 214, 1305.

211. A.M. Maxam and W. Gilbert, 1980, *Meth. Enzymol.*, 65, 499; D.E. Pulleyblank, 1982, *FEBS Letts.*, 139, 276.

212. D.A. Peattie, 1979, *Proc. Natl. Acad. Sci. U.S.A.*, 76, 1760.

213. B. Singer, 1975, *Prog. Nucleic Acid Res. Mol. Biol.*, 15, 219; B. Singer, 1976, *FEBS Lett.*, 63, 85.

214. D.H. Swenson and P.D. Lawley, 1978, *Biochem. J.*, 171, 575; B. Singer, 1976, *Nature*, 264, 333; D.E. Jenson and D.J. Reed, 1978, *Biochemistry*, 17, 5098; D.E. Jenson, 1978, *ibid*, 17, 5108.

215. W.J. Bodell, B. Singer, G.H. Thomas and J.E. Cleaver, 1979, *Nucleic Acids Res.*, 6, 2819.

216. R. Saffhill and P.J. Abbott, 1978, *Nucleic Acids Res.*, 5, 1791; B. Singer, H. Fraenkel-Conrat and J.T. Kusmierek, 1978, *Proc. Natl. Acad. Sci. U.S.A.*, 75, 1722.

217. J.V. Frei, D.H. Swenson, W. Warren and P.D. Lawley, 1978, *Biochem. J.*, 174, 1031.

218. S. Neidle, 1980, *Nature*, 283, 135; D.E. Hathway and G.F. Kolar, 1980, *Chem. Soc. Rev.*, 9, 241.

219. H.B. Gamper, J.C. Bartholomew and M. Calvin, 1980, *Biochemistry*, 19, 3948; W.A. Haseltine, M.K. Lo and A.D. D'Andrea, 1980, *Science*, 209, 929.

220. B.G. Que, K.M. Downey and A.G. So, 1980, *Biochemistry*, 19, 5987.

221. S.A. Lesko, R.J. Lorentzen and P.O.P. Ts'o, 1980, *Biochemistry*, 19, 3023.

222. L. Giloni, M. Takeshita, F. Johnson, C. Iden and A.P. Grollman, 1981, *J. Biol. Chem.*, 256, 8608.

223. D.R. Kearns, 1976, *Prog. Nucleic Acid Res. Mol. Biol.*, 18, 91; D.B. Davies, 1978, *Prog. Nucl. Magn. Reson. Spectrosc.*, 12, 135.

224. 'Structure and Conformation of Nucleic Acids and Protein–Nucleic Acid Interactions', ed. M. Sundaralingam and S.T. Rao, University Park Press, Baltimore, 1975.

225. See, e.g. D.R. Davies and S. Zimmerman, 1980, *Nature*, 283, 11.

226. R.E. Dickerson, H.R. Drew, B.N. Conner, R.M. Wing, A.V. Fratini and M.L. Kopka, 1982, *Science*, 216, 475.

227. 'Basic Principles in Nucleic Acid Chemistry', ed. P.O.P. Ts'o, Academic Press, New York, 1974, Vols. 1 and 2; 'Organic Chemistry of Nucleic Acids', ed. N.K. Kochetkov and E.I. Budowskii, Plenum Press, London and New York, 1971, Parts A and B.

228. K.H. Scheit, 'Nucleotide Analogues: Synthesis and Biological Function', Wiley, New York, 1980.

9 Porphyrins and related compounds

A.H. JACKSON

9.1 Introduction

The macrocyclic pigments of life, haem, chlorophyll and vitamin-B_{12} and the related open-chain counterparts, the bile pigments, play a vital role in biological systems. The chlorophylls, through their light-harvesting capacity, enable plants to use the sun's energy to reduce carbon dioxide to carbohydrates, which are subsequently utilized by the plants themselves, and by mammals and other organisms which feed on the plants. Haem plays a complementary role as the core of a range of respiratory enzymes involved in oxidative metabolism in nearly all living systems, and in transporting oxygen in mammalian systems, thereby releasing the energy originally harvested by the chlorophylls. The algal biliproteins also act as photosynthetic receptors in the red and blue-green algae, whilst the bile pigment, phytochrome, plays a vital role in various aspects of the growth of many plants. Vitamin-B_{12} is now recognized as the prosthetic group of a range of important enzymes which carry out various rearrangement reactions, and *trans*-methylations; it is synthesized by micro-organisms and is an essential factor in the diet of man and animals.

This review also includes a brief discussion of the tripyrrolic bacterial pigments, known as prodigiosins, which display a range of anti-bacterial activity. They are, however, biochemically unrelated to the tetrapyrrolic pigments.

Most of the earlier structural and synthetic work on tetrapyrroles has been reviewed by Hans Fischer,[1] the 'father' of pyrrole chemistry in the three volumes of *Die Chemie des Pyrrols*. During and just after the second world war the pace slackened somewhat but then began to grow again during the 1950s, followed by a rapid and worldwide expansion of activity in the 1960s and 1970s. The main emphasis in recent years has been on structure, synthesis, and biosynthesis, but lately work has entered a new phase as chemists have begun to concern themselves more with studies of the function and mode of action of these vital pigments.

A brief account of some of the early work on haem and chlorophyll has been given in the monograph by Marks;[2] this also included a substantial review of the pharmacological aspects of disorders of porphyrin metabolism (por-phyrias). Falk's original book on porphyrins and metalloporphyrins[3] em-phasized the physico-chemical aspects, although it also included a useful section on the preparative aspects; however, this has now been superseded by a second edition[4] (really an entirely new book) with a much more compre-hensive review of porphyrin and chlorophyll chemistry. A seven-volume treatise[5] on the chemistry and biochemistry of porphyrins has also appeared, followed by a two-volume review[6] devoted to vitamin-B_{12}. Other important general reviews include two books on the chemistry of pyrroles,[7,8] and articles in 'Rodd',[9] and 'Comprehensive Organic Chemistry'.[10]

Two systems of nomenclature are in use for porphyrins, chlorins and bile pigments (Scheme 1). The original numbering system[1] used for porphyrins

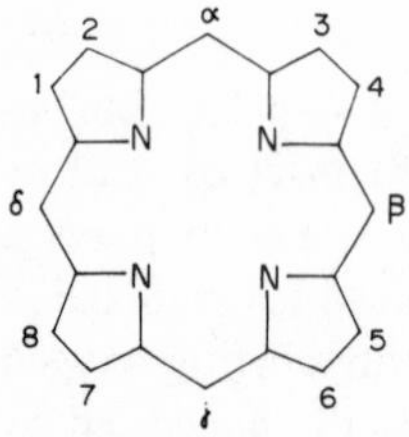

(1a) 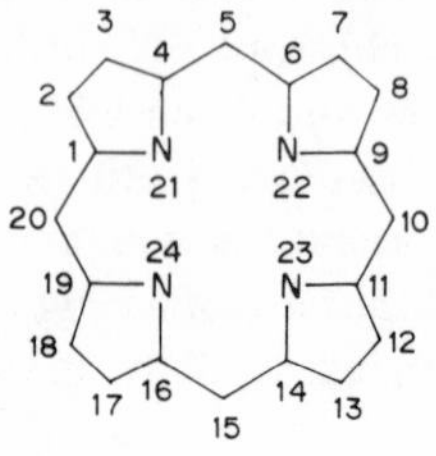   (1b)

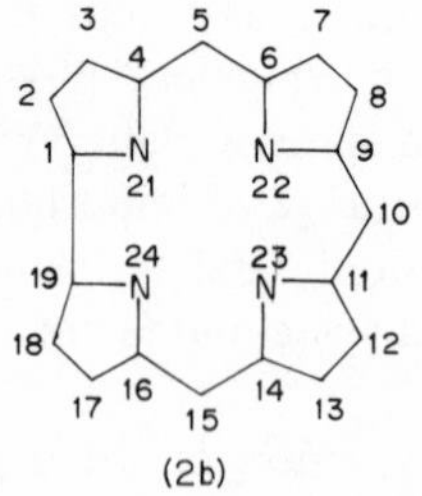

(2b)

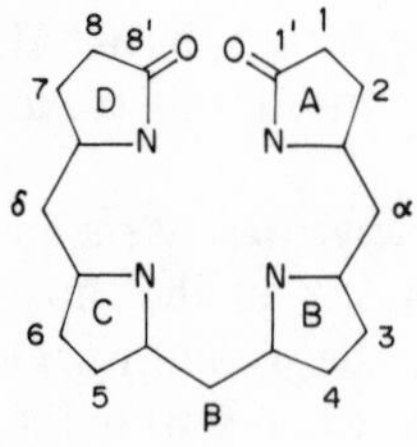

(3a) 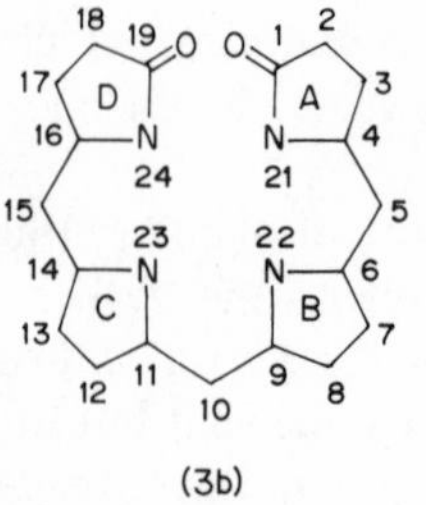   (3b)

Scheme 1 Tetrapyrrole numbering systems

and chlorins by Fischer is shown in structure (1a), the *meso*, or methine, bridge carbon atoms being distinguished by Greek letters. A new numbering system, structure (1b), was recommended by IUPAC[11] in 1960, to allow the same system to be used for the nucleus (2) of vitamin-B_{12}. The corresponding numbering systems for bile pigments are shown in structures (3a) and (3b).

9.2 Porphyrins

Most recent structural studies have relied heavily on spectroscopic and chromatographic comparisons with known derivatives of haem* (4a) (Scheme 2) combined with biosynthetic arguments. Final proof of structure has, however, usually depended on synthesis especially to distinguish between

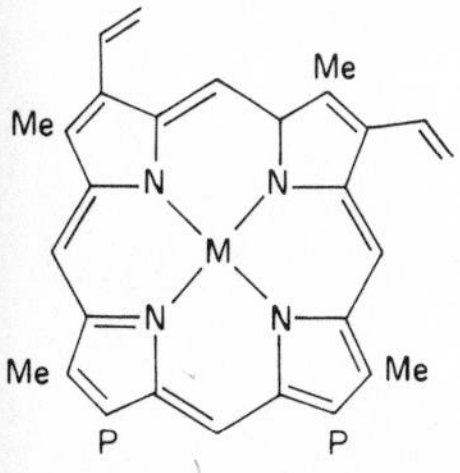

(4) a) M = Fe (II)

b) M = Fe (III) Cl

Scheme 2 Structures of haem and haemin

isomers. Except for the Fischer synthesis from dipyrromethenes[1b] (developed in the 1920s and used in the Nobel prize-winning synthesis of haemin (4b)) most of the porphyrin syntheses now available were developed during the 1960s, i.e. the MacDonald method from dipyrromethanes and dipyr-romethane aldehydes, the *a*- and *b*-oxobilane,[12, 13, 14] the *ac*-biladiene and the *b*-bilene[12, 13, 15] routes. A further variant on the Fischer/*ac*-biladiene route, involving preparation of intermediate tripyrrenes, has now also been de-scribed,[16] and this provides the first truly step-wise method of porphyrin synthesis (see Scheme 10 below for an example).

The key intermediates in haem and chlorophyll formation in nature[17-20] are shown in Scheme 3, and for convenience of reference the ensuing discussion follows the biosynthetic pathway, especially as this has been the raison d'être underlying most of the recent work.

Porphobilinogen (6) was originally synthesized by MacDonald[21] from aliphatic intermediates via a Knorr-type pyrrole synthesis followed by appropriate modifications of the side-chains. A recent variant[22] of this approach (Scheme 4*a*) involved the thallium (III) nitrate oxidative-rearrange-ment of the acetyl pyrrole (12) to the key synthetic intermediate (13); and this

---

*In this and subsequent formulae the abbreviations $A = CH_2CO_2H$, $P = CH_2CH_2CO_2H$, $A^{Me} = CH_2CO_2Me$ and $P^{Me} = CH_2CH_2CO_2Me$ are used for convenience.

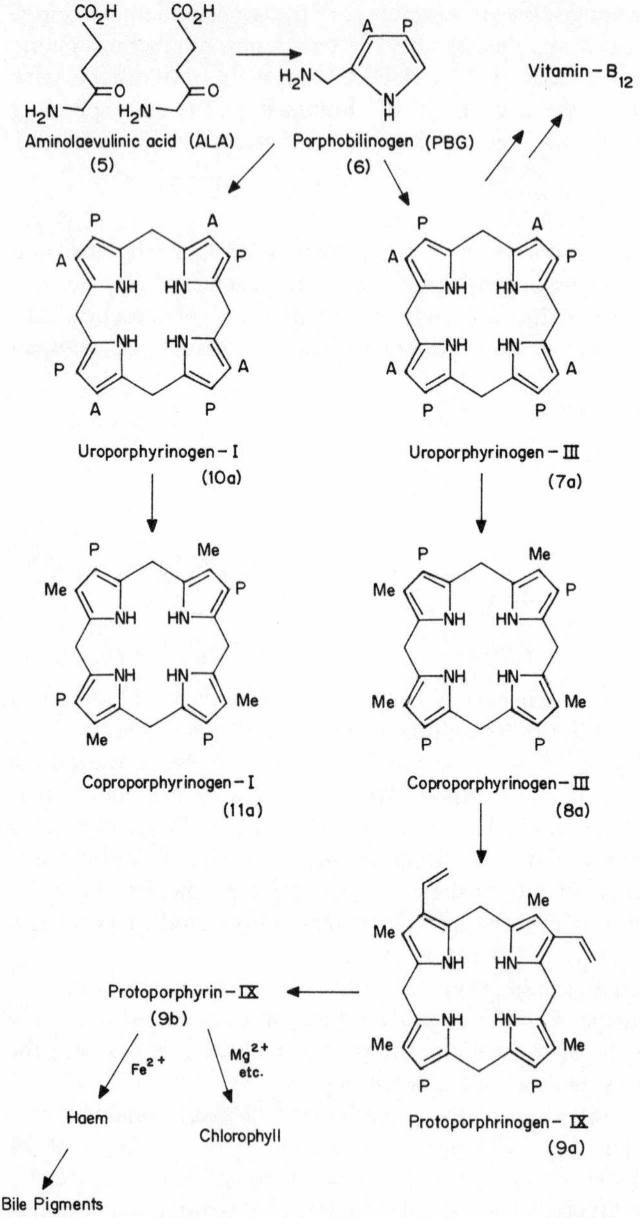

Scheme 3 Biosynthesis of tetrapyrroles. The porphyrinogens in this scheme are denoted by the suffix 'a' after the number; the corresponding porphyrins are denoted in the text by suffix 'b', and the porphyrin methyl esters by the suffix 'c'.

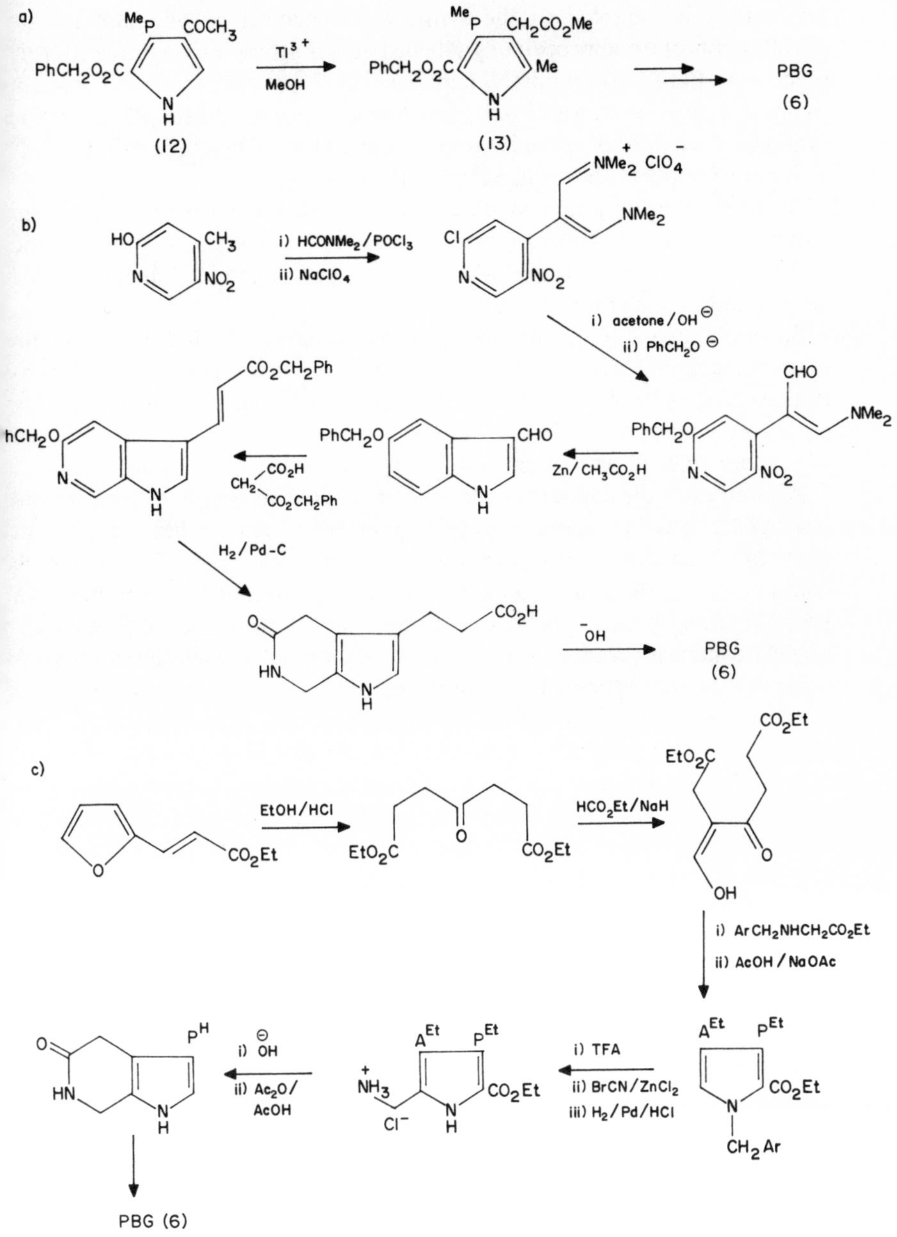

Scheme 4 Recent syntheses of PBG.

can readily be adapted to the synthesis of labelled porphobilinogen by introduction of an appropriately labelled acetyl group into a β-free pyrrole precursor. Earlier Rapoport[23] had developed a novel approach from a pyridine derivative, and this has since been developed and simplified for the synthesis of labelled porphobilinogens required for biosynthetic studies;[24,25] a recent example is shown in Scheme 4b.

Another interesting new synthesis of porphobilinogen (6) has also been reported by Evans,[26] the main novelty being in the synthesis of the aliphatic intermediates required for pyrrole ring construction and introduction of the appropriate side-chain (Scheme 4c).

Porphobilinogen (6) has also been prepared enzymically in 54% yield from δ-aminolaevulinic acid (5) using *Propionibacterium shermanii*.[27] A continuous process involving passage of the acid (5) through a column containing sepharose-linked enzyme has now been developed[28] and this appears to be an even more efficient way of enzymically synthesizing porphobilinogen.

A number of di- and tri-pyrromethanes related to porphobilinogen, and bearing acetic and propionic acid side-chains have been synthesized in recent years in connection with studies of porphyrin biosynthesis,[17-20] but such compounds are difficult to obtain from natural sources as the coupling of porphobilinogen units appears to occur on the enzyme surface of porphobilinogen deaminase (uroporphyrinogen synthetase); only when the synthesis is complete is the open-chain tetrapyrrole liberated, and transformed by uroporphyrinogen-III cosynthetase into uroporphyrinogen-III (7a). This tetrapyrrole was originally thought to be the amino-methyl bilane (14a); indeed ^{13}C-labelled material prepared from a pyrromethane via a tripyrrene and an a,c-biladiene by the Cambridge group[29] was incorporated enzymically into uroporphyrinogen-III (7a) and similar results were also obtained in Stuttgart.[30] More recently, isolation of the intermediate obtained from porphobilinogen and n.m.r. analyses have shown that it is in fact the related hydroxymethylbilane (14b)[31] (also known as 'pre-uroporphyrinogen'). This hydroxymethylbilane is transformed directly into uroporphyrinogen-III by the cosynthetase enzyme, whereas ring closure to uroporphyrinogen-I occurs non-enzymically under mildly acidic conditions, but at a somewhat slower rate.[31] Labelled hydroxymethylbilane was synthesized by coupling a hydroxymethyl pyrromethane (15) with a pyrromethane aldehyde (16) (Scheme 5) followed by hydrolysis of the ester side-chains and careful reduction of the formyl group to hydroxymethyl; subsequent biosynthetic experiments confirmed that it was incorporated intact into uroporphyrinogen-III, and thus that the rearrangement of one of the four porphobilinogen units occurred during the cyclization process.

A number of syntheses of uroporphyrin-III(7b) have been described of recent years, mainly modifications of the MacDonald synthesis, but also

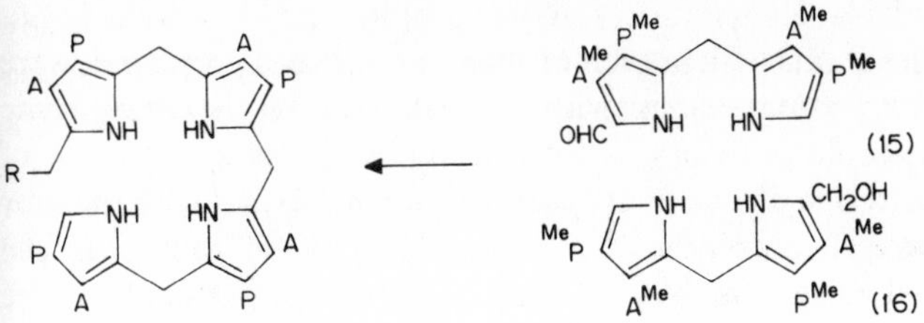

(14) a) R = NH$_2$
      b) R = OH

Scheme 5 Synthesis of the hydroxymethylbilane precursor of uroporphyrinogen-III

including a new approach by the *ac*-biladiene route.[32] The modifications to the MacDonald method enabled specifically labelled uroporphyrinogen-III to be prepared[33-35] for biosynthetic studies of its incorporation into protoporphyrin-IX(**9b**). In normal metabolism only traces of free porphyrins are found in body fluids and excreta, but much larger amounts (several milligrams per day) may be excreted by man (and other mammals) in various types of pathological conditions[36] ('porphyrias'). In porphyria cutanea tarda, which appears to be due to an impairment of uroporphyrinogen de-carboxylase, hepta-, hexa- and penta-carboxylic porphyrins accumulate; as shown by t.l.c. or h.p.l.c. these correspond to the intermediate porphyrinogens between uroporphyrinogen-III(**7a**) and coproporphyrinogen-III(**8a**).[37] The gross structure of the heptacarboxylic porphyrin ('phyriaporphyrin', 'pseudo

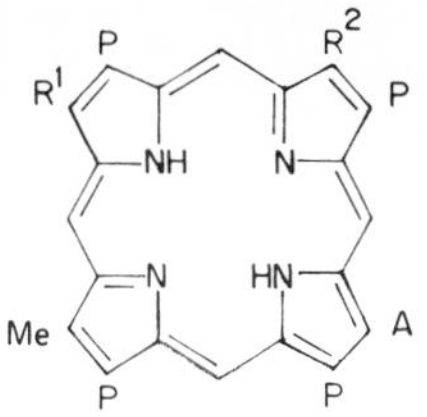

(17) R^1 = R^2 = A

(18) R^1 = Me, R^2 = A

(19) R^1 = R^2 = Me

Scheme 6 Structures of porphyrins related to intermediates between uroporphyrinogen-III and coproporphyrinogen-III

uroporphyrin') was confirmed by mass and n.m.r. spectrometry of its heptamethyl ester; its precise structure (**17**) (see Scheme 6) was subsequently established by synthesis[38-41] of all four isomeric type-III heptacarboxylic porphyrin heptamethyl esters, and by n.m.r. spectroscopic comparisons including titrations with europium shift reagents.[39] (Ester groups on adjacent pyrrole rings undergo bi-dentate complexing with the shift reagent leading to dramatic down-field shifts of the methine proton in between the rings concerned).[42] The hexa- and penta-carboxylic porphyrins[39,40] (18) and (19), excreted in the urine of rats with hexachlorobenzene-induced porphyria, were identified by similar comparisons with the six type-III hexacarboxylic porphyrins (which could in theory be formed following decarboxylation of two of the acetic acid side-chains of uroporphyrinogen-III) and with the four related pentacarboxylic porphyrins.[39-41] The MacDonald route was the method of choice wherever possible, but those porphyrins which could not be synthesized in this way were prepared by the *b*-oxobilane, or *ac*-biladiene routes. The results obtained in this work led to the conclusion that the preferred route from uroporphyrinogen-III to coproporphyrinogen-III involved a 'clockwise' decarboxylation process[39] starting with the D-ring acetic acid of uroporphyrinogen-III (cf. Scheme 3). However, the enzyme is capable of decarboxylating all fourteen hepta-, hexa- and penta-carboxylic porphyrinogens[43] related to urocoproporphyrinogen-III (as well as the other three isomeric uroporphyrinogens[44] and a number of other unnatural porphyrinogens bearing acetic acid side-chains). In subsequent work an interesting modification[45] of the MacDonald method involving condensation of α-formyl-pyrromethan α-carboxylic acids has led to alternative syntheses of uroporphyrin-III, and the hepta-carboxylic porphyrin as their methyl esters (Scheme 7); the method leads to mixtures of products owing to self-condensation of each of the dipyrromethanes, but these can be readily separated by h.p.l.c.

In a rare form of congenital porphyria[46] the uroporphyrinogen cosyn-

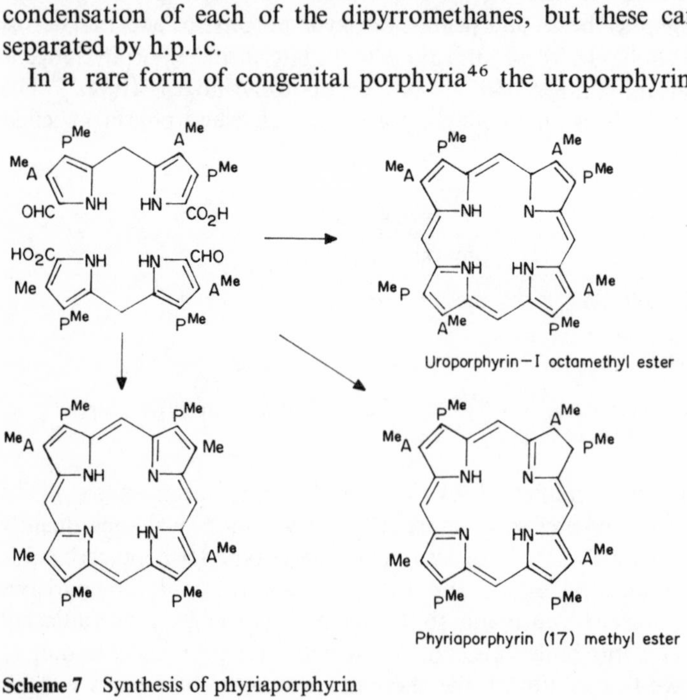

Uroporphyrin—I octamethyl ester

Phyriaporphyrin (17) methyl ester

Scheme 7 Synthesis of phyriaporphyrin

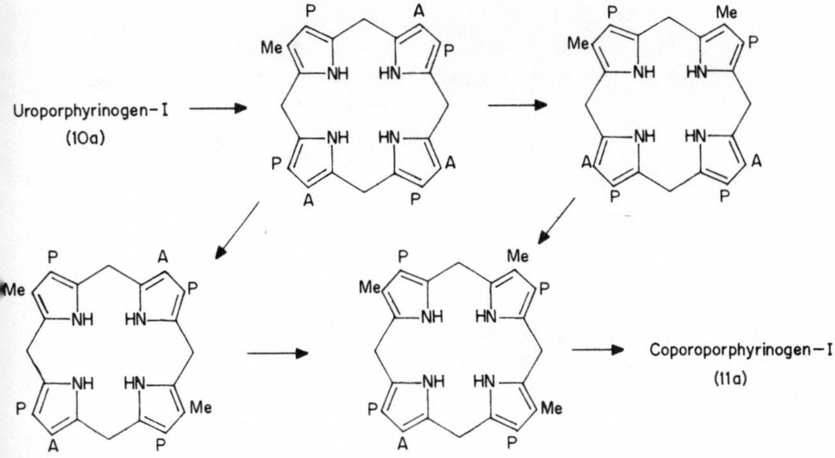

Uroporphyrinogen-I
(10a)

Coporoporphyrinogen-I
(11a)

Scheme 8 Enzymic decarboxylation of uroporphyrinogen-I

thetase enzyme appears to be impaired and substantial amounts of uroporphyrin-I (**10b**) and its decarboxylation products may be formed. The heptacarboxylic porphyrin, the two possible hexacarboxylic porphyrins and the pentacarboxylic porphyrin corresponding to the porphyrinogens shown in Scheme 8 were synthesized as their methyl esters by the *b*-oxobilane and Fischer methods. Comparisons with naturally derived material, together with biosynthetic incorporation studies, showed[47] that the enzymic decarboxylation of uroporphyrinogen-I (**10b**) occurs by both possible routes *in vivo* (Scheme 8).

One of the characteristic features of porphyria cutanea tarda (p.c.t.) is the excretion of a group of four related porphyrins, one of which is isomeric with coproporphyrin-III.[48] Mass and n.m.r. spectroscopic studies of isocoproporphyrin (**20a**) tetramethyl ester, including europium shift reagent titrations (and comparisons with the coproporphyrin isomers I–IV—see Scheme 11 below) enable its structure to be assigned.[42] (Scheme 9). This was subsequently confirmed by total synthesis by the tripyrrene/biladiene,[16] the *b*-

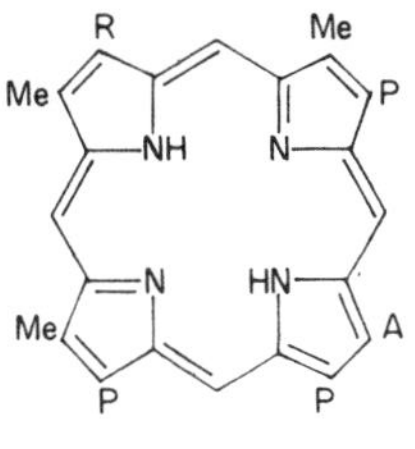

(20) a) R = Et
      b) R = CH = CH$_2$
      c) R = CH(OH)CH$_3$
      d) R = H

Scheme 9 Structures of isocoproporphyrinogen and its congeners

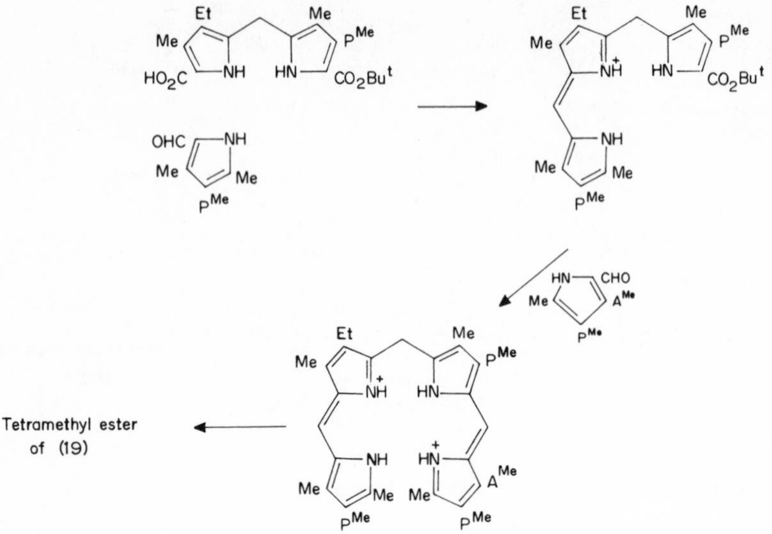

Scheme 10 Tripyrrene biladiene synthesis of isocoproporphyrin

bilene[49] and the b-oxobilane[50] routes. The tripyrrene/biladiene route to isocoproporphyrin is shown in Scheme 10. The intermediate tripyrrene was synthesized by acid-catalysed coupling of an α-formylpyrrole with a dipyr-romethane carboxylic acid (prepared by hydrogenolysis of the corresponding benzyl ester). Deprotection of the tripyrrene t-butyl ester by treatment with trifluoroacetic acid, followed by coupling with a second α-formyl pyr-roleafforded an unsymmetrical ac-biladiene, which was cyclized to isocoprop-orphyrin (20a) tetramethyl ester with copper (II) acetate.[16]

The structures of the other three porphyrins (20b, c and d) were deduced on the basis of spectral comparisons, and they have now also been synthesized by the b-bilene and b-oxobilane routes.[49,50] The formation of isocopropor-phyrin and its congeners is thought to be due to the action of coproporphy-rinogen oxidase on accumulated pentacarboxylic porphyrinogen (cf. 19) formed in p.c.t. prior to decarboxylation of the C-ring acetic acid side-chain.

All four coproporphyrin isomers (Scheme 11) have been synthesized by various methods, both for biochemical experiments with the corresponding porphyrinogens, and for n.m.r. studies. Thus the type-I isomer (11b) was prepared[51] by the Fischer route from the readily available pyrrole (21); further improvements[52] in the yields of this and other related centrosymmetri-cal porphyrins were achieved by heating the intermediate pyrromethene hydrobromide, e.g. (22), with one equivalent of bromine in formic acid (rather than by fusion in molten succinic acid). Coproporphyrin-II has been synthesized by two variants[51] of the MacDonald method, and the III- and IV-

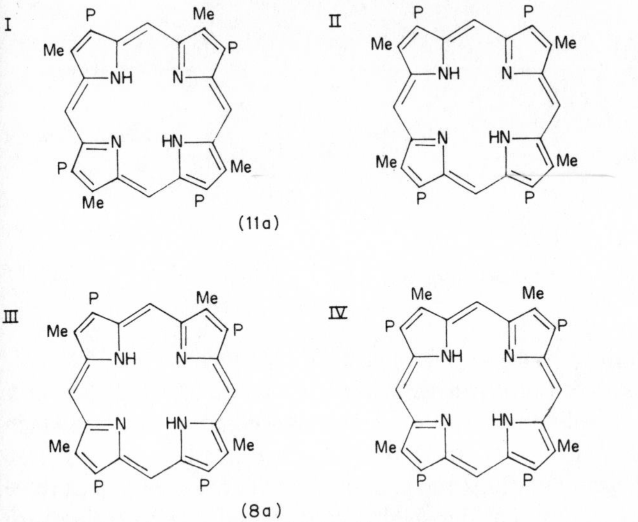

Scheme 11 Structures of coproporphyrin isomers I–IV

isomers by the *a*-oxobilane route.[51] Coproporphyrin-III tetramethyl ester (8c) has also been synthesized [53] from protoporphyrin-IX dimethyl ester (9c) (readily available from haem) by treatment with thallium(III) nitrate in methanol, followed by hydrolysis to the *bis*-aldehyde and reduction with borohydride to the *bis*-hydroxyethyl porphyrin; the latter was then converted via the *bis*-bromoethyl and *bis*-cyanoethyl derivatives to coproporphyrin-III tetramethyl ester (8c) (Scheme 12).

Coproporphyrinogen-III (8a) is degraded *in vivo* via a tricarboxylic

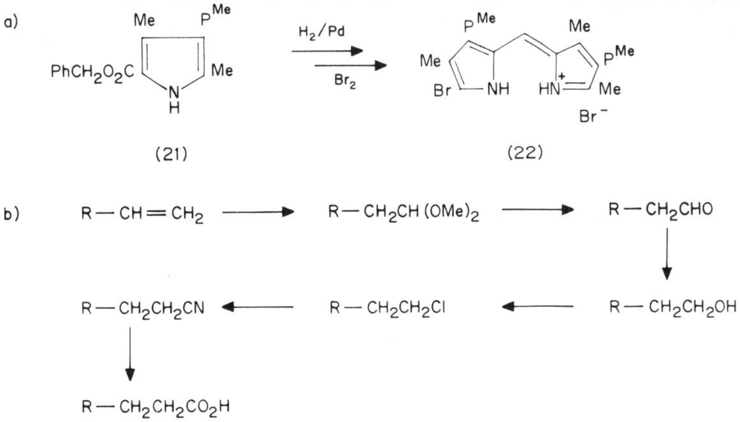

Scheme 12 Intermediates in coproporphyrin synthesis

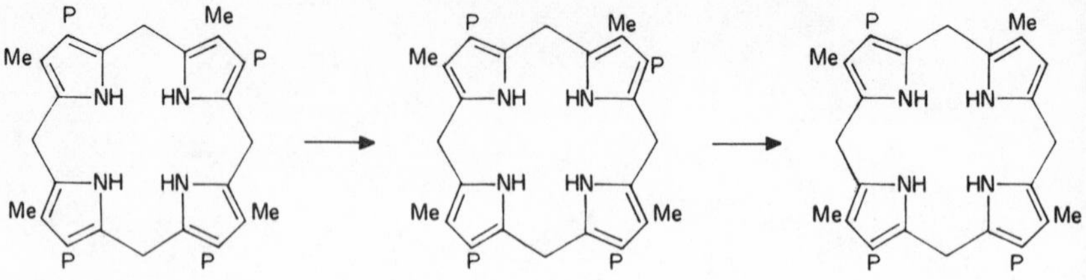

(8a)   (23a)   (9a)

Scheme 13 Formation of protoporphyrinogen-IX

porphyrinogen (23a) to protoporphyrinogen-IX (9a) (Scheme 13). The corresponding tricarboxylic porphyrin (23b) was first isolated[54] (together with copro- and proto-porphyrins) from the Harderian gland of the rat (which occurs at the rear of the eye of this and some other small mammals). This tricarboxylic porphyrin was shown to contain a vinyl group, as well as three propionic acid side-chains and four methyl groups, and its occurrence with copro- and proto-porphyrin pointed to structure (23b) or the isomer with a vinyl group in the B-ring of the porphyrin. Both were synthesized[54,55] by the b-oxobilane route and the trimethyl ester (23c) shown to be identical with that of the natural harderoporphyrin; the vinyl group in each case was fashioned from an acetoxyethyl side-chain carried through from the pyrrole stage. Further syntheses of harderoporphyrins by the MacDonald and b-bilene routes have also been described,[56,57] as well as partial syntheses from protoporphyrin,[58,59] and deuteroporphyrin.[60] Interestingly both harderoporphyrinogen (23a), and its isomer, are substrates for the enzyme coproporphyrinogen oxidase,[58,61] in both plant and mammalian systems, but experiments using labelled coproporphyrinogen precursors together with careful h.p.l.c. studies have shown that the natural pathway to protoporphyrin-IX proceeds regiospecifically *via* harderoporphyrinogen.[61]

An interesting new tetracarboxylic porphyrin was isolated from meconium (the first faeces of the new-born), and given the name 'S411-porphyrin' because of the position of its Soret band in the visible spectrum; visible, n.m.r. and mass spectra showed that it was a dehydrocoproporphyrin, i.e. a monoacrylic acid analogue of coproporphyrin.[62] Biosynthetic arguments favoured either structure (24) or (25), the former being preferred by analogy with harderoporphyrin (23b) (Scheme 14). Both isomers were, therefore, synthesized by the b-oxobilane and MacDonald routes respectively,[63] the acrylic acid side-chains being introduced by formylation of the appropriate precursor with a vacant peripheral position followed by Knoevenagel type condensation with the monomethyl ester of malonic acid (i.e. porphyrin-H→—CHO→—CH=CHCO$_2$Me). Comparisons of both synthetic com-

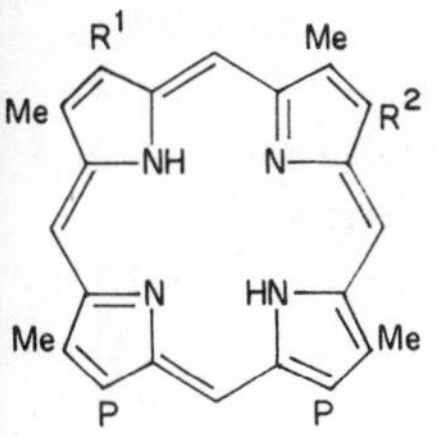

(24) $R^1 = CH \!\!=\!\! CHCO_2H$, $R^2 = P$

(25) $R^1 = P$, $R^2 = CH \!\!=\!\! CHCO_2H$

(26) $R^1 = R^2 = Et$

(27) $R^1 = R^2 = H$

(28) $R^1 = H$, $R^2 = CH \!\!=\!\! CH_2$

(29) $R^1 = CH \!\!=\!\! CH_2$, $R^2 = H$

(30) $R^1 = CHO$, $R^2 = CH \!\!=\!\! CH_2$

(31) $R^1 = CH \!\!=\!\! CH_2$, $R^2 = CHO$

(32) $R^1 = R^2 = CHO$

Scheme 14 Porphyrins related to protoporphyrin-IX

pounds with the natural product confirmed the original prediction. The *b*-bilene synthesis has also been utilized for S-411 synthesis, the acrylic side chain being introduced by reduction and dehydration of a β-keto ester side-chain.[57] A new partial synthesis of S-411 porphyrin (24) has recently been described[64] involving an interesting proto-deacetylation followed by a coupling through a mercury complex with acrylic ester in presence of palladium, i.e. porphyrin-H $\rightarrow$ porphyrin$-$HgCl $\rightarrow$ porphyrin$-$CH$=$CHCO$_2$Me.

Protoporphyrin-IX itself has been synthesized recently in a variety of ways, e.g. by the *a* and *b*-oxobilane,[65] MacDonald,[66] and *ac*-biladiene[67,68] routes; the vinyl groups were either introduced by degradation of preformed acetoxyethyl, or aminoethyl side-chains. The newer methods were also adapted to allow synthesis of various labelled derivatives e.g. trideuteriomethyl analogues[66] (for n.m.r. studies of haemoproteins) and all four *meso*-^{13}C-analogues[69] (for assignments of ^{13}C n.m.r. signals in biosynthetic studies). Clezy[70] has recently synthesized all fifteen protoporphyrin isomers—a substantial achievement reminiscent of Fischer's synthesis of twelve of the fifteen mesoporphyrin isomers.[1]

A number of degradation products of protoporphyrin-IX also occur in small amounts in biological systems, and are formed by reduction, oxidation or removal of one or both vinyl groups. Mesoporphyrin-IX (26) (the diethyl analogue), for example, has been synthesized as its dimethyl ester by the Fischer, *a*- and *b*-oxobilane,[71,72] *a,c*-biladiene routes[73] and *b*-bilane routes[74] although it can also, of course, be obtained by direct reduction of protoporphyrin-IX. Deuteroporphyrin-IX (27) is also readily prepared from protoporphyrin-IX or haemin by fusion with resorcinol; the mechanism of this process has recently been studied by two groups, but an entirely definitive conclusion has not yet been reached.[75]

A new faecal porphyrin, pemptoporphyrin, was identified[76] by n.m.r. and mass spectrometry as a mono-vinyl analogue of protoporphyrin-IX, and its structure (28) subsequently confirmed by total synthesis of the methyl ester and of its isomer (29) by the b-oxobilane,[77] ac-biladiene[78] and b-bilene[79] routes; the vinyl groups were introduced via either acetoxyethyl or aminoethyl side-chains. Pyrrolylacetylenes have also been utilized in a new Russian synthesis of isopemptoporphyrin and protoporphyrin-IX.[80]

Both pempto- and isopempto-porphyrins were converted[77,81,82] into the corresponding formyl-vinyl-porphyrins (30) and (31), chlorocruoro (or Spirographis) porphyrin and its isomer; the iron complex of chlorocruoroporphyrin is the oxygen-carrying pigment of the Mediterranean polychaete worm, Spirographis spallanzanii, and this work confirmed its structure. Both chlorocruoroporphyrin and its isomer have now also been synthesized directly from protoporphyrin-IX (Scheme 15), (i) by perma-

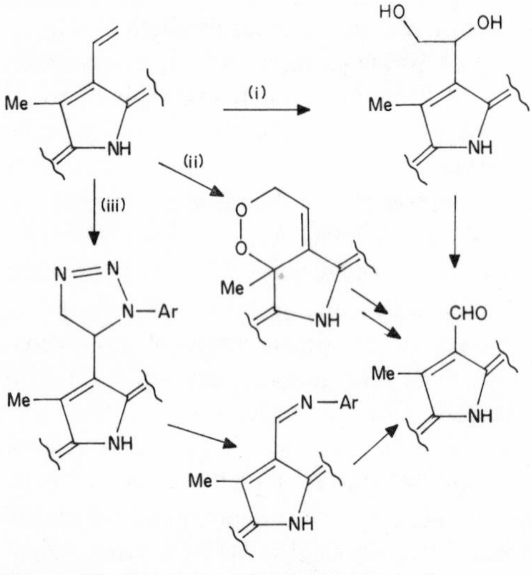

Scheme 15   Routes to chlorocruoroporphyrin

nganate oxidation,[83] (ii) by photo-oxidation and further transformations,[84] and (iii) by treatment with picryl azide.[85] Both the latter processes involve Diels-Alder reactions of the vinyl groups on rings A and B of the protoporphyrin, and in the first and third reactions the use of excess reagent led to the diformylporphyrin (32). The isomeric monoformyl monovinyl porphyrins can readily be separated chromatographically, h.p.l.c. now being the method of choice (cf. ref. 85).

Although most of the haem synthesized in nature is utilized in the form of haemoglobin and myoglobin, a significant amount is also utilized for haemoprotein enzymes e.g. cytochromes, catalases and peroxidases. One of the most important of these is cytochrome oxidase, which can be isolated from beef hearts. Earlier structural work showed that the core of this enzyme, 'haem-a' differed from the haem of haemoglobin by having a formyl group in ring-D (instead of a methyl group) and by the presence of a C_{17}-isoprenoid-derived side-chain in ring-A. Evidence for the structure[86] came from n.m.r. and field desorption mass spectrometric studies of material purified by h.p.l.c., and final proof rested upon total synthesis of porphyrin-a[87] (36) and its hexahydro analogue[88] via the b-bilene route (Scheme 16). The key formyl porphyrin ester (34) was synthesized by successive modifications of the side-chains of the initial porphyrin (33) prepared by oxidative cyclization of a b-bilene. Conversion of the porphyrin ester (34) to acid and then to acid chloride, followed by coupling with the magnesium enolate (37), prepared from E,E-farnesyl bromide and malonic ester, then afforded a β-keto ester, which was demethylated and decarboxylated with lithium iodide to afford the

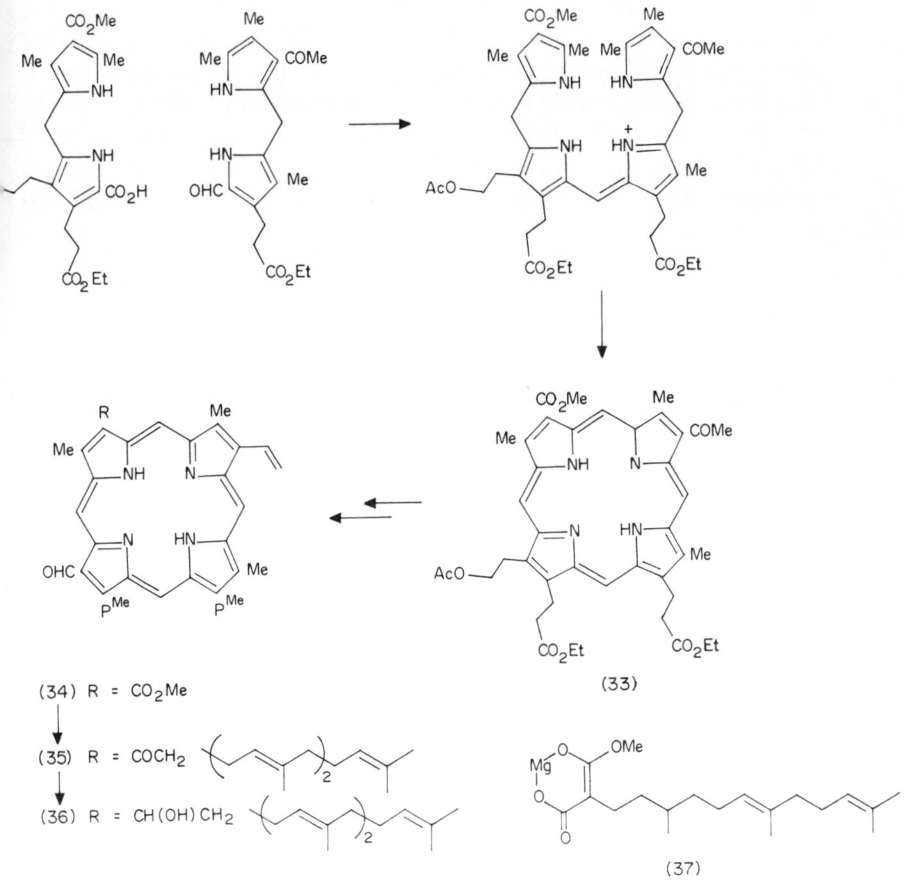

(34) R = CO_2Me

(35) R = $COCH_2$

(36) R = $CH(OH)CH_2$

(33)

(37)

Scheme 16   Structure and synthesis of porphyrin-a (36)

corresponding ketone (35). The formyl group was protected as an acetal whilst the keto group was reduced with borohydride; hydrolysis of the acetal followed by re-esterification then gave (RS)-porphyrin-*a* dimethyl ester identical with the natural product by h.p.l.c., u.v., n.m.r. and 'mixed' n.m.r. using shift reagent. The feasibility of this route to porphyrin-*a* was established by earlier experiments on the synthesis of porphyrin β-keto esters and of a C_{17}-ketoporphyrin; the synthesis of hexahydroporphyrin-*a* was tackled before porphyrin-*a* itself, as well as an analogue (the starting material for which was more readily available). An isomer of porphyrin-*a* was also synthesized, bearing a formyl group on the C rather than the D-ring; it was readily distinguished from porphyrin-*a* both spectroscopically and chromatographically.

Apart from their occurrence in living organisms porphyrins also occur in small amounts in petroleum, mainly as nickel and vanadyl complexes, and traces are found in coal. They are thought to be formed over the course of geological time by degradation of chlorophylls (or haems) and considerable efforts are now being made to separate the mixtures of porphyrins present, and which can only be isolated after lengthy purification procedures.[89] Chromatographic and mass spectrometric studies have revealed that two main series of homologous pigments are present (Scheme 17), one being based

Scheme 17   Structures of aetioporphyrin-III and DPEP

on aetioporphyrin-III (38) (presumably the type-III isomer because of its biological origin), and the other on desoxophylloerythroacetioporphyrin (39) (DPEP) which possesses an isocyclic ring (like the chlorophylls from which it is derived).

Rapoport's original synthesis[90] of DPEP involved a variant of the MacDonald-type synthesis in which one of the two pyrromethanes contained a pre-formed cyclopentanopyrrole unit. More recently Clezy[57b] prepared the key porphyrin (40a) by a variant of the *b*-bilene synthesis (Scheme 18); the ester was hydrolysed, converted to the acid chloride and coupled with the magnesium enolate of the monomethyl ester of malonic acid. The resulting β-

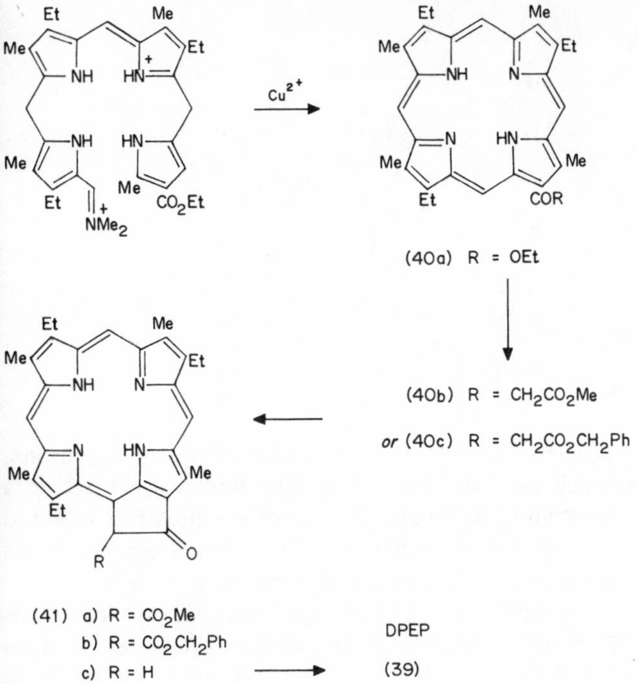

(40a) R = OEt

(40b) R = CH$_2$CO$_2$Me

or (40c) R = CH$_2$CO$_2$CH$_2$Ph

(41) a) R = CO$_2$Me
      b) R = CO$_2$CH$_2$Ph
      c) R = H ⟶ (39)

DPEP

(39)

Scheme 18 Synthesis of DPEP

ketoester (**41b**) was then cyclized by treatment with thallium trifluoroacetate to give the phaeoporphyrin ester (**41a**) which proved surprisingly stable in attempts at hydrolysis and decarboxylation. This was circumvented by preparation of the corresponding benzyl ester (**41b**) (from benzyl hydrogen malonate) followed by hydrogenolysis and decarboxylation; the resulting phylloerythrin (**41c**) was then converted to DPEP (**39**) by Wolf-Kishner reduction. The overall yield from dipyrromethanes was 8%, a marginal improvement over the Rapoport procedure, but the intermediates required were much more readily available.

In addition to the aetioporphyrin and DPEP series, a further type of petroporphyrin having a 'rhodo' rather than an 'aetio' visible spectrum has also been obtained. The rhodo-type spectrum (in which the third band of the four in the visible spectrum is larger than the other three) is usually characteristic of porphyrins bearing peripheral carbonyl substituents, but in the absence of such a group it was suggested that a fused benzene ring could account for the unusual visible spectrum. Clezy[91] has now synthesized such a mono-benzoporphyrin by two different routes, (i) from a tetrahydro indoline

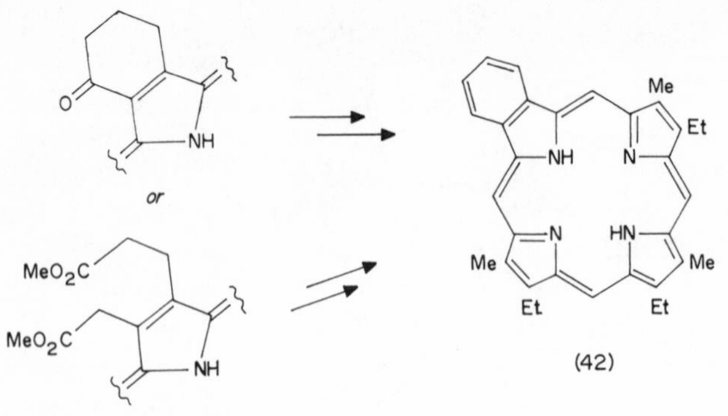

Scheme 19   Synthesis of a benzoporphyrin

by the *b*-bilene route, a dehydrogenation at the end of the synthesis affording the desired benzoporphyrin (**42**), (ii) via a MacDonald synthesis of a porphyrin bearing acetic and propionate ester groups on the A-ring followed by Dieckmann-type cyclization to form an unstable β-keto ester, which was then converted to the same benzoporphyrin (Scheme 19).

A dramatic new development in the chemistry of natural porphyrins has come with the isolation and structure determination of the so called 'green pigments' which are formed in liver after administration of certain drugs known to cause porphyria, e.g. 3,5-diethoxycarbonyl-2,4,6-trimethyl-1,4-dihydro pyridine (DDC), or barbiturates.[92-94] These green pigments are the unstable iron complexes of N-substituted porphyrins, and their basic structure followed from spectral comparisons of the metal-free pigments with simple N-alkyl porphyrins prepared earlier for n.m.r. and mass spectral studies.[95] Perhaps the most important of these green pigments are the N-methyl derivatives which appear to be a mixture of the four isomeric N-monomethyl derivatives of the haem of liver cytochrome-P450; biosynthetic studies with 4-^{14}C-methyl labelled DDC have shown that this is the origin of the N-methyl group in DDC-treated animals. Since N-methylprotoporphyrin has been shown to be a powerful inhibitor of ferrochelatase (the enzyme which inserts iron into protoporphyrin) it has been suggested that this might provide a mechanism for regulating haem synthesis *in vivo*.

The other naturally-occurring green pigments all appear to arise by 'suicide-inactivation' of cytochrome P450 in which the enzyme inserts an oxygen atom into a drug, or other foreign compound, bearing an olefinic or acetylenic substituent, and is itself N-alkylated in the process.[96] Thus rats breathing air containing ethylene gas produced N-hydroxyethyl-protoporphyrins in their livers and even such large molecules as acetylenic steroids, present in certain contraceptive drugs, also give rise to N-steroidally substituted porphyrins *in vivo*. Indeed a whole range of these compounds[96] have now been characterized, mainly by visible and field desorption mass spectrometry, and some by n.m.r. spectroscopy as well, although it is often difficult to isolate sufficient material for full characterization. The precise biological significance of these N-substituted porphyrins is under intensive study at the present time in several laboratories, especially in view of the enormous importance of the role of cytochrome P_{450} as a detoxifying enzyme. Other studies are aimed at elucidation of the mechanisms of the oxygenation process, and the formation of the covalent bond between the nitrogen of the haem and the various substrates. Simple N-alkyl derivatives of protoporphyrin have been prepared by direct alkylation (although di- and trisubstituted compounds are also formed); the isomeric N-methyl compounds have been separated by reversed phase h.p.l.c. and their structures assigned by n.m.r. methods (including n.o.e.).[97]

As mentioned in the introduction there is now a marked trend towards the study of the mode of action of haem enzymes. This has led on the one hand to the synthesis of various substituted derivatives of haem (especially those with deuterium labels in the methyl groups and at the *meso*-positions) to study the n.m.r. spectra of haemoproteins, and on the other hand to the synthesis of a wide variety of haem models. Perhaps the greatest activity in the latter area has come from attempts to prepare model porphyrins,[98] the iron complexes of which it is hoped will show reversible oxygenation characteristics similar to haemoglobin (and which indeed might have practical application as oxygen enrichment devices). These model porphyrins have been described in various ways as 'picket-fence', 'capped', 'strapped', 'basket-handle', 'looping over', 'crowned', etc., but their structures and synthesis fall outside the main scope of this review and the reader is referred elsewhere for details. There has also been considerable interest in *bis* (and even *tris*)metalloporphyrins constructed in such a way that the porphyrin rings are face-to-face.

9.3 Chlorophylls, bacteriochlorophylls and related compounds

Chlorophylls *a* and *b* (**43**) are the major photosynthetic pigments in nature, and their gross structures were determined some forty years ago.[1c] The details were confirmed by later studies, especially in relation to the two additional hydrogens present in the ring system compared with porphyrins, and to their absolute stereochemistry (Scheme 20). Woodward's total synthesis[99] in 1960 of chlorin-e_6 trimethyl ester (**44a**), a degradation product of chlorophyll-*a*, and which had previously been converted back into chlorophyll-*a* by Fischer, was also another major landmark. However, since then a number of interesting new facets of chlorophyll chemistry have been discovered, and other approaches to synthesis of chlorophyll-*a* (**43a**) have also been made.

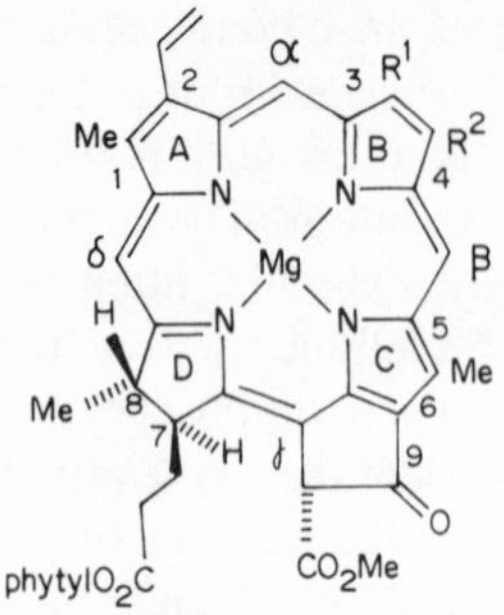

$$(43)$$
a) $R^1 = Me$, $R^2 = Et$

b) $R^1 = CHO$, $R^2 = Et$

c) $R^1 = Me$, $R^2 = CH=CH_2$

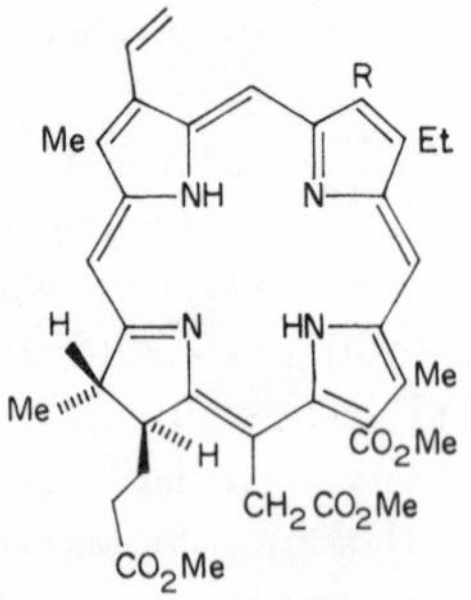

(44) a) R = Me

b) R = CHO

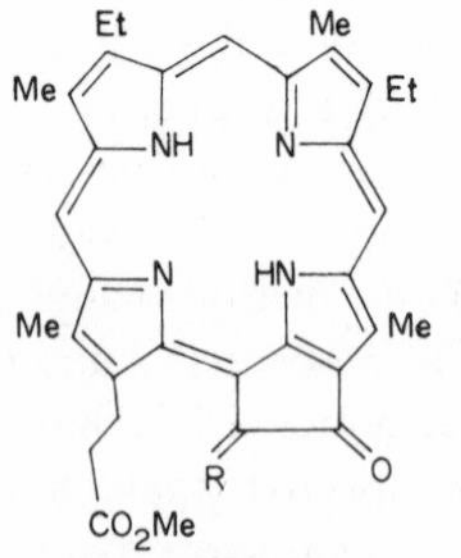

(45) a) R = H$_2$

b) R = O

Scheme 20 Chlorophylls a and b and degradation products

Separation of the chlorophylls from each other and from the carotenoid pigments which occur with them is, of course, the classic example of the use of column chromatography, and not surprisingly chromatographic methods are still of prime importance. However a useful method for larger-scale separation of phaeophytins-a and -b (the demetallated derivatives of the chlorophylls) involved formation of the Girard-T derivative of the formyl group in the phaeophytin-b, followed by column chromatography.[100]

Magnesium can be re-inserted into metal-free chlorins by treatment with Grignard reagents (or alkoxy magnesium derivatives), or by use of magnesium pyridine complexes.[4] However, recent new approaches have involved the use of a hindered magnesium phenolate complex[101] or direct treatment with magnesium acetate in acetone and dimethylsulphoxide.[102]

As well as the lability of the magnesium to mild acid treatment, the chlorophylls are also very susceptible to oxidation, the isocyclic ring being particularly vulnerable in alkaline conditions, when 'allomerization' (oxidation at C-10) occurs; the vinyl group and the δ-*meso*-carbon neighbouring the reduced ring may also undergo oxidation.[103]

The isocyclic ring has long been known to undergo ring opening in alkaline media to afford chlorin-e_6 derivatives, and recently these reactions have been studied kinetically;[104] amines will also cleave the isocyclic ring to give amides derived from chlorine-e_6.[105] On heating in boiling pyridine, or better collidine, the ester group attached to the isocyclic ring is cleaved and the related 'pyro' compounds are formed.[103] Methyl mesopyropheophorbide (45a) (in which the vinyl group is reduced to ethyl) can be oxidized[100] by quinones (e.g. DDQ) to a diketone (45b) (Scheme 21).

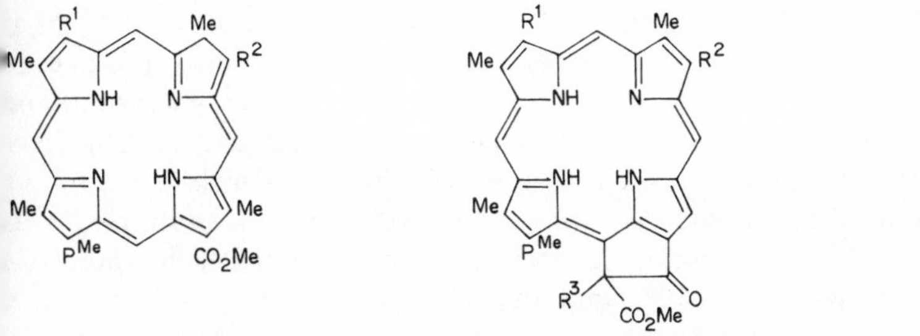

(46) a) $R^1 = R^2 = Et$

b) $R^1 = CH = CH_2$, $R^2 = Et$

c) $R^1 = R^2 = CH = CH_2$

(47) a) $R^1 = R^2 = Et$, $R^3 = OMe$

b) $R^1 = R^2 = Et$, $R^3 = H$

c) $R^1 = CH{=}CH_2$, $R^2 = Et$, $R^3 = H$

Scheme 21 Rhodoporphyrin and phaeoporphyrin derivatives

Both the proton and ^{13}C n.m.r. spectra[106] of chlorophylls-a and b and a wide variety of derivatives have been studied extensively especially in relation to the aromatic ring-current[107] and to the nature of the complexing which occurs in solution.[103] This is of considerable relevance to the way in which the chlorophylls complex *in vivo* and their role in photosynthesis. Visible spectra have also been extensively studied for the same reason in an effort to correlate the spectra observed *in vivo* with those of the species observed *in vitro*. The presence of trace amounts of other ligands, e.g. water, alcohols, ethers, etc. can have a fairly marked effect on the spectra of chlorophylls;[103] this has so far prevented X-ray structure determinations of the chlorophylls themselves although the structure of methyl phaeophorbide-a has been reported by two groups,[108] and more recently the X-ray strucutres of ethyl chlorophyllide-a[109] and ethyl chlorophyllide-b.[110]

The proton n.m.r. spectra of the chlorophylls in pure chloroform show very broad peaks due to aggregation, but on addition of alcohol, or by use of more polar electron-donating solvents well resolved spectra are obtained, which are very similar to those of the corresponding metal free phaeophor bides, but with the addition of resonances due to the phythyl side-chain.[11]

The n.m.r. spectra of the chlorophylls, and their derivatives containing the isocyclic ring and the C-10 ester group, all show small satellite resonance close to many of the main peaks, and these have been assigned to isomer: formed by epimerization of the C-10 ester function via keto-enol tautomerisn in the isocyclic ring.[112] These isomers had been separated in earlier work and denoted as a′ or b′, and it was found that equilibration occurred at different rates in different solvents e.g. the half-life of chlorophyll-a is about 2 hours in polar solvents such as pyridine or tetrahydrofuran, but several times longer in relatively non-polar solvents like benzene. An alternative suggestion that the a′ and b′ isomers are chelated enol forms has now been discounted.[113] It has long been known that the C-10 proton can be readily exchanged for deuterium under very mild conditions (e.g. by treatment with deutero-methanol) and magnesium complexes[113] of the keto-ester system of the isocyclic ring, have been prepared, as well as 9-desoxo-9,10-dehydrochlorophyll.[115]

Although the stereochemistry of the phytyl side-chain was established[103] as long ago as 1959, the absolute stereochemistry of the D ring of chlorophylls-a and b was not established until the late 1960s by correlations with a transformation product of the terpenoid lactone ($-$)-α-santonin, the C-10 ester group having been shown to be $trans$ to the C_7-propionate side chain.[116] These results have also been confirmed by application of Horeau's method.[117]

There has been relatively little synthetic activity in the chlorophyll field since Woodward's synthesis of chlorin-e_6 but a number of preliminary studies have shown the possibility of an approach along biomimetic lines. Thus rhodoporphyrin-XV dimethyl ester (46a) and its mono- and di-vinyl ana logues (46b and c) were synthesized by the b-oxobilane route,[118] the intermediate pyrromethane amides utilized corresponding to the A and B rings of the porphyrins. The vinyl groups were introduced by using acetoxyethyl side-chains as precursors, and the nuclear carboxyl group required in ring C of the rhodoporphyrins was carried through from the pyrrole stage. The nuclear ester group of the rhodoporphyrin-XV diester was then converted into a β-keto ester via reaction of the acid chloride with the magnesium complex of t-butyl methyl malonate, followed by removal of the t-butyl ester group with trifluoroacetic acid.[119] In later work[120] the use of the acylimidazolide and a magnesium enolate of methyl hydrogen malonate was found to provide a more efficient route to the required β-keto ester. Oxidation of the corresponding magnesium porphyrin complex with iodine in methanol in presence of potassium carbonate then effected cyclization[119] to the methoxy phaeoporphyrin derivative (47a); over-oxidation of the desired phaeoporphyrin ester (47b) had evidently occurred in addition to cyclization.

The precise mechanism of this cyclization is uncertain, and it may involve coupling of radicals, a porphyrin radical with the anion of the keto-ester, or a porphyrin dication with the anion; however, it seems very likely that a similar process occurs in the *in-vivo* formation of chlorophyll. A more efficient cyclization[121] of the magnesium β-keto-ester to phaeoporphyrin-a_5 dimethyl ester (47b) was achieved by use of thallium trifluoroacetate as oxidizing agent. 2-Vinyl-2-desethyl-phaeoporphyrin (47c) was also synthesized in the same way, and in a partial synthesis of chlorophyll-*a* the corresponding chlorin (rhodochlorin, obtained by degradation of chlorophyll) was also converted to a β-keto ester, cyclized to chlorin-e_6 and converted to chlorophyll-*a*.[122a] In a different approach a spirocyclic chlorin derivative (49) has been obtained[122] by Vilsmeier cyclization of the porphyrin butyric amide (48), followed by catalytic reduction (Scheme 22); oxidative ring opening of the ketone (49)

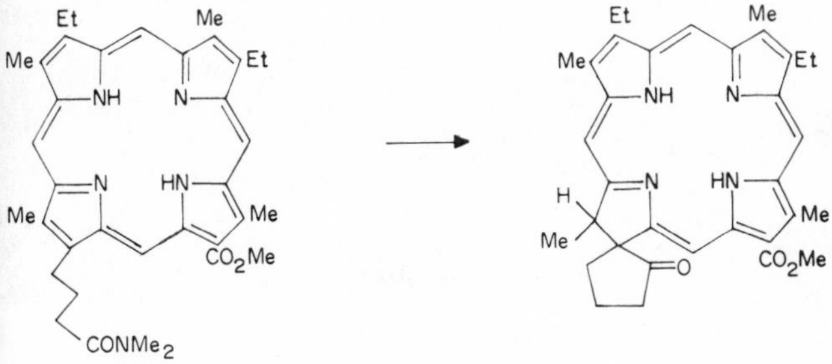

Et   Me
Me—NH   N—Et
Me   N   HN   Me
      CO₂Me
CONMe₂
(48)

⟶

Et   Me
Me—NH   N—Et
H   N   HN   Me
Me   =O   CO₂Me
(49)

Scheme 22 Synthesis of a spirocyclic chlorin

could in principle afford a chlorin with the same functionality as ring D in chlorophyll.

The total synthesis of chlorophyll-*b* is likely to be a much more daunting task than chlorophyll-*a* because of the formyl group in the B-ring, but partial syntheses of rhodin-g_7 (44b) have been carried out from chlorin-e_6 (44a). Thus electrolysis of chlorin-e_6 gave a phlorin-chlorin which then underwent photo-oxidation in dioxan water to a *trans* diol; transformation of the latter into rhodin-g_7 was achieved by acid-catalysed dehydration and rearrangement to a hydroxymethyl derivative which was then oxidized to give the required formyl group.[123] A more direct approach was the direct photo-oxidation of chlorin-e_6 trimethyl ester in carbon tetrachloride which gave rhodin-g_7 ester in low yield, probably via photo-chlorination of the 3-methyl group.[124] Since rhodin-g_7 has been reconverted to chlorophyll-*b* and chlorin-e_6 has been synthesized by Woodward these constitute a formal total synthesis.

Apart from chlorophylls-*a* and *b* a number of other chlorophylls and bacteriochlorophylls have been described in recent years. A new analogue of chlorophyll-a (obtained from a mutant of maize) has recently been characterized by fast atom bombardment, field desorption and 'in beam' electron impact mass spectrometry, as the 4-vinyl-4-desethyl derivative[125] (cf. **43c**). Another new chlorophyll derivative, chlorophyll-c, occurs in a variety of marine algae, dinoflagellates and marine diatoms, often in association with chlorophyll-a. Chlorophyll-c is a mixture of magnesium hexadehydro- and tetradehydro-phaeoporphyrins[126] (**50**); i.e. it is not a chlorin, and moreover, the D-ring acrylic acid group is not esterified as is the corresponding propionate group of all other chlorophylls (Scheme 23). Reduction of the

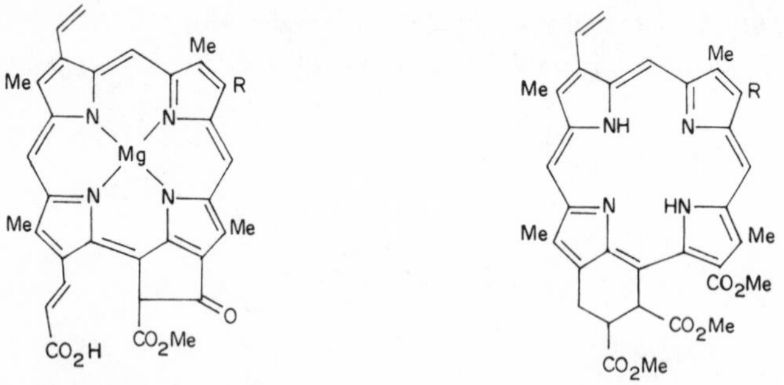

(50)   a) R^1 = CH=CH$_2$
       b) R^1 = Et

(51)   a) R = CH=CH$_2$
       b) R = Et

Scheme 23   Structures of chlorophylls-c and isochloroporphyrins

chlorophyll-c mixture with hydriodic acid afforded phaeoporphyrin-a_5 monomethyl ester (cf. **47b**) which was identical with material derived from chlorophyll-a. Attempts to esterify the free acid in chlorophyll-c with diazomethane led to complex mixtures of products, and prolonged treatment with methanolic hydrogen chloride gave a rearrangement product, 'isochloroporphyrin-c_6 trimethyl ester', (**51**), arising from opening of the isocyclic ring and reclosure onto the acrylic acid residue in the D-ring (Scheme 23). This new material gave a good mass spectrum with molecular ions at m/z 632 and 634, in contrast to chlorophyll-c itself which decomposed in the mass spectrometer. This evidence together with detailed analyses of the n.m.r. spectra and biosynthetic considerations led to the two structures proposed for chlorophylls-c_1 and c_2 (**50a** and **50b**). The two compounds have also been separated chromatographically on polyethylene and the ratio of c_1 and c_2 was about 0.6. Chlorophyll-c is widely distributed in marine organisms and constitutes a major fraction of the world's carbon dioxide fixing capacity.[126]

Both porphyrin-c_1 and c_2 methyl esters (the magnesium-free porphyrin esters corresponding to chlorophylls-c_1 and c_2) have now been synthesized by the *b*-bilene route,[127] the isocyclic ring being introduced by oxidative cyclisation of a β-keto ester with thallium trifluoroacetate; the acrylic acid residue was also produced from a keto-ester side chain by reduction and dehydration.

The main chlorophylls present in purple photosynthetic bacteria are bacteriochlorophylls-*a* and *b* (52) and (53) which both contain the 'bacteriochlorin' (tetrahydroporphyrin) ring system in which opposite pyrrole rings are reduced (Scheme 24).[103] It was originally thought that like chlorophylls-*a* and

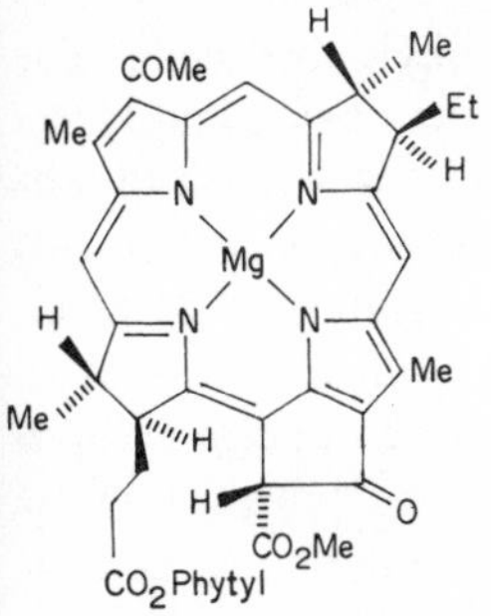

(52) 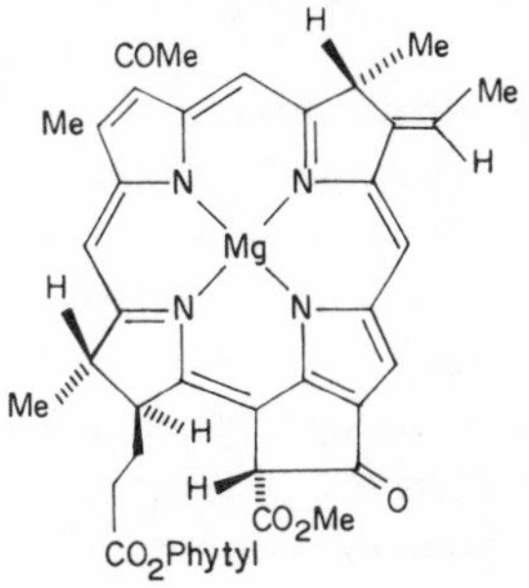   (53)

Scheme 24 Bacteriochlorophylls-*a* and *b*

b the esterifying alcohol on the ring D propionate group of the bacteriochlorophylls-*a* and *b* was also phytol. However, in *Rhodospirillum rubrum* the esterifying group is *all-trans*-geranylgeraniol,[128] and more recently bacteriochlorophylls with straight chain octadecyl, phytyl and geranylgeranyl side-chain ester groups have been found in *Chloroflexus aurantiacus*.[129] Evidence for farnesyl ester in *R. rubrum* has also been obtained,[130] and it seems likely that other polyprenyl esters of chlorophylls may exist in nature, albeit in smaller quantities. The existence of geranyl-geranyl esters is not now surprising in view of recent evidence[131] that the phytyl ester group in chlorophyll arises *in vivo* by reduction of a geranyl-geranyl ester side-chain (rather than by esterification by pre-formed phytol).

Like chlorophylls-*a* and -*b*, bacteriochlorophyll-*a* has also been shown by proton magnetic resonance to exist in solution in equilibrium with a small amount of the a'-epimer. The bacteriochlorophylls also form aggregates in solution like the chlorophylls and it seems likely that they act in a similar fashion in photosynthesis in bacteria.

Bacteriochlorophylls-*c* and -*d* are dihydroporphyrin derivatives produced by the green sulphur bacteria (Chlorobacteriaceae) which are sometimes accompanied by small amounts of bacteriochlorophyll-*a*. These pigments were originally known as *Chlorobium* chlorophylls-660 and -650, respectively, on account of the position of the maximum absorption in their visible spectra. They contain magnesium like other chlorophylls, but in contrast are esterified by farnesol on the D-ring propionate group, and lack the methoxycarbonyl group on the isocyclic ring. Reversed phase partition chromatography showed that the two pigments each consisted of a series of homologous compounds, and structures[132] were assigned following oxidative degradation of the corresponding phaeophorbides and pyrroporphyrins. *Trans*-dihydro-haematinic acid was obtained from the phaeophorbides showing that the D-ring was the same as in chlorophylls-*a* and *b*, whereas the maleic imides obtained from the B- and C-rings contained a range of alkyl substituents. The longer wavelength absorption of the 660 series of pigments was attributed to a δ-*meso*-substituent. Spectroscopic evidence that this was a methyl group was obtained from the more abundant fractions, but it was suggested that two minor components contained *meso*-ethyl groups. The assignment of the *meso*-methyl substituent to the δ-position was questioned[133] on the basis of an analysis of the n.m.r. spectra and it was suggested that this group was at the α-position. However, two phylloporphyrins were synthesized[134] by the *b*-bilene version of the MacDonald route, and the *meso*-methyl porphyrin proved to be identical with a degradation product obtained from fraction 4 of the '660' chlorophylls; however, the *meso*-ethyl analogue was different from any of the naturally derived phylloporphyrins, thus casting considerable doubt on the suggestion that two of the fractions contained ethyl groups.[134] Further proof that the *meso*-methyl groups are in the δ-position (rather than at α) was provided by reductive methylation (hydriodic acid/formic acid) which afforded pyrroles with α-ethyl groups (identified by gas chromatography) derived from the A and D rings of the '660' pigments.[135] Final clarification came only very recently when careful h.p.l.c. separations using reversed phase systems, followed by high field n.m.r. spectroscopy, showed that mixtures of diastereomers were present differing in the configuration of the A-ring hydroxyethyl group[136] (Scheme 25). Thus the compounds previously thought to contain *meso*-ethyl groups were simply diastereomers of other fractions with *meso*-methyl groups.

The structures and absolute configurations[137,138] of bacteriochlorophylls-*d* (55) were obtained by oxidative degradation to imides corresponding to rings B and D of the pigments;[138] the deoxo pigments also gave rise to imides corresponding to ring C. Benzoylation of the ring A hydroxyethyl group, followed by oxidation afforded a maleic imide which on ozonolysis gave (R)-benzoyl lactic acid, thus showing the 2^1-R-configuration[137] of the bacterio-chlorophylls-*d*. H.p.l.c. separations of the pigments and X-ray crystallo-graphy confirmed these deductions.[138] Two new higher homologues of bacteriochlorophylls-*d* bearing neopentyl substituents have also recently been isolated.[139]

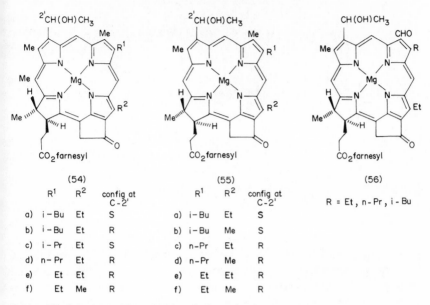

Scheme 25 Structures of bacteriochlorophylls-*c*, -*d* and -*e*

[13]C and [1]H n.m.r. spectra of bacteriochlorophylls-*c* (**54**) derivatives obtained from *Chloropseudomonas ethylicum* and degradative studies showed that four homologous pigments were present,[139] rather than the six obtained from *Chlorobium thiosulphatophilum*. Model chlorins with propyl, propenyl, isobutyl and isobutenyl side-chains were synthesized from rhodin-g_7 trimethyl ester (**42b**) using the formyl group as the basis for elaboration of the alkyl group.[140]

The configuration of the hydroxyethyl side-chain in bacteriochlorophylls-*c*, -*d* and -*e* was also assigned on the basis of n.m.r. spectral analyses of the esters with 2-phenylbutyric acid.[141] Bacteriochlorophyll-*e* (**56**), a new chlorophyll from known species of Chlorobiaciae was shown to be a mixture of three homologous pigments with the structures and absolute configuration shown.[142]

Stereochemically pure epimers of a bacteriochlorophyll-*d* were obtained[143] from the hydroxyethyl analogue of methyl pyrophaeophorbide-*a* by chromatography on a silica column coated with the chiral electron acceptor (+)-TAPA, or by reversed phase chromatography.

Two groups[144,145] have independently synthesized bacteriophaeophorbide-*c* from a derivative of chlorin-e_6-trimethylester (in which the vinyl group was protected) by Vilsmier formylation of the δ-*meso*-position; the δ-formyl group was then reduced via the δ-hydroxymethyl intermediate to a δ-

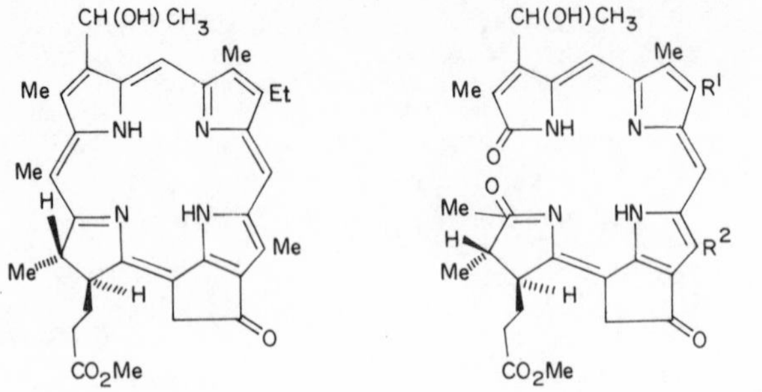

(57)   (58)

Scheme 26 Methyl bacteriophaeophorbide-c, and oxidation products

methyl group, and deprotection of the vinyl group followed by hydration afforded methyl bacteriophaeophorbide-c (57) (which had been obtained previously from band 6 of the bacteriochlorophylls-c). The mixture of epimeric bacteriophaeophorbide-c obtained could be separated (like the corresponding compounds in the d-series) by reversed phase chromatography. During the course of this work the ^{13}C-spectra of a range of chlorins and their derivatives were determined.[106]

An interesting artefact,[229,146] formed by photo-oxidation of one of the bacteriochlorophylls-c during isolation was the open-chain acetyl bilatriene (58); a similar formyl bilatriene had previously been obtained by photo-oxidation of magnesium octaethylporphyrin.[147] Experiments[148] with methyl pyromesophaeophorbide-a using a mixture of [18]O_2 and [16]O_2 have now confirmed that the oxidation involves addition of one molecule of oxygen (presumably singlet oxygen formed by sensitization by the phaeophorbide itself). This result clearly differentiates the ring-opening process from that occurring in haem metabolism to bile pigments (which involves two molecules of oxygen). More importantly, though, reactions of this type may be involved in the breakdown of chlorophyll in the senescent leaf, the precise nature of which is still one of the unsolved mysteries of chlorophyll chemistry and biochemistry.

A novel physiologically active dihydroporphyrin, bonellin[149] (59) (Scheme 27) has recently been isolated from the marine animal *Bonellia viridis*, and the structure elucidated largely by spectroscopic methods and comparisons with known compounds; this has now been confirmed by X-ray crystallographic studies of anhydrobonellin (60), in which the ring C propionate residue has cyclized onto the γ-meso-position. Bonellin is situated mainly in the proboscis of the organism but the body skin is pigmented with neobonellin, an isoleucine conjugate; other conjugates with valine, leucine and alloisoleucine have also been detected.[150] These pigments are unique in

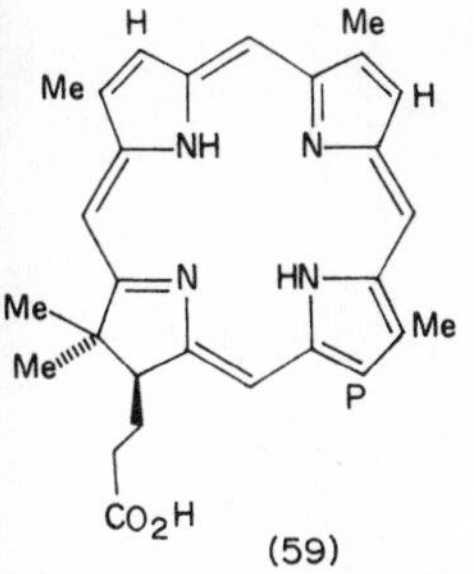

Scheme 27 Bonellin and anhydrobonellin

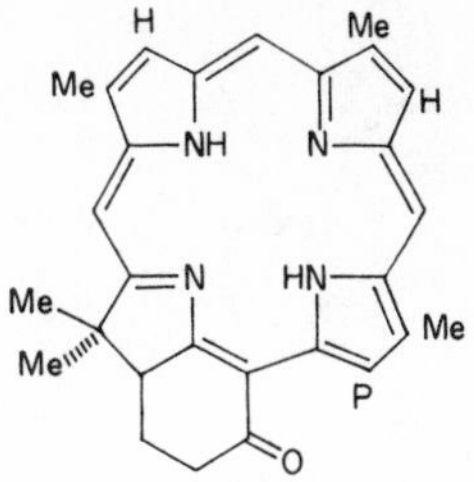

the porphyrin field for although formally bonellin is a methyl substituted dihydro derivative of deuteroporphyrin-IX it is clearly not related either to any of the other known chlorophylls, nor to vitamin-B^{12}.

9.4 Vitamin-B_{12} and related compounds

The early history of vitamin-B_{12} (cyanocobalamin) (61), (Scheme 28) from its recognition as the anti-pernicious anaemia factor to its isolation and structure determination, has recently been recounted by Folkers[151] in a fascinating review. The structure determination of a degradation product of the vitamin by Dorothy Crowfoot-Hodgkin in 1955 is now regarded as one of the milestones in the development of modern X-ray crystallography. The similarities (between the structure of vitamin-B_{12} and those of haem and chlorophyll) were immediately recognized and subsequent work[152] has shown that all three types of structures are formed from the same monocyclic precursor, porphobilinogen, via the tetrapyrrolic uroporphyrinogen-III (Scheme 10). The complexity of the cyanocobalamin structure, its water-solubility and the ability of the central cobalt atom of form metal-alkyl bonds which are such important features of its biological activity, have fascinated chemists and biochemists alike. Many studies of the reactions of the cobalt both in the natural chromophore, and in model systems, have now been carried out in the course of investigations on the mode of action of vitamin B_{12}.[153]

The reactions of the chromophore itself have also been extensively studied,[154] although largely with the cobalt complex itself and related model systems; it is difficult to remove the cobalt without decomposition, and, while simpler model compounds have been synthesized, only small amounts of the cobalt-free corrinoids have been isolated from living systems. The corrin chromophore of vitamin-B_{12} is not aromatic, but shows considerable stability and tends either to remain intact, or to break down completely under

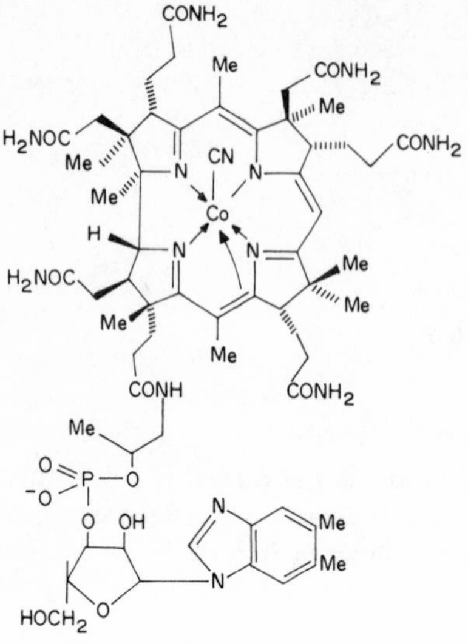

(61)

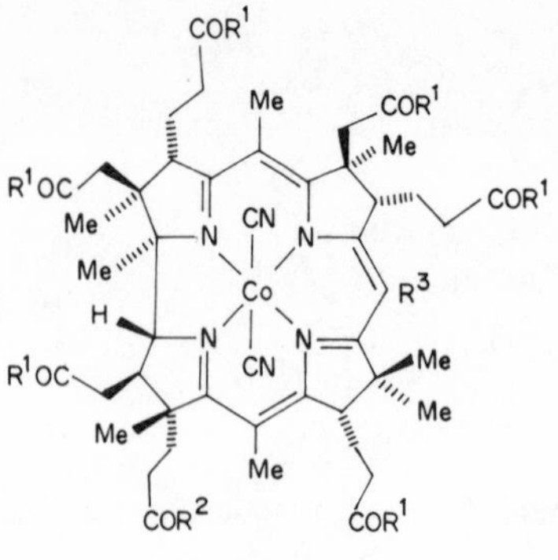

(62)

a) $R^1 = R^2 =$ OMe, $R^3 =$ H

b) $R^1 = R^2 =$ OMe, $R^3 =$ Br

c) $R^1 = NH_2$, $R^2 =$ OH, $R^3 =$ H

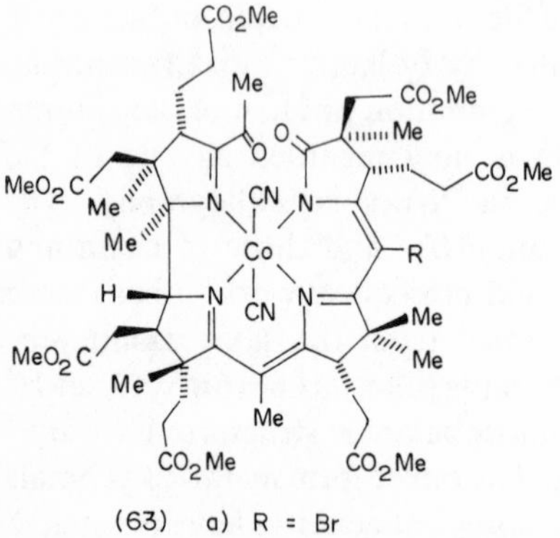

(63)   a) R = Br

        b) R = H

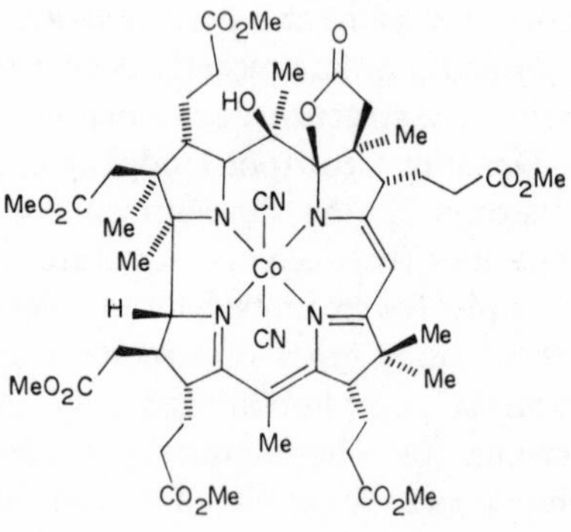

(64)

Scheme 28   Vitamin-B_{12} and degradation products

more vigorous conditions. Thus electrophilic substitution (e.g. deuteration, halogenation, nitrosation etc.) occurs at the β-*meso*-position, although in model systems lacking the methyl groups at the α- and γ-*meso*-positions it is the latter which are preferentially attacked.[154] However, treatment of cobyrinic acid heptamethyl ester (**62a**) (a degradation product of vitamin-B_{12}) with a large excess of chlorine gave a colourless adduct containing chlorine at the α, β, γ- and 'valley' positions;[155] similar products were obtained in reactions with hypochlorous acid.[156]

Oxidizing agents also attack the *meso*-positions preferentially, and in the early degradative studies succinimide derivatives were obtained from the four pyrrole rings.[154] Very recently, however, it has been shown[157] that controlled ozonization of a bromo-cobyrinic ester (62b) with 1.2 mole equivalents of ozone afforded a seco-corrin-dione (63a) (Scheme 28) whereas with two moles of ozone, followed by reductive work-up, cleavage of both the α- and β-bridges occurred with formation of fragments corresponding to the AD and BC rings of the original macrocycle. The bromo-substituent in the secocorrin-dione could be removed either photolytically, or by reduction with borohydride, and subsequent treatment with sulphuric acid effected cyclization back to the corrinoid system (in 18% yield). Photo-oxygenation of the cobyrinic ester affords both the α-seco-corrin dione (63b) and the isomeric γ-seco dione formed by cleavage at the γ-position.[158]

In contrast to the *meso*-positions the β-positions are relatively unreactive although epimerization may occur and also oxidative cyclization of acetic acid side-chains. The *neo*-series of pigments in which the C-ring propionic acid has been epimerized have long been known, because they are formed in highly acidic conditions e.g. in the hydrolysis of vitamin-B_{12} to cobyrinic acid.[159] The mass, n.m.r. and visible spectra of the normal and neo-series differ only slightly from each other but the chiroptical properties of the two series are quite distinctive. Similar epimerizations of the other propionate groups are theoretically possible but the natural configurations of the A, B and D rings appear to be strongly preferred.

The so-called 'stable yellow corrinoids', formed from vitamin-B_{12} derivatives under a variety of oxidizing conditions, are probably hydroxylated at the α-*meso* position (thus interrupting the chromophoric system) and the B-ring acetic acid also cyclizes to form a lactone,[160] e.g. (64). The lactone ring was reductively cleaved by borohydride and the product methylated to afford a mixture of heptamethyl dicyanocobyrinate, and its *meso*-hydroxy derivative. Other similar examples have also been described.[161]

Permanganate oxidations of the *meso*-methyl groups to the corresponding carboxylic acids have been reported, and the latter can be decarboxylated by heating in acetic anhydride/acetic acid.[154] The recent studies of vitamin-B_{12} chemistry have been greatly assisted by extensive use of modern spectroscopic methods, and ^{1}H and ^{13}C n.m.r. spectra of several derivatives have been correlated.[162]

9.4.1 *Synthesis*

Vitamin-B_{12} is now readily available for therapeutic use by microbial fermentation procedures. However, its structural and stereochemical complexity has provided a major challenge to modern synthetic methods in

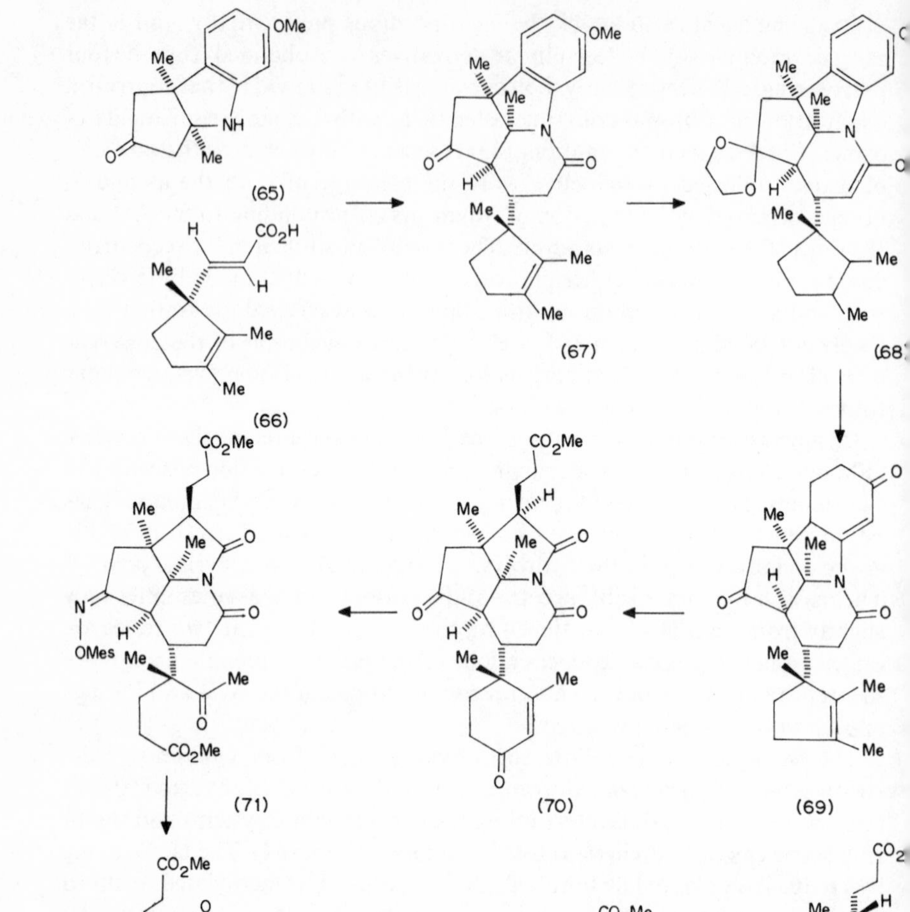

(65)

(66)

(67)

(68)

(69)

(70)

(71)

(72)

(73)

(74)

a) A-ring propionate – α
b) A-ring propionate – β

Scheme 29 Synthesis of the A-D portion of vitamin-B$_{12}$

organic chemistry. The challenge was taken up by teams of chemists at Harvard (led by R.B. Woodward) and Zürich (led by A. Eschenmoser) in a joint venture, which proved to be one of the most outstanding achievements of modern organic chemistry. Over 100 chemists were involved during a fifteen-year period, and the Zürich team also completed a second synthesis in addition to the joint synthesis. Summaries of the work have been published,[13,163] and given by the two team leaders in numerous lectures, but the full account of this mammoth undertaking is now being prepared by Eschenmoser, following the untimely death of Woodward three years ago.

Woodward and Eschenmoser's primary target was cobyric acid (62c), a partial degradation product from vitamin-B_{12}. This had earlier been reconverted to the coenzyme by Bernhauer and his colleagues, and thus its synthesis would constitute a formal total synthesis of vitamin-B_{12}. The synthesis of cobyric acid was indeed a daunting task because there were nine chiral centres in the molecule which were only partially contiguous, but all of which would have to be introduced with the correct relative and absolute stereochemistry. Moreover it was necessary to ensure that the D-ring propionate group could be distinguished from the other six acid side-chains in order to enable the nucleoside side-chain to be attached to complete Bernhauer's original reconstruction of the vitamin from cobyric acid—by no means a trivial undertaking.

The Harvard group undertook the synthesis of the A-D (or western) half of the molecule, and the Zürich group the B-C (or eastern) half; the two parts were then joined firstly between the C and D rings and finally the macrocyclic ring was closed between the A and B rings. In the second synthesis the direct linkage of the A- and D-rings was the final step; many syntheses of model systems were also carried out in Zürich.

The A-D portion of vitamin-B_{12} was the more difficult to synthesize (Scheme 29) partly owing to the 'crowded concatenation of six contiguous asymmetric centres', but the final coupling of this with the B-C unit (Scheme 30) relied heavily upon the experience of the Zürich group in their extensive studies on model corrins, especially the use of the sulphide contraction procedure. In the A-D component the largest groups are all *trans* in the six asymmetric centres, which argued well for 'asymmetric synthesis by induction'. The precursor (65) of the A-ring was constructed from 6-methoxy 2,3-dimethyl indole and propargyl bromide; the stereochemistry of the two five-membered rings in the tricyclic ketone (65) was of necessity *cis*; the latter was resolved using (+)-phenylethyl thiocyanate and the absolute stereochemistries of both enantiomers determined by degradative comparisons with ( )-camphor derivatives.

The D-ring precursor (66) was synthesized from camphor and combined as its acid chloride with the tricyclic ketone (65), and the resulting amide cyclized

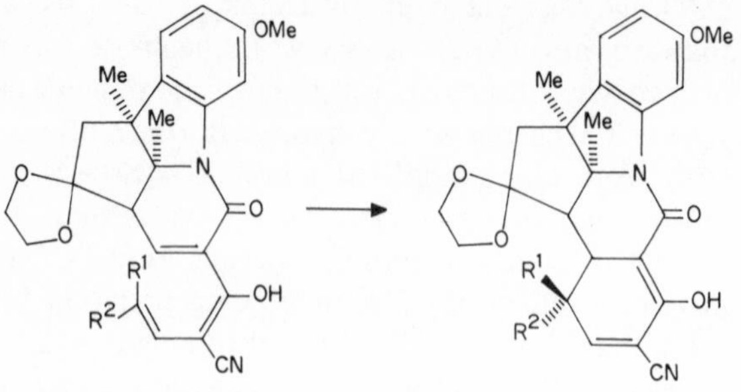

a) $R^1 =$ Me , $R^2 = CO_2Me$
b) $R^1 = CO_2Me$, $R^2 =$ Me

Scheme 30   A redundant route to the A-D portion of vitamin-B_{12} which led to the Woodward-Hofmann rules

with potassium t-butoxide to give the intermediate (**67**) in which five of the desired contiguous asymmetric centres were all of the correct chirality due to asymmetric induction by the large groups at each centre. Following protection of the amide function as a methoxyenamine (**68**) the aromatic ring was degraded by Birch reduction and hydrolysis, although it was not recognized until later that the new asymmetric centre in the cyclohexenone (**69**) was opposite to that desired owing to steric hindrance to protonation in the hydrolysis. After oximination of the five-ring ketone, both the cyclohexenone and cyclopentene rings were oxidized (O_3/KIO_4) to generate the A-ring propionate side-chain and a diketone respectively; the latter was recycled to form a new cyclohexenone (**70**). The oxime of the latter on mesylation and ozonolysis, followed by periodic acid work-up and esterification with diazomethane gave the keto-ester (**71**), containing the potential D-ring nitrogen atom masked as an oxime mesylate. Beckmann rearrangement of the oxime (a key step in the whole strategy of the synthesis) was effected by heating in methanol containing polystyrene sulphonic acid as a catalyst, and afforded the tetracyclic intermediate (**72a**) nicknamed 'α-cornerstone' and a small amount of the β-epimer (**72b**). It was only at this stage that it was recognized that the stereochemistry at the centre bearing the propionate side-chain was incorrect, but by cleaving the A-ring with alkali first, followed by acidification and treatment with diazomethane the β-cornerstone (**72b**), with the *trans* arrangement of groups in ring A, was obtained in 90% yield. Ring opening of the amide to generate the acetic ester side-chain required on ring A was carried out with hydrogen chloride in methanol in presence of thiophenol to afford the ring-opened vinylogous thioimino ether (**73**). Ozonolysis followed by treatment with ammonia and borohydride reduction gave an alcohol containing the D-ring propionate group as an amide; the latter was dehydrated to the nitrile with methane sulphonic anhydride and the mesylate of the alcohol converted to the bromide (74). In this way the propionate residue in the C-ring was differentiated from the other acetic side chain in the form of a nitrile.

Another possible route to the A-D component, also studied at harvard, involved the novel cyclization of the triene (75a). Two products (76a) and (76b) were isolated, and it was subsequently shown that the latter had arisen from a prior isomerization of the original triene to the isomer (75b); thus cyclization of each isomer had occurred in a completely stereospecific manner. It was these observations which led to the realization of the principles of orbital symmetry, and their subsequent enunciation in the Woodward-Hoffmann rules,[165] one of the most important theories of modern organic chemistry.

The B-C component was synthesized from two monocyclic units (Scheme 31), the B-ring being derived from β-methyl-β-acetylacrylic acid (77) and the C-ring from (+)-camphorquinone. The two chiral centres of the B-ring were created in the required relative stereochemistry by utilizing a Diels-Alder cyclo-addition and resolving the product (78). Chromic acid oxidation cleaved the

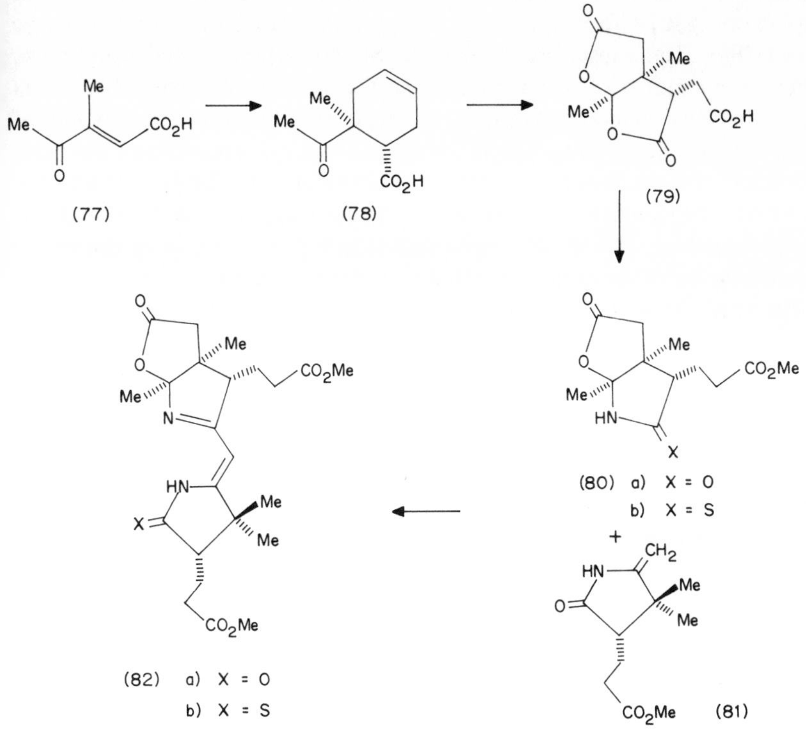

Scheme 31 Synthesis of the B-C portion of vitamin-B$_{12}$

double bond and the intermediate di-acid cyclized with the ketonic carbonyl group to form the bis-lactone acid (79). Arndt-Eistert homologation of the acetate side-chain gave a bis-lactone propionate ester, which was converted into a lactam-lactone with ammonia in methanol, and thence into the thiolactam (80b) by heating with P_2S_5. The synthesis of ring-C essentially followed a route devised by J.W. Cornforth (in unpublished work) directed towards the synthesis of vitamin-B_{12}.

Oxidation of the B-ring precursor with benzoyl peroxide gave a disulphide dimer, which in presence of the C-ring component gave a sulphur-bridged intermediate which underwent 'sulphide contraction' on heating in triethyl phosphite to give a mixture of the desired bicyclic compound (82a) and its B-ring propionate epimer. The corresponding thiolactam (82b) was prepared by O-alkylation of a mercury complex with Meerwein's reagent, followed by treatment with hydrogen sulphide.

The anion derived from this lactam (82b) was then coupled with the A-D bromide (74) to form a thioiminoether which underwent acid-catalysed sulphur contraction to give the tetracyclic A-D-C-B 'corrigenolide' (83a) (Scheme 32). The A-ring lactam was then converted to the thiolactam thiolactone (83b) with P_2S_5 and two routes were devised for cyclization to the corrin (85). The most direct involved S-methylation of the thiolactam, followed by insertion of cobalt to give the dicyano complex (84) which underwent ring closure with DBN. In both cyclization procedures mixtures of eight possible diastereomers were formed due to epimerization of the three propionate side-chains (although the structural formulae drawn in this article only show one isomer). These diastereomers were separated by reversed phase h.p.l.c.—an early use of this technique which provided a very important demonstration of its value in synthetic organic chemistry.

The synthesis was later simplified somewhat by the discovery that iodine oxidation of the mixture afforded a mixture of only four lactones (86), each of which was separated in pure form. The required α- and γ-meso methyl groups were introduced by regio-selective electrophilic benzyloxymethylation, followed by conversion to the bis-phenylthiomethyl adduct and reductive cleavage of the thiophenyl residues by zinc amalgam. The reductive conditions also led to ring opening of the lactones leading to the normal configuration of the propionate residue in the B-ring. Hydrolysis of the D-ring cyanoethyl side-chain to propionamide was achieved by heating with boron trifluoride in acetic acid, and de-amination with dinitrogen tetroxide led to the propionic acid. Conversion of the remaining six ester groups to amide was then effected smoothly by heating with anhydrous ammonia in ethylene glycol, using dry ammonium chloride as catalyst. The cobyric acid (62) was shown to be identical in all respects with naturally-derived material 'most particularly in liquid chromatographic behaviour'.

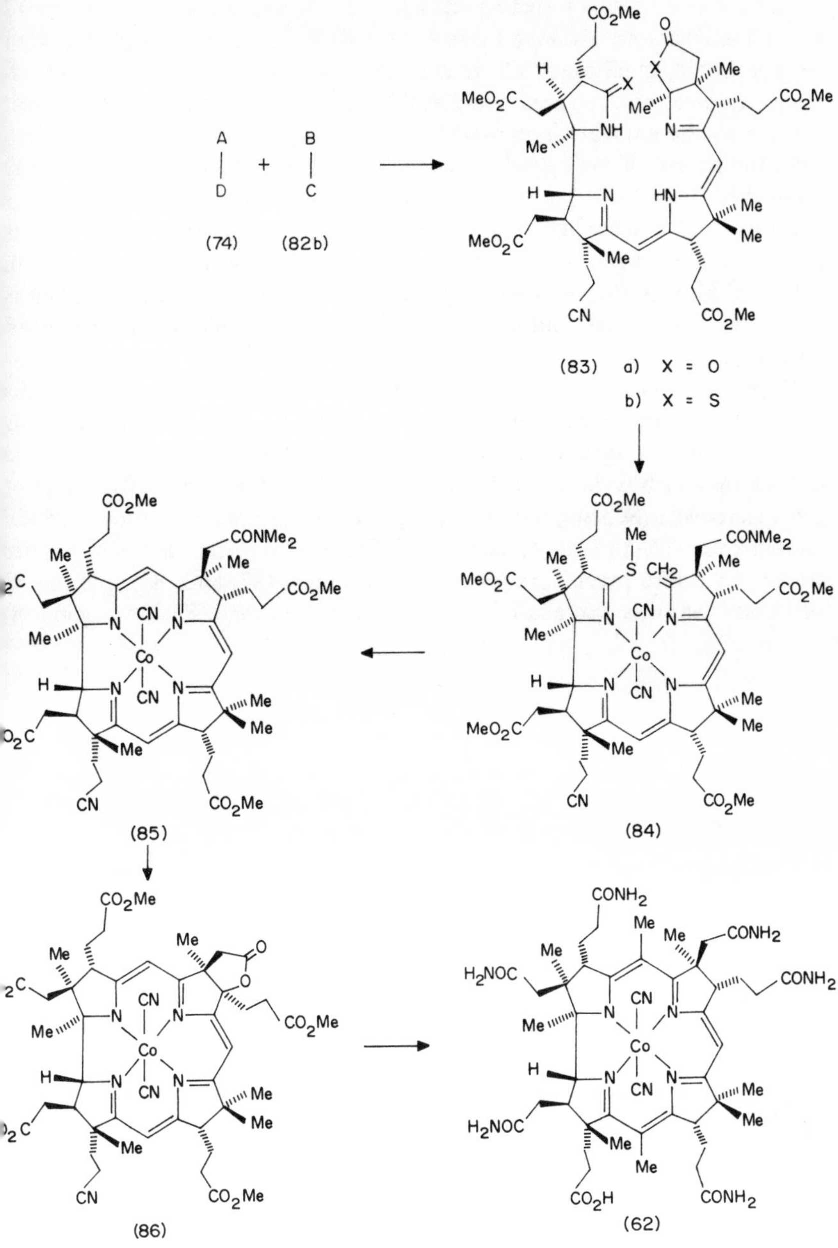

Scheme 32 Synthesis of cobyric acid (mixtures of epimers were formed, but only one is shown for convenience)

Not content with completing the synthesis of cobyric acid, the Harvard group then went on to convert it into vitamin-B_{12} itself by a modification of Bernhauer's original route. The final product proved to be identical spectroscopically (electronic, i.r., c.d. and 270 MHz n.m.r. spectra) and in a biological assay with the natural vitamin-B_{12}.

In the course of their work on model corrins the Zürich group not only developed the sulphide contraction method for joining the individual rings together, but also relied heavily upon activation of amide residues as iminoethers (by means of the hitherto little-used Meerwein trialkyl oxonium salts). In this way Eschenmoser helped to provide both the experimental basis for the joint synthesis, and also laid the foundations for his 'New Road' to corrins.

With the knowledge and experience gained with the earlier model systems Eschenmoser conceived a brilliant new approach to vitamin-B_{12} synthesis based on a photochemical A/D cyclization. Both enantiomers of the cyclohexene carboxylic acid (78) were utilized in syntheses of all four rings of the macrocycle (Scheme 33). The (+) enantiomer was converted into the lactam (80a) as already described (cf. Scheme 31) and this was further transformed into precursors for rings A, B and C (87) and (88); the ring-C precursor (88) was also available from camphorquinone. The (−)-acid (78)

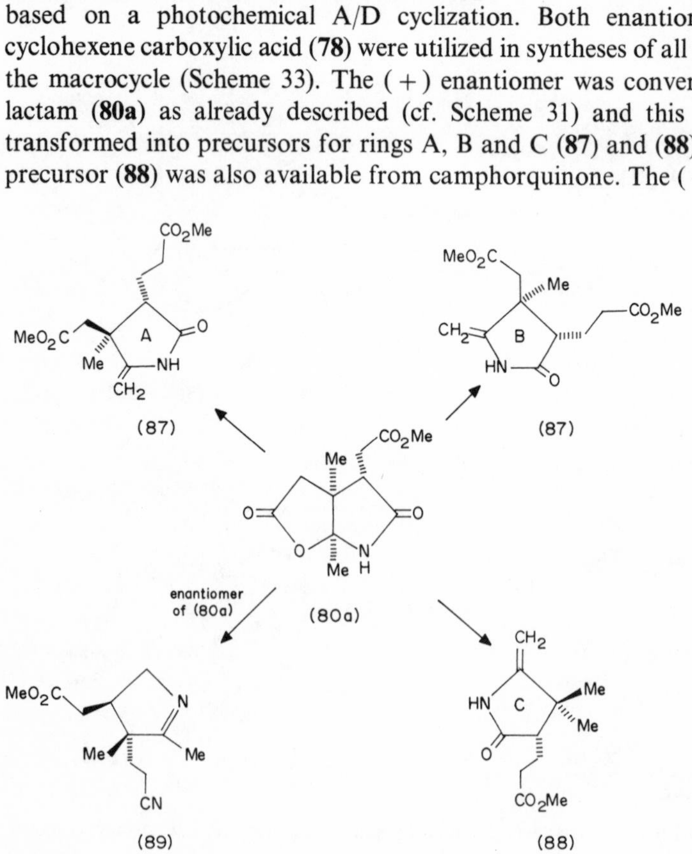

Scheme 33 Intermediates for the Zürich B_{12} synthesis

was converted into the precursor of ring-D, by a process involving an intermediate ketolactam, which was converted into its oxime and subjected to Beckmann fragmentation to give a cyanoethylpyrrolidine (89); the D-ring propionate group was thus differentiated from the other ester side-chains (as in the joint synthesis).

The four rings were then joined by the sulphide contraction route (developed during the model studies) to produce a seco-corrin (90) which was a mixture of eight epimers (owing to epimerization of the propionate side-chains) (Scheme 34). Elimination of HCN (using DBU in sulfolane) followed

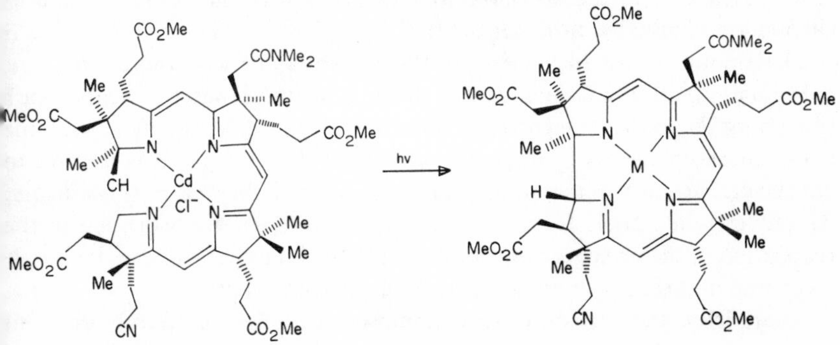

Scheme 34   Photochemical ring closure of a Seco-corrin: the key step in the Zürich B_{12}-synthesis

by photochemical cyclization of the cadmium complex in white light under slightly anaerobic conditions (following conditions developed in the model studies) then afforded the cadmium corrin (91a) in over 90% yield. Removal of the cadmium was readily effected by acid, and insertion of cobalt followed by oxidation led to a mixture of the eight epimeric dicyano cobalt corrins (91b) which could be separated as in the first synthesis and converted to cobyric acid (62) and neocobyric acid.

The remarkable symmetry-allowed antarafacial 1,16-hydrogen shift, followed by A-D cyclization, had been foreshadowed by extensive studies[166] in Zürich with model systems in which it was shown that Li, Na, Mg, Cd, Pd, Pt, Zn and Cd seco-corrins could be cyclized photochemically (but not the Ni, Co, Cu or Mn complexes); the Cd complexes proved to be the best both in terms of the yield and formation of the desired *trans* ring junction. More recently Eschenmoser has shown[166] that the seco-corrin corrin ring closure can also be brought about by electrochemical oxidation of nickel complexes (which do not undergo photochemical cyclization). Variants of the sulphide contraction method for synthesizing seco-corrins have also been investigated, and have led to a 1-oxo A-D secocorrin, irradiation of which afforded a 1-hydroxycorrin, thus showing that a carbonyl group could replace the carbon-carbon double bond in the cyclization process.[166, 168] Furthermore a 19-carboxy-seco-corrin was cyclized to a nickel corrin lacking the carboxyl group (decarboxylation having taken place after cyclization), and a nickel 19-formyl seco-corrin afforded a 19-formyl corrin which could be deformylated under basic conditions.[169] These observations are of considerable significance in relation to vitamin-B_{12} biosynthesis as *meso*-methylation occurs at the δ-position after methylation of the A and B rings of uroporphyrinogen-III, and later during ring contraction to the corrinoid system the δ-*meso* methyl group and δ-carbon are eliminated as *acetic acid*![170]

The monumental achievements of Woodward and Eschenmoser and their colleagues in Harvard and Zürich over a span of nearly twenty years (including the earlier model studies) must surely rank as the most supreme achievement in the art of organic synthesis to date; indeed it is difficult to conceive that such a mammoth undertaking will ever be possible again. Moreover, quite apart from the syntheses themselves, the work led to the realization of the principles of orbital symmetry, which 'constitutes the single most unifying theory in contemporary organic chemistry'.

Apart from the Woodward-Eschenmoser syntheses an entirely different approach to vitamin-B_{12} was conceived by J.W. Cornforth utilizing the previously little-appreciated chemistry of the isoxazole ring system. Isoxazoles may be regarded as a protected form of β-diketones, from which they are commonly prepared, and reductive ring-opening affords amino-ketones. Cornforth proposed to synthesize a tri-oxazole and transform this into a corrin, and his work, carried out in the late 1950s and the 1960s, has been summarized elsewhere.[13]

Independently of Cornforth, Stevens[171] also conceived a similar approach involving isoxazoles. Three routes were used for the synthesis of the isoxazoles (Scheme 35), starting from primary nitro compounds, or oximes. These were transformed into unstable nitrile oxides which reacted *in situ* with alkynes to form isoxazoles (92). In theory the corrin nucleus could be synthesized by

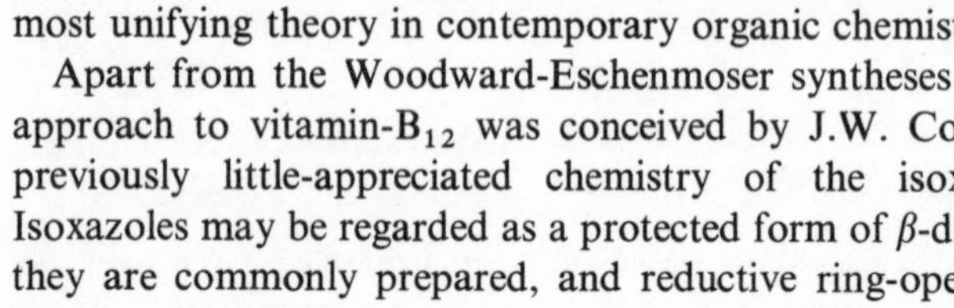

Scheme 35 Syntheses of isoxazoles

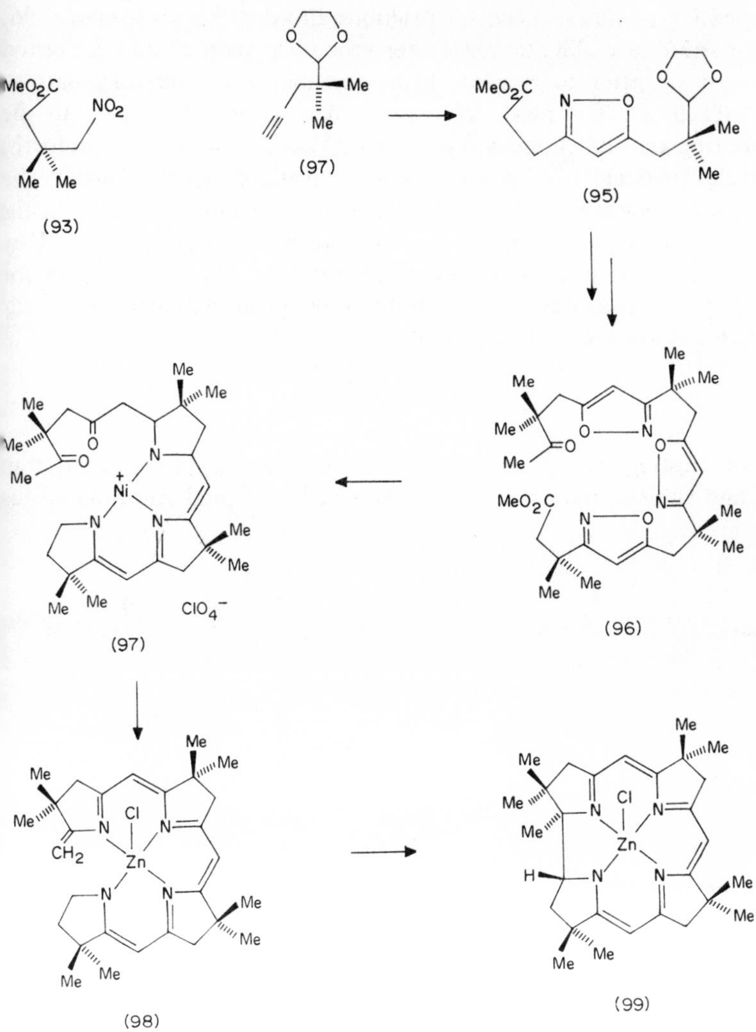

Scheme 36 Synthesis of a model corrin from isoxazole units

combining three isoxazole units in a 'clockwise' or 'anti-clockwise' manner. Model experiments on both approaches have been explored, and for the example, the anti-clockwise route has yielded a corrin as shown in Scheme 36. Thus the nitro-ester (93) was converted into the nitrile oxide and coupled *in situ* with the acetylenic acetal (94) to give the isoxazole (95). The acetal was converted via the corresponding aldehyde to the oxime and nitrile oxide which underwent cycloaddition with another molecule of alkyne. A further repetition of this sequence of reactions afforded the tri-isoxazole (96) in 40% overall yield. The terminal ester group was reduced and converted to the tosylate prior to catalytic hydrogenation of the oxazole units to β-enaminoketones. The newly generated amino groups cyclized onto the neighbouring keto functions (as conceived by both Stevens and Cornforth) and the alkyl toxylate side-chain also cyclized, to afford a tricyclic ligand. The nickel complex (97) of the latter was treated with ammonium acetate to give the tetracyclic ligand, and the nickel was then replaced by zinc, followed by dehydration to the methylene seco-corrin (98). An important reason for exchanging the metal was to facilitate the subsequent photochemical cyclization which gave the *trans*-corrin (99).

9.4.2 Corrin precursors

Several interesting new pigments (Scheme 37) have recently been isolated from nitrite and sulphite reductase enzyme systems[172,173] and from organisms

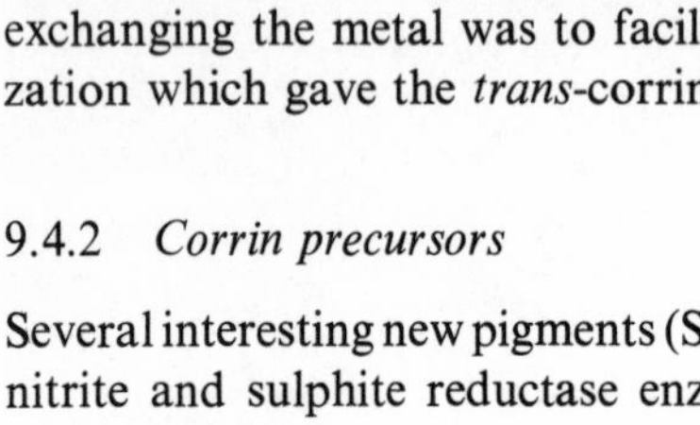

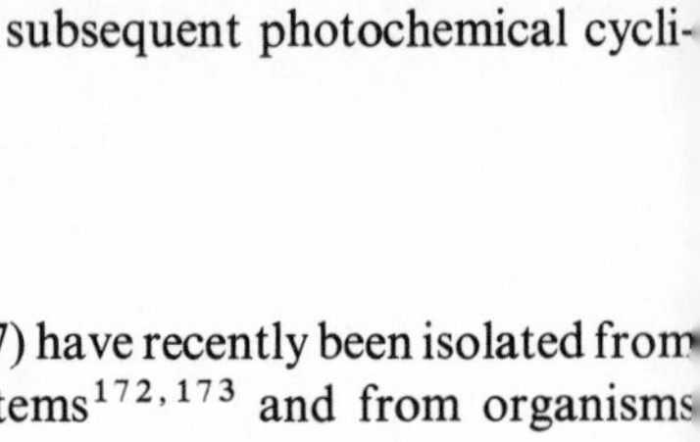

(100) R = H
(104) R = Me

(101)

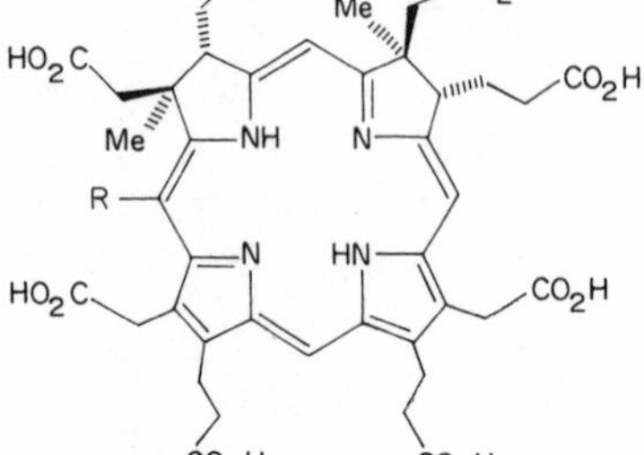

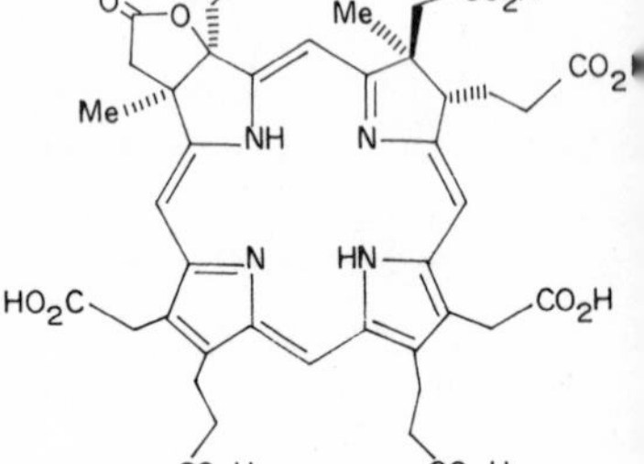

(102) R = H
(105) R = Me

(103)

Scheme 37 Pigments related to biosynthetic precursors of vitamin-B_{12}

producing vitamin B_{12}.[174,175] It is now clear that these are methylated derivatives of uroporphyrinogen-III closely related to intermediates in the biosynthesis of vitamin-B_{12}; their structures were determined by a combination of spectroscopic methods and biosynthetic studies.[152]

A dimethylated pigment ('Faktor II') isolated from *P. shermannii* is now known as sirohydrochlorin (**100**), and its iron complex, sirohaem, is the prosthetic group of sulphite reductases. The visible absorption spectrum clearly indicated that it was an isobacteriochlorin and the gross structure was determined by a combination of mass (field desorption) and n.m.r. spectrometry. The identity of the materials from the different sources was confirmed by the identity of their circular dichroism curves.[176] The positions of the two methyl groups were deduced by biosynthetic experiments with radio-labelled sirohydrochlorin which had itself been previously prepared by incorporation of ^{14}C-methyl-labelled methionine.[177] The cobyrinic acid obtained was then carefully ozonized to give succinimides corresponding to rings B and C and a bicyclic fragment corresponding to rings A and D; approximately 50% of the activity in sirohydrochlorin was found in each of the B and AD fragments (but essentially none in the C-ring fragment) thus rigorously showing that the two methyl groups were in rings A and B of the pigment. These results were also confirmed[178] by incorporation of ^{13}C-labelled aminolaevulinic acid into sirohydrochlorin; the [13]C n.m.r. spectra showed that the lowest field *meso*-carbon atom was that at the γ-position, as it resonated as a triplet in the proton decoupled spectra due to coupling with the neighbouring carbon atoms in the *non-reduced* C and D rings (and each of which like the γ-*meso*-carbon atom is also derived from C-5 of ALA).

Two other related dimethylated pigments have also been isolated from cultures of *P. shermannii*, a monolactone (sirolactone) and a dilactone ('corriphyrin 4') formed by oxidative cyclization of the acetic acid units of rings A and B of sirohydrochlorin by aerial oxygen.[174,176] These compounds, like sirohydrochlorin itself, were isolated as their permethyl esters, and n.m.r. and mass spectrometric studies together with evidence for their interconversion helped to establish their structures, as well as that of sirohydrochlorin itself. A minor component always found during the isolation of sirohydrochlorin was found to be the C-3 epimer,[179] and was probably the same as 'Faktor-IIa'.[180]

A green pigment[175] ('Faktor I') obtained from cell-free enzyme preparations of *C. tetanormorphus* or *P. shermannii* was clearly a chlorin from its visible spectrum, and the mass spectra of its permethyl ester showed that it was a monomethylated macrocycle (**103**) closely related to uroporphyrin-III and to sirohydrochlorin. The chlorin was labelled after incorporation with either labelled ALA or methyl-labelled methionine, and, bearing in mind the structure of sirohydrochlorin, these results led to the conclusion that the new pigment was *mono*-methylated on ring A or ring B of the macrocycle. However, unlike sirohydrochlorin the ^{14}C-methyl-labelled chlorin was not incorporated into cobyrinic acid, unless it had been previously reduced by sodium amalgam, when degradative experiments showed that 96% of the label was carried by the A-D fragment.[175,180] Thus the chlorin Faktor I clearly had structure (103) with the methyl group on the A-ring.

A trimethyl isobacteriochlorin derivative ('Faktor-III') (104) has now also been isolated from enzyme preparations in both Cambridge[181] and Stuttgart.[175] The presence of three methyl groups was established by mass spectrometry of a pigment derived from CD_3-labelled methionine, and by the proton n.m.r. spectra which showed singlets due to two methyl groups like those in sirohydrochlorin, and a third lower-field singlet due to a *meso*-methyl group;[182] detailed n.m.r. studies indicated the surprising result that the latter was at the δ-position and this was confirmed by biosynthetic labelling experiments with 5-^{13}C-ALA and methyl ^{13}C-labelled-methionine.[183,184] The trimethylisobacteriochlorin is also readily incorporated into cobyrinic acid, and the stereochemistry is thus likely to be the same as that of sirohydrochlorin; more importantly, however, the δ-*meso*-methyl group was eliminated together with the *meso*-carbon atom during the metabolic process as acetic acid,[185,186] and thus neither of these two carbon atoms migrate to form the angular methyl group in the natural corrins.

A closely related trimethylated pigment ('corriphyrin-3') was shown to be a hexacarboxylic acid (by the same method as used with most of the other pigments) by comparing the field desorption mass spectra of the premethyl ester (M^+, 956), and the ester (M^+, 974) formed from trideuteriomethanol, which was 18 units higher. The [1]H n.m.r. spectra enabled the structure to be assigned[152] as the dilactone (105) i.e. the *meso*-methyl analogue of corriphyrin-4 (102).

The isolation and structure determination of these pigments was carried out largely on a sub-milligram scale and has provided some novel examples of the combination of spectroscopic methods with biosynthetic studies. The first three stages in the biosynthesis of vitamin-B_{12} from uroporphyrinogen-III involve the corresponding reduced systems at the porphyrinogen level of oxidation, rather than the pigments (102), (100) and (104) themselves; indeed, the 15,23-dihydrosirohydrochlorin has recently been isolated, and its structure determined by ^{1}H and [13]C n.m.r. spectroscopy.[187]

Owing to the sparsity of material available for chemical and biological studies rational routes to isobacteriochlorins are now being developed by ring synthesis.[152,187] An interesting new biomimetic approach is also being explored by Eschenmoser in model systems.[188] This involves the synthesis of hexahydroporphyrins isomeric with porphyrinogens, the interconversion of the porphyrinogen and isobacteriochlorin systems, and studies of pyrrocorphins (substances with a porphyrin skeleton, but a corrin-like chromophore). In a very interesting recent biomimetic study he has shown that a *meso*-methyl substituted dihydropyrrocorphin will undergo a ring contraction to corrin with subsequent elimination of the *meso*-methyl group and *meso*-carbon atom as acetic acid.

9.5 Bile pigments

9.5.1 *Structures and reactions*

The natural bile pigments (Scheme 36) nearly all arise by oxidative ring opening of haem at the α-*meso* position[189] and include the mammalian pigments biliverdin (**106**), bilirubin (**107**) and their bacterial reduction

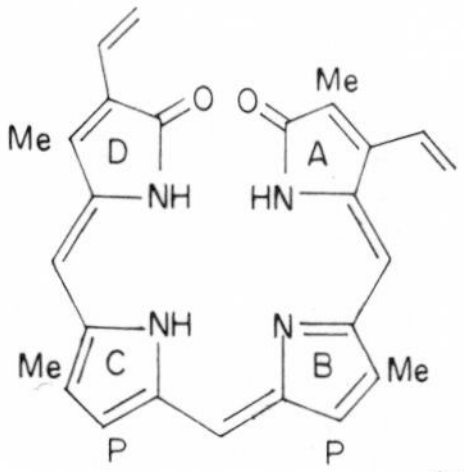

(106)

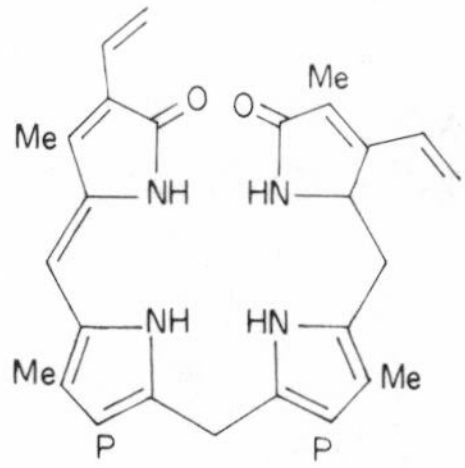

(107)

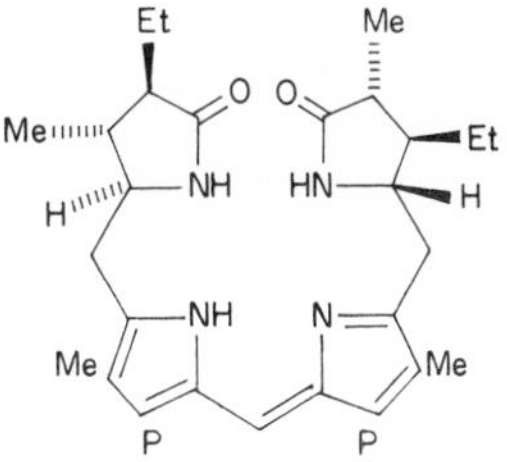

(108)

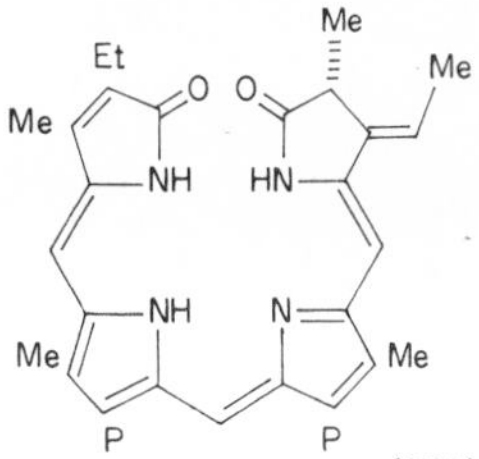

(109)

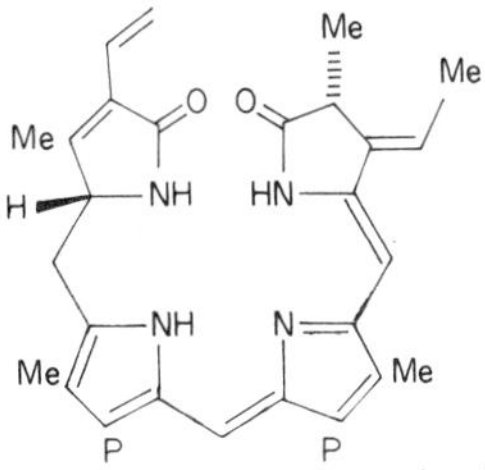

(110)

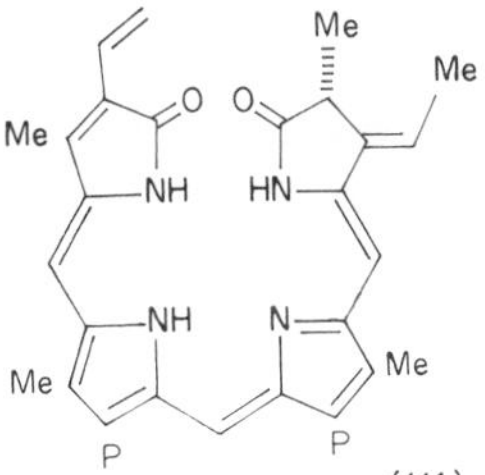

(111)

Scheme 38 Structures of mammalian and plant bile pigments products, the urobilins and stercobilin (**108**), the photosynthetic algal bile pigments[190] phycocyanobilin (**109**) and phycoerythrobilin (**110**) and the plant growth regulator, phytochromobilin (**111**). An interesting new series of bile pigments has, however, recently been isolated from a number of Lepidoptera, and are formed by ring opening of haem at the γ-*meso* position (see below).

The X-ray crystal structure[191] of bilirubin shows that it has an interesting 'ridge-tile' type of structure and this has clarified the reasons for its relative insolubility in aqueous systems; the NH, carbonyl and carboxyl groups are internally hydrogen-bonded and thus not available for solvation by water molecules (see Fig. 9.1). The crystal structure of biliverdin dimethyl ester has

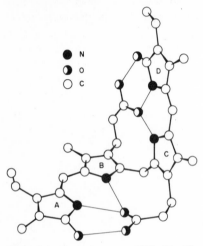

Figure 9.1   X-ray structure of bilirubin. The fine lines show the hydrogen-bonding interaction. Redrawn from Bonnett *et al.*, *Nature* **262**, 326 (1980).

also been determined.[192] The increase in solubility of bilirubin on irradiation with light is largely due to a change in configuration (Z to E) of one of the pyrromethenone units[193–196] and perhaps of cyclization[197] of one of the vinyl groups on to the neighbouring ring. A similar configurational change is thought to account for the well-known changes in the chromophore of phytochrome on irradiation ($P_R \rightleftarrows P_{FR}$) and which is a key feature of its biological activity.[198] Phytochromobilin, like the other algal bile pigments, is covalently bonded to the protein through a sulphur linkage to the ethylidene residue in the A-ring; their spectral properties are profoundly affected by non-covalent interactions with the native protein.[190] A large number of model di-, tri- and tetra-pyrrolic analogues of the bile pigments have been synthesized in recent years and their chemical, stereochemical, physical, spectroscopic and photochemical properties studied in considerable detail in relation to the biological functioning of phytochrome and the algal biliproteins.[199,200]

The photo-oxidation of bilirubin has also received considerable atten-tion[200-202] largely because of the use of phototherapy in treatment of hyperbilirubinaemia, a temporary phenomenon in about 10% of new-born babies, which if untreated can lead to brain damage. The products are mainly maleic imides, the bicyclic propentdyopents, and traces of biliverdin are also found; biliverdin is only very slowly photo-oxidized. The reactions involve oxidation by singlet oxygen formed by self-sensitization by the bilirubin itself.

Bilirubin undergoes reversible isomerization to give mixtures containing the IIIα and XIIIα isomers, and small amounts of these isomers have also been detected by t.l.c. in bilirubin crystallized from bile.[200] Isomerically pure bilirubin-IXα can be prepared by chloranil oxidation of biliverdin-IXα;[203] the older method involving iron (III) chloride oxidation is liable to cause isomerization. The main bilirubin conjugate in bile has been isolated and characterized as the diglucuronide in a careful study involving diazo-coupling reactions and mass spectral analyses of the products.[204] The methylene bridge carbon atom of bilirubin is extruded as formaldehyde after diazo-coupling reactions,[205] and a method has been developed for tritium-labelling the methylene bridge.[206] Bacterial reduction of bilirubin in the gut gives rise to a series of reduced bile pigments [207] and mild chromic acid oxidation of one of these, d-urobilin, has given both methyl ethyl and methyl vinyl maleimide

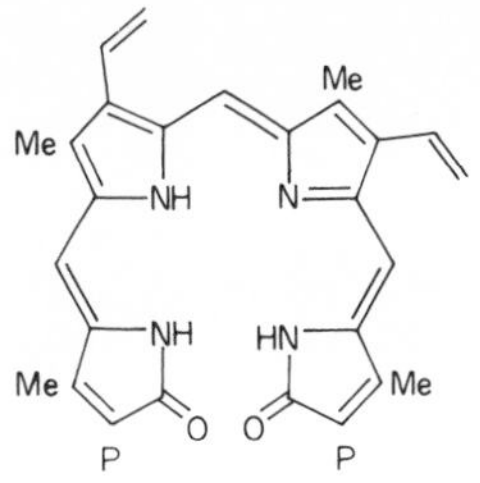

(112)

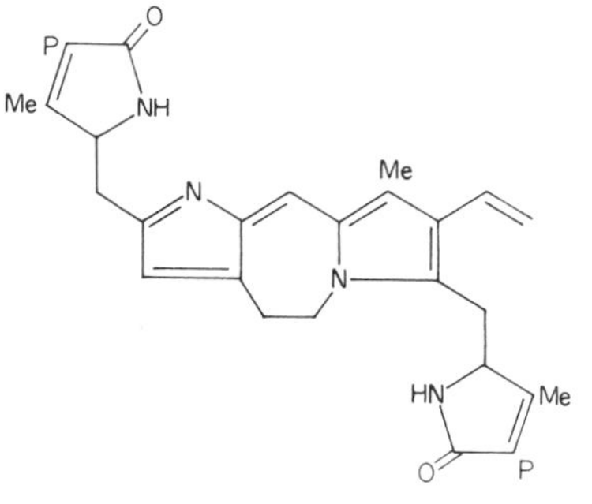

(113)

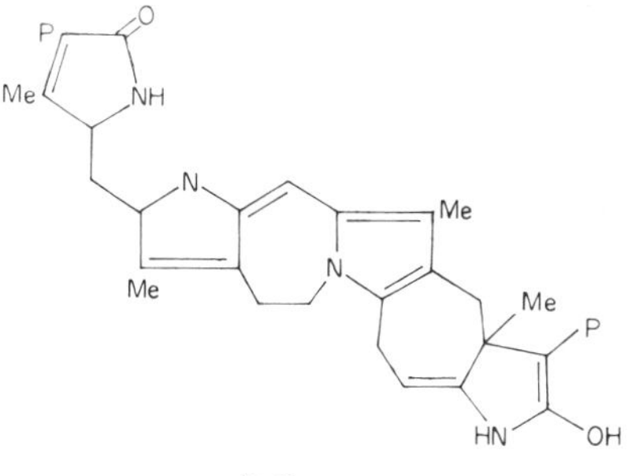

(114)

Scheme 39 Structures of butterfly bile pigments

indicating the existence of a mono-vinyl d-urobilin.[208] A new pigment, uroerythrin, isolated from human urine has a tripyrrolic structure (as shown by spectroscopic methods) corresponding to the A, B and C rings of biliverdin;[209] analogous tripyrrolic compounds have also been obtained by chemical, or photochemical oxidation of model biliverdins.

Bile pigments of the IX-γ series (Scheme 39) have recently been discovered in a wide range of butterflies.[210] The main pigment pterobilin (biliverdin-IXγ) (112) is the most widespread, occurring in over 70% of the 100 species so far screened; it was first isolated from the larvae and pupae of *Pieris brassicae* (the common cabbage white butterfly). More recently phorcabilin (113) and sarpedobilin (114) (which occurs less frequently) have also been isolated and characterized in a number of species.[210] Their structures were assigned largely on the basis of spectroscopic studies, and it is clear that they are intramolecular photocyclization products of biliverdin-IX-γ. Studies of the photocyclization of the (unnatural) biliverdin-IXδ have also been carried out[211] as a model for that of biliverdin-IXγ.

9.5.2   Synthesis

Two approaches to the synthesis of bile pigments have been studied, the oxidative ring opening of metalloporphyrins (by processes mimicking the normal metabolic process) or by direct synthesis from mono- and di- pyrrolic units. Lemberg and Fischer originally showed that oxidation of haems in pyridine solution with molecular oxygen gave biliverdin type pigments after work-up and removal of the iron.[189] Haem itself affords a mixture of all four biliverdins and these can now be separated by h.p.l.c. as their methyl esters.[212] *Meso*-hydroxylated haems are intermediates in these processes and recently it has also been shown that direct oxidation of protoporphyrin-IX itself with benzoyl peroxide affords a mixture of all four *meso*-benzoyloxy-derivatives; the latter can be separated by h.p.l.c., hydrolysed and converted to the oxyhaems, and then oxidized in pyridine solution to form the four biliverdin-IX isomers.[213] Deuterohaem has also been oxidized to give a mixture of all four 'deuterobiliverdin' isomers.[214] The mechanism of these chemical oxidations apears to be the same as that of the *in vivo* process, except for its specificity for the α-position; this is clear from experiments involving incorporation of a mixture of $^{18}O_2$ and $^{16}O_2$ which show that three molecules of oxygen are involved in the overall oxidation process both *in vivo* and *in vitro*.[215] Specific oxidative ring opening of a haem at a particular *meso*-bridge has only been achieved so far with the more symmetrical iron complexes of protoporphyrin-III and -XIII at the α-position.[216] However synthesis[13,14] of the appropriate oxophlorin ('oxyporphyrin') by the b-oxobilane route followed by oxidation of the haem in pyridine, gives the corresponding bile

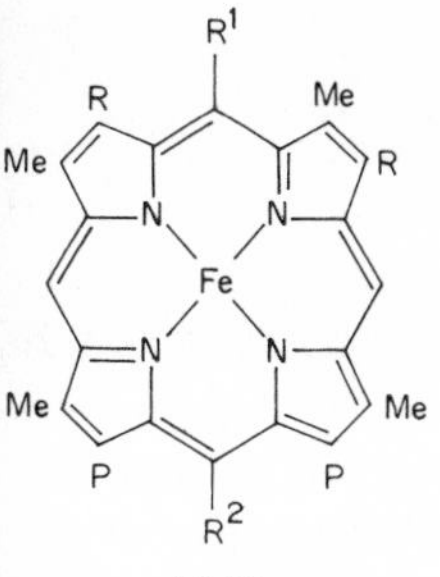

a) R = Et , R^1 = OH , R^2 = H

b) R = vinyl , R^1 = OH , R^2 = H

c) R = vinyl , R^1 = H , R^2 = H

(115)

Scheme 40 Oxyhaems: intermediates in biliverdin formation

pigment. α-Oxymesoporphyrin IX was synthesized in this way[217] and the corresponding haem (115a) oxidized both *in vitro*[217] and *in vivo*[218] to mesobilirubin IXα (Scheme 40). α-Oxyprotoporphyrin-IX has also been synthesized,[219] albeit in very low yield, and the haem (115b) converted to biliverdin-IXα. Similarly in more recent work γ-oxyprotoporphyrin-IX has been prepared by the *b*-oxobilane route, the *meso*-oxygen substituent being protected by an ethoxymethyl group during the generation of the vinyl groups; oxidative ring opening of the γ-oxyhaem (115c) then gave pterobilin (112), the butterfly pigment.[220] Interestingly photo-oxygenation of magnesium (and some other metal) complexes of chlorins affords dihydro-analogues of bile pigments by ring opening at a *meso*-position neighbouring the reduced ring; however, these reactions involve singlet oxygen and are mechanistically different from the oxidative ring opening of haem.[221]

Bilirubin and biliverdin and closely related bile pigments were originally synthesized by Fischer,[1b,13,222] and in a recent adaptation (Scheme 41) of the original methods the α-oxopyrromethenone (116) was condensed with the Mannich base (117) to give a mixture of bilirubin-IXα dimethyl ester (118) (contaminated with a small amount of the XIII α isomer); *meso* ^{14}C-labelled

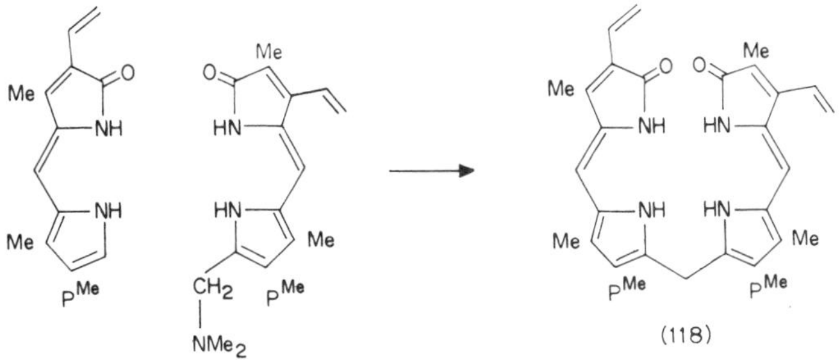

(116)   (117)   (118)

Scheme 41 Synthesis of bilirubin dimethyl ester

bilirubin was prepared in this way, the label being incorporated via Vilsmeier formylation with ^{14}C-dimethylformamide.[223]

During the 1960s substantial effort in Plieninger's laboratory led to the synthesis of natural (−)-stercobilin (108), and of other related bile pigments including the urobilins. More recently Gossauer has made a series of major contributions[222] to the total synthesis of bile pigments including phycocyanobilin (109),[224] phycoerythrobilin (110),[225] mesobilirubin, meso-urobilin analogues and racemic phytochromobilin (111)[226] (all as their dimethyl esters). The strategy adopted was very similar to earlier bile pigment syntheses but a number of new methods were developed (Scheme 42). For example, a diazoketone was

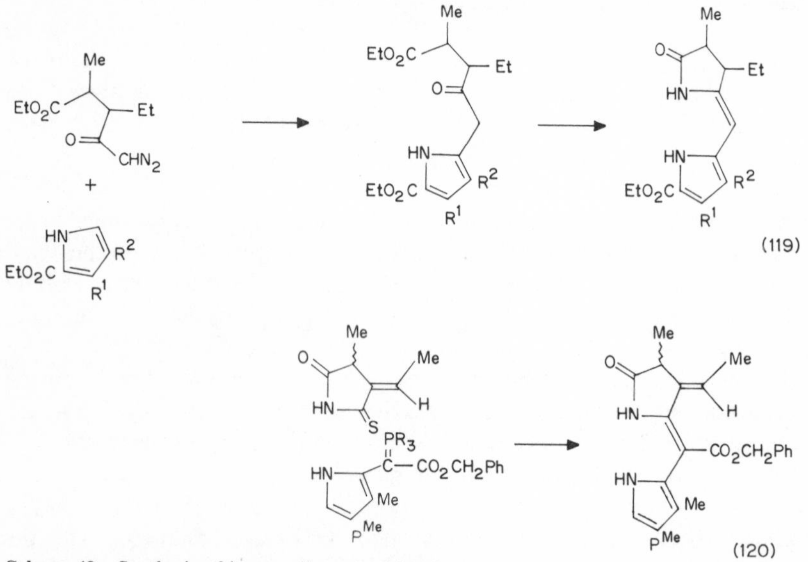

Scheme 42 Synthesis of intermediates for bile pigments

coupled with an α-free pyrrole to give an adduct, which was then cyclized with ammonium acetate to give the oxopyrromethane (119). In the synthesis of phycocyanobilin one of the pyrromethenones (120) was prepared by coupling of the thiosuccinimide with an α-bromomethyl pyrrole ester, followed by 'sulphur contraction' (a method reminiscent of those utilized in the vitamin-B_{12} syntheses); alternatively a phosphorous ylid could be utilized instead of the bromomethyl pyrrole.

Two phycoerythrobilins were synthesized by Gossauer (Scheme 43) by condensing the chiral formylpyrromethenone (121b) with the racemic pyrromethenone (122) in presence of acid; chromatographic separation of the diastereoisomeric products afforded the (+)-(2R),(16R)- and (−)-(2S),(16R)-phycoerythrobilin dimethyl esters (123). The chiral pyrromethenone was

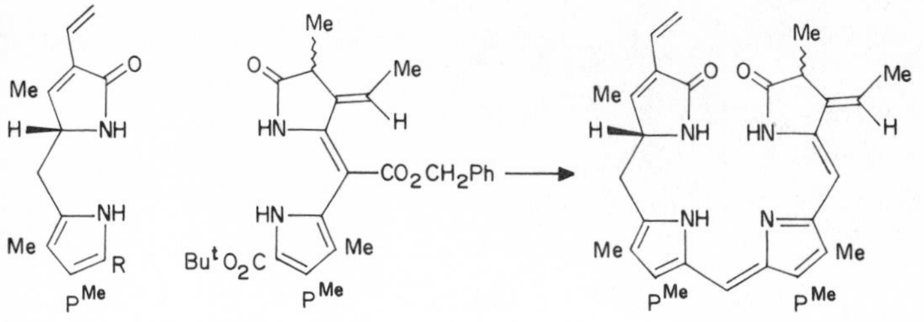

(121)   (122)   (123)

a)   R = CO$_2$ fenchyl
b)   R = CHO

Scheme 43   Synthesis of phycoerythrobilins

prepared via the fenchyl ester (121a) and the absolute configurations of the phycoerythrobilins were determined by oxidative degradations to optically active products, as well as by use of the same chiral intermediate to synthesize an optically active urobilin. It was concluded that the chromophore of the native biliprotein has the α-configuration at both C-2 and C-16.

These methods have now been extended to the synthesis of racemic phytochromobilin dimethyl ester,[227] and its photoisomerization has also been studied. Other work recently carried out in Gossauer's laboratory has involved the total synthesis of two epimeric methanol adducts of phycocyanobilin dimethyl ester and elucidation of their relative configuration.[228] A synthesis of E,Z,Z-biliverdins has also been carried out from appropriate pyrrom-ethenones,[229] whilst Pliéninger has reported a new route to dihydropyr-romethenones required for phytochromobilin model compounds.[230]

9.6   Prodigiosins

Prodigiosin (124a) a tripyrrolic red pigment, possessing considerable antibiotic and antifungal activity, was originally isolated over fifty years ago from *Serratia marcescens*. Several other analogues (Scheme 44) have been discovered more recently, including nonyl prodigiosin (124b)[231] undecyl prodigiosin (124c)[232] metacycloprodigiosin (125),[233] and a cyclic nonyl prodigiosin (126).[234] A cyclic analogue of prodigiosin (cycloprodigiosin), recently isolated from the marine bacterium *Alteromonas rubra*, together with prodigiosin itself, was originally assigned[235] the structure (127); however, this has now been reassigned[236] as (128) on the basis of careful 200 MHz n.m.r. spectral studies of the free base and its hydrochloride.

Prodigiosin itself was synthesized by Rapoport[237] some twenty years ago, the main problem being the efficient coupling of two pyrroles to give the directly

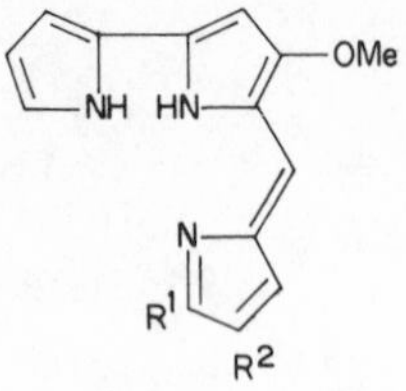

(124)

a) $R^1 = Me$, $R^2 = C_5H_{11}$

b) $R^1 = C_9H_{19}$, $R^2 = H$

c) $R^1 = C_{11}H_{23}$, $R^2 = H$

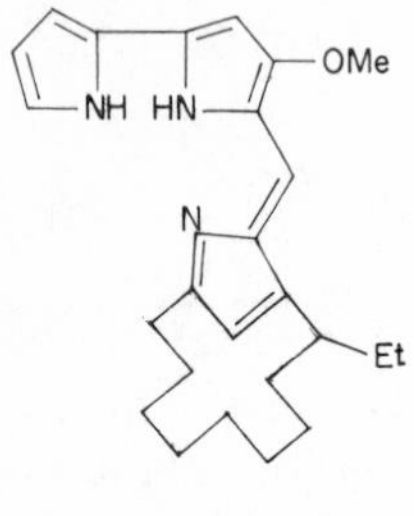

(125)

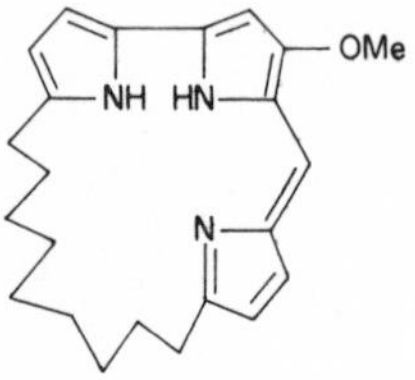

(126)

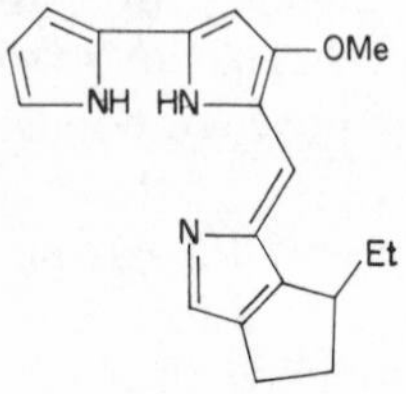

(127)

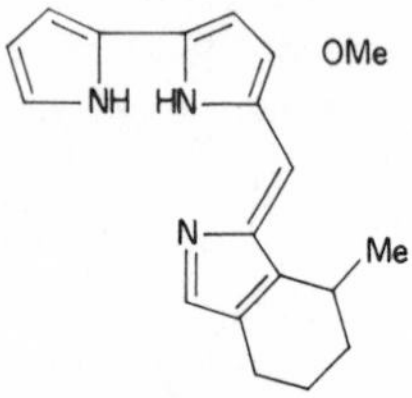

(128)

Scheme 44   Structures of some prodigiosins

linked dipyrrole unit. The undecyl analogue (126) was synthesized in the same way.[232] Metacycloprodigiosin has also been prepared in a similar manner,[233] the major challenge of the synthesis being the construction of the appropriate 'meta'-bridged pyrrole (131). This was achieved in a multi-step sequence (Scheme 45) starting from cyclododecanone in which the ethyl substituent was introduced first α-to the keto group followed by functionalization in the β'-position and conversion to 4-ethyl-cyclododec-2-enone (129); addition of cyanide, followed by conversion to the aldehyde (130) and treatment with ammonium carbonate then afforded the racemic ethylcyclononylpyrrole (131). Acid catalysed condensation with the methoxy-formyl-bipyrrole (132) then afforded metacycloprodigiosin.

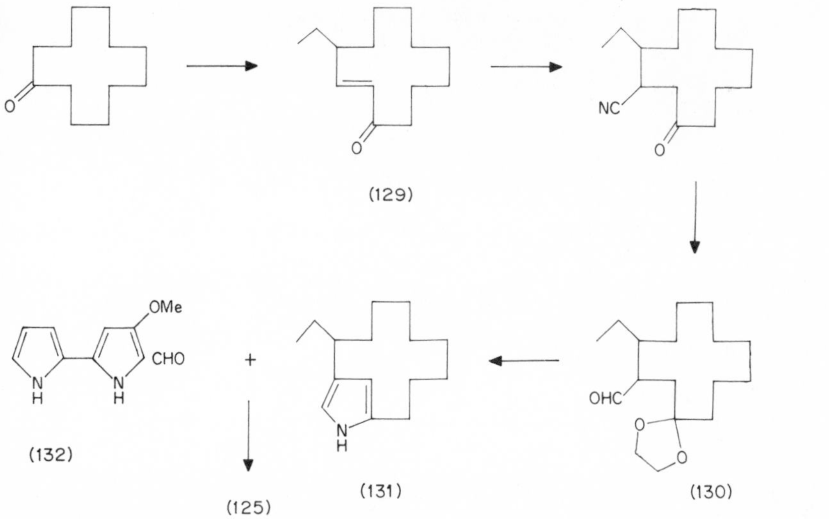

Scheme 45 Synthesis of metacycloprodigiosin

References

1. (a) H. Fischer and H. Orth, 'Die Chemie des Pyrrols', Vol. I, Akademische Verlag, Leipzig 1937.
   (b) H. Fischer and H. Orth, 'Die Chemie des Pyrrols, Vol. II(i), Akademische Verlag, Leipzig, 1937.
   (c) H. Fischer and H. Stern 'Die Chemie des Pyrrols', Vol. II(ii), Akademische-Verlag, Leipzig, 1940.
2. G. Marks, 'Heme and Chlorophyll', Van Nostrand, London, 1962.
3. J.E. Falk, 'Porphyrins and Metalloporphyrins', Elsevier, Amsterdam, 1964.
4. 'Porphyrins and Metalloporphyrins' (ed. K.M. Smith), Elsevier, Amsterdam, 1975; this book also includes a supplementary chapter by K.M. Smith and J.H. Fuhrhop, 'Laboratory Methods in Porphyrin and Metalloporphyrin Research', which is available separately.
5. 'The Porphyrins' ed. D. Dolphin, Academic Press, New York, 1978; Vol. I. Structure and Synthesis Part A; Vol. II Structure and Synthesis Part B; Vol. III Physical Chemistry, Part A, Vol. IV Physical Chemistry, Part B; Vol. V Physical Chemistry Part C; Vol. VI Biochemistry, Part A; Vol. VII Biochemistry, Part B.
6. 'B_{12}' ed. D. Dolphin, Wiley-Interscience, New York, 1982, Vol. 1, Chemistry; Vol. 2, Biochemistry and Medicine.
7. A. Gossauer, 'Die Chemie der Pyrrole', Springer-Verlag, New York, 1974.
8. R.A. Jones and G.P. Bean, 'The Chemistry of Pyrroles', Academic Press, London, 1977.
9. K.M. Smith, 'Pyrrole Pigments', in 'Rodd's Chemistry of Carbon Compounds', 2nd edn., ed. S. Coffey, Elsevier, Amsterdam, 1977, Vol. IV B, p. 237.
10. (a) A.H. Jackson in 'Comprehensive Organic Chemistry', Vol. 4 (ed. P.G. Sammes) Pergamon, Oxford, 1979, 275 (Pyrroles and Bile Pigments).
    (b) K.M. Smith, ibid, p. 321 (Porphyrins, Chlorins and Corrins).
11. IUPAC nomenclature cf. R. Bonnett in ref. 5., Vol. I, p. 1.
12. cf. J.B. Paine III in ref. 5., Vol. 1, p. 101.
13. cf. A.H. Jackson and K.M. Smith in 'The Total Synthesis of Natural Products', (ed. J. ApSimon), Wiley, New York, 1974, p. 144.

448   THE CHEMISTRY OF NATURAL PRODUCTS

14. *cf.* P.S. Clezy and A.H. Jackson in ref. 5, Vol. 1, p. 265.
15. *cf.* A.W. Johnson in ref. 5, Vol. 1, p. 235.
16. J.A.P. Baptista de Almeida, G.W. Kenner, K.M. Smith and M.J. Sutton, 1975, *J. Chem. Soc., Chem. Comm.*, 21, and J.A.P. Baptista de Almeida G.W. Kenner, J. Rimmer and K.M. Smith, 1976, *Tetrahedron*, 32, 1793.
17. A.R. Battersby and E. McDonald in ref. 4, p. 61.
18. D.G. Buckley in 'Ann. Repts. on the Progress of Chemistry', 1977, 74B, 392.
19. L. Bogorad, in ref. 5, Vol. VI, p. 125, and R.B. Frydman, B. Frydman and A. Valasinas in ref. 5, Vol. VI, p. 3.
20. M. Akhtar and P.M. Jordan in 'Comprehensive Organic Chemistry', Vol. 5, ed. E. Haslam, Pergamon Oxford, 1979, 1121.
21. G.P. Arsenault and S.F. MacDonald, 1961, *Can. J. Chem.*, 39, 2043 and preceding papers.
22. (a) G.W. Kenner, J. Rimmer, K.M. Smith and J.F. Unsworth, 1976, *Phil. Trans Roy. Soc. London*, 273B, 255;
    (b) 1977, *idem*, J. Chem. Soc. Perkin I, 332.
23. B. Frydman, S. Reid, W.E. Despuy, and H. Rapoport, 1969, *J. Amer. Chem. Soc.*, 91, 2338 and 1970, 92, 1810.
24. B. Frydman, G. Buldain and J.C. Repetto, 1973, *J. Org. Chem.*, 38, 1824.
25. A.R. Battersby, E. McDonald, H.K.W. Wurziger and K.J. James, 1975, *J. Chem. Soc., Chem. Comm.*, 493.
26. M.I. Jones, C. Fronssidus and D.A. Evans, 1976, *J. Chem. Soc., Chem. Comm.*, 472.
27. G. Muller, 1972, *Z. Naturforsch*, 27b, 473.
28. D. Gwyne and D. Shemin, 1973, *Science*, 180, 1188; and 1976, *Meth. Enzymol.*, 44, 844.
29. A.R. Battersby, E. McDonald, D.C. Williams and H.K.W. Wurziger, 1977, *J. Chem. Soc., Chem. Comm.*, 113.
30. H.O. Dauner, G. Gunzer, I. Hager and G. Müller, 1976, *Z. Physiol. Chem.*, 357, 147.
31. A.R. Battersby, C.J.R. Fookes, K.E. Gustavson-Potter, G.W.J. Matcham and E. McDonald, 1979, *J. Chem. Soc., Chem. Comm.*, 1155.
32. J. Engel and A. Gossauer, 1976, *Annalen*, 1637.
33. B. Franck, D. Guntz, F.P. Montforts and F. Swindtchen, 1972, *Angew. Chem. (Int. Edn.)*, 11, 421.
34. A. Valasinas and B. Frydman, 1976, *J. Org. Chem.*, 48, 2991.
35. A.R. Battersby, M. Ihara, E. McDonald, J. Saunders and P.J. Webby, 1976, *J. Chem. Soc. Perkin I*, 293.
36. U.A. Meyer and R. Schmid in 'The Metabolic Basis of Inherited Disease' (ed. J.B. Stanbury, J.B. Wyngaarden and D.S. Frederickson), 1978, McGraw-Hill, New York, 1166.
37. G.H. Elder, G.B. Lee and J.A. Tovey, 1978, *N. Engl. J. Med.*, 299, 274.
38. A.R. Battersby, E. Hurst, E. McDonald, J.B. Paine, and J. Saunders, 1976, *J. Chem. Soc. Perkin I*, 1008.
39. A.H. Jackson, H.A. Sancovich, A.M. Ferramola, N. Evans, D.E. Games, S.A. Matlin, S.G. Smith and G.H. Elder, 1976, *Phil. Trans. Roy. Soc. Lond. B.*, 273, 191.
40. A.H. Jackson, H.A. Sancovich and A.M. Ferramola de Sancovich, 1980, *Bioorganic Chem.*, 9, 71.
41. P.S. Clezy, T.T. Hai and P.C. Gupta, 1976, *Aust. J. Chem.*, 29, 393.
42. M.S. Stoll, G.H. Elder, D.E. Games, P. O'Hanlon, D.S. Millington and A.H. Jackson, 1973, *Biochem. J.*, 131, 429.
43. A.H. Jackson, A.M. Ferramola, H.A. Sancovich, N. Evans, S.A. Matlin, D.J. Ryder and S.G. Smith, 1976, *Annals Chim. Res.*, 8, 89.
44. D. Mauzerall and S. Granick, 1958, *J. Biol. Chem.*, 232, 1141.
45. S.A. Ali, M. Chakrabarty, A.H. Jackson and G. Philip, 1981, *Heterocycles*, 15, 1199.
46. Y. Nordmann and J.C. Deybach, 1982, *Seminars in Liver Disease*, 2, 154.
47. A.H. Jackson, K.R.N. Rao, D.M. Supphayen and S.G. Smith, 1977, *J. Chem. Soc., Chem. Comm.*, 696.
48. G.H. Elder, 1972, *Biochem. J.*, 126, 877.
49. P.S. Clezy and T.T. Hai, 1976, *Aust. J. Chem.*, 29, 1561.
50. A.H. Jackson, T.D. Lash and D.J. Ryder, unpublished work.

51. A.H. Jackson, G.W. Kenner and J. Wass, 1972, *J. Chem. Soc. Perkin I*, 475.
52. K.M. Smith, 1972, *J. Chem. Soc. Perkin I*, 1471; J.B. Paine, G.K. Chang and D. Dolphin, 1977, *Heterocycles*, 7, 831.
53. G.W. Kenner, S.W. McCombie and K.M. Smith, 1972, *J. Chem. Soc. Chem. Comm.*, 1347.
54. A.H. Jackson, G.W. Kenner, C.J. Suckling and G.Y. Kennedy, 1970, *FEBS Letts.*, 6, 9 and 7, 205.
55. G.W. Kenner, A.H. Jackson, K.M. Smith and C.J. Suckling, 1976, *Tetrahedron*, 32, 2757.
56. J.A.S. Cavaleiro, G.W. Kenner and K.M. Smith, 1974, *J. Chem. Soc. Perkin I*, 1188.
57. (a) I.A. Chaudry, P.S. Clezy and V. Diakiw, 1976, *Aust. J. Chem.*, 30, 879.
    (b) I.A. Chaudry, P.S. Clezy and A.K. Mirza, 1980, *Aust. J. Chem*, 33, 1095.
58. G.W. Kenner, J.M.E. Quirke and K.M. Smith, 1976, *Tetrahedron*, 32, 2753.
59. D. Dolphin and R. Sivasothy, 1981, *Can. J. Chem.*, 59, 779.
60. K.M. Smith and K.C. Langry, 1980, *J. Chem. Soc. Chem. Comm.*, 217.
61. G.H. Elder, J.O. Evans, J.R. Jackson and A.H. Jackson, 1978, *Biochem. J.*, 169, 215; A.H. Jackson, D.M. Jones, G. Philip, T.D. Lash, A.M. del C. Batle and S.G. Smith, 1980, *Int. J. Biochem.*, 12, 681.
62. J.M. French, D.C. Nicholson and C. Rimington, 1966, *Biochem. J.*, 120, 393.
63. P.W. Couch, D.E. Games and A.H. Jackson, 1976, *J. Chem. Soc. Perkin I*, 2492.
64. K.M. Smith and K.C. Langry, 1981, *J. Chem. Soc. Chem. Comm.*, 283, and 1983, *J. Org. Chem.*, 48, 500.
65. R.P. Carr, A.H. Jackson and G.W. Kenner, 1971, *J. Chem. Soc. (C)*, 487.
66. J.A.S. Cavaleiro, A.M. d'A.R. Gonsalves, G.W. Kenner and K.M. Smith, 1977, *J. Chem. Soc. Perkin I*, 1771.
67. R. Grigg, A.W. Johnson and M. Roche, 1970, *J. Chem. Soc. (C)*, 1928.
68. G.V. Ponomarev, S.M. Navarella, A.G. Bybnova and R.P. Evstigneeva, 1973, *Khim. geterosikli Soedin.*, 202.
69. A.R. Battersby, G.L. Hodgson, M. Ihara, E. McDonald and J. Saunders, 1973, *J. Chem. Soc. Perkin I*, 2923.
70. P.S. Clezy and C.J.R. Fookes, 1980, *Aust. J. Chem.*, 33, 557.
71. A.H. Jackson, G.W. Kenner, and G.S. Sach, 1967, *J. Chem. Soc. (C)*, 2045, and A.H. Jackson, G.W. Kenner, G. McGillivray and K.M. Smith, 1968, *J. Chem. Soc. (C)*, 294.
72. P.J. Crook, A.H. Jackson and G.W. Kenner, 1971, *J. Chem. Soc. (C)*, 474.
73. D. Dolphin, A.W. Johnson, J. Leng and P. Van den broek, 1966, *J. Chem. Soc. (C)*, 880.
74. L.I. Fleiderman, A.F. Mironov and R.P. Evstigneeva, 1973, *Zhur. obshch. Khim.*, 43, 886.
75. R. Bonnett, I.H. Campion-Smith and A.J. Page, 1977, *J. Chem. Soc. Perkin I*, 68; R.K. DiNello and D. Dolphin, 1981, *J. Org. Chem.*, 48, 3498.
76. S. Sano, T. Shingu, J.M. French and E. Thonger, 1965, *Biochem. J.*, 97, 250.
77. A.H. Jackson, G.W. Kenner and J. Wass, 1974, *J. Chem. Soc. Perkin I*, 480.
78. P. Barnfield, R. Grigg, R.W. Kenyon and A.W. Johnson, 1968, *J. Chem. Soc. (C)*, 1259.
79. P.S. Clezy, A.J. Liepa and N.W. Webb, 1972, *Aust. J. Chem.*, 25, 1991.
80. A.F. Mironov, A.N. Nizhnik, D.T. Khosich, A.N. Kozyrev and R.P. Evstigneeva, 1981, *Zh. obshch, Khim.*, 51, 699.
81. D.E. Games, P.J. O'Hanlon and A.H. Jackson, 1976, *J. Chem. Soc. Perkin I*, 2501.
82. P.S. Clezy, and V. Diakin, 1975, *Aust. J. Chem.*, 28, 1589.
83. M. Sano and T. Asakura, 1974, *Biochemistry*, 13, 4386.
84. H.H. Inhoffen, G. Bliesener and H. Brockmann, 1969, *Annalen*, 730, 173.
85. A.H. Jackson, S.A. Matlin, A.H. Rees, and R. Towill, 1978, *J. Chem. Soc. Chem. Comm.*, 645.
86. W.S. Caughey, G.A. Smythe, D.H. O'Keeffe, J.E. Maskashy, and M.L. Smith, 1975, *J. Biol. Chem.*, 250, 7602.
87. M. Thompson, J. Barrett, E. McDonald, A.R. Battersby, C.J.R. Fookes, I.A. Chaudry and P.S. Clezy, 1977, *J. Chem. Soc., Chem. Comm.*, 278.
88. P.S. Clezy and C.J.R. Fookes, 1977, *Aust. J. Chem.*, 20, 1799.
89. *cf.* E.W. Baker and S.E. Palmer in ref. 5, Vol. I, p. 486, and J.M.E. Quirke, G. Eglinton and J.R. Maxwell, 1979, *J. Amer. Chem. Soc.*, 101, 7693.
90. M.E. Flaugh and H. Rapoport, 1968, *J. Amer. Chem. Soc.*, 90, 6877.

91. P.S. Clezy, C.J.R. Fookes and A.H. Mirza, 1977, *Aust. J. Chem.*, 30, 1337; P.S. Clezy and A.H. Mirza 1982, *Aust. J. Chem.*, 35, 197.
92. P.R. Ortiz de Montellano, H.S. Beilan and K.L. Kunze, 1981, *Proc. Natl. Acad. Sci. U.S.A.*, 78, 1490.
93. F. de Matteis, A.H. Jackson, A.H. Gibbs, K.R.N. Rao, J. Atton, S. Weerasinghe, and C. Hollands, 1982, *FEBS Letts.*, 142, 44.
94. T.R. Tephly, B.L. Coffmann, G. Ingoll, M.S. Abon Zeit-Han, H.M. Gott, H.D. Tabba and K.M. Smith, 1981, *Arch. Biochem. Biophys.*, 272.
95. A.H. Jackson and G.R. Dearden, 1973, *Annals N.Y. Acad. Sci.*, 206, 151.
96. P.R. Ortiz de Montellano, K.L. Kunze, H.S. Beilan and C. Wheeler, 1982, *Biochemistry*, 21, 1331, and references therein.
97. K.L. Kunze and P.R. Ortiz de Montellano, 1981, *J. Amer. Chem. Soc.*, 103, 4225.
98. T.G. Traylor, 1981, *Acc. Chem. Res.*, 14, 102.
99. R.B. Woodward *et al.*, 1960, *J. Amer. Chem. Soc.*, 82, 3800.
100. G.W. Kenner, S.W. McCombie and K.M. Smith, 1973, *J. Chem. Soc. Perkin I*, 2517.
101. E. Zass, H.P. Isenring, R. Etter and A. Eschenmoser, 1980, *Helv. Chim. Acta*, 63, 1048.
102. M. Strell and T. Crumnow, 1977, *Annalen*, 970.
103. cf. A.H. Jackson in 'Chemistry and Biochemistry of Plant Pigments' (ed. T.W. Goodwin), 2nd edn. Vol. 1, Academic Press, London, 1976, p. 1.
104. B.D. Berezin and O.P. Lapshima, 1976, *Zhur. fiz. Khim.*, 50, 2007.
105. P.H. Hynninen, 1973, *Acta Chem. Scand.*, 27, 1771.
106. H. Scheer and J.J. Katz in ref. 4, p. 399, and T.R. Janson and J.J. Katz in ref. 5, Vol. IV, p. 1; see also V. Wray, U. Jurgens and H. Brockman, 1979, *Tetrahedron*, 35, 2275.
107. R.J. Abraham, K.M. Smith, D.A. Goff and J.J. Lai, 1982, *J. Amer. Chem. Soc.*, 104, 4332.
108. J. Gassman, I. Strell, F. Brand, M. Sturm and W. Hoppe, 1971, *Tetrahedron Lett.*, 4609; M.S. Fischer, D.H. Templeton, A. Zalkin and M. Calvin, 1972, *J. Amer. Chem. Soc.*, 94, 3613.
109. H.C. Chow, R. Serlin and C.E. Strouse, 1975, *J. Amer. Chem. Soc.*, 97, 7230.
110. R. Serlin, H.C. Chow and C.E. Strouse, 1975, *J. Amer. Chem. Soc.*, 97, 7237.
111. G.L. Closs, J.J. Katz, F.C. Pennington, M.R. Thomas and H.H. Strain, 1963, *J. Amer. Chem. Soc.*, 85, 8809.
112. J.J. Katz, H.H. Strain, D.L. Leussing, and R.C. Dougherty, 1968, *J. Amer. Chem. Soc.*, 90, 6841.
113. P.H. Hynninen, M.R. Wasielewski, and J.J. Katz, 1979, *Acta Chem. Scand.*, 33, 637.
114. H. Scheer and J.J. Katz, 1978, *J. Amer. Chem. Soc.*, 100, 561.
115. M.R. Wasielewski and J.F. Thompson, 1978, *Tetrahedron Lett.*, 1043.
116. I. Fleming, 1968, J. Chem. Soc. C, 2765 and H.H. Brockmann, 1968, *Angew. Chem. (Int. Edn.)*, 7, 211.
117. H. Brockmann and J. Bode, 1974, *Annalen*, 1017.
118. T.T. Howarth, A.H. Jackson, J. Judge, G.W. Kenner, and D.J. Newman, 1974, *J. Chem. Soc. Perkin I*, 490.
119. M.T. Cox, T.T. Howarth, A.H. Jackson and G.W. Kenner, 1974, *J. Chem. Soc. Perkin I*, 512.
120. M.T. Cox, A.H. Jackson, G.W. Kenner, S.W. McCombie and K.M. Smith, 1974, *J. Chem. Soc. Perkin I*, 516.
121. G.W. Kenner, S.W. McCombie and K.M. Smith, 1974, *J. Chem. Soc. Perkin I*, 527.
122. (a) K.M. Smith and W.M. Lewis, 1981, *Tetrahedron*, 37, Suppl. 1, 399;
    (b) K.M. Smith, J.A.P. Baptista de Almeida and W.M. Lewis, 1980, *J. Heterocycl. Chem.*, 17, 481.
123. H.H. Inhoffen, P. Jäger and R. Mählhop, 1971, *Annalen*, 749, 109.
124. U. Jurgens and H. Brockmann, 1979, *J. Chem. Res.*, 2379.
125. M.B. Bassaz, C.V. Bradley and R.G. Brereton, 1982, *Tetrahedron Lett.*, 1211.
126. R.C. Dougherty, H.H. Strain, W.A. Svec, R.A. Uphams and J.J. Katz, 1970, *J. Amer. Chem. Soc.*, 92, 2826.
127. P.S. Clezy and C.J.R. Fookes, 1978, *Aust. J. Chem.*, 31, 2491.

128. E. Walter, J. Schreiber, E. Zass and A. Eschenmoser, 1979, *Helv. Chim. Acta*, 62, 899.
129. N. Risch, H. Brockmann and A. Gloe, 1979, *Annalen*, 408.
130. H. Brockmann and R. Knoblock, 1972, *Arch. Mikrobiol.*, 85, 123.
131. W. Rüdiger, P. Hedden, H.-P. Köst and D. Chapman, 1977, *Biochem. Biophys. Res. Comm.*, 1268.
132. A.S. Holt, J.W. Purdie and J.W.F. Wasley, 1966, *Can. J. Chem.*, 44, 88, and references therein; J.L. Archibald, D.M. Walker, K.B. Shaw, A. Markovac and S.F. MacDonald, 1966, *Can. J. Chem.*, 44, 345.
133. J.W. Mathewson, W.R. Richards and H. Rapoport, 1968, *J. Amer. Chem. Soc.*, 85, 364.
134. M.T. Cox, A.H. Jackson and G.W. Kenner, 1979, *J. Chem. Soc. C.*, 1974.
135. R.A. Chapman, M.W. Roomi, T.C. Morton, D.T. Krajcarski and S.F. MacDonald, 1971, *Can. J. Chem.*, 49, 3544.
136. K.M. Smith, L.A. Kehres and H.D. Tabba, 1980, *J. Amer. Chem. Soc.*, 102, 7149.
137. H. Brockmann and R. Tackekarindadian, 1979, *Annalen*, 419.
138. K.M. Smith, D.A. Goff, J. Fajer and K.M. Barkijia, 1982, *J. Amer. Chem. Soc.*, 104, 3747.
139. K.M. Smith, D.A. Goff, J. Fajer and K.M. Barkijia, 1983, *J. Amer. Chem. Soc.*, 105, 1674.
140. K.M. Smith, M.J. Bushell, J. Rimmer and J.F. Unsworth, 1980, *J. Amer. Chem. Soc.*, 102, 2437.
141. H. Risch and H. Brockmann, 1976, *Annalen*, 578.
142. H. Brockmann, A. Gloe, N. Risch and A. Trowitzsch, 1976, *Annalen*, 566.
143. N. Risch and H. Reisch, 1979, *Tetrahedron Lett.*, 4257.
144. K.M. Smith, G.M.F. Bissett and M.J. Bushell, 1980, *J. Org. Chem.*, 45, 2218.
145. U. Jurgens and H. Brockmann, 1982, *Annalen*, 472.
146. H. Brockmann and C. Belter, 1979, *Z. Naturforsch. B*, 34, 127.
147. J.H. Fuhrhop and D. Mauzerall, 1979, *Photochem. Photobiol.*, 13, 453.
148. N. Risch and C. Belter, *Z. Naturforsch. B*, 34, 129; R.F. Troxler, K.M. Smith and S.B. Brown, 1980, *Tetrahedron Lett.*, 491.
149. J.A. Ballantine, A.F. Psaïla, A. Pelter, P. Murray-Rust, V. Ferrito, P. Schembri and V. Jaccarini, 1980, *J. Chem. Soc. Perkin I*, 1080, and refs. therein.
150. A. Pelter, A. Abelamedici, J.A. Ballantine, V. Ferrito, S. Ford, V. Jaccarini and A.F. Psaïla, 1978, *Tetrahedron Lett.*, 2017.
151. K. Folkers, in ref. 6, Vol. 1, p. 1.
152. *cf.* A.R. Battersby and E. MacDonald in ref. 6, Vol. 1, p. 107.
153. J. Halpern in ref. 6, Vol. 1, p. 501, and B.T. Golding in ref. 6, Vol. 1, p. 543.
154. R. Bonnett, in ref. 6, Vol. 1, p. 201.
155. A. Gossauer, K.P. Heise, H. Laas and H.H. Inhoffen, 1976, *Annalen*, 1150.
156. E.M. Sauer, K. Broschinski, L. Ernst and H.H. Inhoffen, 1981, *Helv. Chim. Acta.*, 64, 2257.
157. R.P. Hinze, W. Schaer and H.H. Inhoffen, 1980, *Annalen*, 165; G. Bartels, R.P. Hinze, H.H. Inhoffen and L. Ernst, 1979, *Annalen*, 1440.
158. B. Krautler, 1982, *Helv. Chim. Acta*, 45, 1941.
159. R. Bonnett, J.M. Godfrey and V.B. Math, 1971, *J. Chem. Soc. C*, 3736.
160. A. Gossauer, B. Grüning, L. Ernst, W. Becker and W.S. Sheldrich, 1977, *Angew. Chem. Int. Edn.* 16, 481.
161. G. Schlingmann, B. Dresow, L. Ernst and V.B. Koppenhagen, 1981, *Annalen*, 2061.
162. *cf.* O.D. Hensen, H.A.O. Hill, C.E. McLelland and R.J.P. Williams in ref. 6, Vol. 1, p. 463 and L. Ernst, 1981, *Annalen*, 376.
163. R.V. Stevens in ref. 6, Vol. 1, p. 169.
164. R.B. Woodward, 1973, *Pure Appl. Chem.* 38, 145, and A. Eschenmoser, 1974, *Naturwissen*, 61, 513.
165. R.B. Woodward, and R. Hoffmann, 'The Conservation of Orbital Symmetry', Verlag Chemie, Weinheim, 1970.
166. A. Eschenmoser, 1976, *Chem. Soc. Rev.*, 5, 377.
167. A. Pfaltz, B. Hardegger, P.M. Muller, S. Faroog, B. Krautler and A. Eschenmoser, 1975, *Helv. Chim. Acta*, 58, 1444.
168. E. Gotschi and A. Eschenmoser, 1973, *Angew. Chem.* 85, 952.

169. A. Pfaltz, N. Buhler, R. Neier, K. Hirai and A. Eschenmoser, 1977, *Helv. Chim. Acta*, 60, 2653.
170. L. Mombelli, C. Nussbaumer, H. Weber, G. Müller and D. Arigoni, 1981, *Proc. Natl. Acad. Sci. U.S.A.*, 78,9; A.R. Battersby, M.J. Bushell, C. Jones, N.G. Lewis and A. Pfenninger, 1981, *Proc. Natl., Acad. Sci. U.S.A.* 78, 13.
171. R.V. Stevens, 1976, *Tetrahedron*, 32, 1599.
172. J.R. Postgate, 1956, *J. Gen. Microbiol.* 14, 545; H.E. Jones and G.W. Skyring, 1974, *Aust. J. Biol. Sci.*, 27,7.
173. M.J. Murphy and L.M. Siegel, 1973, *J. Biol. Chem.* 248, 251.
174. A.R. Battersby, K. Jones, E. MacDonald, J.A. Robinson and H. Morris, 1977, *Tetrahedron Lett.*, 2213.
175. R. Deeg, H.P. Kriemler, K.H. Bergmann and G. Müller, 1977, *Hoppe Seyler's Z. Physiol. Chem.* 358, 339; K.H. Bergmann, R. Deeg, K.D. Gneuss, H.P. Kriemler and G. Muller, 1977, *Hoppe Seyler's Z. Physiol. Chem.* 358, 1315.
176. A.R. Battersby, E. McDonald, H. Morris, M. Thompson, D.C. Williams, V. Bykhovsky, N. Zaitseva and V. Bukin, 1977, *Tetrahedron Lett.* 2217.
177. A.R. Battersby, E. McDonald, M. Thompson and V.Y. Bykhovsky, 1978, *J. Chem. Soc. Chem. Comm*, 150.
178. A.I. Scott, A.J. Irwin, L.M. Siegel and J.N. Shoolery, 1978, *J. Amer. Chem. Soc.* 100, 316 and 7987; A.I. Scott, 1978, *Acc. Chem. Res.* 11, 29.
179. A.R. Battersby, E. McDonald, R. Neier and M. Thompson, 1979, *J. Chem. Soc. Chem. Comm.*, 960.
180. K.D. Gneuss, G. Gunzer and H.P. Kriemler, in Vitamin B_{12} (ed. J. Zagalah and W. Friedrich) de Gruyter, Berlin, 1979, p. 279.
181. A.R. Battersby and E. McDonald, 1978, *Bioorg. Chem.* 7, 161.
182. A.R. Battersby, G.W.J. Matcham, E. McDonald, R. Neier, M. Thompson, W.D. Woggon, V.Y. Bykhovsky and H.R. Morris, 1979, *J. Chem. Soc. Chem. Comm.*, 185.
183. N.G. Lewis, R. Neier, G.W.J. Matcham, E. McDonald and A.R. Battersby, 1979, *J. Chem. Soc. Chem. Comm.*, 541.
184. G. Müller, K.D. Gneuss, H.P. Kriemler, A.I. Scott and A.J. Irwin, 1979, *J. Amer. Chem. Soc.*, 101, 3655.
185. L. Mombelli, C. Nussbaumer, H. Weber, G. Muller and D. Arigoni, 1981, *Proc. Natl. Acad. Sci. US.A.*, 78, 9.
186. A.R. Battersby, M.J. Bushell, C. Jones, N.G. Lewis and A. Pfenninger, 1980, *Proc. Natl. Acad. Sci. U.S.A.*, 78, 13.
187. A.R. Battersby, K. Frobel, F. Hammerschmidt and C. Jones, 1982, *J. Chem. Soc. Chem. Comm.*, 455.
188. V. Rasetti, K. Holpert, A. Fassler, A. Pfaltz and A. Eschenmoser, 1981, *Angew. Chem.* 93, 1108, and refs. therein.
189. *cf.* R. Schmid and A.F. McDonagh in ref. 5. Vol. VI p. 258 and A.H. Jackson in 'Iron in Biochemistry and Medicine' (ed. M. Worwood), Academic Press, London, 1974, p. 145.
190. A. Bennett and H. Siegelman, in ref. 5, Vol. VI, p. 493.
191. R. Bonnett, J.E. Davies and M. Hursthouse, 1976, *Nature*, 262, 326.
192. W.S. Sheldrick, 1976, *J. Chem. Soc. Perkin II*, 1457.
193. A.F. McDonagh, L.A. Palmer and D.A. Lightner, 1982, *J. Amer. Chem. Soc.*, 104, 6865 and 6867.
194. S. Onishi, N. Kawade, S. Itoh, K. Isobe, S. Sugiyama, T. Hashimoto and H. Narita, 1981, *Biochem. J.*, 198, 107.
195. M.S. Stoll, N. Vicker, C.H. Gray and R. Bonnett, 1982, *Biochem. J.*, 201, 179.
196. H. Falk, N. Muller, M. Ratzenhofer and K. Winsauer, 1982, *Monats. Chem.*, 113, 1421.
197. S. Onishi, K. Isobe, S. Itoh, N. Kawade and S. Sugiyama, 1980, *Biochem. J.*, 190, 533.
198. T. Brandlmeier, H. Scheer and W. Rüdiger, 1981, *Z. Naturforsch*, 36c, 431; F. Thümmler, T. Brandlmeier and W. Rüdiger, 1981, *ibid.*, 36c, 440.
199. H. Falk, K. Grubmayr, G. Kapl and U. Zrunek, 1982, *Monats. Chem.* 113, 1329, and preceding papers.
200. *cf.* A.F. McDonagh in ref. 5, Vol. VI, p. 293.

201. D.A. Lightner in 'Chemistry and Physiology of the Bile Pigments' (eds. N.I. Berlin, P.D. Berk and C.J. Watson), 1977, Nat. Inst. Health, Bethesda, Maryland, 93.
202. R. Bonnett, 1976, *Biochem. Soc. Trans.*, 4, 222, and refs. therein.
203. P. Manitto and D. Monti, 1979, *Experientia*, 35, 9.
204. E.M. Gordon, C.A. Goresky, T.H. Chang and A.S. Perkin, 1976, *Biochem. J.*, 155, 477.
205. D.W. Hutchinson, B. Johnson and A.J. Knell, 1972, *Biochem. J.*, 127, 907.
206. D.W. Hutchinson, M.N. Willees, and H.Y.N. Au, 1981, *J. Labelled Cpd.Radiopharm.* 18, 1401; F.E. Hancock, D.W. Hutchinson and A.J. Knell, 1976, *Biochem. Soc. Trans.*, 4, 303.
207. *cf.* D.A. Lightner in ref. 5, Vol. VI, p. 521.
208. M. Chedekel, F.A. Bovey, A.I.R. Brewster, Z.J. Petryka, M. Weimer, C.J. Watson, A. Moskowitz and A. Lightner, 1974, *Proc. Natl. Acad. Sci. U.S.A.*, 71, 1599.
209. J. Deruter, J.P. Colombo, and U.P. Schunegger, 1975, *Europ. J. Biochem.*, 56, 239.
210. *cf.* M. Barbier, 1981, *Experientia*, 37, 1060.
211. M. Boischoussy and M. Barbier, 1980, *Helv. Chim. Acta*, 63, 1098.
212. R.D. Rasmussen, W.H. Yokoyama, S.G. Blumenthal, D.E. Bergstrom and R.H. Ruebner, 1980, *Ann. Biochem.*, 101, 66.
213. A.H. Jackson, K.R.N. Rao and M. Wilkins, 1982, *J. Chem. Soc., Chem. Comm.*, 794.
214. K.M. Smith, L.C. Sharkus and J.L. Dallas, 1980, *Biochem. Biophys. Res. Comm.*, 97, 1370.
215. R. Tenhunen, H. Marver, N.R. Pimstone, W.F. Trager, D.Y. Cooper, and R. Schmid, 1972, *Biochemistry*, 11, 1716.
216. R.B. Frydman, J. Awruch, M.L. Tomaro and B. Frydman, 1979, *Biochem. Biophys. Res. Comm.*, 87, 928.
217. P.J. Crook, A.H. Jackson and G.W. Kenner, 1971, *Annalen*, 748, 26.
218. A.H. Jackson, G.W. Kenner, T. Kondo and D.C. Nicholson, 1971, *Biochem. J.*, 121, 601.
219. P.S. Clezy and A.J. Liepa, 1970, *Aust. J. Chem.*, 23, 2477.
220. A.H. Jackson, R.M. Jenkins, D.M. Jones and S.A. Matlin, 1981, *J. Chem. Soc., Chem. Comm.*, 763, and 1983, *Tetrahedron*, 39, 1849.
221. R.F.G.J. King and S.B. Brown, 1978, *Biochem. J.* 174, 103.
222. A. Gossauer and H. Plieninger, in ref. 5, Vol. VI, p. 586.
223. H. Plieminger, F. El-Barkawi, K. Ehl, R.D. Kohler and A.F. McDonagh, 1972, *Annalen*, 758, 195.
224. A. Gossauer and W. Hirsch, 1974, *Annalen*, 1496.
225. A. Gossauer and J.P. Weller, 1978, *J. Amer. Chem. Soc.*, 100, 5928.
226. A. Gossauer and R. Klahr, 1979, *Chem. Ber.*, 12, 2243.
227. A. Gossauer and J.P. Weller, 1980, *Chem. Ber.*, 113, 1608.
228. A. Gossauer, R.P. Hinze and R. Kutschan, 1981, *Chem. Ber.*, 114, 132.
229. A. Gossauer, M. Blacha-Puller, R. Zeisburg and V. Wray, 1981, *Annalen*, 142.
230. H. Plieninger and I. Preuss, 1982, *Tet. Lett.*, 43.
231. N.N. Gerber, 1969, *Appl. Microbiol.* 18, 1.
232. H.H. Wasserman, G.C. Rodgers and D.D. Keith, 1966, *J. Chem. Soc., Chem. Comm.* 825.
233. H.H. Wasserman, D.D. Keith and J. Nadelson, 1969, *J. Amer. Chem. Soc.*, 91, 1264, and 1976 *Tetrahedron*, 32, 867.
234. N.N. Gerber, 1970, *Tet Lett.*, 809.
235. N.N. Gerber and M.J. Gauthier, 1979, *Appl. Environment. Microbiol.*, 37, 1176.
236. H. Laatsch and R.H. Thomson, 1983, *Tetrahedron Lett.*, 2701.
237. H.H. Rapoport and G. Holden, 1962, *J. Amer. Chem. Soc.*, 84, 635.

INDEX